Lecture Notes in Computer Science 4321

Commenced Publication in 1973
Founding and Former Series Editors:
Gerhard Goos, Juris Hartmanis, and Jan van Leeuwen

D0725722

Peter Brusilovsky   Alfred Kobsa
Wolfgang Nejdl (Eds.)

The Adaptive Web

Methods and Strategies of
Web Personalization

 Springer

Volume Editors

Peter Brusilovsky
School of Information Sciences
University of Pittsburgh
Pittsburgh PA 15260, USA
E-mail: peterb@mail.sis.pitt.edu

Alfred Kobsa
Donald Bren School of Information and Computer Science
University of California
Irvine, CA 92697-3425, U.S.A
E-mail: kobsa@uci.edu

Wolfgang Nejdl
L3S Research Center
University of Hannover
Appelstr. 9a, 30167 Hannover, Germany
E-mail: nejdl@l3s.de

Library of Congress Control Number: 2007926322

CR Subject Classification (1998): H.5.4, H.5, H.2-4, C.2, I.2, D.2, J.1, K.4

LNCS Sublibrary: SL 3 – Information Systems and Application, incl. Internet/Web
and HCI

ISSN   0302-9743
ISBN-10   3-540-72078-2 Springer Berlin Heidelberg New York
ISBN-13   978-3-540-72078-2 Springer Berlin Heidelberg New York

Springer is a part of Springer Science+Business Media

springer.com

© Springer-Verlag Berlin Heidelberg 2007
Printed in Germany

Typesetting: Camera-ready by author, data conversion by Scientific Publishing Services, Chennai, India
Printed on acid-free paper   SPIN: 12051878   06/3142   5 4 3 2 1 0

Preface

In the first few years after its inception, the Web was the same for everyone. Web sites presented the same information and the same links to all visitors, regardless of their goals and prior knowledge. A query to a Web search engine or catalog produced the same result for all users, irrespective of their underlying interests and information needs.

With the growth of the available information on the Web, the diversity of its users and the complexity of Web applications, researchers started to question this "one-size-fits-all" approach. Does it make sense for a Web course to present the same learning material to students with widely differing subject knowledge? Do news sites serve clients well when they suggest the very same hot news items to people with different interests? Is it appropriate for health information sites to present identical information to readers with different health problems and different educational backgrounds?

To address these deficits, researchers started developing adaptive Web systems that tailored their appearance and behavior to each individual user or user group. Adaptive systems were designed for different usage contexts and explored different kinds of personalization. For instance, adaptive search systems promoted items in result lists that they deemed more relevant to the user's interests and needs than others. Adaptive hypermedia systems tailored page content to the respective user and pushed recommended links to the fore. Adaptive filtering and recommendation systems, finally, complemented search and browsing based information access by actively recommending items that seem most relevant to users' interests and might otherwise be missed due to information overload. To support these kinds of personalization, adaptive systems collected data about their users by implicitly observing their interaction and explicitly requesting direct input from them, and they built user models (aka "profiles") that enabled them to cater to users' different characteristics.

Year after year, the growing demands on personalization as well as the success of early adaptive Web systems resulted in progressively more advanced systems. Web personalization has grown into a large research field that attracts scientists from different communities such as hypertext, user modeling, machine learning, natural language generation, information retrieval, intelligent tutoring systems, cognitive science, and Web-based education.

Meanwhile, the field of the adaptive Web has reached a certain level of maturity. Adaptive Web systems demonstrated their value in several application areas. A wide range of techniques for user modeling and personalization were developed and evaluated in numerous research projects. The volume of knowledge and experience collected in the field gradually turns the adaptive Web from an area of pure research into an engineering discipline where new adaptive systems can be quickly developed by combining known techniques and ideas. The maturity of the field is demonstrated by a number of review papers focusing on various Web personalization topics, many conference tutorials, and first college courses targeting the new generation of Web practitioners. Yet, there was no book to date that would provide a systematic overview of the ideas and techniques of the adaptive Web and serve as a central source of information for researchers, practitioners, and students. The present volume intends to fill that gap through a comprehensive and carefully planned collection of chapters that map out the most important areas of the adaptive Web, each solicited from the experts and leaders in the field.

To serve the diversity of potential readers, the editors solicited three kinds of chapters. The largest part of the book focuses on personalization techniques and is split into two sections. The first section from Chaps. 1 to 5 discusses the modeling side of personalization, while the second section (Chaps. 6 – 14) focuses on adaptation. Each chapter in these two sections provides a comprehensive review of a specific set of techniques or a specific class of adaptive Web systems. Together, these techniques and system types form the "tool chest" of the adaptive Web that every researcher or practitioner in the field should know. The technique-focused part is complemented by four domain-oriented chapters in the third section of the book (Chaps. 15 – 18). These chapters present a holistic view of personalization from the prospect of four different application areas. Finally, the last section is devoted to recently emerging topics. The ideas and techniques presented in these last six chapters are not yet widespread and are not considered to be part of the "tool chest." Yet the editors believe that the issues covered therein are moving rapidly into the focus of the adaptive Web community and have to be included as a glimpse into a not-so-distant future.

The editors would like to thank all authors for the work they put into writing and rewriting their chapters in several revision cycles. We are grateful to Bettina Berendt, Andreas Dieberger, Jörg Diederich, Daqing He, Judith Masthoff, Filippo Menczer, Martin Svensson and Kalina Yacef who served as external reviewers. We appreciate the input from graduate students who "test rode" chapters of the book in classes by Alfred Kobsa at the Università di Roma Tre and by Peter Brusilovsky at the University of Pittsburgh. Finally we would also like to thank Philipp Kärger who helped assemble the individual chapters into a coherent book.

March 2007

Peter Brusilovsky
Alfred Kobsa
Wolfgang Nejdl

List of Contributors

Liliana Ardissono Dipartimento di
Informatica
Università di Torino Corso Svizzera
185 Torino, Italy
liliana@di.unito.it

Jörg Baus
Saarland University
Germany
baus@uni-sb.de

Daniel Billsus
FX Palo Alto Laboratory, Inc.
3400 Hillview Ave, Bldg. 4
Palo Alto, CA 94304
billsus@fxpal.com

Peter Brusilovsky
DSchool of Information Sciences
University of Pittsburgh,
Pittsburgh PA 15260
peterb@mail.sis.pitt.edu

Andrea Bunt
Department of Computer Science
University of British Columbia
bunt@cs.ubc.ca

Robin Burke
School of Computer Science, Telecommunications and Information Systems
DePaul University
243 S. Wabash Ave.
Chicago, IL 60604
rburke@cs.depaul.edu

Giuseppe Carenini
Department of Computer Science
University of British Columbia
carenini@cs.ubc.ca

Alison Cawsey
School of Mathematical and Computer
Sciences
Heriot Watt University
Edinburgh, UK
alison@macs.hw.ac.uk

Aravind Chandramouli
Electrical Engineering and Computer
Science
Information & Telecommunication
Technology Center
2335 Irving Hill Road, Lawrence
Kansas 66045-7612
aravindc@ittc.ku.edu

Luca Chittaro
HCI Lab, Dept. of Math. and
Computer Science
University of Udine
via delle Scienze 206
33100 Udine, Italy
chittaro@dimi.uniud.it

Cristina Conati Department of
Computer Science
University of British Columbia
conati@cs.ubc.ca

Peter Dolog Department of
Computer Science Aalborg
University Fredrik Bajers Vej 7E
DK-9220 Aalborg, Denmark
dolog@cs.aau.dk

Dan Frankowski Department of
Computer Science University of
Minnesota 4-192 EE/CS Building 200
Union St. SE Minneapolis, MN 55455
dfrankow@cs.umn.edu

Fabio Gasparetti Department of
Computer Science and
Automation Artificial Intelligence
Laboratory Roma Tre University
Via della Vasca Navale, 79 - 00146
Rome, Italy
gaspare@dia.uniroma3.it

Susan Gauch
Electrical Engineering and Computer
Science Information &
Telecommunication Technology
Center 2335 Irving Hill Road,
Lawrence
Kansas 66045-7612
sgauch@ittc.ku.edu

Cristina Gena
Dipartimento di Informatica
Università di Torino
Corso Svizzera 185
Torino, Italy
cgena@di.unito.it

Anna Goy
Dipartimento di Informatica
Università di Torino
Corso Svizzera 185
Torino, Italy
goy@di.unito.it

Floriana Grasso
Department of Computer Science
University of Liverpool Ashton
Building
Ashton Street
Liverpool L69 3BX - (UK)
floriana@csc.liv.ac.uk

Dominik Heckmann DFKI GmbH
Germany
heckmann@dfki.de

Nicola Henze IVS – Semantic
Web Group University of Hannover &
L3S Research Center Appelstr. 4
Hannover, Germany henze@l3s.de

Jon Herlocker
School of Electrical Engineering and
Computer Science
Oregon State University
102 Dearborn Hall
Corvallis, OR 97331
herlock@eecs.oregonstate.edu

Anthony Jameson
DFKI, German Research Center for
Artificial Intelligence,
Germany

Alfred Kobsa
Donald Bren School of Information
and Computer Sciences
University of California, Irvine
Irvine, CA 92697-3440, U.S.A.
kobsa@uci.edu

Antonio Krüger University of
Münster Germany
antonio.krueger@uni-muenster.de

Michael Kruppa DFKI GmbH
Germany kruppa@dfki.de

Mauro Marinilli Department of
Computer Science and
Automation Artificial Intelligence
Laboratory Roma Tre University
Via della Vasca Navale, 79 - 00146
Rome, Italy
marinil@dia.uniroma3.it

Alessandro Micarelli
Department of Computer Science and
Automation
Artificial Intelligence Laboratory
Roma Tre University,
Via della Vasca Navale, 79 00146
Rome, Italy
micarel@dia.uniroma3.it

Eva Millán
University of Malaga
Malaga, Spain
eva@lcc.uma.es

Bamshad Mobasher
Center for Web Intelligence
School of Computer Science, Telecommunication, and Information Systems
DePaul University, Chicago, Illinois, USA
mobasher@cs.depaul.edu

Wolfgang Nejdl
L3S Research Center
University of Hannover
Appelstrasse 9A
30167 Hannover, Germany
nejdl@l3s.de

Cécile Paris
SIRO ICT Centre
Sydney, Australia
Cecile.Paris@csiro.au

Michael J. Pazzani Rutgers
University, ASBIII 3
Rutgers Plaza New Brunswick, NJ
08901 pazzani@rutgers.edu

Giovanna Petrone Dipartimento di
Informatica
Università di Torino Corso Svizzera
185 Torino, Italy
giovanna@di.unito.it

Roberto Ranon IICI Lab,
Dept. of Math. and Computer Science
University of Udine via delle
Scienze 206 33100 Udine, Italy
ranon@dimi.uniud.it

Ben Schafer Department of
Computer Science University of
Northern Iowa Cedar Falls, IA
50614-0507 schafer@cs.uni.edu

Filippo Sciarrone Department of
Computer Science and
Automation Artificial Intelligence
Laboratory Roma Tre University
Via della Vasca Navale, 79 - 00146
Rome, Italy
sciarro@dia.uniroma3.it

Shilad Sen Department of Computer
Science University
of Minnesota 4-192 EE/CS Building
200 Union St. SE Minneapolis, MN
55455 ssen@cs.umn.edu

Barry Smyth The School
of Computer Science and Informatics
University College Dublin
Belfield, Dublin 4, Ireland *also*
ChangingWorlds Ltd. South
County Business Park Leopardstown,
Dublin 18, Ireland.
Barry.Smyth@ucd.ie

Amy Soller Institute for
Defense Analyses 4850 Mark Center
Drive Alexandria, Virginia, USA
asoller@ida.org

Mirco Speretta
Electrical Engineering and Computer
Science
Information & Telecommunication
Technology Center
2335 Irving Hill Road, Lawrence
Kansas 66045-7612
mirco@ittc.ku.edu

Rainer Wasinger
DFKI GmbH
Germany
wasinger@dfki.de

Stephan Weibelzahl
School of Informatics
National College of Ireland
Mayor Street
Dublin, Ireland
sweibelzahl@ncirl.ie

Table of Contents

III. Applications

IV. Challenges

Part I

Modeling Technologies

User Models for Adaptive Hypermedia and Adaptive Educational Systems

Peter Brusilovsky[1] and Eva Millán[2]

[1]School of Information Sciences
University of Pittsburgh, Pittsburgh PA 15260, USA
[2] ETSI Informática
University of Malaga
peterb@pitt.edu, eva@lcc.uma.es

Abstract. One distinctive feature of any adaptive system is the user model that represents essential information about each user. This chapter complements other chapters of this book in reviewing user models and user modeling approaches applied in adaptive Web systems. The presentation is structured along three dimensions: what is being modeled, how it is modeled, and how the models are maintained. After a broad overview of the nature of the information presented in these various user models, the chapter focuses on two groups of approaches to user model representation and maintenance: the overlay approach to user model representation and the uncertainty-based approach to user modeling.

1.1 Introduction

Adaptive hypermedia and other adaptive Web systems (AWS) belong to the class of *user-adaptive software systems* [174]. One distinctive feature of an adaptive system is a *user model*. The user model is a representation of information about an individual user that is essential for an adaptive system to provide *the adaptation effect*, i.e., to behave differently for different users. For example, when the user searches for relevant information, the system can adaptively select and prioritize the most relevant items (see Chapter 6 of this book [125]). When the user navigates from one item to another, the system can manipulate the links (e.g., hide, sort, annotate) to provide adaptive navigation support (see Chapter 8 of this book [21]). When the user reaches a particular page, the system can present the content adaptively (see Chapter 13 of this book [28]). To create and maintain an up-to-date user model, an adaptive system collects data for the user model from various sources that may include implicitly observing user interaction and explicitly requesting direct input from the user. This process is known as *user modeling*. User modeling and adaptation are two sides of the same coin. The amount and the nature of the information represented in the user model depend to a large extent on the kind of adaptation effect that the system has to deliver.

P. Brusilovsky, A. Kobsa, and W. Nejdl (Eds.): The Adaptive Web, LNCS 4321, pp. 3–53, 2007.
© Springer-Verlag Berlin Heidelberg 2007

As mentioned in the introduction, Chapters 1 to 5 of this book are focused mostly on the modeling side of personalization, while the remaining chapters focus mostly on the adaptation side. Chapters 1 and 2 are specifically devoted to user models and user modeling. Beyond this, user modeling issues are discussed at different levels of detail in several other chapters. This chapter attempts to complement the remaining chapters in two ways. First, it provides an overview (a "big picture") of the user modeling side referring readers when necessary to additional information in chapters within this book. Second, it attempts to complement other chapters by presenting aspects that are either not covered in other chapters or covered insufficiently.

To envision the big picture, this chapter follows Sleeman [175] who suggested classifying user models by the nature and form of information contained in the model as well as the methods of working with it. Following his suggestions, we analyze user models along three layers: *what is being modeled* (nature), *how this information is represented* (structure) and *how different kinds of models are maintained* (user modeling approaches). In this book, the overview of user modeling along the first layer, the nature of the represented information, is provided in section 1.2 of this chapter. This section serves as a basis for understanding the user modeling problem as a whole (Fig. 1.1).

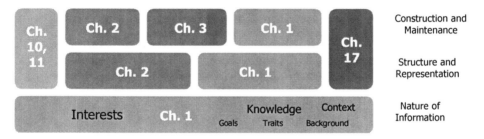

Fig. 1.1. Three layers for the analysis of user modeling approaches and their coverage in different chapters of this book. Horizontal dimension represents user features reflected in the models.

The review of the structure and the representation of information in user models (the second layer) is split between this chapter and Chapter 2 [72]. Together these chapters provide a detailed overview of the two most important and most elaborate types of user models, which were originally developed in the fields of information retrieval [107] and intelligent tutoring systems [158]. Information retrieval and filtering systems attempt to find documents that are most relevant to user interests and then to order them by perceived relevance. The user model that typically powers this kind of systems is known historically as a *user profile* and represents the user's *interests* in terms of keywords or concepts. Intelligent tutoring systems (ITS), strive to select educational activities and deliver individual feedback that is most relevant to the user's level of knowledge. The user model in ITS is known as a *student model* and represents mostly the user's *knowledge* of the subject in relation to expert-level domain knowledge. In this book, section 2.3 in Chapter 2 reviews three main types of user profiles for representing user interests, while section 1.3 (in this chapter) reviews the dominant overlay approach for modeling user knowledge.

Finally, the process of construction and maintenance of user models (the third layer) is discussed in several chapters. The lion's share of this presentation is provided in Chapters 1 and 2 - the main user modeling chapters. These chapters follow their foci on knowledge and interest modeling respectively: section 1.4 of Chapter 1 focuses on Bayesian Networks, the most important approach to knowledge modeling, while section 2.4 of Chapter 2 reviews major interest-modeling approaches. This presentation is complemented by several other chapters, which focus on specific user modeling approaches. Chapter 3 [135] reviews Data Mining approach to user profile constructions. Chapters 9 and 10 [152; 172] review content-based and collaborative user profiling as used for recommendation. Chapter 17 [109] elaborates on context modeling for mobile applications.

The design of this chapter is the result of a compromise between two goals: to provide core content that is determined by the chapter's share in presenting the "big picture" of user modeling while making it useful and meaningful on its own. To achieve the second goal, the chapter was focused on user modeling for Adaptive Hypermedia (AH) and Adaptive Educational Systems (AES). AH and AES are two groups of Web-based systems that extensively employ the user knowledge models covered by the core of this chapter. To better justify this title, the knowledge modeling core of sections 1.3 and 1.4 of this chapter were extended to cover overlay modeling and Bayesian user modeling, which goes "beyond knowledge". The nature of this chapter caused a different balance between the breadth and depth of coverage for the different sections. Section 1.2 attempts to provide a broad coverage while providing relatively few details about specific user models or modeling approaches. It is intended as a good overview of the topic that will be useful for anyone interested in adaptive Web systems. In contrast, sections 1.3 and 1.4 provide more details at the price of a broader coverage. While focusing mainly on the modeling of user knowledge, they also introduce approaches to the modeling of other kinds of information about the user.

1.2 What Is Being Modeled

According to the nature of the information that is being modeled in adaptive Web systems, we can distinguish models that represent features of the user as an individual from models that represent the current context of the user's work. The former are important to all adaptive Web systems while the latter are mostly the concern of mobile and ubiquitous adaptive systems, where context is essential. This section focuses on the five most popular and useful features found when viewing the user as an individual: the user's knowledge, interests, goals, background, and individual traits. We also discuss modeling the context of a user's work. At the end of this section we discuss stereotype-based user modeling that is an alternative to the more popular feature-based modeling.

1.2.1 Knowledge

The *user's knowledge* of the subject being taught or the domain represented in hyperspace appears to be the most important user feature, for existing AES and AHS. In AES, the knowledge is frequently the only user feature being modeled. In AHS, it is used by the majority of systems for both adaptive navigation support and adaptive presentation. The user's knowledge is a changeable feature. The user's knowledge can both increase (learning) or decrease (forgetting) from session to session and even within the same session. This means that an adaptive system relying on user knowledge has to recognize the changes in the user's knowledge state and update the user model accordingly.

The simplest form of a user knowledge model is the scalar model, which estimates the level of user domain knowledge by a single value on some scale – quantitative (for example, a number ranging from 0 to 5) or qualitative (for example, good, average, poor, none). Scalar models, especially qualitative, are quite similar to stereotype models. The difference is that scalar knowledge models focus exclusively on user knowledge and are typically produced by user self-evaluation or objective testing, not by a stereotype-based modeling mechanism. Despite their simplicity, scalar models can be used effectively to support simple adaptation techniques in AHS. A number of AHS's use scalar knowledge models to support adaptive presentation. These systems divide their users into two or three classes according to their knowledge level of the subject (i.e., expert, intermediate, and novice) and serve different versions of the whole page content [63] or page fragments [8; 14; 16] to users with different levels of knowledge.

A good example of adaptive presentation based on a scalar model is the MetaDoc system [14], MetaDoc represents user knowledge of UNIX (which was the domain of the system) as a qualitative scalar value (novice - beginner - intermediate – expert. The scalar model was used to generate an original adaptation based on *stretchtext*. Stretchtext is a special kind of hypertext where clicking on an anchor (hotword) simply "expands" it by inserting a fragment of content *after* or *instead of* the anchor. Another click collapses the expanded content back to the original hotword (a similar approach is used by the well-known Windows Explorer). Each page in MetaDoc is a stretchtext, which may contain many expandable hotwords. The idea of adaptive stretchtext presentation in MetaDoc is to present a requested page with all stretchtext fragments not relevant to the user being collapsed and all fragments relevant to the user being expanded. To achieve this result, an author must first classify expandable text fragments as either an additional explanation or a low-level detail. The user of MetaDoc with an expert level of knowledge of a concept will be presented with additional explanations hidden (collapsed) and low-level details expanded. On the other hand, the user with a beginner's level of knowledge will receive expanded additional explanations in all cases. After its presentation, a stretchtext page can be further adapted by the user who is free to expand and collapse text fragments. The study of MetaDoc demonstrated that this simple technology based on a scalar model can increase the speed and quality of user comprehension of the content [14]. More information about stretchtext and other adaptive presentation techniques can be found in Chapter 13 of this book [28].

The shortcoming of the scalar model is its low precision. User knowledge of any reasonably-sized domain can be quite different for different parts of the domain. For example, in word processing, a user may be an expert in using text annotation, but a novice in formula editing [66]. A scalar model effectively averages the user knowledge of the domain. For any advanced adaptation technique that has to take into account some aspect of user knowledge, the scalar model is not sufficient. For the above reason, AES's that focus on advancing user knowledge and many AHS's use various kinds of *structural models*. The structural models assume that the body of domain knowledge can be divided into certain independent fragments. These models attempt to represent user knowledge of different fragments independently. By the nature of represented knowledge, structural models can be independently classified along two different sub-dimensions, according to:

- the type of represented knowledge (declarative vs. procedural), and
- a comparison of the user's knowledge—represented in the model—to an expert's level of knowledge of the subject, referred to as domain model, expert model, or "ideal student" model.

The most popular form of a structural knowledge model is an *overlay model*. The purpose of the overlay model is to represent an individual user's knowledge as a subset of the *domain model*, which reflects the expert-level knowledge of the subject. For each fragment of domain knowledge, an overlay model stores some estimation of the user's knowledge level of this fragment. The pure overlay model, developed in the field of ITS over 30 years ago [194], assigns a Boolean value, yes or no, to each fragment, indicating whether the user knows or does not know this fragment. In this case, user knowledge is represented at each instant of time as an exact subset or "overlay" of expert knowledge. In its modern form, an overlay model represents the degree to which the user knows such a domain fragment. This can be a qualitative measure (good-average-poor), or a quantitative measure, such as the probability that the user knows the concept.

Since the overlay model represents the user's knowledge as a (weighted) subset of expert knowledge, the nature of the user knowledge reflected in the overlay model depends on the nature of the expert knowledge represented in a specific system. The majority of ITS focused on representing two types of domain knowledge: conceptual knowledge (facts and their relationships) and procedural knowledge (problem-solving skills). Conceptual knowledge is typically represented in the form of a network of concepts. Procedural knowledge is most frequently represented as a set of problem-solving rules. Several knowledge-representation approaches, such as ACT-R [2], or propositional representation [100] allowed one to combine these two types together. More recently, Stellan Ohlsson suggested focusing on a different kind of procedural knowledge – not the knowledge that allows the user to solve the problem, but the knowledge that allows him/her to evaluate the correctness of the solution [144]. This knowledge is typically represented as a set of constraints.

In turn, the nature of knowledge represented in a specific adaptive system is determined by the kind of personalized support it provides. A large class of ITS known as "tutors" focus on helping users solve educational problems and thus rely on the procedural knowledge of either problem solving or evaluation nature. Other types of ITS and IES focus on helping users to select the most relevant piece of educational content and thus rely on conceptual knowledge about the domain. The use of conceptual knowledge is shared by almost all non-educational AHS, which also focus on guiding the user to the most appropriate content. In all these cases, regardless of the nature of the represented expert knowledge, the overlay model can be successfully applied to model individual user knowledge, i.e., it can measure how well the user knows a concept, what is the probability that the user can apply a rule or which of the constraints or propositions are likely mastered.

Overlay models constitute a dramatic step forward from scalar models. Yet, in the field of ITS, overlay models have often been criticized for being "too simple." It has been argued that the state of user knowledge is never an exact subset of expert knowledge. The user may have misconceptions and her knowledge generally progresses to expert-level knowledge not by "filling the gaps," but through a complex process of generalization and refinement. To model user misconceptions, an overlay model was expanded into a *bug model,* representing both correct knowledge and misconceptions (known as buggy knowledge or "bugs"). Bug models were predominantly used to model user procedural problem solving knowledge. The most extensively studied form of bug model is called the *perturbation model.* This model assumes that several incorrect *perturbations* can exist for each element of domain knowledge. Incorrect user behavior may, from the viewpoint of this approach, be caused by the systematic application of one of the perturbations in place of the correct rule. The goal of a system with a bug model is not just to declare that a specific element of domain knowledge is incomplete or missing, but to identify, if possible, specific buggy knowledge that can be used to provide a higher quality adaptation. An even richer model that makes it possible to reflect the development (genesis) of user knowledge from the simple to the complex and from the specific to the general is known as a *genetic model* [74].

While both bug models and genetic models are certainly more powerful than the traditional overlay model, they are also much harder to develop. Research on these models has contributed to the development of the fields of cognitive modeling and ITS [194], but the practical use of these models has been quite limited. Genetic models have never been used in practical systems. Bug models have been used mostly in problem solving ITS created for simple domains, although several well-known systems created by Carnegie Mellon researchers have demonstrated that this approach could work for large-scale practical systems [1; 106]. In the area of Web-based systems the use of bug models is limited to a small subset of Web-based AES that are focused on adaptive problem solving support. Non-educational Web systems do not use bug models since they have no means (and no need) to diagnose misconceptions.

The typical pattern of bug model application in Web-based AES can be demonstrated by such systems as WITS [145], ILESA [116], and Web-PVT [192]. These systems, developed for three different educational domains, provide a personalized analysis of exercise solutions and some form of guidance through problem selection (WITS), problem generation (ILESA) or adaptive navigation support (Web-PVT). The systems use a combination of an overlay model and a bug model. Bug models allow the systems to recognize misconceptions in the user's problem-solving knowledge, distinguish it from random slips, such a typos and calculation errors, and provide a useful personalized explanation. However, the rest of the adaptive functionality (such as problem generation or adaptive navigation support) only considers the balance between correct and incorrect usage of knowledge, as reflected in the overlay model, ignoring the exact nature of the misconceptions represented in the bug model.

In contrast to the powerful but complicated and rarely used bug models, overlay knowledge models are extremely popular in both Web AES and AHS. Almost every

Web AES and a majority of modern AHS are based on some form of overlay models. Multiple projects have demonstrated that an overlay model provides a good balance of simplicity and power. The ability to independently measure user knowledge within different elements of the domain provides a level of power that is sufficient to run the majority of the advanced adaptation techniques. Yet, overlay model are relatively easy to develop, especially for cases with less than 100 domain-knowledge elements. Due to the importance of overlay models for AES and AHS, this chapter provides a special section on the development of overlay models.

1.2.2 Interests

User interests always constituted the most important (and typically the only) part of the user profile in adaptive information retrieval and filtering systems that dealt with large volumes of information. It is also the focus of user models/profiles in Web recommender systems. In contrast, early AES and educational AHS paid no attention to user interests, instead focusing on learning goals when sequencing educational content. As for non-educational AHS, their nature and the small size of their hyperspaces created no demand for interest adaptation. This situation has changed dramatically over the last 10 years. User interests are now competing with user knowledge to become the most important user feature to be modeled in AHS. The change was caused by the rapid growth of the volume of information and the growing popularity of several new kinds of information-oriented AHS such as encyclopedias [84], hypertextual news systems [3], electronic stores [4], museum guides [142], and information kiosks [65] where access to information is mostly interest-driven. Following the pioneer attempts mentioned above, interest modeling was explored in a number of information-oriented AHS. More recently, the abundance of available content and the growing popularity of the interest-driven constructionist approach to education have encouraged more attempts to model user interests in educational AHS.

Starting from the pioneer systems, AHS focused on a new approach for modeling user interests. The predominant representation of user interests at that time was the weighed vector of keywords. This approach was used by nearly every adaptive information retrieval and filtering system and is still the most popular in these areas. More details about this approach can be found in Chapter 2 [72]. In contrast to this keyword-level approach, AHS adopted a concept-level approach to user interest modeling where user interests are represented as a weighed overlay of a concept-level domain model. The concept overlay approach to user interest modeling is very similar to the overlay knowledge modeling approach. Overlay user modeling and specific examples of its application to user interest modeling are presented in more detail in section 1.3 of this chapter.

Concept-level models of user interests are generally more powerful than keyword-level models. Concept-level models allow a more accurate representation of interests. Given a rich domain model, a concept overlay model can separately model different aspects of user interests. For example, a news personalization system can model user interests on distinct topics, based in a specific geographical location, and dealing with specific named entities [96]. An adaptive museum system can separately model interests in the designer, style, or origin of a jewelry item [143]. In addition, semantic links in the domain model allow different kinds of interest propagation to compensate for *sparsity*, a standard problem of large overlay models.

Powerful, concept-level models of user interests in AHS have been enabled by the nature of AHS content that is traditionally manually indexed with domain model concepts. In closed corpus AHS, such as adaptive information kiosks or museum guides, the content was indexed at the time of system creation. In systems with expandable corpus, such as adaptive news systems, new content had to be indexed by a provider at the time of its insertion into the system [3; 96]. The need for manual indexing originally led to the establishment of two distinctive groups or systems that are able to adapt to user interests. Closed corpus systems (mostly AHS) used the concept-level interest model, paying the price of manual indexing. Open corpus systems (mostly information retrieval and filtering systems) used keyword-level models, but were able to work with an unrestricted corpus of document due to their ability to process documents automatically. This dichotomy is no longer clear-cut. A new generation of adaptive information access systems has attempted to combine concept-level interest modeling with automatic document processing. Many of these systems are based on automatic document categorization where each document is automatically assigned to a concept of an existing domain model (such as the Yahoo! directory or the ACM topic ontology). These approaches are presented in detail in Chapter 2 [72] of this book. More recent systems explore automatic multi-concept indexing, where a document can be automatically connected with several domain model concepts [43]. In addition, a new group of hybrid systems attempts to combine concept-level and keyword-level interest models in one system [54]. A more elaborate discussion on bridging the gap between closed corpus AHS and open corpus information retrieval and filtering systems is provided in Chapter 22 of this book [25].

1.2.3 Goals and Tasks

The user's goal or *task* represents the immediate purpose for a user's work within an adaptive system. Depending on the kind of system, it can be the goal of the work (in application systems), an immediate information need (in information access systems), or a learning goal (in educational systems). In all of these cases, the goal is an answer to the question "What does the user actually want to achieve?" The user's goal is the most changeable user feature: it almost always changes from session to session and can often change several times within one session of work. Early research on adaptation to the user's *work goal* was done in the area of adaptive interfaces and intelligent help systems [10; 66; 76]. Adaptation to the user's *learning goal* was explored by instructional planning and sequencing systems [18; 123; 124; 197]. Adaptation to the user's immediate *information need* was explored by adaptive information retrieval systems [17]. Among AHS systems goal modeling techniques were explored in the sub-areas of hypertext-based help systems [63; 78] and adaptive information access systems [88; 97; 121] where the ability to adapt to the current user goal is very important.

Goal modeling in modern AHS and AES mostly follows the approaches suggested in the pioneer research mentioned above. The user's current goal is usually modeled with a *goal catalog* approach, which is somewhat similar to overlay knowledge modeling. The core of this approach is a pre-defined catalogue of possible user goals or tasks that the system can recognize. Frequently this catalogue is simply a small set of independent goals, however, some systems use a more advanced catalog in the form of a goal or task hierarchy, which is inherited from earlier research on adaptive interfaces and instructional planning. In a goal hierarchy, relatively stable higher-level goals are progressively decomposed into subgoals down to the lowest level formed by short-term goals. Most typically, the system assumes that the user has exactly one goal (or one goal at each level of the hierarchy) at any moment of work with the system. The job of the user modeling component is to recognize this goal and to mark it as the current goal in the model. This fires the adaptation rules that refer to possible user goals specified in the catalogue. The adaptation rules can, for example, recommend some pages to the user [23], focus user attention on a subset of the hyperspace [69; 198], or adapt content of the selected page [88].

The goal/task recognition process is difficult and not precise in general. It is especially difficult in AH and other Web-based system where the flow of information from the user (bandwidth) required by user modeling components is thinner than in traditional desktop systems. Over the years, adaptive Web systems explored a number of approaches to fight this problem. To start with, a number of practical systems allow the user to specify the current goal. Typically the user has to select one of the pre-defined goals [69; 88; 151], although some systems are able to gradually learn how to adapt to a completely new goal introduced by the user [97; 121]. A different approach to fight the imprecise goal recognition is to model the user current goal as a probabilistic overlay of the goal catalogue where for each goal the system maintains the probability that this goal is the current goal of the user [63; 126]. Finally, several recent projects explored the use of data mining technologies to identify the current user task in an expected sequence of tasks and to provide personalized task-level support [87; 94]. More information about the use of data mining technologies for Web personalization ns provided is Chapter 3 of this book [135].

A popular example of goal adaptation is provided by the PUSH system [88]. This system has a small catalogue of user goals and adapts the presentation of each selected page to the current goal. Depending on the current goal, some parts of the page can be collapsed using the stretchtext approach explained in subsection 1.2.1 above. The system attempts to deduce the user goal by observing the user's actions and shows the assumed goal to the user in the spirit of a *glass-box adaptation*. The user can also change the current goal by selecting a more appropriate one from the catalog.

A different example of goal modeling is a performance support system ADAPTS [23]. ADAPTS deduces the current goal within a goal hierarchy by following the user's aircraft maintenance operations. Once the goal is recognized, the system generates a page with the list of technical manual fragments that are most relevant to the current goal and user level of knowledge.

1.2.4 Background

The *user's background* is a common name for a set of features related to the user's previous experience outside the *core domain* of a specific Web system (for example, the core domain of a city guide [36] is a specific city and its objects of interest; the core domain for a hospital information system [198] is a specific hospital, its objects and procedures). A range of backgrounds that have been used in adaptive Web systems includes the user's profession, job responsibilities, experience of work in related areas, and even specific view on the domain. For example, medical adaptive hypermedia systems can distinguish two or three categories of users according to their knowledge of medical terminology and adapt content presentation to the user category by selecting either medical terms or everyday language to present the same content [8; 178]. Alternatively these systems can distinguish users by their profession (student, nurse, doctor) which implies both the level of knowledge and responsibilities [53; 198]. More examples for this application area can be found in Chapter 15 of this book [34]. Another example is the categorizing of users by their language ability (i.e., native or non-native speakers), followed by choosing the appropriate version of the content for them [102]. Background information is used most frequently for content adaptation, although there are examples of the use of it within adaptive search [71; 121] and adaptive navigation support [198].

By its nature, user background is similar to the user's knowledge of the subject (i.e., it is also mostly a measure of knowledge beyond the core domain area). However, the representation and handling of user background in adaptive systems is different. Since detailed information about the background is not necessary, the common way to model user background is not an overlay, but a simple stereotype model. In addition, user background typically does not change during work with the system and is nearly impossible to deduce by simply watching the user work. As a result, user background is typically provided explicitly, either by the user herself or by some kind of a superior (a teacher in a college or an administrator at an institution).

1.2.5 Individual Traits

The *user's individual traits* is the aggregate name for user features that together define a user as an individual. Examples are personality traits (e.g., introvert/extravert), cognitive styles (holist/serialist), cognitive factors (e.g., working memory capacity) and learning styles. Similar to user background, individual traits are stable features of a user that either cannot be changed at all, or can be changed only over a long period of time. Unlike user background, however, individual traits are traditionally extracted not from a simple interview, but through specially-designed psychological tests. Many researchers agree on the importance of modeling individual traits and using them for adaptation. While different kinds of user traits are extensively discussed in psychological literature, current work on modeling and using individual traits for personalization focuses mostly on two groups of traits – *cognitive styles* and *leaning styles*. These groups are discussed below. More recently, researchers on adaptive systems started considering individual traits beyond cognitive and learning styles. For example, CUMAPH system [188] made a pioneer attempt to build a user profile from lover level cognitive abilities and apply for adaptively generating page content for the user. Another recent work considered the use of personality factors in the context of adaptive museum guides [77] (more information about adaptive museum guides is presented in Chapter 17 of this book [109]).

Cognitive Styles. By cognitive style, researchers typically mean an individually preferred and habitual approach to organizing and representing information [165]. Research on cognitive style has long attracted attention on researchers in Web personalization and related fields such as human information behavior. Professional literature distinguishes a number of dimensions in which the users cognitive styles may differ: field-dependent/independent, impulsive/reflective, conceptual/inferential, thematic/relational, analytic/global [35; 113]. Most popular among Adaptive Hypermedia researchers are Witkin's field-dependent/independent [205] and Pask's holist/serialist [150]. Since, by its nature, cognitive style influences humans' ability to access information, the work on adaptation to user cognitive style was focused more on the navigational side of AHS and AES. For example, several adaptive hypermedia systems [61; 133; 190; 191] distinguished field-dependent and field-independent users and provided different navigation organization, amount of user control, and navigation support tools for these groups.

A typical scenario for cognitive style adaptation is provided in AES-CS system [190; 191]. AES-CS attempted to adapt to both user knowledge and user cognitive style. A field-dependency test was used to classify users in two groups: field-dependent and field-independent. After that a range of system features were adapted to the identified cognitive style. Field-independent users received an access to the navigation menu to control their navigation. Field-dependent users were only able to proceed through the content sequentially, however they were provided with additional orientation support tools such as a concept map and a path indicator. Depending on their learning style, the users also received different instructions, feedback, and contextual organizers. Evaluated against a static version of content, the AES-CS (with both kinds of adaptation enabled) demonstrated a significant increase of user performance [191].

So far, the research on the use of cognitive style of adaptation is a mixed-success story. On one side, a number of studies confirmed that cognitive style affects both search and browsing behavior [35; 104]. On the other side, few success stories (like AES-CS) on using cognitive styles for adaptation were reported. A number of projects made a similar attempt to distinguish users by their cognitive style and match them with a version of a hypermedia system developed to support this style, but were not able to report any significant differences against a non-adaptive condition. It is interesting that attempts to mismatch user styles and system versions (i.e., to match users to version of the system developed for the alternative style) typically reported significant negative results [133]. Thus, cognitive style remains an important user feature to take into account, but reliable approaches of adaptation to user cognitive styles are yet to be found.

Learning Styles. Learning styles are typically defined as the way people prefer to learn. This group of individual traits is close to cognitive style, but more narrow in scope due to its focus on human learning. The application of learning styles to adaptation is limited to Web-based AES. After a few pioneer works that presented the idea of adapting Web-based AES to learning styles and suggested some ways to implement it [33], this direction of work quickly emerged into arguably the most popular kind of research on individual traits adaptation on the Web. A number of pre-Web approaches or *inventories* to classify and measure learning styles were applied and a number of studies demonstrated differences between users with different styles in Web context. Most of the work on learning style adaptation explored *content-level* adaptation attempting to match users with a specific learning style to content that should be the most appropriate for this style. This adaptation may take different forms, such as selecting the most style-relevant version of content for presentation, ordering content fragments by their relevance to user style or hiding style-irrelevant content. More recently, these popular approaches to learning style adaptation were integrated into several adaptation frameworks and authoring tools such as ACE [179], CAMELEON [111], AHA! [181], and APeLS [47]. For example, a system developed according to APeLS framework [47] matches the user model with content metadata in order to select learning objects that are most relevant to the user's learning style given certain alternatives provided in the pool of resources.

Despite all this work, there are no proven recipes for the application of learning styles in adaptation. It is still unclear which aspects of learning style are worth modeling, and what can be done differently for users with different styles. Dozens of experimental systems that consider different style inventories and suggest different ideas for adaptation were reported, but careful studies are rare and success stories are very few. On the contrary, a number of experimental studies aimed to evaluate the value of treating users with different learning styles differently concluded without finding any significant differences. As a whole, the situation is similar to the situation with cognitive styles: the area of adaptation to individual traits holds a lot of potential, but offer almost no practical suggestions. To progress in this area, we either need to learn more about the relationships between user traits and possible interface settings, or develop trait-agnostic techniques that treat user traits as a black box and attempt adapt to them using case-based and non-symbolic technologies [73].

1.2.6 Context of Work

Adaptation to the context of the user's work is a relatively new research direction within AHS. It was introduced by several pioneer Web-based systems and later expanded into the area of mobile adaptive systems. Early context-adaptive systems explored mostly platform adaptation issues. The growing interest to mobile and ubiquitous systems attracted researchers attention to other dimensions of the context such as user location, physical environment, social context, and affective state. Context modeling is conceptually different from modeling of other user features discussed above. Some information represented in the context models can hardly be considered information about the user in pure sense. However, context modeling and user modeling are tightly interconnected. Many user models include context features; similar techniques are used for context and user modeling; integrated frameworks are being developed for modeling both user context and user features [80; 210]. For all that reason, the overview of context modeling is included in this chapter.

User Platform. Since users of the same server-side Web application may use different equipment at different times, adaptation to the user's platform becomes an important issue. Adaptive hypermedia systems explored a range of techniques that might be used to adapt to such aspects of user platform (computing environment) as hardware, software, and network bandwidth. The largest stream of work focused on adaptation to the screen size by either converting pages designed for viewing on desktop Web browsers to mobile browsers or generating pages differently for desktop and mobile applications. Another stream focused on media presentation capabilities that are a combination of hardware, available software, and bandwidth. Over the last several years, the work on platform adaptation has grown from a pioneer research domain into practice and is now in the focus of W3C [202]. Most recent attempt to standardize the description and use of platform capabilities can be found in [105].

It is important to stress that platform-oriented context models are different from the knowledge, task, and goal models reviewed above. A context is typically described by a potentially long set of name-value pairs where names indicate parameters (i. e, screen width, type of pointing device, or presence of movie player) and values specify parameter values in the current context. This is, however, a *raw model* of the context. While adaptation rules can be written directly addressing the raw parameters, this solution is nowadays neither practical nor scaleable. As a result, we observe an emergence of context adaptation approach that is somewhat similar to stereotype modeling. I.e., a set of all possible combinations of name-value pairs is mapped to a smaller set of stereotypes that are, in turn, used by adaptation rules. For example, an adaptive system can distinguish two or more platforms *types* where each type is formed by a specific range of platform parameters. In this situation, platform recognition mechanism uses parameters of a specific platform to determine the current type. After that, simple adaptation rules check current type and perform different actions for different platform types. For example, if the user accesses the system from a handheld device, the system switches on conversion [112; 115; 206] or generation [15; 149] of presentation for a small screen. If a user platform cannot show color pictures or the bandwidth is low, the system converts the pictures to black-and-white or low resolution [166]. If the platform cannot show movies due to the absence of a movie player or low bandwidth, it can replace the movie with a picture or remove a link to the movie [93; 95]. More advanced technologies can generate considerably different interfaces for different platform types [11; 46; 62] and even use platform limitation to the benefits of user modeling [12]. For example, a Palm Pilot version of an adaptive news system [12] that is characterized by a small screen and low bandwidth requires the user to request pages of news story one by one thus sending each time an implicit relevance feedback to the system.

User Location. Mobile context-adaptive systems naturally focused on adaptation to user location. The modeling and the use of location is slightly different from other context elements. Most frequently the location is used not to fire adaptive presentation rules, but to determine a small subset of nearby objects of interests. This subset defines what should be presented or recommended to the user. This kind of adaptation was explored by early context-adaptive systems in several contexts such as museum guides [142], tourist guides [36], and marine information system [70]. Accordingly, user location is being modeled in a way that supports determining the nearby objects. Depending on the kind of location sensing it is typically a coordinate-based or zone-based representation. More information about location-adaptive systems can be found in Chapter 17 of this book [109].

A Broader View of the Context. More recently, research on mobile and ubiquitous computing has considerably expanded the notion of context [173]. While there is no definite agreement about what should be included into the area of context, most of the work on context adaptation in mobile and ubiquitous computing focuses on a common core that includes *environment* and *human* dimensions [91; 173]. The environment dimension includes spatio-temporal aspect and physical conditions (light, temperature, acceleration, pressure, etc.). The human dimension includes personal context (user pulse, blood pressure, mood, cognitive load, etc.), social context [154], and user task. This may look confusing to the user modeling researchers who consider user tasks as a part of the user model, not context. To avoid confusion, it is important to remember that the research on context modeling is going on in two different research communities that consider context from two different points of view (Fig. 1.2). From the user-centered view employed in the user modeling field, the user task is not a part of the context, while the device itself is. From the device-centered view, which is dominant in the mobile and ubiquitous computing, a range of parameters characterizing the current state of the user are, indeed, a part of the device context. This view is presented in Chapter 17 of this book [109], which provides an extensive discussion of context modeling and reviews its use in several kinds of adaptive systems.

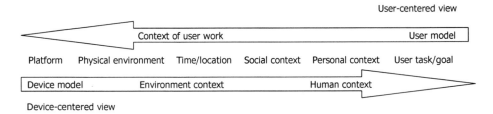

Fig. 1.2. The dimensions of context from user- and device-centered views

To establish a more objective border between user and context modeling, it is useful to observe that user modeling is focused mostly on the longer-term properties of the user that are distilled from observations, while context models attempt to represent the current features of the user and the environment, mostly read from context and physiological sensors [91; 173]. At the moment it is hard to classify the approaches to broader context modeling and adaptation in adaptive Web systems since these systems rarely attempt to model context beyond the platform and location. However, it seems that the emerging stereotype approach for platform adaptation can work within a range of other context parameters. Namely, a system may map a long list of name-value pairs supplied by sensors into a small set of meaningful pre-defined stereotype contexts. Once the current stereotype context is recognized, the system can use adaptation rules that address these stereotypes. For example, a context-aware application on a mobile device may use several parameters to recognize whether a user is rushing somewhere or not (two context stereotypes) and depending on the current context stereotype present information about the user's flight in different ways [137].

Affective State. An important context dimension that deserves to be mentioned separately is the user's affective state. Influenced by the idea of affective computing [156], the research on modeling using the affective state has become very popular in both user modeling and ubiquitous computing areas. A sequence of workshops devoted to "affective modeling" has been held in conjunction with the biannual User Modeling conference series since 1999. The proceedings of the last workshop in the series demonstrate the ranges of issues that are being explored now [30]. The methods used to model user affective state inherited approaches from both research areas mentioned above. While the methods applied by researchers with ubiquitous and pervasive computing backgrounds focus mostly on using various sensor input, researchers with user modeling background explored a range of approaches based on observing user-system interaction. Most important in the context of this chapter is the small stream of works on modeling the user's affective state by observing the user's Web log data. Despite the relatively low bandwidth of this source, a number of researchers have demonstrated that it can be used to detect user motivation, frustration, engagement, and disengagement [7; 9; 38; 193]. It is also interesting to observe that Bayesian networks emerged as the most popular technology to process both sensor input and the user action logs into user affective state [7; 42; 159; 193]. Section 1.4 of this chapter, which focuses on Bayesian approaches to user modeling, provides some further information on Bayesian affective modeling.

1.2.7 Feature-Based Modeling vs. Stereotype Modeling

Feature-based user modeling reviewed above is currently the dominant user modeling approach in adaptive Web systems. Feature-based models attempt to model specific features of individual users such as knowledge, interests goals, etc. During the user's work with the system, these features may change, so the goal of feature-based models is to track and represent an up-to-date state for modeled features. An alternative to the feature-based modeling is stereotype modeling.

Stereotype user modeling is one of the oldest approaches to user modeling. It was developed in the works of Elaine Rich [163; 164] over a quarter of century ago and elaborated in a number of user modeling projects. Stereotype user models attempt to cluster all possible users of an adaptive system into several groups, called stereotypes. All users belonging to the same stereotype are treated in the same way by the adaptation mechanisms. A user in a classical stereotype-based system is represented simply as her current stereotype (i.e., a group she currently belongs to). Naturally, each stereotype groups together users with specific mixture of features. However, stereotype modeling ignores the features and uses the stereotype as a whole. More exactly, the goal of stereotype modeling is to provide mapping from a specific combination of user features to one of the stereotypes. After that, only the user current stereotype is used for adaptation. Any changes in the user's features are responded to by simply re-assigning a user, if necessary, to a different stereotype. Elaine Rich discusses extensively when would be the right moment for this [164].

Stereotypes were extensively used in early adaptive systems between 1989 and 1994. A good overview of this generation of stereotype-based systems is provided in [101]. More recently, stereotype models were overshadowed by feature-based models, yet stereotypes were used in a number of adaptive Web systems [3; 5; 71; 127; 192]. In addition, as was mentioned above, the techniques developed for stereotype modeling and adaptation are used beyond classical stereotype modeling to manage low-granularity feature-based models.

A promising direction for the future application of stereotype models is their use in combination with feature-based models. One of the most popular combinations is the use of stereotypes to initialize an individual feature-based model [3; 5; 192]. This approach allows to avoid a typical "new user" problem in feature-based modeling where effective adaptation to new user is not possible since user modeling is started "from scratch." A good example of this combination is provided by SeAN system [3] for adaptive news presentation. Starting the work with a new user SeAN attempts to map the set of demographic features provided by the user (such as age, education, and type of job) into a set of predefined stereotypes. These stereotypes, in turn, are used to initialize the feature-based model of user knowledge and interests. Another promising direction for stereotype-based modeling is its use in combination with group models. Group models are becoming increasingly popular in Web personalization. More information about group modeling is provided in Chapter 20 of this book [92].

1.3 The Overlay Approach to User Modeling

This section provides more detailed information about an overlay approach to user modeling, the approach that is at the same time most important and most popular for AES and AHS. We start with overlay knowledge modeling, which was the original application of the overlay approach. After that we review the use of overlay approach for modeling user interests and discuss a generalization of the overlay approach for modeling other user features.

1.3.1 Overlay Modeling of User Knowledge

As we mentioned above, the idea of overlay knowledge modeling is to represent an individual user's knowledge as a subset of the domain model that resembles expert knowledge of the subject. Overlay models of user knowledge were introduced and developed in the field of ITS where overlay models were used mainly by systems with task sequencing, curriculum sequencing, and instructional planning functionalities. The popularity of this approach among early AES and AHS systems can be explained by their strong connection with ITS systems. In fact, a number of early AHS were developed in an attempt to extend an ITS system with hypertext functionality [8; 26; 75; 155]. The overlay knowledge models proved to be a good match for the core function of AHS: providing personalized access to information. As a result, within just a few years these models were accepted as de-facto standard by almost all educational and many non-educational adaptive hypermedia systems. This section provides more details on both components required for overlay knowledge modeling: the domain model and the overlay knowledge model.

The Domain Model. The heart of the overlay approach to knowledge modeling is a structured *domain model* that decomposes the body of knowledge about the domain into a set of *domain knowledge elements*. These elements can be named differently in different systems—concepts, knowledge items, topics, knowledge elements, learning objectives, learning outcomes, but in all the cases they denote elementary fragments of domain knowledge or information. In this paper we will be referring to these fragments as *concepts*. Though this name is slightly misleading[1] it is currently the most popular way to name domain knowledge elements. Depending on the domain, the application area, and the choice of the designer, concepts can represent larger or smaller pieces of domain knowledge: from a relatively large chunk of knowledge (i.e., a topic) down to elementary facts [143; 171], rules [167], or constraints [134].

---

[1] The word "concepts" can cause someone to think that concepts can only represent fragments of conceptual knowledge. However, a concept is a general name that can denote a fragment of knowledge of any kind, including procedural knowledge.

The simplest form of domain model is formed by a set of independent (unrelated) concepts. This kind of model is called a *set model* or a *vector model* since the set of concepts has no internal structure [20]. Even this simple form of overlay modeling provides a powerful platform for maintaining a detailed picture of user knowledge that can support a fine grain adaptation in relatively large practical systems [22; 52]. The main problem of vector models is the lack of connection between concepts. A vector model can register user knowledge of a specific concept, but this does not help to model user knowledge of other concepts. As a result, when the number of concepts is large and the number of observations is not sufficient, the system is only able to predict user knowledge for a very small fraction of concepts, related to existing observations.

In a more advanced form of domain model concepts are connected to each other thus allowing some inter-concept inferencing. By its origin, it is possible to distinguish two main types of connected models. The first, relatively rare type is inspired by the ideas of instructional planning and used in a number of AES and educational AHS [32; 110; 117; 176]. The model is formed by a tree of educational objectives where larger objectives (starting with the whole course) are progressively decomposed into smaller objectives.

The second type is both more general and more popular. In this type of domain model, concepts can be connected by different kinds of relationships, thus forming an arbitrarily complex network. This kind of model (known as a *network model*) was inspired by research on semantic networks (Fig. 1.3). Network domain models were used in many AHS and AES, including several development frameworks [6; 24; 50; 83; 148; 160; 169; 183; 189; 199].

Existing systems use network models of various complexity, with one or more types of links. In educational AHS and AES, the most popular links are prerequisite links between concepts, representing the fact that one of the related concepts must be learned before another. Prerequisite links can support several adaptation and user modeling techniques. In many AHS, prerequisite linkage is the only relationship given between concepts [81; 86; 141; 148; 157; 204]. More advanced educational systems as well as adaptive information systems favor classical semantic links such as "is-a" and "part-of" [23; 51; 183; 184; 189; 199].

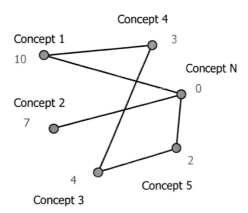

Fig. 1.3. A network domain model with a simple numeric overlay user model

All kinds of links between concepts are used to improve the precision of user modeling. When the user demonstrates a lack of knowledge, links can help to locate the most likely concepts that will remedy the situation. For example, if the user failed to answer a question that is based on the knowledge of several concepts, the concept that has fewer connections with the well-known concepts is the most likely source of problem. When the user demonstrates the presence of knowledge, links between concepts allow the propagation of the presence of knowledge beyond direct observation. For example, observing evidence of knowledge about a concept, the system can deduce the presence of knowledge for its prerequisite concepts. This approach is known as knowledge propagation. Different kinds of links may cause different types of knowledge propagation. An interesting approach for knowledge propagation is used in the AHA! system [48], which allows AHS authors to define new concept relationships types and their corresponding propagation semantics.

Recently, research on network domain models have been taken to the next level by several research teams that have attempted to use more formalized and more elaborated forms of network domain models inspired by the Semantic Web. The core idea here is to use formal domain *ontologies* [56; 58; 103; 120; 185] or domain *topic maps* [55; 146] in place of informal network domain models as a basis for overlay knowledge modeling. More information about the application of Semantic Web approaches in adaptive Web systems can be found in Chapter 23 of this book [57].

The Overlay Knowledge Model. The most important function of the domain model is to provide a framework for the representation of the user's domain knowledge using the overlay knowledge model. The key principle of the overlay model is that for each domain model concept, individual user knowledge model stores some data that is an estimation of the user knowledge level of this concept. The overlay model is powerful and flexible because it can measure independently the user's knowledge of different concepts. In the simplest (and oldest) form it is a binary value (known – not known) that enables the model to represent user's knowledge as an overlay of domain knowledge. This pure form of overlay model was used in several early AHS.

An extension of the pure overlay model is a *weighted overlay* model that can distinguish several levels of user's knowledge about each concept. This approach is used by the majority of modern AHS and AES. There are three popular forms of weighted overlay models: qualitative, simple numeric, and uncertainty-based that correspond to three approaches to user modeling, weight propagation, and adaptation. Qualitative models represent user knowledge of a concept as a qualitative value [22; 148] (for example, good-average-poor). These models are used mostly by systems with rule-based user modeling and adaptation components since qualitative values are very easy to update and use with rules. Simple numeric models use a quantitative value (for example, from 0 to 100) to represent the level of user knowledge [24; 51]. These models are explored by systems with simple algebraic approaches to knowledge modeling and propagation. The uncertainty-based models use different forms of uncertainty management such as Bayesian networks of fuzzy logics to model user knowledge. The user knowledge in these models are represented in the form dictated by the selected approach – most frequently, a probability that the user knows the concept [82; 177] or a probability distribution [3]. Due to its importance, the uncertainty-based approach is presented in detail in section 1.4 of this chapter.

One extension of the weighted overlay knowledge model is a *layered overlay model*. A layered model stores several values to represent user knowledge of each concept. Layered models were suggested to avoid mixing together the estimations of user knowledge obtained from different sources. For example, the system may choose to store separately the levels of user knowledge obtained by direct observation and by weight propagation [189; 203] or levels of user knowledge corresponding to different levels of concept mastery [27] such as Bloom's Taxonomy of Educational Objectives [13]. In these systems, different layers are maintained separately and mixed only in the process of adaptation decision making.

From the technical side, overlay knowledge models of individual users are typically stored as a set of name-value pairs where the names indicate domain model concepts and values represent the level of knowledge according to the representation formalism used in a system. Advanced overlay models such as layered models, fuzzy models, and models with probability distributions are stored as name-aspect-value triples. The details and the structure of the domain model is stored once in the system and is typically not replicated for individual users. As a result, the amount of information stored for each individual user is very small, even in systems with very large domain models.

1.3.2 Overlay Modeling of User Interests

Overlay approach to user interest modeling is very similar to overlay knowledge modeling. User interests are represented as an overlay of a concept-level model of the domain that the system covers. In general, domain models used for interest modeling and those used for knowledge modeling are slightly different in their structure and the size of concepts. However, this is caused mostly by the difference in system types: most of the domain models for knowledge modeling are developed for AES while most of the models for interest modeling are developed for adaptive information systems. This is rather a tradition than a necessity – a number of AES model user interests [85; 146] and many information systems model user knowledge [3; 143]. Conceptually domain models used for interest and knowledge modeling are compatible: most of the models build for knowledge modeling can be used for interest modeling and vice versa. A number of sophisticated AHS take an advantage of this compatibility suggesting modeling both user interests and user knowledge as two separate overlays over the same network of concepts [3; 143; 146; 171].

The variety of known domain models used for interest tracking can be classified into three groups that bear analogy to similar models used for knowledge tracking. First group is formed by systems that track user interests using a vector model: a set of unrelated concepts. In these systems, each information object (a document or a fact) is associated with one or more concepts. A demonstrated interest in an object is modeled by increasing interest in corresponding concept(s). For example interest in a craft object made in a specific style may increase the level of interest in this style [143]. Despite its simplicity, the vector model allows a reliable prediction of user interests for items that are related to concepts with previously registered interests. However, vector models do not allow the prediction of user interests for previously unexplored concepts, which becomes a serious problem as the set of concepts grows. Nowadays vector models are mostly replaced by two kinds of connected models – a taxonomy model and an ontology model. A *taxonomy* model is formed by a classification hierarchy of concepts that are frequently called topics, classes, or categories. Parents and children in this hierarchy are connected by topic-subtopic relationships. This kind of model is preferred by systems with expandable corpus such as Web directories or adaptive news systems [3; 128; 187] as well as similar open corpus systems reviewed in Chapter 2 [72] of this book. A new resource could be easily integrated into a hierarchy of concepts by attributing it to one of the leaf concepts (this classification can be done either manually by content provider or automatically). To increase the power of this model, it is possible to use more than one classification hierarchy to model different aspects of user interests. However, with some exclusion [96], this faceted classification is used only in closed corpus systems since it requires classifying each item along each of the taxonomies.

In an *ontology* model of interests the concepts form a rich network connected by different kinds of links. Most typical links are the same as in a network modeling of knowledge: is-a, part-of, and similarity. This kind of model is preferred by closed corpus adaptive information systems such as museum or tourist guides [171] or store catalogues [4]. The ontology model provides a richer representation of the world behind the information system and allows better interest tracking and propagation. Note that many adaptive systems that claim to use ontology for interest tracking are really using taxonomies that could be considered as a simple case of ontology. In contrast, some systems attempt to use very sophisticated ontologies such as WordNet to track user interests [118].

Both groups of domain models that represent links between concepts allow inter-concept interest propagation that is very important to fight sparsity and to increase the precision of interest tracking. In simple hierarchical domain models interest is typically propagated from child to parent concepts. For example, observed interest in *machine learning* causes increase of user interest in the parent concept *artificial intelligence* [128]. Richer domain models based on concept networks and "true" ontologies allow broader interest propagation. Typically these approaches ignore the nature of the links: if the system registers user interest in a specific concept, it assumes that related concepts in the domain model are also of some interest to the user [170; 187]. Some recent advanced approaches attempt to increase the precision of interest propagation using rules that take into account the structure and the type of connection in the domain model. For example, if a user is interested in a driver of a specific racing team, the user is probably interested in the team itself [43].

1.3.3 Generalized Overlay Models

User features beyond knowledge and interests are not formally modeled as overlays of domain models; however the approaches to modeling such features as goals, backgrounds and even whole stereotypes bear some reasonable similarity to the overlay approach. The framework for modeling these features is based on a space of possible characteristics (a set of possible goals, or a set of possible stereotypes) that is technically similar to a space of all concepts in structured domain models. To reveal this similarity, [19] suggested considering *generalized domain models* and *generalized overlay models*. A generalized domain model is a set of *aspects* where aspects can represent any characteristics of the user such as domain concepts, domain tasks and goals, and possible stereotypes. A generalized overlay user model is a set of pairs "aspect-value" in which the value in each pair can be "true" or "false" (indicating if the user has this characteristic) or some qualitative or quantitative value (measuring an extent to which user has this characteristic).

This generalization turned out to be very useful to find deep similarities between different AHS. Moreover, the ability to draw a similarity between approaches to model a specific user feature (such as background) and the very well explored overlay modeling approach could be instrumental for developing of more powerful and elaborated models. Below we provide some examples of using this similarity by re-considering approaches to goal and stereotype modeling.

Let's start with goal modeling. As reviewed in section 1.2.3, a system that is able to adapt to user goals typically uses a catalogue of user goals or tasks that the system can recognize. A traditional approach to goal modeling attempts to recognize the current goal in the catalogued set and mark it as current in the user model. Now let's consider the set of all goals to be a generalized domain model for goal modeling. As in traditional domain models, the set of goals can be unrelated or can form some connected structure – such as goal-subgoal hierarchy. In this context, the generalized overlay model of individual user can represent a probability that each specific goal in the generalized model is the current user goal. Now the traditional way of goal modeling by marking the current goal becomes simply a very primitive case of this generalized goal model (current goal has overlay value 1 and the others zero). Using this framework in full power allows building more elaborated goal models that can reflect uncertainty in modeling goals. In addition, recognizing the similarity with overlay knowledge modeling allows to re-use powerful uncertainty-based modeling and propagation techniques. Indeed, a few adaptive systems use some elements of this generalized vision to provide a more elaborated goal modeling. For example, some systems maintain a hierarchy of possible goals or tasks [198] and a few pioneer systems use probabilistic overlay modeling where each possible goal in the catalogue is represented by the system as a probability that it is the current user goal [63; 78; 126].

A similar generalization can be used for representing user stereotypes, backgrounds, or individual traits. For example, a stereotype user model is based on a collection of possible user stereotypes. A typical stereotype-based system attempts to recognize current stereotype, mark it as current in the user model and use for adaptation. An application of a generalized overlay approach allows us to treat a set of all possible stereotypes as a generalized domain model. A generalized overlay model now allows handling uncertainty of stereotype recognition as well as to represent more sophisticated cases of stereotype modeling such as the ability of the user to belong to several stereotypes at the same time. With this generalization, stereotype user models can be represented as a set of pairs "stereotype-value," where the value of each pair can either be "true" or "false" (which means that the user belongs or does not belong to the stereotype) or some probabilistic value (which represents the probability that the user belongs to the stereotype). A good case for maintaining a probabilistic overlay for modeling user stereotypes is provided by SeAN system [3] that uses a popular combination of stereotype and overlay modeling. SeAN attempts to map the set of demographic features provided by the user into a set of predefined stereotypes that are used to *seed* the overlay model of user knowledge and interests. Unlike typical stereotype modeling systems, SeAN is not trying to come up with the single, best stereotype for each user, but maintains a probabilistic overlay. This helps the system to achieve a more precise seeding of the overlay model.

The ability to view models of different user characteristics as generalized overlay models allows developers to better understand the similarity between different systems and to reuse more broadly the representations and user modeling approaches developed in the field. For example, as demonstrated in the next section of this chapter, elaborated approaches for knowledge modeling based on Bayesian networks could applied for modeling other user features such as individual traits.

On the practical side, the vision of generalized overlay models is currently supported by some universal authoring frameworks for developing AHS. The best example of such a universal framework is AHA! [49; 50]. AHA! allows the author of an AHS to introduce generalized concepts and relationships between the concepts. The concepts in AHA! can represent different user aspects such as knowledge, goals, interest, etc. For each concept, the system maintains an individual numeric overlay value. The semantics of the concepts, the connections, and the meaning of specific overlay values are encoded by the author into a set of user modeling rules and weight propagation rules. The universal nature of AHA! made it the most popular authoring tool in the field of adaptive hypermedia. So far, AHA! has already been used for modeling such user features as knowledge, interests, and even learning styles [181].

1.4   Uncertainty-Based User Modeling for Adaptive Hypermedia and Adaptive Educational Systems

There is no doubt that in the case of user modeling, there is often the need to deal with information that is uncertain (we are not sure that the available information is absolutely true) and/or imprecise (the values handled are not completely defined). An example of a statement that we need to deal with would be: "the user failed this question, so *most probably* he/she doesn't know concept C," which is *uncertain* information. On the other hand, if we say, "the user has been reading about concept C for *quite* a long time" we are making an *imprecise* observation. Obviously, user modeling is a domain in which there are many different sources of uncertainty and/or imprecision, therefore numerically-approximate reasoning techniques are suitable for this purpose. The two more commonly used in this context are Bayesian Networks and Fuzzy Logic.

Bayesian Networks (BNs) were developed in the eighties by Judea Pearl [153] and since then there has been an increasing interest and enormous progress in the development of new techniques and algorithms, extensions to the model, and applications. BNs are a probabilistic model inspired by causality and provide a graphical model in which each node represents a variable and each link represents a causal influence relationship. Currently they are considered one of the best techniques available for diagnosis and classification problems. The Fuzzy Logic (FL) paradigm was originally proposed by Professor Lofti Zadeh in 1965 [207], and the basic idea is to allow membership functions (in Fuzzy Sets theory) or truth values (in Fuzzy Logic) to take values between 0 and 1, with 0 representing absolute falseness and 1 absolute truth. As in BNs, FL has been the object of intense research, both in the development of models and in its application to real problems, with special emphasis on control problems. Therefore, it is no surprise that in the field of adaptive systems researchers have used such models, since the development of an adaptive system involves diagnosis, classification, and control.

However, in the field of adaptive hypermedia there are few studies that report the use of approximate reasoning techniques (notable exceptions are [82] for BNs and [99] for FL). Probably, the reason is that researchers are devoting much more attention to other relevant aspects such as foundation and core techniques (data mining, meta-data, semantic web, intelligent web agents, web services, etc.) or practical aspects (architectures, privacy, security, usability, evaluation, etc). The situation might be comparable to what happened in other related fields ten years ago (for example, in the field of Intelligent Tutoring Systems only four of the papers presented at ITS'1996 reported the use of such techniques (all of them used BNs), while at ITS'2006, eight papers use BNs and one paper uses FL. Our prediction is that in the future the number of adaptive web-based systems that use BNs or FL will continue to grow.

Nowadays, most researchers who decide to use approximate reasoning techniques in their user models choose the Bayesian network paradigm. The reason for this is the suitability for diagnosis. Also, the fast development of this field over the last twenty years has resulted in an increasing range of techniques and tools for Bayesian reasoning. In this sense, tools like GeNIe (SMILE), HUGIN, or JAVABAYES allow for the easy definition of Bayesian models and seamless integration of learning and updating algorithms. Still, there are also a number of systems that use other approaches like Fuzzy Logic [37; 98; 108] and, more recently, neuro-fuzzy approaches [182]. The interested reader can see [90] for an excellent review of former systems and issues in the field of user modeling, and [99] for some discussion about fuzzy user modeling in the context of adaptive web-based applications.

At the time of this writing, most solid experience in Bayesian user modeling has been accumulated in the area of modeling users that interact with educational systems, also known as student modeling. Student modeling was always the most area for AES and AHS and it was also the area where new technologies were developed and explored before being applied for other types of user modeling. Therefore, the core part of this chapter is focused on Bayesian student models (BSMs), beginning with the modeling of user knowledge. We will then briefly discuss the use of BNs beyond modeling knowledge. Our goal is to explain in detail how to define a Bayesian student model (BSM), which student features can be modeled, and how the parameters of the model can be obtained. The use of BSM in adaptive web-based systems is still quite unusual. Because of this, we will also refer to more relevant models that have been deployed in stand-alone tutoring systems and learning environments. We do this to illustrate how such techniques might be employed in adaptive web-based educational systems following the success of stand-alone implementations.

1.4.1 Basics of BNs

A Bayesian Network can be formally defined as:

- A set V of propositional variables $(X1, ..., Xn)$ (nodes of the network)
- A set E of probabilistic relationships between the variables (arcs of the network)

Such that:

- The graph G = (V,E) is an acyclic directed graph.
- The conditional independence assumptions are satisfied, i.e., each node Xi is conditionally independent from the rest of the nodes (except from its descendants), given the state of its parents.

With respect to nomenclature, the words "node" and "variable" are used interchangeably in literature. The probabilistic relationship between variables is usually referred to as an "arc" or "link."

Though the relationship between the variables does not need to be causal in nature, the BN paradigm is very suitable for that case. Therefore, building a model usually involves putting the information available into a cause-effect schema. Regarding the probability distribution, it can be easily shown that if the conditional independence assumptions are satisfied, then the joint probability distribution is given as the product of the probability distribution of each node given its parents. Because of that, in the case of BN it is enough to provide the conditional probability distribution of each node given its parents (or prior probability distribution for root nodes) to make *any* kind of inference needed. In that sense, a powerful feature of BNs is that they allow for diagnosis (inferences about possible causes of an event) and prediction (future state/evolution of variables given evidence).

To illustrate the concepts presented above, let us consider the following example in the context of student modeling. In order to determine if a student knows a certain domain element K, we can use the result of a certain event E that provides evidence about it. This evidence-bearing event can be an answer to a test item, solution of a problem, teacher's opinion, the number of Web pages relevant to the element K that have been visited, etc. In that case, the nodes of the network are: node K (having knowledge about a domain element K), and node E (result of the evidential event E). In the simplest case, both variables are binary: K can be *known* or *not_known*, while the evidence provided by E can be *positive* or *negative*. To simplify the notation, we will denote the positive states (*known*, *positive*) by 1 and the negative states (*not_known*, *negative*) by 0. There is only one relationship: the state of variable K influences (in this case, causally) the result of the event E (Fig. 1.4).

Fig. 1.4. The simplest BN for knowledge diagnosis

Let's concretize our example. A sample concept could be "Adding natural numbers" and its possible values could be *known* and *not_known*. A possible question could be "3+4" which has the possible states *right* and *wrong* (Figure 1.5).

Fig. 1.5. A simple BN for diagnosing knowledge about adding natural numbers

Regarding the quantitative information (network parameters), we need to provide the conditional probability distribution of each node given its parents (for the nodes without parents, the a-priori probability distribution). So in the simplest general case, three probabilities need to be specified: $P(K=1)$, $P(E=1|K=1)$ and $P(E=1|K=0)$. Let us use the following values for these parameters:

- $P(K=1) = 0.2$, means that, in our experience (for instance, if we are a math teacher), the probability that a random student knows how to add integers is 0.2.
- $P(E=1|K=1) = 0.99$, means that the probability of giving a correct answer to "3+4" is 0.99 for a student who knows how to add integers. Conversely, although the student knows how to add, there is still the probability of 0.01 of giving an incorrect answer, allowing us to model what is called a "slip" in related literature.
- $P(E=1|K=0) = 0.02$, means that the probability of giving a correct answer to "3+4" is 0.02, if the student does not know how to add fractions. Conversely, even if a student does not know how to add there is still a certain probability of guessing the correct answer.

In the related fields of medical diagnosis and statistics such parameters have a clear meaning. For example, in the medical diagnosis domain, if D represents an disease and T a test used to diagnose it, the causal relationship is $D \rightarrow E$, and the parameters needed are: the a priori probability of the disease (which in medicine is called the *prevalence* of the disease), and the conditional probability distribution $P(T/D)$, more concretely $P(T=1/D=0)$, which is the rate of *false positives* (or *type II error* in statistics) and $P(T=0/D=1)$, which is the rate of *false negatives* (or *type I error* in statistics) of test T. Both measures are commonly used as indicators of the quality of the test T or its suitability for the disease D.

Of course, the more complex the structure of the network, the higher the number of parameters needed. In general, if a node N has n parent nodes $P_1, \ldots, P_n$ and each of them has k states, the number of parameters needed for node N is k^n, i.e., the number of parameters needed grows exponentially with the number of parents.

Once the network has been defined, it can be used to make inferences about the domain. Using a BN model allows us to reason both in the diagnostic direction (what are the more probable causes given certain evidences) and in the prediction direction (what is the probability that a given configuration of variables states will happen, given a set of evidence?). To illustrate this, we will add some more nodes to our example (Fig. 1.6).

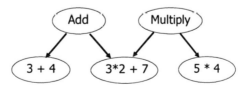

Fig. 1.6. Adding nodes to the basic BN

Let us imagine that our student gave a correct answer to "3+4." We can then reason in the diagnostic direction (i.e., compute the probability that he/she knows concept "Add") and also to predict a future result (i.e., compute the probability that he/she will be able to correctly solve "3*2+7"). This potential for doing both abductive and predictive reasoning is very useful in the case of student modeling. Not only does it allows us to accurately diagnose the student's current state of knowledge, but also to make an informed and adaptive recommendation for the most appropriate next activity (e.g. exercise, test) according to the evidence available about this particular student (previous activities with the system, favorite learning strategy, etc.). For example, for a student that likes to be challenged it might be better to select a "difficult" (lower probability of right answer) exercise while for a student that has low self-confidence it might be better to select an "easy" (higher probability of right answer) exercise.

As pointed out before, BN formalism implements certain independence assumptions amongst variables. Conditional independence modeled in a particular BN model should correspond to independence relationships existing in the real world. In our example, the conditional independence assumptions mean that the knowledge nodes "Add" and "Multiply" are independent a priori (which is reasonable, because if they are not the model should include a link between) and that each evidential node is conditionally independent of the rest of the evidential nodes given the values of its parents (which is also reasonable, because once the value of the corresponding parent nodes, i.e., knowledge nodes, is known, other evidential nodes do not provide information any more). For more information about what the conditional independence assumptions mean, see [140] or [153].

The construction of a student model based on BNs consists of two steps: (a) development of the *qualitative model*, that involves the definition of the structural model (nodes and arcs); and (b) development of the *quantitative model*, that involves the specification of the parameters needed (conditional and prior probability or distributions). There are two ways of obtaining the information needed in the above steps. The first way is to use domain expert's opinion to get the structure/parameters. The second is to infer them from data using BNs learning algorithms [79]. A combination of these two ways is also possible and often employed in practice. As far as learning is concerned, *structural learning* is a process of learning dependencies between the given variables (step a), while *parametric learning* is a process of learning the parameters describing the strength of the dependencies in the model (step b).

If a BN is built by an expert, it is necessary to make sure that the model includes all relevant aspects of the real world, i.e., the model has to account for the relevant variables and relationships between them, and the specified conditional probabilities have to represent the strength of these relationships. On the other hand, if a BN is to be learned from data, the relationships learned on the significance level fixed for the statistical independence tests performed during structural learning. If the significance level is low we might miss some weak relationships while if the significance level is high some hard-to-interpret relationships might be inferred or assumed. A possible solution is to have an expert to specify the structure (or fine-tune the learned structure) and then learn parameters from data.

The structure of the rest of section 1.4 is as follows: In subsection 1.4.2, we will discuss the development of the quantitative model under the assumption that the structure of a BN is to be built by an expert. To this end, we will discuss which aspects of the real world being modeled (in our case, a student) can be taken into account and go through different modeling options[2]. Subsection 1.4.3 gives some clues for defining qualitative modeling, while subsection 1.4.4 discusses the combination of expert knowledge and learning techniques in building BN student models. Section we present some conclusions.

1.4.2 Development of the Qualitative Model

As in any modeling process, the first important step is to specify aspects of the world being modeled which should be represented in the model, in order to achieve the defined goals. In the case of student modeling, the range of features to be modeled is wide and can be grouped into two main categories: knowledge (declarative, procedural, based on skills, competencies, etc.) and user features (learning styles, cognitive and meta-cognitive skills, emotional states, etc). The more user features are represented in the model the more complete it will be. However, the cost of creating and maintaining a more complete model should be carefully balanced against the usefulness of such a model. In the user modeling field, the goal is usually to have a personalized model of the user, in order to provide the basis for adaptation. So we must carefully choose the user features that will be useful to provide adaptation, because if we are going to consider k different variables and each variable can take n different values, then we will need to provide adaptive strategies for the k^n different profiles of students generated by the combination of such values.

To begin with, we will explain how to use BNs to model student's knowledge based on overlay models. Then we will present how to build knowledge models for other approaches such us constraint-based models and misconceptions models. Next, we will discuss how to model features beyond knowledge, such as meta-cognitive skills, personality traits, affective states, and attitudes and perceptions. Finally, we mention some examples where the qualitative model has been developed by experts and some other examples in which learning techniques have been used to infer it from available data. Along the discussion we will present examples of application, which will illustrate the main ideas and concepts.

Modeling Knowledge Based on Overlay Models with Bayesian Networks. As explained before, the first step consists in selecting the variables of interest for our model, which in this case are the knowledge variables and the sources of evidence that will be used to obtain relevant information about them. Each one of them will be modeled as a random variable, and represented as node in the BN. In the case of user modeling, typically the sources of evidence are the results of interaction with the system (e.g. answers to questions or exercises, time spent reading certain content, number of clicks, etc.). Open/scrutable user models allow that either the user or the expert can see and modify this information.

Once the relevant aspects of the real world being modeled have been selected, it is time to translate them to a mathematical model. In the case of BNs, this means that we need to structure this information in a causal relationship schema. To this end, a couple of important aspects related to building BN models should be clarified: first, that each node must represent a propositional value; and second, that the proper

---

[2] A good tutorial about modeling with BNs for troubleshooting problems is [114].

direction of the links needs to be established. We will discuss each of these aspects in more detail.

As aforementioned, each node must represent a *propositional* variable [140], which is defined to be as a variable that takes an *exhaustive* and *mutually exclusive* number of values. This aspect allows us to determine whether a given element of the real world should be modeled as a variable or as a state of a variable. Let us illustrate this using some examples.

Example 1: A Medical Diagnosis Problem

A doctor is trying to diagnose a patient to determine the illness (or illnesses) that he/she is suffering from. Let us define a finite set of illnesses as $\{I_1, ..., I_n\}$. In this case there are two modeling options: (A) to consider a node I with values $I_1, ..., I_n$, and (B) consists in considering n binary nodes $I_1, ..., I_n$. In the case of medical diagnosis, the choice depends on whether or not more than one illness can be present at the same time. If the patient can have one and only one illness, the correct option is A. If the patient can have two or more illnesses at the same time, the correct option is B. It is important to point out that in option A all the possible illnesses must be considered within the set of possible values $\{I_1, ..., I_n\}$, and if this is not possible, one of the states should be labeled "other" to account for any illness not taken into account in the model (thus making the set of states exhaustive). In option B, a node "other" can be included, but it is not necessary.

Example 2: Classification Problems

We have to classify a set of objects as belonging to certain categories $C_1, ..., C_n$. Again, the decision of using a single node C with values $C_1, ..., C_n$ (option A) or n separate binary nodes $C_1, ..., C_n$ (option B). The decision depends on whether or not an object can belong to more than one category. So if we are classifying animals in mammals, amphibians, reptiles, etc. we should choose option A (with the additional value "others" if needed). However, if we are classifying persons according to their educational background into doctors, architects, lawyers, etc. we should use option B.

Example 3. The Use of the Two Different Options in User Modeling

This example illustrates how the two different options presented above have been used in two real systems: KBS Hyperbook [82] and the English tutor I-Peter [161]. In both systems, users are being classified according to their knowledge level K into the categories novice (N), beginner (B), intermediate (I), and advanced (A). But while KBS Hyperbook models student's competence in each knowledge item i by using a node K_i that takes values $\{N, B, I, A\}$, I-Peter models student's competence in English by using four separate binary nodes N, B, I, A. In the second system, the underlying assumption is that a student can belong to more than one category simultaneously. A possible interpretation is that the I-Peter model might be intended to allow smooth transitions between categories. For instance, a possible interpretation of $P(N=Yes)=0.2$ and $P(B=Yes)=0.6$ is that the user is gradually leaving the category *novice* to get to the more advanced state *beginner* (quite like in a Fuzzy Logic model). But other combinations of probability values might be harder to interpret, and therefore estimating the parameters needed might be difficult in I-Peter.

Fig. 1.7. A BN and the equivalent BN with the arc $C_1 \rightarrow$ E reversed

As already mentioned, a second important decision is the more appropriate direction for the arcs in our model, i.e., given two propositional variables X and Y, we need to decide if we will represent the relationship between them as X $\rightarrow$ Y or X $\leftarrow$ Y. This is an especially important aspect, because in spite of the fact that the direction of any arc in a BN can always be reversed [89], BNs are very suitable for the case of causal relationships and therefore the model should stay as close to the notion of causality as possible. It is important to note that if the direction of an arc is reversed new arcs are introduced to the model. This is necessary for the new network to keep the same dependency structure. Fig. 1.7 shows the arc reversal result on a simple network.

Any of these two models would correctly represent the relationships between the two causes and the effect, but the second one (after an arc has been reversed) needs more parameters (which would also be harder to estimate and interpret). So using the proper direction of the arcs saves work and makes the definition of the model easier. To be able to choose the proper direction, we should think about which variables have causal influence in which others. For instance, if X is *type_of_animal* and Y represents a characteristic of an animal like *presence_of_feathers*, it seems clear that the type of animal has *causal influence* on the presence of feathers, and therefore the more adequate option is X $\rightarrow$ Y. However, it seems that humans find it easier to structure the information in terms of IF-THEN rules rather than in terms of causal relationships. Because of that, it is quite common to see the other option, perhaps influenced by the implicit construction in our mental model of the corresponding diagnostic rule (IF it has feathers, THEN it is an animal) instead of the causal relationship (it has feathers BECAUSE it is an animal).

In the case of student modeling, the basic overlay knowledge model usually includes the following relationships: granularity relationships, relationships between knowledge and evidential nodes, and prerequisite relationships. Examples of Bayesian student models that use both modeling options (arcs in both causal and diagnostic directions) for each of such relationships can be easily found in literature and will be discussed in the following paragraphs. We continue using the generic notation introduced in section 1.4.1: K will represent a knowledge node, while E will represent any event that can provide information (evidence) about the student's knowledge K (e.g., a question, test, exercise, task, etc.).

For modeling the relationships between the evidential nodes E and the knowledge node K, the options are:

(o_1) causal direction, i.e., K $\rightarrow$ E; or
(o_2) diagnostic direction, E $\rightarrow$ K.

Option o_2 could be also interpreted as a causal relationship if the variable K is considered to be a measure of how well the student performed in a test. But usually the goal is to evaluate the knowledge of the student and not his/her performance. In fact, the performance can be seen as an effect of the knowledge, together with many other causes such us the emotional state of the student, the difficulty of the evidential task being performed, etc.

Let us see an example in the context of the adding/multiplying problem. The modeling options are as depicted in Fig. 1.8. If we use option (o_1), the nodes "Add" and "Multiply" represent the knowledge that the student has about those two operations. In option (o_2) the nodes "Add" and "Multiply" represent an artificial constructs that measures how well the student did in the test involving such operations.

Though this section is devoted only to qualitative model, thinking about the parameters needed in each option can help in taking a decision. In option (o_1) we need to estimate the prior probability distributions P(Add), P(Multiply), and the conditional probability distributions P((3+4)/Add), P((3*2+7)/Add, Multiply), and P((5*4)/Multiply). All of these parameters have an intuitive meaning: the priors represent the "prevalence" of the two basic operations in the population under study, i.e., how many of our target students know them, while the conditionals represent the probability of a correct answer given that the knowledge required has/has not been acquired. In option (o_2), we would need the priors P(3+4), P(3*2+7) and P(5*4), which can be meaningfully estimated, but also the conditionals P(Add/(3+4), (3*2+7)), P(Multiply/(3*2+7),(3+4)), which seem harder to interpret. For example, what is the probability that the student "performed well" in the test with respect to the operation "Add" given that his/her answer to 3+4 was incorrect and to 3*2+7 was correct?

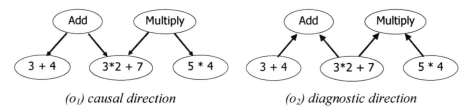

(*o_1*) *causal direction*   (*o_2*) *diagnostic direction*

Fig. 1.8. Options available for modeling relationships between knowledge and evidential nodes: an example

Both modeling options are present in the literature. In [196] evidence is provided by problems that the student needs to solve and knowledge nodes represent the physics rules that the student needs to apply. The conclusion reached in the paper is that if links are defined in the diagnostic direction it "means that the rules are conditionally independent given the evidence, which just isn't true (see [196] for an example). Therefore, the causal direction seems preferable in this context. The same conclusion is drawn in [130] that theoretically compares both options in terms of criteria such as knowledge representation, independence relationships, reasoning process, and knowledge engineering effort required. More recently, [119] performed an empirical study with the goal to compare the accuracy reached with each modeling option in the diagnosis of a group of real students, reporting a maximum accuracy of 0.508 if o_2 (the diagnostic direction) is used (no better than chance, in spite of their efforts of fine-tune the model using data) versus an average test set accuracy of 0.776 of o_1 (causal direction).

With respect to granularity relationships, options available are:

(o_3) the causal direction, i.e., $K_1 \rightarrow K$; or
(o_4) the diagnostic direction $K \rightarrow K_1$.

The first option (o_3) assumes that knowing a component indicates knowledge about the whole, while the second option (o_4) states that knowing the whole has influence in knowing each of its components.

In our example, we can define an aggregated knowledge element "Basic arithmetic," which would be divided into "Add" and "Multiply." These two modeling options are depicted in Fig. 1.9.

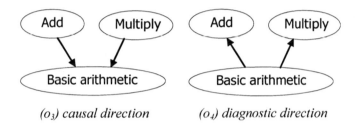

(o_3) causal direction   *(o_4) diagnostic direction*

Fig. 1.9. Two options for modeling granularity relationships

The interpretation of both options follows. The implicit consideration under (o_3) is that if a student knows basic arithmetic, then he/she also knows how to add and how to multiply, while (o_4) assumes that a student has knowledge of basic arithmetic *because* he/she knows how to add and how to multiply.

In the context of granularity relationships, examples of BNs with links defined in the diagnostic direction (o_4) are: [82; 132; 139], while [119; 122; 130; 195; 208] use links defined in the causal direction (o_3). The first theoretical comparison between both options can be found in [90], which discusses some implications but does not explicitly recommend any of them. The causal direction (o_3) is supported by theoretical studies such us [130], which compare both options in terms of the same criteria described above, and empirical studies such as [39], which evaluate three different course hierarchies in the context of adaptive testing, using quality measures such as *test length* and *test coverage*.

Another interesting alternative for the modeling of relationships between knowledge and evidential nodes are dynamic models, which allows for changing the variables over time. This is a desirable characteristic in the context of student modeling since the knowledge and other relevant variables usually change continuously during a student's interaction with the system. In the *Dynamic Bayesian Networks* (DBNs) model [168], a separate BN is constructed for each time slice. This approach is also taken in [119; 122; 130; 209], probably inspired by Reye's work [162]. Reye presents a simple model based on DBN, which captures the dynamic nature of the student's knowledge. In this proposal, for each j = 1, ..., n, ..., the following nodes and relationships between them are defined (Fig. 1.10):

L_j = student's state of knowledge after the j'th interaction with the system
O_j = result of the j'th interaction.

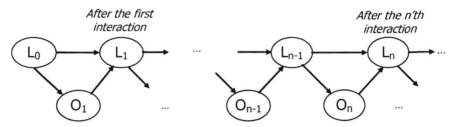

Fig. 1.10. Dynamic BN for student modeling. Adapted from [162]

We can see that at time n, the probability that the student is now in the learned state (L_n) depends on whether the student was already in the learned state at n-1 (L_{n-1}) and on the outcome of the last interaction with the system (O_n). Reye provides formulas for a two-phase updating of this basic model and proves that the knowledge tracing model used in the ACT cognitive tutors [44; 45] can be viewed as a particular case of this more general model (by adding constraints to the general model until the equivalence is shown). Please note that in spite of the fact that this description seems to refer to only one knowledge node, the implementations of this model will contain as many knowledge nodes as needed, resulting in more complex networks. A later study [131] makes the empirical comparison of a static BN with a DBN using simulated students, showing that both models produce a very similar performance in terms of test length and accuracy, although the static model seems to perform slightly better.

In summary, it seems that evidence available (both theoretical and empirical) encourages the use of the causal direction. The arc should then go from variables that cannot be observed directly and need to be estimated (knowledge in the case of student modeling, illness or disease in the case of medical diagnosis, faulty components in the case of trouble-shooting systems) to variables that can be observed (answers to problems, symptoms, or problems, respectively).

With respect to prerequisite relationships, it seems clear that if A is a prerequisite of B, knowing A has *causal influence* on knowing B, so the direction of the link should be A→ B. However, the main difficulty of introducing this new kind of relationship into our model is that, as reported in [31], the meaning of the relationships between the nodes becomes somehow unclear and the specification of the parameters becomes more difficult. To illustrate this, let us consider the following example: in a basic arithmetic course, students are taught how to add and multiply natural numbers (N) and fractions (Q). A basic overlay BN student model for this course (with links in the causal direction) is given in Fig. 1.11 where prerequisite relationships are represented as a dotted line.

In this example, one of the parameters needed is P(Multiply/Mult_N, Mult_Q, Add). But the fact that different types of relationships are mixed in the conditional distribution makes this probability difficult to estimate. Just as an example, we would need to provide the probability of knowing how to multiply, given that "the student knows how to multiply natural numbers and fractions, but does not now how to add," which is quite improbable. As suggested in [31], a possible solution is to disregard that kind of relationship in the model (thus making a simplification of reality, in which prerequisite relationships do exist). Another possible solution is to make a different simplifying assumption: instead of not including prerequisites, consider that both relationships operate at different levels, i.e., to build a multi-layered student model as proposed by [200]. In [31], Carmona et al present an empirical study comparing both alternatives. The study supports a conclusion previously established in [59]: "not considering valid prerequisites relationships does not lead to a wrong assessment of a student's knowledge state, but it renders the assessment less efficient in the sense that more answers than necessary have to be collected." So, if the test length is not an important issue, eliminating prerequisites will not affect the accuracy of our assessment, but if length is important, a possible option would be to use a multilayered model which can improve the performance (in the sense that less questions will be needed to reach the same accuracy for our estimation of the student's knowledge).

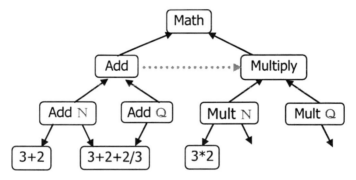

Fig. 1.11. Introducing prerequisites in the BN

Another option for considering prerequisite relationships is presented in [82] and is based on categorizing the knowledge nodes into levels. The first level corresponds to simple concepts, which do not require prerequisites. The successive levels require that knowledge about some of the concepts in the previous levels be understood. In this application, prerequisite relationships are only allowed between the root nodes of each level (recall that this system uses the diagnostic direction (o_4) for the granularity relationships, so the root nodes are the compound concepts). In this way, the problem of confusing different kinds of relationships in the model can be avoided.

Modeling Knowledge Beyond Overlay Models. There are also some proposals to model student knowledge with BNs that are not based on overlay models. Alternative approaches include constraint-based modeling and models based on misconceptions. A good example of constraint-based modeling is the student model of CAPIT [122], a constraint-based tutor for English capitalization and punctuation. In this model each node L_i represents the outcome of the student's last attempt with respect to the violation or not of a certain constraint I. That attempt can take values S (satisfied), V (violated), VFB (violated with feedback) or NR (not relevant). When a new problem is presented, node N_i represents the predicted outcome of this new attempt with respect to the violation of constraint i. The authors decided to include relationships between the constraints, which caused the model to grow more complex but was more realistic and allowed for a better prediction of student performance. The model also includes decision nodes and their utilities, i.e., transforming the BN in an Influence Diagram or Decision Network (DN). This allows for the selection of the next tutorial action by maximizing the expected utility. A similar approach based on DN has also been presented in [138].

As for bug models, a good example is the student model for a tutoring system on decimal numbers presented in [180]. In this case there are several types of nodes: *evidential nodes*, which can be either test items of a decimal comparison test (nodes TI) or variables that account for student behavior during the interaction with the system (nodes B) and *student-type nodes*, which serve the purpose of classification of students according to the type of misconception they have. This classification happens at two levels of granularity: coarse and fine. The basic structure of the network is as represented in Figure 1.12.

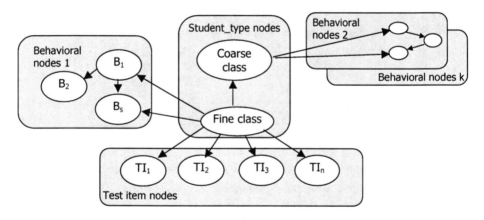

Fig. 1.12. A Bayesian student model based on misconceptions

This model features several groups of behavioral nodes. Those groups can be of different sizes and can have different internal structures (i.e., they can be connected in different ways). They model different instructional episodes. Please note that there is only one group of test item nodes and student type nodes. As it can be seen, in this student model the links follow the causal direction. It is important to note that the performance of this student model has been evaluated by comparing the results of manual expert diagnosis with the automatic system's diagnosis. The comparison took into account the results of more than 2,000 students who took the DCT test and showed an 80-90% agreement rate (the 10% variation was due to varying the values of certain parameters).

Beyond Modeling Knowledge. The basic knowledge models presented above can be enhanced by including information that allows for *model-tracing* (in addition to knowledge-tracing). This can be achieved by adding other variables like student goals, plans, etc. An example of the second kind of model is the Bayesian student model for the physics domain in the tutoring system Andes [40], which divides knowledge nodes into proposition nodes (fact and goal nodes), rule application nodes, and strategy nodes (which allow for modeling the different methods used to solve a problem, and therefore for model tracing). A similar approach was later adopted in [186] for the domain of medical diagnosis. The architecture of the Bayesian student model of the Andes system is depicted in Fig. 1.13.

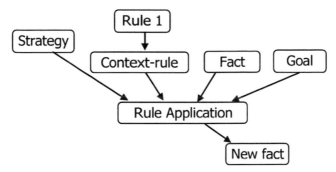

Fig. 1.13. General BN model for facts, rules, strategies, and goals

For example, in the physics domain, if the student knows a certain rule (such as F=m*a) he/she will be able to use it in any context. Then, if he/she also knows certain facts (for example, m=50kg and a=4m/s^2), has a certain goal (compute the force) and a certain strategy (way of solving the problem), then the rule will be applied to the problem and a new fact will be inferred (in this case, that F=200N). In the medical diagnosis example, the idea of "rule" is replaced by "medical concept" that can be applied to derive a hypothesis (equivalent to facts in the physical domain). In both domains (physics and medicine), the application, goal and apply nodes are used in the same way. Strategies are modeled by the fact that students can enumerate the causal hypothesis structure in any order. That's why they are not explicitly introduced as nodes in the model. One difference between these two domains is that causal relationships exist between hypotheses in the medical application but do not exist between facts in the physics tutor.

Recently, there has been increasing interest in widening the range of user characteristics that can be measured and used in adaptive systems. In this sense, BNs have been used to model the student beyond cognitive features (knowledge, goals) towards a richer variety of features, like:

- Meta-cognitive skills, for example, self-explanation [41] and exploration [29]. To support the evaluation of meta-cognitive skills the BN student model includes nodes to represent student's tendency to self-explain, to explore, and also nodes to model student's actions that are indicative of relevant applications of such abilities. Another recent example of modeling a meta-cognitive skill is [136], which presents a computational framework to support learning by using analogical problem solving. To this end, new nodes have been introduced in the Andes Bayesian student model: copy nodes (with values *correct, incorrect, no_copy*) which represent if the student copied the solution from the solved example; similarity nodes (with values *trivial, non_trivial, none*) that measure the degree of similarity of fact nodes between the example solution and the student's solution; analogy nodes (that represent the student's tendency to use the minimum or maximum analogies, i.e., to try to solve the problem on their own and consult the example only when reaching an impasse or to copy as much as possible), and an EBLC node, which allows modeling of the user's tendency to learn by the Explanation Based Learning of Correctness.

- Personality traits. An example is [68], that presented a BN to model learning styles within a web-based education system. To this end, they consider three dimensions of Felder's framework [64], namely perception (sensory/intuitive), processing (active/reflective), and understanding (sequential/global) as unobservable nodes, and several evidential nodes, such as the use of mail, forum, chats, number of examples visited, and exam results. In this case they are using the diagnostic instead of the causal direction for links between variables.

- Affective states. An example is the affective user model in the educational game Prime Climb, a game designed to help children learn about number factorization while being coached by an intelligent agent [42]. This affective user model has nodes such as: goal nodes (that model the objective during game playing, e.g., learn math, have fun, beat partner), action nodes (for both the player and the action), goal satisfaction nodes (to model the degree of satisfaction that each action causes), and emotional nodes that allow for the modeling of six of the 22 emotions described in the OCC theory of emotions [147], namely, joy/distress (states of the node "emotion for the game"), pride/shame (states of the node "emotion for self") and admiration/reproach (states of the node "emotion for agent"). In this case, links are established in the causal direction.

- Attitudes, perceptions. In [7], log-data is used to infer the student's hidden attitudes towards learning, learning gains, and perception of the system. To this end, the student model contains unobservable variables that measure if the students liked the system, learned, seriously tried to learn, wanted to finish quickly, wanted to challenge himself/herself, was concerned with getting help, had a fear of doing the wrong action, etc. Observable variables were: average of hints asked per problem, time between attempts, average seconds per problem, etc. In that model also, links are pointed in the causal direction.

Going beyond modeling the student's knowledge can certainly provide a much better adaptation and therefore better performance in terms of learning gains. However, as explained in this section, the relationship between the cost of building, maintaining and effectively using such enriched student models must be evaluated on an individual basis, to decide whether each one is worth the gained improvement in performance of the system.

Building the Qualitative Model Versus Using Learning Techniques. In the field of student modeling, many researchers have chosen to create structural models, with or without the help of domain experts. Good examples are Andes [40], HYDRIVE [132], Adele [67] and DT-Tutor [138]. However, the relationships between the variables of the model can also be learned from studying the data. To do so, the variables of interest, X_1, X_2, ..., X_n for the domain being modeled, must first be identified. From that list of variables and from a dataset composed of the sets of their values $(x_1, ..., x_n)$, a structural learning algorithm infers the relationships between the variables, according to a desired confidence level. A key point of this approach is to choose a good value for the confidence level. If too low a value is selected, low data-evidence relationships will be inferred. If the value is too high, some important relationships can be missed. A possible solution is having domain experts fine-tune the inferred model.

There are few examples in which the structure has been discovered from studying the data, but the interested reader can find an excellent one in [122], which describe the techniques used and the strengths and weaknesses of this approach.

1.4.3 Development of the Quantitative Model

Once the qualitative part of the model has been defined or learned, the next step is to define the parameters, which in this case are the prior probability distributions of the root nodes and the conditional probability distributions for the rest of the nodes. This task has commonly been cited as one of the main difficulties when building a BN model. The options available are: a) knowledge engineering, i.e., having experts specify the probabilities; b) using pre-existent models to specify part of the probability distributions needed (canonical models, theoretical models, etc.); and c) learning the parameters from available data (cases). A combination of these alternatives can also be used.

Examples of systems in which the parameters have been estimated by experts include those mentioned in subsection 1.4.2 as examples of systems in which experts provided also the qualitative model. However, there are other systems where the structure has been specified or constrained by experts and the parameters are partially or totally derived from data [201], or adjusted to theoretical models, for example, models inspired by Item Response Theory [130] and [90]. Some approaches to simplify parameter specification in the context of student modeling have also been proposed [129]. Finally, an example in which both the structure and parameters have been discovered from data is [122].

Also, some modeling tricks can help to relieve the burden of specifying the numbers. An example of such a trick is grouping the parent nodes into related causal categories, which reduces the number of parameters needed. For a more comprehensive literature guide to tricks and techniques, see [60].

1.4.4 Building Student Models by Combining the Domain Expert's Knowledge and Learning Methods

More recently, researchers have begun to use mixed approaches, in which parts of the model are defined by experts while others are learned from data. Some examples are:

- In the student model for the medical domain by Suebnukarn et al [186], the structure has been specified by experts and then the parameters are mined from the data obtained in transcripts of problem-based learning sessions.
- Stacey et al construct their Bayesian student model based on misconceptions through a combination of elicitation from the domain experts and automated methods [180].
- When developing the knowledge student model for Prime Climb, [119] compares the performance of two different structures (one developed following teacher's suggestions, the other one inspired by causality) and for both of them the parameters are revealed in the data.

- Arroyo and Woolf define the structure of the network, taking into account the knowledge gained during a correlation analysis between variables, and then look for conditional probabilities in the data to evaluate the accuracy of the model by cross validation [7].
- In [29] and [42], Conati and colleagues use an iterative design process: the initial model is developed using the researcher's intuition or the expert's opinions, and then data from real users is collected and used to refine both the parameters and the structure, either by adding new nodes (such as "general exploration" or nodes that describe attitudes towards the intelligent agent, such as "wanting help") or by modifying the direction of the links (for example, between knowledge and exploration nodes or game events and goals).

1.5 Conclusions

A common feature of various adaptive Web systems is the application of user models (also known as profiles) to adapt the systems' behavior to individual users. User models represent the information about users that is essential to support the adaptation functionality of the systems. Adaptive Web systems have investigated a range of approaches to user modeling, exploring how to organize the storage for user information, how to populate it with user data, and how to maintain the current state of the user. The majority of modern adaptive Web systems use feature-based approach to represent and model information about the users. The competing stereotype-based approach, once popular in the pre-Web area of adaptive interfaces, has lost dominance but is still applied, especially in combination with the feature-based approaches.

The most popular features modeled and used by adaptive Web systems are user knowledge, interests, goals, background, individual traits, and context of work. Each individual adaptive system typically uses a subset of this list, as determined by the class of adaptive systems it belongs to and the adaptation needs of this class (Fig. 1.14). Web-based adaptive educational systems (AES) rely mostly on user knowledge and learning goals capitalizing on the modeling and representation techniques established in the field of Intelligent Tutoring Systems (ITS). Adaptive information systems and Web recommenders focus on modeling the user's interests and extend

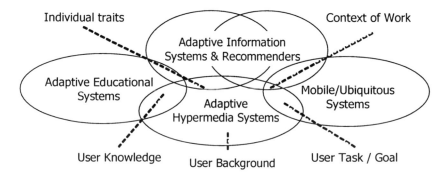

Fig. 1.14. User features typically modeled by different classes of adaptive Web systems

modeling approaches originally developed for adaptive information retrieval systems. Meanwhile, adaptive hypermedia systems attempt to represent and employ an even wider range of user features. In addition to user knowledge and interests, these systems frequently model user goals (following approaches developed in the field of adaptive interfaces), individual traits, and the context of user's work.

Overlay user modeling is currently the leading user modeling approach in AES and AHS. This approach was originally developed in the field of ITS to model user knowledge as an overlay of a concept-level domain model. Currently, the overlay approach has grown to include the modeling of knowledge, interests, goals, and some other features. In the area of adaptive information systems the concept-overlay approach to modeling user interests is now competing with the more traditional keyword-level profiling. This is one example of the convergence that has begun to blur the boundaries between different classes of adaptive Web systems.

One of the main advantages of using a formal method for student modeling is its robustness. Once this model behaves in a stable and theoretically-correct fashion, the evaluation of a system can be focused on other components (such as quality of the learning material, learning strategies used, or adaptation capabilities). The most commonly used formal reasoning technique for student modeling is currently the BN paradigm. Applications of other approaches (like Fuzzy Logic) are less frequent. A very attractive potential use of BNs in the context of adaptive web-based applications would be to employ learning algorithms that would further shape on-the-fly improvements within the model itself. The future adaptation learning algorithm would then be able to process each user's interactive behavior information and simultaneously update the structure of the model. We hope to see this potential use developed in the very near future.

References

1. Anderson, J.R., Corbett, A.T., Koedinger, K.R., Pelletier, R.: Cognitive tutors: Lessons learned. The Journal of the Learning Sciences 4, 2 (1995) 167-207
2. Anderson, J.R., Lebiere, C.: Atomic Components of Thought. Lawrence Erlbaum Associates, Hillsdale, NJ (1998)
3. Ardissono, L., Console, L., Torre, I.: An adaptive system for the personalised access to news. AI Communications 14 (2001) 129-147
4. Ardissono, L., Goy, A.: Tailoring the interaction with users in Web stores. User Modeling and User Adapted Interaction 10, 4 (2000) 251-303
5. Ardissono, L., Goy, A., Meo, R., Petrone, G., Console, L., Lesmo, L., Simone, C., Torasso, P.: A Configurable System for the Construction of Adaptive Virtual Stores. World Wide Web 2, 3 (1999) 143-159
6. Aroyo, L., Dicheva, D.A.: Concept-based approach to support learning in a Web-based support environment. In: Moore, J.D., Redfield, C.L., Johnson, W.L. (eds.) Proc. of AI-ED'2001. IOS Press (2001) 1-12
7. Arroyo, I., Woolf, B.P.: Inferring learning and attitudes from a Bayesian Network of log file data. In: Looi, C.-K., McCalla, G., Bredeweg, B., Breuker, J. (eds.) Proc. of 12th International Conference on Artificial Intelligence in Education, AIED'2005. IOS Press (2005) 33-40
8. Beaumont, I.: User modeling in the interactive anatomy tutoring system ANATOM-TUTOR. User Modeling and User-Adapted Interaction 4, 1 (1994) 21-45

9.   Beck, J.: Engagement tracing: using response times to model student disengagement. In: Looi, C.-K., McCalla, G., Bredeweg, B., Breuker, J. (eds.) Proc. of 12th International Conference on Artificial Intelligence in Education, AIED'2005. IOS Press (2005) 88-95

10. Benyon, D., Murray, D.: Experience with adaptive interfaces. The Computer Journal 31, 5 (1988) 465-473

11. Berti, S., Mori, G., Paternò, F., Santoro, C.: An environment for designing and developing multi-platform interactive applications. In: Ardissono, L., Goy, A. (eds.) Proc. of HCITALY'2003. University of Turin (2003) 7-16

12. Billsus, D., Pazzani, M.J.: A learning agent for wireless news access. In: Lieberman, H. (ed.) Proc. of 2000 International Conference on Intelligent User Interfaces. ACM Press (2000) 94-97

13. Bloom, B.S.: Taxonomy of Educational Objectives, Handbook I: The Cognitive Domain. David McKay Co Inc., New York (1956)

14. Boyle, C., Encarnacion, A.O.: MetaDoc: an adaptive hypertext reading system. User Modeling and User-Adapted Interaction 4, 1 (1994) 1-19

15. Brady, A., Conlan, O., Wade, V.: Dynamic Composition and Personalization of PDA-based eLearning – Personalized mLearning. In: Nall, J., Robson, R. (eds.) Proc. of World Conference on E-Learning, E-Learn 2004. AACE (2004) 234-242

16. Brailsford, T.J., Stewart, C.D., Zakaria, M.R., Moore, A.: Autonavigation, links, and narrative in an adaptive Web-based integrated learning environment. In: Proc. of The 11th International World Wide Web Conference. (2002)

17. Brajnik, G., Guida, G., Tasso, C.: User modeling in intelligent information retrieval. Information Processing and Management 23, 4 (1987) 305-320

18. Brusilovsky, P.: A framework for intelligent knowledge sequencing and task sequencing. In: Frasson, C., Gauthier, G., McCalla, G. (eds.) Proc. of Second International Conference on Intelligent Tutoring Systems, ITS'92. Springer-Verlag (1992) 499-506

19. Brusilovsky, P.: Adaptive hypermedia, an attempt to analyze and generalize. In: Brusilovsky, P., Kommers, P., Streitz, N. (eds.): Multimedia, Hypermedia, and Virtual Reality. Lecture Notes in Computer Science, Vol. 1077. Springer-Verlag, Berlin (1996) 288-304

20. Brusilovsky, P.: Developing Adaptive Educational Hypermedia Systems: From Design Models to Authoring Tools. In: Murray, T., Blessing, S., Ainsworth, S. (eds.): Authoring Tools for Advanced Technology Learning Environments: Toward cost-effective adaptive, interactive, and intelligent educational software. Dordrecht, Kluwer (2003) 377-409

21. Brusilovsky, P.: Adaptive navigation support. In: Brusilovsky, P., Kobsa, A., Neidl, W. (eds.): The Adaptive Web: Methods and Strategies of Web Personalization. Lecture Notes in Computer Science, Vol. 4321. Springer-Verlag, Berlin Heidelberg New York (2007) this volume

22. Brusilovsky, P., Anderson, J.: ACT-R electronic bookshelf: An adaptive system for learning cognitive psychology on the Web. In: Maurer, H., Olson, R.G. (eds.) Proc. of WebNet'98, World Conference of the WWW, Internet, and Intranet. AACE (1998) 92-97

23. Brusilovsky, P., Cooper, D.W.: Domain, Task, and User Models for an Adaptive Hypermedia Performance Support System. In: Gil, Y., Leake, D.B. (eds.) Proc. of 2002 International Conference on Intelligent User Interfaces. ACM Press (2002) 23-30

24. Brusilovsky, P., Eklund, J., Schwarz, E.: Web-based education for all: A tool for developing adaptive courseware. In: Ashman, H., Thistewaite, P. (eds.) Proc. of Seventh International World Wide Web Conference. Vol. 30. Elsevier Science B. V. (1998) 291-300

25. Brusilovsky, P., Henze, N.: Open corpus adaptive educational hypermedia. In: Brusilovsky, P., Kobsa, A., Neidl, W. (eds.): The Adaptive Web: Methods and Strategies of Web Personalization. Lecture Notes in Computer Science, Vol. 4321. Springer-Verlag, Berlin Heidelberg New York (2007) this volume

26. Brusilovsky, P., Schwarz, E., Weber, G.: ELM-ART: An intelligent tutoring system on World Wide Web. In: Frasson, C., Gauthier, G., Lesgold, A. (eds.) Proc. of Third International Conference on Intelligent Tutoring Systems, ITS-96. Lecture Notes in Computer Science, Vol. 1086. Springer Verlag (1996) 261-269

27. Brusilovsky, P., Sosnovsky, S., Yudelson, M.: Ontology-based framework for user model interoperability in distributed learning environments. In: Richards, G. (ed.) Proc. of World Conference on E-Learning, E-Learn 2005. AACE (2005) 2851-2855

28. Bunt, A., Carenini, G., Conati, C.: Adaptive content presentation for the Web. In: Brusilovsky, P., Kobsa, A., Neidl, W. (eds.): The Adaptive Web: Methods and Strategies of Web Personalization. Lecture Notes in Computer Science, Vol. 4321. Springer-Verlag, Berlin Heidelberg New York (2007) this volume

29. Bunt, A., Conati, C.: Probabilistic Student Modelling to Improve Exploratory Behaviour. User Modeling and User-Adapted Interaction 13, 3 (2003) 269-309

30. Carberry, S., de Rosis, F. (eds.): Proceedings of Workshop on Adapting the Interaction Style to Affective Factors hold in conjunction with User Modeling 2005, July 25, 2005. Edinburgh, UK (2005) available online at http://www.di.uniba.it/intint/UM05/WS-UM05.html

31. Carmona, C., Millán, E., Perez de la Cruz, J.-L., Trella, M., Conejo, R.: Introducing Prerequisite Relations in a Multi-layered Bayesian Student Model. In: Proc. of 10th International Conference UM'2005. Vol. 3538. Springer-Verlag (2005) 347-356

32. Carro, R.M., Pulido, E., Rodríguez, P.: Dynamic generation of adaptive Internet-based courses. Journal of Network and Computer Applications 22, 4 (1999) 249-257

33. Carver, C.A., Howard, R.A., Lavelle, E.: Enhancing student learning by incorporating student learning styles into adaptive hypermedia. In: Proc. of ED-MEDIA'96 - World Conference on Educational Multimedia and Hypermedia. AACE (1996) 118-123

34. Cawsey, A., Grasso, F., Paris, C.: Adaptive information for consumers of healthcare. In: Brusilovsky, P., Kobsa, A., Neidl, W. (eds.): The Adaptive Web: Methods and Strategies of Web Personalization. Lecture Notes in Computer Science, Vol. 4321. Springer-Verlag, Berlin Heidelberg New York (2007) this volume

35. Chen, S.Y., Macredie, R.D.: Cognitive styles and hypermedia navigation: Development of a learning model. Journal of the American Society for Information Science and Technology 53, 1 (2002) 3-15

36. Cheverst, K., Davies, N., Mitchell, K., Smith, P.: Providing tailored (context-aware) information to city visitors. In: Brusilovsky, P., Stock, O., Strapparava, C. (eds.) Proc. of Adaptive Hypermedia and Adaptive Web-based Systems. AH'2000. Lecture Notes in Computer Science, Vol. 1892. Springer-Verlag (2000) 73-85

37. Chin, D., Kobsa, A., Wahlster, W.: Modelling what the User Knows in UC. In: Loveland, D.W. (ed.) User Models in Dialog Systems. Symbolic Computation Series, Springer Verlag, Berlin (1989) 74-107

38. Cocea, M., Weibelzahl, S.: Can log files analysis estimate learners' level of motivation? In: Herder, E., Heckmann, D. (eds.) Proc. of 14th Workshop on Adaptivity and User Modeling in Interactive Systems, ABIS 2006. University of Hildesheim (2006) 32-35

39. Collins, J.A., Greer, J.E., Huang, S.H.: Adaptive Assessment Using Granularity Hierarchies and Bayesian Nets. In: Frasson, C., Gauthier, G., Lesgold, A. (eds.) Proc. of 3rd International Conference on Intelligent Tutoring Systems, ITS'96. Lecture Notes in Computer Science, Vol. 1086. Springer-Verlag (1996) 569-577

40. Conati, C., Gertner, A., VanLehn, K.: Using Bayesian Networks to Manage Uncertainty in Student Modeling. User Modeling and User-Adapted Interaction 12, 4 (2002) 371-417

41. Conati, C., Larkin, J., VanLehn, K.: A Computer Framework to Support Self-explanation. In: du Bolay, B., Mizoguchi, R. (eds.) Proc. of 8th World Conference on Artificial Intelligence in Education AIED'97. Knowledge and Media in Learning Systems, IOS Press (1997) 279-286

42. Conati, C., Maclaren, H.: Data-Driven Refinement of a Probabilistic Model of User Affect. In: Proc. of 10th International Conference UM'05. Lecture Notes in Computer Science, Vol. 3538. Springer-Verlag (2005) 40-49

43. Conlan, O., O'Keeffe, I., Tallon, S.: Combining adaptive hypermedia techniques and ontology reasoning to produce dynamic personalized news services. In: Wade, V., Ashman, H., Smyth, B. (eds.) Proc. of 4th International Conference on Adaptive Hypermedia and Adaptive Web-Based Systems (AH'2006). Lecture Notes in Computer Science, Vol. 4018. Springer Verlag (2006) 81-90

44. Corbett, A., Anderson, J.: Student Modelling and Mastery Learning in a Computer-based Programming Tutor. In: Frasson, C., Gauthier, G., McCalla, G.I. (eds.) Proc. of 2nd International Conference on Intelligent Tutoring Systems, ITS'92. Lecture Notes in Computer Science, Vol. 608. Springer-Verlag (1992) 413-420

45. Corbett, A., Anderson, J.R., O'Brien, A.T.: Student Modeling in the ACT Programming Tutor. In: Cognitively Diagnostic Assessment. Erlbaum, Hillsdale, NJ (1995) 19-41

46. Cotter, P., Smyth, B.: WAP-ing the Web: Content personalization for WAP-enabled devices. In: Brusilovsky, P., Stock, O., Strapparava, C. (eds.) Proc. of Adaptive Hypermedia and Adaptive Web-based systens. Lecture Notes in Computer Science, Vol. 1892. Springer-Verlag (2000) 98-108

47. Dagger, D., Conlan, O., Wade, V.P.: An architecture for candidacy in adaptive eLearning systems to facilitate the reuse of learning Resources. In: Rossett, A. (ed.) Proc. of World Conference on E-Learning, E-Learn 2003. AACE (2003) 49-56

48. De Bra, P., Aerts, A., Rousseau, B.: Concept Relationship Types for AHA! 2.0. In: Driscoll, M., Reeves, T.C. (eds.) Proc. of World Conference on E-Learning, E-Learn 2002. AACE (2002) 1386-1389

49. De Bra, P., Aerts, A., Smits, D., Stash, N.: AHA! Version 2.0: More Adaptation Flexibility for Authors. In: Driscoll, M., Reeves, T.C. (eds.) Proc. of World Conference on E-Learning, E-Learn 2002. AACE (2002) 240-246

50. De Bra, P., Calvi, L.: AHA! An open Adaptive Hypermedia Architecture. The New Review of Hypermedia and Multimedia 4 (1998) 115-139

51. De Bra, P., Ruiter, J.-P.: AHA! Adaptive hypermedia for all. In: Fowler, W., Hasebrook, J. (eds.) Proc. of WebNet'2001, World Conference of the WWW and Internet. AACE (2001) 262-268

52. De Bra, P.M.E.: Teaching Hypertext and Hypermedia through the Web. Journal of Universal Computer Science 2, 12 (1996) 797-804

53. de Rosis, F., De Carolis, B., Pizzutilo, S.: User tailored hypermedia explanations. In: Brusilovsky, P., Beaumont, I. (eds.) Proc. of Workshop Adaptive Hypertext and Hypermedia at Fourth International Conference on User Modeling. (1994) http://wwwis.win.tue.nl/ah94/deRosis.html

54. Díaz, A., Gervás, P.: Personalisation in news delivery systems: Item summarization and multi-tier item selection using relevance feedback. Web Intelligence and Agent Systems 3, 3 (2005) 135-154

55. Dichev, C., Dicheva, D., Aroyo, L.: Using Topic Maps for Web-based Education. Advanced Technology for Learning 1, 1 (2004) 1-7

56. Dolog, P., Henze, N., Nejdl, W., Sintek, M.: Personalization in distributed e-learning environments. In: Proc. of The Thirteenth International World Wide Web Conference, WWW 2004 (Alternate track papers and posters). ACM Press (2004) 161-169

57. Dolog, P., Nejdl, W.: Semantic Web Technologies for the Adaptive Web. In: Brusilovsky, P., Kobsa, A., Neidl, W. (eds.): The Adaptive Web: Methods and Strategies of Web Personalization. Lecture Notes in Computer Science, Vol. 4321. Springer-Verlag, Berlin Heidelberg New York (2007) this volume

58. Dolog, P., Schäfer, M.: Learner Modeling on the Semantic Web. In: Proc. of PerSWeb'05, Workshop on Personalization on the Semantic Web at 10th International User Modeling Conference. (2005) http://www.win.tue.nl/persweb/Camera-ready/6-Dolog-full.pdf
59. Dowling, C.E., Hockemeyer, C., Ludwig, A.H.: Adaptive Asessment and Training Using the Neighbourhood of Knowledge States. In: Frasson, C., Gauthier, G., Lesgold, A. (eds.) Proc. of 3rd International Conference on Intelligent Tutoring Systems, ITS'96. Lecture Notes in Computer Science, Vol. 1086. Springer-Verlag (1996) 578-585
60. Druzdzel, M., van der Gaag, L.C.: Building Probabilistic Networks: Where Do the Numbers Come From? - Guest editors' introduction. IEEE Transactions on Knowledge and Data Engineering 12, 4 (2000) 481-486
61. Dufresne, A., Turcotte, S.: Cognitive style and its implications for navigation strategies. In: du Boulay, B., Mizoguchi, R. (eds.) Proc. of AI-ED'97, 8th World Conference on Artificial Intelligence in Education. IOS (1997) 287-293
62. Eisenstein, J., Vanderdonckt, J., Puerta, A.: Applying model-based techniques to the development of UIs for mobile computers. In: Proc. of 6th International Conference on Intelligent User Interfaces. ACM Press (2001) 69-76
63. Encarnação, L.M.: Multi-level user support through adaptive hypermedia: A highly application-independent help component. In: Moore, J., Edmonds, E., Puerta, A. (eds.) Proc. of 1997 International Conference on Intelligent User Interfaces. ACM (1997) 187-194
64. Felder, R.: Learning and teaching styles. Journal of Engineering Education 78, 7 (1988) 674-681
65. Fink, J., Kobsa, A., Nill, A.: Adaptable and adaptive information provision for all users, including disabled and elderly people. The New Review of Hypermedia and Multimedia 4 (1998) 163-188
66. Fischer, G.: User modeling in human-computer interaction. User Modeling and User Adapted Interaction 11, 1-2 (2001) 65-86
67. Ganeshan, R., Johnson, W., Shaw, E., Wood, B.P.: Tutoring Diagnostic Problem Solving. In: Proc. of 7th International Conference on Intelligent Tutoring Systems, ITS'2004. Vol. 1839. Springer-Verlag (2000) 33-42
68. García, F., Amandi, A., Schiaffinoa, S., Campoa, M.: Evaluating Bayesian networks' precision for detecting students' learning styles. Computers & Education (2006) In press
69. Garlatti, S., Iksal, S.: Context filtering and spacial filtering in an adaptive information system. In: Brusilovsky, P., Stock, O., Strapparava, C. (eds.) Proc. of Adaptive Hypermedia and Adaptive Web-based systens. Lecture Notes in Computer Science, Vol. 1892. Springer-Verlag (2000) 315-318
70. Garlatti, S., Iksal, S., Kervella, P.: Adaptive on-line information system by means of a task model and spatial views. In: Brusilovsky, P., Bra, P.D. (eds.) Proc. of Second Workshop on Adaptive Systems and User Modeling on the World Wide Web. (1999) 59-66, also available at http://wwwis.win.tue.nl/asum99/garlatti/garlatti.html
71. Gates, K.F., Lawhead, P.B., Wilkins, D.E.: Toward an adaptive WWW: a case study in customized hypermedia. New Review of Multimedia and Hypermedia 4 (1998) 89-113
72. Gauch, S., Speretta, M., Chandramouli, A., Micarelli, A.: User profiles for personalized information access. In: Brusilovsky, P., Kobsa, A., Neidl, W. (eds.): The Adaptive Web: Methods and Strategies of Web Personalization. Lecture Notes in Computer Science, Vol. 4321. Springer-Verlag, Berlin Heidelberg New York (2007) this volume
73. Gilbert, J.E., Han, C.Y.: Arthur: Adapting Instruction to Accommodate Learning Style. In: Bra, P.D., Leggett, J. (eds.) Proc. of WebNet'99, World Conference of the WWW and Internet. AACE (1999) 433-438
74. Goldstein, I.P.: The genetic graph: a representation for the evolution of procedural knowledge. In: Sleeman, D.H., Brown, J.S. (eds.): Intelligent tutoring systems. Academic press, London (1982) 51-77

75. Gonschorek, M., Herzog, C.: Using hypertext for an adaptive helpsystem in an intelligent tutoring system. In: Greer, J. (ed.) Proc. of AI-ED'95, 7th World Conference on Artificial Intelligence in Education. AACE (1995) 274-281

76. Goodman, B.A., Litman, D.J.: On the interaction between plan recognition and intelligent interfaces. User Modeling and User-Adapted Interaction 2, 1 (1992) 83-115

77. Goren-Bar, D., Graziola, I., Pianesi, F., Zancanaro, M.: The influence of personality factors on visitor attitudes towards adaptivity dimensions for mobile museum guides. User Modeling and User Adapted Interaction 16, 1 (2005) 31-62

78. Grunst, G.: Adaptive hypermedia for support systems. In: Schneider-Hufschmidt, M., Kühme, T., Malinowski, U. (eds.): Adaptive user interfaces: Principles and practice. North-Holland, Amsterdam (1993) 269-283

79. Heckerman, D.: A Tutorial on Learning with Bayesian Networks, Technical Report No. MSR-TR-95-06, Microsoft Research Advanced Technology Division (1995)

80. Heckmann, D., Schwartz, T., Brandherm, B., Schmitz, M., von Wilamowitz-Moellendorff, M.: Gumo - The General User Model Ontology. In: Ardissono, L., Brna, P., Mitrovic, A. (eds.) Proc. of 10th International User Modeling Conference. Lecture Notes in Artificial Intelligence, Vol. 3538. Springer Verlag (2005) 428-432

81. Henze, N., Naceur, K., Nejdl, W., Wolpers, M.: Adaptive hyperbooks for constructivist teaching. Künstliche Intelligenz, 4 (1999) 26-31

82. Henze, N., Nejdl, W.: Student modeling for KBS Hyperbook system using Bayesian networks, Technical report, University of Hannover (1999) available online at http://www.kbs.uni-hannover.de/paper/99/adaptivity.html

83. Henze, N., Nejdl, W.: Adaptation in open corpus hypermedia. International Journal of Artificial Intelligence in Education 12, 4 (2001) 325-350

84. Hirashima, T., Hachiya, K., Kashihara, A., Toyoda, J.i.: Information filtering using user's context on browsing in hypertext. User Modeling and User Adapted Interaction 7, 4 (1997) 239-256

85. Hirashima, T., Matsuda, N., Nomoto, T., Toyoda, J.i.: Context-sensitive filtering for browing in hypertext. In: Proc. of International Conference on Intelligent User Interfaces, IUI'98. ACM Press (1998) 21-28

86. Hockemeyer, C., Held, T., Albert, D.: RATH - A relational adaptive tutoring hypertext WWW-environment based on knowledge space theory. In: Alvegård, C. (ed.) Proc. of CALISCE'98, 4th International conference on Computer Aided Learning and Instruction in Science and Engineering. (1998) 417-423

87. Hollink, V., Someren, M.v., Hage, S.t.: Discovering stages in web navigation. In: Ardissono, L., Brna, P., Mitrovic, A. (eds.) Proc. of 10th International User Modeling Conference. Lecture Notes in Artificial Intelligence, Vol. 3538. Springer Verlag (2005) 473-482

88. Höök, K., Karlgren, J., Wærn, A., Dahlbäck, N., Jansson, C.G., Karlgren, K., Lemaire, B.: A glass box approach to adaptive hypermedia. User Modeling and User-Adapted Interaction 6, 2-3 (1996) 157-184

89. Horvitz, E.J., Breese, J.S., Henrion, M.: Decision Theory in Expert Systems and Artificial Intelligence. International Journal of Approximate Reasoning 2 (1988) 247-302

90. Jameson, A.: Numerical Uncertainty Management in User and Student Modeling: An Overview of Systems and Issues. User Modeling and User-Adapted Interaction 5 (1996) 193-251

91. Jameson, A.: Modeling both the context and the user. Personal Technologies 5, 1 (2001) 29-33

92. Jameson, A., Smyth, B.: Recommendation to groups In: Brusilovsky, P., Kobsa, A., Neidl, W. (eds.): The Adaptive Web: Methods and Strategies of Web Personalization. Lecture Notes in Computer Science, Vol. 4321. Springer-Verlag, Berlin Heidelberg New York (2007) this volume

93. Jantke, K.P., Memmel, M., Rostanin, O., Rudolf, B.: Media and service integration for professional e-learning. In: Nall, J., Robson, R. (eds.) Proc. of World Conference on E-Learning, E-Learn 2004. AACE (2004) 725-732
94. Jin, X., Zhou, Y., Mobasher, B.: Task-Oriented Web User Modeling for Recommendation. In: Ardissono, L., Brna, P., Mitrovic, A. (eds.) Proc. of 10th International User Modeling Conference. Lecture Notes in Artificial Intelligence, Vol. 3538. Springer Verlag (2005) 109-118
95. Joerding, T.: A temporary user modeling approach for adaptive shopping on the Web. In: Brusilovsky, P., De Bra, P. (eds.) Proc. of Second Workshop on Adaptive Systems and User Modeling on the World Wide Web. (1999) 75-79, also available at http://wwwis.win.tue.nl/asum99/joerding/joerding.html
96. Jokela, S., Turnpeinen, M., Kurki, T., Savia, E., Sulonen, R.: The Role of Structured Content in a Personalised News Service. In: Proc. of 34th Hawaii International Conference on System Sciences. (2001) 1-10
97. Kaplan, C., Fenwick, J., Chen, J.: Adaptive hypertext navigation based on user goals and context. User Modeling and User-Adapted Interaction 3, 3 (1993) 193-220
98. Katz, S., Lesgold, A., Eggan, G., Gordin, M.: Modelling the student in SHERLOCK II. In: Greer, J.E., McCalla, G. (eds.): Student Modelling: The Key to Individualized Knowledge-Based Instruction. Series F: Computer and Systems Sciences. NATO ASI Series, Springer Verlag, Berlin Heidelberg (1994) 99-125
99. Kavcic, A.: Fuzzy user modeling for adaptation in educational hypermedia. IEEE Transactions on Systems, Man, and Cybernetics 34, 4 (2004) 439-449
100. Kawai, K., Mizoguchi, R., Kakusho, O., Toyoda, J.: A framework for ICAI systems based on inductive inference and logic programming. New Generation Computing 5 (1987) 115-129
101. Kay, J.: Lies, damned lies and stereotypes: pragmatic approximations of users. In: Kobsa, A., Litman, D. (eds.) Proc. of Fourth International Conference on User Modeling. MITRE (1994) 175-184
102. Kay, J., Kummerfeld, R.J.: An individualised course for the C programming language. In: Proc. of Second International WWW Conference. (1994) http://www.cs.usyd.edu.au/~bob/kay-kummerfeld.html
103. Kay, J., Lum, A.: Ontologies for Scrutable Learner Modeling in Adaptive E-Learning. In: Aroyo, L., Tasso, C. (eds.) Proc. of Workshop on Application of Semantic Web Technologies for Adaptive Educational Hypermedia at the Third International Conference on Adaptive Hypermedia and Adaptive Web-Based Systems (AH'2004). Technische University Eindhoven (2004) 292-301
104. Kim, K.S., Allen, B.: Cognitive and Task Influences on Web Searching Behavior. Journal of the American Society for Information Science and Technology 53, 2 (2002) 109-119
105. Klync, G., Reynolds, F., Woodrow, C., Ohto, H., Johan Hjelm, Butler, M.H., Tran, L.: Composite Capability/Preference Profiles (CC/PP): Structure and Vocabularies 1.0. W3C Recommendation 15 January 2004. (2004) http://www.w3.org/TR/CCPP-struct-vocab/
106. Koedinger, K.R., Anderson, J.R., Hadley, W.H., Mark, M.A.: Intelligent tutoring goes to school in the big city. International Journal of Artificial Intelligence in Education 8 (1997) 30-43
107. Korfhage, R.R.: Information storage and retrieval. Wiley Computer Publishing, N.Y. (1997)
108. Kosba, E., Dimitrova, V., Boyle, R.: Using Fuzzy Techniques to Model Students in Web-Based Learning Environments. International Journal of Artificial Intelligence Tools 13, 2 (2004) 279-297
109. Krüger, A., Baus, J., Heckmann, D., Kruppa, M., Wasinger, R.: Adaptive mobile guides. In: Brusilovsky, P., Kobsa, A., Neidl, W. (eds.): The Adaptive Web: Methods and Strategies of Web Personalization. Lecture Notes in Computer Science, Vol. 4321. Springer-Verlag, Berlin Heidelberg New York (2007) this volume

110. Kumar, A.N.: A Scalable Solution for Adaptive Problem Sequencing and its Evaluation. In: Wade, V., Ashman, H., Smyth, B. (eds.) Proc. of 4th International Conference on Adaptive Hypermedia and Adaptive Web-Based Systems (AH'2006). Lecture Notes in Computer Science, Vol. 4018. Springer Verlag (2006) 161-171

111. Laroussi, M., Benahmed, M.: Providing an adaptive learning through the Web case of CAMELEON: Computer Aided MEdium for LEarning on Networks. In: Alvegård, C. (ed.) Proc. of CALISCE'98, 4th International conference on Computer Aided Learning and Instruction in Science and Engineering. (1998) 411-416

112. Lee, K.B., Grice, R.A.: An Adaptive Viewing Application for the Web on Personal Digital Assistants. In: Proc. of ACM SIGDOC'03. IEEE (2003) 125-132

113. Liu, Y., Ginther, D.: Cognitive styles and distance education. Online Journal of Distance Learning Administration 2, 3 (1999) http://www.westga.edu/~distance/liu23.html

114. Locke, J.: Microsoft Bayesian Networks. Basics of Knowledge Engineering. Microsoft Support Technology (1999) http://freelock.com/files/KE.pdf

115. Lopez, J.F., Szekely, P.: Web page adaptation for universal access. In: Stephanidis, C. (ed.) Proc. of 1st International Conference on Universal Access in Human-Computer Interaction. Lawrence Erlbaum Associates (2001) 690-694

116. López, J.M., Millán, E., Pérez-de-la-Cruz, J.-L., Triguero, F.: ILESA: a Web-based Intelligent Learning Environment for the Simplex Algorithm. In: Alvegård, C. (ed.) Proc. of CALISCE'98, 4th International conference on Computer Aided Learning and Instruction in Science and Engineering. (1998) 399-406

117. Lundgren-Cayrol, K., Paquette, G., Miara, A., Bergeron, F., Rivard, J., Rosca, I.: Explor@ Advisory Agent: Tracing the Student's Trail. In: Fowler, W., Hasebrook, J. (eds.) Proc. of WebNet'2001, World Conference of the WWW and Internet. AACE (2001) 802-808

118. Magnini, B., Strapparava, C.: Improving user modeling with content-based techniques. In: Bauer, M., Gmytrasiewicz, P.J., Vassileva, J. (eds.) Proc. of 8th International Conference on User Modeling, UM 2001. Lecture Notes on Artificial Intelligence, Vol. 2109. Springer-Verlag (2001) 74-83

119. Manske, M., Conati, C.: Modelling Learning in an Educational Game. In: Proc. of 12th World Conference of Artificial Intelligence and Education AIED'05. IOS Press (2005) 411-419

120. Masthoff, J.: Towards an authoring coach for adaptive Web-based instruction. In: De Bra, P., Brusilovsky, P., Conejo, R. (eds.) Proc. of Second International Conference on Adaptive Hypermedia and Adaptive Web-Based Systems (AH'2002). Lecture Notes in Computer Science, Vol. 2347. (2002) 415-418

121. Mathé, N., Chen, J.: User-centered indexing for adaptive information access. User Modeling and User-Adapted Interaction 6, 2-3 (1996) 225-261

122. Mayo, M., Mitrovic, A.: Optimising ITS behaviour with Bayesian networks and decision theory. International Journal of Artificial Intelligence in Education 12 (2001) 124-153

123. McArthur, D., Stasz, C., Hotta, J., Peter, O., Burdorf, C.: Skill-oriented task sequencing in an intelligent tutor for basic algebra. Instructional Science 17, 4 (1988) 281-307

124. McCalla, G., Bunt, R.B., Harms, J.J.: The design of the SCENT automated advisor. Computational Intelligence 2, 2 (1986) 76-91

125. Micarelli, A., Gasparetti, F., Sciarrone, F., Gauch, S.: Personalized search on the World Wide Web. In: Brusilovsky, P., Kobsa, A., Neidl, W. (eds.): The Adaptive Web: Methods and Strategies of Web Personalization. Lecture Notes in Computer Science, Vol. 4321. Springer-Verlag, Berlin Heidelberg New York (2007) this volume

126. Micarelli, A., Sciarrone, F.: A case-based system for adaptive hypermedia navigation. In: Smith, I., Faltings, B. (eds.): Advances in Case-Based Reasoning. Lecture Notes in Artificial Intelligence, Springer-Verlag, Berlin (1996) 266-279

127. Micarelli, A., Sciarrone, F.: Anatomy and empirical evaluation of an adaptive Web-based information filtering system. User Modeling and User Adapted Interaction 14, 159-200 (2004)

128. Middleton, S.E., Shadbolt, N.R., De Roure, D.C.: Ontological User Profiling in Recommender Systems. ACM Transactions on Information Systems 22, 1 (2004) 54-88

129. Millán, E., Agosta, J.M., Perez de la Cruz, J.-L.: Bayesian Student Modelling and the Problem of Parameter Specification. British Journal of Educational Technology 32, 2 (2001) 171-181

130. Millán, E., Perez de la Cruz, J.-L.: A Bayesian Diagnostic Algorithm for Student Modeling. User Modeling and User-Adapted Interaction 12 (2002) 281-330

131. Millán, E., Perez de la Cruz, J.-L., García, F.: Dynamic versus Static Student Models Based on Bayesian Networks: An Empirical Study. In: Proc. of 7th International Conference KES'2003. Lecture Notes in Computer Science, Vol. 2774. Springer-Verlag (2003) 1337-1344

132. Mislevy, R., Gitomer, D.H.: The Role of Probability-Based Inference in an Intelligent Tutoring System. User Modeling and User-Adapted Interaction 5, 3-4 (1996) 253-282

133. Mitchell, T., Chen, S.Y., Macredie, R.: Adapting Hypermedia to cognitive styles: Is it necessary? In: Proc. of Workshop on Individual Differences in Adaptive Hypermedia at the 3rd International Conference on Adaptive Hypermedia and Adaptive Web-based Systems.   (2004) http://www.dcs.bbk.ac.uk/~gmagoulas/AH2004_Workshop/Proceedings.htm

134. Mitrovic, A.: An Intellignet SQL Tutor on the Web. International Journal of Artificial Intelligence in Education 13, 2-4 (2003) 173-197

135. Mobasher, B.: Data mining for Web personalization. In: Brusilovsky, P., Kobsa, A., Neidl, W. (eds.): The Adaptive Web: Methods and Strategies of Web Personalization. Lecture Notes in Computer Science, Vol. 4321. Springer-Verlag, Berlin Heidelberg New York (2007) this volume

136. Muldner, K., Conati, C.: Using Similarity to Infer Meta-Cognitive Behaviours During Analogical Problem Solving. In: Proc. of 10th International Conference UM'05. Vol. 3538. Springer-Verlag (2005) 134-143

137. Müller, C., Großmann-Hutter, B., Jameson, A., Rummer, R., Wittig, F.: Recognizing time pressure and cognitive load on the basis of speech: An Experimental study. In: Bauer, M., Gmytrasiewicz, P.J., Vassileva, J. (eds.) Proc. of 8th International Conference on User Modeling, UM 2001. Lecture Notes on Artificial Intelligence, Vol. 2109. Springer-Verlag (2001) 24-33

138. Murray, R.C., VanLehn, K., Mostow, J.: Looking ahead to select tutorial actions: A decision-theoretic approach. International Journal of Artificial Intelligence in Education 14, 3-4 (2004) 235-279

139. Murray, W.: An Easily Implemented, Linear-time Algorithm for Bayesian Student Modeling in Multi-level Trees. In: Lajoie, S., Vivet, M. (eds.) Proc. of 9th World Conference of Artificial Intelligence and Education AIED'99. IOS Press (1999) 413-420

140. Neapolitan, R.: Probabilistic Reasoning in Expert Systems: Theory and Algorithms. John Wiley & Sons, New York (1990)

141. Neumann, G., Zirvas, J.: SKILL - A scallable internet-based teaching and learning system. In: Maurer, H., Olson, R.G. (eds.) Proc. of WebNet'98, World Conference of the WWW, Internet, and Intranet. AACE (1998) 688-693

142. Not, E., Petrelli, D., Sarini, M., Stock, O., Strapparava, C., Zancanaro, M.: Hypernavigation in the physical space: adapting presentation to the user and to the situational context. New Review of Multimedia and Hypermedia 4 (1998) 33-45

143. Oberlander, J., O'Donell, M., Mellish, C., Knott, A.: Conversation in the museum: experiments in dynamic hypermedia with the intelligent labeling explorer. The New Review of Multimedia and Hypermedia 4 (1998) 11-32

144. Ohlsson, S.: Constraint-based student modeling. Journal of Artificial Intelligence in Education 3, 4 (1992) 429-447

145. Okazaki, Y., Watanabe, K., Kondo, H.: An Implementation of the WWW Based ITS for Guiding Differential Calculations. In: Brusilovsky, P., Nakabayashi, K., Ritter, S. (eds.) Proc. of Workshop "Intelligent Educational Systems on the World Wide Web" at 8th World Conference on Artificial Intelligence in Education. (1997) 18-25, also available at http://www.contrib.andrew.cmu.edu/~plb/AIED97_workshop/Okazaki/Okazaki.html

146. Ong, E., Tay, A.-H., Ong, C.-K., Chan, S.-K.: Personalising Information Assets in Collaborative Learning Environments. In: Looi, C.-K., McCalla, G., Bredeweg, B., Breuker, J. (eds.) Proc. of 12th International Conference on Artificial Intelligence in Education, AIED'2005. IOS Press (2005) 523-530

147. Ortony, A., Clore, G.L., Collins, A.: The Cognitive Structure of Emotions. Cambridge University Press, (1988)

148. Papanikolaou, K.A., Grigoriadou, M., Kornilakis, H., Magoulas, G.D.: Personalising the interaction in a Web-based Educational Hypermedia System: the case of INSPIRE. User Modeling and User Adapted Interaction 13, 3 (2003) 213-267

149. Paris, C., Wan, S., Wilkinson, R., Wu, M.: Generating personal travel guides - and who wants them? In: Bauer, M., Gmytrasiewicz, P.J., Vassileva, J. (eds.) Proc. of 8th International Conference on User Modeling, UM 2001. Lecture Notes on Artificial Intelligence, Vol. 2109. Springer-Verlag (2001) 251-253

150. Pask, G.: A fresh look at cognition and the individual. International Journal on the Man-Machine Studies 4 (1972) 211-216

151. Paternò, F., Paganelli, L.: Intelligent analysis of user interactions with Web applications. In: Gil, Y., Leake, D.B. (eds.) Proc. of 2002 International Conference on Intelligent User Interfaces. ACM Press (2002) 111-118

152. Pazzani, M.J., Billsus, D.: Content-based recommendation systems. In: Brusilovsky, P., Kobsa, A., Neidl, W. (eds.): The Adaptive Web: Methods and Strategies of Web Personalization. Lecture Notes in Computer Science, Vol. 4321. Springer-Verlag, Berlin Heidelberg New York (2007) this volume

153. Pearl, J.: Probabilistic Reasoning in Expert Systems: Networks of Plausible Inference. Morgan Kaufmann Publishers, Inc, San Francisco (1988)

154. Pentland, A.: Socially Aware Computation and Communication. Computer 38, 3 (2005) 33 - 40

155. Pérez, T., Gutiérrez, J., Lopistéguy, P.: An adaptive hypermedia system. In: Greer, J. (ed.) Proc. of AI-ED'95, 7th World Conference on Artificial Intelligence in Education. AACE (1995) 351-358

156. Picard, R.W.: Affective Computing. MIT Press, Cambridge, MA (1997)

157. Pilar da Silva, D., Durm, R.V., Duval, E., Olivié, H.: Concepts and documents for adaptive educational hypermedia: a model and a prototype. In: Brusilovsky, P., De Bra, P. (eds.) Proc. of Second Adaptive Hypertext and Hypermedia Workshop at the Ninth ACM International Hypertext Conference Hypertext'98. Eindhoven University of Technology (1998) 35-43, also available as http://wwwis.win.tue.nl/ah98/Pilar/Pilar.html

158. Polson, M.C., Richardson, J.J. (eds.): Foundations of intelligent tutoring systems. Lawrence Erlbaum Associates, Hillsdale (1988)

159. Prendinger, H., Mori, J., Ishizuka, M.: Recognizing, modeling, and responding to user affective states. In: Ardissono, L., Brna, P., Mitrovic, A. (eds.) Proc. of 10th International User Modeling Conference. Lecture Notes in Artificial Intelligence, Vol. 3538. Springer Verlag (2005) 60-69

160. Prentzas, J., Hatzilygeroudis, I., Garofalakis, J.: A Web-based intelligent tutoring systems using hybrid rules as its representation basis. In: Cerri, S.A., Gouardères, G., Paraguaçu, F. (eds.) Proc. of 6th International Conference on Intelligent Tutoring Systems (ITS'2002). Lecture Notes in Computer Science, Vol. 2363. Springer-Verlag (2002) 119-128

161. Read, T., Bárcena, E., Barros, B., Verdejo, F.: I-PETER: Modelling Personalised Diagnosis and Material Selection for an Online English Course. In: Proc. of 8th Ibero-American Conference IBERAMIA'2002. Lecture Notes in Computer Science, Vol. 2527. Springer-Verlag (2002) 734-744

162. Reye, J.: Two-phase Updating of Student Models Based on Dynamic Belief Networks. In: Proc. of 4th International Conference on Intelligent Tutoring Systems, ITS'98. Lecture Notes in Computer Science, Vol. 1452. Springer-Verlag (1998) 6-15

163. Rich, E.: Building and Exploiting User Models. In: Proc. of Sixth International Joint Conference on Artificial Intelligence. (1979) 720-722

164. Rich, E.A.: Stereotypes and user modeling. In: Kobsa, A., Wahlster, W. (eds.): User models in dialog systems. Vol. 18. Springer-Verlag, Berlin (1989) 35-51

165. Riding, R., Rayner, S.: Cognitive Styles and Learning Strategies: Understanding Style Differences in Learning and Behavior. David Fulton Publisher, London (1998)

166. Rist, T.: A perspective on intelligent information interfaces for mobile users. In: Smith, M., Salvendy, G., Harris, D., Koubek, R.J. (eds.) Proc. of 9th International Conference on Human-Computer Interaction, HCI International'2001. Vol. 1. Lawrence Erlbaum Associates (2001) 154-158

167. Ritter, S.: PAT Online: A Model-tracing tutor on the World-wide Web. In: Brusilovsky, P., Nakabayashi, K., Ritter, S. (eds.) Proc. of Workshop "Intelligent Educational Systems on the World Wide Web" at AI-ED'97, 8th World Conference on Artificial Intelligence in Education. ISIR (1997) 11-17, also available at http://www.contrib.andrew.cmu.edu/~plb/AIED97_workshop/Ritter/Ritter.html

168. Russell, S., Norvig, P.: Artificial Intelligence: A Modern Approach. Prentice Hall, (1995)

169. Sanrach, C., Grandbastien, M.: ECSAIWeb: A Web-based authoring system to create adaptive learning systems. In: Brusilovsky, P., Stock, O., Strapparava, C. (eds.) Proc. of Adaptive Hypermedia and Adaptive Web-based Systems, AH2000. Lecture Notes in Computer Science, Vol. 1892. Springer-Verlag (2000) 214-226

170. Santos Jr., E., Nguyen, H., Zhao, Q., Wang, H.: User modeling for intent prediction in information analysis. In: Proc. of 47th Annual Meeting for the Human Factors and Ergonomics Society (HFES-03). (2003) 1034-1038

171. Sarini, M., Strapparava, C.: Building a User Model for a Museum Exploration and Information-Providing Adaptive System. In: Brusilovsky, P., De Bra, P. (eds.) Proc. of Second Adaptive Hypertext and Hypermedia Workshop at the Ninth ACM International Hypertext Conference Hypertext'98. (1998) 63-68, also available at http://wwwis.win.tue.nl/ah98/Sarini/Sarini.html

172. Schafer, J.B., Frankowski, D., Herlocker, J., Sen, S.: Collaborative filtering recommender systems. In: Brusilovsky, P., Kobsa, A., Neidl, W. (eds.): The Adaptive Web: Methods and Strategies of Web Personalization. Lecture Notes in Computer Science, Vol. 4321. Springer-Verlag, Berlin Heidelberg New York (2007) this volume

173. Schmidt, A., Beigl, M., Gellersen, H.-W.: There is more to context than location. Computers and Graphics 23, 6 (1999) 893-901

174. Schneider-Hufschmidt, M., Kühme, T., Malinowski, U. (eds.): Adaptive user interfaces: Principles and practice. Human Factors in Information Technology, North-Holland, Amsterdam (1993)

175. Sleeman, D.H.: UMFE: a user modeling front end system. International Journal on the Man-Machine Studies 23 (1985) 71-88

176. Sosnovsky, S., Brusilovsky, P.: Layered Evaluation of Topic-Based Adaptation to Student Knowledge. In: Proc. of Fourth Workshop on the Evaluation of Adaptive Systems at 10th International User Modeling Conference, UM 2005. (2005) 47-56

177. Specht, M., Klemke, R.: ALE - Adaptive Learning Environment. In: Fowler, W., Hasebrook, J. (eds.) Proc. of WebNet'2001, World Conference of the WWW and Internet. AACE (2001) 1155-1160

178. Specht, M., Kobsa, A.: Interaction of domain expertise and interface design in adaptive educational hypermedia. In: Brusilovsky, P., De Bra, P. (eds.) Proc. of Second Workshop on Adaptive Systems and User Modeling on the World Wide Web. (1999) 89-93

179. Specht, M., Oppermann, R.: ACE - Adaptive Courseware Environment. The New Review of Hypermedia and Multimedia 4 (1998) 141-161

180. Stacey, K.P., Sonenberg, L., Nicholson, A., Boneh, T., Steinle, V.: A Teaching Model Exploiting Cognitive Conflict Driven by a Bayesian Network. In: Brusilovsky, P., Corbett, A., Rosis, F.d. (eds.) Proc. of 9th International User Modeling Conference. Lecture Notes in Computer Science, Vol. 2702. Springer-Verlag (2003) 352-362

181. Stash, N., Cristea, A., De Bra, P.: Authoring of learning styles in adaptive hypermedia: Problems and solutions. In: Proc. of The 13th International World Wide Web Conference (Alternate track papers and posters). ACM Press (2004) 114-123

182. Stathacopoulou, S., Magoulas, G., Grigoriadou, M., Samarakou, M.: Neuro-fuzzy knowledge processing in intelligent learning environments for improved student diagnosis. Information Sciences 170, 2-4 (2005) 273-307

183. Steinacker, A., Faatz, A., Seeberg, C., Rimac, I., Hörmann, S., Saddik, A.E., Steinmetz, R.: MediBook: Combining semantic networks with metadata for learning resources to build a Web based learning system. In: Proc. of ED-MEDIA'2001 - World Conference on Educational Multimedia, Hypermedia and Telecommunications. AACE (2001) 1790-1795

184. Steinacker, A., Seeberg, C., Rechenberger, K., Fischer, S., Steinmetz, R.: Dynamically generated tables of contents as guided tours in adaptive hypermedia systems. In: Kommers, P., Richards, G. (eds.) Proc. of ED-MEDIA/ED-TELECOM'99 - 11th World Conference on Educational Multimedia and Hypermedia and World Conference on Educational Telecommunications. AACE (1999) 640-645

185. Stojanovic, L., Staab, S., Studer, R.: eLearning based on the Semantic Web. In: Fowler, W., Hasebrook, J. (eds.) Proc. of WebNet'2001, World Conference of the WWW and Internet. AACE (2001) 1774-1783

186. Suebnukarn, S., Haddawy, P.: Modeling Individual and Collaborative Problem Solving in Medical Problem-Based Learning. In: Ardissono, L., Brna, P., Mitrovic, A. (eds.) Proc. of 10th International User Modeling Conference, UM'2005. Lecture Notes in Artificial Intelligence, Vol. 3538. Springer-Verlag (2005) 377-386

187. Tanudjaja, F., Mui, L.: Persona: A contextualized and personalized Web search. In: Proc. of 35th Hawaii International Conference on System Sciences. IEEE (2002) 1232-1240

188. Tarpin-Bernard, F., Habieb-Mammar, H.: Modeling elementary cognitive abilities for adaptive hypermedia presentation. User Modeling and User Adapted Interaction 15, 5 (2005) 459-495

189. Trella, M., Carmona, C., Conejo, R.: MEDEA: an Open Service-Based Learning Platform for Developing Intelligent Educational Systems for the Web. In: Proc. of Workshop on Adaptive Systems for Web-based Education at 12th International Conference on Artificial Intelligence in Education, AIED'2005. IOS Press (2005) 27-34

190. Triantafillou, E., Pomportis, A., Demetriadis, S.: The design and the formative evaluation of an adaptive educational system based on cognitive styles. Computers and Education (2003) 87-103

191. Triantafillou, E., Pomportis, A., Demetriadis, S., Georgiadou, E.: The value of adaptivity based on cognitive style: an empirical study. British Journal of Educational Technology 35, 1 (2004) 95–106

192. Tsiriga, V., Virvou, M.: Modelling the Student to Individualise Tutoring in a Web-Based ICALL. International Journal of Continuing Engineering Education and Lifelong Learning 13, 3-4 (2003) 350-365

193. Ueno, M.: Intelligent LMS with an agent that learns from log data. In: Richards, G. (ed.) Proc. of World Conference on E-Learning, E-Learn 2005. AACE (2005) 2068-2074

194. VanLehn, K.: Student models. In: Polson, M.C., Richardson, J.J. (eds.): Foundations of intelligent tutoring systems. Lawrence Erlbaum Associates, Hillsdale (1988) 55-78
195. VanLehn, K., Martin, J.: Evaluation of an assessment system based on Bayesian Student Modeling. International Journal of Artificial Intelligence in Education 8, 2 (1998) 179-221
196. VanLehn, K., Niu, Z., Siler, S., Gertner, A.S.: Student Modeling from Conventional Test Data: A Bayesian Approach Without Priors. In: Goettl, B., Redfield, C.L., Halff, H.M., Shute, V.J. (eds.) Proc. of 4th International Conference on Intelligent Tutoring Systems, ITS'98. Lecture Notes in Computer Science, Vol. 1452. Springer-Verlag (1998) 434-443
197. Vassileva, J.: An architecture and methodology for creating a domain-independent, plan-based intelligent tutoring system. Educational and Training Technology International 27, 4 (1990) 386-397
198. Vassileva, J.: A task-centered approach for user modeling in a hypermedia office documentation system. User Modeling and User-Adapted Interaction 6, 2-3 (1996) 185-224
199. Vassileva, J.: DCG + GTE: Dynamic Courseware Generation with Teaching Expertise. Instructional Science 26, 3/4 (1998) 317-332
200. Vassileva, J., McCalla, G., Greer, J.: Multi-Agent Multi-User Modeling in I-Help. User Modeling and User-Adapted Interaction 12 (2003) 179-210
201. Vomlel, J.: Bayesian Networks in Educational Testing. International Journal of Uncertainty, Fuzziness and Knowledge-Based System 12 (2004) 83-100
202. W3C: Device Independence: Access to a Unified Web from Any Device in Any Context by Anyone. World Wide Web Consortium (2006) http://www.w3.org/2001/di/
203. Weber, G., Brusilovsky, P.: ELM-ART: An adaptive versatile system for Web-based instruction. International Journal of Artificial Intelligence in Education 12, 4 (2001) 351-384
204. Weber, G., Kuhl, H.-C., Weibelzahl, S.: Developing adaptive internet based courses with the authoring system NetCoach. In: Bra, P.D., Brusilovsky, P., Kobsa, A. (eds.) Proc. of Third workshop on Adaptive Hypertext and Hypermedia. (2001) 35-48, also available at http://wwwis.win.tue.nl/ah2001/papers/GWeber-UM01.pdf
205. Witkin, H.A., Moore, C.A., Goodenough, D.R., Cox, P.W.: Field-dependent and field-independent cognitive styles and their educational implications. Review of Educational Research 47, 1 (1977) 1-64
206. Yin, X., Lee, W.S., Tan, Z.: Personalization of Web Content for Wireless Mobile Device. In: Proc. of Wireless Communications and Networking Conference. IEEE (2004) 2569-2574
207. Zadeh, L.: Fuzzy sets. Information and Control 8 (1965) 338-353
208. Zapata-Rivera, D., Greer, J.: SModel Server: Student Modelling in Distributed Multi-Agent Tutoring Systems. In: Moore, J.D. (ed.) Proc. of 9th World Conference of Artificial Intelligence and Education, AIED'99. IOS Press (2001) 446-455
209. Zapata-Rivera, D., Greer, J.: Inspectable Bayesian student modelling servers in multi-agent tutoring systems. International Journal of Human-Computer Studies 61 (2004) 535-563
210. Zimmermann, A., Specht, M., Lorenz, A.: Personalization and Context Management. User Modeling and User-Adapted Interaction 15, 3-4 (2005) 275 302

User Profiles for Personalized Information Access

Susan Gauch[1], Mirco Speretta[1], Aravind Chandramouli[1], and Alessandro Micarelli[2]

[1] Electrical Engineering and Computer Science
Information & Telecommunication Technology Center
2335 Irving Hill Road, Lawrence Kansas 66045-7612
{sgauch, mirco, aravindc}@ittc.ku.edu
[2] Department of Computer Science and Automation
Artificial Intelligence Laboratory
Roma Tre University,
Via della Vasca Navale, 79 00146 Rome, Italy
micarel@dia.uniroma3.it

Abstract. The amount of information available online is increasing exponentially. While this information is a valuable resource, its sheer volume limits its value. Many research projects and companies are exploring the use of personalized applications that manage this deluge by tailoring the information presented to individual users. These applications all need to gather, and exploit, some information about individuals in order to be effective. This area is broadly called user profiling. This chapter surveys some of the most popular techniques for collecting information about users, representing, and building user profiles. In particular, explicit information techniques are contrasted with implicitly collected user information using browser caches, proxy servers, browser agents, desktop agents, and search logs. We discuss in detail user profiles represented as weighted keywords, semantic networks, and weighted concepts. We review how each of these profiles is constructed and give examples of projects that employ each of these techniques. Finally, a brief discussion of the importance of privacy protection in profiling is presented.

2.1 Introduction

In the modern Web, as the amount of information available causes information overloading, the demand for personalized approaches for information access increases. Personalized systems address the overload problem by building, managing, and representing information customized for individual users. This customization may take the form of filtering out irrelevant information and/or identifying additional information of likely interest for the user. Research into personalization is ongoing in the fields of information retrieval, artificial intelligence, and data mining, among others.

This chapter discusses user profiles specifically designed for providing personalized information access. Other types of profiles, build using different construction techniques, are described elsewhere in this book. In particular, Chapter 4 [40] dis-

P. Brusilovsky, A. Kobsa, and W. Nejdl (Eds.): The Adaptive Web, LNCS 4321, pp. 54–89, 2007.

cusses generic user modeling systems that are broader in scope, not necessarily focused on Internet applications. Related research on collaborative recommender systems, discussed in Chapter 9 of this book [81], combines information from multiple users in order to provide improved information services. Concern over privacy protection is growing in parallel with the demand for personalized features. These two trends seem to be in direct opposition to each other, so privacy protection must be a crucial component of every personalization system. A detailed discussion can be found in Chapter 21 of this book [39].

There are a wide variety of applications to which personalization can be applied and a wide variety of different devices available on which to deliver the personalized information. Early personalization research focused on personalized filtering and/or rating systems for e-mail [49], electronic newspapers [14, 16], Usenet newsgroups [41, 58, 86, 91, 106], and Web documents [4]. More recently, personalization efforts have focused on improving navigation effectiveness by providing browsing assistants [9, 13], and adaptive Web sites [69]. Because search is one of the most common activities performed today, many projects are now focusing on personalized Web search [46, 88, 92] and more details on the subject can be found in Chapter 6 of this book [52]. However, personalized approaches to searching other types of collections, e.g., short stories [76], Java source code [100], and images [14] have also been explored. Commercial products are also adopting personalized features, for example, Yahoo!'s personalized Web portals [110] and Google Lab's personalized search [30].

The aforementioned systems are just a few examples that illustrate the breadth of applications to which personalized approaches are being investigated. Nichols [63] and Oard and Marchionini [64] provide a general overview of some the issues and approaches to personalized rating and filtering and Pretschner [71] describes approximately 45 personalization systems.

Most personalization systems are based on some type of user profile, a data instance of a user model that is applied to adaptive interactive systems. User profiles may include demographic information, e.g., name, age, country, education level, etc, and may also represent the interests or preferences of either a group of users or a single person. Personalization of Web portals, for example, may focus on individual users, for example, displaying news about specifically chosen topics or the market summary of specifically selected stocks, or a groups of users for whom distinctive characteristics where identified, for example, displaying targeted advertising on e-commerce sites.

In order to construct an individual user's profile, information may be collected *explicitly*, through direct user intervention, or *implicitly*, through agents that monitor user activity. Although profiles are typically built only from topics of interest to the user, some projects have explored including information about non-relevant topics in the profile [35, 104]. In these approaches, the system is able to use both kinds of topics to identify relevant documents and discard non-relevant documents at the same time.

Profiles that can be modified or augmented are considered *dynamic*, in contrast to *static* profiles that maintain the same information over time. Dynamic profiles that take time into consideration may differentiate between short-term and long-term interests [37, 93, 103]. *Short-term* profiles represent the user's current interests whereas *long-term* profiles indicate interests that are not subject to frequent changes over time. For example, consider a musician who uses the Web for her daily research. One day, she decides to go on vacation, and she uses the Web to look for hotels, airplane tickets, etc. Her user profile should reflect her music interests as long-term interests, and the vacation-related interests as short-term ones. Once the user returns from her vacation, she will resume her music-related research, and the vacation information in her profile should eventually be forgotten. Because they can change quickly as users change tasks, and less information is collected, short-term user's interests are generally harder to identify and manage than long-term interests. In general, the goal of user profiling is to collect information about the subjects in which a user is interested, and the length of time over which they have exhibited this interest, in order to improve the quality of information access and infer user's intentions.

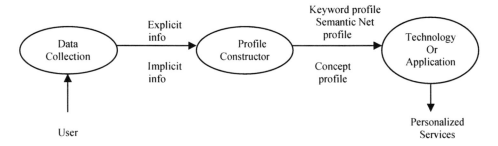

Fig. 2.1. Overview of user-profile-based personalization

As shown in Figure 2.1, the user profiling process generally consists of three main phases. First, an information collection process is used to gather raw information about the user. As described in Section 2.2, depending on the information collection process selected, different types of user data can be extracted. The second phase focuses on user profile construction from the user data. Section 2.3 summarizes a variety of ways in which profiles may be represented and Section 2.4 some of the ways a profile may be constructed. The final phase, in which a technology or application exploits information in the user profile in order to provide personalized services, is discussed in Parts II and III of this book.

2.2 Collecting Information About Users

The first phase of a profiling technique collects information about individual users. A basic requirement of such a system is that it must be able to uniquely identify users. This task is described in more detail in Section 2.2.1. The information collected may be explicitly input by the user or implicitly gathered by a software agent. It may be collected on the user's client machine or gathered by the application server itself. Depending on how the information is collected, different data about the users may be extracted. Several options, and their impacts, are discussed in Section 2.2.2. In general, systems that collect implicit information place little or no burden on the user are more likely to be used and, in practice, perform as well or better than those that require specific software to be installed and/or explicit feedback to be collected.

2.2.1 Methods for User Identification

Although accurate user identification is not a critical issue for systems that construct profiles representing groups of users, it is a crucial ability for any system that constructs profiles that represent individual users. There are five basic approaches to user identification: software agents, logins, enhanced proxy servers, cookies, and session ids. Because they are transparent to the user, and provide cross-session tracking, cookies are widely used and effective. Of these techniques, cookies are the least invasive, requiring no actions on the parts of users. Therefore, these are the easiest and most widely employed. Better accuracy and consistency can be obtained with a login-based system to track users across sessions and between computers, if users can be convinced to register with the system and login each time they visit. A good compromise is to use cookies for current sessions and provide optional logins for users who choose to register with a site.

Web usage mining can also be used to identify users, and these approaches are covered in more detail in Chapter 3 of this book [59]. Many companies rely on data aggregators, such as Acxiom [1], to provide demographic data about customers. This information actually turns out to be more accurate than surveys of customers themselves. Usually, all that is required to get full demographic data is a credit card number or the combination of name and zipcode, information that is often collected during purchase or registration.

The first three techniques are more accurate, but they also require the active participation of the user. Software agents are small programs that reside on the user's computer, collecting their information and sharing this with a server via some protocol. This approach is the most reliable because there is more control over the implementation of the application and the protocol used for identification. However, it requires user-participation in order to install the desktop software. The next most reliable method is based on logins. Because the users identify themselves during login, the identification is generally accurate, and the user can use the same profile from a variety of physical locations. On the other hand, the user must create an account via a registration process, and login and logout each time they visit the site, placing a burden on the user. Enhanced proxy servers can also provide reasonably accurate user identification. However, they have several drawbacks. They require that the user register their computer with a proxy server. Thus, they are generally able to identify users connecting from only one location, unless users bother to register all of the computers they use with the same proxy server.

The final two techniques covered, cookies and session ids, are less invasive methods. The first time that a browser client connects to the system, a new userid is created. This id is stored in a cookie on the user's computer. When they revisit the same site from the same computer, the same userid is used. This places no burden on the user at all. However, if the user uses more than one computer, each location will have a separate cookie, and thus a separate user profile. Also, if the computer is used by more than one user, and all users share the same local user id, they will all share the same, inaccurate profile. Finally, if the user clears their cookies, they will lose their profile altogether, and if users have cookies turned off on their computer, identification and tracking is not possible. Session ids are similar, but there is no storage of the userid between visits – each user begins each session with a blank slate, but their activity during the visit is tracked. In this case, no permanent user profile can be built, but adaptation is possible during the session.

2.2.2 Methods for User Information Collection

User profiles may be based on heterogeneous information associated with an individual user or a group of users who showed similar interests or similar navigational behavior. Broadly, user profile construction techniques can be partitioned by the type of input used to build the profile. In this section we discuss explicit and implicit feedback systems in detail. Hybrid approaches are also possible. Papazoglou [66] uses an automatic component to build a user profile based on user observations, but they also provide a mechanism for explicit relevance feedback in order to better tailor the profiles to user's individual interests.

Explicit User Information Collection. Explicit user information collection methodologies, often called explicit user feedback, rely on personal information input by the users, typically via HTML forms. The data collected may contain demographic information such as birthday, marriage status, job, or personal interests. In addition to simple checkboxes and text fields, a common feedback technique is the one that allows users to express their opinions by selecting a value from a range. All these methodologies have the drawback that they cost the user's time and require the user's willingness to participate. If users do not voluntarily provide personal information, no profile can be built for them.

Commercial systems have been exploring customization for some time. Many sites collect user preferences in order to customize interfaces. This customization can be viewed as the first step to provide personalized services on the Web. Many of the systems described in Section 2.4 rely on explicit user information. The collection of preferences for each user can be seen as a user profile and the services provided by these applications adapt in order to improve information accessibility. For instance, MyYahoo! [110], explicitly ask the user to provide personal information that is stored to create a profile. The Web site content is then automatically organized based on the user's preferences.

More sophisticated personalization projects based on explicit feedback have focused on navigation. One of the earliest, Syskill & Webert [68], recommends interesting Web pages based on explicit feedback. If the user rates some links on a page, Syskill & Webert can recommend other links on the page in which they might be interested. In addition, the system can construct a Lycos query and retrieve pages that might match a user's interest. The Wisconsin Adaptive Web Assistant (WAWA) [84,85] also uses explicit user feedback to train neural networks to assist users during browsing.

One problem with explicit feedback is that it places an additional burden on the user. Because of this, or privacy concerns, the user may not choose to participate. Users may not accurately report their own interests or demographic data, or, since the profile remains static whereas the user's interests may change over time, the profile may become increasingly inaccurate over time. An argument in favor of explicit feedback is that, in some cases, users enjoy providing, and sharing, their feedback. This is most evident in movie rating sites such as NetFlix [62] and sites dedicated to collecting, and sharing, consumer ratings such as ePinions [24].

Implicit User Information Collection. User profiles are often constructed based on implicitly collected information, often called implicit user feedback. The main advantage of this technique is that it does not require any additional intervention by the user during the process of constructing profiles. Kelly and Teevan [36] give an overview of the most popular techniques used to collect implicit feedback, and the type of information about the user that can be inferred from the user's behavior. Table 2.1 summarizes the approaches covered in this chapter, the type of information each approach is able to collect, and the breadth of applicability of the collected information. Because they only require a one time setup, do not require new software to be developed and installed on the user's desktop, and only track browsing activity, proxy servers seem to be a good compromise between easily capturing information and yet not placing a large burden on the user. Capturing activity at the site providing personalized services, for example a search site itself, is also an option in some cases. It requires absolutely no special user activity, but not all personalized sites are used frequently enough by any single user to allow them to create a useful profile.

Table 2.1. Implicit User Information Collection Techniques

Collection Technique	Information Collected	Information Breadth	Pros and Cons	Examples
Browser Cache	Browsing history	Any Web site	**pro**: User need not install anything. **con**: User must upload cache periodically.	OBIWAN [71]
Proxy Servers	Browsing activity	Any Web site	**pro**: User can use regular browser. **con**: User must use proxy server.	OBIWAN [71] Trajkova [99] Barrett et al [6]
Browser Agents	Browsing activity	Any personalized application	**pro**: Agent can collect all Web activity. **con**: Install software and use new application while browsing.	Letizia [43] WebMate [13] Vistabar [50] WebWatcher [58]

Table 2.1 (continued)

Desktop Agents	All user activity	Any personalized application	**pro**: All user files and activity available. **con:** Requires user to install software.	Seruku [83] Surfsaver [94] Haystack [2,17] Google Desktop [29] Stuff I've Seen [22]
Web Logs	Browsing activity	Logged Web site	**pro:** Information about multiple users collected. **con:** May be very little information since only from one site.	Mobasher [59]
Search Logs	Search	Search engine site	**pro:** Collection and use of information all at same site. **con:** Cookies must be turned on and/or login to site. **con:** May be very little information	Misearch [87] Liu et al [45]

Browsing histories are a common source of information from which user interests are extracted. Browsing histories are collected in two main ways: users share their browsing caches on a periodic basis [71]; or users install a proxy server that acts as their gateway to the Internet, thereby capturing all Internet traffic generated by the user [6, 99]. These browsing histories contain the urls visited by the user and the dates and times of the visits. Summary information about the number of visits to a particular url over a variety of time periods can be easily extracted. The time spent on the each page can also be inferred, with some error, as the time between consecutive hyperlink clicks. These browsing histories are typically shared with one particular Web site, allowing that site only to provide personalized services. Another drawback to this approach is that it typically only collects the user's browsing history from a single computer. However, a user could share their browsing caches from multiple computers or install the same proxy server on each computer they use regularly (e.g., home and work). Even if they do not do this, they could, via a login system, use the same user profile in multiple locations, allowing consistent access to personalized services.

Many personalization approaches use agents to collect information interactively, while the user browses. These browser agents are implemented as either a stand-alone application that includes browsing capabilities or a plug-in to an existing browser. Because the browser agents are installed on the user's desktop computer, they are able to capture all of the activities the user performs while browsing. Although not every system collects or uses all available information, this approach allows the system to collect a richer set of information about the user than is available via browsing histories. In addition to the urls visited and accurate information about the amount of time spent on each Web page, the agents can also collect actions performed on the Web page such as bookmarking and downloading to disk. Letizia [43, 44] was one of the first systems to interactively collect and exploit implicit user feedback. Based on previously visited pages and bookmarked pages, it suggests links on the current page that might be of interest. Other browsing assistants based on browsing agents are WebMate [13], Vistabar [50], and Personal WebWatcher [58]. Some literature in this area distinguishes between browsing assistants and browsing agents. Vistabar [50] is a prototypical browsing assistant, a tool that helps users track viewed urls, fill out forms or fetch pages without any specific agenda. In contrast, WebMate [13] and Personal WebWatcher [58] are examples of browsing agents that perform more critical tasks such as highlighting hyperlinks of likely interest to the user, recommending urls, or refining search keywords.

One drawback to this approach is that it requires the user to install a new application on their computer and, in the case of a stand-alone browsing application, it requires them to use a new application during browsing instead of a conventional browser. Another drawback is that this approach requires a large investment in software development and maintenance. In order to capture user information, the personalization system must develop a high quality browsing agent or plug-in, distribute it widely, and maintain and support numerous, widely-deployed versions that would result should the personalized application become successful. A final drawback to this approach is that, since it is resident on a personal computer, the user profile built would typically only be available when the user was using that particular computer. However, this drawback may be offset by the fact that, since it is resident on the user's computer, the user profile could be shared by multiple personalized applications.

There has been a recent surge in the availability of commercial toolbars and browser add-ons that include personalized features. Examples include the Seruku Toolbar [83] and SurfSaver [94], both of which try to help users organize their browsing histories stored in their desktop caches. These products are the direct descendents of the early browser agents developed by the research projects described above. Eventually, these personalized agents may evolve into a fully integrated personalized environment. In such a system, the searches would not be limited to the Web, but they would also include databases to which the user has access, and the user's personal documents. Such search systems are implemented in tools like Google Desktop Search [29] and Stuff I've Seen [22]. Then, the information found in the personal documents and databases could be used to enhance the user profile. The Haystack project [2, 17] presents the infrastructure necessary to create a personalized environment: a general purpose database to store all of the user's documents, the database management system, and the learning module in charge of maintaining the user profiles.

The above approaches all focus on collecting information about the users as they browse or perform other activities. Because they try to capture and share what the user is doing on their computer, they are essentially client-side approaches. All client-side approaches place some burden on the users in order to collect and/or share the log of their activities. In contrast, the final two approaches collect only the activities the user performs while interacting with the site providing the personalized services. Although they have access to less information than client-side approaches, they place no burden on the user at all, and can silently collect the information via cookies, logins, and/or session ids. There are two main sources of information for server-side personalization, browsing activity on the site and search interactions. Web logs capture the browsing histories for individual users at a given website. This information can be used to create Web sites that adapt their organization based on the user's behavior. Since web log mining is covered in detail in Chapter 3 of this book [59], it will not be discussed further here. However, search histories are discussed in some detail below.

Recently, search histories have been explored as a source of information for user profiling that can then be exploited to provide personalized search. Search histories contain information about the queries submitted by a particular user and the dates and times of those queries. The personalization system can also cache the urls and snippets of the result sets for each user's queries simultaneously with formatting that information for presentation to the user. If the personalization system wraps the presented results appropriately, the user clicks on particular results can also be collected. The personalization system could also download the complete Web pages for the visited urls. However, the network delays for this process are such that this cannot be done quickly enough to provide acceptable interactivity. Although downloading could be done as an offline process, this source of information is rarely used. As mentioned previously, this approach has the advantage that user does not need to install a desktop application or plug-in to collect their activities and/or upload their information to the personalized service. The service that is providing the personalized search collects the user activities as the user interacts directly with the site. If the site requires a login process, the same profile can be used whenever they visit the site regardless of the particular computer they are using. The disadvantage is that because only the activities at the search site itself are tracked, much less information is available. Also, the amount of representative text collected per interaction, i.e., the queries and/or snippets, is much less than the full text of Web pages typically collected for browsing-based profiles. However, several projects [45, 88] have been able to successfully provide personalized search by building user profiles based on this information.

Comparing Implicit and Explicit User Information Collection. Only recently have researchers begun to investigate the most effective source of information on which to build profiles. In 2000, Quiroga and Mostafa [73] compared systems using explicit feedback, implicit feedback, and a combination of the two by studying 18 users searching a collection of 6,000 health records classified into 15 different topics. Each user used the system for 15 sessions, and the highest precision of approximately 68% was achieved with profiles build from combined feedback. In contrast, explicit feedback alone produced a maximum precision of around 63% and the implicit feedback alone produced a maximum precision of around 58%. These differences were found to be statistically significant, suggesting that systems using the explicitly created profile or a profile built from a combination of explicit and implicit feedback produced better results than a system that made use of an implicitly created profile alone.

However, in contrast to the above findings, White et al. [102] did not find significant differences between profiles constructed using implicit and explicit feedback. They developed a system that used both implicit and explicit feedback to improve search on the Web. To compare these systems, they performed experiments with 16 users who searched the Web to answer specific questions on four topics. The successful completion of the task, the amount of time, and the number of result pages viewed to perform the task were used as metrics to evaluate the systems. The users who used the implicitly constructed profile were able to complete 61 out of the 64 tasks, while the users who used the explicitly created profile were able to complete only 57 tasks. Also, the average time per task for users with the implicit profile was 372 seconds, while the users with the explicit profile spent on 437 seconds on average. However, users with implicitly created profiles viewed approximately 3.3 results pages per task, more than the 2.5 pages viewed by users with explicitly created profiles. Since none of these differences was statistically significant, the authors concluded that implicit and explicit feedback were somewhat interchangeable.

In 2004 Wærn [100] studied the effect of user intervention on automatic filtering. The author compared the effectiveness of user profiles that were partially or completely built with automatic means. The study showed that although user intervention during profile construction can be useful, were not able to judge the quality of filtering and, furthermore, they were not able to improve the filters that were performing adequately.

Most recently, in 2005, Teevan et al. [98] evaluated a variety of information sources available to a client-side profiling agent, i.e., the Web pages visited, emails exchanged, calendar items, and all other documents stored on the client machine. Different rules, generating different collections, were used to gather information about the user, for example, recent documents only, Web pages only, documents only, and combinations of sources. In addition, two "lighter-weight" profiles were created: one constructed from search histories (queries issued in the past) and another from a list of all domains visited while browsing. They found that the richer the amount of information available, the better the profile performed. In particular, they found that the user profile built from the user's entire desktop index (the set of all information created, copied, or viewed by the user) was most accurate, followed by the profile built from recent information only, then that based on Web pages only. The least accurate profile was built from user-submitted queries only, but even it outperformed non-personalized search. They were also able to show that the profiles built from text collected implicitly from the user's desktop index could perform better than profiles built from explicit relevance feedback, a very promising result for future personalization systems.

These three studies, taken together, show that there is no clear answer on whether implicitly created profiles are more or less accurate than explicitly created profiles. However, the trend seems to be that the earlier study found explicit feedback better, the next study that the two forms of feedback were comparable, and the most recent study that implicit feedback was superior. This may indicate that, as experience with ways to collect and use implicit feedback has grown, the quality of the profiles constructed from this type of information improved. Since implicit feedback places less burden on the user, and it automatically updates as the user interacts with the system, it seems to be the preferable method of collecting information about users. One drawback to implicit feedback techniques is that they can typically only capture positive feedback. When a user clicks on an item or views a page, it seems reasonable to assume that this indicates some user interest in the item. However, it is not as clear, when a user fails to examine some data item, that this is an indication of disinterest. Thus, in general, implicit feedback techniques do not collect negative feedback.

2.3 User Profile Representations

User profiles are generally represented as sets of weighted keywords, semantic networks, or weighted concepts, or association rules. Because association rules are primarily used in the field of Web log mining, the subject of Chapter 3 of this book [59], they will not be discussed further here. Keyword profiles are the simplest to build, but because they fundamentally have to capture and represent all (or most) words by which interests may be discussed in future documents, they require a large amount of user feedback in order to learn the terminology by which a topic might be discussed. This problem is also shared by most semantic network-based profiles – they must learn the terminology with which concepts are discussed. Concept profiles, in contrast, are trained on examples for each concept *a priori*, and thus begin with an existing mapping between vocabulary and concepts. Thus, they can build profiles that are robust to variations in terminology with less user feedback. Many of the approaches described in this section rely on extracting, and weighting, keywords from documents and comparing documents to each other. The reader is referred to Chapter 5 of this book [54] on document representations for discussions of term weighting, vector representations of documents, and document similarity calculations.

2.3.1 Keyword Profiles

The most common representation for user profiles is sets of keywords. These can be automatically extracted from Web documents or directly provided by the user. Weights, which are usually associated with keywords, are numerical representations of user's interests. Each keyword can represent a topic of interest or keywords can be grouped in categories to reflect a more standard representation of user's interests. An example of a weighted keyword-based user profile is shown below in Figure 2.2.

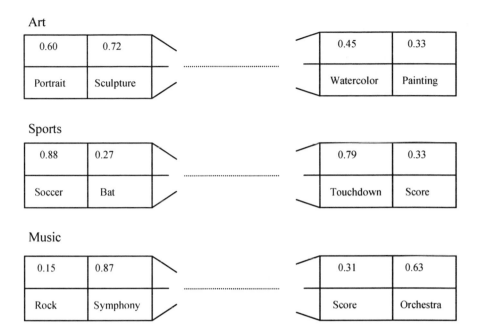

Fig. 2.2. A keyword-based user profile

Profiles represented in this way were among the first to be explored. The keywords in the profile are extracted from documents visited by the user during browsing, Web pages bookmarked or saved by the user, or the keywords were explicitly provided by the user. Each keyword is usually associated with a numerical weight representing its importance in the profile. *Amalthaea* [61] is one of many systems that creates keyword profiles by extracting keywords from Web pages. They weight the keywords with the widely used tf*idf weighting scheme from information retrieval [80]. Each profile is represented in the form of a keyword vector, and the documents that are retrieved by the system in response to a search are converted to similar weighted keyword vector. These vectors are then compared to the profile using the cosine formula [80], and only the corresponding documents for those vectors that are closest to the profile are then passed on to the user. The system also provides the user with the option of explicitly specifying their profile, which is weighted higher than the profile built by the system. This project is somewhat unique in that it employs a learning algorithm based on genetic algorithms to adapt and expand the user profiles. Weighted keyword vectors have also been used in *Anatagonomy* [78], a personalized online newspaper, *Fab* [5], a Web page recommender, and *Letizia* [43], a browsing assistant, and *Syskill & Webert* [68] a recommender system.

PEA [60] is also a personalized Web search assistant that builds keyword-based profiles using terms extracted from the user's bookmarked Web pages. However, it differs from the other approaches in that, rather than creating a single profile for the user, the user is represented as a set of keyword/weight vectors, one per bookmark. The rationale behind this extension is that, if a user is interested in two topics, combining the keywords from both topics in a single vector results in a profile that points halfway between them. In contrast, representing each area of interest, as indicated by a bookmark, as a separate vector is likely to provide a more accurate profile. As the user browses, additional pages are recommended to user when the vector for a potential new page is similar to a vector for an existing bookmark. *WebMate* [13] also builds user profiles containing one keyword vector per user's area of interest whereas *Alipes* [103] expands upon this approach by representing each interest with three keyword vectors, i.e., a long-term descriptor and two short-term descriptors, one positive and one negative.

PSUN [91], a personalized system for reading Usenet news, improves on the keyword vector representation by representing user profiles using weighted word sequences. The profiles are thus made up of weighted n-grams, i.e., word sequences of length n. Each n-gram has an associated weight that estimates the likelihood of the words in the n-gram co-occurring in a document and a strength that represents the importance of that n-gram relative to all other n-grams in the profile. One of the main drawbacks to keyword-based profiles is that many words have multiple meanings. Because of this polysemy, the keywords in the user profile are ambiguous, making the profile inaccurate. By focusing on word sequences, which are essentially statistically derived phrases, the contexts of the individual words are constrained. They report that profiles built from n-grams of length 2, i.e., word pairs, are more accurate than profiles built from individual keywords, but no formal analysis is presented.

More details about personalization based on keyword profiles can be found in Chapter 10 of this book [67].

2.3.2 Semantic Network Profiles

In order to address the polysemy problem inherent with keyword-based profiles, the profiles may be represented by a weighted semantic network in which each node represents a concept. Minio and Tasso [56] explore an approach based on this in which each node contains a particular word found in the corpus and arcs are created based upon co-occurrences of the two words in the connected nodes. Their user model is further enhanced by the inclusion of a set of attribute-value pairs corresponding to the structured part of the documents, e.g., host, size, number of images, etc., that have previously been of interest to the user [4]. The SiteIF project also uses a word-based semantic to represent user profiles [92]. However, they found that representing individual words as nodes in the semantic network was not accurate enough to discriminate word meanings. Instead, they used information inherent in WordNet to group related words together in concepts called "synonym sets," or synsets. They represent a user profile as a semantic network in which the nodes are synsets, the arcs are co-occurrences of the synset members within a document of interest to the user, and the node and arc weights represent the user's level of interest.

InfoWeb [28], a filtering system for online digital libraries documents, also builds semantic network based profiles that represent long-term user interests. Each user profile is represented as a semantic network of concepts. Initially, each semantic network contains a collection of unlinked nodes in which each node represents a concept. Concept nodes, called *planets*, contain a single, representative weighted term for that concept. As more information about the user is gathered, the profile is enriched to include additional weighted keywords associated with the concepts. These keywords are stored in subsidiary nodes, called *satellites*, linked to their associated concept nodes (planets). Links are also added between planets representing associations between concepts. Figure 2.3 shows an example excerpt of a user model based on this representation.

This representation was extended in WIFS [53], a filtering interface for personalizing results from the AltaVista [3] search engine. In this system, user profiles consist of three components: a header, including the user's personal data, a set of stereotypes, and a list of interests. A stereotype, or prototypical user, comprises a set of interests, represented by a frame of slots. Each slot contains three facets: *domain, topic*, and *weight*. The domain identifies an area of interest for the user, the topic is the specific term used by the user to identify the interest, and the weight indicates the user's degree of interest in the topic. The user model is represented as a frame containing the facets *semantic links* and *justification links*, as well as *domain, topic*, and *weight*. Figure 2.4 shows a sample profile based on this representation.

The semantic links include lists of keywords co-occurring in a document associated with the slot and having a degree of affinity with the topic. In this case, the profile is seen as a set of semantic networks, for which a slot is a planet and semantic links are the *satellites*. Figure 2.5 offers a simple example of just such a semantic network.

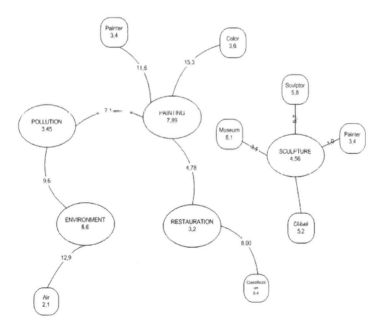

Fig. 2.3. An excerpt of a user profile based on semantic networks

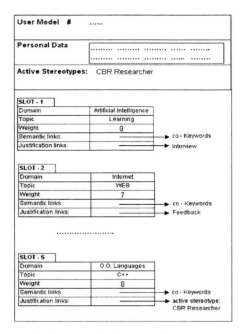

Fig. 2.4. An excerpt of user profile based on frames and semantic networks

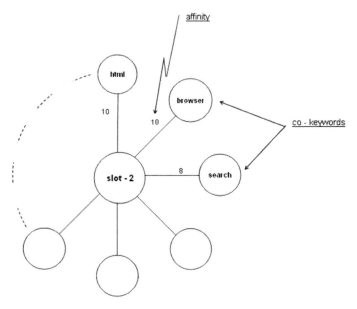

Fig. 2.5. An example of a semantic network

The justification links track down the reason why the slot to which they belong was inserted into the model. Their use is described in Section 2.4.2.

The system described in [25, 26] creates semantic network-based profiles that are used to model the interaction between users and information sources. Their Search of Associative Memory model tries to represent human memory, taking into consideration both the structure and the processes operating within it. The profile is organized into two main components, called the Long-Term Store (LTS) and the Short-Term Store (STS). Thus, each profile essentially consists of two keyword vectors, one that represents the long-term interests of the user (LTS) and another that represents the user's short-term interests. (STS). In particular, the STS identifies volatile information that is a subset of the LTS components. Links are created between words and context and each link is assigned a strength value that is used both in the learning and in the retrieval process. During the learning process, this value is calculated based on the amount of time each pair is temporarily stored in the STS.

2.3.3 Concept Profiles

Concept-based profiles are similar to semantic network-based profile in the sense that both are represented by conceptual nodes and relationships between those nodes. However, in concept-based profiles, the nodes represent abstract topics considered interesting to the user, rather than specific words or sets of related words. Concept profiles are also similar to keyword profiles in that often they are represented as vectors of weighted features, but the features represent concepts rather than words or sets of words. Various mechanisms are applied to express how much the user is interested in each topic. The simplest technique is a numerical value, or weight, associated with each topic.

Bloedorn et al. [8] suggest using hierarchical concepts, rather than a flat set concepts, because this enables the system to make generalizations. The levels in the concept hierarchy can be fixed [99], or they can change dynamically according to the user's interests [15]. The simplest concept hierarchy based profiles are constructed from a reference taxonomy or thesaurus. More complex profiles may be constructed from reference ontologies. In the latter case, relationships between concepts are explicitly specified and the resulting profile may include richer information and a wide variety of relationship types.

Concept hierarchies were initially used to represent the content of Web pages [31, 42] but have more recently been used to represent user profiles. Most systems are based on a reference concept hierarchy, or taxonomy, from which a subset of the concepts and relationships are extracted and weighted to form a user profile. Because creating a broad and deep concept hierarchy is an expensive, mostly manual process, profiles are typically based on subsets of existing concept hierarchies. Conceptual search projects have used the *Sensus* ontology [31, 38], a taxonomy of approximately 70,000 nodes, and a subset of the *Yahoo!* directory [42, 111] as their reference conceptual hierarchies.

When using an existing directory as a source of concepts, certain transformations must take place to turn directory's contents into a concept hierarchy. Because the directory is designed to enable end-user browsing, not all parent-child links are conceptual. Some topics are split into children alphabetically, merely to partition the content. Others are split geographically. Some topics have dozens or hundreds of children whereas others may have few or none. Finally, some topics may have many Web pages linked to that subject whereas others may have little or no associated content. The profiling project must take these issues into consideration and decide which of the directory's subjects to include in the concept hierarchy. The more levels used, the more specific the user profile representation can become. However, if too many levels are used, general areas of interest may be lost. Often, non-conceptual parent-child subjects are removed and also those topics which have too few associated Web pages to act as examples for the profiling algorithm.

One of the first projects to build concept-based user profiles was the OBIWAN project [72]. Initially, they used a reference concept hierarchy containing 4,417 topics from the top four levels of the Magellan site. After the Magellan site ceased to exist, the group experimented with subject hierarchies downloaded from Yahoo! [111] and Lycos [48], eventually selecting the Open Directory Project (ODP) as a replacement, primarily because their directory is open source [65]. Initially, they represented profiles using 1,869 concepts from the top three levels of the ODP concept hierarchy [9] but, because the ODP has grown, they have used as many as 2,991 concepts from the top three levels [99]. Figure 2.6 shows an example of a conceptual user profile built from user's browsing histories by the OBIWAN project using the top three levels of the ODP [9,72,99].

Figure 2.7 shows the Web display of a particular user's profile in the misearch system [88]. This system builds the user profiles from implicit feedback collected via search engine queries and clicked results. Users may view their top-weighted concepts with percentages that convey the relative weights of the concepts in the profile.

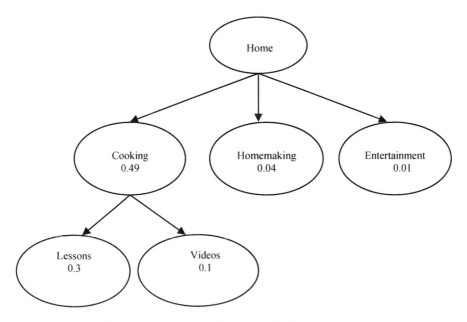

Fig. 2.6. An excerpt of a user profile based on concepts

Fig. 2.7. A conceptual user profile representing a user interested in cooking, builtfrom the top 3 levels of the ODP and the user's search history

The ODP is also used as the reference concept hierarchy in *Persona* [97]. However, because they use all concepts at any level in the ODP, they build more specific user profiles. These profiles contain only those concepts containing associated urls actually visited by the users, keeping the profile size scalable. On the other hand, because the Outride Personalized Search System [70] uses only 1,000 concepts from the Open Directory Project directory, their profiles are somewhat smaller than those used in OBIWAN and misearch, and they focus on capturing broad trends.

Although many of the previous projects may refer to their concept hierarchy as an ontology, the only relationship expressed is a parent-child relationship which generally represents an *is-a* and/or *has-a* relationship. The Semantic Web initiative is focusing on the creation and use of richer ontologies that can capture a wider variety of relationship types [7]. These ontologies are modeled using ontology representation languages such as *SHOE* [34, 47], *Extensible Markup Language* (XML) [106], the *Resource Description Framework* (RDF) [74], *RDF Schema* [75], *DAML+OIL* [19], or the *Web Ontology Language* (OWL) [101]. Some recent projects are exploring the use of these richer ontologies for improved search results [32, 112]. User profiles based on these richer ontologies may not be far away, however there remain serious roadblocks in the way, primarily due to scalability issues in creating large, diverse ontologies and exploiting them for searching large, distributed document collections. A comprehensive discussion of Semantic Web technologies for personalization can be found in Chapter 23 of this book [21].

2.4 User Profile Construction

User profiles are constructed from information sources using a variety of construction techniques based on machine learning or information retrieval. Depending on the user profile representation desired, different techniques may be appropriate. Techniques commonly used to construct keyword profiles are described in Section 2.4.1, whereas Section 2.4.2 and 2.4.3 describe construction techniques appropriate for semantic network profiles and concept profiles respectively. Profiles may be constructed manually by the users or experts, however, this is difficult and time consuming for most users and would be a barrier to widespread adoption of a personalized service. Techniques which automatically construct the profiles from user feedback are much more popular. Although some approaches use genetic algorithms or neural networks to learn the profiles, simpler, more efficient approaches based on probabilities or the vector space model are widely used and have been found to be effective in many applications.

No matter which construction method is chosen, the profile must be kept current to reflect the user's preferences accurately; this has proven to be a very challenging task [89]. Profile updating can be done automatically and/or manually. Automatic methods are preferred because it is less intrusive to the end user. Some authors warn against fully automatic profile updates, advising that user feedback, which requires minimal effort, should be used [90]. However, the results of experiments on fully automatic profile updating are promising [11, 12, 18, 72, 93].

2.4.1 Building Keyword Profiles

Keyword-based profiles are initially created by extracting keywords from Web pages collected from some information source, e.g., the user's browsing history or bookmarks. Some form of keyword weighting is done to identify the most important keywords from a given Web page, and often the number of words extracted from a single page is capped so that only the top N most highly weighted terms from any page contribute to the profile.

The simplest type of profiling construction technique produces a single keyword profile for each user. *Amalthaea* [61] is one of many systems that creates profiles by extracting keywords from Web pages. They weight the keywords with the widely used tf*idf weighting scheme from information retrieval [80]. This project is somewhat unique in that it employs a learning algorithm based on genetic algorithms to adapt and expand the user profiles. In addition to the tf*idf weighting scheme, other projects have explored using Latent Semantic Indexing (LSI) [20] and Linear Least Squares Fit (LLSF) [45] for creating the keyword-based feature vectors.

Building multiple keyword profiles for each user, one per interest area, creates a more accurate picture of the user. Consider a user interested in Sports and Cooking. A single keyword vector will point towards the middle of these two topics, creating a picture of a user fascinated in athletes who cook, or people who cook for Superbowl parties. In contrast, by using a pair of vectors, the user profile more accurately represents the user's two independent interests.

Table 2.2. Keyword Profile Construction Techniques

Profile Representation	Information Source	Construction Technique	Example
Single Keyword Vector	Web pages Implicit, positive feedback	Extract top-weighted keywords	Amalthaea [61]
One Keyword Vector per Interest	Web pages Explicit, positive feedback	Create document vector Compare interest vectors Merge closest interest vectors	WebMate [13]
Multiple Keyword Vectors per Interest	Web pages Explicit, positive and negative feedback	Create document vector Compare to interest vectors Add to closest match	Alipes [103]

WebMate [13] is an example of a system that builds user profiles that contain multiple keyword vectors, one per interest. Users provide explicit feedback on Web pages they view as they browse. Document vectors are created by extracting keywords from the Web pages that receive positive feedback. Stop words, very common words such as 'and' and 'or', are removed and light stemming, removal of common word suffixes, is done to decrease the vocabulary size. Words are weighted using the tf*idf method common in vector space approaches. Title and heading words are specifically identified and weighted more highly. Unlike other systems that require the user to explicitly label interesting documents with their area of interest, *WebMate* automatically learns the interest areas. The learning algorithm is supplied with a fixed number of desired interests, N. The first N positive examples are each assumed to be a unique interest, and the vector for each document is used as an interest vector. Once there are more than N positive examples supplied, the two most similar interest vectors, as determined by the cosine similarity metric [80], are combined into a single interest vector. Figure 2.8 shows the creation of an user profile in *WebMate*.

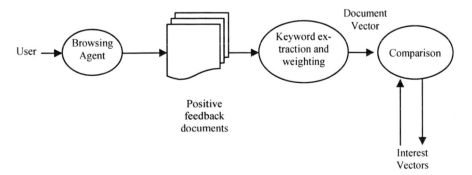

Fig. 2.8. Creation of keyword-based user profile in *WebMate*

Alipes [103] also creates user profiles that are based upon interest vectors, however they use multiple vectors per interest. In their case, each interest is modeled by three keyword vectors: long-term; short-term (positive), and short-term (negative). They consider negative feedback in addition to positive feedback, and the learning rate is affected by the strength of the user's preference. Like *WebMate*, they also automatically learn the user's interests, however, they base the creation of new interests on a similarity threshold rather than on a fixed number of desired interests. When a document vector is added to the user profile, it is compared to each of the three vectors for each interest using the cosine similarity metric. If the similarity exceeds a threshold, the document vector is added to the best matching interest. The strength of the user's feedback affects the amount of contribution the new document makes to the short-term vector, but the contribution to the long-term vector is determined by the number of example documents that have been learned so far, with the contribution factor declining over time. If, however, there is no match of sufficient strength between the document vector and the existing interest vectors, then a new interest is created and seeded with the document vector.

2.4.2 Building Semantic Network Profiles

Semantic network-based profiles are typically built by collecting explicit positive and/or negative feedback from users. Similar to keyword vector profile construction techniques, keywords are extracted from the user-rated pages. The techniques differ from those in the previous section because, rather than adding the extracted keywords to a vector, the keywords are added to a network of nodes. The nodes may represent individual words or, in more sophisticated approaches, a particular concept and its associated words. The terms "concepts" and "interests" are often used interchangeably in the literature. In this section, concept refers to a specific fine-grained idea and a collection of associated words, e.g., *dog* and its synonyms, whereas interest refers to higher level topics of interest to a user, e.g., *Animal Rights*, which in turn may be represented by a collection of associated concepts.

Semantic user profiles have an advantage over keyword-based profiles because they can explicitly model the relationship between particular words and higher-level concepts. Thus, they can deal more effectively with the inherent ambiguity and synonymy of natural language. However, this also places a barrier to the ease of constructing such system. They must either exploit an existing mapping between words and concepts, for example WordNet used by SiteIF, or they must build this through a learning mechanism as done by ifWeb [4], PIN [96], and InfoWeb [28], or they must build this manually, as is done in WIFS [53].

In the simplest systems, each user is represented by a single semantic network in which each node contains a single keyword. The ifWeb system [4] initially builds this type of profile by presenting the user with a pre-determined small set of documents (4-6) and collecting positive and negative feedback on these documents. The profile is then refined as the user browses via a browsing agent and provides further feedback on Web pages proposed by ifWeb. Keywords are extracted from each of the pages receiving user feedback. These keywords undergo standard preprocessing, i.e., segmentation, stopword removal, stemming, and weighting. Moreover, keywords that occur too few times in a document, compared to a given threshold, are excluded.

Table 2.3. Semantic Network Profile Construction Techniques

Profile Representation	Information Source	Construction Technique	Example
Single Semantic Network: One Node per Word	Sample documents Web pages Explicit positive and negative feedback	Extract top weighted words Create one node in semantic network per word Link nodes when the words they contain co-occur in documents	ifWeb [4]
Single Semantic Network: One Node per Concept	Web pages Implicit positive feedback	Extract top weighted words Map words into concepts using WordNet	SiteIF [92]
Single Semantic Network: One Node per Concept	Web pages Explicit positive feedback	Extract nouns Learn concepts using neural networks	PIN [96]
Single Semantic Network: One Planet per Concept, One Satellite per Word	Collection of stereotype documents Explicit positive and negative feedback Direct user refinement	Create concept nodes from explicit feedback Add keyword nodes and arcs by refinement	InfoWeb [28]
One Semantic Network per Interest: One Planet per Interest, One Satellite per Word	Collection of stereotype documents User Interview Explicit User Feedback, Direct Manipulation	Create concept nodes and keyword nodes using human experts Add concept nodes, keyword nodes, and arcs by refinement	WIFS [53]

These keywords are then submitted to the IFTool subsytem [56] that is in charge of updating the semantic network representing the user profile. Keywords are added to the semantic network in which each node represents a keyword and the arcs represent the co-occurrence of the keywords within a document. If the keyword is already present in the semantic network, that node's score is increased or decreased, according to user feedback. If the keyword does not already appear, then a new node is created. Finally, the set of keywords extracted from the document are used to update the weights on the co-occurrence arcs. The IFTool Linguistic Processor is used both for user profile construction and for document evaluation. When the document is to be evaluated, IFTool extracts the information about its structure and its content which is used in order to build the document internal representation. The ifWeb system is able to consider different formats of documents such as HTML, PDF, plain text and post-script documents. The analysis of these documents is performed by a syntax-directed parser for each document format and modules for segmentation, stopword removal, stemming, and contextual weighting. By comparing a browsed document to the user model using user-settable criteria, the ifWeb system classifies the document in one of the three categories 'interesting,' 'not-interesting,' or 'indifferent.'

Because the same concept can be expressed using many different words, semantic network profiles in which the nodes represent concepts, rather than individual words, are likely to be more accurate. The SiteIF [92] system builds this type of semantic network-based profile from implicit user feedback. Essentially, the nodes are created by extracting concepts from a large, pre-existing collection of concepts, WordNet [105]. As the user browses the Web, representative keywords are extracted from documents using the same process as the ifWeb system. These keywords are mapped into concepts using WordNet, a collection of 100,000 word forms organized into 80,000 *synsets*. Polysemous words are then disambiguated by analyzing their synsets to identify the most likely sense given the other words in the document. Finally, the synsets for the disambiguated representative keywords are combined to yield a user profile that is a semantic net whose nodes are concepts (synsets) and whose arcs represent the co-occurrence of two concepts within a document. Every node and every arc has a weight that represents the user's level of interest. In order to capture a shift in the user's interests over time, the weights in the network are periodically reconsidered and possibly lowered, depending on the time passed from the last update. Also, nodes and arcs that are no longer useful may be removed from the net.

When building user profiles based on semantic networks of concepts, the personalized system may not be able to afford the time and space overhead necessary to incorporate a large concept network. In other situations, there may not exist appropriate online collections of concepts from which nodes can be created. This may happen when the user's interests are domain specific, or based on recently created topics. PIN [96] is a personalized news system that also builds semantic network based user profiles in which the nodes represent concepts. Each node consists of an interest term, essentially the name of the concept, and a set of keywords related to that concept. Explicit user feedback is collected via user-supplied ratings of news articles. Because news stories, by definition, often cover previously unknown topics, PIN learns its concepts from the examples rather than mapping its examples to existing concepts. The profile is initially instantiated by the user supplying one or more concepts, consisting of an interest term and one or more keywords. For each positively rated article, a morphological analyzer is used to identify the part-of-speech of each word. To reduce complexity by reducing the number of features supplied to the neural network, only nouns are extracted. These are reduced to their stem, and the profile is searched to see if that keyword already appears in the list of keywords for the existing interest. If the keyword is found, the count for that keyword is incremented. The user feedback also sends the positively rated article to ARAM [95], a neural network based learning system. Through a refinement process, ARAM [95] learns new concepts that are not explicitly mentioned by the user, enhancing the user profile. The learned concepts are weighted according to the rating given to the Web page by the user.

InfoWeb [28], a query expansion and document filtering system for an online digital library, also builds user profiles represented as a semantic network of concepts. The profiles are built based on explicit user feedback collected by a browsing agent. One problem with explicit feedback techniques is the time and effort required from the users. By carefully selecting the documents on which the user feedback is collected, rather than presenting the user with random documents, they are able to collect representative information with less user effort. In order to identify documents that represent the scope of the digital library, they cluster the documents in the collection into a pre-determined set of k possible categories. To ensure that the categories created by the clustering algorithm are semantically meaningful, a domain expert chooses k documents that are representative of the k semantic categories into which the expert divides the collection. These documents are then used as the seeds for the clustering algorithm. After the clustering, the document closest to the centroid of the cluster, is selected to act as the representative document for the cluster, called the *stereotype*.

To build their initial profile, users are asked to give explicit relevance feedback (both positive and negative) about the stereotypes. Once the feedback is collected, the rated documents are processed to extract the highest weighted keywords, creating a single semantic network representation of the user profile. Initially, each extracted keyword from the rated stereotype is represented as a planet. Implicit feedback from users, or direct user manipulation of the profile, is used to maintain and refine the semantic network. As the user interacts with the system, feedback documents are matched to the profile based on a linear combination of the individual terms in the document and the user's semantic network, similar to that used in Rocchio classification [79]. This feedback is used to enrich the profile – when the weights of the links exceed a certain threshold – by adding satellite nodes linked to the appropriate planet node. User feedback is also used to create links between concepts, viewed as creating links between planets. These links are added and updated according to a distance metric calculated between the terms in the documents.

In the aforesaid approaches, a single semantic network is built to represent the user's interests. This approach suffers from the same lack of accuracy of systems that build a single keyword vector per user. It is generally preferable to produce a more fine-grained representation of user profiles as collections of interests, each represented by a semantic network. This approach is used by WIFS [53], an extension to InfoWeb [28] applied to searching the World Wide Web. Since this system is applied to Web search, it is not as easy to create the initial set of topics from which user interests can be selected. The Web is far too large to cluster its contents so as to determine all possible topics. Instead, the WIFS system features a preliminary work led by human experts, who identified the set of terms, stored in a Terms Data Base, deemed most relevant for each specific field of interest. Besides, these experts set a basic level of knowledge of stereotypes, each one representing the prototype user's information needs. The first time a user starts a working session, he is interviewed by the system to obtain a first set of information needs. This information is used to determine the stereotype(s) (named *active stereotypes*) that best approximate his information needs.

The user model is initialized with the information provided in the interview and with the data inherited from the active stereotypes.

As the user interacts with WIFS [53], his relevance feedback is used to refine the profile by adding, updating, or removing planets, satellites and affinities. There are five different ways to update the user model, four automatic and one manual. Firstly, the user's current interest, chosen from their existing interests, is updated according to the match between keywords in the semantic networks for each interest and keywords extracted from the currently viewed Web page and the query used to locate that page. Secondly, occurrences of keywords in the Web page that are already contained in the semantic network for the current interest are used to modify the affinity value of the arc between the satellite node, for the keyword, and the planet node, for the interest. The increase is proportional to the feedback value supplied by the user. If the new value does not fall within a predefined range, then the keyword node is removed from the network. Thirdly, new keywords extracted from the Web page are used to add new satellite nodes to the semantic network for the current interest. The value of the user's feedback on the page is used to set the weights on the arcs (affinity value) linking the new satellite nodes to the planet node whose topic pertains to the document; in the way the system maintains the term co-occurrence information. Fourthly, if the feedback value on the Web page exceeds a threshold, keywords extracted from the Web page can be used to add a new topic with the term's name and the domain = "filler". Finally, users are also able to explicitly modify their personal profile through direct manipulation. The consistency of the model is maintained by a simple justification-based truth maintenance system (JTMS) that uses the justification links described in Figure 2.4. They help provide explanations of the reason why the slot was inserted into the model and the evaluation of its weight. Such links maintain the consistency of the model, should the cause be removed. For example, if a shift in the user's interests occurs during the course of an interaction, thereby changing the active stereotype, the slots previously justified by the older stereotype will be eliminated.

2.4.3 Building Concept Profiles

This section describes three representative systems that build user profiles represented as weighted concept hierarchies. Although each uses a different construction methodology, they each use a reference taxonomy as the basis of the profile. These profiles differ from semantic network profiles because they describe the profiles in terms of pre-existing concepts, rather than modeling the concepts as part of the user profile itself. Thus, they all require some way of determining which concepts a user is interested in based on their feedback. Although some systems collect feedback on pre-classified documents, many collect feedback on a wide variety of documents then do text classification to identify the concepts exemplified by each document. Many research projects in this section refer to their concept hierarchies as *ontologies*. However, in this section, we use the term concept hierarchy when the ontology contains only "is-a" links, and restrict the use of the word ontology to (future) systems that support a rich variety of relationships between the concepts, including logical propositions that formally describe the relationship.

Table 2.4. Concept Profile Construction Techniques

Reference Taxonomy	Information Source	Construction Technique	Example
Open Directory Project All Concepts	Explicit positive feedback on pre-classified Web pages	Tree-coloring	Persona [97]
Yahoo!	Implicit positive feedback on Web pages and search results	Clustering	ARCH [86]
CORA 97 Concepts	Implicit and explicit positive feedback on pre-classified research papers	Tree-coloring Propagation to parent concepts	Foxtrot [55]
Open Directory Project ~2,000 Concepts	Implicit positive feedback on any Web page[1] or queries and search results[2]	Text classification to identify concepts	OBIWAN [72] Misearch [87]
Open Directory Project 619 Concepts	Implicit positive feedback using queries, search results; explicit positive feedback on categories	Text classification to identify concepts Expand classifier training based on feedback	Liu et al [45]
Open Directory Project 55 Concepts	Implicit positive feedback on any Web page	Text classification to identify concepts Taxonomy adapts to add/remove concepts	PVA [15]
ACM Topic Hierarchy 1,287 Concepts	Implicit feedback via bibliography contents, queries Explicit feedback via profile manipulation	Tree coloring Direct manipulation Recommendations	Bibster [33]

Persona [97] is exploring personalized search that exploits user profiles represented as a collection of weighted concepts based upon the Open Directory Project's concept hierarchy. The system builds a taxonomy of user interest and disinterest using a tree coloring method. As the user searches the collection of pre-classified documents in the ODP, they are asked to provide explicit feedback on the resulting pages. This feedback is then used to update their profile. Since the pages are already manually mapped into the ODP concepts, the user profile can be easily updated by keeping a count of the number of times a given concept was visited, i.e., had a page viewed by the user, and the number of positive and negative feedbacks the node received, and the set of urls associated with the node. Because the system uses pre-classified documents, the profile is able to contain any or all concepts in the ODP and the mapping of visited pages to concepts is very accurate. One difficulty with this approach is that, because the ODP hierarchy is so deep and contains so many concepts, the profile can become very large and contain many very narrow concepts. When using this profile to provide personalized services, matching may need to be done using the parent, grandparent, or higher level ancestors of colored concepts, and deciding the level at which to perform matching remains to be investigated.

The ARCH system [86] is a hybrid approach combining keyword vector based user profiles with a concept hierarchy. The system collects implicit user feedback in the form of browsing and search activity in order to identify a set of documents in which the user has shown interest. These documents are clustered to identify their areas of interest and the centroid for each cluster is calculated, producing a weighted keyword vector representing that interest. The authors expand upon the keyword vector approaches described in Section 2.4.1, however, by mapping between user interests and concepts in the Yahoo! concept hierarchy. For each Yahoo! concept, the system calculates the centroid of a set of training documents. When a the user enters a query, they identify the most similar interest vector, then calculate the similarity between that interest vector and the concept vectors to find the most similar concept. Terms from the top-matching concept are then used for query expansion. Although this is an interesting approach, the other projects that explicitly model user profiles as collections of weighted concepts are computationally more efficient and likely to be as accurate as this hybrid.

Persona builds its profiles from manually classified documents and ARCH employs clustering to identify user interests. All of the other systems in this section rely on text classification in order to map the information collected about the user into the appropriate concept(s) in a concept hierarchy. This approach seems quite robust, but hinges on the quality of the information used to train the text classifier, the match between the feedback documents and the training documents, and the accuracy of the classifier. Text classification is a supervised approach that attempts to assign documents to the best matching concept(s) from a predefined set of concepts. It is comprised of two phases: learning and classification. In the learning phase, the system is given a series of documents classified by hand, and it attempts to acquire enough information from them in order to classify a new document. In the classification phase, the system receives a new document and assigns it a concept label based on its match with the training data. Several methods for text classification have been developed, each with a different approach for comparing the new documents to the reference set. These include comparisons between a variety of frequently-used vector representations of the documents (Support Vector Machines, k-nearest neighbor, linear least-squares fit, tf * idf); the use of the joint probabilities of the words being in the same document (Naive Bayesian); decision trees; and neural networks. A very complete survey and comparison of such methods is presented in [108], and more are discussed in [67, 77, 80]. Recent approaches focus on extensions of traditional classification approaches to hierarchical concept hierarchies [23, 57].

The Foxtrot recommender system for a digital library of computer science papers [55] takes a similar approach to Persona. The authors organize the library contents into a concept hierarchy of 97 classes [51], and manually provide 5-10 example documents for each concept. They employ text classification techniques to automatically classify the remainder of the papers in the library. As users interact with an online digital library of pre-classified research papers, implicit (browsing activity) and explicit feedback (relevance judgments) is collected. The concepts associated with the documents are used to update the concept profile, with explicit feedback contributing 10 times more to a concept's weight than implicit feedback. The system also includes a linear time-delay factor so that, as days go by, previously contributed papers contribute less and less to the weight of the concepts in the profile. This may not be a necessary enhancement to the algorithm since, as more feedback is collected, concepts that are no longer of interest will cease to grow whereas concepts for current interests will continue to grow and the relative weights of the past and current interests will shift. The time-delay factor merely accelerates this process. The system also propagates the 50% of the weight of low-level concepts to their parent concepts. This is an important enhancement, allowing the profile to represent the fact that a user interested in, for example, "Machine Learning" and "Data Mining" is also interested in the parent concept, "Artificial Intelligence."

The OBIWAN project represents user profiles as a weighted concept hierarchy built from a reference concept hierarchy. The profile creation process in OBIWAN is shown in Figure 2.9. Initially, the Magellan directory was used [71], but more recently the Open Directory Project has been adopted [9]. The main difference between this approach and Persona is that the system is not restricted to building the user profiles from pre-classified documents. Any source of representative text may be automatically classified by the system to find the best matching concepts from the ODP, and then those concepts have their weights increased. The project focuses on personalized search and navigation as a way to validate the quality of the profiles produced. The authors have built the profiles based upon browsing histories submitted as browser caches [27], collected by proxy servers [98], or captured from desktop screens [10]. Most recently, this approach has been used in the misearch project [87] that builds the profiles by collecting and classifying the user search histories rather than the user's browsing history. Unlike Persona, which requires explicit feedback from the user, the profiles are built from implicit information, the queries submitted and snippets of search results clicked on by the user. These profiles, regardless of which information source has been used, have been shown to be able to statistically significantly improve personalized search.

Similar to the OBIWAN project, Liu et al [45] construct user profiles based on ODP categories using text classification. The authors use only the top 2 levels in their user profile, creating a broader overview of user interests. The system trains the classifier for the ODP categories using words extracted from the ODP associated documents,

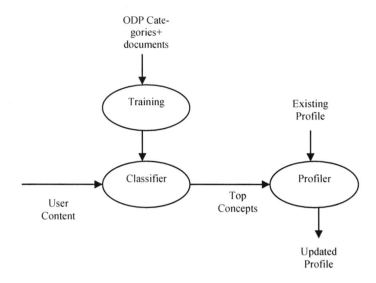

Fig. 2.9. User profile creation in OBIWAN

but, because they enhance the classifier with terms extracted from user-supplied feed-back, their profiles will be less sensitive to the terminology contained in the ODP training documents. Because the system requires users to explicitly indicate the cate-gories of interest, this approach is not entirely based on implicit feedback.

PVA [15] also builds a user profile represented as a weighted concept hierarchy from implicit feedback, in their case, information collected from a proxy server. Similar to OBIWAN, they automatically classify the representative documents for a user into a pre-existing concept hierarchy. Currently, the system uses a three-level concept hierarchy containing 55 concepts used by the Yam [107] search site. Because the authors are focusing a personalized recommendation of news articles, a fast changing domain, the system retrains the classifier daily with manually-classified new content collected from three news agencies. PVA differs from previous projects, however, in that it does not apply a variation of a tree coloring algorithm. Rather, the system uses the concepts returned by the classifier to dynamically grow and shrink the user profile. Initially, the user profile is represented just as the collection of top level concepts in the hierarchy. The results of the classification are therefore used to in-crease the weight, which they call Energy, of the appropriate top-level concept, and the full classification path for each classified document. For example, if a document is classified into "/Sports/Basketball/NBA", the Energy for "/Sports" will be in-creased. As documents are classified over a period of time, the Energy value for a concept may increase beyond a threshold. When this occurs, the system then splits that concept into two concepts based upon the classification paths of the contained documents. Thus, the profile grows as more information is collected about a given user. The authors also incorporate an aging process in their user profiling method. As time passes, the contribution a document makes to its associated concept's Energy value decreases. If the total Energy for a concept falls below a threshold, that concept is merged back into its parent concept. Thus, user profiles can grow and shrink to better reflect the user's interests.

As user profiles begin to move out from the research world and into use, the need to keep them accurate over time increases. PVA represents an entirely automatic approach to profile adaptation. In contrast, Bibster [33] employs collaborative feedback from multiple users with similar interests in order to recommend profile changes to users with similar users. Although the authors discuss users as having ontologies, essentially each user is represented by a concept profile, a weighted set of concepts selected from the ACM Topic Hierarchy. Users manually manipulate their profiles by adding/removing concepts and implicit feedback from the user's personal bibliography of documents (which are pre-categorized with respect to the concept hierarchy) and as they search. The unique feature of this system is that, as one user manipulates/updates their profile, these changes are also suggested to other users with similar profiles. The advantage of a collaborative approach is that, when many similar people are providing feedback, less feedback per individual is needed in order to construct and maintain an accurate profile. The drawback is that, when many people's feedback is combined, the fit between the profile and a particular individual may not be as good. The authors circumvent this process by presenting prospective changes to the profile as recommendations that are subject to user review before being adopted in their personal profile.

2.5 Conclusions

In conclusion, there is a tremendous growth in the approaches taken to represent, construct, and employ user profiles. These enabling technologies are key to providing users with accurate, personalized information services. There are a variety of techniques being investigated, but implicitly-created profiles place less burden on the user and, in several instances, seem to be able to adequately capture the user's interests. As these technologies mature, we see a move from simple keyword vectors to richer, conceptual representations. In future, profiles will also need to incorporate temporal and contextual information such as: What is the user doing now? What information has the user already seen? Where is the user located? However, personalized services are becoming a reality as user profiles move from the laboratory to the Internet.

References

1. The Acxiom Corporation, http://www.acxiom.com/ (last access on February 2006)
2. Adar, E., Karger, D.: Haystack: Per-User Information Environments. In: Proceedings of the 8th International Conference on Information Knowledge Management (CIKM), Kansas City, Missouri, November 2-6 (1999) 413-422
3. Altavista search engine, http://www.altavista.com/ (last access on October 2005)
4. Asnicar, F., Tasso, C.: ifWeb: A Prototype of User Model-Based Intelligent Agent for Documentation Filtering and Navigation in the World Wide Web. In: Proceedings of the 6th International Conference on User Modeling, Chia Laguna, Sardinia, Italy, June 2-5 (1997) 3-11
5. Balabanovic, M., Shoham, Y.: Fab: Content-Based Collaborative Recommendation. In: Communications of the ACM, 40(3), March (1997), 66-72

6. Barrett, R., Maglio, P., Kellem, D.C.: How to Personalize the Web. In: Proceedings of the SIGCHI conference on Human factors in computing systems, Atlanta, March 22-27 (1997) 75-82
7. Berners-Lee, T., Hendler, J., Lassila, O.: The Semantic Web. Scientific American, 284(5) May (2001) 34–43
8. Bloedorn, E., Mani, I., MacMillan, T.R.: Machine Learning of User Profiles: Representational Issues. In: Proceedings of AAAI 96, IAAA 96, Portland, Oregon, August 4-8, (1) (1996) 433-438
9. Chaffee, J., Gauch, S.: Personal Ontologies for Web Navigation. In: Proceedings of the 9th International Conference On Information Knowledge Management (CIKM), Washington, DC, November 6-11 (2000) 227-234
10. Challam, V., Gauch, S.: Contextual Information Retrieval Using Ontology Based User Profiles. In: ACM Transactions on Internet Technologies (pending)
11. Chan, K.P.: A Non-Invasive Learning Approach to Building User Web Profiles. In: Proceedings of the KDD-99 Workshop on Web Usage Analysis and User Profiling, San Diego, August 15-18 (1999) 39-55 (last access on October 2006) http://citeseer.ist.psu.edu/chan99noninvasive.html
12. Chan, K.P.: Constructing Web User Profiles: A Non-Invasive Learning Approach. In: Web Usage Analysis and User Profiling, LNAI 1836, Springer- Verlag (2000) 39-55
13. Chen, L., Sycara, K.: A Personal Agent for Browsing and Searching. In: Proceedings of the 2nd International Conference on Autonomous Agents, Minneapolis/St. Paul, May 9-13, (1998) 132-139
14. Chen, Y-S., Shahabi, C.: Automatically improving the accuracy of user profiles with genetic algorithm. In: Proceedings of IASTED International Conference on Artificial Intelligence and Soft Computing, Cancun, Mexico, May 21-24 (2001) 283-288
15. Chen, C., Chen, M., Sun, Y.: PVA: A self-adaptive Personal View Agent. Journal of Intelligent Information Systems, 18 (2-3), March-May (2002) 173-194
16. Chesnais, P., Mucklo, M., Sheena, J.: The Fishwrap Personalized News System. In: Proceedings of IEEE 2nd International Workshop on Community Networking: Integrating Multimedia Services to the Home, Princeton, NJ, June 20-22 (1995) 275-282
17. Chien, W.: Learning Query Behavior In the Haystack System. Master's thesis, MIT, June (2000)
18. Crabtree, B., Soltysiak, S.: Identifying and Tracking Changing Interests. International Journal on Digital Libraries, 2(1), (1998), 38-53
19. The DARPA Agent Markup Language Homepage, http://www.daml.org/ (last access on October 2005)
20. Deerwester, S., Dumais, S., Furnas, G., Landauer, T.K., Harshman, R.: Indexing by Latent Semantic Analysis. Journal of the American Society for Information Science, 41(6) (1990) 391-407
21. Dolog, P., and Nejdl, W.: Semantic Web Technologies for the Adaptive Web. In: Brusilovsky, P., Kobsa, A., Nejdl, W. (eds.): The Adaptive Web: Methods and Strategies of Web Personalization, Lecture Notes in Computer Science, Vol. 4321. Springer-Verlag, Berlin Heidelberg New York (2007) this volume
22. Dumais, S., Cutrell, E., Cadiz, J.J., Jancke, G., Sarin, R., Robbins, D.C.: Stuff I've seen: a system for personal information retrieval and re-use. In: Proceedings of the 26th annual international ACM SIGIR conference on Research and development in information retrieval, Toronto, Canada, July 28 - August 01, (2003) 72-79
23. Dumais, S., Chen, H.: Hierarchical classification of Web content. In: Proceedings of the 23rd ACM International Conference on Research and Development in Information Retrieval, (2000) 256-263
24. Epinions website, http://www.epinions.com/ (last access on February, 2006)

25. Gasparetti, F.: Adaptive Web Search: User Modeling based on Associative Memory and Multi-Agent Focused Crawling. PhD Thesis, University of Roma Tre, 2005
26. Gasparetti, F., Micarelli, A.: User Profile Generation Based on a Memory Retrieval Theory. In: The 1st International Workshop on Web Personalization, Recommender Systems and Intelligent User Interfaces (WPRSIUI 2005), Reading, UK, October 3-7 (2005) http://citeseer.ist.psu.edu/gasparetti05user.html
27. Gauch, S., Chaffee, J., Pretschner, A.: Ontology-Based User Profiles for Search and Browsing. In: Web Intelligence and Agent Systems 1(3-4) (2003) 219-234
28. Gentili, G., Micarelli, A., Sciarrone, F.: Infoweb: An Adaptive Information Filtering System for the Cultural Heritage Domain. Applied Artificial Intelligence 17(8-9) (2003) 715-744
29. Google Desktop, http://desktop.google.com/ (last access on October 2005)
30. Google Personalized Search, https://www.google.com/psearch/ (last access on September 2005)
31. Guarino, N., Masolo, C., Vetere, G.: OntoSeek: Content-Based Access to the Web. IEEE Intelligent Systems, May 14(3) (1999) 70-80
32. Guha, R., McCool, R., Miller, E.: Semantic Search. In: Proceedings of the WWW2003, Budapest, Hungary, May 20-24 (2003) 700-709
33. Haase, P., Hotho, A., Schmidt-Thieme, L., Sure, Y.: Collaborative and Usage-driven evolution of personalized ontologies. In: Proceedings of the 2nd European Semantic Web Conference, Heraklion, Greece, May 29-June 1 (2005) 486-499
34. Heflin, J., Hendler, J., Luke S.: SHOE: A Knowledge Representation Language for Internet Applications. Technical Report CS-TR-4078 (UMIACS TR-99-71), University of Maryland at College Park (1999) (last access on September 2005) http://www.cs.umd.edu/projects/plus/SHOE/pubs/techrpt99.pdf
35. Hoashi, K., Matsumoto, K., Inoue, N., Hashimoto, K.: Document Filtering method using non-relevant information profile. In: Proceedings of the 23rd Annual International ACM SIGIR Conference on Research and Development in Information Retrieval, Athens, Greece, July 24-28 (2000) 176-183
36. Kelly, D., Teevan, J.: Implicit feedback for inferring user preference: a bibliography. ACM SIGIR Forum 37(2) (2003) 18-28
37. Kim, H., Chan, P.: Learning implicit user interest hierarchy for context in personalization. In: Proceedings of IUI' 03, Miami, Florida, January 12-15 (2003) 101-108
38. Knight, K., Luk, S.: Building a Large Knowledge Base for Machine Translation. In: Proceedings of American Association of Artificial Intelligence Conference (AAAI), Orlando, Florida, July 18–22 (1999) 773- 778
39. Kobsa, A.: Privacy-Enhanced Web Personalization. In Brusilovsky, P., Kobsa, A., Nejdl, W. (eds.): The Adaptive Web: Methods and Strategies of Web Personalization, Lecture Notes in Computer Science, Vol. 4321. Springer-Verlag, Berlin Heidelberg New York (2007) this volume
40. Kobsa, A.: Generic User Modeling Systems. In Brusilovsky, P., Kobsa, A., Nejdl, W. (eds.): The Adaptive Web: Methods and Strategies of Web Personalization, Lecture Notes in Computer Science, Vol. 4321. Springer-Verlag, Berlin Heidelberg New York (2007) this volume
41. Konstan, J., Miller, B., Maltz, D., Herlocker, J., Gordon, L., Riedl, J.: GroupLens: Applying Collaborative Filtering To Usenet News. Communications of the ACM 40(3) (1997) 77-87
42. Labrou, Y., Finin, T. : Yahoo! As An Ontology – Using Yahoo! Categories To Describe Documents. In: Proceedings of the 8th International Conference On Information Knowledge Management (CIKM), Kansas City, Missouri, November 2-6 (1999) 180-187
43. Lieberman, H.: Letizia: An Agent That Assists Web Browsing. In: Proceedings of the 14th International Joint Conference On Artificial Intelligence, Montreal, Canada, August (1995) 924-929

44. Lieberman, H.: Autonomous Interface Agents. In: Proceedings of the ACM Conference on Computers and Human Interaction (CHI'97) Atlanta, Georgia, March 22-27 (1997) 67 - 74
45. Liu, F., Yu, C., Meng, W.: Personalized web search by mapping user queries to categories. In: Proceedings CIKM'02, Mclean, Virginia, November 4-9 (2002) 558-565
46. Liu, F., Yu, C., Meng, W.: Personalized Web Search For Improving Retrieval Effectiveness. In: IEEE Transactions on Knowledge and Data Engineering, 16(1), January (2004) 28-40
47. Luke, S., Spector, L., Rager, D., Hendler, J.: Ontology-Based Web Agents. In: Proceedings of the First International Conference on Autonomous Agents (AA'97) Association for Computing Machinery, California, February 5-8 (1997) 59-66
48. Lycos, http://www.lycos.com (last access on September 2005)
49. Malone, T., Grant, K., Turbak, F., Brobst, S., Cohen, M.: Intelligent Information Sharing Systems. Communications of the ACM 30(5) (1987) 390-402
50. Marais, H., Bharat, K.: Supporting cooperative and personal surfing with a desktop assistant. In: Proceedings of ACM UIST'97, Banff, Alberta, Canada, October 14-17 (1997) 129-138
51. McCallum, A.K., Nigam, K., Rennie, J., Seymore, K.: Automating the Construction of Internet Portals with Machine Learning. In: Information Retrieval 3(2) (2000) 127-163
52. Micarelli, A., Gasparetti, F., Sciarrone, F., and Gauch S.: Personalized Search on the World Wide Web. In Brusilovsky, P., Kobsa, A., Nejdl, W. (eds.): The Adaptive Web: Methods and Strategies of Web Personalization, Lecture Notes in Computer Science, Vol. 4321. Springer-Verlag, Berlin Heidelberg New York (2007) this volume
53. Micarelli, A., Sciarrone, F.: Anatomy and Empirical Evaluation of an Adaptive Web-Based Information Filtering System. User Modeling and User-Adapted Interaction 14(2-3) June (2004) 159-200
54. Micarelli, A., Sciarronne, F., Marinilli, M.: Web Document Modeling. In Brusilovsky, P., Kobsa, A., Nejdl, W. (eds.): The Adaptive Web: Methods and Strategies of Web Personalization, Lecture Notes in Computer Science, Vol. 4321. Springer-Verlag, Berlin Heidelberg New York (2007) this volume
55. Middleton, S.E., Shadbolt, N.R., De Roure, D.C.: Capturing interest through inference and visualization: Ontological user profiling in recommender systems. In: International Conference on Knowledge Capture, K-CAP 2003, Sanibel Island, Florida, September (2003) 62-69 http://portal.acm.org/citation.cfm?id=945657
56. Minio, M., Tasso, C.: User Modeling for Information Filtering on INTERNET Services: Exploiting an Extended Version of the UMT Shell. In: UM96 Workshop on User Modeling for Information Filtering on the WWW; Kailua-Kona, Hawaii, January 2-5 (1996) http://ten.dimi.uniud.it/~tasso/UM-96UMT.html
57. Mladenic, D.: Turning Yahoo into an Automatic Web-Page Classifier. In: Proceedings of the 13th European Conference on Aritficial Intelligence ECAI (1998) 473-474
58. Mladenić, D.: Personal WebWatcher: Design and Implementation. Technical Report IJS-DP-7472, J. Stefan Institute, Department for Intelligent Systems, Ljubljana, Slovenia (1998) (last access on October 2006) http://www-ai.ijs.si/DunjaMladenic/papers/PWW/pwwTR.ps.Z
59. Mobasher, B.: Data Mining for Web Personalization. In Brusilovsky, P., Kobsa, A., Nejdl, W. (eds.): The Adaptive Web: Methods and Strategies of Web Personalization, Lecture Notes in Computer Science, Vol. 4321. Springer-Verlag, Berlin Heidelberg New York (2007) this volume
60. Montebello, M., Gray, W., Hurley, S.: A Personal Evolvable Advisor for WWW Knowledge-Based Systems. In: Proceedings of the 1998 International Database Engineering and Application Symposium (IDEAS'98), Cardiff, Wales, U.K, July 8-10 (1998) 224-233
61. Moukas, A.: Amalthaea: Information Discovery And Filtering Using A Multiagent Evolving Ecosystem. In: Applied Artificial Intelligence 11(5) (1997) 437-457

62. Netflix Website, http://www.netflix.com/ (last access on February, 2006)
63. Nichols, D.: Implicit Rating and Filtering. In: Proceedings of the 5th DELOS Workshop on Filtering and Collaborative Filtering, Budapest, November 10-12 (1998) 31-36 (last access on October 2006) http://citeseer.ist.psu.edu/nichols98implicit.html
64. Oard, D., Marchionini, G.: A Conceptual Framework for Text Filtering. Technical Report EE-TR-96-25 CAR-TR-830 CLIS-TR-9602 CS-TR-3643. University of Maryland, May (1996)
65. The Open Directory Project (ODP), http://dmoz.org (last access on September 2005)
66. Papazoglou, M.: Agent-oriented technology in support of e-business. In: Communications of the ACM, 44(4), April (2001) 71-77
67. Pazzani, M., Billsus, D.: Content-based recommendation systems.In Brusilovsky, P., Kobsa, A., Nejdl, W. (eds.): The Adaptive Web: Methods and Strategies of Web Personalization, Lecture Notes in Computer Science, Vol. 4321. Springer-Verlag, Berlin Heidelberg New York (2007) this volume
68. Pazzani, M., Muramatsu, J., Billsus, D.: Syskill & Webert: Identifying Interesting Web Sites. In: Proceedings of the 13th National Conference On Artificial Intelligence Portland, Oregon, August 4–8 (1996) 54-61
69. Perkowitz, M., Etzioni, O.: Adaptive Web Sites: Automatically Synthesizing Web Pages. AAAI, Madison, Wisconsin, July 26–30 (1998) 727-732
70. Pitkow, J., Schütze, H., Cass T. et all.: Personalized search. CACM 45(9) (2002) 50-55
71. Pretschner, A.: Ontology Based Personalized Search. Master's thesis. University of Kansas, June (1999)
72. Pretschner, A., Gauch, S.: Ontology Based Personalized Search. In: Proceedings of the 11th IEEE International Conference on Tools with Artificial Intelligence (ICTAI) November 8-10 (1999) 391-398
73. Quiroga, L., Mostafa, J.: Empirical evaluation of explicit versus implicit acquisition of user profiles in information filtering systems. In: D. H. Kraft (Ed.), Proceedings of the 63rd annual meeting of the American Society for Information Science and Technology, Medford, NJ: Information Today 37 (2000) 4-13
74. Resource Description Framework, http://www.w3.org/RDF/ (last access on October 2005)
75. Resource Description Framework Schema, http://www.w3.org/TR/rdf-schema/ (last access on October 2005)
76. Rich, E.: Users are Individuals: Individualizing User Models. In: International Journal of Man-Machine Studies 18 (1983) 199–214
77. Ruiz, M., Srinivasan, P.: Hierarchical Neural Networks For Text Categorization. In: Proceedings of the 22nd Annual International ACM SIGIR Conference on Research and Development in Information Retrieval, Berkeley, California, August 15-19 (1999) 281-282
78. Sakagami, H., Kamba, T.: Learning Personal Preferences on Online Newspaper Articles From User Behaviors. In: Proceedings of the 6th International WWW Conference, Santa Clara, California, April 7-11 (1997) 291-300
79. Salton, G. Developments in automatic text retrieval. Science. Vol.253. Pages 974-979, 1991
80. Salton, G., McGill, M.: Introduction to Modern Information Retrieval. McGraw-Hill (1983)
81. Schafer, J.B., Frankowski, D., Herlocker, J., and Sen, S.: Collaborative Filtering Recommender Systems. In Brusilovsky, P., Kobsa, A., Nejdl, W. (eds.): The Adaptive Web: Methods and Strategies of Web Personalization, Lecture Notes in Computer Science, Vol. 4321. Springer-Verlag, Berlin Heidelberg New York (2007) this volume
82. Sebastiani, F.: Machine Learning in Automated Text Categorization. ACM Computing Surveys 34(1) (2002) 1–47
83. Seruku Toolbar, http://www.seruku.com/index.html (last access on October 2005)

84. Shavlik, J., Eliassi-Rad, T.: Intelligent Agents for Web-Based Tasks: An Advice-Taking Approach. In: Working Notes of the AAAI/ICML-98 Workshop on Learning for text categorization. Madison, WI, July 26-27 (1998) (last access on October 2006) http://citeseer.ist.psu.edu/shavlik98intelligent.html
85. Shavlik, J., Calcari, S., Eliassi-Rad, T., Solock, J.: An Instructable, Adaptive Interface for Discovering and Monitoring Information on the World Wide Web. In: Proceedings of the 1999 International Conference on Intelligent User Interfaces. Redondo Beach, California, January 5-8 (1999) 157-160
86. Sheth, B.: A Learning Approach to Personalized Information Filtering. Master's thesis. Massachusetts Institute of Technology (1994)
87. Sieg, A., Mobasher, B., Burke, R.: Inferring users information context: Integrating user profiles and concept hierarchies. In: 2004 Meeting of the International Federation of Classification Societies, IFCS, Chicago, July (2004) http://maya.cs.depaul.edu/~mobasher/p-pers/arch-ifcs2004.pdf
88. Speretta, M., Gauch, S.: Personalized Search based on User Search Histories. In: IEEE/WIC/ACM International Conference on Web Intelligence (WI'05). Compiegne University of Technology, France September 19-22 (2005) 622-628
89. Stadnyk, I., Kass, R.: Modeling User's Interests in Information Filters. Communications of the ACM, 35(12) December (1992), 49-50
90. Soltysiak, S.J., Crabtree, I.B.: Automatic Learning Of User Profiles - Towards the Personalization of Agent Services. BT Technology Journal 16 (3), July (1998) 110-117
91. Sorensen, H., McElligott, M.: PSUN: A Profiling System for Usenet News. In: Proceedings of CIKM'95 Workshop on Intelligent Information Agents, Baltimore Maryland, December 1-2 (1995)
92. Stefani, A., Strappavara, C.: Personalizing Access to Web Sites: The SiteIF Project. In: Proceedings of the 2nd Workshop on Adaptive Hypertext and Hypermedia HYPERTEXT'98 Pittsburgh, June 20-24 (1998) http://www.contrib.andrew.cmu.edu/~plb/HT98_workshop/Stefani/Stefani.html
93. Sugiyama, K., Hatano, K., Yoshikawa, M.: Adaptive web search based on user profile constructed without any effort from users. In: Proceedings 13th International Conference on World Wide Web, New York, May 17-22 (2004) 675-684
94. SurfSaver, http://www.surfsaver.com/ (last access on October 2005)
95. Tan, A.: Adaptive Resonance Associative Map. Neural Networks, 8(3) (1995) 437-446
96. Tan, A., Teo, C.: Learning user profiles for personalized information dissemination. In: Proceedings of 1998 IEEE International Joint Conference on Neural Networks, Alaska, May 4-9 (1998) 183-188
97. Tanudjaja, F., Mui, L.: Persona: A Contextualized and Personalized Web Search. In: Proc 35th Hawaii International Conference on System Sciences, Big Island, Hawaii, January (2002) 53
98. Teevan, J., Dumais, S., Horvitz, E.: Personalizing Search via Automated Analysis of Interests and Activities. In: Proceedings of 28th Annual International ACM SIGIR Conference on Research and Development in Information Retrieval, Salvador, Brazil, August 15-19 (2005) 449-456
99. Trajkova, J., Gauch, S.: Improving Ontology-Based User Profiles. In: Proceedings of RIAO 2004, University of Avignon (Vaucluse), France, April 26-28 (2004) 380-389
100. Wærn, A.: User Involvement in Automatic Filtering: An Experimental Study. In User Modeling and User-Adaptive Interaction, 14 (2-3), June (2004) 201-237
101. Web-Ontology(WebOnt) Working Group, http://www.w3.org/2001/sw/WebOnt/ (last access on February 2004)

102. White, R.W., Jose, J.M., Ruthven, I.: Comparing explicit and implicit feedback techniques for Web retrieval: TREC-10 interactive track report. In: Proceedings of the Tenth Text Retrieval Conference (TREC 2001, Gaithersburg, MD) 534-538
http://trec.nist.gov/pubs/trec10/papers/glasgow.pdf
103. Widyantoro, D.H., Yin, J., El Nasr, M., Yang, L., Zacchi, A., Yen, J.: Alipes: A Swift Messenger In Cyberspace. In: Proc. 1999 AAAI Spring Symposium Workshop on Intelligent Agents in Cyberspace, Stanford, March 22-24 (1999) 62-67
http://citeseer.ist.psu.edu/widyantoro99alipes.html
104. Widyantoro, D.H., Ioerger, T.R., Yen, J.: Learning User Interest Dynamics with Three-Descriptor Representation. Journal of the American Society of Information Science and Technology (JASIST) 52(3) February (2001) 212-225
105. The Wordnet Website, http://wordnet.princeton.edu/ (last access on February 2006)
106. eXtensible Markup Language, http://www.xml.com (last access on October 2005)
107. Yam search engine, http://www.yam.com (last access on October 2005)
108. Yan, T., García-Molina, H.: SIFT – A Tool for Wide-Area Information Dissemination. In: Proceedings of USENIX Technical Conference, New Orleans, Louisiana, January 16-20 (1995) 177-186
109. Yang, Y., Liu, X.: A Re-Examination Of Text Categorization Methods. In: Proceedings of the 22nd Annual International ACM SIGIR Conference on Research and Development in Information Retrieval, California, August 15-19 (1999) 42-49
110. Yahoo Personalized Portal, http://my.yahoo.com/ (last access on September 2005)
111. Yahoo Directory, http://dir.yahoo.com/ (last access on October 2005)
112. Zhu, H., Zhong, J., Li, J., Yu, Y.: An Approach for Semantic Search by Matching RDF Graphs. In: Proceedings of the Fifteenth International Florida Artificial Intelligence Research Society Conference, Pensacola Beach, Florida, May 14-16 (2002) 450-454

3

Data Mining for Web Personalization

Bamshad Mobasher

Center for Web Intelligence
School of Computer Science, Telecommunication, and Information Systems
DePaul University, Chicago, Illinois, USA
mobasher@cs.depaul.edu

Abstract. In this chapter we present an overview of Web personalization process viewed as an application of data mining requiring support for all the phases of a typical data mining cycle. These phases include data collection and preprocessing, pattern discovery and evaluation, and finally applying the discovered knowledge in real-time to mediate between the user and the Web. This view of the personalization process provides added flexibility in leveraging multiple data sources and in effectively using the discovered models in an automatic personalization system. The chapter provides a detailed discussion of a host of activities and techniques used at different stages of this cycle, including the preprocessing and integration of data from multiple sources, as well as pattern discovery techniques that are typically applied to this data. We consider a number of classes of data mining algorithms used particularly for Web personalization, including techniques based on clustering, association rule discovery, sequential pattern mining, Markov models, and probabilistic mixture and hidden (latent) variable models. Finally, we discuss hybrid data mining frameworks that leverage data from a variety of channels to provide more effective personalization solutions.

3.1 Introduction

The ultimate goal of any user-adaptive system is to provide users with what they need without them asking for it explicitly [89]. Automatic personalization, therefore, is a central technology used in such systems. In the context of the Web, personalization implies the delivery of dynamic content, such as textual elements, links, advertisement, product recommendations, etc., that are tailored to needs or interests of a particular user or a segment of users.

We distinguish between "automatic personalization" and what is sometimes referred to as "customization". Both customization and personalization refer to the delivery of content tailored to a particular user. What separates these two notions is who controls the creation of user profiles as well as the presentation of interface elements to the user. In customization, the users are in control of (often manually) specifying their preferences or requirements, based on which the interface elements are created. Examples of customization on the Web include customized Web sites, such as MyYahoo (www.yahoo.com), and a variety of e-commerce Web sites (such as www.dell.com)

P. Brusilovsky, A. Kobsa, and W. Nejdl (Eds.): The Adaptive Web, LNCS 4321, pp. 90–135, 2007.

that allow for manual configurations of systems or services before purchase. Automatic personalization, on the other hand, implies that the user profiles are created, and potentially updated, automatically by the system with minimal explicit control by the user. Examples of automatic personalization in commercial systems include Amazon.com's personalized recommendations, music or playlist recommenders such as Mystrand.com, and a variety of news filtering agents available today.

Traditional approaches to automatic personalization have included content-based, collaborative, and rule-based filtering systems. Each of these approaches is distinguished by the specific type of data collected to construct user profiles, and by the specific type of algorithmic approach used to provide personalized content. Generally, the process of personalization consists of a data collection phase in which the information pertaining to user interests is obtained and a learning phase in which user profiles are constructed from the data collected. Learning from data can be classified into memory based (also known as lazy) learning and model based (or eager) learning depending on whether the learning is done online while the system is performing the personalization tasks or offline using training data.

Standard user-based collaborative filtering and most content based filtering systems that use lazy learning algorithms are examples of the memory-based approach to personalization, while item-based and other collaborative filtering approaches that learn models prior to deployment are examples of model-based personalization systems.

Memory based systems simply memorize all the data and generalize from it at the time of generating recommendations. They are therefore more susceptible to scalability issues. Model-based approaches, that perform the computationally expensive learning phase offline, generally tend to scale better than memory based systems during the online deployment stage. On the other hand, as more data is collected, memory based systems are generally better at adapting to changes in user interests compared to model based techniques in which model must either be incremental or be rebuilt in order to account for the new data. These advantages and shortcomings have led to an extensive body of research and practice comprised of a variety of personalization or recommender systems that generally fall into the aforementioned categories.

Our goal in this chapter is not to provide an overview of automatic personalization, in general. Rather, we focus more specifically on *Web personalization* where the recommended objects come from a repository of Web objects (items or pages) browseable through navigation of links between the objects, usually in a particular Web site. Furthermore, we are particularly interested in a data mining approach to personalization where the goal is to leverage all available information about users of the Web site to deliver a personal experience.

Kohavi et al. [62] suggest five desiderata for success in data mining applications:

- data rich with descriptions to enable search for patterns beyond simple correlations;
- large volume of data to allow for building reliable models;
- controlled and reliable (automated) data collection;
- the ability to evaluate results; and
- ease of integration with existing processes (to build systems that can effectively take advantage of the mined knowledge).

Seldom are all these criteria satisfied in a typical data mining application. Personalization on the Web, and more specifically in e-commerce, has been considered the "killer app" for data mining, in part because many of these elements are indeed present. However, to be able to take full advantage of the flexibility provided by the data, and to effectively use the discovered models in an automatic personalization system, the process of personalization must be viewed as an application of data mining requiring support for all the phases of a typical data mining cycle [27], including data collection, pre-processing, pattern discovery and evaluation, in an off-line mode, and finally the deployment of the knowledge in real-time to mediate between the user and the Web.

The advantages and flexibilities afforded by the data mining approach to personalization come precisely from the fact that personalization is viewed as a holistic process rather than as individual algorithms or specific data types. Indeed, many of the traditional algorithms used for personalization can also be placed within the context of this process.

In this chapter we present a comprehensive view of the data mining approaches to personalization. We focus primarily on Web usage mining where the goal is to leverage data collected as a result of user interactions with the Web in order to learn user models and to use these models for personalization. We provide a detailed discussion of a host of data mining activities necessary for this process, including the preprocessing and integration of data from multiple sources, common pattern discovery techniques that are applied to this data in order to derive aggregate user models, and recommendation algorithms for combining the discovered knowledge with the current status of a user's activity in a Web site to provide personalized content to a user.

The remainder of this chapter is organized as follows. In Section 3.2 we provide a brief background on traditional approaches to automatic personalization and methods for profile generation based on different types of data. This discussion motivates our focus on the data mining approach. In Section 3.3, we discuss the essential data modeling and representation issues relevant to the personalization tasks, and in particular, provide a detailed discussion of the preprocessing and integration stage of the data mining cycle in the context of Web usage mining. Section 3.4, we consider a number of classes of data mining algorithms used particularly for Web personalization, and for each class, we present a number of specific approaches used in the literature. In this Section, we also discuss some of the shortcomings of the pure usage-based approaches and show how hybrid data mining frameworks, that leverage data from a variety of sources, can provide potential solutions to these shortcomings. Finally, in Section 3.5, we provide an overview dimensions along which personalization models can be evaluated an discuss some of commonly used evaluation metrics.

3.2 Automatic Personalization and Data Mining

The ability of a personalization system to tailor content and recommend items implies that it must be able to infer what a user requires based on previous or current interactions with that user, and possibly other users. The personalization task can therefore be viewed as a prediction problem: the system must attempt to predict the user's level of interest in, or the utility of, specific content categories, pages, or items, and rank these according to their predicted values. Furthermore, the task of delivering personalized content is often framed in terms of a recommendation task in which the system recommends items with the highest predicted interest values or utilities to an active user. In general, a personalization system can be viewed as a mapping of users and items to a set of "interest values". The view of personalization function as a prediction task comes from the fact that this mapping is not, in general, defined on the whole domain of user-item pairs, and thus requires the system to estimate the interest values for some elements of the domain.

Automatic personalization systems, generally, differ in the type of data and the method used to create user profiles, and in the type of algorithmic approaches used to make predictions. We will briefly describe each of these two dimensions below and provide an overview of the data mining approach to personalization which will guide our discussion in the remainder of this chapter.

3.2.1 Approaches to Personalization

From an architectural and algorithmic point of view personalization systems fall into three basic categories: Rule-based systems, content-filtering systems, and collaborative filtering systems. Our primary focus in this chapter is on model-based approaches to collaborative filtering in which models are learned through a variety of data mining techniques. However, we provide brief descriptions of each of these categories below. Additional details on traditional (e.g., memory-based) collaborative filtering techniques and content-based filtering algorithms can be found in Chapters 9 [117] and 10 [103] of this book, respectively. Furthermore, a great deal of work has focused on creating hybrid systems that combine various elements of these algorithms. A detailed characterization of hybrid recommender systems can be found in Chapter 12 [22].

Rule-Based Personalization Systems. Rule-based filtering systems rely on manually or automatically generated decision rules that are used to recommend items to users. Many existing e-commerce Web sites that employ personalization or recommendation technologies use manual rule-based systems. Such systems allow Web site administrators to specify rules, often based on demographic, psychographic, or other personal characteristics of users. In some cases, the rules may be highly domain dependent and reflect particular business objectives of the Web site. The rules are used to affect the content served to a user whose profile satisfies one or more rule conditions. Like most rule-based systems, this type of personalization relies heavily on knowledge engineering by system designers to construct a rule base in accordance to the specific characteristics of the domain or market research. The user profiles are generally obtained through explicit interactions with users. Some research has focused on machine learning techniques for classifying users into one of several categories based on their demographic attributes, and therefore, automatically derive decision rules that can be used for personalization [101].

The primary drawbacks of rule-based filtering techniques, in addition to the usual knowledge engineering bottleneck problem, emanate from the methods used for the generation of user profiles. The input is usually the subjective description of users or their interests by the users themselves, and thus is prone to bias. Furthermore, the profiles are often static, and thus the system performance degrades over time as the profiles age.

Content-Based Filtering Systems. In Content-based filtering systems, a user profile represent the content descriptions of items in which that user has previously expressed interest. The content descriptions of items are represented by a set of features or attributes that characterize that item. The recommendation generation task in such systems usually involves the comparison of extracted features from unseen or unrated items with content descriptions in the user profile. Items that are considered sufficiently similar to the user profile are recommended to the user.

In most content-based filtering systems, particularly those used on the Web and in e-commerce applications, the content descriptions are textual features extracted from Web pages or product descriptions. As such, these systems often rely on well-known document modeling techniques with roots in information retrieval [112] and information filtering [11] research. Both user profiles, as well as, items themselves, as represented as weighted term vectors (e.g., based on TF.IDF term-weighting model [112]). Predictions of user interest in a particular item can be derived based on the computation of vector similarities (e.g., using the Cosine similarity measure) or using probabilistic approaches such as Bayesian classification. Furthermore, in contrast with approaches based on collaborative filtering, the profiles are individual in nature, built only from features associated with items previously seen or rated by the active user. Chapter 5 of this book [76] provides a more detailed discussion of various approaches used in Web document modeling.

Examples of early personalized agents using this approach include Letizia [70], NewsWeeder [68], Personal WebWatcher [79], InfoFinder [66], Syskill and Webert [102], and the naïve Bayes nearest neighbor approach used by Schwab et al. [120]. A survey of the commonly used text-learning techniques in the context of content-based filtering can be found in [80].

The primary drawback of content-based filtering systems is their tendency to overspecialize the item selection since profiles are solely based on the user's previous rating of items. User studies have shown that users find online recommenders most useful when they recommend unexpected items [124], suggesting that using content similarity alone may result in missing important "pragmatic" relationships among Web objects such as their common or complementary utility in the context of a particular task. Furthermore, content-based filtering requires that items can be represented effectively using extracted textual features which is not alway practical given the heterogeneous nature of Web data.

A more detailed discussion of content-based filtering systems is provided in Chapter 10 [103].

Collaborative Filtering Systems. Collaborative filtering [64, 49] has tried to address some of the shortcomings of other approaches mentioned above. Particularly, in the context of e-commerce, recommender systems based on collaborative filtering have achieved notable successes [118]. These techniques generally involve matching the ratings of a current user for objects (e.g., movies or products) with those of similar users (nearest neighbors) in order to produce recommendations for objects not yet rated or seen by an active user. Traditionally, the primary technique used to accomplish this task is the standard memory-based k-Nearest-Neighbor (kNN) classification approach which compares a target user's profile with the historical profiles of other users in order to find the top k users who have similar tastes or interests.

However, collaborative filtering techniques have their own potentially serious limitations. The most important of these limitations is their lack of scalability. Essentially, kNN requires that the neighborhood formation phase be performed as an online process (i.e., the modeling phase is performed in real-time, in contrast to model-based approaches in which model learning is performed off-line from training data). As the numbers of users and items increase, this approach may lead to unacceptable latency for providing recommendations or dynamic content during user interaction.

Another limitation of kNN-based techniques emanates from the sparce nature of the dataset. As the number of items in the database increases, the density of each user record with respect to these items will decrease. This, in turn, will decrease the likelihood of a significant overlap of visited or rated items among pairs of users resulting in less reliable computed correlations. Furthermore, collaborative filtering usually performs best when explicit non-binary user ratings for similar objects are available. In many Web sites, however, it may be desirable to integrate the personalization actions throughout the site involving different types of objects, including navigational and content pages, as well as implicit product-oriented user events such as shopping cart changes, or product information requests.

A number of optimization strategies have been proposed and employed to remedy these shortcomings [2, 116, 140, 143]. These strategies include similarity indexing and dimensionality reduction to reduce real-time search costs and remedy the sparsity problems, as well as offline clustering of user records, allowing the online component of the system to search only within a matching cluster. A model-based variant of collaborative filtering is known as *item-based* collaborative filtering [114] in which, starting from the same user-rating profile databases, an item-item similarity matrix is built offline, and used in the prediction phase to generate recommendations. Rather than basing item similarity on content descriptions of the items, similarity between items is based on user ratings of these items. Each item is represented by a vector, and the similarities are computed using metrics such as cosine similarity and correlation-based similarity. The recommendation process predicts the rating for items not previously seen or rated by an active user using a weighted sum of the ratings, by that user, of items in the item neighborhood of the target item. Evaluation of the item-based collaborative filtering approach [35] has shown that item-based collaborative filtering can provide recommendations that are, in general, of similar quality when compared to memory-based collaborative approach.

Most data mining approaches to personalization can be viewed as extensions of collaborative filtering. In these approaches the pattern discovery algorithms take as input the historical rating or navigational profiles of past users and generate aggregate user models. The user models, in turn, can be used, in conjunction with the profile of an active user, to predict future user behavior or generate recommendations. This viewpoint will guide our presentation through the subsequent sections of this chapter.

A more detailed discussion of collaborative filtering systems is provided in Chapter 9 [117].

3.2.2 Approaches to User Profiling

All approaches to personalization, and to a greater degree, personalization based on data mining, require the collection of data that accurately reflect the interests of the users and their interactions with applications and items. Personalized systems differ, not only in the algorithms used to generate recommendations or make prediction, but also in the manner in which user profiles are built using this underlying data.

Rule-based and content-based personalization systems generally build an individual model of user interests and use this profile to tailor future interactions with only that user. As noted earlier, the content-based filtering systems require content features of items extracted from item descriptions, or relational attributes associates with items in the backend databases. In such systems the process of building a profile for a user requires two stages. First, the system must determine the level of user interest in a subset of items. This task may be accomplished implicitly by passively observing the user and using various heuristics to classify items as interesting or non-interesting [70, 79], or it can be based on explicit user judgment assigning ratings to items or manually identifying positive and negative examples [68, 102]. The transformation of each item (usually a Web page or document) into a bag or words (vector) representation, with each token being assigned a weight using methods such as *TF.IDF* [112] or minimum description length [109]. The profile is then used to recommend other similar items to the user. A major disadvantage of approaches based on an individual profiles is the lack of serendipity as recommendations are very focused on the user's previous interests. Also, the system depends on the availability of content descriptions of the items being recommended.

In the case of rule-based systems, particularly those based on demographic filtering, each user profile may be represented by a vector of personal and demographic attributes, sometimes called a *fingerprint*. In e-commerce and Web analytics applications, the visitor fingerprints may also include such computed attributes as total amount spent as well as the recency and frequency of purchase or visit. Few systems use demographic data within the recommendation process. This is due to the fact that such data is more difficult to collect on the Web and, when collected, tends to be of poor quality. Also, recommendations purely based on demographic data have been shown to be less accurate than those based on the item content and user behavior [101]. In Lifestyle Finder [65], externally procured demographic data (Claritas's PRIZM) was used to enhance demographic attributes obtained from the user, through an iterative process where the system only requests information pertinent to classifying the user into one of 62 demographic clusters defined within the PRIZM classification. Once classified, objects most relevant to that demographic cluster are recommended to the user.

In collaborative filtering, the system not only uses the profile for the active user but also maintains a database of other users' profiles. In contrast to content-based filtering in which item-to-item similarities form the basis for recommendation generation, collaborative systems rely on user-to-user similarities. Profiles are generally represented as a vector or set of ratings providing the user's preferences on a subset of items. An active user's profile is used to find other users with similar preferences, referred to as the active user's neighborhood. Note that as opposed to content-based filtering, the actual content descriptions of items are not part of the profile.

While traditional collaborative filtering only uses rating data, hybrid collaborative approaches that utilize both content and user rating data have also been proposed [6, 28, 75]. Furthermore, both in the case of collaborative and content-based filtering, various approaches have been explored to integrate ontological domain knowledge with user profiles [77, 45, 122, 145, 139]. In the presence of a domain ontology, the user profiles may actually reflect the structure of the domain, and thus may require a more complex representation than the flat representations used in standard approaches.

Regardless of the algorithmic approach to personalization, the data for user profiling can be collected implicitly or explicitly. Explicit collection usually requires the user's active participation. In systems that rely on demographic or personal information user interaction may take the form of participating in online surveys at the time of registration or providing personal and financial information during a purchase (which can then be combined with offline demographic data available through a variety of data aggregation services). Similarly, as noted above, content-based filtering systems can also use either implicit or explicit user feedback to determine the level of user interest in items. Traditional collaborative filtering systems used in e-commerce generally use explicit user feedback in the form of ratings on individual items. However, many collaborative systems, particularly Web personalization systems that use clickstream or other types of behavioral data, attempt to measure user interest in individual or groups of items based on heuristic indicators (such as time spent viewing the item, whether the item is purchased, etc.). Many e-commerce systems, such as Amazon.com, monitor each customer's purchase and activity history and use information as part of the user profiles.

The advantage of using implicit feedback for user profiling is that it removes the burden associated with providing personal information from the user. The system collects relevant data, based on users' observed behavior, and infers user-specific information. Implicit profiling implies that the system must be able to track and monitor user behavior in order to identify browsing or buying patterns. Implicit data could be collected on the client or on the server side. Approaches to personalization can be classified based on whether these approaches have been developed to run on the client side or on the server-side. The key distinction between these personalization approaches is the breadth of data that are available to the personalization system. On the client side, data is only available about the individual user and hence the only approach possible on the client side is *individual*. On the server side, the business has the ability to collect data on all its visitors and hence both individual and collaborative approaches can be applied. On the other hand, server side approaches generally only have access to interactions of users with content on their Web site while client side approaches can access data on the individual's interaction with multiple Web sites.

Most client side applications are content-based systems aimed at personalized search across the Web or multiple repositories [99, 122, 26, 138]. The lack of common domain ontologies across Web sites, the unstructured nature of the Web, and the sparseness of available behavioral data currently reduce the possibilities for personalization of navigational as opposed to search based interactions with the Web as whole.

Collaborative personalization systems based on Web usage mining, which are the primary focus of the remainder of this Chapter, rely on clickstream and navigational data automatically collected by Web and application servers and stored in server log file. Another source of customer data are transaction databases, pre-sale and after-sale support data, or demographic information. Such data could be dynamically collected by a Web site or purchased from third parties. In many cases data is stored in different formats in multiple, disparate databases.

We focus primarily on profiles built from implicit user feedback, collected automatically by monitoring users' activity histories, generally on the server-side. Our discussion is mainly centered around the application of data mining methodology and machine learning techniques that attempt to learn group profiles and generate user models that can be used to tailor a Web site's interactions with future users.

For a detailed discussion of various approaches to Web user profiling see Chapter 2 of this book [40].

3.2.3 Data Mining Approach to Personalization

The foregoing background motivates our focus on data mining (and more specifically, Web usage mining) as an approach to personalization. What makes the data mining approach to Web personalization different from the other approaches discussed above, is that Web usage mining is not a specific algorithm, but rather it follows the typical data mining cycle. As such, it provides a great deal of flexibility for leveraging different data channels in a comprehensive manner, and allows for the personalization tasks to be better integrated with other existing applications. Furthermore, because of the focus of data mining on efficient model-based pattern discovery algorithms, personalized systems based on data mining tend to be more scalable than those based on traditional approaches such as standard collaborative filtering.

Web usage mining [31, 130, 81] can be defined as the automatic discovery and analysis of patterns in clickstream and associated data collected or generated as a result of user interactions with Web resources on one or more Web sites. The goal of Web usage mining is to capture, model, and analyze the behavioral patterns and profiles of users interacting with a Web site. The discovered patterns are usually represented as collections of pages, objects, or resources that are frequently accessed by groups of users with common needs or interests.

Traditionally, the goal of Web usage mining has been to support the decision making processes by Web site operators in gaining better understanding of their visitors, create a more efficient or useful organization for the Web sites, and to do more effective marketing. However, these models can also be used by adaptive systems automatically in order to achieve various personalization functions.

The overall process of Web personalization based on Web usage mining consists of three phases: data preparation and transformation, pattern discovery, and recommendation. Of these, only the latter phase is performed in real-time.

The data preparation phase transforms raw Web log files into user profile or Web transaction data that can be processed by data mining tasks. This phase also includes data integration from multiple sources, such as backend databases, application servers, and site content. A variety of data mining techniques can be applied to this data in the pattern discovery phase, such as clustering, association rule mining, sequential pattern discovery, and probabilistic modeling. The results of the mining phase are transformed into aggregate user models, suitable for use in the recommendation phase. The recommendation engine considers the active user's profile in conjunction with the discovered patterns to provide personalized content.

In the following sections, we provide a detailed overview of the techniques and algorithms used in each of these phases.

3.3 Data Collection, Preprocessing, and Modeling

Viewing personalization as a data mining application, the aim is to create a set of user-centric data models (user profiles), representing the interests and activities of all users, that can be used as input to a variety machine learning algorithms for pattern discovery. The output from these algorithms, i.e., the patterns discovered, can then be used for predicting future interests of users. The exact representations of these user models differ based on the approach taken to achieve personalization and the granularity of the information available. The pattern discovery tasks would therefore differ in complexity based on the expressiveness of the user profile representation chosen and the data available.

3.3.1 Data Modeling and Representation

For the purposes of our discussion, we assume the existence of a set of m users, $U = \{u_1, u_2, \cdots, u_m\}$ and a set of n items, $I = \{i_1, i_2, \cdots, i_n\}$. We represent the profile for a user $u \in U$ as an n-dimensional vector of ordered pairs,

$$u^{(n)} = \langle (i_1, s_u(i_1)), (i_2, s_u(i_2)), \cdots, (i_n, s_u(i_n)) \rangle, \qquad (3.1)$$

where i_j's $\in I$ and s_u is a function for user u assigning (possibly null) interest scores to items.

In a typical data mining approach, such profiles are collected over time and stored for all users interacting with the system. Conceptually, the database of all user profiles can be represented as the $m \times n$ matrix, $UP = [s_{u_k}(i_j)]_{m \times n}$, where $s_{u_k}(i_j)$ is the degree of interest in item i_j by a user u_k.

Formally, a *personalization system* can be viewed as a mapping $PS : \mathcal{P}(UP) \times U \times I \rightarrow R \cup \{\texttt{null}\}$, assigning interest values to pairs of users and items, according to a set of user profiles. Because the mapping PS is not, in general, defined on the whole domain of user-item pairs, the system must estimate or predict the interest scores of a given user for elements of the domain. Depending on the prediction algorithm used, the system may not be able to an interest score for a particular user-item pair, in which case the PS mapping produces a $\texttt{null}$ value. In other words, the task of a personalization system can be viewed as one of predicting, for a given *target user* $u_k \in U$ and a *target item* $i_j \in I$, and the databases of user profiles UP, $PS(UP, u_k, i_j) = s_{u_k}(i_j)$.

Indeed, most of the approaches to personalization and user profiling, discussed in Sections 3.2.1 and 3.2.2 can be placed within this general framework. In content-based and some rule-based approaches, the user profile databases UP contains only a single profile, that of the target user, u_k, and the prediction of interest score, $s_{u_k}(i_j)$ for the target item i_j is based on its similarity to the user profile or based on the demographic or other personal attributes of the user. On the other hand, in the standard collaborative filtering context, the interest scores usually represent rating values from an ordered but discrete scale, and UP contains the past ratings of all users of the system. It that case, the prediction or estimation of the interest score for the target user is based, usually, on the similarity of that user's profile to other profiles in UP.

In the data mining approach to personalization, a variety of machine learning techniques are applied to UP in order to discover aggregate user models based on which a prediction is made for the target user. More specifically, in the context of personalization based on Web usage mining, our main focus in the remainder of this chapter, UP generally contains user transaction records representing their online activity (including clickthroughs or purchase transactions) in one or more sessions. The items are data abstractions representing pages, content categories, or products available on the Web site, and the interest scores are usually derived based on implicit observation of user activity on the Web site, such as time spent on a page, the purchase or selection of a product, etc.

Based on the above discussion, there are two important questions that must be answered before any type of pattern discovery or prediction can be performed: (a) what elements constitute the items in I, and (b) how is the function s_{u_k} defined for each user u_k? The answers to these questions, of course, depend on the type of approaches used for personalization and user profiling, the underlying application domain, and the types and sources of data available. In the knowledge discovery framework, the generation of the user-centric data representation is achieved through the application of several (often domain-specific) data collection, manipulation, and transformation operations. Collectively, we call the application of these operations the *data preprocessing* stage.

The goal of the preprocessing stage is to transform the raw data into a set of data abstractions that can be used in the above general framework. This includes the extraction and transformation of features or attributes that can be used to represent each item, as well as the extraction and transformation of explicit or implicit user attributes that are used to determine users' interest in various items (i.e., the functions $s_{u_k}(\cdot)$). As noted earlier, the extraction and transformation tasks vary depending on the application domain and context of personalization. Because our primary focus is on Web personalization, i.e., personalization of the Web users' navigational experience, in the following discussion we focus primarily on data preprocessing tasks for Web usage mining.

3.3.2 Data Sources for Web Usage Mining

The primary data sources used in Web usage mining are the server log files, which include Web server access logs and application server logs. Additional data sources that are also essential for both data preparation and pattern discovery include the site files and meta-data (including content features and structural elements of pages), operational databases, application templates, and domain knowledge [32, 130, 82]. In some cases and for some users, additional data may be available due to client-side or proxy-level (Internet Service Provider) data collection, as well as from external clickstream or demographic data sources (e.g., ComScore, NetRatings, MediaMetrix, and Acxiom).

The most important of these sources for Web usage mining is the clickstream data recorded automatically by the Web and application servers in log files. This data represents the fine-grained navigational behavior of visitors. Each hit against the server, corresponding to an HTTP request, generates a single entry in the server access logs. Each log entry (depending on the log format) may contain fields identifying the time and date of the request, the IP address of the client, the resource requested, possible parameters used in invoking a Web application, status of the request, HTTP method used, the user agent (browser and operating system type and version), the referring Web resource, and, if available, client-side cookies which uniquely identify a repeat visitor.

Depending on the goals of the analysis, this data needs to be transformed and aggregated at different levels of abstraction. In Web usage mining, the most basic level of data abstraction is that of a *pageview*. A pageview is an aggregate representation of a collection of Web objects contributing to the display on a user's browser resulting from a single user action (such as a click-through). Conceptually, each pageview can be viewed as a collection of Web objects or resources representing a specific "user event", e.g., reading an article, viewing a product page, or adding a product to the shopping cart. At the user level, the most basic level of behavioral abstraction is that of a *session*. A session is a sequence of pageviews by a single user during a single visit. The notion of a session can be further abstracted by selecting a subset of pageviews in the session that are significant or relevant for the analysis tasks at hand. A session can be used directly as the user profile (as described in the formal representation given in 3.1). However, if the goal of analysis is to capture the behavior of users over time (i.e., over multiple sessions), all sessions belonging to a user can be combined and aggregated to create the profile for that user.

The content data in a site is the collection of objects and relationships that are conveyed to the user. For the most part, this data is comprised of combinations of textual material and images. The data sources used to deliver or generate this data include static HTML/XML pages, multimedia files, dynamically generated page segments from scripts, and collections of records from the operational databases. The site content data also includes semantic or structural meta-data embedded within the site or individual pages, such as descriptive keywords, document attributes, semantic tags, or HTTP variables. The underlying domain ontology for the site is also considered part of the content data. Domain ontologies may include conceptual hierarchies over page contents, such as product categories, explicit representations of semantic content and relationships via an ontology language such as RDF, or a database schema over the data contained in the operational databases.

The structure data represents the designer's view of the content organization within the site. This organization is captured via the inter-page linkage structure among pages, as reflected through hyperlinks. The structure data also includes the intra-page structure of the content within a page. For example, both HTML and XML documents can be represented as tree structures over the space of tags in the page. The hyperlink structure for a site is normally captured by an automatically generated "site map", usually represented as a directed graph. A site mapping tool must have the capability to capture and represent the inter- and intra-pageview relationships. For dynamically generated pages, the site mapping tools must either incorporate intrinsic knowledge of the underlying applications and scripts, or must have the ability to generate content segments using a sampling of parameters passed to such applications or scripts.

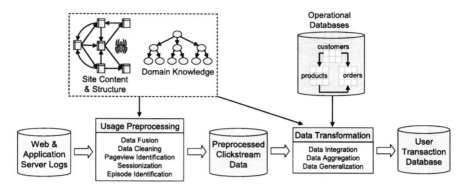

Fig. 3.1. Summary of the primary tasks and elements in usage data preprocessing.

Finally, the operational databases for the site may include additional information about user and items. Such data may include demographic information about registered users, user ratings on various objects such as products or movies, past purchase or visit histories of users, as well as other explicit or implicit representations of a user's interests. Product databases or content management systems may also include additional content descriptors and relational attributes that can be used as part of the representation of content information for items. Some of this data can be captured anonymously as long as there is the ability to distinguish among different users.

3.3.3 Data Preprocessing for Web Usage Mining

The goal of the preprocessing stage in Web usage mining is to transform the raw click-stream data into a set of user profiles (as described in the formal representation given in equation 3.1). From a navigational point of view each such profile captures a delimited sequence or a set of pageviews representing a user session. This sessionized data can be used as the input for a variety of data mining algorithms or further transformed and abstracted. Web usage data preprocessing presents a number of unique challenges which have led to a variety of algorithms and heuristic techniques for preprocessing tasks such as data fusion and cleaning, user and session identification, pageview identification [32]. The successful application of data mining techniques to Web usage data is highly dependent on the correct application of the preprocessing tasks.

Figure 3.1 provides a summary of the primary tasks and elements in usage data preprocessing. We provide a brief discussion of each of these elements below.

Data fusion refers to the merging of log files from several Web and application servers. This may require global synchronization across these servers. In the absence of shared embedded session ids, heuristic methods based on the "referrer" field in server log entries along with various sessionization and user identification methods (see below) can be used to perform the merging. This step is essential in "inter-site" Web usage mining where the analysis of user behavior is performed over the log files for multiple related Web sites [137].

Data cleaning involves tasks such as, removing extraneous references to embedded objects, style files, graphics, or sound files, and removing references due to spider navigations. The latter task can be performed by maintaining a list of known spiders, using heuristics, or using classification algorithms to build models of spider and Web robot navigations [135]. Also, not all client page requests are recorded in server access logs. Client-side or proxy-side caching can often result in missing references to those pages or objects that have been cached. Most of these missing references can be heuristically inferred through a process called *path completion* which relies on the knowledge of site structure and referrer information from server logs [32]. In the case of dynamically generated pages, form-based applications using the HTTP POST method result in all or part of the user input parameter not being appended to the URL accessed by the user, and thus not appear in server log entries (though, in the latter case, it is possible to re-capture the user input through packet sniffers on the server side).

Pageview identification is the process of aggregating a collection of objects or pages that should be considered an atomic unit for the purpose of analysis. This process is heavily dependent on the linkage structure of the site, as well as on the site contents. The level of abstraction captured in a pageview is also determined, in part, by the the underlying site domain knowledge and by the type of analysis required. In the simplest case, each HTML file has a one-to-one correlation with a pageview. In multi-framed sites, several files may make up a given pageview. In addition, it may be desirable to consider pageviews at a higher level of aggregation, where each pageview represents a collection of pages or objects, for examples pages related to the same concept category. In order to provide a flexible framework for a variety of data mining activities a number of attributes must be recorded with each pageview. These attributes include the pageview id (normally a URL uniquely representing the pageview), static pageview type (e.g., information page, product view, category view, or index page), and other meta-data, such as content attributes (e.g., keywords or product attributes). In the context of our discussion, the pageviews represent the abstract items $i_j \in I$ (in equation 3.1) that are objects of personalization.

In Web usage mining it is necessary to distinguish between the activities of different users. In the absence of an authentication mechanism, a common approach to distinguishing among unique visitors is the use of client-side cookies. Not all sites, however, employ cookies, and due to privacy concerns, client-side cookies are sometimes disabled by users. IP addresses, alone, are not generally sufficient for mapping log entries onto the set of unique visitors. This is mainly due the proliferation of ISP proxy servers which assign rotating IP addresses to clients as they browse the Web. In such cases, it is possible to more accurately identify unique users through combinations of IP addresses and other information such as the user agents and referrers [32].

Assuming that unique user records can be identified, we refer to the sequence of logged activities belonging to the same user as the *user activity log*. *Sessionization* is the process of segmenting the user activity log of each user into sessions, each representing a single visit to the site. Web sites without the benefit of additional authentication information from users and without mechanisms such as embedded session ids must rely on heuristics methods for sessionization. The goal of a sessionization heuristic is the reconstruction, from the clickstream data, of the actual sequence of actions performed by one user during one visit to the site. Generally, sessionization heuristics fall into two basic categories: time-oriented or structure oriented. Time-oriented heuristics apply either global or local time-out estimates to distinguish between consecutive sessions, while structure-oriented heuristics use either the static site structure or the implicit linkage structure captured in the referrer fields of the server logs. Various heuristics for session-ization have been identified and studied [32]. A formal framework for measuring the effectiveness of such heuristics has been proposed [129], and the impact of different heuristics on various Web usage mining tasks has been analyzed in [12].

An *Episode* is a subset or subsequence of a session comprised of semantically or functionally related pageviews. Episode identification can be performed as a final step in preprocessing of the clickstream data in order to focus on the relevant subsets of pageviews in each user session. This task may require the automatic or semi-automatic classification of pageviews into different functional types or into concept classes according to a domain ontology or concept hierarchy. In highly dynamic sites, it may also be necessary to map pageviews within each session into "service-based" classes according to a concept hierarchy over the space of possible parameters passed to script or database queries [13]. For example, the episode may ignore the quantity and attributes of an item added to the shopping cart, and focus only on the action of adding the item to the cart.

The above preprocessing tasks ultimately result in a set of n pageviews, $P = \{p_1, p_2, \cdots, p_n\}$, and a set of v user transactions, $T = \{t_1, t_2, \cdots, t_v\}$, where each $t_i \in T$ is an l-length sequence of ordered pairs:

$$t = \langle (p_1^t, w(p_1^t)), (p_2^t, w(p_2^t)), \cdots, (p_l^t, w(p_l^t)) \rangle,$$

where each $p_i^t = p_j$ for some $j \in \{1, \cdots, n\}$, and $w(p_i^t)$ is the weight associated with pageview p_i^t in the transaction t.

Each items $i_j \in I$ in the general framework of Section 3.3.1 (see Equation 3.1) can each represent a pageview. Note that a pageview in this context is not just a Web page, but as noted above, an abstraction which may represent a conceptual or functional entity in the application domain (e.g., a Web page, a product view or purchase, a task, or a content category). The notion of *user transaction* introduced above is meant to capture the activity of a user vis-a-vis these pageviews within the site during a partic-ular session (thus, sometimes we refer to these transactions as sessions). The weights can be determined in a number of ways, in part based on the type of analysis or the intended personalization tasks. For example, in a standard collaborative filtering ap-plication, weights may be determined based on user ratings of items. In most Web usage mining tasks, the focus is generally on anonymous user navigational activity in which case the weights are either binary, representing the existence or non-existence of a pageview in the transaction; or a function of the duration of the pageview in the user's session.

Finally, one or more transactions or sessions associated with a given user u_k can be aggregated to form the final profile for that user resulting in user profile representa-tion of Section 3.3.1, in which each item i_j is a pageview and the value of the interest function, $s_{u_k}(i_j)$, is determined as a function of the weight of the associated pageview, $w(p_j^t)$. If the profile is constructed from a single session, then it represents the short-term interests of that user during a single visit, while the aggregation of multiple ses-sions results in profiles that capture the user's long-term interests. The collection of these profiles will comprise the $m \times n$ matrix UP that can be used to perform various data mining tasks. For example, similarity computations can be performed among the profile vectors for clustering and kNN neighborhood formation tasks, or an association rule discovery algorithm can be applied (with pageviews as items) to find frequent itemsets of pageviews.

After the basic clickstream preprocessing steps, data from a variety of other sources must be integrated. The integration of content, structure and user data in various phases of the Web usage mining process may be essential in providing the ability to further analyze and reason about the discovered patterns. For example, the integration of semantic knowledge from the site content or semantic attributes of products can be used by personalization systems to provide more useful recommendations [33, 41, 59]. In e-commerce applications, the integration of both customer and product data (e.g., demographics, ratings, purchase histories) from operational databases with usage data can allow for the discovery of important business intelligence metrics such as customer conversion ratios and lifetime values [20, 61]. The use of structure data is necessary during preprocessing (for example in pageview identification, sessionization, and path completion). But, it can also be used to improve the results of model-based personalization techniques [90, 69].

One direct source of semantic knowledge that can be integrated into the mining process is the collection of content features associated with items or pageviews on a Web site. These features include keywords, phrases, category names, or specific attributes associated with items or products, such as price, brand, etc. Content preprocessing involves the extraction of relevant features from text and meta-data.

Extending the general framework of Section 3.3.1, each item i_j can be represented as a k- dimensional feature (or attribute) vector, where k is the total number of extracted features. Each dimension in a feature vector represents the corresponding feature weight associated with the item. Thus, the feature vector for an item i_j is given by:

$$ i_j = \langle fw_j(f_1), fw_j(f_2), \cdots, fw_j(f_k) \rangle, $$

where $fw_j(f_d)$, is the weight of the dth feature in $i_j \in I$, for $1 \le d \le k$. For features extracted from textual content of pages, the feature weight is usually the normalized tf.idf value for the term. In order to combine feature weights from meta-data (specified externally) and feature weights from the text content, proper normalization of those weights must be performed as part of preprocessing.

Further preprocessing on content features can be performed by applying text mining techniques. For example, classification of content features based on a concept hierarchy can be used to limit the discovered usage patterns to those containing pageviews about a certain subject or class of products. Performing clustering or association rule mining on the feature space can lead to composite features representing concept categories. In many Web sites it may be possible and beneficial to classify pageviews into functional categories representing identifiable "tasks" (such as completing an online loan application) [56]. The mapping of pageviews onto a set of concepts or tasks allows for the analysis of user sessions at different levels of abstraction according to a concept hierarchy or according to the types of activity performed by users [94, 37].

3.4 Pattern Discovery for Predictive Web User Modeling

As noted earlier, model-based collaborative techniques, including those used in the pattern discovery phase of Web usage mining, use a two stage process for recommendation generation. The first stage is carried out offline, where user behavioral data collected during previous interactions is mined and an explicit model generated for use in future online interactions. The second stage, is carried out in real-time as a new visitor begins an interaction with the Web site. Data from the current user session is scored using the models generated offline, and recommendations generated based on this scoring. The application of these models are generally computationally inexpensive compared to memory-based approaches such as traditional collaborative filtering, aiding scalability of the real-time component of the recommender system.

Model generation can be applied to explicitly and implicitly obtained user behavioral data. While the most commonly used implicit data is Web usage data, data pertaining to the structure and content are also often used. A number of data mining algorithms have been used for offline model building including Clustering, Classification, Association Rule Discovery, Sequential pattern Discovery, Markov models, and hidden (latent) variable models. In this section we briefly describe these approaches.

3.4.1 Personalization Approaches Based on Clustering

Clustering aims to divide a data set into groups or clusters where inter-cluster similarities are minimized while the similarities within each cluster are maximized. Generally speaking, clustering methods can be divided into three categories [47]:

- Partitioning methods, that create k partitions of a given data set, where each partition represents a cluster. The most widely used partitioning method is the k-means algorithm.
- Hierarchical methods either using a top-down approach (divisive) or a bottom-up approach (agglomerative) to create a hierarchy of clusters. Divisive methods start from the whole data set of items as a single cluster and recursively partition this data, while agglomerative methods start from individual items as clusters and iteratively combine smaller clusters.
- Model-based methods, that discover the best fit between data points given a mathematical model, usually specified as a probability distribution.

Various clustering algorithms have been used in standard collaborative filtering applications where interest scores for items are generally explicit ratings. Most of these approaches, however, generalize easily to the context of Web usage mining where the items are pageviews and interest scores are normally based on the characteristics of user behavior (such as pageview duration). In this context, clustering is usually used in one of two ways: to cluster users or to cluster items. In user-based clustering, users are grouped together based on the similarity of their user profiles in matrix UP (see Section 3.2.2). In item based clustering, items are clustered based on the similarity of the interest scores for these items across all users, or based on similarity of their content features or attributes. Some of the past methods used in this context include partitioning algorithms such as, K-means for item and user-based clustering [140], ROCK [95] for item-based clustering, agglomerative hierarchical clustering [95] for item-based clustering, divisive hierarchical clustering for user-based and item-based clustering [63], mixture resolving algorithms such as EM [34] to cluster users based on their item ratings [19] and Gibbs Sampling [19].

As noted earlier, the primary motivation behind the use of clustering (and more generally, model-based algorithms) in collaborative filtering and Web usage mining is to improve the efficiency and scalability of the real-time personalization tasks. For example, both user-based clustering and item-based clustering have been used as an integrated part of a Web personalization framework based on Web usage mining [88, 82]. Motivated by reducing the sparseness of the rating matrix, O'Connor and Herlocker proposed the use of item clustering as a means for reducing the dimensionality of the rating matrix [95]. Column vectors from the ratings matrix were clustered based on their similarity, measured using Pearson's correlation coefficient, in user ratings. The clustering resulted in the partitioning of the universe of items and each partition was treated as a separate, smaller ratings matrix. Predictions were then made by using traditional collaborative filtering algorithms independently on each of the ratings matrices. While some statistical methods such as sampling, as well as clustering, can mitigate the online computational complexity of collaborative filtering, these methods often result in reduced recommendation quality [72]. However, in the context of Web usage mining it has been shown that proper preprocessing of the usage data can help the clustering approach achieve prediction accuracy in par with standard k-nearest-neighbor approach [84].

A typical user-based clustering starts with the matrix UP of user profiles and partitions this multi-dimensional space into k groups of profiles (or Web transactions) that are close to each other based on a measure of distance or similarity among the vectors (such as the Pearson's correlation coefficient). Clusters obtained in this way can represent user or visitor segments based on their common navigational behavior or interest shown in various items. The discovered user segments are then employed in the user-based neighborhood formation task, rather than individual profiles [88].

In order to determine similarity between a target user and a user segment, the centroid vector corresponding to each cluster is computed and used as the aggregate representation of the user segment. For each cluster C_k, the centroid vector $\mathbf{v_k}$ is computed as: $\mathbf{v_k} = \frac{1}{|C_k|} \sum \mathbf{u_n}$, where $\mathbf{u_n}$ is the vector in UP for a user profile $u_n \in C_k$.

To make a recommendation for a target user u and target item i, a neighborhood of user segments that have a ratings or interest scores for i and whose aggregate profile v_k is most similar to u are selected. This neighborhood represents the set of user segments of which the target user is most likely to be a member. Given that the aggregate profile of a user segment contains the average interest scores for each item within the segment, a prediction can be made for item i in the same manner as in standard collaborative filtering using k-nearest-neighbor [49]. For example, the predicted score for a target item i and target user u can be computed as:

$$p_{u,i} = \bar{s}_u + \frac{\sum\limits_{v \in V} sim(u,v)(s_v(i) - \bar{s}_v)}{\sum\limits_{v \in V} |sim(u,v)|} \qquad (3.2)$$

where V is the set of k most similar segments; $s_v(i)$ is the weight (average interest score) of i in the neighbor segment v; $\bar{s}_u$ and $\bar{s}_v$ are the average interest scores over all items for user u and segment v, respectively; and $sim(u,v)$ is the similarity between u and segment v.

As noted above, many other approaches based on user-based or item-based clustering have been used in the context of personalization based on Web usage mining. For example, an algorithm called *PageGather* has been used to discover significant groups of pages based on user access patterns [105, 106]. This algorithm uses, as its basis, clustering of pages using the graph-based Clique (complete link) clustering technique. The resulting clusters are used to automatically synthesize alternative static index pages for a site, each reflecting possible interests of one user segment. In PageGather an edge is added between two nodes (pages) if the corresponding pages co-occur in more than a certain number of sessions. Clusters are then generated by finding connected components or cliques within this graph. A new index page for the Web site is created from each cluster with hyperlinks to all the pages in that cluster. One advantage of this approach is that it creates overlapping clusters. However, the problem of finding (maximal cliques) in a graph is generally not computationally feasible for large graphs (i.e., for sites with many pages).

Because in Web usage mining it is often desirable to group users into multiple categories, a number of approaches based on fuzzy clustering have been explored. For example, a fuzzy clustering approach is proposed in [57] for clustering user sessions. The Web site hyperlink structure is used as a bias in computing the similarity between sessions by taking into account the relative position of pages within sessions in the site tree. The clustering algorithm used are variants of the Fuzzy C-means (FCM) clustering method [14] which allows one piece of data to belong to two or more clusters. Similarly, Nasraoui et al. [91] proposed an unsupervised relational clustering algorithm based on the competitive agglomeration algorithm to discover aggregate user models. This approach was later extended with fuzzy clustering algorithms such as Relational Fuzzy C-Maximal Density Estimator (RFC-MDE) and Fuzzy C Medoids algorithm (FCMdd), both of which are again based on FCM [92].

Most distance-based approaches, such as those described above, do not consider the sequential ordering inherent in Web transactions. Clustering can also be applied to Web transactions viewed as sequences rather than as vectors. For distance-based clustering algorithms to handle this type of data, a measure of distance (or similarity) which takes ordering among items into account is necessary. Some clustering approaches have integrated sequential representation of user session data and defined pairwise distance functions between user sessions [7, 132]. For example in [7] a graph-based algorithm was introduced to cluster Web transactions based on a function of longest common subsequences. The novel similarity metric used for clustering takes into account both the time spent on pages as well as a significance weight assigned to pages.

Model-based clustering algorithms also have the advantage of not requiring an explicit distance measure. Therefore, despite their potential high computational cost, they are often applicable in a more general context. For example, Cadez et al. [23] used the Expectation-Maximization (EM) algorithm [34] on a mixture of Markov models for clustering user sessions. Each Markov model in this framework captures the behavior of a particular subgroup of users according to their navigational activities. The algorithm was used as the basis of a tool called WebCANVAS, designed to visualize user navigation paths in each cluster. The EM algorithm was also used by Anderson et al. [5] for discovering predictive Web usage models. The user navigation sessions were assumed to belong to one or more clusters, and the EM algorithm was used to compute the model parameters for each cluster. The probability of visiting a certain page is estimated by calculating its conditional probability for each cluster. The standard Markovian assumption is made that occurrences of pages in a particular session are independent given the cluster, resulting in a Naive Bayesiam mixture model.

Another approach that has been used effectively in item-based clustering of Web usage data is *Association Rule Hypergraph partitioning* (ARHP) [46]. In this approach, first association rule mining (see Section 3.4.2) is used to discover a set E of frequent itemsets among the items (pageviews) in the set of all items I. These itemsets are used as hyperedges to form a hypergraph $H = \langle V, E \rangle$, where $V \subseteq I$. A hypergraph is an extension of a graph in the sense that each hyperedge can connect more than two vertices. The weights associated with each hyperedge can be computed based on a variety of criteria such as the average confidence of the association rules involving the items in the frequent itemset. The hypergraph H is recursively partitioned until a stopping criterion for each partition is reached resulting in a set of clusters. A connectivity measure for a vertex (a pageview appearing in the frequent itemset) with respect to a cluster is defined based on the weights of hyperedges connecting it to other vertices in the cluster. The vertices with connectivity measure greater than a given threshold value are considered to belong to the partition, and the remaining vertices are dropped from the partition. This approach has also been used in the context of Web personalization [88].

3.4.2 Personalization Using Association Discovery

Association rule discovery techniques, such as the Apriori algorithm [3], were initially developed as techniques for mining supermarket basket data but have since been used in various domains including Web mining [10]. Association rule discovery on usage data results in finding groups of items or pages that are commonly accessed or purchased together. This, in turn, enables Web sites to organize the site content more efficiently, or to provide effective cross-sale product recommendations. For example, a high-confidence rule such as

$$\{\text{special-offers/, /products/software/}\} \Rightarrow \{\text{shopping-cart/}\}$$

might provide some indication that a promotional campaign on software products is positively affecting online sales. Such rules can also be used to optimize the structure of the site. For example, if a site does not provide direct linkage between two pages A and B, the discovery of a rule $\{A\} \Rightarrow \{B\}$ would indicate that providing a direct hyperlink might aid users in finding the intended information.

The discovery of association rules from transaction data consists of two main parts: the discovery of frequent itemsets (i.e., itemsets which satisfy a minimum *support* threshold) and the discovery of association rules from these frequent itemsets which satisfy a minimum *confidence* threshold.

Given a set of transactions T and a set $I = \{I_1, I_2, \ldots, I_k\}$ of itemsets over T. The *support* of an itemset $I_i \in I$ is defined as

$$\sigma(I_i) = \frac{|\{t \in T : I_i \subseteq t\}|}{|T|}$$

An association rule, r, is an expression of the form $X \Rightarrow Y$ (σ_r, α_r), where X and Y are itemsets, $\sigma_r = \sigma(X \cup Y)$ is the support of $X \cup Y$ representing the probability that X and Y occur together in a transaction. The confidence for the rule r, α_r, is given by $\sigma(X \cup Y)/\sigma(X)$ and represents the conditional probability that Y occurs in a transaction given that X has occurred in that transaction. Additional metrics have been proposed in literature that aim to quantify the interestingness of a rule [97, 123, 136], however support and confidence as these are the most commonly used metrics when using association and sequence based approaches to personalization.

Although not as widely used as clustering for Web personalization, the results of association rule mining on the user profile and items space can result in models that, in conjunction with the activity or profile of a target user, can be used for recommendation generation [39, 71, 83, 115]. For example, in the collaborative filtering context, Sarwar, et al. [115], used association rules in the context of a *top-N* recommender systems for e-commerce. The preferences of the target user are matched against the items in the antecedent X of each rule, and the items on the right hand side of the matching rules are sorted according to the confidence values. Then the top N ranked items from this list are recommended to the target user.

One problem for association rule recommendation systems is that a system cannot give any recommendations when the dataset is sparse (which is often the case in Web usage mining and collaborative filtering applications), and hence larger itemsets often do not meet the minimum support constraint. Sarwar, et al. [115] rely on some standard dimensionality reduction techniques to alleviate this problem. Fu et al. [39] propose two potential solutions to this problem. The first solution is to rank all discovered rules calculated by the degree of intersection between the left-hand-side of rule and a user's active session and then to generate the top k recommendations. The second solution is to utilize collaborative filtering: the system finds "close neighbors" who have similar interest to a target user and makes recommendations based on the close neighbor's history.

In [71] a collaborative recommendation system was presented using association rules. The proposed mining algorithm finds an appropriate number of rules for each target user by automatically selecting the minimum support. The recommendation engine generates association rules for each user, among both users and items. If a user minimum support is greater than a threshold, the system generates recommendations based on user association, else it uses item associations.

Because it is difficult to find matching rule antecedents with a full user profile (e.g., a full session in Web usage mining context), association-based recommendation algorithms typically use a sliding window w over the target user's active profile or session. The size of this window is iteratively decreased until an exact match with the antecedent of a rule is found. A problem with the naive approach to this algorithm is that it requires repeated search through the rule-base. However, efficient trie-based data structures can be used to store the discovered itemsets and allow for efficient generation of recommendations without the need to generate all association rules from frequent itemsets. Such data structures are commonly used for string or sequence searching applications. In the context of association rule mining, the frequent itemsets are stored in a directed acyclic graph. The Frequent Itemset Graph is an extension of the lexicographic tree used in the "tree projection algorithm" [1]. The graph is organized into levels from 0 to k, where k is the maximum size among all frequent itemsets. Each node at depth d in the graph corresponds to an itemset, I, of size d and is linked to itemsets of size $d+1$ that contain I at level $d+1$. The single root node at level 0 corresponds to the empty itemset. To be able to match different orderings of an active session with frequent itemsets, all itemsets are sorted in lexicographic order before being inserted into the graph. The user's active session is also sorted in the same manner before matching with patterns.

Using this general framework, the recommendation engine matches the current user session window with the previously discovered frequent itemsets to find candidate items (pages) for recommendation. Given an active session window w and a group of frequent itemsets, the algorithm considers all the frequent itemsets of size $|w| + 1$ containing the current session window by performing a depth-first search of the Frequent Itemset Graph is performed to level $|w|$. The recommendation value of each candidate is based on the confidence of the corresponding association rule whose consequent is the singleton containing the page to be recommended. If a match is found, then the children of the matching node n containing w are used to generate candidate recommendations. The details of this general recommendation algorithm [83] are given in Figure 3.2.

Association rules have also been used in conjunction with other data mining algorithms, such as clustering, in personalization based on Web usage mining (as well as other applications). In Section 3.4.1, we already described the item-based clustering approach used in [88] in which frequent itemsets of pageviews are organized in an *Association Rule Hypergraph*, and the resulting hypergraph is partitioned into pageview clusters.

Another approach that combines clustering and association rule mining is a two-level model-based collaborative filtering technique described in [133]. In the first level, a fuzzy C-Means clustering algorithm called Relational Fuzzy Subtractive Clustering (RFSC) is used to cluster the user sessions. Then the clusters are *defuzzified* by assigning the sessions to a cluster to with highest membership. This defuzzification process removes the noise and reveals the real structure in the data. In the second level, single-consequent association rules are discovered from within each cluster. For an active profile (a session) of a target user, the algorithm first finds the nearest cluster prototype, and then matches the profile with the antecedent of each rule in that cluster to find the matching score for each rule. The matching scores are weighted with the confidence of each rule to obtain the complete recommendation score of the item (page) appearing is the consequent of the rule.

Input: an active session window $w = \{p_1, p_2, \ldots, p_n\}$ in lexicographic order
                Minimum confidence threshold α
Output: Recommendation set REC

$REC = \emptyset$
$Node = root; depth = 0;$
repeat
    $depth$++;
    if $Node.children \neq \emptyset$
           and $\exists X \in Node.children$ with $X.itemset \subseteq \{p_1, \ldots, p_{depth}\}$
      then $Node = X;$
    else
        $Node$ = NULL; Break;
until $depth > |w|$
if $Node \neq$ NULL then
    for each node $N \in Node.children$ do
      let $c = N.support/Node.support;$
      if $c \geq \alpha$
        $p = (N.itemset - w)$
        $p.rec\_score = c$
        $REC = REC \cup \{p\}$
    end if
    end for
end if

Fig. 3.2. Recommendation Algorithm Based on Association Rules

3.4.3 Personalization Using Sequential Modeling

As with association rule discovery, Sequence rule discovery techniques [4] were also initially developed as techniques for mining supermarket basket data. The key difference between these algorithms is that while association rule discovery algorithms do not take into account the order in which items have been accessed, sequential pattern discovery algorithms do consider the order when discovering frequently occurring itemsets. Hence, given a user transaction $\{i_1, i_2, i_3\}$, the transaction supports the association rules $i_1 \Rightarrow i_2$ and $i_2 \Rightarrow i_1$ but not the sequential pattern $i_2 \Rightarrow i_1$.

When discovering sequential patterns from Web logs, two types of sequences are identified: Contiguous or Closed Sequences and Open Sequences [10]. Contiguous sequences require that items appearing in a sequence rule appear contiguously in transactions that support the sequence. Hence the contiguous sequence pattern $i_1, i_2 \Rightarrow i_3$ is satisfied by the transaction $\{i_1, i_2, i_3\}$ but not by the transaction $\{i_1, i_2, i_4, i_3\}$, as i_4 appears in the transaction between the items appearing in the sequence pattern. On the other hand, both transactions support the rule if it were an open sequence rule.

Given a transaction set T and a set $S = \{S_1, S_2, \ldots, S_n\}$ of frequent (contiguous) sequential patterns over T, the support of each S_i is defined as follows:

$$\sigma(S_i) = \frac{|\{t \in T : S_i \text{ is (contiguous) subsequence of } t\}|}{|T|}$$

The confidence of the rule $X \Rightarrow Y$, where X and Y are (contiguous) sequential patterns, is defined as

$$\alpha(X \Rightarrow Y) = \frac{\sigma(X \circ Y)}{\sigma(X)},$$

where $\circ$ denotes the concatenation operator. The Apriori algorithm used in association rule mining can also be adopted to discover open and contiguous sequential patterns. This is normally accomplished by changing the definition of support to be based on the frequency of occurrences of subsequences of items rather than subsets of items [4].

In the context of Web usage mining, contiguous sequential patterns can be used to capture *frequent navigational paths* among user trails [128, 119]. In contrast, items appearing in open sequential patterns, while preserving the underlying ordering, need not be adjacent, and thus they represent more general navigational patterns within the site. Frequent item sets, discovered as part of association rule mining, represent the least restrictive type of navigational patterns, since they focus on the presence of items rather than the order in which they occur within user session.

An approach for efficiently representing contiguous navigational sequences is to insert each sequence into a trie structure. A well-known example of this approach is the notion of *aggregate tree* introduced as part of the WUM (Web Utilization Miner) system [128]. The aggregation service of WUM extracts the transactions from a collection of Web logs, transforms them into sequences, and merges those sequences with the same prefix into the aggregate tree (a trie structure). Each node in the tree represents a navigational subsequence from the root (an empty node) to a page and is annotated by the frequency of occurrences of that subsequence in the transaction data (and possibly other information such as markers to distinguish among repeat occurences of the corresponding page in the subsequence). WUM uses a powerful SQL-like mining query language, called MINT, to discover generalized navigational patterns from this trie structure. MINT includes mechanism to specify sophisticated constraints on pattern templates, such as wildcards with user-specified boundaries, as well as other statistical thresholds such as support and confidence.

It is also possible to insert frequent sequences (after or during sequential pattern mining) into a trie structure [104, 86]. In the context of personalization, sequential patterns are typically stored in a single trie structure with each node representing an item and the root representing the empty sequence. Recommendation generation can be achieved in $O(s)$ by traversing the tree, where s is the length of the current user transaction deemed to be useful in recommending the next set of items. Mobasher et al. [86] use a fixed size sliding window, w, over the current transaction for recommendation generation. Hence the maximum depth of the tree required to be generated is $|w| + 1$. The size of the trees generated during the offline mining can be controlled by setting different minimum support and confidence thresholds. Thus, a similar general algorithm used in Section 3.4.2 (see Figure 3.2 for generating recommendations from frequent item-sets, can easily be adopted in the context of open and contiguous sequential patterns.

An empirical evaluation of association and sequential pattern based recommendation showed that site characteristics such as site topology and degree of connectivity can have a significant impact on the usefulness of sequential patterns over non-sequential (association) patterns [90]. Additionally, it has also been shown that contiguous sequential patterns are particularly restrictive and hence are more valuable in page prefetching applications (were the intent is to predict the immediate next page to be accessed) rather than in the more general context of recommendation generation [86].

Another type of approach for sequential modeling is based on stochastic methods that from the sequences of pageviews in user sessions learn probabilistic models that can used for predicting subsequent visits. One such approach is to model the navigational activity in the Web site as a Markov chain. A Markov model is represented by the 3-tuple $\langle A, S, T \rangle$ where A is a set of possible actions, S is the set of n states for which the model is built and T is the Transition Probability Matrix that stores the probability of performing an action $a \in A$ when the process is in a state $s \in S$. Specifically, $T = [p_{i,j}]_{n \times n}$, where $p_{i,j}$ represents the probability of a transition from state s_i to state s_j. The *order* of the Markov model corresponds to the number of prior events used in predicting a future event. So, a kth-order Markov model predicts the probability of the next event by looking at the past k events. Given a set of all paths R, the probability of reaching a state s_j from a state s_i via a (non-cyclic) path $r \in R$ is given by: $p(r) = \prod p_{k,k+1}$, where k ranges from i to $j - 1$. The probability of reaching s_j from s_i is the sum over all paths: $p(j|i) = \sum_{r \in R} p(r)$.

In the context of recommendation systems, A is the set of items and S is the visitor's navigation history, defined as a k-tuple of items visited, where k is the order of the Markov model. In Web usage analysis, they have been proposed as the underlying modeling machinery for Web prefetching applications or to minimize system latencies [36, 98, 107, 113]. Such systems are designed to predict the *next* user action based on a user's previous surfing behavior. On the other hand, Markov models can also be used to discover high-probability user navigational paths in a Web site. For example, Borges and Levene [17] modeled user sessions as a hypertext probabilistic grammar (or alternatively, an absorbing Markov chain) whose higher probability paths correspond to the user's preferred trails. An algorithm is provided to efficiently mine such trails from the model.

As the order of the Markov model increases, so does the size of the state space, S. On the other hand the coverage of that space reduces, leading to an inaccurate transition probability matrix. To counter the reduction in coverage, various Markov models of differing order can be trained and used to make predictions. The resulting model is referred to as the *All-Kth-Order* Markov model [107]. The downside of using the All-Kth-Order Markov model is the large number of states. Selective Markov models that only store some of the states within the model have been proposed as a solution to this problem [36]. A post pruning approach is used to prune out states that cannot be expected to be accurate predictors. Three pruning approaches based on the support, confidence and estimated error were proposed.

Rather than pruning states as a post process, sequence rule discovery and association rule discovery algorithms actively prune the state space during the discovery process using support. A further post pruning, based on confidence of the discovered rules, is also carried out. Hence the Selective Markov model is analogous to sequence rule discovery algorithms. Note however that the actual pruning process based on confidence proposed by Deshpande and Karypis [36] is not the same as that carried out during sequence rule discovery. Evaluation of Selective Markov models has shown that up to 90% of states can be pruned without a reduction in accuracy.

Other types of stochastic methods include various mixture models [23, 100, 141]that have been used to model navigational patterns. We have already discussed some of these approaches and their use in clustering approach to personalization (see Section 3.4.1). Recent work in this area has shown that mixture models are able to capture more complex, dynamic user behavior. This is in part because the observation data (i.e., the user-item space) in some applications (such as large and very dynamic Web sites) may be too complicated to be modeled by basic probability distributions such as a normal or a multinomial distribution. In particular, each user may exhibit different "types" of behavior corresponding to different tasks, and common behaviors may each be reflected in a different distribution within the data.

The general idea behind mixture models (such as a mixture of Markov models) is as follow. We assume there exist k types of user behavior (or k user clusters) within the data, and each user session is assumed to be generated via a generative process which models the probability distributions of observed variables and hidden variables. First, a user cluster is chosen with some probability; then the user session is generated from a Markov model with parameters specific to that user cluster. Next, the probabilities associated with the user cluster are estimated (usually via the EM [34] algorithm), as well as the parameters of each mixture component. Mixture-based user models can provide a great deal of flexibility. For example, a mixture of first-order Markov models [23] can not only probabilistically cluster user sessions based on similarities in navigation behavior, but also characterize each type of user behavior using a first-order Markov model, thus capturing popular navigation paths or characteristics of each user cluster. New user sessions can be easily fit into the model, and dynamic predictions or recommendations can be generated based on the probablitiy of association of the target user profile to various clusters. However, mixture models tend to fall victim to overfitting problems, largely due to their naive data generation assumptions. A more detailed discussion of mixture models is provided in the next section.

3.4.4 Approaches Based on Latent Variable Models

Latent variable models (LVMs) [8, 38] have recently become popular modeling approaches in data mining related fields and, particularly, in Web usage mining. By introducing latent variables as hidden factors underlying observation data, LVMs use probabilistic approaches to effectively discover the structural and semantic relationships within the data.

Two commonly used latent variable models are *Factor Analysis* (FA) models and *Finite Mixture Models* (FMM). By learning a low dimensional latent space from a high dimensional observation space, FA models aim at summarizing and explaining the complex dependency relationship among the observation data. Factor analysis models have a long history of successful applications in many domains, including in patterns recognition. Only recently, however, they have been effectively used the context of collaborative filtering [25] and personalization based on Web usage mining [144]. However, in the context of Web user modeling, FA models, as in most clustering approaches, generally ignore the sequential information conveyed in user sessions.

FMMs, on the other hand, use a finite number of components to model the observation data. Theoretically, the component models can be any probability distribution. In FMMs, one generally assumes the existence of k components (each component is a probability distribution) that account for all the observation data. Each single observation (e.g., a user's rating for an item, or a pageview in a user session) is assumed to be generated by the following process: first, a component with a certain probability is chosen, and then the chosen component is used to generate the observations. As noted earlier, the EM algorithm is usually used to fit the model and estimate the parameters associated with each component.

For example, a mixture of multinomial models is proposed in [24] to analyze the e-commerce transaction data. Transactions generated by individual users are probabilistically clustered into k groups, where each cluster is modeled by a multinomial distribution. The experiments show that the mixture model distinctly outperforms non-mixture techniques (a single multinomial model) in predicting out-of-sample individual behavior. As we noted in Section 3.4.3, in [23], a mixture of first-order Markov models was proposed to cluster Web users in which each component was modeled as a first-order Markov model. It was formally shown that the mixture of first-order Markov models is not first-order Markov model, and that it can model much more complex user behavior. A mixture of hidden Markov models [141] is also proposed for modeling clickstreams of Web surfers. In addition to user-based clustering, this approach can also be used for automatically page categorization.

Mixture models tend to have their own shortcomings. From the data generation perspective, each individual observation (such as a user session) is generated from one and only one component model. The probability assignment to each component only measures the uncertainty about this assignment. This assumption limits this model's ability of capturing complex user behavior, and more seriously, may result in overfitting problems [110].

Probabilistic Latent Semantic Analysis (PLSA) [52] provides a reasonable solution to some of these problems. PLSA adopts a totally different data generation idea. In the context of Web user navigation, each observation (a user visiting a page) is assumed to be generated as follows. First a user is selected with a certain probability. Next, conditioned on the selected user, a hidden variable is selected. Finally, the page to visit is slected conditioned on the chosen hidden variable. Since each user usually visits multiple pages, this data generation process ensures that each user is explicitly associated with multiple hidden variables, thus eliminating the overfitting problem associated with above mixture models. The PLSA model also uses the EM algorithm to estimate the parameters which probabilistically characterize the hidden variables underlying the co-occurrence observation data, and measure the relationship among hidden variables and observed variables. Due to its great flexibility, the PLSA model has been successfully used in a variety of application domains, including information retrieval [51], text learning [18, 60], and co-citation analysis [29, 30].

Another type of hidden variable model is the *Latent Dirichlet Allocation* (LDA) model [16]. The LDA model uses two levels of hidden variables. Each observation is assumed to be a multinomial distribution of k hidden variables, and each multinomial distribution is further constrained by a global variable with dirichlet distribution. The two levels of hidden variables are used to ensure that training observations and non-training observations can be generated via the same process. A side effect of having two levels of hidden variables is that exact inference for the LDA model is not feasible. Methods such as Variational Bayes [16], Markov Chain Monte Carlo [44] and Expectation-Propagation (EP) [78] are proposed to learn the model. Recently, the LDA model has been used in text mining [16], author-topic analysis [131], and collaborative filtering [73]. Although PLSA and LDA seem to be quite different in terms of parameter learning, research has shown that they are essentially equivalent in terms of modeling method, and PLSA is just a *Maximum A Posterior* (MAP) estimation of LDA model [42]. LDA introduces an extra set of hidden variables and is able to naturally fit in new data. However this also makes the learning of the LDA model computationally more expensive than PLSA.

In order to see how a hidden variable modeling approach, such as PLSA, can be used in the context of personalization, we provide more detail on a general recommendation algorithm, based on PLSA, that can be adopted both in the context of standard collaborative filtering [53], as well as, in Web usage mining [55, 56].

As in the approaches based on clustering, the PLSA-based approach begins with the discovery of user segments with similar behavior: Given a set of n user profiles, $UP = \{u_1, u_2, \cdots, u_n\}$, and a set of m items, $I = \{i_1, i_2, \cdots, i_m\}$ the PLSA model associates an unobserved factor variable $Z = \{z_1, z_2, \cdots, z_l\}$ with observations in the data. Each observation corresponds to the interest score $s_{u_k}(i_j)$ for an item i_j in the user profile for a user u_k (i.e., a rating or a weight associated with a pageview).

For a target user u and a target item i, the following joint probability can be defined:

$$Pr(u, i) = \sum_{k=1}^{l} Pr(z_k) \bullet Pr(u|z_k) \bullet Pr(i|z_k)$$

In order to explain the observations in (UP, I), we need to estimate the parameters $Pr(z_k)$, $Pr(u|z_k)$, and $Pr(i|z_k)$, while maximizing the following likelihood $L(UP, I)$ of the observation data:

$$L(UP, I) = \sum_{u \in UP} \sum_{i \in I} s_u(i) \bullet \log Pr(u, i)$$

where $s_u(i)$ is the interest score (e.g., rating) of user u for item i.

The Expectation-Maximization (EM) algorithm [34] is used to perform maximum likelihood parameter estimation. Based on initial values of $Pr(z_k)$, $Pr(u|z_k)$, and $Pr(i|z_k)$, the algorithm alternates between an expectation step and maximization step. In the expectation step, posterior probabilities are computed for latent variables based on current estimates, and in the maximization step, Lagrange multipliers [52] are used to obtain the re-estimated parameters. Iterating the expectation and maximization steps monotonically increases the total likelihood of the observed data $L(UP, I)$, until a locally optimal solution is reached.

Next the segments of user profiles that have similar underlying interests are identified. For each latent variable z_k, a user segment C_k is created and all user profiles having probability $Pr(u|z_k)$ exceeding a certain threshold μ are selected. If a user profile's probability does not exceed the threshold for any latent variable, it is associated with the user segment of highest probability. Thus, every user profile will be associated with at least one user segment, but may be associated with multiple segments. This allows authoritative users to have broader influence over predictions, without adversely affecting coverage in sparse rating data.

For each user segment C_k, the associated user profiles are aggregated into a weighted profile vector $\mathbf{v_k}$, computed as the mean vector or centroid of all $u_i \in C_k$. This the aggregate profile for a user segment to be represented in the original n-dimentioal space of items. To make a recommendation for a target user u and target item i, a neighborhood of user segments is selected that have defined interest scores for i and whose aggregate profile v_k is most similar to u. This neighborhood represents the set of user segments of which the target user is most likely to be a member, based on a measure of similarity (such as the Pearson's correlation coefficient which is usually used with rating data). A prediction for item i can now be derived using equation 3.2, used earlier in Section 3.4.1 in the context of the clustering approach.

3.4.5 Hybrid Models for Web Personalization

Pure usage-based approaches to personalization have some important drawbacks. The recommendation process relies on the existing user transactions or rating data, thus items or pages added to a site recently cannot be recommended. This is commonly referred to as the "new item problem". Furthermore, because such systems do not take into account the semantic or structural knowledge inherent in the underlying domain, they generally lack the ability to recommend complex objects or concepts based in their semantic attributes or based on other information channels available in the particular application domain. This limitation also hampers the ability of these systems to explain or reason about the discovered user models or recommendations.

In traditional collaborative filtering a number of hybrid approaches have been proposed. The most common form of hybrid recommender combines content-based and collaborative filtering [28, 75]. Other approaches have also incorporated other information sources such as user demographics [101, 127]. A detailed examination of different approaches to create hybrid recommender systems is presented in [21] (see also Chapter 12 of this book [22]). In the following we focus primarily on the data mining approaches to personalization, and particularly those based on Web usage mining, in which various information channels have been integrated in the knowledge discovery and recommendation generation processes.

Integration of Content Features with Usage-Based Models. A common approach to resolving the "new item problem" is to integrate content characteristics of pages with the user-based data (i.e., navigational or rating data). Generally, in these approaches, keywords are extracted from the content on the Web site and are used to either index pages by content or classify pages into various content categories. In Web personalization, this approach would allow the system to recommend pages to a user, not only based on similar users, but also (or alternatively) based on the content similarity of these pages to the pages user has already visited. The semantic information extracted as keyword-based features can be leveraged at various steps in the knowledge discovery process, namely in the preprocessing phase, in the mining phase, or during the post-processing of the discovered patterns.

A direct approach for the integration of content and usage data for Web personalization is to transform each user profile in UP (see Section 3.3.1), into a "content enhanced" profile containing the semantic features of the underlying items. This process, performed as part of data preprocessing, involves mapping each item or page in a user profile to one or more content features extracted from items, or a set of concepts (for example, from an externally available concept hierarchy). The the range of this mapping can be the full feature space or the concept space obtained as described above. Conceptually, the transformation can be viewed as the multiplication of the user-item matrix UP by the item-feature or an item-concept matrix. The result is a new matrix $UF = \{t'_1, t'_2, \ldots, t'_m\}$, where each t'_i is a k-dimensional vector over the feature (or concept) space. Thus, a user profile can be represented as a concept vector, reflecting that user's interests in particular concepts or topics. A variety of data mining algorithms can then be applied to this transformed user data.

For example, in [94], usage mining is enhanced by mapping user navigational data to concepts in an ontology underlying a particular Web site. The semantic annotation of the Web content is assumed to have been performed a priori. In order to mine interesting patterns, first user transactions are semantically enriched with concept labels, and then the transformed transaction space is mined to extract patterns reflecting users' changing interest in terms of concepts.

Following a similar approach, in [37] Web usage logs are enriched with semantics derived from the content features extracted from of the Web site's pages. The extraction of the keywords that describe each Web page is performed using standard information retrieval based techniques. These keywords are then mapped to the categories of a predefined concept hierarchy. The enhanced Web logs are then used as input to the Web mining process. The output consists of patterns representing users' navigational behavior in the form of clusters or association rules. This set of patterns is then used as the recommendation basis for each user or group of users, resulting in a broader yet semantically focused set of recommendations.

Haase et al. create semantic user profiles from usage and content information to provide personalized access to bibliographic information on a Peer-to-Peer bibliographic network [45]. The semantic user profile consists of the expertise, recent queries, recent relevant instances and a set of weights for the similarity function.

The integration of content features with user models can also be performed during or after the mining phase. In this case, patterns are discovered independently from the user profile data and the content data, and then combined in the recommendation generation process. For example, the results of user-based clustering can be combined with "content profiles" derived from the clustering of content features in pages [87]. The feature clustering is accomplished by applying a clustering algorithm to the transpose of the item-feature matrix UF, described above. This approach treats each feature as a vector over the space of items. Thus the centroid of a feature cluster can be viewed as a set (or vector) of items with associated weights. This representation is similar to that of aggregate models described in Section 3.4.1, however, in this case the weight of an item in the aggregate model represents the prominence of the features of the item that are associated with the corresponding cluster. The combined set of aggregate content and usage models can then be used seamlessly to generate recommendations.

Such approaches have also been useful in the context of e-commerce recommender systems. For example, Niu et al. [93] build customer profiles based on a product hierarchy in order to learn customer preferences. Ghani and Fano [41] proposed a recommender system based on a custom-built knowledge base of product semantics. The focus is on generating "soft" attributes from the online marketing text, describing the products browsed, and using them to generate cross category recommendations.

One type of integration approach is that for each user, one builds a local prediction model using algorithms such as naïve Bayes or k-Nearest Neighbor based on content data. Then all the individual models are integrated to form a global model via approaches such as linear combinations or probabilistic combination. An example of such integration is shown in [142], where a combined recommendation model is proposed. For each user, a probabilistic SVM (Support Vector Machine) model is built only based on the content information of this user's interested items. These individual models enable the system to make predictions for unvisited/unrated items only based on the content information of these items. Then all the individual models were combined under a hierarchical Bayesian framework, and the final prediction is the result of combining predictions from all individual models.

Finally, a number of approaches have attempted to integrate content and usage data based on hidden variable and mixture models (see Section 3.4.4). For example, in [108], an extension of the PLSA model was used to handle three-way co-occurrence data including users, items, and content features. The proposed extended PLSA model is used to discover the hidden relationships among users, items and attributes. A limitation of this approach is that, since the three-way observation data does not exist, and is generated subjectively from other observation data, it may not be consistent with the original navigational or content data.

Jin et al. [54] proposed a more robust approach based on hidden variable models in which users' navigational data and the content features associated with items are seamlessly integrated using a maximum entropy approach [111]. The goal of a maximum entropy model is to find a probability distribution which satisfies all the constraints in the observed data while maintaining maximum entropy. One of the advantages of such a model is that it enables the unification of information from multiple knowledge sources in one framework. First, probabilistic user models are discovered from the usage data, based on the PLSA approach, and used one set of constraints for the maximum entropy framework. Secondly, for content information, Latent Dirichlet Allocation (LDA) [16] is used to discover the hidden semantic relationships among visited items and specify another set of constraints based on these item association patterns. These two set of constraints are used in a unifying maximum entropy framework to generate recommendations.

Integration of Structured Semantic Knowledge and Usage-Based Models. The integration of content features with usage-based personalization is desirable when we are dealing with sites where text descriptions are dominant and other structural relationships in the data are not easy to obtain, e.g., news sites or online help systems, etc. Keyword-based approaches, however, are incapable of capturing more complex relationships among objects at a deeper semantic level based on the inherent properties associated with these objects. For example, potentially valuable relational structures among objects such as relationships between movies, directors, and actors, or between students, courses, and instructors, may be missed if one can only rely on the description of these entities using sets of keywords.

To be able to recommend different types of complex objects using their underlying properties and attributes, the system must be able to rely on the characterization of user segments and objects, not just based on keywords, but at a deeper semantic level using the domain ontologies for the objects. For instance, in a traditional personalization system on a university Web site might recommend courses in Java to a student, simply because that student has previously taken or shown interest in Java courses. On the other hand, a system that has knowledge of the underlying domain ontology, might recognize that the student should first satisfy the prerequisite requirements for a recommended course, or be able to recommend the best instructors for Java course, and so on.

This observation has led to a number of efforts that attempt to use "ontological user profiles" for personalization. For example, Middleton et al. [77] use an ontological profile for a user within their research paper recommendation system, QuickStep. The profile is based on a topic hierarchy alone. They also attempt to use externally available ontologies based on personnel records and user publications to address the cold-start problem for their recommendations system. The existence of such additional knowledge, while applicable in their specific application domain, cannot however be assumed in a general e-tailer scenario.

Dai and Mobasher [33] provide a general framework for integrating domain knowledge with Web usage mining for user based personalization. The primary focus of the proposed approach is to transform aggregate user models, that are the results of pattern discovery (clustering) on Web transaction data, into ontology enhanced aggregate models. In the initial discovery phase, each "page-level" aggregate models, m, is represented as a vector of item-weight pairs. Specifically, given a session cluster c, the aggregate model m as a set of pageview-weight pairs obtained by computing the centroid of c. Thus, m can be viewed as vector over the n-dimensional space if items (pages): $m = \langle w_m(p_1), w_m(p_2), \cdots, w_m(p_n) \rangle$ where $w_m(p_i)$ is the average weight of p_i across all sessions in the cluster c. Using the domain ontology, objects instances of ontology classes are extracted from each page p_i, and m is transformed into an object-level model $om = \{\langle o_1, w_{o_1} \rangle, \langle o_2, w_{o_2} \rangle, \cdots, \langle o_k, w_{o_k} \rangle\}$ in which each o_i is an object instance in the underlying domain ontology and w_{o_i} represents o_i's significance.

Objects that belong to the same class are combined to form an aggregated pseudo object belonging to that class. In this aggregation process, attribute values for objects of the same class are combined using aggregation functions for different attributes, defined in the domain ontology. The transformed aggregate model represents a set of objects accessed together frequently by a group of users in the same cluster. This new object-space is then used as input to additional data mining algorithm and for generating recommendations for pages that are similar at the object level. An important benefit of aggregation is that the pattern volume is significantly reduced, thus relieving the computation burden for the recommendation engine.

Kearny et al. [59] also investigate how Web usage data may be combined with semantic domain knowledge to provide a deeper understanding of user behavior. In particular, an "impact" measure is introduced based on information theory that captures the influence of a given concept from the domain ontology on user behavior. The impact measures for each of the concepts within the ontology are then combined to create an ontological profile for each user. This approach also begins by mapping each page within user sessions onto the concepts in the ontology. Then the specific instances are generalized to an Ontological Profile (OP). Thus, each page can be represented as a vector over the set of concepts where each dimension measures the degree to which the page belongs to the corresponding concept. In a similar manner as in [33], a composite distance measure based specific domain characteristics of each concept is defined and used as part of the mining and recommendation generation process.

Using Linkage Structure for Model Learning and Selection. Aside from the content features associated with items or pages, there are other information channels and knowledge sources that can be leveraged in the data mining approach to personalization. These include structured semantic information such as that available from domain ontologies or relational databases, and, in the context of Web personalization, the hyperlink structure of the Web site. We discuss the integration of ontological information in the next section. Here, we focus our attention to approaches that have used linkage information as part of the mining and recommendations processes.

Based on their study on the impact of site characteristics on the usefulness of sequential patterns over non-sequential (association) patterns, Nakagawa and Mobasher [90] proposed a hybrid recommendation system that switched between different recommendation algorithms based on the degree of connectivity in the site and the current location of the user within the site. The study showed that the performance of each recommendation model depends, in part, on the structural characteristics of the Web site. For example, in a highly connected Web site with short navigational paths, non-sequential models perform well by achieving higher overall precision and recall than sequential pattern models. In this hybrid approach, a measure of localized connectivity (LCM) is defined with respect to the current page being visited by the user. A logistic regression function is then learned from a set of training user profiles based on the LCM values of pages within the profiles and the best recommendations achieved for each user. This function is then used as a switching criterion to select the best recommendation model for the target user. Evaluation of this approach revealed that the hybrid model outperformed the base recommendation models in both precision and recall.

In [92], the site's hierarchical linkage structure is treated as an implicit concept hierarchy that is exploited in computing the similarity between pages. This similarity function allows for a more robust comparison of sessions that contain pages that are different but structurally related.

Lin and Zaiane [69] proposed a hybrid Web recommender system that combines access history and the content of visited pages, as well as the connectivity between the pages on a Web site, in order to model users' concurrent information needs and generate navigational patterns. These simultaneous goals of users are called "missions". A mission is a sub-session with a consistent goal as determined based on the content similarity of the pages within the session. These missions are in turn clustered to generate navigational patterns, and augmented with their linked neighborhood and ranked according to their authority determined based on site connectivity. These new clusters (i.e., augmented navigational patterns) are provided to the recommendation engine. When a visitor starts a new session, the session is matched with these clusters to generate a recommendation list.

3.5 Evaluating Personalization Models

As in any data mining application, before the discovered models can be deployed as part of a personalization framework, it is essential to evaluate their accuracy and effectiveness. The evaluation of personalization models, however, is an inherently challenging task for several reasons. First, the various modeling approaches and recommendation algorithms, such as those described in the previous section, may require different evaluation metrics. Secondly, the required personalization actions may be quite different depending on the underlying domain, intended application, and the data gathered for personalization. Finally, there is a lack of consensus among researchers and practitioners as to what factors most affect quality of service in personalized systems. Ultimately, the goal of evaluation in this context is to judge the "quality" of recommendations (or personalized content) generated by the system. The factors mentioned above, however, affect how this notion of quality is defined in different settings and according to the personalization task.

Herlocker et al. [50] have identified several types of personalization tasks performed by typical systems. These tasks include providing annotations in context (i.e., annotating items or existing content with prediction scores), finding (some or all) "good" items, recommending a sequence of items, providing decision support for browsing or making purchases, and providing credible recommendations. While evaluating the performance of a personalization system vis-a-vis these tasks generally requires measuring the *accuracy* of recommendations, the aforementioned study indicates that some accuracy metrics are more appropriate for a given task than others. Here, we briefly discuss some of the most commonly used accuracy metrics, and then we consider some of the other factors that impact the quality of recommendations.

The most common approach to evaluation in collaborative filtering systems is to measure the effectiveness of the system's predictive accuracy. Such metrics measures how close the recommender system's predicted ratings are to the actual user ratings. Particularly when dealing with user ratings of items, a frequently used metric is the *Mean Absolute Error* (MAE) [121, 49], which measures the average absolute deviation between a predicted rating and the user's actual rating. Several related accuracy metrics have been proposed for the prediction task with numeric ratings, including root mean squared error and mean squared error, that implicitly assign a greater weight to predictions with larger errors, and the normalized mean squared error [43] that aims to normalize MAE across datasets with varying rating scales.

Massa and Avesani suggest another variant of MAE called the mean absolute user error that calculates the mean absolute error for each user and then averages over all users [74]. This was based on their observation that recommender systems tend to have lower errors when predicting ratings by prolific raters rather than less frequent ones.

This metric is particularly useful when the number of items in the test set varies for each user. For example, this metric may be appropriate if the number of items in the test set is based on a percentage of items rated by a user.

While the MAE and its variants are useful in measuring the accuracy of predictions, they may not provide a complete picture of how good the recommendations are. These metrics may be less appropriate for tasks such as finding "good" items [50], where a ranked result is returned to the user. In such systems the target users usually only view items at the top of the ranking, and thus the accuracy of predictions for items of no interest to the user is not a determinant factor.

Classification metrics, on the other hand, measure the frequency with which a recommender system makes correct or incorrect decisions about recommending an item. Two commonly used metrics in this context are *Precision* and *Recall* which are standard metrics used in evaluating information retrieval effectiveness, but have also been adopted to evaluate ranked ordering of recommended items in personalization [58, 116, 15]. While precision measures the probability that a recommended item is relevant, recall measures the probability that a relevant item is recommended. In order to compute precision and recall in recommender systems, it is necessary to distinguish between the item set that is returned to the user (i.e., selected or recommended), and the item set that is not. One approach in doing so is to determine the set of top N recommended items for a fixed N and consider the remaining items as not recommended.

One advantage of metrics such as precision and recall is that they can be used in the evaluation of personalization systems in which the underlying user preferences are not determined by numeric ratings. This, of course, is the case when dealing with navigational data in which an item is either visited or it is not. In the context of numeric ratings on a continuous scale, it would be necessary to first transform ratings into a binary scale. For example, a rating scale of 1–5 may be transformed into a binary scale by converting every rating of 4 or 5 to "relevant" and all ratings between 1 and 3 to "nonrelevant". The determination of which items are relevant and which are not poses its own unique challenges when dealing with Web navigation data. A recorded visit to a particular page on a Web site, cannot necessarily be taken as an indication of relevance or interest. One approach that can address this problem is to record the amount of time spent on each page during a session (the pageview duration). To accurately convert this data into a binary scale, it is usually necessary to standardize the pageview durations with respect to the mean duration for that page. In this way, pageviews that last significantly less than the mean duration can be removed from the relevant list.

It should also be noted that there is often a trade-off between precision and recall, so it is important to consider both of these metrics for a given system. Some metrics attempt to combine precision and recall into a single number. One such metric is the The F1 measure which is computed as the harmonic mean of precision and recall [9, 85]. A more general form of the F1 measure can be devised which allows for weighting one of these metrics more than the other, depending on their relative importance in a particular application domain.

A measure that provides and alternative to precision and recall is the Receiver Operating Characteristic (ROC) which has roots in signal detection theory [48]. The ROC metric attempts to measure the extent to the system can successfully distinguish between signal (relevance) and noise. It assumes that the information system will assign a predicted level of relevance to every potential item. The ROC-curve is a plot of the systems sensitivity (the probability of signal, or, in the context of recommendation, the true positive rate) by the complement of its specificity (the probability of noise, or, in the recommendation context, the complement of the true negative rate). Generally, to compare the recommendation accuracy in two systems, the size of the area under the ROC-curve is measured with a larger value indicating better performance.

As noted earlier, the focus of the aforementioned metrics is generally the evaluation of recommendation accuracy. However, the research and practice in personalization technologies has led to the emerging consensus that measuring accuracy alone may not paint a complete picture of how users view the recommendations. The recommendations, or more generally the personalized content generated by the system, must also be "useful" to users. For example, a system that only recommend highly popular items (such as best seller books, or in the context of Web usage, highly visited pages in a site), may be quite accurate based on the above measures, but one can argue that such items are not particularly useful for the users of the system.

Recent user studies have found that a number of issues can affect the perceived usefulness of personalization systems including, trust in the system, transparency of the underlying recommendation algorithm, ability for a user to refine the system generated profile, and diversity of recommendations [134, 125, 145]. Therefore, the evaluation of personalization systems needs to be carried out along a number of dimensions, in addition to accuracy, some of which are better understood that others and have well established metrics available. The key dimensions along which personalization systems can be evaluated (aside from accuracy) include the coverage, utility, explainability, robustness, scalability, and user satisfaction.

Coverage measures the percentage of the universe of items that the recommendation system is capable of producing. For the prediction task it is calculated as the ratio of items for which the system can provide recommendations to all available items. Since it may not be practical to compute predictions for all user-item pairs in the system, this metric is usually estimated by selecting a random sample of user-item pairs, attempting to generate a prediction for each pair, and measuring the percentage for which a prediction was provided. An alternative is to calculate coverage as a percentage of items of interest to a user rather than considering the complete universe of items [50]. If the predictive accuracy is computed by withholding a selection of ratings and then predicting those ratings, the coverage can be measured as the percentage of withheld items for which a prediction is obtained.

The notion of "usefulness" suggests that measuring the utility of a recommendation for a user may be required. Breese et al. [19] suggested a metric based on the expected utility of the recommendation list. The utility of each item is calculated by the difference in vote for the item and a "neutral" weight. The metric is then calculated as the weighted sum of the utilities of all items in the list where the weight signifies the probability that an item in the ranked list will be viewed or selected by the user. This likelihood that a user will view or select each successive item is defined by an exponential decay function, where the decay factor is described by a half-life parameter. The basic, and rather strong, assumption behind this metric is that the true utility (in terms of cost/benefit analysis) rapidly (exponentially) drops as the search length increases.

The utility of recommendations or personalized content produced by the system can also be viewed in terms of their novelty. If the system only produces obvious recommendations, even if accurate, the recommendation may not be perceived as useful by the users. Clearly, the novelty of recommendations is not only user-specific, but also domain dependent and, therefore, measuring it would require domain specific metrics. For example, in the context of Web navigation, several metrics have been proposed that measure utility based on the distance of the recommended item from the current page (referred to as navigation distance) [5]. Although novelty may be an important consideration, it should be noted that several studies have found that there is, in fact value in providing user with some "obvious" recommendations [134]. Such recommendations tend to increase user confidence in the system leading to the a user perception that the system does generate credible recommendation; an important factor in the success of the personalization system.

A number of metrics have been proposed in literature for evaluating the robustness of a recommender system. Such metrics attempt to provide a quantitative measure of the extent to which an attack can affect a recommender system. Stability of prediction [96] measures the percentage of unrated (user,items) pairs that have a prediction shift less that a predefined constant. Power of an attack [96] on the other hand measures the average change in the gap between the predicted and target rating for the target item. The target item is the item that the attack is attempting to push or nuke. The power of attack metric assumes that the goal of the attack is to force item ratings to a target rating value. Noting that the effect of an attack on an items current rating is not necessarily going to affect its ability to be recommended, Lam and Herlocker [67] proposed an alternative metric called the Change in Expected change in top-N occupancy. It is calculated as the average expected occurrence of the target items in the top-N recommendation list of users.

The performance and scalability dimension aims to measure the response time of a given recommendation algorithm and how easily it can scale to handle a large number of concurrent requests for recommendations. Typically, these systems need to be able to handle large volumes of recommendation requests without significantly adding to the response time of the Web site that they have been deployed on.

Finally, attempts to measure user satisfaction range from using business metrics for customer loyalty such as RFM and life-time value through to more simplistic measures such as recommendation uptake. For example, the físchlár video recommendation system [126] implicitly obtains a measure of user satisfaction by checking is the recommended items were played or recorded.

3.6 Conclusions

In this chapter we have presented a comprehensive discussion the Web personalization process viewed as an application of data mining which must therefore be supported during the various phases of a typical data mining cycle. We have discussed a host of activities and techniques used at different stages of this cycle, including the preprocessing and integration of data from multiple sources, and pattern discovery techniques that are applied to this data. We have also presented a number of specific recommendation algorithms for combining the discovered knowledge with the current status of a user's activity in a Web site to provide personalized content to a user. The approaches we have detailed show how pattern discovery techniques such as clustering, association rule mining, and sequential pattern discovery, and probabilistic models performed on Web usage collaborative data, can be leveraged effectively as an integrated part of a Web personalization system.

While a research into personalization has led to a number of effective algorithms and commercial success stories, a number of challenges and open questions still remain.

A key part of the personalization process is the generation of user models. The most commonly used user models are still rather simplistic, representing the user as a vector of ratings or using a set of keywords. Even where more multi-dimensional or ontological information has been available, the data is generally mapped onto a single user-item table which is more amenable for most data mining and machine learning techniques. To provide the most useful and effective recommendations, personalization systems need to incorporate more expressive models. Some of the discussion on the integration of semantic knowledge and ontologies in the mining process suggests that some strides have been made in this direction. However, most of this work has not, as of yet, resulted in true and tested approaches that can become the basis of the next generation personalization systems.

Another important and difficult of challenge is the modeling of user context. In particular profiles commonly used today lack in their ability to model user context and dynamics. Users access different items for different reasons and under different contexts. The modeling of context and its use within recommendation generation needs to be explored further. Also, user interests and needs change with time. Identifying these changes and adapting to them is a key goal of personalization. However, very little research effort has been expended the evolution of user patterns over time and their impact on recommendations. This is in part due to the trade-offs between expressiveness of the profiles and scalability with respect to the number of active users.

Solutions to these important challenges are likely to lead to the creation of the next-generation of more effective and useful Web personalization and recommender systems that can be deployed in increasingly more complex Web-based environments.

References

1. Agarwal, R., Aggarwal, C., Prasad, V.: A tree projection algorithm for generation of frequent itemsets. Journal of Parallel and Distributed Computing **61**(3) (2001) 350–371
2. Aggarwal, C.C., Wolf, J.L., Yu, P.S.: A new method for similarity indexing for market data. In: Proceedings of the 1999 ACM SIGMOD Conference, Philadelphia, PA (June 1999) 407–418
3. Agrawal, R., Srikant, R.: Fast algorithms for mining association rules. In: Proceedings of the 20th International Conference on Very Large Data Bases (VLDB'94), Santiago, Chile (September 1994) 487–499

4. Agrawal, R., Srikant, R.: Mining sequential patterns. In: Proceedings of the International Conference on Data Engineering (ICDE'95), Taipei, Taiwan (March 1995) 3–14
5. Anderson, C., Domingos, P., Weld, D.: Adaptive web navigation for wireless devices. In: Proceedings of the 17th International Joint Conference on Artificial Intelligence, Seattle, Washington (August 2001) 879–884
6. Balabanovic, M., Shohan, Y.: Fab: Content-based, collaborative recommendation. Communications of the ACM **40**(3) (1997) 66–72
7. Banerjee, A., Ghosh, J.: Clickstream clustering using weighted longest common subsequences. In: Proceedings of the Web Mining Workshop at the 1st SIAM Conference on Data Mining, Chicago, Illinois (April 2001)
8. Bartholomem, D., Knott, M.: Latent Variable Models and Factor Analysis. Oxford University Press, New York, USA (1999)
9. Basu, C., Hirsh, H., Cohen, W.: Recommendation as classification: Using social and content-based information in recommendation. In: Proceedings of the Recommender System Workshop at AAAI 98, Madison, Wisconsin (July 1998) 11–15
10. Baumgarten, M., Büchner, A.G., Anand, S.S., Mulvenna, M.D., Hughes, J.: User-driven navigation pattern discovery from internet data. web usage analysis and user profiling. In Masand, B., Spiliopoulou, M., eds.: Web Usage Analysis and User Profiling: Proceedings of the WEBKDD'99 Workshop. Lecture Notes in Computer Science 1836. Springer-Verlag (2000) 74–91
11. Belkin, N., Croft, B.: Information filtering and information retrieval. Communications of ACM **35**(12) (2001) 29–37
12. Berendt, B., Mobasher, B., Nakagawa, M., Spiliopoulou, M.: The impact of site structure and user environment on session reconstruction in web usage analysis. In Zaïane, O.R., Srivastava, J., Spiliopoulou, M., Masand, B., eds.: Proceedings of WEBKDD 2002 - Mining Web Data for Discovering Usage Patterns and Profiles. Volume 2703 of LNCS. Springer Berlin / Heidelberg (2003) 159–179
13. Berendt, B., Spiliopoulou, M.: Analysis of navigation behaviour in web sites integrating multiple information systems. VLDB Journal, Special Issue on Databases and the Web **9**(1) (2000) 56–75
14. Bezdek, J.C.: Pattern Recognition with Fuzzy Objective Function Algorithms. Plenum Press, New York (1981)
15. Billsus, D., Pazzani, M.J.: Learning collaborative information filters. In: Proceedings of the 15th International Conference on Machine Learning (ICML'98), Madison, Wisconsin (July 1998) 46–53
16. Blei, D., Ng, A., Jordan, M.: Latent dirichlet allocation. Journal of Machine Learning Research **3** (2003) 993–1022
17. Borges, J., Levene, M.: Data mining of user navigation patterns. In Masand, B., Spiliopoulou, M., eds.: Web Usage Analysis and User Profiling: Proceedings of the WEBKDD'99 Workshop. LNAI 1836. Springer-Verlag (1999) 92–111
18. Brants, T., Chen, F., Tsochantaridis, I.: Topic-based document segmentation with probabilistic latent semantic analysis. In: Proceedings of the Eleventh International Conference on Information and Knowledge Management, Washington D.C. (November 2002) 211–218
19. Breese, J.S., Heckerman, D., Kadie, C.: Empirical analysis of predictive algorithms for collaborative filtering. In: Proceedings of the Fourteenth Annual Conference on Uncertainty in Artificial Intelligence, Madison, Wisconsin (July 1998) 43–52
20. Büchner, A., Mulvenna, M.D.: Discovering internet marketing intelligence through online analytical web usage mining. SIGMOD Record **4**(27) (1998) 54–61
21. Burke, R.: Hybrid systems for personalized recommendations. In Mobasher, B., Anand, S.S., eds.: Intelligent Techniques in Web Personalisation. LNAI 3169. Springer-Verlag (2005) 133–152

22. Burke, R.: Hybrid web recommender systems. In Brusilovsky, P., Kobsa, A., Nejdl, W., eds.: The Adaptive Web: Methods and Strategies of Web Personalization. Volume 4321 of Lecture Notes in Computer Science. Springer-Verlag, Berlin Heidelberg New York (2007) This Volume

23. Cadez, I., Heckerman, D., Meek, C., Smyth, P., White, S.: Model-based clustering and visualization of navigation patterns on a web site. Journal of Data Mining and Knowledge Discovery 7(4) (2003) 399–424

24. Cadez, I., Smyth, P., Ip, E., Mannila, H.: Predictive profiles for transaction data using finite mixture models. Technical Report Technical Report No. 01-67, Information and Computer Science Department, University of California, Irvine, Irvine, CA (2001)

25. Canny, J.: Collaborative filtering with privacy via factor analysis. In: Proceedings of the 25th Annual International ACM SIGIR Conference on Research and Development in Information Retrieval, Tampere, Finland (August 2002) 238–245

26. Cassel, L., Wolz, U.: Client side personalization. In: Proceedings of the Second DELOS Network of Excellence Workshop on Personalization and Recommender Systems in Digital Libraries, Dublin, Ireland (June 2001)

27. Chapman, P., Clinton, J., Kerber, R., Khabaza, T., Reinartz, T., Shearer, C., Wirth, R.: Crisp-dm 1.0: Step-by-step data mining guide. http://www.crisp-dm.org (2000)

28. Claypool, M., Gokhale, A., Miranda, T., Murnikov, P., Netes, D., Sartin, M.: Combining content-based and collaborative filters in an online newspaper. In: Proceedings of the ACM SIGIR '99 Workshop on Recommender Systems: Algorithms and Evaluation, Berkeley, California (August 1999)

29. Cohn, D., Chang, H.: Probabilistically identifying authoritative documents. In: Proceedings of the Seventeenth International Conference on Machine Learning, Stanford, CA (June 2000) 167–174

30. Cohn, D., Hofmann, T.: The missing link: A probabilistic model of document content and hypertext connectivity. In Todd K. Leen, T.G.D., Tresp, V., eds.: Advances in Neural Information Processing Systems 13. MIT Press, Vancouver, Canada (2001) 430–436

31. Cooley, R., Mobasher, B., Srivastava, J.: Web mining: Information and pattern discovery on the world wide web. In: Proceedings of the 9th IEEE International Conference on Tools with Artificial Intelligence (ICTAI'97), Newport Beach, CA (November 1997) 558–567

32. Cooley, R., Mobasher, B., Srivastava, J.: Data preparation for mining world wide web browsing patterns. Journal of Knowledge and Information Systems 1(1) (1999) 5–32

33. Dai, H., Mobasher, B.: A road map to more effective web personalization: Integrating domain knowledge with web usage mining. In: Proceedings of the International Conference on Internet Computing, IC03, Las Vegas (June 2003) 58–64

34. Dempster, A., Laird, N., Rubin, D.: Maximum likelihood from incomplete data via the em algorithm. Journal of Royal Statistical Society B(39) (1977) 1–38

35. Deshpande, M., Karypis, G.: Item-based top-n recommendation algorithms. ACM Transactions on Information Systems 22(1) (2004) 1–34

36. Deshpande, M., Karypis, G.: Selective markov models for predicting web-page accesses. ACM Transactions on Internet Technology 4(2) (2004) 163–184

37. Eirinaki, M., Vazirgiannis, M., Varlamis, I.: Sewep: Using site semantics and a taxonomy to enhance the web personalization process. In: Proceedings of the 9th SIGKDD International Conference on Data Mining and Knowledge Discovery (KDD'03), Washington, DC (August 2003) 99–108

38. Eveitt, B.: An Introduction to Latent Variable Models. Champman and Hall, New York, USA (1984)

39. Fu, X., Budzik, J., Hammond, K.J.: Mining navigation history for recommendation. In: Proceedings of the 2000 International Conference on Intelligent User Interfaces, New Orleans, LA, ACM Press (January 2000) 106–112

40. Gauch, S., Speretta, M., Chandramouli, A., Micarelli, A.: User profiles for personalized information access. In Brusilovsky, P., Kobsa, A., Nejdl, W., eds.: The Adaptive Web: Methods and Strategies of Web Personalization. Volume 4321 of Lecture Notes in Computer Science. Springer-Verlag, Berlin Heidelberg New York (2007) This Volume

41. Ghani, R., Fano, A.: Building recommender systems using a knowledge base of product semantics. In: Proceedings of the Workshop on Recommendation and Personalization in E-Commerce, at the 2nd Int'l Conf. on Adaptive Hypermedia and Adaptive Web Based Systems, Malaga, Spain (May 2002)

42. Girolami, M., Kaban, A.: On an equivalence between plsi and lda. In: Proceedings of the 26th Annual International ACM SIGIR Conference (SIGIR'03), Toronto, Canada (July 2003) 433–434

43. Goldberg, K., Roeder, T., Gupta, D., Perkins, C.: Eigentaste: A constant time collaborative filtering algorithm. Information Retrieval 4(2) (2001) 133–151

44. Griffiths, T., Steyvers, M.: Finding scientific topics. Proceedings of the National Academy of Sciences, PNAS 2004 101 (April 2004) 5228–5235

45. Haase, P., Ehrig, M., Hotho, A., Schnizler, B.: Personalized information access in a bibliographic peer-to-peer system. In: Proceedings of the AAAI Workshop on Semantic Web Personalization, AAAI Workshop Technical Report (2004) 1–12

46. Han, E., Karypis, G., Kumar, V., Mobasher, B.: Hypergraph based clustering in high-dimensional data sets: A summary of results. IEEE Data Engineering Bulletin 21(1) (March 1998) 15–22

47. Han, J., Kamber, M.: Data Mining: Concepts and Techniques. Morgan Kaufmann, San Francisco, CA (2001)

48. Hanley, J.A., McNeil, B.J.: The meaning and use of the area under a receiver operating characteristic (roc) curve. Radiology (143) (1982) 29–36

49. Herlocker, J., Konstan, J., Borchers, A., Riedl, J.: An algorithmic framework for performing collaborative filtering. In: Proceedings of the 22nd ACM Conference on Research and Development in Information Retrieval (SIGIR'99), Berkeley, CA (August 1999) 230–237

50. Herlocker, J.L., Konstan, J.A., Terveen, L.G., Riedl, J.: Evaluating collaborative filtering recommender systems. ACM Transactions on Information Systems 22(1) (2004) 5–53

51. Hofmann, T.: Probabilistic latent semantic indexing. In: Proceedings of the 22nd International Conference on Research and Development in Information Retrieval, Berkeley, CA (August 1999) 50–57

52. Hofmann, T.: Unsupervised learning by probabilistic latent semantic analysis. Machine Learning Journal 42(1) (2001) 177–196

53. Hofmann, T.: Latent semantic models for collaborative filtering. ACM Transactions on Information Systems 22(1) (2004) 89–115

54. Jin, X., Zhou, Y., Mobasher, B.: A unified approach to personalization based on probabilistic latent semantic models of web usage and content. In: Proceedings of the AAAI 2004 Workshop on Semantic Web Personalization (SWP'04), San Jose, CA (2004)

55. Jin, X., Zhou, Y., Mobasher, B.: Web usage mining based on probabilistic latent semantic analysis. In: Proceedings of the ACM SIGKDD Conference on Knowledge Discovery and Data Mining (KDD04), Seattle, WA (August 2004) 197–205

56. Jin, X., Zhou, Y., Mobasher, B.: Task-oriented web user modeling for recommendation. In: Proceedings of the 10th International Conference on User Modeling (UM'05), Edinburgh, UK (July 2005) 109–118

57. Joshi, A., Krishnapuram, R.: On mining web access logs. In: Proceedings of the ACM SIGMOD Workshop on Research Issues in Data Mining and Knowledge Discovery (DMKD 2000), Dallas, Texas (May 2000)

58. Karypis, G.: Evaluation of item-based top-n recommendation algorithms. In: Proceedings of the tenth International conference on Information and knowledge management (CIKM'01), Atlanta, Georgia (October 2001) 247–254
59. Kearney, P., Anand, S.S., Shapcott, M.: Employing a domain ontology to gain insights into user behaviour. In: Proceedings of the 3rd Workshop on Intelligent Techniques for Web Personalization, at IJCAI 2005, Edinburgh, Scotland (August 2005)
60. Kim, Y., Chang, J., Zhang, B.: a empirical study on dimensionality optimization in text mining for linguistic knowledge acquisition. In: Proceedings of the Seventh Pacific-Asia Conference on Knowledge Discovery and Data Mining (PAKDD-03), Seol, Korea (April 2003) 111–116
61. Kohavi, R., Mason, L., Parekh, R., Zheng, Z.: Lessons and challenges from mining retail e-commerce data. Machine Learning 57(1–2) (2004) 83–113
62. Kohavi, R., Provost, F.: Applications of data mining to electronic commerce. Data Mining and Knowledge Discovery 5(1–2) (2001) 5–10
63. Kohrs, A., Mérialdo, B.: Clustering for collaborative filtering applications. In: Proceedings of the International Conference on Computational Intelligence for Modelling, Control & Automation (CIMCA'99), Vienna, Austria (February 1999)
64. Konstan, J., Miller, B., Maltz, D., Herlocker, J., Gordon, L., Riedl, J.: Grouplens: Applying collaborative filtering to usenet news. Communications of the ACM 40(3) (1997) 77–87
65. Krulwich, B.: Lifestyle finder: Intelligent user profiling using large-scale demographic data. AI Magazine 18(2) (1997) 37–45
66. Krulwich, B., Burkey, C.: Learning user information interests through extraction of semantically significant phrases. In: Proceedings of the AAAI Spring Symposium on Machine Learning in Information Access, Stanford, California (March 1996)
67. Lam, S.K., Riedl, J.: Shilling recommender systems for fun and profit. In: Proceedings of the 13th international World Wide Web conference (WWW'04), New York, NY (May 2004) 393–402
68. Lang, K.: Newsweeder: Learning to filter netnews. In: Proceedings of the 12th International Conference on Machine Learning, Tahoe City, California (July 1995) 331–339
69. Li, J., Zaiane, O.: Using distinctive information channels for mission-based recommender systems. In: Proceedings of the sixth WEBKDD workshop: Webmining and Web Usage Analysis (WEBKDD04), in conjunction with the 10th ACM SIGKDD conference (KDD'04), Seattle, Washington (August 2004)
70. Lieberman, H.: Letizia: An agent that assists web browsing. In: Proceedings of the 14th International Joint Conference in Artificial Intelligence (IJCAI'95), Montreal, Quebec, Canada (August 1995) 924–929
71. Lin, W., Alvarez, S.A., Ruiz, C.: Efficient adaptive-support association rule mining for recommender systems. Data Mining and Knowledge Discovery 6 (2002) 83–105
72. Linden, G., Smith, B., York, J.: Amazon.com recommendations: Item-to-item collaborative filtering. IEEE Internet Computing 7(1) (2003) 76–80
73. Marlin, B.: Modeling user rating profiles for collaborative filtering. In: Proceedings of the 17th Annual Conference on Neural Information Processing System (NIPS'03), Vancouver, B.C., Canada (December 2003)
74. Massa, P., Avesani, P.: Trust-aware collaborative filtering for recommender systems. In: Proceedings of International Conference on Cooperative Information Systems, Larnaca, Cyprus (October 2004) 492–508
75. Melville, P., Mooney, R., Nagarajan, R.: Content-boosted collaborative filtering. In: Proceedings of the SIGIR2001 Workshop on Recommender Systems, New Orleans, LA (September 2001)

76. Micarelli, A., Sciarrone, F., Marinilli, M.: Web document modeling. In Brusilovsky, P., Kobsa, A., Nejdl, W., eds.: The Adaptive Web: Methods and Strategies of Web Personalization. Volume 4321 of Lecture Notes in Computer Science. Springer-Verlag, Berlin Heidelberg New York (2007) This Volume

77. Middleton, S.E., Shadbolt, N.R., Roure, D.C.D.: Ontological user profiling in recommender systems. ACM Transactions on Information Systems **22**(1) (2004) 54–88

78. Minka, T., Lafferty, J.: Expectation-propagation for the generative aspect model. In: Proceedings of the 18th Conference on Uncertainty in Artificial Intelligence, Edmonton, Alberta, Canada (August 2002) 352–359

79. Mladenic, D.: Personal web watcher: Implementation and design. Technical Report IJS-DP-7472, Department of Intelligent Systems, J. Stefan Institute, Slovenia (1996)

80. Mladenic, D.: Text-learning and related intelligent agents: A survey. IEEE Intelligent Systems **14**(4) (July/August 1999) 44–54

81. Mobasher, B.: Web usage mining. In Wong, J., ed.: Encyclopedia of Data Warehousing and Data Mining. Idea Group Publishing (2005) 1216–1220

82. Mobasher, B.: Web usage mining and personalization. In Singh, M.P., ed.: Practical Handbook of Internet Computing. CRC Press (2005)

83. Mobasher, B., Dai, H., Luo, T., Nakagawa, M.: Effective personalization based on association rule discovery from web usage data. In: Proceedings of the 3rd ACM Workshop on Web Information and Data Management (WIDM01), Atlanta, Georgia (November 2001)

84. Mobasher, B., Dai, H., Luo, T., Nakagawa, M.: Improving the effectiveness of collaborative filtering on anonymous web usage data. In: Proceedings of the IJCAI 2001 Workshop on Intelligent Techniques for Web Personalization (ITWP01), Seattle, WA (August 2001)

85. Mobasher, B., Dai, H., Luo, T., Nakagawa, M.: Discovery and evaluation of aggregate usage profiles for web personalization. Data Mining and Knowledge Discovery **6**(1) (2002) 61–82

86. Mobasher, B., Dai, H., Luo, T., Nakagawa, M.: Using sequential and non-sequential patterns for predictive web usage mining tasks. In: Proceedings of the IEEE International Conference on Data Mining, Maebashi City, Japan (December 2002) 669–672

87. Mobasher, B., Dai, H., Luo, T., Sun, Y., Zhu, J.: Integrating web usage and content mining for more effective personalization. In: E-Commerce and Web Technologies: Proceedings of the EC-WEB 2000 Conference. Lecture Notes in Computer Science (LNCS) 1875, Springer (September 2000) 165–176

88. Mobasher, B., Dai, H., T. Luo, M.N.: Discovery and evaluation of aggregate usage profiles for web personalization. Data Mining and Knowledge Discovery **6** (2002) 61–82

89. Mulvenna, M., Anand, S.S., Büchner, A.G.: Personalization on the net using web mining. Communication of ACM **43**(8) (2000) 122–125

90. Nakagawa, M., Mobasher, B.: A hybrid web personalization model based on site connectivity. In: Proceedings of the WebKDD 2003 Workshop, at the ACM-SIGKDD Conference on Knowledge Discovery in Databases (KDD'2003), Washington, DC (August 2003)

91. Nasraoui, O., Frigui, H., Krishnapuram, R., Joshi, A.: Extracting web user profiles using relational competitive fuzzy clustering. International Journal on Artificial Intelligence Tools **9**(4) (2000) 509–526

92. Nasraoui, O., Krishnapuram, R., Joshi, A., Kamdar, T.: Automatic web user profiling and personalization using robust fuzzy relational clustering. In Segovia, J., Szczepaniak, P., Niedzwiedzinski, M., eds.: Studies in Fuzziness and Soft Computing. Volume 105. Springer-Verlag, Heidelberg (2002) 233–261

93. Niu, L., Yan, X., , Zhang, C., , Zhang, S.: Product hierarchy-based customer profiles for electronic commerce recommendation. In: Proceedings of the 1st International Conference on Machine Learning and Cybernetics. (2002) 1075–1080

94. Oberle, D., Berendt, B., Hotho, A., Gonzalez, J.: Conceptual user tracking. In: Proceedings of the Atlantic Web Intelligence Conference (AWIC'03), Madrid, Spain (May 2003) 155–164

95. O'Connor, M., Herlocker, J.: Clustering items for collaborative filtering. In: Proceedings of ACM SIGIR'99 Workshop on Recommender Systems: Algorithms and Evaluation, Berkeley, California (August 1999)

96. O'Mahony, M., Hurley, N., Kushmerick, N., Silverstre, G.: Collaborative recommendations: A robustness analysis. ACM Transactions on Internet Technologies 4(4) (2004) 344–377

97. Padmanabhan, B., Tuzhilin, A.: Unexpectedness as a measure of interestingness in knowledge discovery. Decision Support Systems 27(3) (1999) 303–318

98. Palpanas, T., Mendelzon, A.: Web prefetching using partial match prediction. In: Proceedings of the 4th International Web Caching Workshop (WCW99), San Diego, CA (March 1999)

99. Parent, S., Mobasher, B., Lytinen, S.: An adaptive agent for web exploration based on concept hierarchies. In: Proceedings of the 9th International Conference on Human Computer Interaction, New Orleans (August 2001) 903–907

100. Pavlov, D.: Sequence modeling with mixtures of conditional maximum entropy distributions. In: Proceedings of the Third IEEE International Conference on Data Mining (ICDM'03), Melbourne, Florida (November 2003) 251–258

101. Pazzani, M.: A framework for collaborative, content-based and demographic filtering. Artificial Intelligence Review 13(5-6) (1999) 393–408

102. Pazzani, M., Billsus, D.: Learning and revising user profiles: The identification of interesting web sites. Machine Learning 27 (1997) 313–331

103. Pazzani, M.J., Billsus, D.: Content-based recommendation systems. In Brusilovsky, P., Kobsa, A., Nejdl, W., eds.: The Adaptive Web: Methods and Strategies of Web Personalization. Volume 4321 of Lecture Notes in Computer Science. Springer-Verlag, Berlin Heidelberg New York (2007) This Volume

104. Pei, J., Han, J., Mortazavi-Asl, B., Zhu, H.: Mining access patterns efficiently from web logs. In: Proceedings of the 4th Pacific-Asia Conference on Knowledge Discovery and Data Mining (PAKDD'00), Kyoto, Japan (April 2000) 396–407

105. Perkowitz, M., Etzioni, O.: Adaptive web sites: Automatically synthesizing web pages. In: Proceedings of the 15th National Conference on Artificial Intelligence, Madison, WI (July 1998) 727–732

106. Perkowitz, M., Etzioni, O.: Adaptive web sites. Communications of ACM 43(8) (2000) 152–158

107. Pitkow, J., Pirolli, P.: Mining longest repeating subsequences to predict www surfing. In: Proceedings of the 2nd USENIX Symposium on Internet Technologies and Systems, Boulder, Colorado (October 1999)

108. Popescul, A., Ungar, L., Pennock, D., Lawrence, S.: Probabilistic models for unified collaborative and content-based recommendation in sparse-data environments. In: Proceedings of 17th Conference in Uncertainty in Artificial Intelligence, Seattle, WA (August 2001) 437–444

109. Rissanen, J.: Modelling by shortest data description. Automatica 14 (1978) 465–471

110. Rivasseau, J.: Understanding and applying lda model to first-order markov chains. Univ. of british columbia, canada, technical report, Univ. of British Columbia, Canada, Canada (2003)

111. Rosenfeld, R.: Adaptive statistical language modeling: A maximum entropy approach. Phd dissertation, CMU (1994)

112. Salton, G., McGill, M.: Introduction to Modern Information Retrieval. McGraw-Hill, New York, NY (1983)

113. Sarukkai, R.: Link prediction and path analysis using markov chains. In: Proceedings of the 9th International World Wide Web Conference, Amsterdam (May 2000)

114. Sarwar, B., Karypis, G., Konstan, J., Riedl, J.: Item-based collaborative filtering recommendation algorithms. In: Proceedings of the 10th International WWW Conference, Hong Kong (May 2001) 285–295

115. Sarwar, B.M., Karypis, G., Konstan, J., Riedl, J.: Analysis of recommender algorithms for e-commerce. In: Proceedings of the 2nd ACM E-Commerce Conference (EC'00), Minneapolis, MN (October 2000) 158–167

116. Sarwar, B.M., Karypis, G., Konstan, J.A., Riedl, J.: Application of dimensionality reduction in recommender system - a case study. In: Proceedings of the WebKDD 2000 Web Mining for E-Commerce Workshop at ACM SIGKDD 2000, Boston (August 2000)

117. Schafer, J.B., Frankowski, D., Herlocker, J.L., Sen, S.: Collaborative filtering recommender systems. In Brusilovsky, P., Kobsa, A., Nejdl, W., eds.: The Adaptive Web: Methods and Strategies of Web Personalization. Volume 4321 of Lecture Notes in Computer Science. Springer-Verlag, Berlin Heidelberg New York (2006) This Volume

118. Schafer, J., Konstan, J., , Riedl, J.: Recommender systems in e-commerce. In: Proceedings of the ACM Conference on Electronic Commerce, Denver, Colorado (November 1999) 158–166

119. Schechter, S., Krishnan, M., Smith, M.D.: Using path profiles to predict http requests. In: Proceedings of the 7th International World Wide Web Conference, Brisbane, Australia (April 1998)

120. Schwab, I., Kobsa, A., Koychev, I.: Learning about users from observation. In: Adaptive User Interfaces: Papers from the 2000 AAAI Spring Symposium, Menlo Park, CA, AAAI Press (2000)

121. Shardanand, U., Maes, P.: Social information filtering: Algorithms for automating word of mouth. In: Proceedings of the 1995 ACM Conference on Human Factors in Computing Systems (CHI'95), Denver, Colorado (May 1995) 210–217

122. Sieg, A., Mobasher, B., Burke, R.: Inferring user's information context from user profiles and concept hierarchies. In: Proceedings of the 2004 Meeting of the International Federation of Classification Societies, IFCS 2004, Chicago, IL (July 2004) 563–574

123. Silberschatz, A., Tuzhilin, A.: What makes patterns interesting in knowledge discovery systems. IEEE Transactions on Knowledge and Data Engineering **8**(6) (1996) 970–974

124. Sinha, R., Swearingen, K.: Comparing recommendaions made by online systems and friends. In: Proceedings of Delos-NSF Workshop on Personalisation and Recommender Systems in Digital Libraries. (June 2001)

125. Sinha, R., Swearingen, K.: The role of transaprency in recommender systems. In: CHI '02 extended abstracts on Human factors in computing systems. (2002) 830–831

126. Smeaton, A., Murphy, N., O'Connor, N.E., Marlow, S., Lee, H., McDonald, K., Browne, P., Ye, J.: The físchlár digital video system: a digital library of broadcast tv programmes. In: Proceedings of the 1st ACM/IEEE-CS Joint Conference on Digital Libraries, Roanoke, Virginia (June 2001) 312–313

127. Smyth, P.: Probabilistic model-based clustering of multivariate and sequential data. In Heckerman, D., Whittaker, J., eds.: Proceedings of the Seventh International Workshop on AI and Statistics, Los Gatos, CA, Morgan Kaufmann (January 1999)

128. Spiliopoulou, M., Faulstich, L.: Wum: A tool for web utilization analysis. In: Proceedings of EDBT Workshop at WebDB'98. LNCS 1590, Springer Verlag (1999) 184–203

129. Spiliopoulou, M., Mobasher, B., Berendt, B., Nakagawa, M.: A framework for the evaluation of session reconstruction heuristics in web usage analysis. INFORMS Journal of Computing - Special Issue on Mining Web-Based Data for E-Business Applications **15**(2) (2003)

130. Srivastava, J., Cooley, R., Deshpande, M., Tan, P.: Web usage mining: Discovery and applications of usage patterns from web data. SIGKDD Explorations **1**(2) (2000) 12–23
131. Steyvers, M., Smyth, P., Rosen-Zvi, M., Griffiths, T.: Probabilistic author-topic models for information discovery. In: Proceedings of the International Conference on Knowledge Discovery and Data Mining (KDD'04), Seattle, Washington (August 2004) 306–315
132. Strehl, A., Ghosh, J.: Relationship-based clustering and visualization for high-dimensional data mining. INFORMS Journal Of Computing, Special Issue on Web Mining, (A. Tuzhilin and L. Rashid, guest Eds.) **15**(2) (2003) 208–230
133. Suryavanshi, B.S., Shiri, N., Mudur, S.P.: Improving the effectiveness of model based recommender systems for highly sparse and noisy web usage data. In: Proceedings of the IEEE/WIC/ACM International Conference on Web Intelligence (WI'05), Compiegne, France (September 2005) 618–621
134. Swearingen, K., Sinha, R.: Beyond algorithms: An hci perspective on recommender systems. In: Proceedings of the ACM SIGIR Workshop on Recommender Systems, New Orleans. LA (September 2001)
135. Tan, P., Kumar, V.: Discovery of web robot sessions based on their navigational patterns. Data Mining and Knowledge Discovery **6** (2002) 9–35
136. Tan, P., Kumar, V., Srivastava, J.: Selecting the right objective measure for association analysis. Information Systems **29**(4) (2004) 293–313
137. Tanasa, D., Trousse, B.: Advanced data preprocessing for intersite web usage mining. IEEE Intelligent Systems **19**(2) (2004) 59–65
138. Teevan, J., Dumais, S.T., Horvitz, E.: Personalizing search via automated analysis of interests and activities. In: Proceedings of 28th ACM SIGIR Conference on Research and Development in Information Retrieval, Salvador, Brazil (August 2005) 449–456
139. Trajkova, J., Gauch, S.: Improving ontology-based user profiles. In: Proceedings of the Recherche d'Information Assiste par Ordinateur, RIAO 2004, University of Avignon (Vaucluse), France (April 2004) 380–389
140. Ungar, L., Foster, D.P.: Clustering methods for collaborative filtering. In: Proceedings of the AAAI98 Workshop on Recommendation Systems, Madison Wisconsin (July 1998)
141. Ypma, A., Heskes, T.: Categorization of web pages and user clustering with mixtures of hidden markov models. In: Proceedings of the WEBKDD 2002 Workshop: Web Mining for Usage Patterns and User Profiles, at SIGKDD 2002), Edmonton, Alberta, Canada (July 2002)
142. Yu, K., Schwaighofer, A., Tresp, V., Ma, W., Zhang, H.: Collaborative ensembling learning. Combining collaborative and content-based information filtering. In: Proceedings of the 19th Conference on Uncertainty in Artificial Intelligence (UAI'03), Acapulco, Mexico (August 2003) 616–623
143. Yu, P.S.: Data mining and personalization technologies. In: Proceedings of the International Conference on Database Systems for Advanced Applications (DASFAA99), Hsinchu, Taiwan (April 1999) 6–13
144. Zhou, Y., Jin, X., Mobasher, B.: A recommendation model based on latent principle factors in web navigation data. In: Proceedings of the 3rd International Workshop on Web Dynamics at WWW 2004 Conference, New York (2004)
145. Ziegler, C., McNee, S.M., Konstan, J.A., Lausen, G.: Improving recommendation lists through topic diversification. In: Proceedings of the 14th international World Wide Web conference, Chiba, Japan (May 2005) 22–32

Generic User Modeling Systems

Alfred Kobsa

Donald Bren School of Information and Computer Sciences
University of California, Irvine
Irvine, CA 92697-3440, U.S.A.
kobsa@uci.edu
http://www.ics.uci.edu/~kobsa

Abstract. This chapter reviews research results in the field of Generic User Modeling Systems. It describes the purposes of such systems, their services within user-adaptive systems, and the different design requirements for research prototypes and commercial deployments. It discusses the architectures that have been explored so far, namely shell systems that form part of the application, central server systems that communicate with several applications, and possible future agent-based user modeling systems. Major implemented research prototypes and commercial systems are briefly described.

4.1 User Modeling Shell Systems

4.1.1 Historical Development

User modeling is usually traced back to the works of Allen, Cohen and Perrault (see, e.g., [1, 16, 74]) and Elaine Rich [79, 80]. Inspired by their seminal research, numerous application systems in various application areas were subsequently developed that collected different kinds of information about the current user, and adapted to the user in different ways. Several publications from this time [66, 69, 96] offer comprehensive reviews of first-generation user-adaptive applications.

In this early work, all user modeling was performed by the application system. In most cases, there was no clear distinction between system components that served user modeling purposes and components that performed other tasks. From the mid-eighties onwards, such a separation was increasingly made (e.g., in [2, 38, 46, 86]), but no efforts are reported to make the user modeling component reusable for the development of other user-adaptive systems.

In 1986, Tim Finin published his "General User Modeling System" GUMS [20, 21]. This software allows programmers of user-adaptive applications the definition of simple stereotype hierarchies (see Chapter 2 of this book [26]). For each stereotype, one can define the Prolog facts describing stereotype members and the rules prescribing the system's reasoning about them. At runtime, GUMS accepts and stores new facts about the user which are provided by the application system, verifies the consistency of a new fact with currently held assumptions, informs the application about recognized inconsistencies, and answers queries of the application concerning the currently held assumptions about the user.

Albeit GUMS was never used together with an application system, it set the stage for future generic (i.e., application-independent) user modeling systems. At the same time, GUMS also defined their basic functionality, namely the provisioning of selected user modeling services at runtime that can be configured during development time. When filled by the developer with application-specific user modeling knowledge at the time of development, these systems would serve as a separate user modeling component in an application system at runtime. Early systems usually included a representation system for expressing the contents of the user model (such as some logic formalism, rules, or simple attribute-value pairs) and a reasoning mechanisms for deriving assumptions about the user from existing ones and for detecting inconsistencies in the user model.

Kobsa [48] seems to be the first author who used the term "user modeling shell system" for such kinds of software tools. The term "shell system", or "shell" for short, had been borrowed from the field of Expert Systems. There, van Melle [93] and Buchanan and Shortliffe [12] had condensed the experiences made with the medical expert system MYCIN [85] into EMYCIN ("Essential" MYCIN), an "empty" expert system that had to be filled with domain-specific rules for deployment as a "real" expert system. Commercial expert system shells like Knowledge Craft [45], KEE [36] and ART [15] became very popular in the late seventies and early eighties. User modeling "shells" had similar purposes as expert system shells, but the general underlying aims, namely software decomposition and abstraction to support modifiability and reusability, is of course much older than expert system shells.

4.1.2 Example Systems

A number of user modeling shell systems were developed after GUMS, which comprised different representation mechanisms for user models as well as associated inference processes. Below we list four representative examples.[1]

UMT [6] allows the user model developer the definition of hierarchically ordered user stereotypes, and of rules for user model inferences and contradiction detection. Information about the user that is received from the application is classified as invariable premises or (later still retractable) assumptions. When new information is received, stereotypes may become activated and their contents (which describe the respective user subgroups) added to the user model. UMT then applies inference rules (including contradiction detection rules) to the set of premises and assumptions, and records the inferential dependencies. After the firing of all applicable inference rules and the activation of all applicable stereotypes, contradictions between assumptions are sought and various resolution strategies applied ("truth maintenance").

---

[1] Also see [5, 35, 59, 65] for additional systems.

PROTUM [95] represents user model content as a list of constants, each with associated type (i.e., observed, derived from stereotype, default) and confidence factor. It is related to UMT except that it possesses more sophisticated stereotype retraction mechanisms than UMT.

TAGUS [71] represents assumptions about the user in first-order formulas, with meta-operators expressing the different assumption types (namely users' beliefs, goals, problem solving capabilities and problem solving strategies). The system allows for the definition of a stereotype hierarchy and contains an inference mechanism, a truth maintenance system (with different strengths of endorsements for assumptions about the user), and a diagnostic subsystem including a library of misconceptions. It also supports powerful update and evaluation requests by the application, including a simulation of the user (i.e., forward-directed inferences on the basis of the user model) and the diagnosis of unexpected user behavior.

um [40, 41] is a toolkit[2] for user modeling that represents assumptions about the user's knowledge, beliefs, preferences, and other user characteristics in attribute-value pairs. Each piece of information is accompanied by a list of evidence for its truth and its falsehood. The source of each piece of evidence, its type (observation, stereotype activation, rule invocation, user input, told to the user) and a time stamp is also recorded. Explanations for components, sources of evidence, and types of evidence sources may be entered as well. At runtime, competing specialized inference processes (the so-called "resolvers") interpret the available evidence and conclude the value of a component. Applications have to decide which resolvers to employ. Users can inspect and edit their user models [42].

4.2   User Modeling Servers

4.2.1 Characteristics

The purpose of user modeling servers, like that of user modeling shells, is to separate user modeling functionality from user-adaptive application systems. In contrast to user modeling shell systems, user modeling servers are not a part of an application system but rather independent from it (i.e., they are not functionally integrated into the application but communicate with the application through inter-process communication). User modeling servers may reside on the same platform as the application system and only serve one instance of this application [54]. Much more commonly however, they will be part of a local area network or a wide area network and serve more than one application instance at a time (possibly even several 100,000 simultaneous instances in the case of personalized e-commerce websites, cf. Chapter 16 of this book [27]). They communicate with application systems through protocols that both sides support, such as LDAP, ODBC, remote procedure calls, or plain TCP/IP.

---

[2] From the point of view of the application system, um was more a library of user modeling functions than an independent user modeling component. It therefore is not a user modeling shell in a strict sense.

A client-server based architecture provides a number of advantages in comparison to embedded user modeling components that were described in the previous section (see [24] and [4] for more comprehensive discussions):

- All information about the user is maintained in a repository with clearly defined points of access (usually one single access point).

- User information is at the disposal of more than one application at a time.

- User information acquired by one application can be employed by other applications, and vice versa.

- Information about users is stored in a non-redundant manner.

- The consistency and coherence of information gathered by different applications can be more easily ascertained.

- Information about user groups, either available a priori as stereotypes (e.g., [79-82]) or dynamically calculated as user group models (e.g., [70, 73]), can be maintained with low redundancy.

- Methods and tools for system security, identification, authentication, access control and encryption can be applied for protecting user models in user modeling servers [57, 84].

- Complementary user information that is dispersed across the enterprise (e.g., demographic data from client databases, past purchase data from transactional systems, user segmentations from marketing research) can be integrated more easily with the information in the user model repository.

User modeling servers may be "centralized" (e.g., reside on a single platform only). This facilitates their implementation, but exposes them to the typical downsides of centralization such as the need for a permanent network connection and the jeopardy of a single point of failure and potential bottleneck. Most modern user modeling servers, specifically the commercial ones, are therefore distributed across several platforms to increase their performance and availability (typically through CORBA [77]). Most commercial servers also allow the virtual integration of heterogeneous "outside" resources of user information, sometimes to the point that transparent read and write access to these outside resources is possible dynamically at runtime. Increasingly, user modeling servers also allow for the (partial) replication of user modeling server entries, which mitigates the need for a reliable permanent network connection.

A large number of user modeling servers have been developed over the past 15 years, ranging from academic prototypes to commercial systems. In the next section, we briefly describe examples of major research prototypes, and thereafter the currently available commercial systems.

4.2.2 Examples of Research Prototypes of User Modeling Servers

BGP-MS [55, 76] allows assumptions about the user and stereotypical assumptions about user groups to be represented in a first-order predicate logic. A subset of these assumptions is stored in a terminological logic. Different assumption types, such as (nested) beliefs and goals as well as stereotypes, are represented in different partitions that can be hierarchically ordered to exploit inheritance of partition contents (a partition together with all its direct and indirect ancestor partitions thereby establishes a so-called *view* of the full user model). Inferences across different assumption types (i.e. partitions) can be defined in a first-order modal logic. The BGP-MS system can be used as a network server with multi-user and multi-application capabilities.

DOPPELGÄNGER [70] accepts information about the user from hardware and software sensors. Techniques for generalizing and extrapolating data from the sensors (such as beta distributions, linear prediction, Markov models) are put at the disposal of user model developers. Unsupervised clustering (see Chapter 3 of this book [68]) is available for collecting individual user models into so-called 'communities' whose information serves the purpose of stereotypes. In contrast to all other user modeling shell systems, membership in a stereotype is probabilistic rather than definite. The different representations of DOPPELGÄNGER are quite heterogeneous. As is the case for um (see Section 4.1.2), users can inspect and edit their user models.

CUMULATE [9] is designed to provide user modeling functionality to a student-adaptive educational system (see Chapters 1 and 22 of this book [10, 31]). It collects evidence (events) about a student's learning from multiple servers that interact with the student. It stores students' activities and infers their learning characteristics, which form the basis for individual adaptation to them. In this vein, external and internal inference agents process the flow of events and update the values in the inference model of the server. Each inference agent is responsible for maintaining a specific property in the inference model, such as the current motivation level of the student or the student's current level of knowledge for each course topic. Brusilovsky et al. [11] describe the interaction of CUMULATE with an ontology server, which stores the ontological structures of the taught domain and provides the platform for the exchange between different user model servers of higher-level information about students' knowledge.

Personis [43] and a simplified version of it, PersonisLite [14], have the same representational foundations as their predecessor um that was described in Section 4.1.2. The components from um form objects in Personis that reside in an object layer over Berkeley DB, a near-relational database system. The object database structures user models into hierarchically ordered contexts similar to the partitions of BGP-MS (see Section 4.1.2). It also holds objects defining the views that include components from all levels of the user model context hierarchy. The authors distinguish two basic operations upon this representation: accretion, which involves the collection of uninterpreted evidence about the user, and resolution, the interpretation of the current collection of evidence (cf. the resolvers in um).

UMS [23, 52] is a user modeling server that is based on the Lightweight Directory Access Protocol (LDAP). Its Directory Component allows for the representation of different models, such as user and usage profiles as well as system and service models. "Pluggable" User Modeling Components are internal clients of the Directory Component. They can access these models and perform dedicated user modeling tasks, such as collaborative filtering, domain-based inferences, etc. The stored models can also be accessed by External Clients, such as user-adaptive applications or tools for user model inspection, visualization and statistical analysis. The use of LDAP makes it possible to base the storage component of user modeling servers on international industry-adopted standards, to distribute user information across a network and replicate and loosely synchronize such information (both often increases the performance, scalability, availability and reliability of a service), and to realize a "virtually centralized distributed architecture" for user models that is internally distributed but provides a common point of access to all clients.

4.2.3 Examples of Commercial User Modeling Servers

Group Lens [92] originally employed various collaborative filtering algorithms [7, 32] for predicting users' interests, based on explicitly provided users ratings, implicit ratings derived from users' navigation, and transaction histories (e.g., shopping basket operations, purchases). GroupLens stored all user ratings in a database, but kept a correlation matrix of all ratings in cache memory during runtime. This created memory problems and huge performance problems on the largest sites. They were temporarily solved by statistically selecting reduced-size models (with careful sampling, the reduced-size models did not show much quality degradation). The commercial version of Group Lens eventually moved to item-item models, which can be truncated substantially without much loss in quality [67].

ATG Adaptive Scenario Engine [3] allows for the definition of rules that assign individual users to one or more user groups (e.g., customer segments) based on their demographics, their system usage, and their software, hardware and network environments. Rules can also be defined for inferring individual assumptions about the user from his or her navigation behavior, and for personalizing the content of web pages. The operation of Personalization Server thus follows very much the "stereotype approach" from classical user modeling research (see Chapter 2 of this book [26]). Customer data from legacy databases can be integrated via SQL, XML and Web Services.

enQuire™ Identity Server [19] is a multi-functional server product with an embedded virtual directory engine. It supports the development of user modeling servers by introducing a flexible virtualization layer between multiple repositories and applications that provide user data via LDAP, ODBC or an API. The enQuire Server component stores information about user data sources and their structure in an enterprise, enforces security policies and rule-based access control, federates user data from connected information sources, applies rules to filter and transform user data sets, and presents consolidated user data in a standard format. The results of the federation process can be stored in a persistent cache, which eliminates the dependency on source-specific data structures. enQuire supports the assignment of users to static or dynamically constructed user groups. enQuire plug-ins are customization components that enable developers to describe actions to be executed under specific circumstances or at desired points in the request execution process.

Other Identity Management and User Provisioning Systems. In addition to the commercial systems mentioned above, a large number of so-called "Identity Management Systems" or "User Provisioning Systems" are commercially available which to some extent provide important functionality of a user modeling server (see, e.g., [17, 61, 89, 91]). Such functionality includes one or more of the following: integration of disparate user data into a single centralized repository, federated provisioning of disparate user data, account linking, policy-based access control, user account management, and support for privacy and security audits. These systems however lack other essential functionality, such as inference capabilities on the basis of user data (including the assignment of users to user groups), or triggers for personalization methods. They can therefore not yet be regarded as user modeling servers.

4.3   Required Services and Characteristics of Generic User Modeling Systems

4.3.1   From Ingredients to Services

Developers of user modeling shell systems (see Section 4.1) and early user modeling servers (see Section 4.2) aimed at condensing basic structures and processes into these systems that they deemed important for user-adaptive application systems (e.g., certain knowledge representation systems, inference mechanisms, truth maintenance systems, and tell/ask interfaces). For identifying such important structures and processes, developers mostly relied on their intuitions and/or their experience through prior work on user-adaptive systems. Efforts to put these decisions on more empirical grounds were seemingly only made by Kleiber [44] and Pohl [75, 76]. Even these authors however merely identified individual user-adaptive application systems in the literature that would have profited from the functionality of their own shell system, rather than conducting a comprehensive review of prior user-adaptive systems, and determining current as well as predicting future needs of user-adaptive application systems.

In an attempt to extend the de facto definition of user modeling shells introduced by GUMS and to avoid characterizing user modeling shell systems via internal structures and processes, Kobsa [49] listed the following frequently found *services* of such systems:

- "the representation of assumptions about one or more types of user characteristics in models of individual users (e.g. assumptions about their knowledge, misconceptions, goals, plans, preferences, tasks, and abilities);

- the representation of relevant common characteristics of users pertaining to specific user subgroups of the application system (the so-called stereotypes);

- the classification of users as belonging to one or more of these subgroups, and the integration of the typical characteristics of these subgroups into the current individual user model;

- the recording of users' behavior, particularly their past interaction with the system;
- the formation of assumptions about the user based on the interaction history;
- the generalization of the interaction histories of many users into stereotypes;
- the drawing of additional assumptions about the current user based on initial ones;
- consistency maintenance in the user model;
- the provision of the current assumptions about the user, as well as justifications for these assumptions;
- the evaluation of the entries in the current user model, and the comparison with given standards."

This list is of course subject to changes when new forms of adaptation to the user require new services from generic user modeling systems. It is surprisingly stable though, i.e. it is by and large still valid today. The only important addition today would be services to secure users' data and to protect users' privacy (such services are discussed in Chapter 21 of this book [51]). In the light of recent progress in the field of recommender systems (see Chapters 9 and 12 of this book [13, 83]), services that compare users with other users might also be useful (e.g., delivering a list of nearest neighbors of a given user). In early years researchers tacitly strived for "universal" generic user modeling systems that would ideally perform all the important user modeling services. Today, however, a typical generic user modeling system only delivers a small portion of the services listed above, and it is unlikely that this will change very much in the future (see Section 4.4.5).

4.3.2 Required Characteristics of Generic User Modeling Systems

Several characteristics of generic user modeling systems have been regarded as very important over the years. We will discuss some of them in the following. The first two requirements were already proposed very early in the history of generic user modeling systems, while the other requirements became important only a few years ago when commercial generic user modeling systems were developed for use in web site personalization.

Generality, Including Domain Independence. This requirement states that user modeling systems should be usable in as many domains as possible, and within these domains for as many user modeling tasks as possible. While this requirement seemed very important in earlier years, it is less so to date. "Subclasses" of generic user modeling systems have already evolved, most prominently one for student-adaptive tutoring systems that impose very specific requirements on generic student modeling systems (see, e.g., [8, 35, 59, 65, 72] and Chapter 1 of this book [10]). Such generic student modeling systems are expected to be usable for teaching different subject matters, but not for additional applications besides educational ones.

Expressiveness and Strong Inferential Capabilities. Shell systems and early user modeling servers were expected to be able to express many different types of assumptions about the user at a time. These included beliefs, goals, plans and preferences of the user, as well as various reflexive assumptions regarding the user and the system (see [47, 90]), and moreover uncertainty and vagueness in these assumptions. Generic user modeling systems were also expected to perform all sorts of reasoning, such as reasoning in a first-order predicate logic, complex modal reasoning (e.g., reasoning about types of modalities), reasoning with uncertainty, plausible reasoning when full information is not available, and to perform conflict resolution when contradictory assumptions are detected.

The rationale for assigning so much importance to expressiveness and strong inferential capabilities lies in the affinity of user modeling research of those days to artificial intelligence, natural-language dialog [58], and intelligent tutoring [39, 87]. User modeling shells were expected to support the complex assumptions and complex reasoning about the user that had been identified in these domains, and additionally to be usable in a wide range of other domains as well. When in the mid-nineties user-adaptive application systems shifted towards different domains with less demanding user modeling requirements (like user-adaptive learning environments that are described in Chapters 1 and 22 of this book [10, 31], as well as personalized web sites [53]), such highly expressive user modeling and powerful reasoning capabilities became largely redundant.

Support for Quick Adaptation. In order to bond first-term visitors with web stores, adaptations should already take place during their (usually relatively short) initial interaction. Several commercial user modeling servers can therefore select between more than one modeling and personalization methods with different degrees of complexity, depending on the amount of data that is already available about the user.

Extensibility. Current user modeling servers support a number of user model acquisition and personalization methods, but companies may want to integrate their own methods or third-party tools. Application Programmer Interfaces (APIs) and interfaces that allow for the (possibly bi-directional) exchange of user information between user-modeling tools are therefore required.

Import of External User-Related Information. Many businesses already own customer and marketing data, and usually want to integrate these into user modeling systems when starting with personalized e-commerce. To access external data, ODBC interfaces or native support for a wide variety of databases are required. Due to legacy business processes and software, external user-related information often continues to be updated in parallel to the e-commerce application, and therefore needs to be continually integrated at reasonable cost and without impairing the response time.

Management of Distributed Information. The ability of a generic user modeling system to manage distributed user models is becoming more and more important. Commercial personalized websites often utilize several sources of user information, such as user profiles, purchase records from legacy systems, and customer segmentations from marketing research. Current commercial user modeling servers integrate these information sources already today to a greater or lesser extent [24]. In the future, support for the management of distributed information will also facilitate the integration of mobile user models and of user models in smart appliances (see Sections 4.4.1 and 4.4.2).

Support for Open Standards. Adherence to open standards in the design of generic user modeling systems is decisive since it fosters their interoperability. There already exist efforts in some subfields of user modeling to come up with standards, e.g. for the exchange of user models [28, 29, 56, 60] and for a common user modeling ontology [29, 30, 64, 78], but without much progress so far. Of particular importance for the openness of generic user modeling systems would be the adoption of LDAP [33, 34, 63] which is based on IETF standards. LDAP directories constitute a widely adopted and supported industry standard for storing and retrieving various kinds of people-related information including names, phone numbers, salaries, photographs, digital certificates, passwords, preferences, and even mobile "user agents". Moreover, they support the representation of information about organizations, groups (e.g. administrators) and devices (e.g. printers). Pre-defined ontologies ("schemas") exist for these information types that are easily extensible.

Load Balancing. Under real-world conditions, user model servers will experience dramatic changes in their average load. Noticeable response delays or denials of requests should only occur in emergency situations. User modeling servers should be able to react to load increases through load distribution (ideally with CORBA-based components that can be distributed across a network of computers) and possibly by resorting to less thorough (and thereby less time-consuming) user model analyses.

Failover Strategies. Centralized architectures need to provide fallback mechanisms in case of a breakdown.

Transactional Consistency. Parallel read/write on the user model and abnormal process termination can lead to inconsistencies that must be avoided by carefully selected transaction management strategies [22]. For instance, if a consistent state cannot be reached, the user model has to be reset to the original state.

Privacy Support. Users' privacy concerns, company privacy policies, industry privacy norms, and national and international privacy legislation have a considerable impact on what user-adaptive application may do. Moreover, numerous privacy-enhancing software tools and Internet services are emerging. Generic user modeling systems should also facilitate the compliance with such regulations and user preferences, as well as support privacy-enhancing services. Chapter 21 [51] discusses privacy-related issues in more detail.

4.4 Future Trends

It goes without saying that predictions concerning the future of user modeling systems are fairly speculative, due to the rapidly changing nature of application scenarios for personalization and of computing devices that are thereby used. Since personalization has already been demonstrated to benefit both the users and the providers of computer systems and since personalization is therefore likely to stay, it is practically certain that generic tool systems that allow for the easy development and maintenance of personalized systems will be equally necessary in the years to come. The exact shape that future user modeling systems will assume is however likely to be very much contingent on many characteristics of future system usage that are difficult to predict. Here are a few considerations on likely future avenues.

4.4.1 Mobile User Models

Computing is becoming increasingly mobile and ubiquitous, yet interaction with the user should nevertheless still be personalized. In ubiquitous scenarios, users' handheld devices as well as sensor-equipped environments around them are supposed to provide personalized services wherever users go. In the near future at least, the availability and reliability of mobile networks (and possibly also their bandwidths) is however unlikely to meet the demands of today's client-server architectures for user modeling systems, which require permanent reliable connectivity. Hence mobile user models seem to be worth considering for such scenarios. These user model agents may reside on the server side and be replicated at the beginning of each interaction (see e.g. UMS [23, 52] which is based on LDAP that supports partial or full replication). Or, they may be truly mobile and stay with the user all the time, either on his or her computing device or on an item that the user always wears (like in a wristwatch or jewelry with a wireless connection [25]).

4.4.2 User Models for Smart Appliances

Personalization has so far been almost exclusively confined to computing systems. Recently, however, appliances are being offered that feature limited but very useful personalization. Examples include car radios with a chip card that contains an authorization code and also stores the driver's preferences concerning pre-set stations, sound volume and tone, and traffic news. Electronic car keys exist that adjust the driver seat, the mirrors and the GPS system to the driver's individual preferences when plugged into the ignition lock. While these are proprietary solutions with proprietary minuscule user models, it is likely that we will see far more examples of personalizable appliances in the future. Since people will not want to carry a small user model gadget for each and every personalized appliance, standardized solutions and hence the need for generic tool systems will soon arise. A particular challenge will be the integration of such local minuscule user models into larger models (e.g. to make them accessible to all cars that the user drives). Generic user modeling systems that support distributed models will have a decisive advantage in this regard.

4.4.3 Agent-Based User Modeling Systems

The internal structure of user modeling servers (which constitute the currently predominant form of generic user modeling systems) is not transparent to the outside. As was discussed in Section 4.2.1, various realizations are possible which range from central-repository approaches such as in BGP-MS [55, 76] and Personis [14, 43] to servers that integrate heterogeneous resources of user information (e.g., ATG Adaptive Scenario Engine [3] and enQuire™ Identity Server [19]) and to servers that support the (partial) replication of user modeling server entries (UMS, [23, 52]). These forms of decentralization all help increase the performance and failure tolerance of user modeling servers and their ability to integrate into existing environments. Such architectures proved to be able to support personalization even at the largest current websites.

Despite their internal heterogeneity, user modeling servers have however two characteristics in common. For one, they have clearly defined points of access (usually one), at which information about the user and requests for user model entries can be submitted and where answers or spontaneous notifications of the user modeling server can be received. Moreover, even when user modeling servers are decentralized, the interaction of their parts (specifically the flow of control and information) is predetermined by design. Individual parts of user modeling servers are hardly autonomous, i.e. cannot exist without the other parts, and only rarely exhibit "initiative" of their own.

Attempts have been made in the past few years to develop new types of architectures for user modeling purposes that do not exhibit these two characteristics. Ideally, such user modeling systems would be conglomerates of independent and autonomous services (or "agents") which collaborate with each other as the need arises (those services may have even been developed by different parties). Decisions on collaboration would not be predetermined by design, but be made dynamically at runtime. A need for such novel kinds of user modeling architectures arose particularly in the field of ubiquitous computing where numerous small sensors and devices acquire limited information about users and are supposed to exchange it for the provision of personalized services (see Section 4.4.1). A second area is the support of communities of computer users in which individual users need to contact others who have desired characteristics, e.g. specific knowledge and skills.

Within this broad agent-based paradigm, major differences exist however in the ways in which user models are stored in current research prototypes. In the expert finder system DEMOIR [98], every user possesses a unique user agent. User agents monitor their users and store assumptions about their knowledge and skills. When users need to find experts with certain knowledge and skills, their user agents broadcast the desired profile to all other user agents, collect their responses, and determine the most similar experts. In a variant architecture, all agents report (summaries of) their user model contents to a central repository, and contact this repository first when external expertise is sought.

I-Help [94] also features one personal agent per user, which contains a coarse user model that describes predominantly the users' preferences for helping others. I-Help also has other agents, such as diagnostic agents which develop models of the user's knowledge in various domains, and application agents which characterize the services of the applications to which they pertain. Matchmaker agents facilitate locating agents that possess information resources or represent users who are knowledgeable on certain topics. For this purpose, they maintain profiles of the knowledge and some other characteristics of users and applications. When users seek help, their personal agent acts as their representative and negotiates with other agents. In order to do this effectively, user agents also create models of the other agents' "character" and priorities, and collect references to other agents who keep information about their users, for example, diagnostic agents. User model fragments pertaining to a given user may therefore be maintained by several agents in I-Help (in a field study, the authors found up to 20 fragmented models of the same user). These fragments may partially overlap and may even be globally inconsistent.

Even more radical than the mentioned approaches is the work of Lorenz [62] and Specht et al. [88], who envision to break the one-to-one relationship between users and their representing user agents and to replace these agents by a network of small active entities on the client side. In the spirit of self-organizing networks, the authors propose distributed active user modeling agents with extremely limited functionality (which the authors categorize into sensoring, modeling, controlling and actuating). The agents would be able to act out of their own initiative and would self-organize appropriately to form "modeling networks". For the communication between the potentially countless tiny agents, the authors foresee a mix of the blackboard and the messaging approaches from multi-agent systems [37, 97]. Agents register their services and requests with blackboards provided by local brokers, and brokers broadcast the requests between each other. Communication between different applications would be facilitated by a common user model exchange language (e.g., UserML [28, 29]) and a common user modeling ontology (such as GUMO [29, 30] or that of Razmerita [78]).

Agent-Based User Modeling Systems are currently the object of intensive research and discussion (see, e.g., [18]). One can expect that the same push towards abstraction and more generic architectures that lead to the development of current generic user modeling systems will take place as soon as these architectures become more homogeneous and more widely adopted, and the requirements from the application side more clearly specified.

4.4.4 Multiple-Purpose Usage

Information about the characteristics of individual users may not only be interesting for personalization purposes. Other possible applications include organizational directory services, skill inventory systems, and in-house or global expert-finding applications. It seems worthwhile and possible to develop generic user model systems that can support all these different usage purposes for people-related information. This would not only create more powerful generic systems, but could also improve them both from a theoretical and practical perspective. For instance, these new applications may shed new light on the pros and cons of a user model server architecture versus an agent based architecture (see [98]). Or from a more practical perspective, basing generic user modeling systems on LDAP that was developed for directory services (see [23, 52]) would also help elevate these systems from proprietary developments to the realm of open industry standards.

4.4.5 Diverse Generic User Modeling Systems will Co-exist

A tacit assumption in the 1980s and early 1990s was that only few different user modeling shell systems (and later server systems) would be needed to support the complete range of possible personalized applications. The vast increase of possible scenarios for personalized services with their inherent different demands and constraints has made the likelihood of a very limited number of "user modeling pearl systems" quite unlikely. Instead, we will likely see a variety of generic user modeling systems in the future, each of which is going to support only few of the very different future manifestations of personalization and other uses of information about users. Privacy requirements, the need to include user information from legacy systems, and the need to exchange user information across different systems will however enforce some standardization, at least at the communication level.

Acknowledgment. The preparation of this paper (which is a radically revised and extended version of [50]) was supported by an Alexander von Humboldt Research Prize.

References

1. Allen, J. F.: A Plan-Based Approach to Speech Act Recognition. Dept. of Computer Science, University of Toronto, Canada, Technical Report 131/79, (1979).
2. Allgayer, J., Harbusch, K., Kobsa, A., Reddig, C., Reithinger, N., and Schmauks, D.: XTRA: A Natural-Language Access System to Expert Systems. International Journal of Man-Machine Studies 31, (1989) 161-195, DOI 10.1016/0020-7373(89)90026-6.
3. ATG Adaptive Scenario Engine. (2006), http://www.atg.com/en/products/engine/
4. Billsus, D. and Pazzani, M. J.: User Modeling for Adaptive News Access. User Modeling and User-Adapted Interaction: The Journal of Personalization Research 10, (2000) 147-180, DOI 10.1023/A:1026501525781.
5. Blank, K.: Benutzermodellierung für adaptive interaktive Systeme: Architektur, Methoden, Werkzeuge und Anwendungen. Berlin, Germany: Academic Publishing Society (infix) (1996).
6. Brajnik, G. and Tasso, C.: A Shell for Developing Non-monotonic User Modeling Systems. International Journal of Human-Computer Studies 40, (1994) 31-62, DOI 10.1006/ijhc.1994.1003.
7. Breese, J., Heckerman, D., and Kadie, C.: Empirical Analysis of Predictive Algorithms for Collaborative Filtering. Proceedings of the Fourteenth Annual Conference on Uncertainty in Artificial Intelligence (UAI-98), San Francisco (1998) 43-52. ftp://ftp.research.microsoft.com/pub/tr/tr-98-12.pdf.
8. Brusilivsky, P and Maybury, M.: From Adaptive Hypermedia to the Adaptive Web. Communications of the ACM 45, (2002) 31-33, DOI 10.1145/506218.506239.
9. Brusilovsky, P.: KnowledgeTree: A Distributed Architecture for Adaptive E-Learning. Thirteenth International World Wide Web Conference, WWW 2004 (Alternate track papers and posters), New York, NY (2004) 104-113, 10.1145/1013367.1013386.
10. Brusilovsky, P. and Millán, E.: User Models for Adaptive Hypermedia and Adaptive Educational Systems. In: The Adaptive Web: Methods and Strategies of Web Personalization, Brusilovsky, P., Kobsa, A., and Nejdl, W., Eds. Berlin Heidelberg New York: Springer Verlag (2007) this volume.
11. Brusilovsky, P., Sosnovsky, S., and Yudelson, M.: Ontology-based Framework for User Model Interoperability in Distributed Learning Environments. World Conference on E-Learning, E-Learn 2005, Vancouver, Canada (2005) 2851-2855, http://www2.sis.pitt.edu/~peterb/papers/eLearn2005-adapt.pdf.
12. Buchanan, B. G. and Shortliffe, E. H.: Rule-Based Expert Systems: The MYCIN Experiments of the Stanford Heuristic Programming Project. Reading, MA: Addison-Wesley (1984).
13. Burke, R.: Hybrid Web Recommender Systems. In: The Adaptive Web: Methods and Strategies of Web Personalization, Brusilovsky, P., Kobsa, A., and Nejdl, W., Eds. Berlin Heidelberg New York: Springer Verlag (2007) this volume.

14. Carmichael, D. J., Kay, J., and Kummerfeld, B.: Consistent Modelling of Users, Devices and Sensors in a Ubiquitous Computing Environment. User Modeling and User-Adapted Interaction: The Journal of Personalization Research 15, (2005) 197-234, DOI 10.1007/s11257-005-0001-z.
15. Clayton, B. D.: ART Programming Tutorial, Version 1.0. Inference Corporation, Los Angeles, CA. (1985).
16. Cohen, P. R. and Perrault, C. R.: Elements of a Plan-Based Theory of Speech Acts. Cognitive Science 3, (1979) 177-212, DOI 10.1016/S0364-0213(79)80006-3.
17. Critical Path: Identity Management. (2006), http://www.criticalpath.net/en/22/identitymanagement/
18. Dolog, P. and Vassileva, J., Eds.: Workshop on Decentralized, Agent Based and Social Approaches to User Modelling (DASUM): 9th International Conference on User Modelling, Edinburgh, Scotland (2005), http://www.l3s.de/~dolog/dasum/DASUM-proceedings.pdf.
19. enQuire Identity Server. Persistent Systems Pvt. Ltd. (2006), http://www.persistentsys.com/products/enquire/enquire.htm
20. Finin, T. W.: GUMS: A General User Modeling Shell. In: User Models in Dialog Systems, Kobsa, A. and Wahlster, W., Eds. Berlin, Heidelberg: Springer-Verlag (1989) 411-430.
21. Finin, T. W. and Drager, D.: GUMS1: A General User Modeling System. Sixth Canadian Conference on Artificial Intelligence, Montreal, Canada (1986) 24-29.
22. Fink, J.: Transactional Consistency in User Modeling Systems. In: UM99 User Modeling: Proceedings of the Seventh International Conference, Kay, J., Ed. Wien New York: Springer-Verlag (1999) 191-200, http://bistrica.usask.ca/UM/UM99/Proc/fink.pdf.
23. Fink, J.: User Modeling Servers - Requirements, Design, and Evaluation. Amsterdam, Netherlands: IOS Press (2004), http://books.google.com/books?q=isbn: 1586034057.
24. Fink, J. and Kobsa, A.: A Review and Analysis of Commercial User Modeling Servers for Personalization on the World Wide Web. User Modeling and User-Adapted Interaction: The Journal of Personalization Research 10, (2000) 209-249, DOI 10.1023/A:1026597308943.
25. Fink, J., Kobsa, A., and Jaceniak, I.: Individualisierung von Benutzerschnittstellen mit Hilfe von Datenchips für Personalisierungsinformation. GMD-Spiegel 1/1997, (1997) 16-17, http://www.ics.uci.edu/~kobsa/papers/1997-GMD-kobsa.pdf.
26. Gauch, S., Speretta, M., Chandramouli, A., and Micarelli, A.: User Profiles for Personalized Information Access. In: The Adaptive Web: Methods and Strategies of Web Personalization, Brusilovsky, P., Kobsa, A., and Nejdl, W., Eds. Berlin Heidelberg New York: Springer Verlag (2007) this volume.
27. Goy, A., Ardissono, L., and Petrone, G.: Personalization in E-Commerce Applications. In: The Adaptive Web: Methods and Strategies of Web Personalization, Brusilovsky, P., Kobsa, A., and Nejdl, W., Eds. Berlin Heidelberg New York: Springer Verlag (2007) this volume.
28. Heckmann, D. and Krüger, A.: A User Modeling Markup Language (UserML) for Ubiquitous Computing. In: User Modeling 2003: 9th International Conference, UM 2003, Brusilovsky, P., Corbett, A., and de Rosis, F., Eds. Heidelberg: Springer Verlag (2003) 403-407, http://springerlink.metapress.com/link.asp?id=vnjcm5xr8jhjhvaj.
29. Heckmann, D., Schwartz, T., Brandherm, B., and Kröner, A.: Decentralized User Modeling with UserML and GUMO. Workshop on Decentralized, Agent Based and Social Approaches to User Modelling (DASUM), 9th Intl Conference on User Modeling, Edinburgh, Scotland (2005) 61-64, http://www.l3s.de/~dolog/dasum/DASUM-proceedings.pdf.
30. Heckmann, D., Schwartz, T., Brandherm, B., Schmitz, M., and von Wilamowitz-Moellendorff, M.: GUMO: The General User Model Ontology. In: User Modeling 2005: 10th International Conference, UM 2005, Edinburgh, Scotland., Ardissono, L., Brna, P., and Mitrovic, A., Eds. (2005) 428-432, DOI 10.1007/11527886_58.

31. Henze, N. and Brusilivsky, P.: Open Corpus Adaptive Educational Hypermedia. In: The Adaptive Web: Methods and Strategies of Web Personalization, Brusilovsky, P., Kobsa, A., and Nejdl, W., Eds. Berlin Heidelberg New York: Springer Verlag (2007) this volume.

32. Herlocker, J., Konstan, J., Borchers, A., and Riedl, J.: An Algorithmic Framework for Performing Collaborative Filtering. Proc. of the 22nd Annual International ACM SIGIR Conference on Research and Development in Information Retrieval., New York (1999) 230-237, DOI 10.1145/312624.312682.

33. Howes, T., Smith, M., and Good, G.: Understanding and Deploying LDAP Directory Services. Indianapolis, IN: Macmillan (1999).

34. Howes, T. A. and Smith, M.: LDAP: Programming Directory-Enabled Applications with Lightweight Directory Access Protocol. Indianapolis, IN: Macmillan (1997).

35. Huang, X., McCalla, G. I., Greer, J. E., and Neufeld, E.: Revising Deductive Knowledge and Stereotypical Knowledge in a Student Model. User Modeling and User-Adapted Interaction: The Journal of Personalization Research 1, (1991) 87-115, DOI 10.1007/BF00158953.

36. Intellicorp. http://www.intellicorp.com

37. Jennings, N. R. and Wooldridge, M. J., Eds.: Agent Technology: Foundations, Applications, and Markets. Heidelberg, Germany: Springer Verlag (2002).

38. Kass, R.: Acquiring a Model of the User's Beliefs from a Cooperative Advisory Dialog. Ph.D. Thesis, Dept. of Information and Computer Science. Philadelphia, PA: University of Pennsylvania (1988).

39. Kass, R.: Student Modeling in Intelligent Tutoring Systems -- Implications for User Modeling. In: User Models in Dialog Systems, Kobsa, A. and Wahlster, W., Eds. Berlin, Heidelberg: Springer-Verlag (1989) 386-410.

40. Kay, J.: UM: A Toolkit for User Modelling. In: Second International Workshop on User Modeling. Honolulu, HI (1990) 1-11.

41. Kay, J.: The um Toolkit for Reusable, Long Term User Models. User Modeling and User-Adapted Interaction: The Journal of Personalization Research 4, (1995) 149-196, DOI 10.1007/BF01100243.

42. Kay, J.: A Scrutable User Modelling Shell for User-Adapted Interaction. In Basser Department of Computer Science: University of Sydney, Australia (1999), http://www.cs.usyd.edu.au/~judy/Homec/Pubs/thesis.bz2.

43. Kay, J., Kummerfeld, B., and Lauder, P.: Personis: A Server for User Models. In: Adaptive Hypermedia and Adaptive Web-Based Systems: Second International Conference, AH 2002, De Bra, P., Brusilovsky, P., and Conejo, R., Eds. Berlin Heidelberg: Springer (2002) 203–212, http://springerlink.metapress.com/link.asp? id=2l54yrgc0p8n2d5g.

44. Kleiber, U.: Erklärung in interaktiven Systemen und Unterstützungsmöglichkeiten durch das System BGP-MS. WG Knowledge-Based Information Systems, Department of Information Science, University of Konstanz, Germany, WIS Memo 6, (1994).

45. Knowledge Craft 3.2 Edition. Carnegie Group, Inc., Pittsburgh, PA (1988).

46. Kobsa, A.: Benutzermodellierung in Dialogsystemen. Berlin, Heidelberg: Springer Verlag (1985).

47. Kobsa, A.: A Taxonomy of Beliefs and Goals for User Models in Dialog Systems. In: User Models in Dialog Systems, Kobsa, A. and Wahlster, W., Eds. Berlin, Heidelberg: Springer-Verlag (1989) 52-68.

48. Kobsa, A.: Modeling The User's Conceptual Knowledge in BGP-MS, a User Modeling Shell System. Computational Intelligence 6, (1990) 193-208.

49. Kobsa, A.: Editorial. User Modeling and User-Adapted Interaction 4, Special Issue on User Modeling Shell Systems (1995) iii-v. DOI 10.1007/BF01099427

50. Kobsa, A.: Generic User Modeling Systems. User Modeling and User-Adapted Interaction: The Journal of Personalization Research 11, (2001) 49-63, DOI 10.1023/A:1011187500863.

51. Kobsa, A.: Privacy-Enhanced Web Personalization. In: The Adaptive Web: Methods and Strategies of Web Personalization, Brusilovsky, P., Kobsa, A., and Nejdl, W., Eds. Berlin Heidelberg New York: Springer Verlag (2007) this volume.
52. Kobsa, A. and Fink, J.: An LDAP-Based User Modeling Server and its Evaluation. User Modeling and User-Adapted Interaction: The Journal of Personalization Research 16, (2006) 129-169, DOI 10.1007/s11257-006-9006-5.
53. Kobsa, A., Koenemann, J., and Pohl, W.: Personalized Hypermedia Presentation Techniques for Improving Customer Relationships. The Knowledge Engineering Review 16, (2001) 111-155, DOI 10.1017/S0269888901000108.
54. Kobsa, A., Müller, D., and Nill, A.: KN-AHS: An Adaptive Hypertext Client of the User Modeling System BGP-MS. Proc. of the Fourth International Conference on User Modeling, Hyannis, MA (1994) 99-105, Reprinted in M. Maybury and W. Wahlster, eds. (1998). Readings in Intelligent User Interfaces. San Mateo, CA: Morgan Kaufman, 372-378. http://www.ics.uci.edu/~kobsa/papers/1994-UM94-kobsa.pdf.
55. Kobsa, A. and Pohl, W.: The BGP-MS User Modeling System. User Modeling and User-Adapted Interaction: The Journal of Personalization Research 4, (1995) 59-106, DOI 10.1007/BF01099428.
56. Kobsa, A., Pohl, W., and Fink, J.: A Standard for the Performatives in the Communication between Applications and User Modeling Systems (Draft). (1996), http://www.ics.uci.edu/~kobsa/papers/1996-kobsa-pohl-fink-rfc.pdf
57. Kobsa, A. and Schreck, J.: Privacy through Pseudonymity in User-Adaptive Systems. ACM Trans. on Internet Technology 3, (2003) 149–183, DOI 10.1145/767193.767196.
58. Kobsa, A. and Wahlster, W.: User Models in Dialog Systems. Berlin: Springer-Verlag (1989).
59. Kono, Y., Ikeda, M., and Mizoguchi, R.: THEMIS: A Nonmonotonic Inductive Student Modeling System. Journal of Artificial Intelligence in Education 5, (1994) 371-413.
60. Kummerfeld, R. and Kay, J.: Remote Access Protocols for User Modelling. Proceedings and Resource Kit for Workshop User Models in the Real World, Chia Laguna, Sardinia (1997) 12-15, http://www.cs.usyd.edu.au/~judy/Homec/Pubs/1997_umnet.html.
61. Liberty Alliance Project: Digital Identity Defined. (2006), http://www.projectliberty.org/
62. Lorenz, A.: A Specification for Agent-Based Distributed User Modelling in Ubiquitous Computing. Workshop on Decentralized, Agent Based and Social Approaches to User Modelling (DASUM), 9th International Conference on User Modeling, Edinburgh, Scotland (2005) 31-40, http://www.l3s.de/~dolog/dasum/DASUM-proceedings.pdf.
63. Loshin, P.: Big Book of Lightweight Directory Access Protocol (LDAP) RFCs. San Diego, CA: Morgan Kaufmann (2000).
64. LTSC: Learning Technology Standards Committee. (2006), http://ieeeltsc.org/
65. Machado, I., Martins, A., and Paiva, A.: One for All and All in One: A Learner Modelling Server in a Multi-Agent Platform. In: UM99 User Modeling: Proceedings of the Seventh International Conference, Kay, J., Ed. Wien, New York: Springer-Verlag (1999) 211-221.
66. McTear, M., Ed.: Special Issue on User Modeling. vol. 7 (1993).
67. Miller, B. N., Konstan, J. A., and Riedl, J.: PocketLens: Toward a Personal Recommender System. ACM Transactions on Information Systems 22, (2004) 437-476, DOI 10.1145/1010614.1010618.
68. Mobasher, B.: Data Mining for Web Personalization. In: The Adaptive Web: Methods and Strategies of Web Personalization, Brusilovsky, P., Kobsa, A., and Nejdl, W., Eds. Berlin Heidelberg New York: Springer Verlag (2007) this volume.
69. Morik, K.: Überzeugungssysteme der Künstlichen Intelligenz: Validierung vor dem Hintergrund linguistischer Theorien über implizite Äußerungen. Tübingen, Germany: Niemeyer (1982).

70. Orwant, J.: Heterogenous Learning in the Doppelänger User Modeling System. User Modeling and User-Adapted Interaction: The Journal of Personalization Research 4, (1995) 107-130, DOI 10.1007/BF01099429.
71. Paiva, A. and Self, J.: TAGUS: A User and Learner Modeling System. In: Proc. of the Fourth International Conference on User Modeling. Hyannis, MA (1994) 43-49.
72. Paiva, A. and Self, J.: TAGUS -- A User and Learner Modeling Workbench. User Modeling and User-Adapted Interaction: The Journal of Personalization Research 4, (1995) 197-226, DOI 10.1007/BF01100244.
73. Paliouras, G., Karkaletsis, V., Papatheodorou, C., and Spyropoulos, C.: Exploiting Learning Techniques for the Acquisition of User Stereotypes and Communities. In: UM99 User Modeling: Proceedings of the Seventh International Conference, Kay, J., Ed. Wien, New York: Springer-Verlag (1999) 169-178.
74. Perrault, C. R., Allen, J. F., and Cohen, P. R.: Speech Acts as a Basis for Understanding Dialogue Coherence. Department of Computer Science, University of Toronto, Canada, Report 78-5, (1978).
75. Pohl, W.: Logic-Based Representation and Reasoning for User Modeling Shell Systems. Sankt Augustin, Germany: infix (1998).
76. Pohl, W.: Logic-Based Representation and Reasoning for User Modeling Shell Systems. User Modeling and User-Adapted Interaction: The Journal of Personalization Research 9, (1999) 217-282, DOI 10.1023/A:1008325713804.
77. Pope, A.: The CORBA Reference Guide: Understanding the Common Object Request Broker Architecture. Sydney, Australia: Addison-Wesley (1997).
78. Razmerita, L., Angehrn, A., and Maedche, A.: Ontology-Based User Modeling for Knowledge Management Systems. In: User Modeling 2003: 9th International Conference, UM 2003, Brusilovsky, P., Corbett, A., and De Rosis, F., Eds. Heidelberg, Germany: Springer (2003) 213-217, http://springerlink.metapress.com/link.asp?id=thw9rmvmvklx9hac.
79. Rich, E.: Building and Exploiting User Models. PhD. Thesis, Department of Computer Science. Pittsburgh, PA: Carnegie-Mellon University (1979).
80. Rich, E.: User Modeling via Stereotypes. Cognitive Science 3, (1979) 329-354.
81. Rich, E.: Users are Individuals: Individualizing User Models. International Journal of Man-Machine Studies 18, (1983) 199-214.
82. Rich, E.: Stereotypes and User Modeling. In: User Models in Dialog Systems, Kobsa, A. and Wahlster, W., Eds. Berlin, Heidelberg: Springer (1989) 35-51.
83. Schafer, J. B., Frankowski, D., Herlocker, J., and Sen, S.: Collaborative Filtering Recommender Systems. In: The Adaptive Web: Methods and Strategies of Web Personalization, Brusilovsky, P., Kobsa, A., and Nejdl, W., Eds. Berlin Heidelberg New York: Springer Verlag (2007) this volume.
84. Schreck, J.: Security and Privacy in User Modeling. Dordrecht, Netherlands: Kluwer Academic Publishers (2003), http://www.security-and-privacy-in-user-modeling.info.
85. Shortliffe, E. H.: Computer-Based Medical Consultations: MYCIN. New York: North-Holland (1976).
86. Sleeman, D.: UMFE. A User Modelling Front-End Subsystem. International Journal of Man-Machine Studies 23, (1985) 71-88.
87. Sleeman, D. and Brown, J. S.: Intelligent Tutoring Systems. New York: Academic Press (1982).
88. Specht, M., Lorenz, A., and Zimmermann, A.: Towards a Framework for Distributed User Modelling for Ubiquitous Computing. Workshop on Decentralized, Agent Based and Social Approaches to User Modelling (DASUM), 9th International Conference on User Modeling, Edinburgh, Scotland (2005) 31-40, http://www.l3s.de/~dolog/dasum/ DASUM-proceedings.pdf.
89. Sun: Sun Java System Identity Manager. (2006), http://www.sun.com/software/products/identity_mgr/.

90. Taylor, J. A., Carletta, J., and Mellish, C.: Requirements for Belief Models in Cooperative Dialogue. User Modeling and User-Adapted Interaction: The Journal of Personalization Research 6, (1996) 23-68, DOI 10.1007/BF00126653.
91. Tivoli: IBM Tivoli Identity Manager. (2006), http://www-306.ibm.com/software/tivoli/products/identity-mgr/
92. Tornago: Net Perceptions. (2006), http://www.tornago.com
93. van Melle, W.: System Aids in Constructing Consultation Programs: EMYCIN. Ann Arbor, MI: UMI Research Press (1982).
94. Vassileva, J., McCalla, G., and Greer, J.: Multi-Agent Multi-User Modeling in I-Help. User Modeling and User-Adapted Interaction: The Journal of Personalization Research 13, (2003) 179-210, DOI 10.1023/A:1024072706526.
95. Vergara, H.: PROTUM: A Prolog Based Tool for User Modeling. WG Knowledge-Based Information Systems, Department of Information Science, University of Konstanz, Germany, WIS-Report 10, (1994).
96. Wahlster, W. and Kobsa, A.: User Models in Dialog Systems. In: User Models in Dialog Systems, Kobsa, A. and Wahlster, W., Eds. Heidelberg: Springer (1989).
97. Wooldridge, M. and Jennings, N.: Intelligent Agents: Theory and Practice. The Knowledge Engineering Review 10, (1995) 115-152.
98. Yimam, D. and Kobsa, A.: Expert Finding Systems for Organizations: Problem and Domain Analysis and the DEMOIR Approach. In: Beyond Knowledge Management: Sharing Expertise, Ackerman, M., Cohen, A., Pipek, V., and Wulf, V., Eds. Cambridge, MA: MIT Press (2003), http://www.ics.uci.edu/~kobsa/papers/2003-JOCEC-kobsa.pdf.

5

Web Document Modeling

Alessandro Micarelli, Filippo Sciarrone, and Mauro Marinilli

Department of Computer Science and Automation
Artificial Intelligence Laboratory
Roma Tre University
Via della Vasca Navale, 79 - 00146 Rome, Italy
{micarel, sciarro, marinil}@dia.uniroma3.it

Abstract. A very common issue of adaptive Web-Based systems is the modeling of documents. Such documents represent domain-specific information for a number of purposes. Application areas such as Information Search, Focused Crawling and Content Adaptation (among many others) benefit from several techniques and approaches to model documents effectively. For example, a document usually needs preliminary processing in order to obtain the relevant information in an effective and useful format, so as to be automatically processed by the system. The objective of this chapter is to support other chapters, providing a basic overview of the most common and useful techniques and approaches related with document modeling. This chapter describes high-level techniques to model Web documents, such as the Vector Space Model and a number of AI approaches, such as Semantic Networks, Neural Networks and Bayesian Networks. This chapter is not meant to act as a substitute of more comprehensive discussions about the topics presented. Rather, it provides a brief and informal introduction to the main concepts of document modeling, also focusing on the systems that are presented in the rest of the book as concrete examples of the related concepts.

5.1 Introduction

The Web document, in its various representation forms, is the focal point of this chapter, which aims at illustrating the most common techniques employed in Web Document Modeling[1] literature, with particular attention given to those used in Adaptive Web-Based systems. The purpose of this chapter is to offer a support for the other chapters in this volume dealing with the various adaptive systems in greater detail.[2] The present chapter is therefore not to be considered a comprehensive guide to the discussed topics, nor a substitute for more specialized literature. Readers are encouraged to consult the provided references in order to broaden their grasp of the discussed topics or equivalent literature.

---

[1] In this context, with the term *modeling* we mean the construction of an abstract representation of the document, useful for all applications aimed at processing information automatically.

[2] Fig. 5.1 shows the structure of this chapter together with the links to the other chapters of this book.

P. Brusilovsky, A. Kobsa, and W. Nejdl (Eds.): The Adaptive Web, LNCS 4321, pp. 155–192, 2007.
© Springer-Verlag Berlin Heidelberg 2007

It is well known that, owing to the Internet, each one of us can benefit from a large quantity of information, available on-line in several, essentially standard, formats, such as HTML, XML and XHTML text pages and *jpeg* and *tif* graphic ones. More complex formats, particularly multimedia (audio and/or video) usually require a longer search and interaction with the Web, as well as more sophisticated fruition tools installed on user clients, now widely accessible.

The quantity of information available on the World Wide Web is increasing exponentially, and this boom has paved the way for a new era, creating new opportunities in many different fields, such as e-business, e-commerce, e-marketing, e-finance and e-learning, just to mention some of the most interesting ones. *Web Intelligence* [90, 91] and *Wisdom Web* [93, 48, 92] are other examples of new disciplines born from the development of the World Wide Web. All this proceeds from the birth of HTTP, *HyperText Transfer Protocol* [80, 39] and of HTML, *HyperText Markup Language*, a subset of SGML, *Standard Generalized Markup Language* [29], which made the Internet enjoyable for anyone who had a computer with a browser on-board.[3] With time, the number of Web surfers increased, and so did the available Web documents, especially HTML pages. This eventually required the need to gather the information to be supplied in more structured containers, enabling more thorough, personalized searches, directly correlated to the semantics of the document itself, for a more intelligent fruition and to provide more sophisticated search-tools systems.

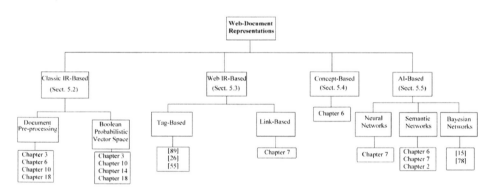

Fig. 5.1. The structure of this chapter with the references to other chapters of this volume

All the aforementioned techniques are the ones used concretely to represent documents on various Web sites, and may therefore be called *Layout Representations*. For automatic systems that operate on the Web, however, e.g. adaptive Web-Based systems, such documents often need to be *pre-processed*, that is, recasting them in representations different from primitive ones and more apt to undergo a particular elaboration. Think of a personalized search system in which the calculation of the relevance of a

---

[3] The reader can visit the w3c Web site: http://www.w3c.org for further reading on Web standards and protocols.

particular HTML document, as retrieved by a single user, needs to be automatically processed. Only a suitably recast representation can ensure a thorough evaluation of the document itself. Some of these representation methods are inspired by the classic Information Retrieval (IR) - an area of research conceived when the Internet still had not dawned - that enjoys a long tradition in the modeling and treatment of documents [74, 76]. Nonetheless, in literature, especially since the Web was born, other IR techniques have been proposed, capable of exploiting the hypertext features of the Web. While some proposals focused on the enhancement of classic IR systems, others searched for new retrieval strategies, exploiting the Web's hypertext features, such as page hyperlinks and HTML general tags [14, 46, 55]. The result was the advent of *Web Information Retrieval* (Web-IR), one of the new IR techniques used to retrieve documents from the Web [2, 1]. Finally, there are also modeling techniques that are based on document-concept representations and on the models and methods of Knowledge Representation, typical of Artificial Intelligence.

This chapter is therefore subdivided into four main parts: the first part (Section 5.2) describes the classic IR methods and techniques for representing documents, with particular examples based on real Web pages; the second part (Section 5.3) describes the Web-IR methods and techniques, with particular reference to the typical hypertext features of the majority of Web documents; the third part (Section 5.4) illustrates document representation methods based on concept approaches, where documents are modeled by means of concept sets. The last part (Section 5.5), deals with document modeling methods based on Artificial Intelligence techniques. More precisely, Section 5.2 illustrates the approach based on classic IR techniques to represent documents. In particular, Subsection 5.2.1 illustrates the typical characteristics of the document pre-processing techniques, including the weighting of the various terms that appear within it. Subsection 5.2.2 describes the *Boolean Model* document modeling technique, characterized by its simplicity and easiness of implementation. Subsection 5.2.3 presents the *Probabilistic Model* document modeling technique, which ranks the query-document similarity according to the probability theory. Subsection 5.2.4 deals with the *Vector Space Model*, the most common one in classic IR systems: it models a document as a point in an n-dimensional space, and calculates the document-query similarity following geometric rules. Section 5.3 illustrates the modern Web-IR techniques, based on HTML tags and hyperlinks. Subsection 5.3.1 shows three examples of *Tag-Based* document modeling techniques where an HTML document is modeled taking into account its HTML structure. Finally, in Subsection 5.3.2, we discuss two algorithms, *HITS* and *PageRank*, that model documents exploiting the hypertext structure of the Web. In Section 5.4 we discuss a concept approach to document modeling, called *Latent Semantic Indexing*, where a document is modeled by a set of concepts. Section 5.5, studies in-depth some document representation methods based on Artificial Intelligence techniques. Subsection 5.5.1 illustrates the characteristics of document modeling via Artificial Neural Networks; Subsection 5.5.2 presents a brief description of the characteristics of document modeling based on Semantic Networks ; finally, Subsection 5.5.3 illustrates how Bayesian Networks are used in diverse adaptive systems to model documents. Our conclusions are drawn in Section 5.6.

5.2 Classic IR-Based Document Representations

This section illustrates the classic IR document modeling techniques, namely the document modeling techniques born from the classic works of Salton [74], of Salton and McGill [76] and of Van Rijsbergen [67], suitable for the retrieval of relevant documents from a collection, and proposed when the Internet was only just dawning. Many adaptive systems on the Web employ such IR techniques to represent Web documents (see for example [58, 43, 78, 11, 36]), obtaining good results. Obviously, this description cannot be fully exhaustive, considering the vastness of the topic; rather, it is to be considered a starting point for possible in-depth studies quoted in the relevant bibliography.

Subsection 5.2.1 deals with *Document Pre-processing*, the process used to remove from the original document all information deemed non-relevant for the semantics of the very document, such as HTML tags and *stopwords*, thus leaving relevant terms only. In fact, in a classic IR environment, a document is simply a *bag of words*, i.e., an unstructured set of words or terms appearing in it, each one having an associated relevance weight. This approach does not take into account typical features of human language, such as homonymy and polysemy: for example, the term *mouse* is represented in the same way, both when it indicates an animal and a computer device; the terms *home* and *house* are represented as two uncorrelated terms. Nonetheless, this document modeling approach, which is uniquely based on large-scale statistics, is currently the most widespread, mainly because of its simplicity and of its suitability for the automatic processing of documents (see for example [77]).

The *Boolean Model* is illustrated in Subsection 5.2.2. This document modeling technique, which can be implemented very easily, is based on the Theory of Sets and on Boolean Algebra.

The *Probabilistic Model* is illustrated in Subsection 5.2.3. This type of document modeling follows the probability theory in calculating the relevance of a document following a given query.

Finally, the *Vector Space Model* is illustrated in Subsection 5.2.4. This technique is based on vector space for the representation of queries and documents. However, such a technique requires large-scale statistics calculated on the entire document collection from which the document to be modeled is retrieved. Many IR systems employ this technique to retrieve relevant documents through query-document similarity metrics, expressed in terms of geometric distance.

5.2.1 Document Pre-processing

A document may be published on the Web in several formats: HTML, PDF, DOC, TEXT, etc, but in all cases, the information it contains is to be *enriched* with particular character sequences, not intended for the user but for the computer visualization driver, such as the browser in the case of HTML Web pages. In order to keep the explanation as simple as possible, we shall henceforth deal only with HTML documents, without distorting conclusions whatsoever, unless exclusive details of implementation come about. As a starting example, we use the Web page illustrated in Fig. 5.2: it represents a Web page as the user sees it through his/her browser.[4] We can say this is the *user's point of*

---

[4] Web site: http://www.springeronline.com.

view of the document. Fig. 5.3, on the other hand, shows a fragment of the same Web page, as the browser *sees* it: we can say that this is the *browser's point of view* of the same document. They are obviously two very different viewpoints of the same thing: the user does not, and must not, see HTML formatting tags for the browser, while the browser must process everything included in the HTML Web page. In order to better comprehend this detail, crucial to get a grasp of the issue, take a look at the following HTML code lines, included in the aforesaid Web page:

```
<link rel=''stylesheet" href=''include/style\_0.css"
type=''text/css"> <script language=''JavaScript"
src=''include/item.js"></script>
<script language=''JavaScript"
src=''include/fw\_menu.js"></script>
<span class=''bodytext">

Benefit from attractive savings on Springer books
by signing up for Springer's free new book e-mail
notification service. New title info, news and
special announcements: with Springer Alerts it
pays to be informed.

</span>
```

In classic IR approach, none of the terms included in the first five lines offer significant information to the user, since they don't refer to document content, but to the layout, whereas the real information content is to be found in the following four lines, i.e., *bodytext* lines. A traditional IR system should therefore somehow extract the four significant lines for the user in order to select the correct terms to represent the Web page. This term-selection process is a part of the document pre-processing task which is broken down into four phases, explained in the following paragraphs.

1. HTML Tag Removal Phase. This phase consists in removing all HTML instructions (i.e., tags) from an HTML page. The simplest approach excludes all terms belonging to HTML tags. In the case in point, only this text would remain:

```
Benefit from attractive savings on Springer books
by signing up for Springer's free new book e-mail
notification service. New title info, news and
special announcements: with Springer Alerts it
pays to be informed.
```

2. Stopwords Removal Phase. Not all terms of a document are necessarily relevant. Some frequently used terms, within the document itself, tend to be removed: these terms are known as *Stopwords* (i.e., "a", "the", "in", "to"; or pronouns: "I", "he", "she", "it"). Stopwords obviously vary according to the language. It is possible to download an

Fig. 5.2. The HTML example page from the user point of view

```
<!DOCTYPE HTML PUBLIC "-//W3C//DTD HTML 4.0 Transitional//EN">
<!-- saved from url=(0049)http://www.springer.com/uk/home?SGWID=3-102-0-0-0 -->
<?xml version="1.0" encoding="UTF-8" ?><HTML lang=en-UK xml:lang="en"
xmlns="http://www.w3.org/1999/xhtml">
<HEAD>
<TITLE>
Springer UK - Academic Journals, Books and Online Media
</TITLE>
<!-- FA1.1 --><!-- Revision: 12435 -->
<META http-equiv=content-type content="text/html; charset=utf-8">
<META
content="Buy academic journals, books and online media at Springer. Choose from thousands
name=description>
<META
content="Springer, Publishing, Springer Online, Springer-Verlag, Books, Journals, Publisher
name=keywords>
<META http-equiv=imagetoolbar content=no>
<META content=noindex,nofollow name=robots>
<META content=all name=audience>
<META content=global name=distribution><LINK title="ICRA labels"
href="/sgw/labels.rdf" type=application/rdf+xml rel=Meta><LINK
href="springer_new_file/springer.css" type=text/css rel=stylesheet><LINK
href="/favicon.ico" type=image/ico rel=icon><LINK href='/favicon.ico"
type=image/x-icon rel="shortcut icon">
<SCRIPT language=JavaScript>
```

Fig. 5.3. The HTML example page from the browser point of view

example of *stoplist*, i.e., a set of standard stopwords, of about 500 stopwords, from the *http://bll.epnet.com* Web site.[5] By exploiting this list for our text fragment, we obtain:

```
Benefit  attractive savings  Springer books  signing
Springer book e-mail notification service  title info
special announcements Springer Alerts  pays informed
```

---

[5] http://bll.epnet.com/help/ehost/Stop_Words.htm.

3. Stemming Phase. The goal of this phase is to reduce a term to its morphologic root, in order to recognize morphologic variations of the word itself. For example, the root *comput* is the reduced version of "comput-er", "comput-ational", "comput-ation" and "compute". The morphologic analysis must be specific for every language, and can be extremely complex. The simplest stemming systems just identify and remove suffixes and prefixes. Considering its widespread use, it is worthwhile mentioning Porter's Stemmer [65]. It is a simple procedure, which cyclically recognizes and removes known suffixes and prefixes without having to use a dictionary. This algorithm can also generate terms which are not language words, as is to be seen in the previous example: "computer", "computational", "computation" all become "comput". Terms which can actually be different are unified, and the algorithm doesn't recognize morphologic variations. The reader can find more detailed information on this algorithm on the Web site: *http://www.mozart-oz.org*.[6] After having processed our example text fragment by Porter's stemmer,[7] we get the mapped text shown in Tab. 5.1. In Chapter 10 [59] the reader can find a relevant example of such a technique in the case of a newspaper article.

Table 5.1. The mapping between stemmed and not stemmed terms of the example text after the Porter's stemming process

Original Terms	Stemmed Terms
benefit	benefit
attractive	attract
savings	save
springer	springer
books	book
signing	sign
springer	springer
book	book
email	email
notification	notif
service	servic
title	titl
info	info
special	special
announcements	announc
alerts	alert
pays	pai
informed	inform

4. Term Weighting Phase. By indicating our example document with d, we obtain a first representation of our original HTML document as a simple set of its terms:

---

[6] http://www.mozart-oz.org/mogul/doc/lager/porter-stemmer/.

[7] We submitted our text fragment to the Web site:
http://maya.cs.depaul.edu/ classes/ds575/porter.html.

$d \equiv \{t_1, t_2, \ldots, t_{16}\}$, i.e., $d \equiv \{$benefit, attract, save, springer, book, sign, email, notif, servic, titl, info, special, announc, alert, pai, inform$\}$. Each term t_k of the document d belonging to a document collection D, is named *feature* or *index term* and its relevance within the document d is measured by means of an associated numeric *weight* w_k. In classic IR, weight w_k is calculated by taking into account the whole collection D of documents from which the document d has been retrieved [76, 5]: the more w_k is high the more term t_k is important for the discerning of the document d. A first simple weight, also used as a starting point to build more sophisticated weights proposed in the literature, is $w = Term\ Frequency\ TF$, namely, the number of times term t appears in the document d and indicated by $TF(t, d)$. This leads to the representation of a document d by a vector of pairs $d \equiv \{(t_1, w_1), (t_2, w_2) \ldots (t_n, w_n)\}$, with $w_k = TF(t_k, d)$. In the case in point, the representation of our example document d becomes that of Tab. 5.2.

Table 5.2. A simple representation of the example document through Term Frequency

Index Terms	TF(t,d)
benefit	1
attract	1
save	1
springer	3
book	2
sign	1
email	1
notif	1
servic	1
titl	1
info	1
special	1
announc	1
alert	1
pai	1
inform	1

Generally speaking, we may assert that a weight w is mostly built as a function of the frequency of term t in document d, as expressed by Eq. 5.1.

$$w = f[TF(t, d)] \tag{5.1}$$

For every document of the collection D, all terms t are extracted and the *Index Term Database* (ITD) is built: it consists of all index terms of all documents belonging to the collection D. However, not all the terms in a document d have the same relevance in discerning the document d itself for a correct representation and retrieval. Tab. 5.3 shows the generic *Term-Document Frequency Matrix*, i.e., a matrix A whose elements a_{ij} represent the weights w_{ij} of the generic index terms t_i in the document d_j, for a collection D composed of n documents and with $m = |ITD|$.

Table 5.3. Term-Document Frequency Matrix

ITD	d_1	d_2	...	d_n
t_1	w_{11}	w_{12}	...	w_{1n}
t_2	w_{21}	w_{22}	...	w_{2n}
...	...	...	...	...
t_m	w_{m1}	w_{m2}	...	w_{mn}

A term's weight must ensure the needed discerning power for a correct representation and retrieval of documents containing it; thus, a term's relevance must follow some specific guidelines, namely [68, 76]:

- The more a term t appears in a document d, the more this term can characterize the topic dealt with by the document itself.
- A term t that appears in almost all documents of the collection D does not entail a relevant information content for the characterization of the topic of a particular document.

Hence, Term Weighting involves at least two components: the frequency of a term t within a document d and the frequency of term t within the whole collection D. These general guidelines provided the first calculation methods for index terms weighting, i.e., Salton and Buckley Weighting Schema [74, 75].

It is now possible to resort to more complicated approaches based on function f defined in Eq. 5.1, defining a new variable, called *Document Frequency $DF(t)$*, namely, the number of documents in which term t appears at least once. A high value of $DF(t)$ should reduce the importance of term t, that is, the reduction of its weight w. The following items show the most common calculation methods in literature, and the ones used and shown in other chapters of this book.

- *Boolean Weighting*. It represents the simplest version of the calculation of weights w_i. The calculation formula is as follows:

$$w_i = 1 \Leftrightarrow t_i \in d \tag{5.2}$$

$$w_i = 0 \Leftrightarrow t_i \notin d \tag{5.3}$$

In the case of a generic collection D of documents, we could turn to Tab. 5.4.

Table 5.4. Example of Boolean weights

ITD	d_1	d_2	...	d_n
t_1	0	1	...	0
t_2	1	0	...	1
...	...	...	...	...
t_m	1	1	...	1

A practical example of this weighting method is to be seen in the Subsection on the Boolean Model, Subsection 5.2.2, whereas application examples can be studied further in [58].

- *TFxIDF Weighting.* The weight w_i of the term t_i is calculated in such a way to be proportional to the frequency of the term t_i in the document d, and inversely proportional to the number of documents in the collection D in which t_i appears. Given that $|D|$ is the number of documents in the collection D, the weight w_i is calculated by the following formula:

$$w_i = TF(t_i, d) \log \frac{|D|}{DT(t_i)} \tag{5.4}$$

being $DT(t_i)$ the number of documents of the collection D that include the term t_i. *TFxIDF* weighting is a very common technique, owing to its simplicity and effectiveness and many IR systems employ it in literature. In the Subsection on Vector Space Model, the reader will find a clear implementation example and a link to other chapters of the book that use it. Besides, several proposals have been made to adapt *TFxIDF* to hypertext links on the Web [41], leading to good results.

An in-depth example of this weighting technique is shown in Chapter 10 [59]. In Chapter 3 [54] the reader can find an example of the use of TFxIDF in Content-Based Filtering Systems while in Chapter 6 [51] is shown the WATSON system [18] that exploits this weighting method to create the contextual query. Finally, another important application of TFxIDF is illustrated in Chapter 18 [12]: here this weighting technique is used in order to learn user models for news access.

- *Okapi BM25 Weighting.* It is worthwhile illustrating the Okapi weighting scheme [84] employed with good results on the benchmark TREC[8] [70, 69].
This weighting technique is part of the probabilistic models for the calculation of term relevance within a document. This approach computes a term weight according to the probability of its appearance in a relevant document, and to the probability of it appearing in a non-relevant document in a collection D. A simplified version of the Okapi BM25 formula to assess the relevance of a term t of a document d belonging to a collection D, is the following [41]:

$$w_i = TF(t_i, d) \frac{\log \frac{(|D| - TF(t_i, d) + 0,5)}{(|D| + 0,5)}}{k_1 \cdot ((1 - b) + b \cdot \frac{|d|}{\bar{d}}) + TF(t_i, d)} \tag{5.5}$$

with $\bar{d}$ being the average length of a document included in the collection D, $|D|$ the number of documents in the collection D, k_1 and b two constants to be determined experimentally.[9]

---

[8] See http://www.soi.city.ac.uk/~mg/okapi-pack/old_mg_bak/okapi-pack.html.
[9] The reader can visit http://www.soi.city.ac.uk/organisation/is/research/cisr/ for an in-depth reading of the Okapi project.

- *Entropy weighting.* This approach is based on ideas of the Information Theory. Some studies [31] proved its efficiency and performance compared to other methods. In this case, the weight is given by:

$$w_i = \log(TF(t_i, d) + 1)(1 + \frac{1}{\log(|D|)} \sum_{j=1}^{|D|} [\frac{TF(t_i, d_j)}{DF(t_i)} \log \frac{TF(t_i, d_j)}{DF(t_i)}]) \quad (5.6)$$

where the value of:

$$\frac{1}{\log(|D|)} \sum_{j=1}^{|D|} [\frac{TF(t_i, d_j)}{DF(t_i)} \log \frac{TF(t_i, d_j)}{DF(t_i)}] \quad (5.7)$$

is equal to the *Entropy* of term t_i, which is equal to -1 if t_i is equally distributed in all documents, and equal to 0 if t_i is present in only one document.

- *Genetic Programming Weighting.* Another interesting modern approach to the calculation of weights is the one based on the Genetic Programming Theory [47]. In the work of Cummins and O'Riordan [25] term weighting schemes are automatically determined by genetic evolution and then tested on standard test collections. Finally they are compared to the traditional TFxIDF weighting scheme and to the BM25 weighting scheme using standard IR performance metrics.

For further reading on Term Weighting techniques, the reader can find other methods and representations for example in the works of Park *et al.* [57] where a new and interesting method for ranking documents, based on discrete wavelet transform, is proposed.

5.2.2 Boolean Model

In the Boolean Model, documents are represented by sets of keywords, extracted manually and/or automatically from all the documents of a collection D [76, 67]. It is a very simple model, easy to understand and implement but with some effectiveness problems. This model is based on Set Theory and on Boolean Algebra: it ascribes a binary value to the weight w_i of a term t_i accordingly to its appearance (or non-appearance) in a document $d_k \in D$. In this way, the document d_k is represented by a vector $\vec{d_k}$:

$$\vec{d_k} \equiv \{(t_1, w_{1k}), (t_2, w_{2k}), \ldots, (t_m, w_{mk})\} \quad (5.8)$$

where $w_{ik} = 1$ iff $\Leftrightarrow t_i \in d_k$, $w_{ik} = 0$ *otherwise* and where terms t_i are all the index terms belonging to the ITD of the collection D. In this model, also the query is represented through Boolean expressions such as [5]:

$$q = k_\alpha \wedge (k_\beta \vee k_\gamma) \quad (5.9)$$

where terms k_i could be present or absent in the documents of the collection D.

   We reckon it could be interesting, by this stage, for the reader, to apply such a representation model to a concrete example, in order to better illustrate the concepts described in this Subsection. Starting from the example of the HTML page illustrated in Fig.5.2, imagine a collection D consisting of only 3 documents, $D \equiv (d_1, d_2, d_3)^{10}$ and of an ITD T composed of 16 terms, $ITD \equiv (t_1, t_2, \ldots, t_{16})$, represented in the Boolean Model shown in Tab. 5.5.

Table 5.5. An example of a document collection D composed of 3 documents and of an ITD of 16 index terms in the Boolean Model

ITD	Terms	d_1	d_2	d_3
t_1	benefit	1	0	0
t_2	attract	1	1	0
t_3	save	1	1	0
t_4	springer	1	1	1
t_5	book	1	0	1
t_6	sign	0	0	1
t_7	email	1	1	0
t_8	titl	0	1	1
t_9	info	0	0	1
t_{10}	special	0	0	1
t_{11}	announc	0	0	1
t_{12}	alert	0	0	1
t_{13}	pai	0	1	0
t_{14}	inform	1	0	1
t_{15}	notif	0	1	1
t_{16}	servic	1	0	1

Suppose we have the following query:

$$q = springer \wedge (inform \vee info) \tag{5.10}$$

In this case the system will retrieve documents d_1 and d_3 from the collection D. Concluding, the Boolean Model is a very simple retrieval model based on Boolean Algebra and very easy to implement. This model only retrieves exact matches: in the example above, the document d_3 matches more keywords than the other retrieved document d_1 but nevertheless this retrieval system returns both d_1 and d_3 at the same level of relevance. Put into practice, this model appears to retrieve too much, or too little. This is also partially due to the way users write their queries. In fact, users are often not familiar with Boolean queries and tend to write too relaxed or too constraining queries that affect the search effectiveness. Finally, there is no weighting of terms in a document or in the query expression, hence all terms are equally important in this model. In the above example the term *Springer* appears in all documents of our collection, thus being somewhat less *useful* than other terms regarding the retrieval effectiveness, as highlighted in the guidelines of Subsection 5.2.1. Nevertheless the Boolean model is quite appealing for its simplicity and computational performance.

---

[10] After the pre-processing phases.

5.2.3 Probabilistic Model

In the Probabilistic Model [68, 44], a document (and even a query) is modeled through a binary weight vector, as is done in the Boolean Model. The difference is to be seen in the model for the calculation of the query-document similarity function. Indeed, the probabilistic model tries to answer the following Basic Question [44]:

What is the probability that a certain document is relevant to a certain query?

Furthermore, the objective of asking the Basic Question is to rank documents according to their probability of relevance. This maximizes the system effectiveness, as retrieved documents are ranked by decreasing probability of relevance. The user inspects the ranked list of documents and assess their relevance by him/herself. Assuming that terms are distributed differently in relevant and non-relevant documents, one could base the representation and retrieval of documents on term distribution. Both for query q and for the document d_j, index terms are represented by binary weights: $w_{ij} \in \{0, 1\}$ and $w_{iq} \in \{0, 1\}$ and the *similarity function* $sim(d_j, q)$, i.e., the function that calculates the query-document similarity, is the following:

$$sim(d_j, q) = \frac{P(R/d_j)}{P(\overline{R}/d_j)} \qquad (5.11)$$

where R is the set of documents known as relevant documents while the set $\overline{R}$ is the complement of set R, namely the set of non-relevant documents. $P(R|d_j)$ is the probability that document d_j will be relevant for the query q, and $P(\overline{R}|d_j)$ is the probability that d_j will not be relevant for q.

Using the Bayes theorem [30]:

$$P(a/b) = \frac{P(b/a)P(a)}{P(b)} \qquad (5.12)$$

and using odds rather than probabilities, defined as:

$$O(z) = \frac{P(z)}{P(\overline{z})} = \frac{P(z)}{1 - P(z)} \qquad (5.13)$$

we can calculate Eq. 5.11 as follows:

$$sim(d_j, q) = \frac{P(d_j/R)P(R)}{P(d_j|\overline{R})P(\overline{R})} \qquad (5.14)$$

Assuming that $P(R)$ and $P(\overline{R})$ are constant for each document we get:

$$sim(d_j, q) \sim \frac{P(d_j/R)}{P(d_j/\overline{R})} \qquad (5.15)$$

Assuming that terms occur independently of each other and switching to logarithms (skipping some steps for brevity) we then have:

$$sim(d_j, q) \sim \sum_{i=1}^{n} log \frac{P(t_i/R)P(\overline{t_i}/\overline{R})}{P(t_i/\overline{R})P(\overline{t_i}/R)} \tag{5.16}$$

Having thought of d_j as a vector of n binary independent term occurrences:

$$P(d_j/R) = \prod_{t=1}^{n} P(t_i/R) \tag{5.17}$$

where $P(t_i/R)$ is the probability that term t_i appears in a document randomly selected from set R, and $P(\overline{t_i}/R)$ the probability that term t_i is not present in a document randomly selected from set R. Eq. 5.16 is not usable at start up, when there are no retrieved documents. In this case a number of simplifying assumptions can be done as follows:

$$P(t_i/R) = 0.5 \tag{5.18}$$

$$P(t_i/\overline{R}) = n_i/N \tag{5.19}$$

with $N = |D|$ and $n_i = |ITD|$. Let's see now a practical example of this simple probabilistic model at work for our imaginary collection D, formed by 3 documents. Tab. 5.6 illustrates, in the last column, the value calculated by the Eq. 5.19 of $P(t_i/\overline{R})$, used to start up the system. By this technique, the initial values calculated through Eq. 5.18 and Eq. 5.19 are modified taking into account the distribution of terms t_i in the retrieved documents, considering the non retrieved documents as non-relevant.

Table 5.6. Example of a document collection D consisting of 3 documents, represented by the Probabilistic Model

ITD	Terms	d_1	d_2	d_3	$P(t_i/\overline{R})$
t_1	benef	1	0	0	0.33
t_2	attract	1	1	0	0.66
t_3	sav	1	1	0	0.66
t_4	springer	1	1	1	1
t_5	book	1	0	1	0.66
t_6	sign	0	1	0	0.33
t_7	email	1	1	0	0.66
t_8	titl	0	1	1	0.66
t_9	info	0	0	1	0.33
t_{10}	special	0	0	1	0.33
t_{11}	announc	0	0	1	0.33
t_{12}	alert	0	0	1	0.33
t_{13}	pai	0	1	0	0.33
t_{14}	inform	1	0	1	0.66
t_{15}	notif	0	1	1	0.66
t_{16}	servic	1	0	1	0.66

The final calculation formula for $P(t_i/R)$ and $P(t_i/\overline{R})$ is played down to Eq. 5.20 and 5.21 [5].

$$P(t_i/R) = \frac{V_i + \frac{n_i}{N}}{V + 1} \qquad (5.20)$$

$$P(t_i/\overline{R}) = \frac{n_i - V_i + \frac{n_i}{N}}{N - V + 1} \qquad (5.21)$$

where $N = |D|$, $n_i = DT(t_i)$, V is the overall number of documents currently retrieved and V_i is the number of such documents containing the term t_i.

The main asset of the probabilistic model is that retrieved documents may be ranked in a descending order, according to their relevance probability, while the main drawbacks of this model are the following:

- Division of the set of documents into relevant and non-relevant documents.
- Index Terms. This approach does not take into account the frequency of index terms in documents: all weights are binary.
- Assumption that terms are all independent from each other: this assumption enables to devise a formula of the calculation of Eq. 5.11 actually computable.

5.2.4 Vector Space Model

In this Subsection, we introduce the *Vector Space* representation of a document [74]. This model provides a technique to retrieve relevant documents from any set of documents. Still taking cue from the Web page previously used (www.springeronline.com), with a collection D consisting of 3 HTML pages taken from this web site, firstly submit each one of them to the pre-processing phases illustrated in Subsection 5.2.1. Secondly, build the Term-Document Frequency Matrix A, as defined in Subsection 5.2.1 and shown in Tab. 5.7. Thirdly, compute the TFxIDF values for each index term of each document included in the collection D.[11]

Table 5.7. Representation of our example collection of documents by Term-Document Frequency Matrix

ITD	Terms	d_1	d_2	$d3$
t_1	benef	1.0	0.0	3.0
t_2	attract	1.0	2.0	0.0
t_3	sav	1.0	0.0	0.0
t_4	springer	1.0	3.0	0.0
t_5	book	1.0	0.0	1.0

[11] The exposed values of Tab. 5.7 are purely indicative, considering the calculation simplicity and clarity.

The last step is to consider a document as a vector of real numbers in an m-dimensional space, namely, considering documents as points in an m-dimensional space, with $m = |ITD|$, where every term t_i represents a dimension. Such an m dimension could be very high, even several thousands, depending on the number, on the length of documents in the collection D and on the semantic domain from where collection D was gathered. To avoid computational problems in computing the Vector Space Model, due to high-dimension [67],[12] it is possible to resort to a *Dimension Reduction* process where, starting from the original m-dimension high space, a new n-dimension space is built, with $n \ll m$. Generally a document d is a very sparse vector, i.e., it contains a very small subset of the ITD. Every term $t_i \in d_j$ has a weight w_{ij}, so that: $w_{ij} > 0$ iff $t_i \in d_j$ and t_i does not belong to all documents of the collection D. The value of w_{ij} is thus the coordinate calculated according to the correspondent index term, as shown in Eq. 5.22. Obviously such weights represent the relevance of the associated terms in the very document.

$$\vec{d_j} = \{w_{1j}, w_{2j}, \ldots w_{mj}\} \tag{5.22}$$

In the case in point, by enforcing the rules of formation of weights TFxIDF (see Subsection 5.2.1) we have the three vectors represented in the rows of Tab. 5.8.

Table 5.8. Representation of our example collection of documents by the Space Vector Model with TFxIDF weighting

Document	w_1	w_2	w_3	w_4	w_5
$\vec{d_1}$	0.176	0.176	0.417	0.176	0.176
$\vec{d_2}$	0.000	0.350	0.000	0.528	0.000
$\vec{d_3}$	0.528	0.000	0.000	0.000	0.176

Each direction of the vector space corresponds to a unique index term in the document collection, while the component of a document vector along a given direction corresponds to the importance of that term to the document. In such a geometric representation, versors $\vec{T_k}$ are assumed to be orthonormal (namely, index terms are assumed to appear independently from each other in documents). The importance of this type of document representation is the fact that it enables, in a simple way, even the representation of a query q, since, in this space, both query and documents can be represented through vectors of weights. In particular, even query q is transformed as if it were a document, following the same rules: an m-dimensional vector representation is obtained, as expressed in Eq. 5.23. Obviously, the q vector will be a very sparse one.

$$\vec{q} = \{w_{1q}, w_{2q}, \ldots w_{mq}\} \tag{5.23}$$

The advantage of this document representation is that of enabling to retrieve relevant documents through a very simple document-query similarity function. In fact, the similarity function is fulfilled through a function that also calculates the similarity ranking

---

[12] In the IR literature such a problem is known as the so-called *Curse of Dimensionality*.

of single documents $\vec{d}$ with respect to the query $\vec{q}$. The most employed similarity function in the literature is the cosine one:

$$sim(d_j, q) = \cos(\vec{d_j}, \vec{q}) = \frac{\vec{d_j} \bullet \vec{q}}{|\vec{d_j}||\vec{q}|} \tag{5.24}$$

Indeed, this measure is equal to the cosine of the angle formed by the two vectors $\vec{d_j}$ and $\vec{q}$ in the m-dimension vector space. Some of the main benefits of this model include:

- Term weighting that enhances response quality.
- Since partial matching between documents and queries may occur, it is possible to obtain responses by approximating the user's requests. In fact, the cosine angle formula allows to rank documents according to similarity document-query.

We may assert that the main drawback of this representation is the assumption that terms are independent from each other, although there is no proof this actually is a drawback in real IR systems. In Chapter 3, [54], Chapter 10 [59] and Chapter 18 [12], the reader can find some examples of the use of the Vector Space Model to represent documents.

More complex text analyses dealing with morphologic-syntactic analysis, semantics analysis and structure terminology analysis are not included in this chapter.[13] Should the reader want to study more in-depth what is illustrated in this Section, s/he can download *IRTools*, a software toolkit intended for IR research.[14] Finally, in Chapter 14 [21] the reader can find other methods of document modeling apt for representing 3D graphical objects, such as X3D-based Web documents.

5.3 Web-IR Document Representations

The representations of documents illustrated in the previous Section come from the traditional IR research field, whose original goal was to retrieve relevant documents matching user needs, expressed through queries. With the advent of the Web and of the HTML language, used to write the vast majority of web documents, other document representations have been suggested, not only based on the single terms that made them up, as occurs for classic IR, but also on other features typical of the hypertext environment, expressed through hyperlinks and/or HTML tags. Indeed, the Web can be considered a huge hypertext and hence interesting for the *Hypertext Information Retrieval* (HIR) research field [2, 24, 33, 23]. That's how the Web-IR came about. It expresses the union between classic IR and HIR for the Web, namely the enhancement of the classic pre-Web IR systems. It exploits the characteristics of hypertext languages of documents [1]: a term t of a Web document d is given a weight also according to the HTML tags between which it is. For example, if a term t appears between the two HTML tags <TITLE> and < /TITLE>, it means it is part of the document title, and must thus be significantly weighted, as it could be correlated to the semantics of the

---

[13] Links to several IR resources, including test collections, IRS lists and text-analysis tools can be found on the http://www.dcs.gla.ac.uk/idom/ir_resources Web site.

[14] Web site: http://sourceforge.net/projects/irtools.

very document, unlike a term included between the two HTML tags <BODY> and < /BODY>, a part of a plain text, where its importance could be relative. An example is to be seen in Fig. 5.3, which highlights the terms between the two HTML tags <TITLE> and < /TITLE>:

```
<TITLE>

Springer UK - Academic Journals, Books and Online Media

</TITLE>
```

there are only a few terms, but they are all significant and correlated to the general semantics of the Web page, while in the following text fragment, taken from the same source:

```
<span class="bodytext">

Benefit from attractive savings on Springer books
by signing up for Springer's free new book e-mail
notification service. New title info, news and
special announcements: with Springer Alerts it
pays to be informed.


</span>
```

there are some terms not all strictly correlated to the semantics of the page in which the text fragment is included (e.g.,"new" and "free").

Some important algorithms adopted by Web search engines, such as *HITS* [46] and *PageRank* [14], mainly exploit hypertext links between pages, considerably improving retrieval processes. In order to better guide the reader in an in-depth analysis of these modeling techniques, this Section has been broken down into two parts: the first part shows document modeling techniques mainly based on HTML tags, with simple, explicative examples. The second part illustrates modeling techniques based on hypertext links, and clarifies *HITS* and *PageRank*, two algorithms which are currently the most commonly used search ones on the Web.

For in-depth studies on semi-structured document modeling examples on the Web, the reader can refer for example to INEX conferences, Initiative for the Evaluation of XML Retrieval [37].[15]

5.3.1 Tag-Based Document Models

As mentioned in the introduction of this section, many proposals in literature have tried to enhance the performance of classic IR systems, mostly based on Vector Space Model, taking into account the possibilities offered by the HTML language. The idea is that of modifying the document pre-processing process, illustrated in Subsection 5.2.1, taking

---

[15] Visit http://www.informatik.uni-trier.de/~ ley/db/conf/inex/index.html.

into account the HTML structure of the Web page as well. Clear examples of document modeling based on HTML tags are to be found in [27, 55, 87, 88]. Now we describe three relevant examples on the use of such techniques.

The IRIS System. In [87] we have a Vector Space representation of the document, where the IRIS system [89] calculates the weight of index terms even according to the tags between which they are comprised, such as heading texts (i.e., terms between $<$Hn$>$ and $</$Hn$>$tags), titles, and meta-keywords. We now illustrate the document modeling phase and the term-weighting phase, in order to better understand the retrieval and modeling process.

- *Document Modeling Phase.* The document is modeled into three main parts:
  - – Body text terms.
  - – Heading text terms.
  - – A combination of the two.

  In the case in point, we have the scheme illustrated in Fig. 5.4, where, in order to make things more straightforward, the single terms have been left as they appear in the fragment of the Web page used as an example, without undergoing the stopword and stemming elimination phases of Subsection 5.2.1. After having subdivided the document into three parts, as shown in Fig. 5.4, the IRIS system, depending on the query length, uses them as index term sources.

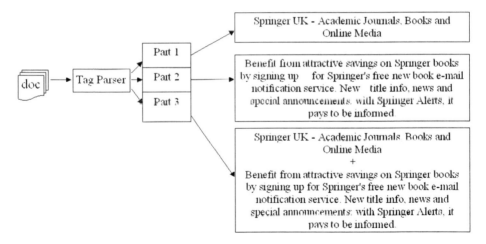

Fig. 5.4. The document modeling process used by the IRIS system

- *Term Weighting Phase.* The single terms are weighted differently, even depending on their position within the document itself. For example, the frequency of single terms comprised in header texts is multiplied by 10. Subsequently, one of the classic weighting system technique, the *SMART lnu* weighting schema with slope 0.3 [17, 16], is adopted.

The example document thus provides three different frequency tables, to be used to weight terms.

An Indexing HTML Model. Another interesting example of tag-based document modeling is to be seen in [55, 62]. In this case, the layout structure of an HTML document is taken into consideration: a *Tag Ranking* weights the several index terms as illustrated in Tab. 5.9.

Table 5.9. The tag classes hierarchy used by the Indexing Model in [62]

Rank	CLASS NAME	CLASSIFIED TAGS/PARAMETERS
1	Title	TITLE, META keyword
2	Header 1	H1, FONT SIZE=7
3	Header 2	H2, FONT SIZE=6
4	Header 3	H3, FONT SIZE=5
5	Linking	HREF
6	Emphasized	EM, STRONG, B, U, I, STRIKE, S, BLINK, ALT
7	Lists	UL, OL, DL, MENU, DIR
8	Emphasized 2	BLOCK QUOTE, CITE, BIG, PRE, CENTER, TH, TT
9	Header 4	H4, CAPTION, CENTER, FONT SIZE = 4
10	Header 5	F5, FONT SIZE = 3
11	Header 6	H6, FONT SIZE=2
12	Delimiters	P,DT,....

- *Document Modeling Phase.* After the HTML parsing stage, the document is shown as modeled in the 12 classes of terms illustrated in Tab. 5.9. Basically, 12 classes of terms are formed, which together make up the document model after the parsing phase, as illustrated in Fig. 5.5. The overall document is then represented by a vector $\vec{S} \equiv \{S_1, S_2, \ldots S_{12}\}$, where $S_i \geq 0$ represents the number of terms in the tag class C_i.

- *Term Weighting Phase.* The frequency distribution of a term t is represented by a vector $\vec{T} \equiv \{T_1, T_2, \ldots T_n\}$, where $T_i \geq 0$ represents the frequency of term t in the tag class C_i and $n = |ITD|$. Firstly, a weight w_i is assigned to each tag class C_i. The weight to be associated with the single term t, $F(d, t)$, is then calculated as the weighted average, with weight w_i, of the several normalized frequencies of the term in all the tag classes, F_{ctag_i}, multiplied by the inverse of the term frequency in the documents collection, IDF_t, as expressed by the Eq. 5.25, similar to the classic formula TFxIDF (see Section 5.2.1):

$$F(d, t) = \sum_{i=1}^{n} w_i F_{ctag_i}(d, t)(IDF_t) \tag{5.25}$$

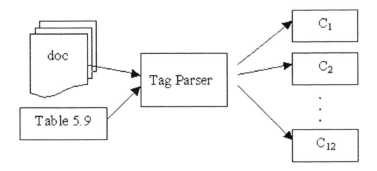

Fig. 5.5. The document modeling process used in [55]

The WEBOR System. Another useful example for the reader is to be seen in [27, 26], where the tag hierarchy is formed only by the six classes illustrated in Tab. 5.10. The IR system employed is *WEBOR*, WEB-based search tool for Organization Retrieval. [16]

Table 5.10. The tag class hierarchy used by the WEBOR system

Class Name	HTML Tags
Anchor	A
H1-H2	H1-H2
H3-H6	H3, H4, H5, H6
Strong	STRONG, B, EM, I, U, DL, OL, UL
Title	TITLE
Plain Text	None of the above

- *Document Modeling Phase.* The document modeling method consists of two stages:
  - Parsing HTML. The document is broken down into 6 parts, each one formed by the groups of terms of the document contained between the tags as shown in Tab. 5.10.
  - Vector Space Modeling. In this stage, the document is modeled with the classic Vector Space Model, to lead the query-similarity operation following the Eq. 5.24.
- *Term Weighting Phase.* Firstly, a vector $\overrightarrow{CIV}$, *Class Importance Vector*, $\overrightarrow{CIV} \equiv \{civ_1, civ_2, \dots civ_6\}$ is formed, where $civ_i \geq 0$ represents a weight determined experimentally by means of genetic algorithms. Such a weight is the contribution of the $i - th$ tag class to the formation of the overall weight of term t in document d. Another vector is then formed, $\overrightarrow{TFV}$, *Term Frequency Vector*, which contains the occurrence frequency of term t in document d for each tag class of Tab. 5.10: $\overrightarrow{TFV} \equiv \{tif_1, tif_2, \dots tif_6\}$. The last step is the calculation of weight w, to be associated with term t by the following formula:

---

[16] See http://nexus.data.binghamton.edu/~ yungming/webor.html.

$$w_t = (\overrightarrow{CIV} \bullet \overrightarrow{TFV})idf \qquad (5.26)$$

where idf is the inverse document frequency of term t in the documents of the collection D.

Finally, the document vector $\overrightarrow{d} \equiv \{w_1, w_2, \ldots, w_n\}$ is built. It contains the weight of each term of the ITD with respect to the document, to be used to carry out the query-similarity operation, as illustrated in Subsection 5.2.1.

5.3.2 Link-Based Web Document Models

The previous Subsection described the document modeling techniques that use HTML features to strengthen classic IR systems for the Web. This section ends with the illustration of document modeling examples based on HTML hypertext links, and on the two algorithms *HITS* [46] and *PageRank* [14] which utilize them. As previously mentioned, the Web can be considered a huge hypertext $G = (V, E)$, formed by documents V connected through links E. In this context, an HTML page can be modeled as a hypertext node featuring hypertext in-links (i.e., anchors) and outgoing links.

HITS. The first step in order to run a hypertextual algorithm, such as HITS, is to build the underlying hyperlink-directed graph from the set of available pages $G = (V, E)$. It consists of a set of nodes V and of a set of edges E, where each edge is an ordered pair of nodes. The *in-degree* and the *out-degree* of a node u are the number of nodes v such that $(v, u) \in E$ and the number of nodes v, such that $(u, v) \in E$ respectively. Of course, each node corresponds to a page and an edge is a hypertextual link between two pages.

Following the idea of backlink count that assigns a rank to a page according to its popularity, that is, the number of pages pointing to it, the in-degree can also be employed in more sophisticated algorithms where the importance of pointing pages is recursively taken into consideration.

Regarding scholarly publications, where it is possible to recognize a class of papers related to surveys and reviews on a particular topic that cite many significant research works, Kleinberg identifies two classes of Web pages. *Authorities*, that have relevant content about a particular topic, and *Hubs*, which contain several links to relevant authoritative pages [46].

A recursive algorithm has been conceived to identify highly relevant authority and hub pages, a recursive algorithm has been conceived. A query-dependent subgraph of the Web is chosen for the analysis, retrieving the first results of a *broad-topic query* from a search engine. This kind of query is characterized by an appreciable number of relevant pages that can overload user search activity. The initial root set of highest-ranked pages is then expanded considering all the pages that point to it and all the pages pointed by it, by means of a search engine that allows this kind of query on the link structure among pages. The obtained set is the input of the HITS iterative algorithm that assigns each page the two measures, authority and hubness, following this method:

$$a_p \leftarrow \sum_{q:(q,p)\in E} h_q \qquad (5.27)$$

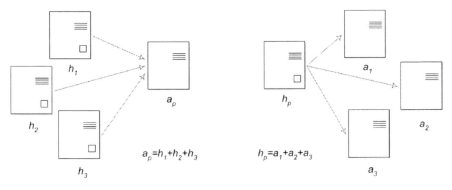

Fig. 5.6. An example of how the authority a and hubness h measures for a page p are drawn

$$h_p \leftarrow \sum_{q:(p,q)\in E} a_q \qquad (5.28)$$

where a and h are the authority and hubness of pages respectively, and E is the link set (see also Fig. 5.6). After a certain number of iterations, the equilibrium is reached and the two measures are assigned to each page. The basic steps of the HITS algorithm is itemized in Algorithm 1. An interesting way to analyze the output is to make two copies of each page and visualize the graph as bipartited, where the hubs point to the authorities. Details on the algorithm and the related matrix formulation, the time and the conditions of convergence are to be seen in [46].

Algorithm 1. Pseudo-code of the HITS algorithm, where authorities a_p and hubs h_p of the pages are stored in the vectors a and h.

$V \leftarrow$ collection of n pages
$N \leftarrow$ number of iterations
$z \leftarrow (1, 1, ..., 1) \in \Re^{|V|}$
$a_0 \leftarrow z$
$h_0 \leftarrow z$
for $i = 0$ to N **do**
    {apply Eq. 5.27 to $(a_{i-1}; h_{i-1})$ and draw the new authority vector $\hat{a}_i$}
    {apply Eq. 5.28 to $(\hat{a}_i; h_{i-1})$ and draw the new hub vector $\hat{h}_i$}
    {normalize both $\hat{a}_i$ and $\hat{h}_i$ to 1}
    $a_i \leftarrow \hat{a}_i$
    $h_i \leftarrow \hat{h}_i$
end for

The output of the HITS algorithm consists of two lists: the pages with the highest hubness, and the ones with the highest authority. The latter can be used to better rank the pages of the initial set built from the results of broad-topic queries, a process named *Topic Distillation*. The hub list is useful for users or for crawling systems that need valid starting points to begin their seeking activities. The pages included in that list point to high authoritative pages, and are therefore useful to quickly find good resources. The HITS has also been used to identify and analyze fine-grained hyperlinked communities, and the related topics of interest [20]. Another method to discover hub pages through link analysis is discussed in [56].

Several enhancements of the HITS algorithm that achieve improvements in terms of precision, and fix some potential problems on the original formulation, such as the *Topic Drift* phenomenon, are described in [10]. Amento *et al.* [3] offers a comparison of hyperlink-based metrics, such as HITS and PageRank, with judgments given by humans about the quality of pages.

PageRank. While Kleinberg proposes a two-level weight propagation scheme where authorities and hubs are the two measures taken into consideration, Page and Brin suggest to assign a single measure to the pages, with value being high if the sum of the measures of its backlinks is high [14]. Therefore, it is possible to see the related PageRank algorithm as a direct enhancement of the classic backlink count.

One of the assets of PageRank over HITS is that we do not need to restrict the calculation to a subgraph of the Web relevant to a given query. For a general-purpose search engine, hypertext-based algorithms, such as the described ones, are computational, intensive processes, that cannot be performed for each submitted query. The only way to include those kinds of relevance measures is to periodically run the algorithm off-line, keeping all sets of values up-to-date. Focused crawlers tailor their search to a subset of topics and retrieve much less documents, and for this reason it is possible to run algorithms such as HITS on-line, during the crawling.

Assuming the Web as a strongly connected graph, the PageRank can be described through the *Random surfer model*. A surfer in that model is able to randomly click on one of the links contained in a page p with equal probability $1/N_p$, where N_p is the number of links in p. A simplified formula to compute the PageRank is:

$$rank(p) = c \sum_{q:(q,p)\in E} \frac{rank(q)}{N_q} \tag{5.29}$$

where c is a normalization constant less than 1. Each page q contributes with a quantity that is proportional to its rank $rank(q)$ but inversely proportional to the number of links N_q. So the PageRank flows from one page to another, decreasing its value if the outdegree of a page is high. Like the HITS algorithm, the equation is recursive and it must be computed until convergence. In the random surfer model the rank could be seen as the probability that the surfer is currently browsing a given page.

The Web is not strongly connected thus there cannot be a situation in which two pages are linked to each other and do not have any outgoing links. In this scenario, the two pages will increase their rank without transferring the value to other documents outside the loop. For this reason, the previous formula has been enhanced:

$$rank(p) = c \sum_{q:(q,p)\in E} \frac{rank(q)}{N_q} + \frac{(1-c)}{N} \tag{5.30}$$

where N is some vector that gives the source of the rank for each page in the corpus.

The second term denotes the event that the imaginary surfer who is randomly clicking on links will eventually stop clicking and start visiting a different page. It is generally assumed that this factor assumes the constant value 0.15.

As in the case of HITS, PageRank can be periodically calculated to keep the ranks associated with the retrieved pages in a given collection up-to-date. The best ranked pages can be further analyzed by focused crawlers, for example, by extracting the outgoing links and putting them at the top of the URL QUEUE. In order to tackle the Webspam issue, an enhanced version of PageRank named *TrustRank* has been proposed [40]. In Chapter 7 [50] the reader can find several examples on the use of a link-based document representation for Focused Crawling.

5.4 Concept-Based Document Modeling: Latent Semantic Indexing

With the Latent Semantic Indexing (LSI) technique, the reader may complete the general survey on the most common document modeling techniques employed in IR literature. This Section illustrates a further extension of the document model, toward a representation based on concepts and semantic relations between index terms.

LSI [32, 28] is a document representation technique that assumes there is some hidden structure in using the terms included in a collection of documents: the topic dealt with by a text is more associated with the concepts that are used to describe it rather than with the terms actually used; hence, the idea is to represent a document through concepts, rather than through index terms. In order to do so, the high dimensional space $ITD \equiv \{t_1, t_2, \ldots, t_m\}$, formed by all the m index terms of a document collection, is mapped by means of Linear Algebra techniques into a lower dimensionality space S^n, with $n << m$, where every component s_j represents a *concept*. This can be obtained by clustering the terms t_i into sets s_j, to form a sort of *association by concepts*. This technique automatically solves synonymity problems, since the terms that appear most frequently together are grouped in the same concept. The fundamental LSI techniques chiefly focus on the correlations between documents and terms. Once such nexuses are found, the goal becomes that of understanding such relations and highlighting the most relevant ones, even when a linguistic nexus is not available. However, it should be pointed out that, broadly speaking, the result of LSI techniques cannot be interpreted from a linguistic viewpoint. Such a result has a purely mathematical value.

In order to better understand this document modeling technique, the reader can refer to the example of the set of documents used in Subsection 5.2.4. The starting point is the term-document matrix A_{ij} that, in our example, is the matrix shown in Tab. 5.11. It is obtained with $D \equiv \{d_1, d_2, d_3\}$, $n = 3$ documents, for a total of $m = 7$ index terms, i.e., $|ITD| = 7$. Thus, we still have to decide which is the best technique to reveal these hidden relations within the example matrix illustrated in Tab. 5.11. In order to do so, we can resort to a mathematical technique called *Singular Value Decomposition* (SVD).

SVD is a widespread technique in the solving of problems such as matrix rank estimation and the canonical analysis of correlations [38, 9]. Given a matrix $A_{n,m}$, without losing in generality, we can suppose $m \geq n$ and $rank(A) = r$. The SVD of matrix A, indicated as $SVD(A)$, is defined as follows:

Table 5.11. Term-Document Matrix A

ITD	d_1	d_2	d_3
new	1	1	0
benefit	1	1	0
attractive	0	1	0
service	1	0	0
springer	0	0	1
info	0	1	1
special	0	0	1

$$A = U \cdot \Sigma \cdot V^T \tag{5.31}$$

where U is a matrix $m \times r$ orthonormal ($U^T \cdot U = I_r$), V is a matrix $n \times r$ orthonormal ($V^T \cdot V = I_r$), $\Sigma = \text{diag}(\sigma_1, \sigma_2, \ldots, \sigma_n)$, with $\sigma_i > 0$, $1 \leq i \leq r$ and $\sigma_i > \sigma_{i+1}$.
The matrix calculus that reduces the space is represented through the following notation:

$$A_k = U_k \cdot \Sigma_k \cdot V_k^T \tag{5.32}$$

where U_k is a matrix $m \times k$ obtained by taking the first k columns from U, V_k is a matrix $n \times k$ obtained by taking the first columns from V, Σ_k is a matrix $k \times k$ obtained from the first k values of the diagonal of Σ.

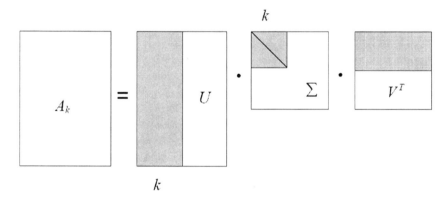

Fig. 5.7. Singular Value Decomposition

Fig. 5.7 reflects the reduction (to rank k) of matrix A, which is used by LSI to get hold of the semantic structure of the used index terms. If $k = r$ then LSI executes the similarity of literal matching (since $A_k = A$, thus two documents are similar only if they are identical). If $k = 1$ all the documents are associated with the same concept. Intuitively, k represents the measure of the overall number of concepts that are to be found in the several documents. By using the above mentioned reduction process, lesser terminology discrepancies are ignored.

Going back to the matrix in the previous example, its rank is $r = 3$, and Σ is given by:

$$\Sigma = diag(2.41, 1.73, 1.10) \tag{5.33}$$

By choosing, for example, $k = 2$, we get:

$$\Sigma = diag(2.41, 1.73) \tag{5.34}$$

The resulting matrix A_k is shown in Tab. 5.12.

Table 5.12. The resulting Matrix A_k

-1.3457	1.0762	1.6799
-1.3518	-0.0000	-1.0829

In this matrix, every column represents one document in the new compressed space. In Chapter 6 [51] the reader can find an example of an SVD application to model a click-through element.

5.5 AI-Based Document Representations

The previous Sections illustrated document modeling techniques based on classic IR, on Web-IR methods and on concept sets. We saw that such methods have the advantage of being simple, allowing to develop automatic IR systems offering good performances. For these reasons, they are widely used in real systems. Nonetheless, literature offers several other document representation methods, such as the ones based on Artificial Intelligence techniques, on which this very paragraph focuses. A typical feature of these methods is that of modeling a document through a richer and more complex knowledge representation of the domain, even though it sometimes entails a higher computational effort. The following Subsections illustrate some of these techniques used is several adaptive Web-Based systems: Artificial Neural Networks, Semantic Networks and Bayesian Networks.

5.5.1 Artificial Neural Networks

Artificial Neural Networks (ANN) provide a method for the automatic understanding of classification and regression functions [42, 13, 71]. An ANN comprises of a certain number of nodes or *neurons* connected by arcs or *synapses*, each one associated to a real value w called *weight*. Each neuron is characterized by an *Activation State*, as determined by the input values and by the weight of the corresponding connections via an *Activation Function*. In literature there are many types of ANN [71]: from the classical *Multi-Layer Perceptron* (MLP), consisting of an input layer, one or more hidden layers and an output layer, with feed-forward synapses and supervised learning, to *Self Organizing Maps* (SOM) networks, consisting of only two layers, of the feed-forward type, but with a unsupervised learning. The latter are very commonly used in Web document clustering (see for example [86, 83, 73]). Other types of networks, such as ART networks, entail more complex architectures, which lie outside the purposes illustrated in this chapters. Several IR systems use ANN to model documents, queries and, in the case of adaptive systems, even users. However, the goal is still that of retrieving the document that mostly fits the user's query.

Generally, a multi-layer system is built to model a document with an ANN: the query's terms are associated with the input layer of neurons, the terms of a set of documents with an intermediate layer, and each document with an output neuron. Following a query, the network is *trained* to provide, in output, a ranking of the documents that are most similar to that query.

An example of an ANN used for document modeling is the one illustrated in [85, 5], where the network actually forms an IR system, whereas a document, through its terms, forms a layer. In this model, an ANN is formed by the terms of the query, input neurons, the terms of all the documents of the collection, neurons of the hidden layer, and by the documents themselves, neurons of the output layer. The reader may find interesting a brief description of the system's operating mechanism on the whole, in order to better comprehend the use of an ANN for document modeling. Consider a collection D of documents, $D \equiv \{d_1, d_2, d_3\}$, consisting of the following three documents, after the pre-processing phases (see Subsection 5.2.1): $d_1 \equiv \{$new, benefit, service$\}$, $d_2 \equiv \{$new, benefit, attractive, info$\}$ and $d_3 \equiv \{$springer, info, special$\}$ built from three Web pages taken from the Web site shown as an example in Section 5.2. The ITD is given by:

$$T \equiv \{\text{new, benefit, attractive, service, springer, info, special}\}$$

Finally, consider a query q:

$$q \equiv \{\text{benefit, service, info}\}$$

According to this model, all the terms of set T form an intermediate layer of neurons, each one bi-directionally connected to the document or documents containing it. The inputting of query q activates only the neurons corresponding to the neuron-terms in the set of the intermediate layer, belonging to the query, highlighted in Fig. 5.8; subsequently, an iterative process activates the last layer's neurons, which represent the documents to be retrieved, as illustrated in Fig. 5.8. The connections or synapsis between the generic term t_i of the intermediate layer and the generic documents d_j that contain it, is indicated by w_{ij} and calculated with the formula of a normalized TFxIDF [85]:

$$w_{ij} = \frac{w_{ij}^*}{\sqrt{\sum_{j=1}^{n} d_{ij}^2}}$$

being w_{ij}^* the TFxIDF weight and $n = |ITD|$. If the sum of the signals received by the terms activated by the query exceeds a certain threshold, then the neuron-document emits a signal that is input, weighted by synapsis, into all the nodes of the terms that form it, and that's the path it follows until the activated node-documents are stable. Each document is thus represented as a set of neurons. In this way, documents are selected according to the query and ranked on the basis of the query-document similarity. This

ANN features a mixed topology: the first connections are feed-forward while the others are bidirectional, as shown in Fig. 5.8.

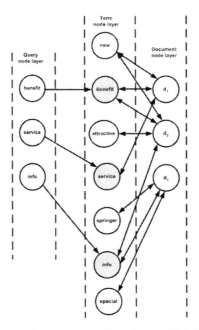

Fig. 5.8. An Example of an IR System Based on Artificial Neural Networks

Another interesting type of document modeling through ANN can be seen in [45]. Here, the basic idea is to use a set of neural networks, represented by three MLP, to perform a score representing the correlation among terms and documents as shown in Fig. 5.9. Two ANN are trained starting from the pairs (term, document) taken as input, while the term's absence or presence in the document is taken as output: if the term t_i belongs to the document d_j, the score is high; low on the contrary. Starting from a TFxIDF weighing for document d_j and from a *one-hot* representation for term t_i, two more representations of terms and documents are built, respectively t_i^* and d_j^*. In particular, the d_j^* representation is an enriched representation of the document through the neural network, that takes into account the probability distribution of single terms within the set of documents [8]. In this way, given a term and a document, it is possible to perform a term-document score, to be used for retrieval.

The reader can find another interesting example of an ANN application to model small documents such as Web links in Chapter 7 [50] while an example to e-mail routing can be found in [22]. In this routing system, named *LINGER*, incoming e-mails that are first pre-processed (see Subsection 5.2.1) form the input layer of a MLP. Every output node of the network represents a single predefined category.

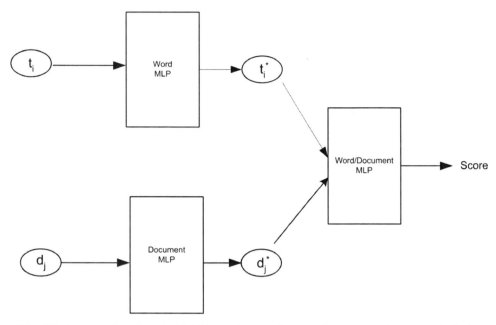

Fig. 5.9. A system based on Artificial Neural Networks to perform a term-document similarity score

5.5.2 Semantic Networks

Semantic Networks (SN) are useful for representing conceptual knowledge and, in particular, the relationship between concepts [66, 79]. In general, a SN is formed by a directed graph, whose nodes are organized in hierarchic structures, while the arcs connecting them represent the binary relations between them, such as relations *is-a* and *part-of* (see for example [72]).

An example of a system using SN for document modeling is to be seen in [6]. This system presents a conceptual indexing method based on WORDNET, a large lexical database, organized as a freely available semantic network,[17] which has received a lot of attention within the computational linguistics community. Nouns, verbs, adjectives and adverbs are organized into synonym sets (i.e., *synsets*), each one representing an underlying lexical concept. Synsets are linked by different semantic relations and organized in hierarchies. The synsets identified in WORDNET derive from a thorough lexicographic work and many one-grained sense distinctions are made [53] (there are 99, 642 synsets in version 1.6). In Baziz's system the document is mapped on the SN WORDNET and converted from a set of terms to a set of concepts (*Concept Detection* phase). The extracted concepts (single or multi-words) are then weighted as in the classical index term case, according to a kind of TFxIDF weighting technique which is also a variant of the OKAPI system. By this stage, the document, having been transformed from a set of terms to a set of concepts, is treated by the system.

---

[17] http://wordnet.princeton.edu/.

Another example is the one shown in [49], where the *SiteIF* system is illustrated: a personal agent for a bilingual news web site that learns user's interests from the requested pages. Even in this case, the system utilizes WORDNET to suggest a word meaning-based document representation, used subsequently to build the user model, along with the extension WORDNET DOMAINS where each term of the lexical database is also labeled with one or more semantic domains, to which it belongs. A document is treated to extract its semantic contents, and is eventually represented through a list of synsets that are relevant for a certain domain. The obtained list is further treated to form a new SN, called *Word Sense Document Representation*, which is the starting point to build the user model.

The last example of such a representation is the WIFS system illustrated in Chapter 6 [51] and in Chapter 2 [35] of this volume. Therein, a document is represented by a set of terms and co-occurrences between them [52]. Fig. 6.6b, in Chapter 6, shows an example of document modeling through simple SN: the document terms that also belong to a database of terms, known to the system, represent the *planets* of the networks, whereas all other terms form the so-called *satellites*, namely the secondary nodes, linked to all the planets with which they co-occur. Even the frequency in the document is calculated for each one of the aforesaid terms. This representation is used to match the document with the user model, in order to assess the document's relevance for that specific user.

5.5.3 Bayesian Networks

A Bayesian Network (BN) is a probabilistic model for Knowledge Representation [60, 61]. It consists of an acyclic direct graph with nodes and arcs linking them. Each node is labeled with a random variable. For example, as illustrated in Fig. 5.10, node o, named *antecedent node* or *parent node*, is connected to node p, called *consequent node* or *child node*. Each arc is associated with a causal relation, expressed by a probability matrix which, for every pair of values V_p and V_q of random variables associated with node p and q respectively, indicates the probability that value V_q occurs once value V_p has occurred, i.e., $P(V_q/V_p)$. The *Evaluation* of the network requires the calculation of the probability of consequent nodes, following a probability distribution of the antecedent nodes. BN are commonly used in adaptive filtering [7, 19, 4] and in IR [5, 15, 64, 63]. In some cases BN turned out to be very useful in improving the retrieval performance, as shown for example in [82].

A document is represented through a network with nodes distributed on two levels (document network): the document is a consequent node, whose previous nodes are the terms contained in the document itself. The weight associated with arcs is calculated through the TFxIDF technique. An example of document representation through this technique is presented in [34], which shows an IR system entirely based on BN. Through the network, the system ranks the document-query similarity. BN are built following these rules:

1. Build the set of terms included in the collection D of all documents.
2. Build the set of possible topics with which the query can be input.
3. Build a BN associated with each possible topic with which the query can be input by the user. Fig. 5.11 illustrates this network. The node *Topic T_i* is associated with

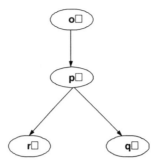

Fig. 5.10. An example of a simple Bayesian Network

event: *the document is relevant for topic* T_i, whereas node t_{ij} is associated with the event *term* t_{ij} *is present in the document*. In order to build the network, each arc (T_i, t_{ij}) is associated with a probability, in automatic or in manual manner.

Once the network is built, the document is represented by the document nodes that form it. Bayes' theorem is used to calculate the query-document similarity:

$$P(T_i/t_{1i}, t_{2i} \ldots, t_{ni}) = \frac{P(T_i)P(t_{1i}, t_{2i} \ldots, t_{ni}/T_i)}{P(t_{1i}, t_{2i} \ldots, t_{ni})} \qquad (5.35)$$

In order to simplify the calculation, it is possible to resort to a linear simplification, using function g [75]:

$$g(T_i/t_{1i}, t_{2i} \ldots, t_{ni}) = \sum_k P(t_{ik})w(t_{ik}, T_i) \qquad (5.36)$$

where $P(t_{ik}) = 1$ if term $t_{ik} \in d$, otherwise $P(t_{ik}) = 0$ and $w(t_{ik}, T_i)$ equivalent to a weight associated with each term. The query is thus represented as a topic T, and the document as a set of BN nodes. The calculation of the BN through the simplified function g gives the query-document similarity ranking.

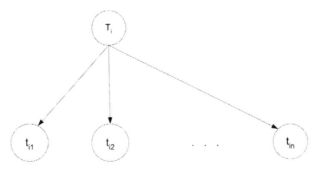

Fig. 5.11. An Example of a Two-Level Bayesian Network Model of IR

Another example of a document representation through BN can be seen in [15], where the *Inquery* system is used, that is an IR system utilizing a document modeling technique based on BN.

Finally, a recent work of B. Piwowarski shows the use of a BN for an IR system on XML-based documents, tested on set INEX of documents [63], whereas in [81] the reader may find another relevant example of Web document representation through a BN.

5.6 Conclusions

In this chapter we discussed the various high level approaches to document modeling and representation. The first approaches to be illustrated derived from the classic IR field, such as the Vector Space Model and its variants, called pre-Web modeling techniques. Another group of techniques followed: they were conceived with the Web, just like those based on HTML tags and hypertext links. The concept modeling was illustrated in the third part, while the final one presented several other approaches inspired by AI techniques, such as Neural Networks, Semantic Networks and Bayesian Networks. The reader will surely have noticed that the Vector Space Model is still very popular, even for the Web, owing to its simplicity and efficacy. These techniques are applied by the various systems described in the rest of the book and by a number of application domains such as Focused Crawling, Content Adaptation, Intelligent Information Search and others. In conclusion, this chapter introduced some of the basic issues related with document modeling in complex and adaptive Web-based systems and applications, providing references to actual uses throughout the rest of the book.

References

1. Agosti, M., Melucci, M.: Information retrieval on the web. In Agosti, M., Crestani, F., Pasi, G. (eds.): ESSIR, Lecture Notes in Computer Science, Vol. 1980. Springer (2000) 242–285
2. In Agosti, M., Smeaton, A.F. (eds.): Information Retrieval and Hypertext. Kluwer Academic Publishers, Dordrecht, NL (1997)
3. Amento, B., Terveen, L., Hill, W.: Does authority mean quality? predicting expert quality ratings of web documents. In: SIGIR '00: Proceedings of the 23rd annual international ACM SIGIR conference on Research and development in information retrieval, New York, NY, USA, ACM Press (2000) 296–303
4. Asnicar, F., Tasso, C.: ifWeb: a prototype of user models based intelligent agent for document filtering and navigation in the World Wide Web. In P.Brusilovsky, Fink, J., Kay, J. (eds.): Proceedings of Workshop Adaptive Systems and User Modeling on the World Wide Web at Sixth International Conference on User Modeling, UM97, Chia Laguna, Sardinia, Italy (June 2 1997) 3–11
5. Baeza-Yates, R., Ribeiro-Neto, B.: Modern Information Retrieval. Addison-Wesley (1999)
6. Baziz, M., Boughanem, M., Traboulsi, S.: A concept-based approach for indexing documents in IR. In: Actes du XXIII ème Congrès INFORSID, Grenoble, Grenoble, INFORSID (May 24–27 2005) 489–504
7. Belkin, N.J., Croft, W.B.: Information filtering and information retrieval: Two sides of the same coin? Communications of the ACM **35**(12) (1992) 29–38

8. Bengio, Y., Ducharme, R., Vincent, P., Jauvin, C.: A neural probabilistic language model. Journal of Machine Learning Research **3** (2003) 1137–1155
9. Berry, M.W.: Large-scale sparse singular value computations. International Journal of Supercomputer Applications **6**(1) (Spring 1992) 13–49
10. Bharat, K., Henzinger, M.R.: Improved algorithms for topic distillation in a hyperlinked environment. In: SIGIR '98: Proceedings of the 21st annual international ACM SIGIR conference on Research and development in information retrieval, New York, NY, USA, ACM Press (1998) 104–111
11. Billsus, D., Pazzani, M.J.: User modeling for adaptive news access. User Modeling and User-Adapted Interaction **10**(2-3) (2000) 147–180
12. Billsus, D., Pazzani, M.J.: Adaptive news access. In Brusilovsky, P., Kobsa, A., Nejdl, W. (eds.): The Adaptive Web: Methods and Strategies of Web Personalization, Lecture Notes in Computer Science, Vol. 4321. Springer-Verlag, Berlin Heidelberg New York (2007) this volume
13. Bishop, C.M.: Neural Networks for Pattern Recognition. Clarendon Press, Oxford (1995)
14. Brin, S., Page, L.: The anatomy of a large-scale hypertextual Web search engine. Computer Networks and ISDN Systems **30**(1-7) (1998) 107–117
15. Broglio, J., Callan, J.P., Croft, W.B., Nachbar, D.W.: Document retrieval and routing using the INQUERY system. In Text REtrieval Conference (TREC) TREC-3 Proceedings, Department of Commerce, National Institute of Standards and Technology (1994) 29–38 NIST Special Publication 500-226: Overview of the Third Text REtrieval Conference (TREC-3).
16. Buckley, C., Singhal, A., Mitra, M.: Using query zoning and correlation within SMART: TREC 5. In Text REtrieval Conference (TREC) TREC-5 Proceedings, Department of Commerce, National Institute of Standards and Technology (1996) NIST Special Publication 500-238: The Fifth Text REtrieval Conference (TREC-5).
17. Buckley, C., Singhal, A., Mitra, M., Salton, G.: New retrieval approaches using SMART: TREC 4. In Harman, D. (ed.): NIST Special Publication 500-236: The Fourth Text REtrieval Conference (TREC-4), Department of Commerce, National Institute of Standards and Technology (November 1995)
18. Budzik, J., Hammond, K.J., Birnbaum, L.: Information access in context. Knowl.-Based Syst. **14**(1-2) (2001) 37–53
19. Callan, J.: Document filtering with inference networks. In Frei, H.P., Harman, D., Schäuble, P., Wilkinson, R. (eds.): Proceedings of the Nineteenth Annual International ACM SIGIR Conference on Research and Development in Information Retrieval, New York, ACM Press (August 18–22 1996) 262–269
20. Chakrabarti, S., Dom, B.E., Kumar, S.R., Raghavan, P., Rajagopalan, S., Tomkins, A., Gibson, D., Kleinberg, J.: Mining the web's link structure. Computer **32**(8) (1999) 60–67
21. Chittaro, L., Ranon, R.: Adaptive 3d web sites. In Brusilovsky, P., Kobsa, A., Nejdl, W. (eds.): The Adaptive Web: Methods and Strategies of Web Personalization, Lecture Notes in Computer Science, Vol. 4321. Springer-Verlag, Berlin Heidelberg New York (2007) this volume
22. Clark, J., Koprinska, I., Poon, J.: A neural network based approach to automated E-mail classification. In: IEEE/WIC International Conference on Web Intelligence (WI'03), IEEE Computer Society (2003) 702–705
23. Croft, W.B., Belkin, N.J., Bruandet, M.F., Kuhlen, R.: Hypertext and information retrieval: What are the fundamental concepts? (panel). In: ECHT. (1990) 362–366
24. Croft, W.B., Turtle, H.R.: Retrieval strategies for hypertext. Information Processing & Management **29**(3) (1993) 313–324
25. Cummins, R., O'Riordan, C.: Evolving local and global weighting schemes in information retrieval. Information Retrieval **9**(3) (June 2006) 311–330

26. Cutler, M., Deng, H., Maniccam, S., Meng, W.: A new study on using HTML structures to improve retrieval. In: Proceedings of the 11th IEEE International Conference on Tools with Artificial Intelligence (ICTAI'99), Chicago, Illinois, USA, IEEE Computer Society (8-10 November 1999) 406–409

27. Cutler, M., Shih, Y., Meng, W.: Using the structure of HTML documents to improve retrieval. In: USENIX Symposium on Internet Technologies and Systems. (1997) 241–252

28. Deerwester, S., Dumais, S.T., Furnas, G.W., Landauer, T.K., Harshman, R.: Indexing by latent semantic analysis. Journal of the American Society for Information Science **41** (1990) 391–407

29. DeRose, S.J.: The SGML FAQ Book : Understanding the Foundation of HTML and XML. Kluwer Academic Publications (1997)

30. Devore, J.L.: Probability and Statistics for Engineering and the Sciences. 3rd edn. Brooks/Cole (1991)

31. Dumais, S.T.: Improving the retrieval of information from external sources. Behavior Research Methods, Instruments and Computers **23** (1991) 229–236

32. Dumais, S.T.: Latent semantic indexing (LSI) and TREC-2. In: Text REtrieval Conference (TREC) TREC-2 Proceedings, Department of Commerce, National Institute of Standards and Technology (1993) 105–116 NIST Special Publication 500-215: The Second Text REtrieval Conference (TREC 2).

33. Frei, H.P., Stieger, D.: Making use of hypertext links when retrieving information. In: Proceedings of the Fourth ACM Conference on Hypertext. Information Retrieval (1992) 102–111

34. Fung, R., Del Favero, B.: Applying Bayesian networks to information retrieval. Communications of the ACM **38**(3) (March 1995) 42–48

35. Gauch, S., Speretta, M., Chandramouli, A., Micarelli, A.: User profiles for personalized information access. In Brusilovsky, P., Kobsa, A., Nejdl, W. (eds.): The Adaptive Web: Methods and Strategies of Web Personalization, Lecture Notes in Computer Science, Vol. 4321. Springer-Verlag, Berlin Heidelberg New York (2007) this volume

36. Gentili, G., Micarelli, A., Sciarrone, F.: Infoweb: An adaptive information filtering system for the cultural heritage domain. Applied Artificial Intelligence **17**(8-9) (2003) 715–744

37. Geva, S., Sahama, T.: The NLP task at INEX 2004. SIGIR Forum **39**(1) (2005) 50–53

38. Golub, G.H., Loan, C.F.V.: Matrix Computations. second edn. The Johns Hopkins University Press, Baltimore, MD, USA (1989)

39. Gourley, D., Totty, B.: HTTP: the definitive guide. First edn. O'Reilly Media (September 2002)

40. Gyöngyi, Z., Garcia-Molina, H., Pedersen, J.: Combating web spam with TrustRank. In: Proceedings of the 30th International Conference on Very Large Databases, Morgan Kaufmann (2004) 576–587

41. Hawking, D., Upstill, T., Craswell, N.: Toward better weighting of anchors. In: Proceedings of the 27th Annual International ACM SIGIR Conference on Research and Development in Information Retrieval. Posters (2004) 512–513

42. Haykin, S.: Neural Networks: A Comprehensive Introduction. Prentice-Hall International Editions (1999)

43. Joachims, T., Freitag, D., Mitchell, T.M.: Webwatcher: A tour guide for the world wide web. In: Proceedings of the 15h International Conference on Artificial Intelligence (IJCAI1997). (1997) 770–777

44. Jones, K.S., Walker, S., Robertson, S.E.: A probabilistic model of information retrieval: development and comparative experiments - part 2. Information Processing & Management **36**(6) (2000) 809–840

45. Keller, M., Bengio, S.: A neural network for text representation. In: Proceedings of the 15th International Conference on Artificial Neural Networks: Biological Inspirations, ICANN, Lecture Notes in Computer Science, volume LNCS 3697. Springer-Verlag, 2005. (2005) 667–672

46. Kleinberg, J.: Authoritative sources in a hyperlinked environment. Journal of the ACM **46**(5) (November 1999) 604–632

47. Koza, J.R.: Genetic Programming: On the Programming of Computers by Means of Natural Selection. MIT Press, Cambridge, MA, USA (1992)

48. Liu, J., Zhong, N., Yao, Y., W.Ras, Z.: The wisdom web: New challenges for web intelligence (WI). Journal of Intelligent Information Systems **20**(1) (2003) 5–9

49. Magnini, B., Strapparava, C.: User modelling for news web sites with word sense based techniques. User Modeling User-Adapted Interaction **14**(2-3) (2004) 239–257

50. Micarelli, A., Gasparetti, F.: Adaptive focused crawling. In Brusilovsky, P., Kobsa, A., Nejdl, W. (eds.): The Adaptive Web: Methods and Strategies of Web Personalization, Lecture Notes in Computer Science, Vol. 4321. Springer-Verlag, Berlin Heidelberg New York (2007) this volume

51. Micarelli, A., Gasparetti, F., Sciarrone, F., Gauch, S.: Personalized search on the world wide web. In Brusilovsky, P., Kobsa, A., Nejdl, W. (eds.): The Adaptive Web: Methods and Strategies of Web Personalization, Lecture Notes in Computer Science, Vol. 4321. Springer-Verlag, Berlin Heidelberg New York (2007) this volume

52. Micarelli, A., Sciarrone, F.: Anatomy and empirical evaluation of an adaptive web-based information filtering system. User Modeling and User-Adapted Interaction **14**(2-3) (2004) 159–200

53. Miller, G.A.: WordNet: A lexical database for English. Communications of the ACM **38**(11) (1995) 39–41

54. Mobasher, B.: Data mining for web personalization. In Brusilovsky, P., Kobsa, A., Nejdl, W. (eds.): The Adaptive Web: Methods and Strategies of Web Personalization, Lecture Notes in Computer Science, Vol. 4321. Springer-Verlag, Berlin Heidelberg New York (2007) this volume

55. Molinari, A., Pereira, R.A.M., Pasi, G.: An indexing model of HTML documents. In: Proceedings of the 2003 ACM Symposium on Applied Computing (SAC), March 9-12, 2003, Melbourne, FL, USA, ACM (2003) 834–840

56. Pant, G., Menczer, F.: Topical crawling for business intelligence. In Koch, T., Sølvberg, I. (eds.): Research and Advanced Technology for Digital Libraries, 7th European Conference, ECDL 2003, Trondheim, Norway, August 17-22, 2003, Proceedings. Volume 2769 of Lecture Notes in Computer Science., Springer (2003) 233–244

57. Park, L.A.F., Ramamohanarao, K., Palaniswami, M.: A novel document retrieval method using the discrete wavelet transform. ACM Transactions on Information Systems **23**(3) (July 2005) 267–298

58. Pazzani, M.J., Billsus, D.: Learning and revising user profiles: The identification of interesting web sites. Machine Learning **27** (1997) 313–331

59. Pazzani, M.J., Billsus, D.: Content-based recommendation systems. In Brusilovsky, P., Kobsa, A., Nejdl, W. (eds.): The Adaptive Web: Methods and Strategies of Web Personalization, Lecture Notes in Computer Science, Vol. 4321. Springer-Verlag, Berlin Heidelberg New York (2007) this volume

60. Pearl, J.: Fusion, propagation, and structuring in belief networks. Artificial Intelligence **29**(3) (1986) 241–288

61. Pearl, J.: Probabilistic Reasoning in Intelligent Systems-Second Edition. Morgan Kauffmann, Los Altos, CA (1988)

62. Pereira, R.A.M., Molinari, A., Pasi, G.: Contextual weighted representations and indexing ieee computer society models for the retrieval of HTML documents. Soft Computing **9**(7) (2005) 481–492
63. Piwowarski, B., Gallinari, P.: A bayesian network for XML information retrieval: Searching and learning with the INEX collection. Information Retrieval **8**(4) (December 2005) 655–681
64. Piwowarski, B., Vu, T., Gallinari, P.: Bayesian networks for structured information retrieval. In: Learning Methods for Text Understanding and Mining, Grenoble, France (January 26–29 2004)
65. Porter, M.F.: An algorithm for suffix stripping. Program **14**(3) (1980) 130–137
66. Quillian, M. In: Semantic Memory. MIT Press (1968)
67. Rijsbergen, C.J.V.: Information Retrieval. Second edn. Department of Computer Science, University of Glasgow (1979)
68. Robertson, S.E., Jones, K.S.: Relevance weighting of search terms. Journal of the American Society for Information Science **27** (1976) 129–146
69. Robertson, S.E., Walker, S.: Okapi/keenbow at TREC-8. In: Text REtrieval Conference (TREC) TREC-8 Proceedings, Department of Commerce, National Institute of Standards and Technology (1999) 151–162 NIST Special Publication 500-246: The Eighth Text REtrieval Conference (TREC 8).
70. Robertson, S.E., Walker, S., Hancock-Beaulieu, M.: Experimentation as a way of life: Okapi at TREC. Information Processing and Management **36**(1) (2000) 95–108
71. Rumelhart, D.E., McClelland, J.L.: Parallel Distributed Processing, Volume 1:Foundations, (ed. w/ PDP Research Group). MIT Press Cambridge, Massachusett (1986)
72. Russel, S., Norvig, P.: Artificial Intelligence: a modern approach. Prentice Hall International (1998)
73. Salem, A.B.M., Syiam, M.M., Ayad, A.F.: Unsupervised artificial neural networks for clustering of document collections. Egyptian Computer Science Journal **26**(1) (2004)
74. Salton, G.: The Smart Retrieval System. Experiments in Automatic Document Processing. First edn. Prentice Hall, Englewood Cliffs (1971)
75. Salton, G., Buckley, C.: Term-weighting approaches in automatic text retrieval. Information Processing and Management **24**(5) (1988) 513–523
76. Salton, G., McGill, M.J.: An Introduction to modern information retrieval. Mc-Graw Hill (1983)
77. Sebastiani, F.: Machine learning in automated text categorization. ACM Computing Surveys **34**(1) (2002) 1–47
78. Segal, R.B., Kephart, J.O.: MailCat: an intelligent assistant for organizing e-mail. In Etzioni, O., Müller, J.P., Bradshaw, J.M. (eds.): Proceedings of the Third International Conference on Autonomous Agents (Agents'99), Seattle, WA, USA, ACM Press (1999) 276–282
79. Shastri, L.: Why semantic networks. In Sowa, J.F., ed.: Principles of Semantic Networks: Explorations in the Representation of Knowledge. Morgan Kaufmann, San Mateo, CA (1991) 108–136
80. Thomas, S.: HTTP Essentials: Protocols for Secure, Scalable Web Sites. Wiley (2001)
81. Tsikrika, T., Lalmas, M.: Combining evidence for web retrieval using the inference network model: an experimental study. Information Processing & Management **40**(5) (2004) 751–772
82. Turtle, H.R., Croft, W.B.: Evaluation of an inference network-based retrieval model. ACM Transactions On Information Systems **9**(3) (1991) 187–222
83. Vlajic, N., Card, H.C.: An adaptive neural network approach to hypertext clustering. In: IEEE International Conference on Neural Networks (IJCNN'99). Volume VI., Washington DC, IEEE (July 1999) 3722–3726
84. Walker, S.: The Okapi online catalogue research projects. In Hildreth, C., ed.: The online catalogue. Research and directions. Library Association, London, UK (1989) 84–106

85. Wilkinson, R., Hingston, P.: Using the cosine measure in a neural network for document retrieval. In Bookstein, A.; Chiaramella, Y.; Salton, G.; Raghavan, V.V. (eds.): Proceedings of the 14th Annual International ACM/SIGIR Conference on Research and Development in Information Retrieval, Chicago, Ill., USA, ACM Press (October 1991) 202–210
86. Yang, C.C., Chen, H., Hong, K.: Visualization of large category map for Internet browsing. Decision Support Systems 35(1) (2003) 89–102
87. Yang, K.: Combining text and link-based retrieval methods for web IR. In Voorhees, E., Harman, D. (eds.): The Ninth Text REtrieval Conference (TREC 9) (2001) 609–618
88. Yang, K., Albertson, D.E.: WIDIT in TREC-2003 web track. In: Text REtrieval Conference (TREC) TREC 2003 Proceedings. (2003) 328–336
89. Yang, K., Maglaughlin, K.L.: IRIS at TREC-8. In: Text REtrieval Conference (TREC) TREC-8 Proceedings, Department of Commerce, National Institute of Standards and Technology (1999) 645–656 NIST Special Publication 500-246: The Eighth Text REtrieval Conference (TREC 8).
90. Yao, Y., Zhong, N., Liu, J., Ohsuga, S.: Web intelligence (WI). Lecture Notes in Computer Science 2198 (2001) 1–17
91. Yao, Y., Zhong, N., Liu, J., Ohsuga, S.: Web intelligence: exploring structures, semantics, and knowledge of the web. Knowledge-Based Systems 17(5-6) (2004) 175–177
92. Zhong, N., Liu, J., Yao, Y.: In search of the wisdom web. IEEE Computer 35(11) (2002) 27–31
93. Zhong, N., Liu, J., Yao, Y.: A New Paradigm for Developing the Wisdom Web and Social Network Intelligence. In: Web Intelligence. Springer-Verlag, Berlin Heidelberg (2003) 1–15

Part II

Adaptation Technologies

6

Personalized Search on the World Wide Web

Alessandro Micarelli[1], Fabio Gasparetti[1],
Filippo Sciarrone[1], and Susan Gauch[2]

[1] Department of Computer Science and Automation
Artificial Intelligence Laboratory
Roma Tre University
Via della Vasca Navale, 79 - 00146 Rome, Italy
{micarel, gaspare, sciarro}@dia.uniroma3.it
[2] Information & Telecommunication Technology Center
University of Kansas
2335 Irving Hill Road, Lawrence Kansas 66045-7612
sgauch@ittc.ku.edu

Abstract. With the exponential growth of the available information on the World Wide Web, a traditional search engine, even if based on sophisticated document indexing algorithms, has difficulty meeting efficiency and effectiveness performance demanded by users searching for relevant information. Users surfing the Web in search of resources to satisfy their information needs have less and less time and patience to formulate queries, wait for the results and sift through them. Consequently, it is vital in many applications - for example in an e-commerce Web site or in a scientific one - for the search system to find the right information very quickly. Personalized Web environments that build models of short-term and long-term user needs based on user actions, browsed documents or past queries are playing an increasingly crucial role: they form a winning combination, able to satisfy the user better than unpersonalized search engines based on traditional Information Retrieval (IR) techniques. Several important user personalization approaches and techniques developed for the Web search domain are illustrated in this chapter, along with examples of real systems currently being used on the Internet.

6.1 Introduction

Recently, several search tools for the Web have been developed to tackle the information overload problem, that is, the over-abundance of resources that prevent the user from retrieving information solely by navigating through the hypertextual space. Some make use of effective personalization, adapting the results according to each user's information needs. This contrasts with traditional search engines that return the same result list for the same query, regardless of who submitted the query, in spite of the fact that different users usually have different needs. In order to incorporate personalization into full-scale Web search tools, we must study the behavior of the users as they interact with information sources.

P. Brusilovsky, A. Kobsa, and W. Nejdl (Eds.): The Adaptive Web, LNCS 4321, pp. 195–230, 2007.

There are three information access paradigms that users undertake each time they need to meet particular information needs on the Web hypertextual environment: *searching by surfing* (or *browsing*), *searching by query* and *recommendation*. Recommendation-based systems suggest items, such as movies, music or products, analyzing what the users with similar tastes have chosen in the past [67, 58], see Chapter 12 of this book [12] for details.

In searching by surfing, users analyze Web pages one at a time, surfing through them sequentially, following hyperlinks. This is a useful approach to reading and exploring the contents of a hypertext, but it is not suitable for locating a specific piece of information. Even the most detailed and organized catalogs of Web sites, such as YAHOO! DIRECTORY[1] and the OPEN DIRECTORY PROJECT[2], do not always allow users to quickly locate the pages of interest. The larger the hypertextual environment is, the more difficulty a user will have finding what he is looking for.

The other dominant information access paradigm involves querying a search engine, an effective approach that directly retrieves documents from an index of millions of documents in a fraction of a second. This approach is based on an classic Information Retrieval (IR) model [71] wherein documents and information needs are processed and converted into ad-hoc representations. These representations are then used as the inputs to some similarity function that produces the document result list. Further details about this basic approach can be found in Chapter 5 [55] and 2 [29] [55] of this book.

Information Retrieval has always been characterized by relatively stable information sources and sequences of possibly unrelated user queries. It is usually considered distinct from the Information Filtering (IF) process [59], where the user needs are stable and there are large volumes of dynamically generated collections of documents. The user's interests in IF change relatively slowly with respect to the rate at which information sources become available. The Web is a highly dynamic environment, with information constantly being added, updated and removed, therefore IF prototypes seem to be the most appropriate choice on which to build Web search systems. Nevertheless, IF mostly employs complex representations of user needs and the time needed to perform the retrieval process, that is, matching the incoming stream of information with the model of user's interests, is quite long. This slow response is one of the reasons why IF prototypes have not become a widespread tool to retrieve information from the Web. For a closer examination of the most important user modeling techniques developed for IF, see Chapter 2 of this book [29].

In the last few years, attention has focused on the adaptation of traditional IR system to the Web environment, and related implementations of personalization techniques. The former task is accomplished by periodically collecting newly-created documents through re-crawling, keeping the search system's internal document index updated. This chapter discusses the second topic, personalization techniques and their implementation in real systems.

The two paradigms, searching by query and browsing, coexist: most of the times, browsing is useful when the user does not know beforehand the search domain keywords. Often, the user actually learns appropriate query vocabulary while surfing. Be-

---

[1] http://dir.yahoo.com
[2] http://dmoz.org

cause searching by query allows users to quickly identify pages containing specific information, it is the most popular way that users begin seeking information [35, 74], making the relevance of this paradigm paramount. For this reason, sophisticated search techniques are required, enabling search engines to operate more accurately for the specific user, abandoning the "one-size-fits-all" method. *Personalized search* aims to build systems that provide individualized collections of pages to the user, based on some form of model representing their needs and the context of their activities. Depending on the searcher, one topic will be more relevant than others. Given a particular need, e.g., a query, the results are tailored to the preferences, tastes, backgrounds and knowledge of the user who expressed it.

In spite of the fact that search engines are the principal tool by which users locate information on the Web, only a few search engines provide tools that adapt to user interaction. Moreover, users often judge these tools as not easy to personalize. In particular, the accessibility of these approaches is low since, as the personalization level increases, the users have more difficulty using these features [41]. There could be several reasons for this phenomenon. First, most personalization techniques are based on user profiles that incorporate information about the user, such as their information needs, interests, and preferences. Users may be uncomfortable with having their personal information stored on an external search system, see Chapter 21 [43]. Second, the personalization of Web search results is a computationally-intensive procedure. A typical search engine usually performs hundreds of queries per second and serves millions of users. Thus, the requirement to provide tailored results in a fraction of a second is not easy to accomplish. Finally, while users are familiar with the current search engines' interface, if the personalization is provided by some sort of new feature, users may find it difficult to understand and profitably use.

This chapter focuses on personalization approaches, techniques and systems developed for search activities, that is, when the user is actively looking for a particular piece of information on the Web. A strongly related topic is *Focused Crawling*, where the search is performed by specific information systems that autonomously crawl the Web collecting pages related to a given set of topics, reducing the network and computational resources. Chapter 7 of this book [53] provides a wide overview on this related topic, with a bias toward approaches which are able to dynamically adapt their behavior during the search according to the alterations of the environment or the given topics of interest.

The most common personalization approaches presented in literature are discussed in the next sections. Related techniques and prototypes are included for each discussed approach. The chapter is organized as follow: Sect. 6.2 provides a brief overview of the personalized search approaches, providing the reader with a broad description of the various methods and techniques proposed in the literature (some of which are fully treated in other chapters of the present volume, e.g., Chapters 2 [29], 9 [75] and 20 [36] of this book). Further details on the above-mentioned approaches are provided in the other sections.

The collection of implicit feedback from the current activity's context or search histories is reported in Sect. 6.3 and Sect. 6.4 respectively. Approaches in which complex and rich representations of user needs are built from user feedback are reported in Sect. 6.5. Section 6.6 discusses collaborative search approaches while personalized clustering of the results are summarized in Sect. 6.7. Section 6.8 explores how hyperlink-based algorithms can be used to adapt the search engine's result lists to the user needs. Hybrid approaches to personalization are discussed in Section 6.9 and, finally, conclusions are presented in Sect. 6.10.

6.2 A Short Overview on Personalized Search

After a brief introduction of the motivations and goals of the personalized search, it is interesting to examine the personalization approaches and tools proposed to achieve this goal.

We begin with a preliminary taxonomy based on content and collaborative-based distinction. We then move on to how user profiles are implemented in the personalized systems and the typical sources employed to recognize user needs. An overview of the different personalized search approaches, which are discussed in depth later on in this chapter, closes this section.

6.2.1 Content and Collaborative-Based Personalization

Many techniques on which search engines are based on originated from the IR field, e.g., Vector Space Model (VSM) [72, 70], mostly *content-based* techniques, wherein each user is assumed to operate independently. The content of documents is used to build a particular representation that is exploited by the system to suggest results to the user in response to ad-hoc queries (Chapter 5 [55] provides details on document representations). The searching-by-query paradigm is definitely quicker when the user is aware of the problem domain and knows the appropriate discerning words to type in the query [60]. However, analyzing search behavior, it is possible to see that many users are not able to accurately express their needs in exact query terms. The average query contains only 2 to 3 terms [50, 78].

Due to *polysemy*, the existence of multiple meanings for a single word, and *synonymy*, for the existence of multiple words with the same meaning, the keyword search approach suffers from the so-called *vocabulary problem* [27]. This phenomenon causes mismatches between the query space and the document space, because a few keywords are unlikely to select the right pages to retrieve from sets of billions [26]. Synonymy causes relevant information to be missed if the query does not contain the exact keywords occurring in the documents, inducing a recall reduction. Polysemy causes irrelevant documents to appear in the result lists, affecting negatively the system precision. For these reasons, users face a difficult battle when searching for the exact documents and products that match their needs. Understanding the meaning of Web content and, more importantly, how it relates to the real meaning of the user's query, is a crucial step in the retrieval process. Figure 6.1 shows the principal content-based personalization approaches, discussed later in this section.

When the algorithm used to build the result list also takes into account models of different users, the approach is usually named *collaborative* [32, 66]. The basic idea

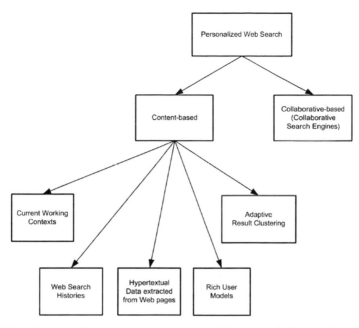

Fig. 6.1. Principal personalization approaches arranged by content/collaborative-based distinction.

behind collaborative-based approaches is that users with similar interests are likely to find the same resources interesting for similar information needs. *Social navigation* is the word coined by Dieberger *et al.* [21] to refer to software that allows people to leave useful traces on Web sites, such as reviews, comments, or votes, used by other people during browsing and searching-by-query.

Because most of the collaborative systems do not employ any search technology, this chapter does not cover them. These systems are discussed in Chapter 20 [36] and Chapter 9 [75] of this book. Two exceptions are EUREKSTER and I-SPY search engines, described in Sect. 6.6, which employ the collaborative or community-based approach to suggest pages that other users who submitted the same query selected frequently. Figure 6.1 shows a taxonomy of collaborative and content-based personalization approaches discussed later in this section.

6.2.2 User Modeling in Personalized Systems

Tracking what pages the user has chosen to visit and their submitted queries is a type of *user modeling* or *profiling* technique, from which important features of users are learned and then used to get more relevant information. Most of the personalized search systems discussed in this chapter employ a user modeling component that occurs during the information retrieval or filtering. Basically, this is the major component needed to provide tailored results that satisfy the particular needs of single users.

In the simplest cases, user models consist of a registration form or a questionnaire, with an explicit declaration of interest by the user. In more complex and extended cases, a user model consists of dynamic information structures that take into account background information, such as educational level and the familiarity with the area of interest, or how the user behaves over time. For example, the *ifWeb* prototype [6] makes use of user models based on semantic networks [64, 18] in order to create a representation of the available topics of interests. It supports users during Web surfing, acting as hypermedia search assistant (see Sect. 6.5.1 for details).

As an example of a very simple personalized search tool, GOOGLE's *Alerts* is an agent that automatically sends emails to the user each time new results for given query terms become available, both from the Web and News sites. GOOGLE's Alerts builds user models using an *explicit approach* where users explicitly construct the model by describing the information in which they are interested in. In this particular case, the user suggests a set of keywords, sometimes called *routing query*, which must appear in the retrieved documents, thus filtering the information stream. As soon as new information is published on the Web, the system evaluates it according to the stored profile, *alerting* the user of such new and potentially interesting contents. The obtained profiles are relatively simple and act as standard queries. Since the routing queries are suggested by users and the results are never adapted by the system to particular needs or tasks, the system's personalization is really limited.

A further tool named GOOGLE's *Personalized Search* used to deliver customized search results based on user profiles overcomes some of the Alert's problems. The results were instantly rearranged by dragging a series of sliders that define the personalization level concerning pre-defined sets of topics. Basically, while indexing, the engine categorizes pages collected from the Web according to a topic taxonomy. When users submit a query, the system looks through pages associated with their interests, that is, the selected topics, to find matches affecting the search results. Due to the kind of feedback employed to build the profiles, the user is still required to point the system to the information that is considered most interesting or, in some cases, suggesting data to be ignored in the future. For this reason this tool has been replaced with a new technology discussed in Sect. 6.4.

In personalized search systems the user modeling component can affect the search in three distinct phases, showed in Fig. 6.2:

- *part of retrieval process*: the ranking is a unified process wherein user profiles are employed to score Web contents.
- *re-ranking*: user profiles take part in a second step, after evaluating the corpus ranked via non-personalized scores.
- *query modification*: user profiles affect the submitted representation of the information needs, e.g., query, modifying or augmenting it.

The first technique is more likely to provide quick query response, because the traditional ranking system can be directly adapted to include personalization, avoiding repeated or superfluous computation. However, since the personalization process usually takes a long time compared with traditional non-personalized IR techniques, most search engines do not employ any personalization at all. Time constraints that force the system to provide result lists in less than a second cannot be met for all users.

On the other hand, re-ranking documents as suggested by an external system, such as a search engine, allows the user to selectively employ personalization approaches

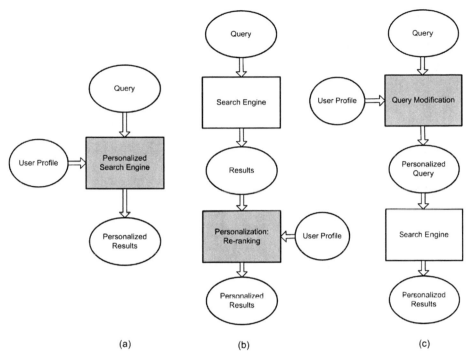

Fig. 6.2. Personalization process where the user profile occurs during the retrieval process *(a)*, in a distinct re-ranking activity *(b)* or in a pre-processing of the user query *(c)*.

able to increase precision. Many systems implement this approach on the client-side, e.g., [62, 54, 77], where the software connects to a search engine, retrieving query results that are then analyzed locally. In order to avoid spending time downloading each document that appears in the result list, the analysis is usually only applied to the top ranked resources in the list, or it considers only the snippets associated with each result returned by the search engine.

Because of the time needed to access a search engine and retrieve the pages to be evaluated, the re-ranking approach implemented via client-side software can be considerably slow. Nevertheless, complex representations of user needs can be employed, considerably improving the personalization performances (see Sect. 6.5).

Finally, profiles can modify the representations of the user needs before that retrieval takes place. For instance, if the user needs are represented by queries, the profile may transform them by adding or changing some keywords to better represent the needs in the current profile. Short queries can be augmented with additional words in order to reduce the vocabulary problem, namely, polysemy and synonymy, which often occur in this kind of keyword-based interaction. Alternatively, if the query retrieves a small number of resources, it is possible to expand it using words or phrases with a similar meaning or some other statistical relations to the set of relevant documents (see *query expansion* technique [7]). The major advantage of this approach is that the amount of work required to retrieve the results is the same as in the unpersonalized scenarios.

Nevertheless, user profiles affect the ranking only by altering the query representations. Unlike ranking that takes place in the retrieval process, the query modification approach is less likely to affect the result lists, because it does not have access to all the ranking process and its internal structures.

6.2.3 Sources of Personalization

The acquisition of user knowledge and preferences is one of the most important problems to be tackled in order to provide effective personalized assistance. Some approaches employ data mining techniques on browsing histories or search engine logs (see Chapter 3 [57]), while others use machine learning [87] to analyze *user data*, that is, information about personal characteristics of the user, in order to learn the knowledge needed to provide effective assistance. The user data usually differs from *usage data*. The latter are related to a user's behavior while interacting with the system. Examples of sources of user data are: personal data, e.g., name, address, phone number, age, sex, education; or geographic data, e.g., city and country.

Techniques such as *relevance feedback* and *query expansion* introduced in the IR field [72, 3] can be employed in the personalization domain in order to update the profile created by users. Basically, to improve ranking quality, the system automatically expands the user query with certain words that bring relevant documents not literally matching the original query. These words are usually extracted from pages in a previously retrieved list of ranked documents that have been explicitly judged interesting by the user through relevance feedback.

Besides considering important synonyms of the original queries' keywords that are able to retrieve additional documents, expansion helps users to disambiguate queries. For example, if the user submits the query '*Jaguar*', the result list will include information on the animal, the car manufacturer, the operating system, etc.. Following relevance feedback on a subset of documents relating to the meaning of interest to the user, the query is updated with words that help the system filtering out the irrelevant pages. Using a lexicon, it is also possible to expand queries such as '*IR*' to '*information retrieval*', increasing the chance of retrieving useful pages.

Even though these techniques have been shown to improve retrieval performance, some studies have found that explicit relevance feedback is not able to considerably improve the user model especially if a good interface is not provided to manage the model and clearly represent the contained information [86]. Users are usually unwilling to spend extra effort to explicitly specify their needs or refine them by means of feedback [5], and they are often not able to use those techniques effectively [79, 85], or they find them confusing and unpredictable [44].

Moreover, studies show that users often start browsing from pages identified by less precise but more easily constructed queries, instead of spending time to fully specify their search goals [84]. Aside from requiring additional time during the seeking processes, the burden on the users is high and the benefits are not always clear (see for example [88]), therefore the effectiveness of explicit techniques may be limited.

Because users typically do not understand how the matching process works, the information they provide is likely to miss the best query keywords, i.e., the words that

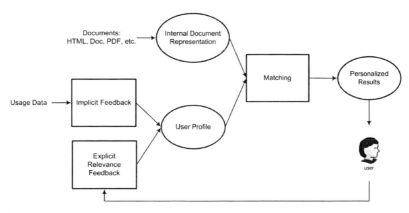

Fig. 6.3. Implicit and Explicit Feedback are used to learn and keep updated the profile of the user used during the personalization.

identify documents meeting their information needs. Moreover, part of the user's available time must be employed for subordinate tasks that do not coincide with their main goal. Instead of requiring user's needs to be explicitly specified by queries or manually updated by the user feedback, an alternative approach to personalize search results is to develop algorithms that infer those needs implicitly.

Basically, *implicit feedback* techniques unobtrusively draw usage data by tracking and monitoring user behavior without an explicit involvement (see Fig. 6.3). Personalized systems can collect usage data on the server-side, e.g., server access logs or query and browsing histories, and/or on the client-side, such as cookies and mouse/keyboard tracking. For a closer examination on implicit feedback techniques see for example [40, 17, 13] and Chapter 21 of this book [43] for the related privacy concerns.

For example, Bharat *et al.* [8] proposes monitoring some current user activity and implicitly building a user profile to provide *content personalization* in Web-based newspaper domain. They suggest that events such as scrolling or selecting a particular article reflect the user's current interest in the given topic. Each event adds a score to the current article, and if it exceeds a certain threshold, the global score increases and the change is reflected in the user profile. Basically, a subset of keywords are extracted from the article and included in the profile with a certain weight that will be updated if the same keyword appears in other articles browsed by the user. When a new article which includes some of the keywords contained in the profile appears, it gets a high score and is included in the personalized newspaper.

The system is somewhat unique in that it allows the user to control the system behavior by controlling the amount of personalization they wish be applied to their results. Sometimes, the user just wants to have an overview on general news, which may require a low-level of personalization since the personalization feature usually filters out information that is judged unrelated to the topics recognized in previous articles read by the user. For this reason, a control bar reduces the effect of personalization, allowing suggestions of popular articles unrelated with the past interests.

Table 6.1. Several important types of personalized search arranged by the type of feedback, implicit or explicit, used to learn the user profile and keep it updated, and the typical data related to the user given as input to the algorithms, such as, resources selected by the user from the results of a search engine, or subsets of pages browsed so far.

Sect.	Personalization based on:	Implicit/Explicit	Typical Input Data
6.3	Current Context	implicit	Word docs, emails, browser's Web pages...
6.4	Search History	implicit	past queries and browsed pages, selected results
6.5	Rich User Models	both	user feedback on results, past queries...
6.6	Collaborative approaches	both	past queries, selected results, user ratings
6.7	Result Clustering	explicit	selected clusters in taxonomies
6.8	Hypertextual Data	both	queries, selected pages...

As a further source of personalization, several desktop search systems, e.g., Copernic, Google Desktop Search, Mac Tiger, Windows Desktop Search, Yahoo! Desktop Search or X1, and several Search Toolbars provide simple access to indexes of information created, copied, or viewed by users. Microsoft's Stuff I've Seen (SIS) project [24] and the associated personalization technique [83, 85] provide personalization by exploiting this type of information. SIS does not involve the retrieval of new information, rather the re-use of what has been previously seen, providing a search interface based on an index of all personal information, such as emails, Web pages, documents, appointments, etc.. The ability to quickly retrieve such data has been proven to be very useful for the user. Essentially, the personalization technique re-ranks the search engine results as a function of a simplified user model based on the keywords occurring in the documents that the user has seen before. This kind of approach is able to use implicit feedback to build and update the user profile which can be used to disambiguate queries. Some of the advantages of this approach to personalization, which is also used by several Web-based personalized systems, is described further on in this chapter.

6.2.4 An Overview on Personalization Approaches

Personalized Search on the Web is a recent research area with a variety of approaches, sometimes tough to arrange in a framework where it is able to identify basic principles and techniques. A possible organization is shown in Table 6.1, where the personalization approaches are arranged by the type of feedback used to build user profiles, and the typical data related to the user given as input during the profiling. Obviously, it is possible to develop systems where more than one search approach is properly combined and implemented (see Sect. 6.9).

The first two approaches, discussed in Sect. 6.3 and Sect 6.4, are based on implicit feedback techniques, where users do not have to explicitly state their preferences or needs. Client-side software captures user interaction and information, such as browsed pages, edited documents or emails stored on hard drives, which are impossible for the search engine itself to collect and analyze. These pieces of data are very useful to understand the user's current working context, that can in turn be employed for query refinement or as an implicit source of evidence on the user's interests. Personalization based on the *Current Context* exploits this information to recognize the current user needs, which are used to retrieve documents related to the user activities.

If the personalization is limited to the Web *Search History*, we distinguish the related personalized systems from the previous category. The reason is that search engines are able to access this information for each user, with no client-side software requirements. User query histories, resources returned in the search results, documents selected by the user, plus information such as anchor text, topics related to a given Web page, or data such as click through rate, browsing pattern and number of page visits, are easily collected and mined server-side. Moreover, the personalization process can be done during the traditional retrieval process, obtaining a faster response than a distinct post-ranking activity.

Nevertheless, usage data are sometimes not available or they contain too much noise to be successfully exploited by implicit feedback techniques. In that case, explicit user feedback may be the only viable way to learn the user profile and keep it updated.

Most of the time, explicit feedback corresponds to a preferential vote assigned to a subset of the retrieved results. This kind of technique, called relevance feedback is really helpful whenever the user is not able to correctly specify a query, because he can submit a vague query and then analyze the query's results and select the documents that are mostly related to what he is searching. In spite of the negative features of explicit relevance feedback previously discussed, the information collected usually allows the system to build *Rich Representations of User Needs*, composed of more than just Boolean sets or bag-of-words models. Examples of this approach are described in Sect. 6.4.

In environments where a large amount of low-quality items are present, such as the Web, the concept of social filtering is that users help each other to distinguish between high and low quality items by providing ratings for items they have analyzed. All the ratings are collected and can then be used by other users to find the best-rated items.

Delivering relevant resources based on previous ratings by users with similar tastes and preferences is a form of personalized recommendation that can also be applied in the Web search domain, following a *Collaborative approach*. Moreover, since the filtering does not depend on the content of the objects, social filtering is able to provide recommendation for objects such as movies and music, that are usually hard to represent and manage in information systems. Section. 6.6 introduces a few of these collaborative-based systems.

In many cases, search engines retrieve hundreds or thousands of links to Web sites in response to a single query. Although the user may find the material he is looking for in the result list, or at least find Web pages from which the browsing process may begin, the sheer vastness of the results list can make sifting through the retrieved information an impossible task.

One idea to help the users during their search is to group the query results into several *clusters*, each one containing all the pages related to the same topic. In this way, an overview of the retrieved document set is available to the user and interesting documents can be found more easily. Typically, the pages might be clustered either into an exhaustive partition or into a hierarchical tree structure. The clusters are matched against the given query, and the best ones are returned as a result, possibly sorted by score. A retrieval system that organizes the results into clusters can be considered personalized because the user is able to customize the set of shown results navigating through the clusters driven by their search needs. This kind of *Adaptive Result Clustering* is shortly investigated in Sect. 6.7, while Chapter 13 of this book [11] provides an extensive dissertation on the Adaptive content presentation.

In the same scenario of queries that retrieve a large number of documents, following a given content-based matching function, search engines might assign the same ranks to several resources that share similar content. This is why some search engines include additional factors in their ranking algorithm on top of the query/document content similarity used by traditional information retrieval techniques. These factors may take into account the popularity and/or the authority of a page, in order to assign more accurate scores to the resources during the ranking process. Usually those measures are based on the Web hypertextual data, which is possible to collect by analyzing the set of links between pages. For a closer examination of these measures, such as the PageRank or HITS's authority and hubness, see Chapter 5 of this book [55].

Both the ranking techniques, the traditional IR's and the hyperlink-based algorithms, compute rank values based on page content as a single and global value for each Web page, ignoring any form of personalization based on the user's preferences regarding the quality for an individual page. Recent work aims to extend hyperlink-based algorithms by considering different notions of importance for different users, queries and domains. In other words, the idea is to create personalized views of the Web by redefining the importance assigned by the hypertextual algorithms according to the implicitly expressed user preferences, for example, through previously submitted queries, or explicitly, via a subset of bookmarks or categories in a given taxonomy. Obviously, the query results that match the user-selected topics will be ranked higher by the search engine, providing tailored output for each user. Section 6.8 introduces this personalization based on *Hypertextual data*.

6.3 Contextual Search

Rhodes [68] proposes a new approach for the search named Just-in-Time IR (JITIR) where the information system proactively suggests information based on a person's working context. Basically, the system continuously monitors the user's interaction with the software, such as typing in a word processor or surfing with Internet browsers, in a non-intrusive manner, automatically identifying their information needs and retrieving useful documents without requiring any action by the user. The retrieval process can exploit a variety of data sources, i.e., any number of pre-indexed databases of documents, such as e-mails or commercial databases of articles.

The JITIR approach combines the alerting approach of Google Alert, briefly described in Sect. 6.2.2, with personalization based on the events inside the user's local working context. Alerting pushes information related to predefined sets of topics toward the user regardless of his current activity, usually requiring a sudden change of user attention. By means of a dynamic user profile kept updated according to changes of the local working contexts, JITIR provides the information tailored to the current user activity.

Describing the JITIR approach, the author suggests three different implementations based on agents. The *Remembrance Agent* presents a list of documents that are related to what the user is typing or reading. *Margin Notes* follows an adaptive hypermedia approach, automatically rewriting Web pages as they are loaded, adding hyperlinks to related local files. The third agent, *Jimminy*, provides information related to the user's physical environment, e.g., spatial location, time of day, subject of conversation, etc., by means of a wearable computer that includes different ways to sense the outside world.

Each of the agents in the JITIR approach share the same back-end system, called *Savant*. It consists of a client-resident search engine that is queried by the agents as the user interacts with the system. The search engine index usually stores public corpora as newspapers or journal articles, and/or personal sources such as e-mail and notes. In order to extract the data needed to build the index, Savant is able to recognize, parse, and index a variety of document formats. During retrieval, the fields extracted from the current document and the ones from the stored documents are compared sequentially, and an overall similarity score is calculated using a linear combination of those similarities. Kulyukin's *MetaCenter* [47] shares many features of the previous prototype, performing automatic query generation according to the current resources the user is working on, e.g., browsed pages or Word documents. The queries are submitted to search engines that operate on online collections of documents to which the user subscribes.

A further instance of the JITIR approach is *Watson* [9, 10]. It monitors the user's actions and the files that he is currently working on to predict the user's needs and offer them related resources. The Watson agent works in a separate window and can track the user across different applications, such as Internet Explorer and Mozilla browsers, and the Microsoft Office suite. As the user's work goes on, Watson looks for related information, following a different context for every open window it is tracking. Relying on the contextual information learned by the agent from the current active window, it generates its own queries to several sources of information and presents them after a result aggregation process. It is also possible to submit explicit queries, which are added to the contextual query while the result post-processing takes place to aggregate results from the different sources.

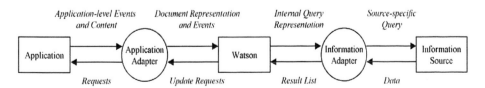

Fig. 6.4. Watson monitors the user activity and sends ad-hoc queries to specific information sources.

According to the authors, Watson uses several sources of information, such as the search engines ALTAVISTA, YAHOO! and DOGPILE, news sources such as Reuters and the New York Times, Blog sites and e-commerce sites. *Application adapters* are used to gain access to internal representations and events generated by user interactions with a specific application, as shown in Fig. 6.4. For example, if the user edits a Word document, the keyboard events trigger Watson to request an updated representation of it. This representation is translated into a query to be submitted to an appropriate source by means of an *Information adapter*. The current task affects the choice of the information source to query, e.g., if the user is editing a medical document, Watson might choose a specialized search engine on this topic.

The user can access the top ranked results from all the relevant sources or filter them by resource type, i.e., Web, news, etc., a capability provided by recent local search systems, such as Apple Tiger's Spotlight and Google Desktop Search. The *TFxIDF* technique [73] is used to create the contextual query based on the currently active window (see Chapter 5 [55] for further details). The bag of words representation in which any document is treated as a set of words regardless of the relations that may exist among them, enhanced with additional formatting information, is used to create a list of term-weight pairs. The top 20 weighted pairs are sorted in their original order of appearance in the document and are used to create the query that is submitted to the information sources. Many heuristics have been considered in order to increase the performance, such as removing *stop words* (common words, such as I, the, when, etc.)and giving more weight to the words that appear at the beginning of documents and those that are emphasized via specialized formatting.

6.4 Personalization Based on Search Histories

User queries are undoubtedly an important source in recognizing the information needs and personalizing the human-computer interaction. A search engine is able to access and process all this information in a non-invasive way, i.e., without installing external proxy servers or client desktop bots, therefore it can tailor the query results based on the previous requests and interests [49]. Simple log-in forms and cookies can be employed in order to identify the user and the related click streams data instead of complex heuristics based on IPs, last access times or user agents data, which cannot be considered entirely accurate [61].

As already noticed, if the user submits a short query, such as *Visa*, it is not clear if he is looking for the credit card company, the policy and procedures to travel to foreign countries, the procedures to change the citizenship, a last name, etc.. The browsing/query history could be a way to weight the different alternatives for example. If the user has recently searched for a flight to a foreign country, a Visa query is more likely to be related to bureaucratic procedures.

Approaches based on search history can be organized in two groups. Offline approaches exploit history information in a distinct pre-processing step, usually analyzing relationships between queries and documents visited by users. Online approaches capture these data as soon as they are available, affecting user models and providing personalized results taking into consideration the last interactions of the user. Even though the latter approaches provide updated suggestions, an offline approach can implement more complex algorithms because there are usually less urgent time constraints than in an online one.

6.4.1 Online Approaches

Following the first personalization attempt briefly described in Sect. 6.2, Google Labs released an enhanced version of Personalized Search that builds the user profile by means of implicit feedback techniques. In particular, the system records a trail of all queries and the Web sites the user has selected from the results, as shown in Fig. 6.5, building an internal representation of his needs. During the search process, the search engine adapts the results according to needs of each user, assigning a higher score to the resources related to what the user has seen in the past. Unfortunately, no details or evaluations are available on the algorithms exploited for that re-ranking process at present except that contained in the patent application filed in 2004 [92]. Nevertheless, the developers claim they can produce more relevant search results based on what the system learns from the search history, especially when the history contains enough data to be analyzed.

Raghavan and Sever [65] use a database of past queries that is matched with the current user query. If a significant similarity with a past query is found, the past results associated with the query are proposed to the user. The research focuses on the similarity measure used to calculate the query-to-query similarity. This cannot be based on traditional word-to-word IR matching functions, such as the cosine measure, because the short nature of queries makes them particularly susceptible to the vocabulary problems of polysemy and synonymy.

Speretta and Gauch developed the *misearch* system [77], which improves search accuracy by creating user profiles from their query histories and/or examined search results. These profiles are used to re-rank the results returned by an external search service by giving more importance to the documents related to topics contained in their user profile.

In their approach, user profiles are represented as weighted concept hierarchies. The OPEN DIRECTORY PROJECT (ODP) is used as the reference concept hierarchy for the profiles. GOOGLE has been chosen as the search engine to personalize through a software wrapper that anonymously monitors all search activities. For each individual user, two different types of information are collected: the submitted queries for which at least one result was visited, and the *snippets*, i.e., titles and textual summaries, of the results selected by the user. Afterward, a classifier trained on the ODP's hierarchy, chooses the concepts most related to the collected information, assigning higher weights to them. In the current implementation, for comparison purposes, the query and the snippet data are kept distinct and therefore two different profiles are built.

After a query is submitted to the wrapper, the search result snippets are classified into the same reference concept hierarchy. A matching function calculates the degree of similarity between each of the concepts associated with result snippet j and the user profile i:

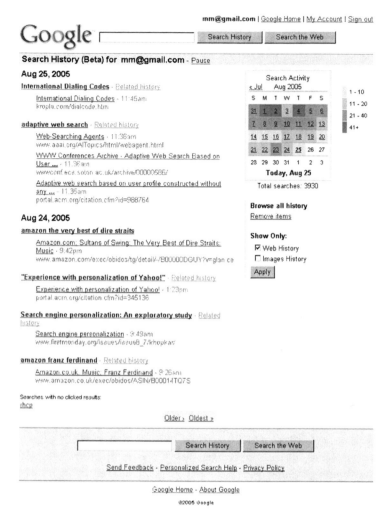

Fig. 6.5. The Search History feature of the Google Labs' Personalized Search records the history of searches and the search results on which the user has clicked. This information is exploited to personalize search results by ranking resources related to what the user has seen in the past higher. *(Reproduced with permission of Google)*

$$sim(user_i, doc_j) = \sum_{k=1}^{N} wp_{i,k} \cdot wd_{j,k} \tag{6.1}$$

where $wp_{i,k}$ is the weight of the concept k in the user profile i, $wd_{j,k}$ is the weight of the concept k in the document j, and N is the number of concepts.

The final weight of the document used for reordering - so that the results that best match the user's interests are ranked higher in the list - is calculated by combining the previous degree of similarity with GOOGLE's original rank, using the following weighting scheme:

$$match(user_i, doc_j) = \alpha \cdot sim(user_i, doc_j) + (1 - \alpha) \cdot googlerank(doc_j) \quad (6.2)$$

where α gets values between 0 and 1. When α is 0, conceptual rank is not given any weight, and the $match$ is equivalent to the original rank assigned by GOOGLE. If α has a value of 1, the search engine ranking is ignored and pure conceptual match is considered. Obviously, the conceptual and search engine-based rankings can be blended in different proportions by varying the value of α.

A thorough evaluation has been done in order to investigate the effectiveness of user profiles built out of queries and snippets. The accuracy of such profiles is analyzed comparing, for user-selected results, GOOGLE's original rank with the conceptual rank based on the profile. The evaluation employed 6 users. Using a profile built from 30 queries, the performance measured in terms of the rank of the user-selected result improves of 33%. A user profile built from snippets of 30 user-selected results showed an improvement of 34% (see [77] for details). Therefore, it is possible to assert that, even though the text a user submits to the search engine is quite short, it is enough to provide more accurate, personalized results.

The ability to recognize user interests in a completely non-invasive way, without installing software or using proxy servers, and the accuracy obtained from the personalized results, are some of the major advantages of this approach. Moreover, result-ordering does not exclusively depend on a global relevance measure, where the computed rank for the whole population is deemed relevant for each individual, but it is tailored to a personal relevance where the rank is computed according with each user within the context of their interactions.

Liu and Yu [51] take a similar approach to personalization, where user profiles are built by analyzing the search history, both queries and selected result documents, comparing them to the first 3 levels of the OPEN DIRECTORY PROJECT category hierarchy. Basically, for each query, the most appropriate categories are deduced and used along with the query as current query context. Because queries are short, they are often ambiguous, so they are likely to match multiple categories in the ODP. The system can automatically use the top-matching category for query expansion, or the user can reduce the ambiguity, by explicitly choosing one of the three top-ranked categories provided by the categorization algorithm.

Each category in a user profile is represented with a weighted term vector, where a highly-weighted keyword indicates a high degree of association between that keyword and the category. The system updates the user profile after a query, when the user clicks on a document, and there is a reasonable duration before the next click, or the user decides to save or print it.

Koutrika and Ioannidis [45] proposed an online approach where user needs are represented by a combination of terms connected through logical operators, e.g., conjunction, disjunction, negation, substitution. These operators are used to transform the queries in personalized versions to be submitted to the search engines. The content of the documents for which the user has performed explicit feedback is used to build the user profile. An evaluation shows that when this personalization approach is applied, the users satisfy their needs faster compared with a traditional search engine, improving the number of relevant documents found among the top results.

Quickstep system [56] follows a quasi-online approach and shares some features with the previous systems. A proxy server monitors browsed research papers and a nearest neighbor classifier assigns OPD categories to them overnight. Sets of recommendations based on the correlations between the user profile and research paper topics are drawn on a daily basis. The user can provide feedback in the form of new training examples or adjustments in the classification outcomes. The user profile consists of a set of topics and the related items, computed following the number of browsed research papers about the given topic, while the Vector Space Model is employed to represent the documents.

6.4.2 Offline Approaches

An innovative personalized search algorithm is the *CubeSVD* algorithm, introduced by Sun *et al.* [80] based on the *click-through data* analysis [38]. This technique is suitable for the typical scenario of Web searching, where the user submits a query to the search engine, the search engine returns a ranked list of the retrieved Web pages, and finally the user clicks on pages of interest. After a period of usage, the system will have recorded useful click-through data represented as triples: c·

$$\langle user,\ query,\ visited\ page \rangle$$

that could be assumed to reflect users' interests. The proposed algorithm aims to model the users' information needs by exploiting such data. It addresses two typical challenges of Web search. The first concerns the study of the complex relationship between user, the query, and the visited Web pages: given a user and her/his query, how to recommend the right Web page to visit? The authors propose a framework for capturing the latent associations among the aforesaid objects.

The second challenge faces the problem of data sparseness: a user generally submits a small number of queries compared with the size of the query set submitted by all the users, and visits few pages. In this case, recognizing relationships among the data becomes a hard task to carry out.

The authors develop a unified framework to model a click-through element as a 3-order tensor, that is, a higher order generalization of a vector (first order tensor) and a matrix (second order tensor), on which 3-mode analysis is performed using the *Singular Value Decomposition* (SVD) technique [33], generalized to HOSVD, Higher-Order SVD [48]. The tensor element measures the preference of a $\langle user, query \rangle$ pair on a given Web page.

Indeed, the CubeSVD algorithm takes the click-though data set as input and outputs a *reconstructed* tensor $\hat{A}$. The tensor measures the degree of relationship among users, queries, and Web pages. The output is represented by a quadruple of the type:

$$\langle user,\ query,\ visited\ page,\ w \rangle$$

representing w the probability that the *user*, after having submitted a given *query*, would be interested in visiting a particular *page*. In this way, relevant Web pages can be recommended to the user by the system. Users are not consulted on the relevance of the visited Web pages during the search process, and the system records and analyzes their clicks as in other implicit feedback based approaches.

An evaluation on a 44.7 million record click-through data set showed that CubeSVD, thanks to high order associations identified by the algorithm, achieves better accuracy compared with collaborative filtering and LSI-based approaches [20]. Although the whole computation is remarkably time-consuming, it is part of an offline process that does not affect the runtime activity. Nevertheless, the algorithm has to be periodically run in order to take into consideration new click-through data.

Further offline approaches exploiting data mining techniques are discussed in Chapter 3 of this book [57].

6.5 Personalization Based on Rich Representations of User Needs

This section presents three prototypes of personalized search systems based on complex representations of user needs constructed using explicit feedback: *ifWeb*, *Wifs* and *InfoWeb*. They are mostly based on frames and semantic networks, two AI structures developed in order to represent concepts in a given domain, and the related relationships between them. Even though these prototypes share some features, the mechanisms employed to build the profiles and the way the needs are represented are fairly different. Therefore, we prefer to discuss them in distinct sections. Complex user modeling techniques applied to the Web personalization are exhaustively discussed in Chapter 2 of this book [29].

6.5.1 ifWeb

ifWeb [6] is a user model-based intelligent agent capable of supporting the user in Web navigation, retrieval, and filtering of documents taking into account specific information needs expressed by the user with keywords, free-text descriptions, and Web document examples. The *ifWeb* system exploits semantic networks in order to create the user profile.

More specifically, the user profile is represented as a weighted semantic network whose nodes correspond to terms (concepts) found in documents and textual descriptions given by the user as positive or negative examples, i.e., relevance feedback. Network's arcs link pairs of terms that co-occurred in some document. The use of the semantic network and of the co-occurrence relationships allows *ifWeb* to overcome the limitations of simple keyword matching, particularly polysemy.

The *ifWeb* prototype also performs autonomous focused crawling (see Chapter 7 [53] for details), collecting and classifying interesting documents. From specific documents pointed out by the user or identified through search engines, the system autonomously performs an extended opportunistic navigation of the Web, then retrieves and classifies documents relevant to the user profile. As a result, the system shows the user the documents that have been classified as the most relevant ones, in decreasing order of probable interest.

The user profile is updated and refined by explicit relevance feedback provided by the user: *ifWeb* presents a collection of documents to the user (usually no more than ten for each feedback session), who then explicitly selects the ones that meet his needs. Then, *ifWeb* autonomously extracts the information necessary to update the user profile from the documents on which the user expressed some positive feedback. Moreover, the prototype includes a mechanism for temporal decay called *rent*, which lowers the weights associated with concepts in the profile that have not been reinforced by the relevance feedback mechanism for a long period of time. This technique allows the profile to be kept updated so that it always represents the current interests of the user.

6.5.2 Wifs

The *Wifs* system described in [54] is capable of filtering HTML or text documents retrieved by the search engine ALTAVISTA[3] in response to a query input by the user. This system evaluates and reorders page links returned by the search engine, taking into account the user model of the user who typed in the query. The user can provide feedback on the viewed documents, and the system uses that feedback to update the user model accordingly.

In short, the user model consists of a frame whose *slots* contain terms (topics), each one associated with other terms (co-keywords) which form a simple semantic network. Slot terms, that is, the topics, must be selected from those contained in a *Terms Data Base* (TDB), created *a priori* by experts who select the terms deemed most relevant for the pertinent domain. Figure 6.6a illustrates a simplified description of a hypothetical user model.

The filtering system is based on a content-based approach, where the documents retrieved by ALTAVISTA are assessed solely according to their contents. The document modeling is not based on traditional IR techniques, such as the Vector Space Model, due to the high variability of Web information sources.

The abstract representation of the document may be seen as described in Fig. 6.6b, where active terms, or *planets*, $T_1, T_2, ..., T_n$ are the ones contained both in the document and TDB, whereas the *satellite* terms $t_1, t_2, ..., t_m$ are the terms included in the document, but not in the TDB, but which co-occur with T_i's. It is evident that the structure is similar to the user model one, but there are no affinity values between the planets and the satellites. For each of these terms, however, document occurrence is calculated. The occurrence value of a term t appearing in a retrieved document is given by the following formula:

$$Occ(t) = c_1 * freq_{body}(t) + c_2 * freq_{title}(t) \tag{6.3}$$

where $freq_{body}(t)$ is the frequency with which term t appears in the body, while $freq_{title}(t)$ is the frequency with which term t appears in the document title, and c_1 and c_2 are two constants.

For the document evaluation, the $\overrightarrow{Rel}$ vector is built, where the element Rel_i represents a relevant value of term t_i compared to user information needs. The user model, the query, and the TDB are taken into account to draw the relevance.

This calculation is done as follows:

---

[3] http://www.altavista.com

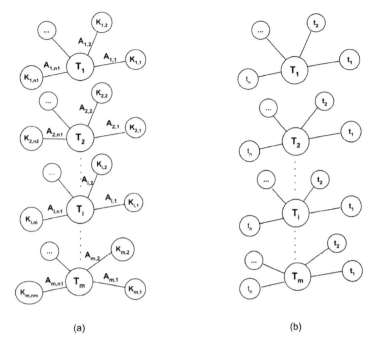

(a)   (b)

Fig. 6.6. Representations of the User model *(a)* and Document model *(b)*

- *Step 1*. The term t's relevance $Rel_{new}(t)$ (where the term t belongs to the document and user model, as a slot topic) is calculated by intensifying the old relevance value, $Rel_{old}(t)$, through the following formula:

$$Rel_{new}(t) = Rel_{old}(t) + c_3 * \sum_j w_j, \forall w_j : t \in slot_j \qquad (6.4)$$

where c_3 is a constant whose value is 2, calibrated experimentally, and w_j the weight associated to slot j containing term t as a topic. In a few words, the new relevance value of term t is obtained from the old value plus the sum of all semantic network weights of the user model containing term t as topic.
- *Step 2*. If the term, as well as belonging to the user model and document, also belongs to the q query input by the user, then the term relevance value is further strengthened, through the following formula:

$$Rel_{new}(t) = Rel_{old}(t) * w_{slot} \qquad (6.5)$$

where w_{slot} is the weight associated with topic t. This way, query q, which represents the user's immediate needs, is used to effectively filter the result set to locate documents of interest.
- *Step 3*. If term t belongs to query q, to document d, and to the TDB, but is not included in the user model, then the only contribution to relevance is given by the following formula:

$$Rel_{new}(t) = Rel_{old}(t) + c_3 \tag{6.6}$$

- *Step 4*. If term t is a topic for the $slot_j$, then this step is considers the contributions given by co-keywords. This is where the true semantic network contributes: all the co-keywords K connected to topic t give a contribution, even if previously unknown to the system, i.e., not currently belonging to the user model, nor to the TDB, but only to the document.

$$Rel_{new}(t) = Rel_{old}(t) + w_j * \sum_i A_{j,i} \tag{6.7}$$

$\forall co - keyword\ L_i \in slot_j : K_i \in doc\ \forall slot_j : topic_j \in d$

In this stage, the system calculates the final relevance score assigned to the document as follows:

$$Score(Doc) = \overrightarrow{Occ} \cdot \overrightarrow{Rel} = \sum_{\forall t \in Doc} Occ(t) * Rel(t) \tag{6.8}$$

where $\overrightarrow{Occ}$ is the vector consisting of elements Occ_i, and $\overrightarrow{Rel}$ is the vector consisting of elements Rel_i, evaluated in the previous steps.

This system is capable of dynamically updating the user model upon receipt of relevance feedback on the viewed documents provided by the user. In addition, the system uses a *renting* mechanism to decrease the weights of the terms appearing in the model that do not receive positive feedback after a period of time. Further details on the user model updates in *Wifs* are described in Chapter 2 of this book [29].

The *Wifs* system has been evaluated to determine the effectiveness of the user profile in providing personalized reordering of the documents retrieved by ALTAVISTA. Considering the whole set of documents retrieved by the search engine following the query, three relevance sorting structures are taken into account based on results provided by ALTAVISTA, *Wifs* and the user. The metrics defined in [39, 90] have been employed, in the perfect ranking hypothesis, to measure the gaps between *user-AltaVista* and *user-system* sorting. By means of a non parametric test, it was shown that the two distributions are different, with the *user-system* variable giving lower values, which shows that the alternative hypothesis is real. Hence, the system sorts sets of documents in a more relevant way for user needs. The evaluation considered 15 working sessions (where for each session a query was submitted) and 24 users. The ordering of the first 30 results was considered. It shows that the system provides roughly a 34% improvement when compared to the search engine's non-personalized results (see [54] for details).

Another interesting experiment showed that the system is capable of responding quickly to the user's sudden interest changes, through the aforementioned dynamic update mechanism, activated by relevance feedback supplied directly by the user.

6.5.3 InfoWeb

A further approach to personalization is taken by *InfoWeb* [30], an interactive system developed for adaptive content-based retrieval of documents belonging to Web digital libraries. The distinctive characteristic of *InfoWeb* is its mechanism for the creation and management of a stereotype knowledge base, and its use for user modeling. A *stereotype* [69] contains the vector representation of the most significant document belonging to a specific category of users, initially defined by a domain expert. The system helps the domain expert build the stereotypes through a *k-means* clustering technique [52], which is applied to the whole document collection in an off-line phase. The clustering starts with specific documents as initial seeds, each one acting as a representative centroid for a class of users. *InfoWeb* uses the stereotypes exclusively for the construction of the initial user model. The user's profile evolves over time in accordance to the user's information needs, formulated through queries, using an explicit relevance feedback algorithm that allows the user to provide an assessment of the documents retrieved by the system.

The filtering system extends the traditional one based on the Vector Space model because it also takes into account the co-occurrences of terms in the computation of document relevance and involves user profiles to perform query expansion. The final document evaluation process involves the representation of the documents, of the user model, and of the expanded query. The results of the experiments are promising, both in terms of performance and in the ability to adapt to the user' shifting interests.

The *InfoWeb* prototype is specifically designed for digital libraries with an established document collection and the presence of a domain expert. Nevertheless, some of the proposed techniques, e.g., stereotypes and automatic query expansion, can be also adapted to vast and dynamic environments, such as the Web.

6.6 Collaborative Search Engines

The EUREKSTER[4] search engine includes a proprietary module named *SearchParty* based on collaborative filtering to help users find the best pages related to a given query. EUREKSTER implements social filtering by storing all the results selected by the users for each query submitted to the search engine. Those results will be shared among the community of users interested in the same topics.

In addition to the social filtering module, the EUREKSTER search engine stores all the queries submitted by a user and the resources on which he clicks. If a certain amount of time is spent on a particular resource, when the user re-submits the same query later on, the previously clicked pages are ranked higher in the result list. Thus, the user does not have to wade through a long list of search results again in order to find a previously selected page.

A social adaptive navigation system called *Knowledge Sea* [89] exploits both the traditional IR approach, where documents and queries are represented through the Vector Space Model, and social navigation based on past usage history and user annotations.

Users can search socially, referring to other users' behavior and opinions, by examining the color lightness and exploring icons next to each result, which respectively provide users with information about the popularity of the page and allow the user to view any available annotations. For example, a dark background means that a document

---

[4] http://www.eurekster.com

Fig. 6.7. After having selected a particular topic, such as "Personalization of the Internet", EU-REKSTER is able to suggest results that other people have previously found useful. In this example, the first two results are proposed by social filtering. *(Reproduced with permission of Eurekster.)*

is popular or it has many annotations, while a light foreground color suggests that the users have chosen to view the document less frequently than most. Even though the results of the search are not socially re-ranked, every result is annotated with social visual cue according with the other users' past searches.

In order for a search engine to employ a collaborative approach, it is important to calculate similarity measures among user needs, which could be identified through queries, and selected documents in result lists. Glance [31] states that the measure of relatedness among two queries should not depend on the actual terms in them, but on the documents returned by the queries. Two queries could be considered synonymous, even though they contain no terms in common, such as 'handheld devices' and 'mobile computers', by looking at the relationship between the documents returned by each. If the search engine produces many common results for two syntactically different queries, they should be considered semantically correlated. Zhao *et al.* [94] present a framework where the similarity among queries is extended by analyzing the temporal characteristics of the historical click-through, that is, the timestamps of the log data.

In the I-SPY collaborative search engine [76], the queries are considered sets of unique terms on which the Jaccard measure is used to compute the similarity measure:

$$Sim(q_1, q_2) = \frac{|q_1 \cap q_2|}{|q_1 \cup q_2|}$$

Two queries are considered similar if the value computed by the aforesaid formula exceeds a given similarity threshold. For example 'modem adsl' and 'modem usb' are considered to be duplicates above a 0.25 similarity threshold but not above a 0.5 threshold.

Based on the idea that specialized search engines, that is, engines focused on a particular topic, attract communities of users with similar information needs, it is possible to build a statistical model of query-page relevance based on the probability that a page p is selected by a user when returned as a result for a given query q. In practice, I-Spy improves result lists from a traditional general purpose search engine analyzing the interests of communities of users. A community may be identified by a query log of a search box located on specialized Web sites.

This model allows the search engine to personalize search results without relying on content-analysis techniques, but on the relative frequency with which a page has been selected in the past in response to a given query. Results frequently selected by users are promoted ahead of other results returned by a traditional search engines by means of the following relevance:

$$Relevance(p_j, q_i) = \frac{H_{ij}}{\sum_{\forall j} H_{ij}}$$

where H_{ij} indicates the number of users that have selected a page p_j given the query q_i so far. The H matrix represents the statistical model of query-page relevance built with data extracted from a specialized search engine, therefore different matrices are used for different communities of users. For a closer examination of further group recommendation approaches see Chapter 20 of this book [36].

Compass Filter [46] follows a similar collaborative approach, but it is based on Web communities, that is, sets of Web documents that are highly inter-connected. A pre-processing step identifies these communities analyzing the Web hyperlink structure, similarly to the HITS algorithm [42]. If the user has frequently visited documents in a particular community X, when he submits a query about X, all the results that fall into the same community are boosted by the collaborative service. Instead of performing a re-ranking process, a different approach uses Web communities in order to find contextualization cues to be combined with the queries [4]. Claypool *at al.* [16] explores a possible combination of collaborative and content-based approaches by basing the interest prediction of a document on a weighted average adapted to the individual user. An evaluation has shown good results in the on-line newspaper domain.

6.7 Adaptive Result Clustering

Traditional search engines show the query results in long lists ranked by the similarity between query and page content. Users usually sift through the list sequentially, examining the titles and the textual snippets extracted from the pages, in order to find the information matching their needs. Obviously, this activity might take a long time, especially if the user is not able to clearly formulate and submit to the search engine a textual representation of what he is looking for.

Several Web search engines organize results into folders by grouping pages about the same topics together, for example CLUSTY[5] and KARTOO[6]. The former is based on the VIVÍSIMO[7] clustering engine that arranges results in the style of folders and sub-folders. In addition to the traditional HTML layout, the meta search engine KARTOO organizes the returned resources on a graphic interactive map. When the user moves the pointer over those resources, a brief description of the site appears. The size of the icons corresponds to the relevance of the site to the given query. As previously noticed, search systems that arrange results into clusters can be considered personalized because the users are able to customize the results by navigating and choosing selected clusters based on their needs.

In the Web domain, clustering is usually performed after the retrieval of the query results, therefore the whole process must be fast enough to be computed interactively, while the user waits for results. For this reason, the clustering algorithms usually take document snippets instead of whole documents as a representation of page contents. Since, unlike classification, clustering does not require pre-defined categories, the number and the organization of the clusters should be chosen so that the user can navigate easily through them. Finally, clustering should provide concise and accurate cluster descriptions that allow the user to find the most useful ones, even in case of polysemous or misleading queries. For a brief overview of clustering techniques, see Chapter 5 [55].

Further clustering systems are described in the literature, e.g., [91, 93]. The SnakeT meta-search engine [25] includes an innovative hierarchical clustering algorithm with reduced time complexity. It allows the users to select a subset of the clusters that are more likely to satisfy their needs. Then, the system performs a query refinement, building and submitting a new query that incorporates keywords extracted by the system from the selected clusters.

Scatter/Gather [19] uses a similar approach, where the user is able to select one or more clusters for further analysis. The system gathers together all the selected groups and applies the clustering again, scattering the Web sites into a small number of clusters, which are again presented to the user. After a sequence of iterations, the clusters become small enough and the resources are shown to the user.

---

[5] http://www.clusty.com/
[6] http://www.kartoo.com/
[7] http://www.vivisimo.com/

6.8 Hyperlink-Based Personalization

Based on one of the enhanced versions of the PageRank algorithm [37], Chirita *et al.* [15] proposed a personal ranking platform called *PROS* that provides personalized ranking of Web pages according to user profiles built automatically, using user bookmarks or frequently-visited page sets.

In short, the PageRank (PR) is a vote assigned to a page A collected from all the pages $T_1..T_n$ on the Web that point to it. It represents the importance of the page pointed to, where a link to a page counts as a vote of support. The PageRank of a page A is given as follows:

$$PR(A) = (1 - d) + d \left[\frac{PR(T_1)}{C(T_1)} + ... + \frac{PR(T_n)}{C(T_n)} \right]$$

where the parameter d is the damping factor that can be set between 0 and 1 and $C(T_n)$ is defined as the number of links in the page T_n. The PR scores provide *a priori* importance estimates for all of the pages on the Web, independent of the search query. At query time, these importance scores are combined with traditional IR scores to rank the query results.

Briefly, in PROS, the pages judged more interesting for the user are given to the *HubFinder* module that collects hub pages related to the user topics, that is, pages that contain many links to high-quality resources. That module analyzes just the link structure of the Web, running a customized version of Kleinberg's HITS algorithm [42]. A further algorithm, called *HubRank*, combines the PR value with the hub value of Web pages in order to further extend the result set of HubFinder. The final page set is given to the personalized version of PageRank [37] that re-ranks the result pages each time the user submits a query.

The two algorithms, HubFinder and HubRank use the Web link structure to find topic-related pages and to rank the Web pages needed to build the user profile for the Personalized PageRank algorithm. The pages judged more interesting are collected and the expanded sets are built automatically, using bookmarks and the most visited pages. The process does not require explicit activity by the user.

In order to enable "topic sensitive" Web searches, in [34], the importance for each page is calculated by tailoring the PageRank scores for each topic. Thus, pages considered important in some subject domains may not be considered important in others. For this reason, the algorithm computes 16 topic-sensitive PageRank sets of values, each based on URLs from the top-level categories of the OPEN DIRECTORY PROJECT. Each time a query is submitted, it is matched to each of these topics and, instead of using a single global PageRank value, a linear combination of the topic-sensitive ranks are drawn, weighted using the similarities of the query to the topics. Since all the link-based computation are performed off-line, the time spent for the process is comparable to the original PageRank algorithm. Experiments led on this system concluded that the use of topic-specific PageRank scores can improve Web search accuracy.

Qui and Cho [63] extends the Topic-Sensitive PageRank computing multiple ranks, one for each OPD topic. When a query is submitted, the most suitable rank is selected (that is, the rank of the topic that most closely matches the given query) and used for ranking. A personalized version of PageRank based on DNS domains is proposed in

[2], while a personalized system named *Persona* based on the ODP taxonomy, and on an improved version of the HITS algorithm [14] that incorporates user feedback is discussed in [81].

6.9 Combined Approaches to Personalization

Some prototypes provide personalized search combining more than one adaptive approach. For example Outride uses both the browsing history and the current context in order to perform personalization, in the form of query modification and result re-ranking. A second system, named *infoFACTORY*, uses an alerting approach trained according to the categories explicitly selected by the user.

6.9.1 Outride

Outride Inc., an information retrieval technology company acquired by GOOGLE in 2001, introduced a contextual computing system for the personalization of search engine results [62]. *Contextualization* and *individualization* are the two different computational techniques used to perform the personalization. The former is related to the "interrelated conditions that occur within an activity", e.g., the kind of information available, the applications in use, and the documents currently examined, while individualization refers to the "characteristics that distinguishes an individual", such as his goals, knowledge and behaviors assumed during the search.

Adomavicius and Tuzhilin [1] stress this division from the user profiling point of view, identifying two components: *behavioral* and *factual*. The latter corresponds to the output of the above-mentioned individualization process. In contrast, the behavioral component contains information about the on-line activities of the user. For instance, a common representation is based on association rules, where interesting associations or correlation relationships among large set of usage data are extracted, e.g., when shopping on Friday, user X usually spends more than $20 on DVDs. Further details can be found in Chapter 3 of this book [57].

Outride's user model includes both the contextualization and individualization technique, aiming at determining a measure of importance that differs from the traditional relevance measures based, for example, on citation and hyperlink approaches. Those measures are characterized by values that affect the results for the entire user population, without taking into account any contextual or individual information on the user, the change in the user's interests and knowledge over time, or the documents he deems relevant.

In practice, the Outride client is integrated into the sidebar of the Internet Explorer browser. Its user model is based on the hierarchical taxonomy of the OPEN DIRECTORY PROJECT, where a subset of categories are weighted according to the current user needs. These weights are initially set by looking through links suggested by the user, and they are kept updated each time the user clicks on a document, while a surfing history stores the last 1,000 selected links. Therefore, both the explicit and implicit feedback have been utilized.

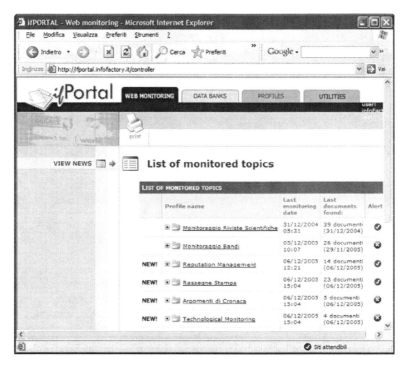

Fig. 6.8. The main page of the *infoFACTORY* monitoring service. From this page, the user can access documents classified according to custom, user-defined categories (see the folder icons). New recently discovered updates, are labeled with a *NEW!* icon. Users may have several profiles, one per topic. A round green icon with a check mark inside indicates that the notification service is enabled for that profile.

The user model is used both for query augmentation and result re-ranking. In the first case, information from the selected categories from the ODP and the Web document currently viewed are compared to the query. If they are similar, the submitted query is related to topics the user has previously seen, an the system can improve it with similar terms in order to disambiguate the query and suggest synonyms. The results from a search engine are re-ranked according to the content of the user model and the current user context, extracting textual information from the pages, e.g., titles and other metadata, and comparing them with a VSM-based representation of the profile. An evaluation of the time spent completing a given set of tasks shows that both novice and expert users are able to find information more quickly using the Outride client than using traditional tools.

6.9.2 infoFACTORY

Finally, it is worth mentioning that *infoFACTORY* [82] contains a large set of integrated Web tools and services that are able to evaluate and classify documents retrieved following a user profile. This system suggests new, potentially interesting contents as soon as it is published on the Web. Thus, it is an application of personalized information provided by means of *push* technology, instead of the traditional *pull* technology employed by search engines.

The *infoMONITOR* component of the system automatically and periodically monitors a selected set of Web resources in order to discover and notify the user about new and interesting documents. Examples of monitored resources include Web sites, portals devoted to a particular topic, daily news sites, journals and magazines, UseNet news, and search engines. Users are able to define their own custom categories, each one represented by a topic-specific profile. Documents are collected and classified into these user-defined categories, which are then used to display the new information, as shown in Fig. 6.8. The user can customize the notification service, requesting e-mail and/or SMS alerting.

6.10 Conclusions

Personalized search on the Web is a research field that has been recently gaining interest, since it is a possible solution to the information overload problem. The reason is quite simple: information plays a crucial role for each user, and users are constantly challenged to take charge of the information they need to achieve both their personal and professional goals. The ability to filter and create a personalized collection of resources simplifies the whole searching process, increasing search engine accuracy and reducing the time the user has to spend to sift through the results for a given query. The same personalization techniques could also be employed to provide advertisements tailored to the current user activity or to proactively collect information on behalf of a user. This chapter provides an introduction to that field, focusing on some of the most interesting and promising approaches and techniques. Some of these researches have been employed in real information systems, while others remain under exploration in research labs. The novelty and liveliness of the personalization field suggests that, over the next few years, new and interesting algorithms and approaches will be proposed and probably transferred to the information systems with which users interact in every day use, such as, search engines or desktop search tools. Ontologies and the Semantic Web[8] are two important research fields that are beginning to receive attention in this context. Gauch *et al.* [28] are investigating techniques that build ontology-based user profiles without user interaction, automatically monitoring the user's browsing habits. Dolog *et al.* [22] are studying mechanisms based on logical mapping rules and description logics, which allow metadata and ontology concepts to be mapped to concepts stored in user profiles. This logical characterization formally enables the personalization techniques in a common language, such as *FOL*, and the reasoning over the Semantic Web (for a closer examination see Chapter 23 [23]).

If the user is working to achieve specific goals, successful systems should recognize those goals and predict aspects of their future behavior. Since the system has expectations about the next user actions, if it is flexible it can adapt itself to the users, thus it should be possible to considerably speed up human-computer interaction.

---

[8] http://www.w3.org/2001/sw/

The *plan-recognition* techniques applied during personalization usually attempt to recognize patterns in user behavior, finding in the set of past actions the ones that are likely to be taken next. For example, some statistical models based on random variables make assumptions about unknown parameters, extrapolating them from observed sample results. These parameters could represent aspects of a user's future behavior, such as their goals, allowing the system to predict their forthcoming actions [95].

Language semantic analysis to understand the meaning of Web content and - more importantly - how it relates to a user's query is a further important field of research in the personalization domain. Language Modeling and Question Answering are two important Natural Language Processing (NLP) research areas that could lead to breakthroughs in the development of personalized search systems. New search engines based on these technologies may be able to understand the users' intention through the analysis of user-supplied natural language questions. They may be able to better understand keywords in the queries by recognizing various sentence types, analyze syntax, and disambiguate word senses in context. As a result, search results will be more accurate, satisfactory, and reliable.

References

1. Adomavicius, G., Tuzhilin, A.: User profiling in personalization applications through rule discovery and validation. In: KDD '99: Proceedings of the fifth ACM SIGKDD international conference on Knowledge discovery and data mining, New York, NY, USA, ACM Press (1999) 377–381
2. Aktas, M.S., Nacar, M.A., Menczer, F.: Personalizing pagerank based on domain profiles. In Mobasher, B., Liu, B., b.Masand, Nasraoui, O., eds.: Proceedings of the sixt WEBKDD workshop Web Mining and Web Usage Analysis (WEBKDD'04), Seattle, Washington, USA (2004) 83–90
3. Allan, J.: Incremental relevance feedback for information filtering. In: Proceedings of the 19th Annual International ACM SIGIR Conference on Research and Development in Information Retrieval, Zurich, Switzerland (1996) 270–278
4. Almeida, R.B., Almeida, V.A.F.: A community-aware search engine. In: WWW '04: Proceedings of the 13th international conference on World Wide Web, New York, NY, USA, ACM Press (2004) 413–421
5. Anick, P.: Using terminological feedback for web search refinement: a log-based study. In: SIGIR '03: Proceedings of the 26th annual international ACM SIGIR conference on Research and development in informaion retrieval, New York, NY, USA, ACM Press (2003) 88–95
6. Asnicar, F.A., Tasso, C.: ifWeb: a prototype of user model-based intelligent agent for document filtering and navigation in the world wide web. In: Proceedings of Workshop Adaptive Systems and User Modeling on the World Wide Web (UM97), Sardinia, Italy (1997) 3–12
7. Baeza-Yates, R., Ribeiro-Neto, B.: Modern Information Retrieval. Addison-Wesley (1999)
8. Bharat, K., Kamba, T., Albers, M.: Personalized, interactive news on the web. Multimedia Syst. 6(5) (1998) 349–358
9. Budzik, J., Hammond, K.J.: User interactions with everyday applications as context for just-in-time information access. In: IUI '00: Proceedings of the 5th international conference on Intelligent user interfaces, New York, NY, USA, ACM Press (2000) 44–51
10. Budzik, J., Hammond, K.J., Birnbaum, L.: Information access in context. Knowledge-Based Systems 14(1-2) (2001) 37–53

11. Bunt, A., Carenini, G., Conati, C.: Adaptive content presentation for the web. In Brusilovsky, P., Kobsa, A., Nejdl, W., eds.: The Adaptive Web: Methods and Strategies of Web Personalization. Volume 4321 of Lecture Notes in Computer Science. Springer-Verlag, Berlin Heidelberg New York (2007) this volume

12. Burke, R.: Hybrid web recommender systems. In Brusilovsky, P., Kobsa, A., Nejdl, W., eds.: The Adaptive Web: Methods and Strategies of Web Personalization. Volume 4321 of Lecture Notes in Computer Science. Springer-Verlag, Berlin Heidelberg New York (2007) this volume

13. Chan, P.K.: Constructing web user profiles: A non-invasive learning approach. In: WEBKDD '99: Revised Papers from the International Workshop on Web Usage Analysis and User Profiling, London, UK, Springer-Verlag (2000) 39–55

14. Chang, H., Cohn, D., McCallum, A.: Learning to create customized authority lists. In: ICML '00: Proceedings of the Seventeenth International Conference on Machine Learning, San Francisco, CA, USA, Morgan Kaufmann Publishers Inc. (2000) 127–134

15. Chirita, P.A., Olmedilla, D., Nejdl, W.: Pros: A personalized ranking platform for web search. In: 3rd International Conference Adaptive Hypermedia and Adaptive Web-Based Systems (AH 2004). Volume 3137 of Lecture Notes in Computer Science., Eindhoven, The Netherlands, Springer (aug 2004) 34–43

16. Claypool, M., Gokhale, A., Miranda, T., Murnikov, P., Netes, D., Sartin, M.: Combining content-based and collaborative filters in an online newspaper. In: ACM SIGIR Workshop on Recommender Systems - Implementation and Evaluation, ACM Press (1999) http://www.csee.umbc.edu/~ian/sigir99-rec/.

17. Claypool, M., Le, P., Wased, M., Brown, D.: Implicit interest indicators. In: IUI '01: Proceedings of the 6th international conference on Intelligent user interfaces, New York, NY, USA, ACM Press (2001) 33–40

18. Collins, A.M., Quillian, R.M.: Retrieval time from semantic memory. Journal of Learning and Verbal Behavior **8** (1969) 240–247

19. Cutting, D.R., Karger, D.R., Pedersen, J.O., Tukey, J.W.: Scatter/gather: a cluster-based approach to browsing large document collections. In: SIGIR '92: Proceedings of the 15th annual international ACM SIGIR conference on Research and development in information retrieval, New York, NY, USA, ACM Press (1992) 318–329

20. Deerwester, S.C., Dumais, S.T., Landauer, T.K., Furnas, G.W., Harshman, R.A.: Indexing by latent semantic analysis. Journal of the American Society of Information Science **41**(6) (1990) 391–407

21. Dieberger, A., Dourish, P., Höök, K., Resnick, P., Wexelblat, A.: Social navigation: techniques for building more usable systems. Interactions **7**(6) (2000) 36–45

22. Dolog, P., Henze, N., Nejdl, W., Sintek, M.: Towards the adaptive semantic web. In Bry, F., Henze, N., Maluszynski, J., eds.: Principles and Practice of Semantic Web Reasoning, International Workshop, PPSWR 2003, Mumbai, India, December 8, 2003, Proceedings. Volume 2901 of Lecture Notes in Computer Science., Springer (2003) 51–68

23. Dolog, P., Nejdl, W.: Semantic web technologies for personalized information access on the web. In Brusilovsky, P., Kobsa, A., Nejdl, W., eds.: The Adaptive Web: Methods and Strategies of Web Personalization. Volume 4321 of Lecture Notes in Computer Science. Springer-Verlag, Berlin, Heidelberg, and New York (2007) this volume

24. Dumais, S., Cutrell, E., Cadiz, J., Jancke, G., Sarin, R., Robbins, D.C.: Stuff i've seen: a system for personal information retrieval and re-use. In: SIGIR '03: Proceedings of the 26th annual international ACM SIGIR conference on Research and development in informaion retrieval, New York, NY, USA, ACM Press (2003) 72–79

25. Ferragina, P., Gulli, A.: A personalized search engine based on web-snippet hierarchical clustering. In: WWW '05: Special interest tracks and posters of the 14th international conference on World Wide Web, New York, NY, USA, ACM Press (2005) 801–810

26. Freyne, J., Smyth, B.: An experiment in social search. In Bra, P.D., Nejdl, W., eds.: Adaptive Hypermedia and Adaptive Web-Based Systems, Third International Conference, AH 2004, Eindhoven, The Netherlands, August 23-26, 2004, Proceedings. Volume 3137 of Lecture Notes in Computer Science., Springer (2004) 95–103

27. Furnas, G.W., Landauer, T.K., Gomez, L.M., Dumais, S.T.: The vocabulary problem in human-system communication. Commun. ACM **30**(11) (1987) 964–971

28. Gauch, S., Chaffee, J., Pretschner, A.: Ontology-based personalized search and browsing. Web Intelligence and Agent System **1**(3-4) (2003) 219–234

29. Gauch, S., Speretta, M., Chandramouli, A., Micarelli, A.: User profiles for personalized information access. In Brusilovsky, P., Kobsa, A., Nejdl, W., eds.: The Adaptive Web: Methods and Strategies of Web Personalization. Volume 4321 of Lecture Notes in Computer Science. Springer-Verlag, Berlin Heidelberg New York (2007) this volume

30. Gentili, G., Micarelli, A., Sciarrone, F.: Infoweb: An adaptive information filtering system for the cultural heritage domain. Applied Artificial Intelligence **17**(8-9) (2003) 715–744

31. Glance, N.S.: Community search assistant. In: IUI '01: Proceedings of the 6th international conference on Intelligent user interfaces, New York, NY, USA, ACM Press (2001) 91–96

32. Goldberg, D., Nichols, D., Oki, B.M., Terry, D.: Using collaborative filtering to weave an information tapestry. Commun. ACM **35**(12) (1992) 61–70

33. Golub, G.H., Loan, C.F.V.: Matrix computations (3rd ed.). Johns Hopkins University Press, Baltimore, MD, USA (1996)

34. Haveliwala, T.H.: Topic-sensitive pagerank. In: WWW '02: Proceedings of the 11th international conference on World Wide Web, New York, NY, USA, ACM Press (2002) 517–526

35. Höscher, C., Strube, G.: Web search behavior of internet experts and newbies. In: Proceedings of the 9th World Wide Web Conference (WWW9), Amsterdam, Netherlands (2000) 337–346

36. Jameson, A., Smyth, B.: Recommending to groups. In Brusilovsky, P., Kobsa, A., Nejdl, W., eds.: The Adaptive Web: Methods and Strategies of Web Personalization. Volume 4321 of Lecture Notes in Computer Science. Springer-Verlag, Berlin Heidelberg New York (2007) this volume

37. Jeh, G., Widom, J.: Scaling personalized web search. In: WWW '03: Proceedings of the 12th international conference on World Wide Web, New York, NY, USA, ACM Press (2003) 271–279

38. Joachims, T.: Optimizing search engines using clickthrough data. In: Proceedings of 8th ACM SIGKDD International Conference on Knowledge Discovery and Data Mining, ACM Press (2002) 133–142

39. John Kemeny, J.L.S.: Mathematical Models in the Social Sciences. MIT Press, New York (1962)

40. Kelly, D., Teevan, J.: Implicit feedback for inferring user preference: a bibliography. SIGIR Forum **37**(2) (2003) 18–28

41. Khopkar, Y., Spink, A., Giles, C.L., Shah, P., Debnath, S.: Search engine personalization: An exploratory study. First Monday **8**(7) (2003) http://www.firstmonday.org/issues/issue8_7/khopkar/index.html..

42. Kleinberg, J.: Authoritative sources in a hyperlinked environment. In: Proceedings of the 9th annual ACM-SIAM symposium on Discrete algorithms, San Francisco, CA, USA (1998) 668–677

43. Kobsa, A.: Privacy-enhanced web personalization. In Brusilovsky, P., Kobsa, A., Nejdl, W., eds.: The Adaptive Web: Methods and Strategies of Web Personalization. Volume 4321 of Lecture Notes in Computer Science. Springer-Verlag, Berlin Heidelberg New York (2007) this volume

44. Koenemann, J., Belkin, N.J.: A case for interaction: a study of interactive information retrieval behavior and effectiveness. In: CHI '96: Proceedings of the SIGCHI conference on Human factors in computing systems, New York, NY, USA, ACM Press (1996) 205–212
45. Koutrika, G., Ioannidis, Y.: A unified user profile framework for query disambiguation and personalization. In: Proceedings of the Workshop on New Technologies for Personalized Information Access (PIA2005), Edinburgh, Scotland, UK (2005) 44–53 http://irgroup.cs.uni-magdeburg.de/pia2005/docs/KouIoa05.pdf.
46. Kritikopoulos, A., Sideri, M.: The compass filter: Search engine result personalization using web communities. In Mobasher, B., Anand, S.S., eds.: Intelligent Techniques for Web Personalization, IJCAI 2003 Workshop, ITWP 2003, Acapulco, Mexico, August 11, 2003, Revised Selected Papers. Volume 3169 of Lecture Notes in Computer Science., Springer (2003) 229–240
47. Kulyukin, V.A.: Application-embedded retrieval from distributed free-text collections. In: AAAI/IAAI. (1999) 447–452
48. Lathauwer, L.D., Moor, B.D., Vandewalle, J.: A multilinear singular value decomposition. SIAM J. Matrix Anal. Appl. **21**(4) (2000) 1253–1278
49. Lawrence, S.: Context in web search. IEEE Data Eng. Bull. **23**(3) (2000) 25–32
50. Lawrence, S., Giles, C.L.: Context and page analysis for improved web search. IEEE Internet Computing **2**(4) (1998) 38–46
51. Liu, F., Yu, C., Meng, W.: Personalized web search for improving retrieval effectiveness. IEEE Transactions on Knowledge and Data Engineering **16**(1) (2004) 28–40
52. MacQueen, J.B.: Some methods for classification and analysis of multivariate observations. In: Proceedings of 5-th Berkeley Symposium on Mathematical Statistics and Probability. Volume 1., University of California Press (1967) 281–297
53. Micarelli, A., Gasparetti, F.: Adaptive focused crawling. In Brusilovsky, P., Kobsa, A., Nejdl, W., eds.: The Adaptive Web: Methods and Strategies of Web Personalization. Volume 4321 of Lecture Notes in Computer Science. Springer-Verlag, Berlin Heidelberg New York (2007) this volume
54. Micarelli, A., Sciarrone, F.: Anatomy and empirical evaluation of an adaptive web-based information filtering system. User Modeling and User-Adapted Interaction **14**(2-3) (2004) 159–200
55. Micarelli, A., Sciarrone, F., Marinilli, M.: Web document modeling. In Brusilovsky, P., Kobsa, A., Nejdl, W., eds.: The Adaptive Web: Methods and Strategies of Web Personalization. Volume 4321 of Lecture Notes in Computer Science. Springer-Verlag, Berlin Heidelberg New York (2007) this volume
56. Middleton, S.E., Roure, D.C.D., Shadbolt, N.R.: Capturing knowledge of user preferences: ontologies in recommender systems. In: K-CAP '01: Proceedings of the 1st international conference on Knowledge capture, New York, NY, USA, ACM Press (2001) 100–107
57. Mobasher, B.: Data mining for web personalization. In Brusilovsky, P., Kobsa, A., Nejdl, W., eds.: The Adaptive Web: Methods and Strategies of Web Personalization. Volume 4321 of Lecture Notes in Computer Science. Springer-Verlag, Berlin Heidelberg New York (2007) this volume
58. Montaner, M., Lopez, B., Rosa, J.L.D.L.: A taxonomy of recommender agents on the internet. Artificial Intelligence Review **19** (2003) 285–330
59. Oard, D.W.: The state of the art in text filtering. User Modeling and User-Adapted Interaction **7**(3) (1997) 141–178
60. Olston, C., Chi, E.H.: ScentTrails: Integrating browsing and searching on the web. ACM Transactions on Computer-Human Interaction **10**(3) (2003) 177–197
61. Pirolli, P.L.T., Pitkow, J.E.: Distributions of surfers' paths through the world wide web: Empirical characterizations. World Wide Web **2**(1-2) (1999) 29–45

62. Pitkow, J., Schütze, H., Cass, T., Cooley, R., Turnbull, D., Edmonds, A., Adar, E., Breuel, T.: Personalized search. Commun. ACM **45**(9) (2002) 50–55
63. Qiu, F., Cho, J.: Automatic identification of user interest for personalized search. In: WWW '06: Proceedings of the 15th international conference on World Wide Web, New York, NY, USA, ACM Press (2006) 727–736
64. Quillian, R.M.: Semantic memory. In Minsky, M., ed.: Semantic information processing. The MIT Press, Cambridge, MA, USA (1968) 216–270
    `http://citeseer.ist.psu.edu/ambrosini97hybrid.html`.
65. Raghavan, V.V., Sever, H.: On the reuse of past optimal queries. In: Research and Development in Information Retrieval. (1995)
    344–350 `http://citeseer.ist.psu.edu/raghavan95reuse.html`.
66. Resnick, P., Iacovou, N., Suchak, M., Bergstrom, P., Riedl, J.: Grouplens: an open architecture for collaborative filtering of netnews. In: CSCW '94: Proceedings of the 1994 ACM conference on Computer supported cooperative work, New York, NY, USA, ACM Press (1994) 175–186
67. Resnick, P., Varian, H.R.: Recommender systems. Commun. ACM **40**(3) (1997) 56–58
68. Rhodes, B.J.: Just-In-Time Information Retrieval. PhD thesis, MIT Media Laboratory, Cambridge, MA (May 2000)
    `http://citeseer.ist.psu.edu/rhodes00justtime.html`.
69. Rich, E.: User modeling via stereotypes. In: Readings in intelligent user interfaces. Morgan Kaufmann Publishers Inc., San Francisco, CA, USA (1998) 329–342
70. Rijsbergen, C.J.V.: Information Retrieval. Butterworth-Heinemann, Newton, MA, USA (1979)
71. Robertson, S.E.: Theories and models in information retrieval. Journal of Documentation **33**(2) (1977) 126–148
72. Salton, G., McGill, M.: An Introduction to modern information retrieval. Mc-Graw-Hill, New York, NY (1983)
73. Salton, G., Wong, A., Yang, C.: A vector space model for automatic indexing. Commun. ACM **18**(11) (1975) 613–620
74. Savoy, J., Picard, J.: Retrieval effectiveness on the web. Information Processing & Management **37**(4) (2001) 543–569
75. Schafer, J.B., Frankowski, D., Herlocker, J.L., Sen, S.: Collaborative filtering recommender systems. In Brusilovsky, P., Kobsa, A., Nejdl, W., eds.: The Adaptive Web: Methods and Strategies of Web Personalization. Volume 4321 of Lecture Notes in Computer Science. Springer-Verlag, Berlin Heidelberg New York (2007) this volume
76. Smyth, B., Balfe, E., Freyne, J., Briggs, P., Coyle, M., Boydell, O.: Exploiting query repetition and regularity in an adaptive community-based web search engine. User Modeling and User-Adapted Interaction **14**(5) (2005) 383–423
77. Speretta, M., Gauch, S.: Personalized search based on user search histories. In: Web Intelligence (WI2005), France, IEEE Computer Society (2005)
    622–628 `http://dx.doi.org/10.1109/WI.2005.114`.
78. Spink, A., Jansen, B.J.: A study of web search trends. Webology **1**(2) (2004) 4
    `http://www.webology.ir/2004/v1n2/a4.html`.
79. Spink, A., Jansen, B.J., Ozmultu, H.C.: Use of query reformulation and relevance feedback by excite users. Internet Research: Electronic Networking Applications and Policy **10**(4) (2000) 317–328 `http://citeseer.ist.psu.edu/spink00use.html`.
80. Sun, J.T., Zeng, H.J., Liu, H., Lu, Y., Chen, Z.: Cubesvd: a novel approach to personalized web search. In: WWW '05: Proceedings of the 14th international conference on World Wide Web, New York, NY, USA, ACM Press (2005) 382–390

81. Tanudjaja, F., Mui, L.: Persona: A contextualized and personalized web search. In: HICSS '02: Proceedings of the 35th Annual Hawaii International Conference on System Sciences (HICSS'02)-Volume 3, Washington, DC, USA, IEEE Computer Society (2002) 67

82. Tasso, C., Omero, P.: La Personalizzazione dei contenuti Web: e-commerce, i-access, e-government. Franco Angeli (2002)

83. Teevan, J.: Seesaw: Personalized web search. Student Workshop for Information Retrieval and Language (SWIRL '04)
(November 2004) http://ciir.cs.umass.edu/~hema/swirl/swirl.htm.

84. Teevan, J., Alvarado, C., Ackerman, M.S., Karger, D.R.: The perfect search engine is not enough: a study of orienteering behavior in directed search. In: CHI '04: Proceedings of the SIGCHI conference on Human factors in computing systems, New York, NY, USA, ACM Press (2004) 415–422

85. Teevan, J., Dumais, S.T., Horvitz, E.: Personalizing search via automated analysis of interests and activities. In: SIGIR '05: Proceedings of the 28th annual international ACM SIGIR conference on Research and development in information retrieval, New York, NY, USA, ACM Press (2005) 449–456

86. Wærn, A.: User involvement in automatic filtering: An experimental study. User Modeling and User-Adapted Interaction 14(2-3) (2004) 201–237

87. Webb, G.I., Pazzani, M., Billsus, D.: Machine learning for user modeling. User Modeling and User-Adapted Interaction 11(1-2) (2001) 19–29

88. White, R., Jose, J.M., Ruthven, I.:   Comparing explicit and implicit feedback techniques for web retrieval: Trec-10 interactive track report.   In: TREC. (2001) http://trec.nist.gov/pubs/trec10/papers/glasgow.pdf.

89. wook Ahn, J., Brusilovsky, P., Farzan, R.: Investigating users' needs and behavior for social search. In: Proc. of Workshop on New Technologies for Personalized Information Access at 10th International User Modeling Conference, UM 2005. (2005) 1–12 http://irgroup.cs.uni-magdeburg.de/pia2005/docs/AhnBruFar05.pdf.

90. Yao, Y.: Measuring retrieval effectiveness based on user preference of documents. Journal of the American Society for Information Science 46(2) (1995) 133–145

91. Zamir, O., Etzioni, O.: Web document clustering: a feasibility demonstration. In: SIGIR '98: Proceedings of the 21st annual international ACM SIGIR conference on Research and development in information retrieval, New York, NY, USA, ACM Press (1998) 46–54

92. Zamir, O.E., Korn, J.L., Fikes, A.B., Lawrence, S.R.: Us patent application #0050240580: Personalization of placed content ordering in search results (July 2004)

93. Zeng, H.J., He, Q.C., Chen, Z., Ma, W.Y., Ma, J.: Learning to cluster web search results. In: SIGIR '04: Proceedings of the 27th annual international ACM SIGIR conference on Research and development in information retrieval, New York, NY, USA, ACM Press (2004) 210–217

94. Zhao, Q., Hoi, S.C.H., Liu, T.Y., Bhowmick, S.S., Lyu, M.R., Ma, W.Y.: Time-dependent semantic similarity measure of queries using historical click-through data. In: WWW '06: Proceedings of the 15th international conference on World Wide Web, New York, NY, USA, ACM Press (2006) 543–552

95. Zukerman, I., Albrecht, D.W.: Predictive statistical models for user modeling. User Modeling and User-Adapted Interaction 11(1-2) (2001) 5–18

7

Adaptive Focused Crawling

Alessandro Micarelli and Fabio Gasparetti

Department of Computer Science and Automation
Artificial Intelligence Laboratory
Roma Tre University
Via della Vasca Navale, 79 – 00146 Rome, Italy
{micarel, gaspare}@dia.uniroma3.it

Abstract. The large amount of available information on the Web makes it hard for users to locate resources about particular topics of interest. Traditional search tools, e.g., search engines, do not always successfully cope with this problem, that is, helping users to seek the right information. In the personalized search domain, focused crawlers are receiving increasing attention, as a well-founded alternative to search the Web. Unlike a standard crawler, which traverses the Web downloading all the documents it comes across, a focused crawler is developed to retrieve documents related to a given topic of interest, reducing the network and computational resources. This chapter presents an overview of the focused crawling domain and, in particular, of the approaches that include a sort of adaptivity. That feature makes it possible to change the system behavior according to the particular environment and its relationships with the given input parameters during the search.

7.1 Introduction

Traditional search engines allow users to submit a query suggesting, as output, an ordered list of pages ranked according to a particular matching algorithm. The underlying Information Retrieval model's goal is to allow users to find those documents that will best help them meet information needs and make it easier to accomplish their information-seeking activities. The query can be considered the user's textual description of the particular information request. If the engine works in the Web domain, a software system usually named *crawler* traverses it, collecting HTML pages or other kinds of resources [69, 83]. It exploits the hyperlink structure in order to find all the destination anchors (or targets) reachable from a given starting set of pages through the outgoing links. The crawler also keeps the snapshot and the internal index of the search engines up-to-date, periodically recrawling and updating the pages with fresh images.

General-purpose search engines employ crawlers to collect pages covering different topics. At query time, the engine retrieves subsets of pages that are more likely to be related to the user current needs expressed by means of sets of keywords. Models of the user needs are usually not employed, therefore the search results are not personalized

P. Brusilovsky, A. Kobsa, and W. Nejdl (Eds.): The Adaptive Web, LNCS 4321, pp. 231–262, 2007.
© Springer-Verlag Berlin Heidelberg 2007

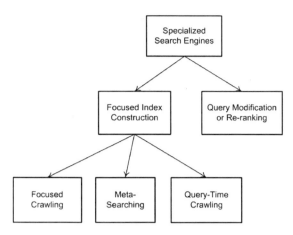

Fig. 7.1. A taxonomy of approaches to build specialized search engines, as shown in [80].

for the user. Basically, two users, with different interests, knowledge and preferences, obtain the same results after having submitted the same query.

A first step toward a better search tool is developing *specialized search engines*, which provide tailored information on particular topics or categories for focused groups of people or individual users. The heavy constraints of a general-purpose search engine, i.e., indexing billions of pages and processing thousands of queries per second, are no longer required for these kinds of tools. New techniques can be included to represent Web pages and to match these representations against the interests of users, e.g., algorithms based on Natural Language Processing (NLP), usually avoided due to the computational resources needed.

There are several approaches to build specialized search engines [80], as shown in Fig. 7.1. Query modification and re-ranking exploit traditional search tools, filtering their content by augmenting user queries with keywords, or re-ordering the results, removing irrelevant resources. These techniques are widely discussed in Chapter 6 of this book [61]. Specialized search engines are also based on focused indexes, which contain only the documents related to the given topics of interest. To retrieve and index those documents, it is possible to meta-search specialized databases, or perform an autonomous search at query time. The most interesting technique is to perform focused crawling on the Web. It concerns the development of particular crawlers able to seek out and collect subsets of Web pages that satisfy some specific requirements. In particular, if the goal is to collect pages related to a given topic chosen by the user, the crawlers are usually named *focused* or *topical* [20, 17] (see Fig. 7.2). Focused crawlers are also employed in different domains from specialized IR-based search engines, but usually related to the retrieval and monitoring of useful hypertextual information, as shown later on in this chapter.

The focused crawling approach entails several advantages in comparison with the other approaches employed in specialized search engines. Performing an autonomous search at query time considerably delays the retrieval of result lists. Meta-searching provides results from existing general-purpose indexes that often contain outdated versions

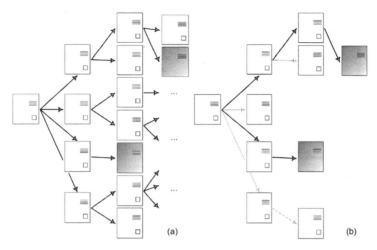

Fig. 7.2. Focused crawling attempts to find out and download only pages related to given topics (b), while standard crawlers employs traditional strategies (a), e.g., breadth-first search or further techniques to balance the network traffic on Web servers.

of the available documents. Due to the reduced storage requirements, focused crawling can employ techniques to crawl part of the deep Web (dynamically generated Web pages not accessible by search engines, see Sect. 7.2) or to keep the stored information fresh updating it frequently.

User queries, which usually consist of 2-3 keywords for traditional search engines [51, 77, 7], can be enriched by further information, such as subsets of taxonomy classes judged interesting by users, e.g. categories in the Open Directory Project[1] (ODP) or Yahoo! Directory[2], or by means of relevance feedback [75], see also Chapter 2 [42] and 6 [61] of this book for details. Better representations of the user needs help the focused crawlers to find out interesting pages, obtaining more accurate results.

If a focused crawler includes learning methods to adapt its behavior to the particular environment and its relationships with the given input parameters, e.g. the set of retrieved pages and the user-defined topic, the crawler is named *adaptive*. Non-adaptive crawlers are usually based on classifiers whose learning phase ends before the searching process starts. Even though some of them employ hypertextual algorithms, such as HITS - which lead to better results, making more information available - the adaptability is actually not manifest. Sometimes adaptive crawlers provide a sort of feedback mechanism where useful information is extracted from the analyzed documents, and the internal classifier updated consequently [18]. Other approaches can explicitly model the set of pages around the topical ones. Such models capture important features that appear in valuable pages, and describe the content of the pages that are frequently associated with relevant ones. The searching process may benefit from those models, obtaining better overall crawl performance [72, 32, 1].

---

[1] www.dmoz.org
[2] www.yahoo.com

Adaptive focused crawlers are key elements in personalizing the human-computer interaction. Traditional non-adaptive focused crawlers are suitable for communities of users with shared interests and goals that do not change with time. In this case, it is easy to recognize the requested topics and start a focused crawl to retrieve resources from the Web. The adaptive crawler's advantage is the ability to learn and be responsive to potential alterations of the representations of user needs. This could happen when users do not know exactly what they are looking for, or if they decide to refine the query during the execution if the results are not deemed interesting. Therefore, adaptive focused crawlers are more suitable for personalized search systems that include a better model of the information needs, which keeps track of user's interests, goals, preferences, etc.. As a consequence, adaptive crawlers are usually trained for single users and not for communities of people.

A further advantage of adaptive crawlers is the sensitivity to potential alterations in the environment. Web pages are constantly being updated, as well as the related hyperlink structure. Therefore, a focused crawler should be able to monitor any change in order to look for new and interesting information.

The related domains that can really benefit from the focused crawling are the so-called vertical portals and studies on Web evolution. The former domain is related to Web sites that provide information on a relatively narrow range of goods and services for communities of users. Retrieving valuable, reliable and up-to-date information resources for the user is a typical purpose of these portals, as well as offering personalized electronic newspapers [10], personal shopping agents [34] or conference monitoring services [48]. Research activities on the study of the evolution of Web hypertextual environment [54, 66] usually requires a constant monitoring of specific portions of Web and of related changes during a given time interval, e.g., bursty communities of blogs [49]. Both domains could really benefit from employing focused crawling systems to retrieve information that meets the given input constraints, discovering properties that combine the topical content of pages and the linkage relationship between them.

The remainder of the chapter is structured as follows: research on the WWW and issues related to crawlers' development are discussed in Sect. 7.2. Section 7.3 includes references to focused crawlers where the adaptivity feature is not always explicitly included. Section 7.3.1 briefly introduces the topical locality phenomenon, used by focused crawlers to look for pages related a given topic. Section 7.4 converges on the adaptive focused crawling approaches explicitly based on AI agent or multi-agent systems, and particularly on genetic algorithms and Ant-paradigm based approaches. Section 7.5 presents an overview of further focused crawling approaches that aim at extracting important features from the crawled pages in order to adapt the search behavior. Finally, Sect. 7.6 gives a brief account on the methodologies chosen so far to evaluate focused crawlers.

7.2 Crawlers and the World Wide Web

Analyzing a general search engine's architecture, as pictured in Fig 7.3, it is possible to recognize some sub-systems. An internal repository stores a copy of the retrieved documents, which will be used to extract links and convert the plain text into an *ad hoc*

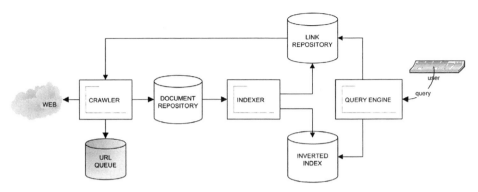

Fig. 7.3. A general overview of a search engine architecture [2].

representation, called *Inverted Index*, used to provide quick responses for a given query. This index makes it possible to build lists of documents containing a specific keyword in a very short time. The global result list is drawn simply by merging all these sublists, and by ranking each resource with a given weighting algorithm.

The crawler is a very important module of a search engine. It is a high-performance system, expected to download hundreds of millions of pages over several weeks, minimizing the time spent for crash recoveries. Therefore, its design presents many challenges, such as I/O and network efficiency, robustness and manageability [15, 45, 64, 76, 33]. Moreover, it is meant to ensure a *politeness policy* that sets an interval between two connections to the same Web server, in order to avoid excessive traffic that could cause network congestions or software crashes.

Due to the peculiar characteristics of the Web domain, its vastness, heterogeneity and owing to the authors quickly changing the page contents, many different research fields focusing on the characteristics of this environment have been drawing a lot of interest over the past years. Some of these works are important to understand and develop focused crawlers, for this reason we report some references in the following sections.

7.2.1 Growth and Size of the Web

Estimating the size of the Web is a tough task. Central repositories of the Web content are missing, therefore it is impossible to draw statistics of the size and growth. A recent study revealed that at the beginning of 2005, the part of the Web considered potentially indexable by major engines was at least 11.5 billion pages [43]. Previous studies came to the conclusion that the size of the Web doubled in less than two years [52, 53]. A further consideration that has been widely accepted is that the network speed has improved less than storage capacities, from the single user to big organizations.

Other works try to estimate how often the Web pages change during a given period. For example, after having monitored a corpus of more than 0.7 million pages on a daily basis for four months, it has been estimated that 40% of the page set changed within a week, and 23% of the pages that fell into the .com domain changed daily [26]. An extended analysis on 150 million pages can be found in [36].

In order to keep results as fresh as possible, the search engines' local copies of remote Web pages must be kept up-to-date. It is possible to develop refresh policies that guarantee a low consumption of computational and network resources by estimating the change frequency of individual pages. Such estimation techniques are usually based on analysis and models that predict the change frequency of pages, monitoring how often it is updated over a given period and the last date of change [23, 24, 26]. Freshness of the index and the size of the indexable Web are two interrelated parameters to take under consideration when developing and tuning crawlers. A crawl may last for weeks, so at some point crawlers should begin revisiting previously retrieved pages, to check for changes. This strategy means that some pages could never be retrieved.

An attempt to automatically notify the search engine when particular pages change or a new page is added to a Web site is Google Sitemaps[3]. The webmaster has to keep up-to-date a simple XML file that lists the pages on the Web site and how often they change. Pushing this information from the Web site to search engines reduces the time they spend to find resources and adapts the re-crawling process optimizing the network resources. This approach is particularly suitable for large Web sites of dynamically generated pages, where the XML report can be periodically drawn up by the system.

Distributed Web crawling is another important research area that aims to develop infrastructures where single computational units collaborate and distribute the crawling in order to speed up the download and let the whole system tolerate possible crashes [76, 11]. An important subtopic of Web crawling architectures is that of parallelization policies. They regard the assignment of different sets of URLs to single crawling processes in order to maximize the download rate, avoiding multiple downloads of the same resources and minimizing the overhead from parallelization [25].

Nevertheless, after taking into consideration the estimate results on the size and the growth of the Web, it is becoming a common opinion that crawling the Web is not trivial because it is growing faster than our search engine technology. The frequent update of some Web pages, which forces the crawler to re-download a great amount of data in a short time is not the only matter.

7.2.2 Reaching the Web: Hypertextual Connectivity and Deep Web

Bailey *et al.* named *dark matter* all the information that a search engine is not able to access [3]. The possible reasons range from the simple robot exclusion protocol[4] (a simple textual file stored on the Web server used by page authors to avoid undesired visits by the search engines' crawlers), to the lack of connectivity in the link graph [16].

Figure 7.4 shows a diagram where 200 million pages and 1.5 billions of links in a 1999 crawl were analyzed discovering four distinct page sets. The Strongly Connected Component (*SCC*) is the larger component, accounting for 27%, consisting of pages that can reach one another along directed links. The second and third sets are called *IN* and *OUT*, each accounting for 21% approximately, and consist of pages that can reach the SCC but cannot be reached from it, or that are accessible from the *SCC* but do not link back to it respectively. The rest of the pages, a further 29%, cannot be reached and

---

[3] http://www.google.com/webmasters/sitemaps/

[4] http://www.robotstxt.org/

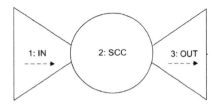

Fig. 7.4. A study on the Web's hypertextual connectivity shows three large sets, where from the IN set a user can reach a page in OUT passing through SCC. The complete bow-tie diagram can be found in [16].

do not reach *SCC*. It is obvious how some parts of the Web are not easily reachable. If the crawler starts its navigation from the *OUT* set, it can reach just a fifth of whole Web.

Another important restriction that reduces the part of the Web available for crawlers is due to the existence of dynamic page generators. For example, commercial Web sites usually offer a text box where the user can submit a query in order to retrieve a list of products. This phenomenon appears each time a page is generated by user interaction with a Web server, an interaction which is more than a simple click of a link. Usually the pages are built by querying an internal database where the information is stored. *Deep Web* is used to indicate the set of all those pages, but sometimes the meaning of the deep Web is expanded to include non-textual files, such as audio, movies, etc., for which a traditional search engine would have problems assigning a content representation. It has been estimated that public information on the deep Web is currently up to 550 times larger than the normally accessible Web [8]. Crawlers can collect those product pages in two different ways: generating and submitting a set of keywords capable of retrieving most of the dynamically generated pages [71, 30, 67], and simulating the user's surfing behavior through agents [50].

Studies on search engine coverage confirm how the indexes contain only a small fraction of the Web. In January 2005, among the top 4 search engines, the best was Google, reaching more than 76% of the indexable Web, while Ask/Teoma reached approximately 58% approximately [43]. A previous study [53] assigned less than 40% of the indexable Web to the best search engine. According to most of the search engine experts, dynamically-generated Web pages and the related techniques used to uncover that information during the crawls are currently ignored by search engines, keeping them from being part of the query results. In other words, most of the huge amount of information on the Web is not accessible to the user through search engines.

The vastness of the Web, and in particular of the Deep Web, which is not indexed by traditional search tools, and the low freshness and coverage of search engines suggest how different kind of crawlers have the chance to increase the performance and the accuracy of the search results. Before analyzing focused crawlers in detail, we shall briefly introduce some of the strategies evaluated on the Web domain to partially overcome the aforesaid issues.

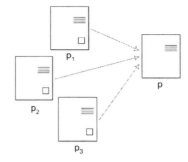

Fig. 7.5. The backlink count metric assigns an importance to a page p based on the number of Web pages pointing to it.

7.2.3 Crawling Strategies

Assuming that a search engine is unable to keep its index fresh due to the extent and growth of the Web, research is focusing on the best strategies to prioritize the downloads. In other words, the crawlers can retrieve only a fraction of the Web pages within a given period, so it becomes important to download the most important pages first, that is, the pages that satisfy most of the queries submitted by users.

The implementation of a crawler includes a queue of URLs to be retrieved, fed during the surfing with URLs extracted from pages, the URL QUEUE in Fig. 7.3. The download strategy used by the crawler affects the queue ordering, moving ahead the resources to be crawled first. In abstract terms, the strategy assigns a score to each resource that denotes an estimate of the importance for the current crawl. Of course, at any given moment that score is based only on the information gathered by the crawler until then, that is, the part of the Web that has been crawled.

Cho *et al.* [27] compared three ordering metrics on a corpus of 179,000 pages collected from an university Web site. The metrics are: Breadth-first search, Backlink count and PageRank. The first metric coincides with the well-known search algorithm. If the Web is represented by a graph of pages and a hyperlink structure of edges, the neighbors of a starting edge are considered before analyzing the neighbors of the visited edges, and so forth. Obviously, the edges are never visited twice.

The *backlink count* metric assigns an importance to each page to be downloaded as a function of the number of the crawled pages that point to it, see Fig. 7.5. Intuitively, if many Web authors decide to put a link to a given site, that site should be judged more important than one that is seldom referenced. This simple metric derives from bibliometry research, that is, the analysis of the citation graph of scholarly articles, and from social network studies. The same research influenced HITS [47] and PageRank [68], two iterative algorithms employed in the Web domain to assign relevance measures to each resource, taking into account the hyperlink structures among pages. In huge collections of documents, query results are often too numerous to be analyzed by users, therefore it becomes important to accurately rank the most important pages. HITS and PageRank try to cope with these problems exploiting the presence of the link structure between documents to unearth the new information used during the ranking process. A better description of these algorithms is available in Chapter 5 of this book [62].

The evaluation in [27] shows that PageRank outperforms the backlink ordering if the goal is to crawl the pages with the highest in-degree, that is, the most popular. The authors show that backlink is more biased by the choice of the starting points, usually reaching locally interesting pages instead of globally interesting ones, focusing more on single clusters. Its behavior does not improve as the crawl proceeds and more information becomes available. If the goal is to crawl pages with content matching a given topic, breadth-first search ensures better performance. It is possible to explain such a result by observing how pages usually point to other pages that share the same topic.

The assets of the breadth-first search are discussed also in [65] where an extended corpus of 328 million of pages has been examined. This strategy is able to discover the highest quality pages assessed by means of the PageRank metric. The authors speculate that there are usually many hosts pointing to high quality pages, and for this reason, regardless of the host or page from which the crawl originates, they can be found sooner during the search.

7.3 Focused Crawling

This section provides an overview of the most important references to focused crawling systems. Even though some of them do not include any adaptivity form, it is useful to analyze the proposed techniques and algorithms before introducing other approaches.

One of the features that characterizes a focused crawler is the way it exploits hypertextual information. Traditional crawlers convert a Web page into plain text extracting the contained links, which will be used to crawl other pages. Focused crawlers exploit additional information from Web pages, such as anchors or text surrounding the links. This information is used to predict the benefit of downloading a given page. Because we do not know anything about the page's content, this prediction avoids to waste CPU and network resources to analyze irrelevant documents.

Basically, all the focused crawlers are based on the topical locality phenomenon. According to this phenomenon, Web pages on a given topic are usually clustered together, with many links that connect one page to another. Once a good page is found, the crawler can analyze its cluster to retrieve pages on the same topic.

A brief description of the topical locality phenomenon and how the hypertextual information is exploited by crawlers in the search activity opens this section.

7.3.1 Exploiting the Hypertextual Information

The Web consists of hypertextual resources, typically Web pages, connected by means of links. Traditional IR content-based approaches take into consideration the textual information inside each page in order to assign representations and match them with the user needs. The *hyper information* [56], that is, the information related to a given page that takes into account also the Web structure it is part of, is usually ignored. Besides

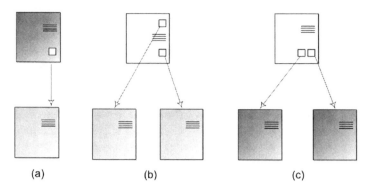

(a)   (b)   (c)

Fig. 7.6. Some phenomena occurring in the hypertextual Web environment: A page is significantly more likely to be topically related to the pages which point to it, (a), and sibling pages are more similar when the links from the parent are located closer together (c) than far away (b).

improving the performance of traditional search engines [15], the information extracted from the link structure is also employed by focused crawlers to navigate among pages in order to find interesting pages. The topical locality phenomenon, the anchors and two hypertextual algorithms, HITS and PageRank, are without doubt important topics to examine in this context.

Topical Locality and Anchors. An important phenomenon that lays the foundations of almost all the focused crawlers is *Topical locality*. Davison makes a deep empirical evaluation along with a study on the *proximal cues* or *residues* extracted from links [29], while Menczer extends the study, also considering the link-cluster conjecture, that is, pages about the same topic clustered together [57]. Topical locality occurs each time a page is linked to others with related content, usually because Web authors put a link to another page, to allow users to see further related information or services.

Focused crawlers exploit this phenomenon crawling clusters of pages each time they find an interesting one. A cluster consists of pages that can be reached by the current page, extracting its links. The crawl usually stops when pages not correlated with selected topics are discovered.

Empirical evaluations acknowledged that this phenomenon is largely common on the Web (see also [28]). A page is much more likely to be topically related to the pages which point to it, as opposed to other pages selected at random or other pages that surround it. Further results show how sibling pages, namely those that share a page that points to both of them, are more similar when the links from the parent are closer together [29], as shown in Fig. 7.6.

The notion of proximal cues or residues corresponds to the imperfect information at intermediate locations exploited by users to decide what paths to follow in order to reach a target piece of information [37, 70]. In the Web environment, text snippets[5], anchor text or icons are usually the imperfect information related to a certain distant content.

---

[5] Snippets usually correspond with the short textual description that search engines associate with each query results.

Davison [29] shows how anchor text is often very informative about the contents of documents pointed to by the corresponding URL, therefore it may be useful in discriminating unseen child pages. That phenomenon has been exploited for several tasks, such as inferring user information needs [41, 22], document summarization [31] and query translation [55].

Some search engines associate the text of a link with the page the link points to [15], including a partial representation of the linked page in the index, even though that page has not been crawled, nor indexed yet, and despite it does not contain any textual information to be indexed, e.g., Macromedia Flash-based homepages. Moreover, the chance to include text usually written by different authors often provides more accurate descriptions of Web pages and increases the retrieval performance; for details see the related vocabulary problem [38].

HITS and PageRank. The wide research activity on citation analysis started by Garfield in the 1950s is the foundation of two famous hypertextual algorithms, PageRank [68] and HITS [47]. They all aim to assign measures of relevance relying only on the structure of links extracted from sets of Web pages, ignoring the textual content. Their effectiveness has been proved when coupled with traditional IR systems, e.g., [15]. They are employed in more than one crawling algorithm to assign a hypertextual-based rank to the resources to be crawled, and finding new *Seed sets*, i.e., initial pages where the search starts [27, 20, 74].

In general terms, when a huge collection of hypertextual documents needs to be analyzed to retrieve a small set of resources that satisfy a given query, the ranking becomes a very important process. The user usually scans a small number of documents. In a large study on queries submitted to a search engine, it has been noticed that 28.6% of users examine only one page of results, and 19% look at two pages only, therefore half of the users do not check more than 20 URLs [78]. If there are many documents that satisfy a query, the system may assign the same rank to the first results. For this reason, further relevance measures that take into account the popularity, the prestige or the authority of a page must be included in the ranking process.

These kinds of hypertextual algorithms can be successfully employed in focused crawling systems. One of the goals of this type of crawler is to optimize the computation resource to analyze a few pages, ignoring the part of the Web that is not interesting or hardly relevant for the user. Moreover, unlike general-purpose search engines, the number of retrieved pages per query is not an important parameter. Short result lists of highly ranked documents are undoubtedly better than long lists of documents that force the users to sift through them in order to find the most valuable information.

A description of the two link analysis algorithms, HITS and PageRank, is to be found in Chapter 5 of this book [62].

7.3.2 An Overview of the Focused Crawling Approaches

Although one of the first focused crawlers dates back to 1994, we had to wait many years before seeing other approaches in the literature. It is worth mentioning the *Fish-search* algorithm [14, 13] because of its key principle that enhances the traditional crawling algorithms.

The algorithm takes as input one or more starting URLs and the user's query. The starting pages could be collected from the first results of search engines or user's bookmarks, and correspond to the seed URLs. The crawler's queue of pages to be downloaded becomes a priority list initialized with the seed URLs. At each step, the first URL is popped from the queue and downloaded. When a text page becomes available, it is analyzed by a scoring component, evaluating whether it is relevant or irrelevant with respect to the search query. Based on that score, a heuristic decides whether to pursue the exploration in the related direction or not. If it is not relevant, its links will be ignored by further downloads. See Fig. 7.2 as an example of a focused crawl compared with a crawl based on a breadth-first search.

In the Fish search, whenever a document is fetched, it is scanned for links. A depth value is assigned to each linked page. If the parent is relevant, the depth of the children is set to some predefined value. Otherwise, the depth is set to be one less than the parent's depth. When the depth reaches zero, the direction is dropped and none of its children are inserted into the queue.

It is possible to limit the search to either a fixed period of time or to the retrieval of a certain number of relevant pages. Although the authors provided a number of heuristics to optimize the search, Web server operators noted that Fish-search is not far from an exhaustive search. If all users were to employ a fish-search for every search, the overall bandwidth demand would impose a heavy burden on Web servers.

Further details of the Fish algorithm plus the Shark-search description, which uses a better measure of relevance to draw the similarity between documents and the given query and several other improvements, can be found in [14, 13, 44].

This section also includes other important non-adaptive focused crawling approaches proposed in the literature. A first approach exploits Web taxonomies, such as ODP or Yahoo!, and the HITS algorithm to predict the importance of the pages to be crawled. Tunneling and Contextual crawling aim at discovering important features to understand where the topical pages are located, following the paths they belong to. Finally, a further approach targets the Semantic Web environment.

Taxonomies and Distillation. A popular focused crawler approach is proposed by Chakrabarti et al. [20, 19]. This focused crawling system makes use of two hypertext mining programs that guide the crawl: a *classifier* that evaluates the relevance of hypertext documents regarding the chosen topics, and a *distiller* that identifies hypertext nodes that are considered great access points to many relevant pages through a few links, by means of the HITS algorithm.

After having collected a set of interesting pages, the user selects the best matching nodes in the category tree of a classifier trained on a given taxonomy, e.g., Yahoo!. The hierarchy helps filtering the correct pages during the crawl. Basically, after having retrieved a document, an algorithm finds the leaf node of that tree with the highest probability of similarity. If some ancestor of these nodes has been marked as interesting (see Fig. 7.7), the URLs found in the document will be crawled in future, otherwise the crawl is pruned at the document or a low priority is assigned to them.

Tunneling. Focused crawlers heavily rely on the topical locality phenomenon discussed in Sect.7.3.1. A page about a given topic is more likely to point to other pages

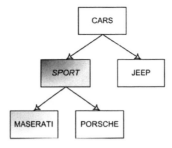

Fig. 7.7. A link contained in a page on `Maserati` cars is visited if the user has selected the `Maserati` topic, or one of its ancestors in the taxonomy, such as `Sport`, during the training of the classifier.

about the same topic. Even though very useful and effective during the crawl, such a phenomenon might generate some drawbacks. Sometimes, pages of the same topic do not point directly one anther and therefore it is necessary to go through several off-topic pages to get to the next relevant one. Bergmark *et al.* [9] suggest to allow the crawl to follow a limited number of bad pages in order to reach the good ones, naming this technique *tunneling*.

The authors use the terms *nugget* and *dud* to indicate pages correlated with at least one of the given topic collections judged interesting, and the pages that do not match any of the collections; 500,000 pages were downloaded and analyzed in order to find patterns of document-collection correlations along link paths. The idea is to statistically estimate an adaptive cutoff for the tunneling strategy, that is, how far the crawler is allowed to go when starting from an interesting page. Once the cutoff is reached, the crawl in that direction is halted.

The evaluation results show how a focused crawl with adaptive cutoff downloads resources at a higher rate, therefore, exploring the Web faster than a fixed cutoff crawler. At the same time, the adaptive version can limit the amount of irrelevant material it has to crawl through. Briefly, the tunneling technique seems to improve the effectiveness of searching by recognizing and pruning paths which look hopeless.

Contextual Crawling. Diligenti *et al.* [32] try a different approach address the very problem of assigning the right importance to different documents along a crawl path following a different approach. Given a set of interesting documents, by querying some search engines they are able to build a representation of the pages that occur within a certain link distance of those documents. In particular, the Web is *backcrawled*; the analyzed pages are the ones that point to the set of interesting documents. All the retrieved information is stored in a structure named *context graph* which maintains for each page the related distance, defined as the minimum number of links necessary to traverse in order to reach a page in the initial set.

The context graph is passed to a set of Naïve Bayes classifiers that identify the categories according to the expected link distance from the page to the target documents. In this way, given a generic document, the crawler predicts how many steps away from a target document it is likely to be. Of course, documents that are expected to lead more quickly to the interesting documents are crawled sooner. The only relevant drawback of the context-focused crawling approach is the need for search engines providing the reverse link information needed to build the context graph.

Semantic Web. A further focused crawling approach [35] is related to the Semantic Web[6], the current effort to enhance the traditional Web presentation languages such as HTML and help grasp the meaning of data, and let autonomous agents understand and interact with information sources. Ontologies are one of the basic elements needed to describe static domain knowledge in terms of common domain structures to be reused by different systems. Basically, the crawling approach aims to define a relevance measure to map a Web page content with an existing ontology provided by users. Textual matches against the ontology's lexicon and taxonomic relationships between super and sub-concepts for calculating accurate relevance scores are undoubtedly the core elements of that approach. The comparison between standard focused crawling approaches and simple keyword matching shows relevant improvements in global performance.

7.4 Agent-Based Adaptive Focused Crawling

In recent years, the research community has tried to propose new original approaches in order to build focused crawlers. Some of them speculate on the Web and the analogy with huge environments where single autonomous units live and keep moving, looking for interesting resources. From this point of view, the AI field has developed a wide range of architectures, models and algorithms providing a solid foundation where it has been possible to build interesting prototypes. This section provides information on two algorithms, a genetic-based one and an ant paradigm-based one, including the related focused crawling techniques.

7.4.1 Genetic-Based Crawling

One of the most renowned adaptive focused crawlers is *InfoSpiders*, also known as *ARACHNID* (Adaptive Retrieval Agents Choosing Heuristic Neighborhoods for Information Discovery) [58] based on genetic algorithms. Even though the prototype is developed using reinforcement learning [81], which will be discussed in Sect. 7.5.2, we decided to discuss it in this section because it is also based on an agent-based approach.

Genetic algorithms have been introduced in order to find approximate solutions to hard-to-solve combinatorial optimization problems. Those kinds of algorithms are inspired by evolutionary biology studies. Basically, a population of chromosomes encoded by means of a particular data structure evolves towards a potential solution through a set of genetic operators, such as inheritance, mutation, crossover, etc.. The chromosomes that get closer to best solutions have more chances to live and reproduce, while the ones that are ill-suited for the environment die out. The initial set of chromosomes is usually generated randomly and for this reason the first solutions are more likely to be rather poor.

---

[6] http://www.w3.org/2001/sw/

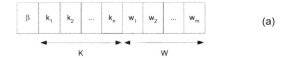

(a)

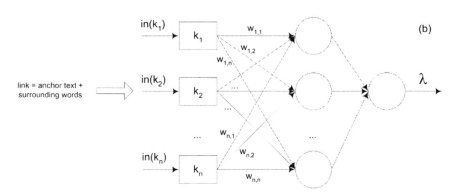

(b)

Fig. 7.8. An Infospiders's genotype (a) composed of a set of keywords K and weights W of a neural network used to evaluate a link, that is, the anchor text plus the surrounding words, contained in a Web page (b).

The random mutations between two chromosomes is an important feature of these algorithms in preventing solution sets from converging to sub-optimal or local solutions. For the same reason, even pairs of organisms that do not receive high fitness, namely, not close to good solutions, have a chance to breed.

As in a real environment, if a lot of changes occur all of a sudden, many individuals of the population risk dying out. On the contrary, if those changes take place gradually, the species can gradually evolve along with it. The whole evolution process is repeated until a solution that is judged fairly close to the optimal solution is found.

In InfoSpiders an evolving population of intelligent agents browses the Web driven by user queries, imitating the humans during their browsing activities. The agents are able to assess the relevance of resources with a given query and to reason autonomously about future actions regarding the next pages to be retrieved, with little or no interaction among them.

Each agent is built on top of a *genotype* (shown in Fig. 7.8), basically a set of chromosomes, which determines its searching behavior. The same genotype takes part in the offspring generation process. The first component of an agent's genotype is a parameter β that represents the extent to which an agent trusts the textual description about the outgoing links contained in a given page. The other two components are: a set of keywords K initialized with the query terms and a vector of real-valued weights W, initialized randomly with uniform distribution. The latter component corresponds to the information stored in a feed-forward neural network used for judging what keywords in the first set best discriminate the documents relevant to users. Basically, for each link in a page, an agent extracts the set of keywords that appear around it and that are also included in the genotype set. The set is the input to the neural network. The function *in()*, within the network, weights each word, counting their occurrences and calculating their positions on the anchor text. More importance is given to terms that are closer to the link.

The adaptivity is both supervised and unsupervised. If the user has previously submitted a relevance feedback on the document chosen by the agent, the related content is used to evaluate the retrieved documents (supervised learning). Otherwise, the relevance assessments are employed for training the neural network's weights W, comparing the outcome of the agent's behavior with the initial query terms corresponding to the profile of user interests. The network's weights are updated through the backpropagation of error [73] according to the agents' actions, e.g., irrelevant pages are retrieved, in order to alter the subsequent behavior of the agent.

Two functions, *benefit()* and *cost()* are employed to determine the energy gain related to a link selection. The former gets high values if the keywords K are contained in the visited document, while the *cost* function is correlated to resources used during the retrieving of documents, e.g., network load or document size. The whole crawling process is summarized in Algorithm 1.

Algorithm 1. Pseudo-code of the InfoSpiders algorithm.

```
{initialize each agent's genotype, energy and starting page}
PAGES ← maximum number of pages to visit
while number of visited pages < PAGES do
    while for each agent a do
        {pick and visit an out-link from the current agent's page}
        {update the energy estimating benefit() − cost()}
        {update the genotype as a function of the current benefit}
        if agent's energy > THRESHOLD then
            {apply the genetic operators to produce offspring}
        else
            {kill the agent}
        end if
    end while
end while
```

The second form of adaptivity to the environment is achieved by mutations and crossovers among agents. They guarantee the inclusion of relevant resources, they spawn new offspring and adapt agents to the environment. Each single agent learns and exploits important local features of the environment. The multi-agent system as a whole captures more general features, which represent resources that are relevant to users.

The keyword vector K is subjected to mutation during the reproduction of the agent population. A stochastic approach takes into consideration both local context, selecting a word that describes documents that led to the energy increase, and global context, selecting words from documents where the user has submitted relevance feedback. According to the authors, that form of keyword evolution, operated through local selection, mutation and crossover, implements a kind of selective query expansion. The neural net is mutated by adding some random noise to the set of weights W.

Basically, a value corresponding to the agent's energy is assigned at the beginning of the search, and it is updated according to the relevance of the pages visited by that agent. Both neural networks and genetic algorithms select terms from those documents that are considered relevant. They are also responsible for using agents' experience to modify the behavior of the agents, and for identifying new links leading to the energy increase. The energy determines which agents are selected for reproduction and the ones to be killed amidst the population.

The agent architecture based on a set of computational units facilitates a multi-process scalable implementation that can be adapted to the current resources, in terms of available network bandwidth and CPU.

Chen *et al.* [21] compared a genetic-based focused crawler, called *Itsy Bitsy spider*, with a Best-first search approach, where a function assigns a score to each outgoing link from the collection of retrieved pages, corresponding to the priority to visit the link. The links with the highest score are the ones the crawler visits first.

Starting from a set of pages related to the current user information needs, the genetic-based crawler extracts words and links from that set. By means of a particular similarity measure among sets, it is possible to compare two different search space states and evaluate the outcome of a system's action, that is, the download of a particular link. The similarity measure is the same one used as a priority function in the best-first search approach.

The mutation operator is implemented performing a search on the YAHOO! database in order to suggest new, promising unknown pages for further explorations. The crossover operates on sets of pages connected to the seed set. The probability to perform one of the two operators during the crawl is a system's parameter. In this way, depending on whether the user prefers to locally crawl a part of the Web or aims at the whole Web, the probability to use the mutation operator gets lower or higher values with respect to crossover.

During the evaluation, the genetic approach does not outperform the best first search. The recall values of the former are significantly higher than the best first search, but precision is not statistically different. Nevertheless the mutation operator includes resources that would be hardly retrievable through conventional search processes. This kind of generation of new configurations is also considered in the Yang and Chen's crawler [82], where a hybrid version of simulated annealing is employed instead of genetic algorithms. Simulated annealing is a generalization of the Monte Carlo method for combinatorial problems inspired by the annealing process of liquids and metals [60]. A further evaluation, where an improved version of InfoSpiders is compared against the best-first search is discussed in [59].

It should be pointed out that the agent architecture and the adaptive representation of InfoSpiders, consisting of a set of computation units with single genotypes, differ considerably from the Itsy Bitsy spider approach. The neural network and the other components of the genotype make it possible for a single agent to autonomously determine and adapt effectively its behavior according to both the local context of the search and the personal preferences of the user. That feature is basically missing in the Itsy Bitsy spider.

Balabanović [6, 5] combines the collaborative and content-based approaches in a single multi-agent recommender system named *Fab*. *Collection* agents search the Web according to profiles representing their current topics. The users are grouped into clusters according to the similarity of their profiles, which do not usually correspond to the collection agents' profile. The topic of a set of search agents can be of interest to many users, while one user can be interested in many topics. The retrieved pages are forwarded to those users whose profiles they match beyond some threshold. If a page is highly rated by one user, it is passed to the users with similar profiles (user's cluster).

Collection agents are based on topical locality, and perform a best-first search of the Web. An agent follows the links in a given page if it is pertinent to the given topic. Other kinds of agents construct queries to be submitted to various search engines and randomly collect links from human-suggested *cool* Web sites.

The population of collection agents evolves by adapting to the population of users. The agents that retrieve pages rarely seen or that receive low feedback scores by users are killed. If the agent is able to retrieve good pages, it is given the chance to reproduce itself. Hence, the population dynamically adapts to the user requests, to the environment and to the available computing resources.

7.4.2 Ant-Based Crawling

Research on ant foraging behaviors inspired a different approach for building focused crawlers. In [39, 40] an adaptive and scalable Web search system is described, based on a multi-agent reactive architecture, stemming from a model of social insect collective behavior [12]. That model has been created by biologists and ethologists to understand how blind animals, such as ants, are able to find out the shortest ways from their nest to feeding sources and back. This phenomenon can be easily explained, since ants can release a hormonal substance to mark the ground, the *pheromone*, leaving a trail along the followed path. This pheromone allows other ants to follow such trails, reinforcing the released substance with their own.

Intuitively, the first ants returning to their nest from the feeding sources are those which have chosen the shortest paths. The back-and-forth trip (from the nest to the source and back) allows them to release pheromone on the path twice. The following ants leaving the nest are attracted by this chemical substance. If they have to select a path, they will prefer the one which has been frequently covered by the other ants that followed it previously. This is the reason why they will direct themselves towards the shortest paths, which are the first to be marked (see Fig. 7.9).

The decision-making of each computational unit is realized through a set of simple behaviors, which allows the agents to wander from one document to another, choosing the most promising paths, i.e., those leading to resources relevant for users. The outcome of the agent exploration is indirectly handed out to other agents, which can exploit it to improve future explorations. This form of social ability ensures the indispensable adaptivity needed to to work in complex and dynamic environments, such as the Web.

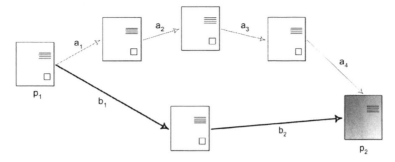

Fig. 7.9. If two paths lead to an interesting page, the first agents that reach that page are the ones that have followed the shortest path $\langle b_1, b_2 \rangle$, and therefore they are the first to release the pheromone that attracts the following agents to the same path.

Aside from the topical locality phenomenon (Sect. 7.3.1), a further observation widely discussed in [63] is taken into consideration: to reach the current page, Web page authors suppose that surfers have followed a special path and, therefore, have visited and read certain pages too. Therefore, the contents of Web pages are often not self-contained: they do not always contain all the information required to completely understand and explain what they deal with. In order to satisfy user information needs, the sequences of connected pages should be considered as a virtual information unit that could be suggested to users [46]. For this reason, during an exploration, the pages on the path through which a surfer reaches the page under examination, i.e., the current *context* of the exploration, are not to be ignored [9].

Each agent corresponds to a virtual ant that has the chance to move from the hypertextual resource where it is currently located url_i, to another url_j, if there is a link in url_i that points to url_j. At the end of each exploration, the pheromone trail is released on the agent's route, that is, sequence of URLs of the followed path. When the agent is located in a certain resource, it can match the related content with the user's needs, and measure the amount of pheromone on the paths corresponding to the outgoing links.

The pheromone trails allows ants to make better local decisions with limited local knowledge both on environment and group behavior. The ants employ them to communicate the exploration results one another: the more interesting resources an ant finds, the more pheromone trail it leaves on the followed path. As long as a path carries relevant resources, the corresponding trail will be reinforced and the number of attracted ants will increase.

The system execution is divided into cycles; in each one of them, the ants make a sequence of moves among hypertextual resources. The maximum number of allowable moves depend proportionally on the value of the current cycle, that is, $maxmoves = k \cdot currentcycle$ where k is a constant. At the end of a cycle, the ants update the pheromone intensity values of the followed path as a function of the retrieved resource scores.

This increases the number of the allowed movements and, consequently, the number of the visited resources, as the number of cycles increases too. During the first cycles, the trails are not so meaningful, due to the small set of crawled resources, so the ex-

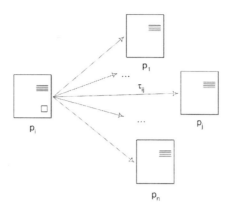

Fig. 7.10. When an agent selects a link to follow, it draws the transition probabilities P_{ij} of all the out-going links $i \rightarrow 1, i \rightarrow 2, \ldots, i \rightarrow n$ from the current page p_i, i.e., $\forall j : (i,j) \in E$, according with the pheromone trails τ_{ij}, as in Eq. 7.1.

ploration is basically characterized by a random behavior. But after a certain number of cycles, the paths that permit to find interesting resources will be privileged with a major intensity of pheromone, therefore, a great number of ants will have the chance to follow them. In other words, the process is characterized by a positive feedback loop where the probability to follow a particular path depends on the number of ants that followed it during the previous cycles.

It is possible to describe the operations that the ant-agents perform with these *task accomplishing behaviors*:

1. at the end of the cycle, the agent updates the pheromone trails of the followed path and places itself in one of the start resources
2. if an ant trail exists, the agent decides to follow it with a probability which is function of the respective pheromone intensity
3. if the agent does not have any available information, it moves randomly

Intelligent behavior emerges from the interaction of these simple behaviors, and from the interaction that agents establish with their environment. The resources from which the exploration starts can be collected from the first results of a search engine, or the user's bookmarks, and correspond to the seed URLs.

To select a particular link to be followed, a generic ant located on the resource url_i at the cycle t, draws the *transition probability* value $P_{ij}(t)$ for every link contained in url_i that connects url_i to url_j. The $P_{ij}(t)$ is considered by the formula:

$$P_{ij}(t) = \frac{\tau_{ij}(t)}{\sum_{l:(i,l)\in E} \tau_{il}(t)} \tag{7.1}$$

where $\tau_{ij}(t)$ corresponds to the pheromone trail between url_i and url_j, and $(i,l) \in E$ indicates the presence of a link from url_i to url_l (see Fig. 7.10).

To keep the ants from following circular paths, and to encourage page exploration, each ant stores a list L containing the visited URLs. A probability related to the path from url_i to url_j, if the url_j belongs to L, is 0. At the end of every cycle, the list is emptied out.

When the limit of moves per cycle is reached, the ants start the *trail updating process*. The *updating rule* for the pheromone variation of the k-ant corresponds to the mean of the visited resource scores:

$$\Delta\tau^{(k)} = \frac{\sum_{j=1}^{|p^{(k)}|} score(p^{(k)}[j])}{|p^{(k)}|} \tag{7.2}$$

where $p^{(k)}$ is the ordered set of pages visited by the k-ant, $p^{(k)}[i]$ is the i-th element of $p^{(k)}$, and $score(p)$ is the function that, for each page p, returns the similarity measure with the current information needs: a $[0, 1]$ value, where 1 is the highest similarity.

The process is completed with the τ value updates. The τ_{ij} trail of the generic path from url_i to url_j at the cycle $t+1$ is affected by the ant's pheromone updating process, through the computed $\Delta\tau^{(k)}$ values:

$$\tau_{ij}(t+1) = \rho \cdot \tau_{ij}(t) + \sum_{k=1}^{M} \Delta\tau^{(k)} \tag{7.3}$$

where ρ is the trail evaporation coefficient. It must be set to a positive value less than 1 to avoid unlimited accumulation of substance caused by the repeated positive feedback. The summation widens to a subset of the N ants living in the environment. In order to avoid the implementation of the third accomplishing behavior, at the beginning of the execution, all the $\tau_{ij}(0)$ values are set to a small constant τ_0. Algorithm 2 shows each step of the crawling process.

Algorithm 2. Pseudo-code of the Ant-based crawler

```
{initialize each agent's starting page}
PAGES ← maximum number of pages to visit
cycle ← 1
t ← 1
while number of visited pages < PAGES do
    while for each agent a do
        for move = 0 to cycle do
            {calculate the probabilities P_ij(t) of the out-going links as in Eq. 7.1}
            {select the next page to visit for the agent a}
        end for
    end while
    {update all the pheromone trails}
    {initialize each agent's starting page}
    cycle ← cycle + 1
    t ← t + 1
end while
```

The developed architecture and, in particular, the trail updating process, permit to take into consideration two empirical observations: Web pages on a given topic are more likely to link to those on the same topic, and the content of Web pages is often not self-contained. It is easy to notice how the pheromone attraction bears the ants' exploration toward pages linked to the most interesting ones. Moreover, if a path leads to a page that is not deemed very interesting, the probabilities to follow its outgoing links could be relevant just the same, because the pheromone trail on a link is function of all the pages' scores in the path. In this way, the context of a page, that is the content of the pages from which it is possible to reach a page, is also inspected.

As for the adaptability, there is a twofold instance: the first concerns the opportunity for users to refine queries during the execution when the results are not so satisfactory or, in general, when users do not know how to input a query to express what they want. The second form regards the possibility of analyzed hypertextual resource changes due to environment instability. Of course, this adaptability does not directly concern the user and variations of his/her needs, but it is related to the environment's dynamics. These two types of adaptivity are possible because at every cycle the value of pheromone intensities $\tau_{ij}(t)$ is updated according to the visited resource scores.

Once the query, or the content of the visited resources changes, the similarity measure changes affects the $\Delta\tau_{ij}(t)$ variation which influences, in its turn, $\tau_{ij}(t)$ (see the pheromone updating process), that is, the results of past explorations. Moreover, as cycles go by, the current values of pheromone intensities tend to become ideal, i.e. function of the current environmental status, with the pheromone evaporation effects.

For instance, given a particular path P, in case of a negative variation of the $\Delta\tau_{ij}^{k}(t)$ values (due to, for instance, a change of the user query), where the path from url_i to url_j belongs to P, the trails $\tau_{ij}(t+1)$ are subjected to a feedback which is reduced if it is compared to the one in the former cycles. For this reason, a smaller number of ants will be attracted, therefore, the $\Delta\tau_{ij}^{k}(t+2)$ increments are still further reduced, and so forth. In other words, each change in the environment causes a feedback consequently modifying the system behavior in order to adapt itself to the new conditions.

The major drawback of this approach is that agents need to start the crawl from the same set of pages at the end of each cycle. As the crawl widens, many pages are re-crawled many times before discovering new resources on the frontier, wasting network and computational resources. Periodically updating the seed set with new pages is a possible way to tackle this problem.

The proposed architecture is a reactive one, where the system's rational behavior emerges from the interaction and the cooperation of an agent population; therefore, it must be attempted to completely understand the relationship between local and global behavior. In particular, it is useful to be aware of the changes of the global behavior in function of some parameter variations, such as the initial values and the decay of the pheromone intensities, or the number of agents employed. For this purpose, the agents' behavior has to be analyzed, i.e. how agents choose paths in function of interesting resource positions, to see if they behave properly under all the possible circumstances.

7.5 Machine Learning-Based Adaptive Focused Crawling

In the last section, two approaches based on genetic algorithms and animal foraging models have been described. Other adaptive focused crawling approaches exploit different techniques in order to represent and recognize features that can drive the search toward the interesting resources. In this section, some of these approaches based on machine learning are briefly introduced.

7.5.1 Intelligent Crawling Statistical Model

Aggarwal *et al.* [1] introduce an interesting adaptive crawling framework, called *Intelligent Crawling*. It aims to statistically learn the characteristics of the Web's linkage structure while performing the search. In particular, given an unseen page, a set of customizable features named *predicates* (e.g., the content of the pages which are known to link to the unseen URL, or tokens in the unseen page's URL), are taken into consideration. Those features are analyzed in order to understand how they are connected to the probability that the related unseen page satisfies the information needs.

The feature set and the linkage structure of that portion of the Web which has already been crawled are the input of the statistical model. At the beginning of the search, no statistical information is available and each single feature gets the same importance, therefore the crawler virtually behaves randomly. As soon as an interesting page is downloaded, its features are analyzed in order to find any possible correlation. For example, if the fraction of pages about music composer *Bach* is 0.3%, and it is found that 10% of the pages that point to them contain the word *Bach* in their content, following the terminology in [1], we may as well assert that:

$$P(C|E) > P(C) \tag{7.4}$$

the particular knowledge E about a candidate URL, e.g., the 10% of pages that contain the word *Bach*, increases the prior probability $P(C)$ that a given Web page will be related to user needs. In order to understand if a given feature, namely, knowledge E, is favorable to probability $P(C)$, the *interest ratio* function is defined:

$$I(C, E) = \frac{P(C|E)}{P(C)} = \frac{P(C \cap E)}{P(C)P(E)} \tag{7.5}$$

An approximation of $P(C \cap E)$ and $P(E)$ can be obtained using the information that the crawler has retrieved so far, as well as $P(C)$, fraction of retrieved pages that satisfy the user defined predicate. The interest ratio gets values greater than 1 if a feature E is favorable to the probability that the unseen page satisfies the user needs. Otherwise, if event E is unfavorable, the interest ratio will be within the range $(0, 1)$.

For example, the interest ratio for the event corresponding to the word *Bach* occurring in the in-linking page is $0.1/0.03 = 3.33$. Once those ratios identify the best features to be taken into consideration, the crawler uses them to decide the candidate pages that are more likely to satisfy the user's needs.

The interesting point of this framework is that it does not need any collection of topical examples for training. At the beginning, users specify their needs by means of some predicates, e.g., the page content or the title must contain a given set of keywords, and the crawler adapts its behavior learning correlations among the given features. Nevertheless, choosing the right set of predicates and/or adding new predicates besides the ones proposed by authors might not be an easy task to accomplish.

7.5.2 Reinforcement Learning-Based Approaches

The idea of using textual information contained in pages that point to the ones to be evaluated during the download is exploited also by Chakrabarti et al. [18]. Their work starts from a traditional focused crawler [20], where a classifier evaluates the relevance of a hypertextual document with respect to the chosen topics. The training of the classifier is done at the beginning of the crawl, where the user selects the best matching nodes in the category tree of a hierarchical classifier. The enhancement of the crawler concerns the inclusion of a second classifier called *apprentice*, which assigns priorities to unvisited URLs in the crawl frontier. In order to draw these priorities, the classifier extracts some features from a particular representation of structured documents, named Document Object Model (DOM)[7], of the pages that point to unvisited URLs.

The original focused crawler's classifier, whose role was to assign a similarity measure to crawled pages given some user needs, now trains instances for the apprentice. Basically, for each retrieved page v, the apprentice is trained on information from the original classifier and on some features around the link extracted from crawled pages that point to v. Those predictions are then used to calculate if it is worth traversing a particular URL, and therefore order the queue of URLs to be visited. The evaluation shows how the false positives decrease significantly between 30% and 90%.

In [72] reinforcement learning techniques are employed to map what actions have generated a benefit during the crawling. That map becomes important each time the system decides the next action to undertake, evaluating the future reward expected from executing it.

The interesting documents found in the environment are the rewards, while following a particular link corresponds to an action. The authors make some simplifying assumptions concerning the representation of an environment state, that is, the remaining set of interesting documents to be found, and the set of links that have been discovered, in order to deal with the problem. Basically, the idea is to learn during the crawl the text in the neighborhood of hyperlinks that are most likely to point to relevant pages. For this reason, each link is replaced with the surrounding words, so it is possible to generalize across different hyperlinks by comparing the related text. A collection of Naïve Bayes text classifiers perform the mapping by casting this regression problem as classification.

---

[7] The Document Object Model is an interface that allows programs and scripts to dynamically access and process the document's content and structure, see http://www.w3.org/DOM/ for details.

7.6 Evaluation Methodologies

Defining an evaluation methodology for a *standard* crawler does not require a great effort. Once a subset of the Web is available, it is possible to run an instance of the crawl on a workstation and monitor the most important parameters to measure its effectiveness, such as: computation time to complete the crawl, or the number of downloaded resources per time unit. Since the Web corpus is fixed, different crawling strategies and the related results can be directly compared. Some technical issues must be addressed in order to construct a collection that is a good sample of the Web [4]. Many important elements should be considered when defining an evaluation methodology regarding a *focused* crawling system. In this section, we briefly mention some of these elements, referring to literature for a deeper analysis.

One of the first evaluation parameters to take into consideration is the soundness of the retrieved documents. The traditional crawlers' goal is to download as many resources as possible, whereas a focused crawler should be able to filter out the documents that are not deemed related to the given topics of interest. All the research in the information retrieval domain can help us define measures of soundness for the retrieved results, but some of them, such as precision and recall, become meaningless in the Web domain.

In order to evaluate the retrieval effectiveness, precision P_r corresponds to the fraction of top r ranked documents that are relevant to the query over the total number of retrieved documents, interesting and not:

$$P_r = \frac{found}{found + false\ alarm} \tag{7.6}$$

while recall R_r is the proportion of the total number of relevant documents retrieved in the top r over the total number of relevant documents available in the environment:

$$R_r = \frac{found}{found + miss} \tag{7.7}$$

see also the diagram in Fig. 7.11.

As pointed out in [20], the recall indicator is hard to measure because it is impossible to clearly derive the total number of relevant documents present on the Web.

The Information Retrieval community has identified many approaches based on very large collections to provide objective testbeds for evaluations. A standard collection guarantees the *reliability* propriety, essential to obtain the same results when tests are repeated under the same initial conditions and the chance to compare the results among different systems. Moreover, it is possible to clearly point out the sets of relevant documents and therefore calculate the traditional IR evaluation measures. Nevertheless, focused crawlers should access the Web directly, to avoid that during the search some paths be ignored because of some pages not being included in the collection. For this reason, standard collections are rarely employed for evaluations of focused crawlers, and it is almost impossible to make comparisons from results obtained by the different algorithms.

In order to evaluate the goodness of the retrieved resources, many evaluations use the same measures employed in the focused crawler's algorithms. For example, the

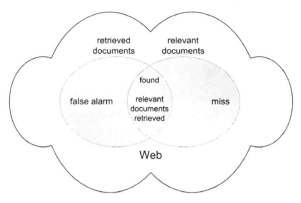

Fig. 7.11. The Web and the two sets, retrieved and relevant documents, which sometimes do not correspond. Precision and recall aim at measuring this deviation.

classifiers' outcome in [20], the VSM-based similarity measures in [39], and the importance metrics in [27] are employed both to guide the search of crawlers and to evaluate the relevance of the retrieved resources. That approach may appear improper, but actually the measures are defined without considering the document set to be evaluated, therefore, the results do not affect the definition of evaluation measures.

A further evaluation measure often employed during the evaluation is the percentage of important pages retrieved over the progress of the crawl. The goal is to select relevant pages only, avoiding to spend time crawling the uninteresting regions of the Web. If the percentage is high, the crawler is able to save its computational resources and retrieve a good page set. Chakrabarti *et al.* [20] consider the average quality measures of the retrieved pages for different time slices of the crawl. In this way, it is possible to see if crawlers get lost during the search or if they are able to constantly keep the crawl over the relevant documents.

The time spent for crawling and analyzing the documents is another important element to measure. A focused crawler usually collects thousands of URLs per hour. Of course, computational resources, such as the available network bandwidth, memory and CPU, required for the processing, affect the crawl time. Some algorithms, such as HITS, based on the link graph, need plenty of time to draw their relevance measures, especially for complex graphs. Focused crawlers based on these algorithms, e.g., [27, 20, 74], are expected to reduce the rate of page downloads as the crawl goes on.

In [79] an extended analysis of the literature about focused crawling suggests a general framework for crawler evaluation research. That framework considers important issues, such as the choice of seed pages where the crawls start. Of course, if seed pages are related to the topic of the crawl, the search for related pages is much easier because of the topical locality phenomenon (see Sect.7.3.1). The framework also provides a mechanism to control the level of difficulty of the crawl task by means of the hypertextual structure among pages.

At present, focused crawling evaluations that also include adaptivity analysis are not available. One of the reasons could be the difficulty to measure the reaction of crawlers to user needs refinements, or alterations of sub-sets of Web pages. How long does it take to adapt the crawl to a user relevance feedback and provide new interesting documents? How many environment alterations are tolerable before the crawling performance falls below a given threshold? Standard methodologies to assess those features, thus allowing to compare results, are yet to be developed.

7.7 Conclusions

Focused crawling has become an interesting alternative to the current Web search tools. It concerns the development of particular crawlers able to seek out and collect subsets of Web pages related to given topics. Many domains may benefit from this research, such as vertical portals and personalized search systems, which provide interesting information to communities of users. Some of the most interesting approaches have been described, along with important hypertextual algorithms, such as HITS and PageRank.

Particular attention has been given to adaptive focused crawlers, where learning methods are able to adapt the system behavior to a particular environment and input parameters during the search. Evaluation results show how the whole searching process may benefit from those techniques, enhancing the crawling performance. Adaptivity is a must if search systems are to be personalized according to user needs, in particular if such needs change during the human-computer interaction.

Besides new crawling strategies that take into consideration peculiar characteristics of the Web, such as topical locality, we can expect future research to head in several directions. Some techniques used to look into the dark matter (hidden Web), and Natural Language Processing (NLP) analysis can help understand the content of Web pages and identify user needs. In this way, the effectiveness of crawlers can be improved both in terms of precision and recall.

Extending the focused crawlers' range to consider resources available on the Semantic Web is a further research topic, which will gain importance as this new technology that allows data to be shared and reused across applications becomes more popular in the Web community. Current prototypes of focused crawlers do not consider the reuse of past experience - namely, what crawlers have discovered and analyzed from past crawls - in new explorations of the Web. The chance to exploit such data in the crawling process could help retrieving information more quickly, reducing the network and computational resources.

References

1. Aggarwal, C.C., Al-Garawi, F., Yu, P.S.: Intelligent crawling on the world wide web with arbitrary predicates. In: Proceedings of the 10th World Wide Web Conference (WWW10), Hong Kong (2001) 96–105 http://www10.org/cdrom/papers/110/.
2. Arasu, A., Cho, J., Garcia-Molina, H., Paepcke, A., Raghavan, S.: Searching the web. ACM Transactions on Internet Technology (TOIT) 1(1) (2001) 2 43
3. Bailey, P., Craswell, N., Hawking, D.: Dark matter on the web. In: Poster Proceedings of the 9th World Wide Web Conference(WWW9), Amsterdam, Netherlands (2000) http://www9.org/final-posters/poster30.html.

4. Bailey, P., Craswell, N., Hawking, D.: Engineering a multi-purpose test collection for web retrieval experiments. Information Processing and Management **39**(6) (2003) 853–871
5. Balabanović, M.: Exploring versus exploiting when learning user models for text recommendation. User Modeling and User-Adapted Interaction **8**(1-2) (1998) 71–102
6. Balabanović, M., Shoham, Y.: Fab: content-based, collaborative recommendation. Communications of the ACM **40**(3) (1997) 66–72
7. Beitzel, S.M., Jensen, E.C., Chowdhury, A., Grossman, D., Frieder, O.: Hourly analysis of a very large topically categorized web query log. In: SIGIR '04: Proceedings of the 27th annual international conference on Research and development in information retrieval, New York, NY, USA, ACM Press (2004) 321–328
8. Bergman, M.K.: The deep web: Surfacing hidden value. The Journal of Electronic Publishing **7**(1) (August 2001)
9. Bergmark, D., Lagoze, C., Sbityakov, A.: Focused crawls, tunneling, and digital libraries. In: ECDL '02: Proceedings of the 6th European Conference on Research and Advanced Technology for Digital Libraries, London, UK, Springer-Verlag (2002) 91–106 http://citeseer.ist.psu.edu/bergmark02focused.html.
10. Bharat, K., Kamba, T., Albers, M.: Personalized, interactive news on the web. Multimedia Systems **6**(5) (1998) 349–358
11. Boldi, P., Codenotti, B., Santini, M., Vigna, S.: Ubicrawler: a scalable fully distributed web crawler. Software, Practice and Experience **34**(8) (2004) 711–726
12. Bonabeau, E., Dorigo, M., Theraulaz, G.: Inspiration for optimization from social insect behavior. Nature **406** (2000) 39–42
13. Bra, P.D., Houben, G.J., Kornatzky, Y.: Information retrieval in distributed hypertexts. In: Proceedings of the 4th RIAO, Intelligent Multimedia, Information Retrieval Systems and Management, New York, NY, USA (1994) 481–491 http://citeseer.ist.psu.edu/debra94information.html.
14. Bra, P.D., Post, R.: Searching for arbitrary information in the www: The fish-search for mosaic. In: Proceedings of the 2nd World Wide Web Conference (WWW2), Chicago, USA (1994) http://citeseer.ist.psu.edu/172936.html.
15. Brin, S., Page, L.: The anatomy of a large-scale hypertextual web search engine. Computer Networks and ISDN Systems **30** (1998) 107–117
16. Broder, A., Kumar, R., Maghoul, F., Raghavan, P., Rajagopalan, S., Stata, R., Tomkins, A., Wiener, J.: Graph structure in the web. In: Proceedings of the 9th World Wide Web Conference (WWW9), Amsterdam, Netherlands (2000) 309–320 http://www9.org/w9cdrom/160/160.html.
17. Chakrabarti, S.: Recent results in automatic web resource discovery. ACM Computing Surveys **31**(4es) (1999) 17
18. Chakrabarti, S., Punera, K., Subramanyam, M.: Accelerated focused crawling through online relevance feedback. In: WWW '02: Proceedings of the 11th international conference on World Wide Web, New York, NY, USA, ACM Press (2002) 148–159 http://www2002.org/CDROM/refereed/336/.
19. Chakrabarti, S., van den Berg, M., Dom, B.: Distributed hypertext resource discovery through examples. In: VLDB '99: Proceedings of the 25th International Conference on Very Large Data Bases, San Francisco, CA, USA, Morgan Kaufmann Publishers Inc. (1999) 375–386 http://www.vldb.org/conf/1999/P37.pdf.
20. Chakrabarti, S., van den Berg, M., Dom, B.: Focused crawling: A new approach to topic-specific web resource discovery. In: Proceedings of the 8th World Wide Web Conference (WWW8), Toronto, Canada (1999) 1623–1640 http://www8.org/w8-papers/5a-search-query/crawling/index.html.
21. Chen, H., Chung, Y.M., Ramsey, M., Yang, C.C.: A smart itsy bitsy spider for the web. Journal of the American Society for Information Science **49**(7) (1998) 604–618

22. Chi, E.H., Pirolli, P., Chen, K., Pitkow, J.: Using information scent to model user information needs and actions on the web. In: Proceedings of the ACM Conference on Human Factors in Computing Systems (CHI2001), Seattle, WA, USA (2001) 490–497
23. Cho, J., Garcia-Molina, H.: The evolution of the web and implications for an incremental crawler. In: VLDB '00: Proceedings of the 26th International Conference on Very Large Data Bases, San Francisco, CA, USA, Morgan Kaufmann Publishers Inc. (2000) 200–209
24. Cho, J., Garcia-Molina, H.: Synchronizing a database to improve freshness. In: SIGMOD '00: Proceedings of the 2000 ACM SIGMOD international conference on Management of data, New York, NY, USA, ACM Press (2000) 117–128
25. Cho, J., Garcia-Molina, H.: Parallel crawlers. In: WWW '02: Proceedings of the 11th international conference on World Wide Web, New York, NY, USA, ACM Press (2002) 124–135 http://www2002.org/CDROM/refereed/108/index.html.
26. Cho, J., Garcia-Molina, H.: Estimating frequency of change. ACM Transactions on Internet Technology (TOIT) **3**(3) (2003) 256–290
27. Cho, J., Garcia-Molina, H., Page, L.: Efficient crawling through url ordering. Computer Networks and ISDN Systems **30**(1–7) (1998) 161–172
28. Chung, C., Clarke, C.L.A.: Topic-oriented collaborative crawling. In: CIKM '02: Proceedings of the eleventh international conference on Information and knowledge management, New York, NY, USA, ACM Press (2002) 34–42
29. Davison, B.D.: Topical locality in the web. In: SIGIR '00: Proceedings of the 23rd annual international ACM SIGIR conference on Research and development in information retrieval, New York, NY, USA, ACM Press (2000) 272–279
30. de Carvalho Fontes, A., Silva, F.S.: Smartcrawl: a new strategy for the exploration of the hidden web. In: WIDM '04: Proceedings of the 6th annual ACM international workshop on Web information and data management, New York, NY, USA, ACM Press (2004) 9–15
31. Delort, J.Y., Bouchon-Meunier, B., Rifqi, M.: Enhanced web document summarization using hyperlinks. In: HYPERTEXT '03: Proceedings of the fourteenth ACM conference on Hypertext and hypermedia, New York, NY, USA, ACM Press (2003) 208–215
32. Diligenti, M., Coetzee, F., Lawrence, S., Giles, C.L., Gori, M.: Focused crawling using context graphs. In: VLDB '00: Proceedings of the 26th International Conference on Very Large Data Bases, San Francisco, CA, USA, Morgan Kaufmann Publishers Inc. (2000) 527–534 http://www.vldb.org/conf/2000/P527.pdf.
33. Diligenti, M., Maggini, M., Pucci, F.M., Scarselli, F.: Design of a crawler with bounded bandwidth. In: WWW Alt. '04: Proceedings of the 13th international World Wide Web conference on Alternate track papers & posters, New York, NY, USA, ACM Press (2004) 292–293 http://www2004.org/proceedings/docs/2p292.pdf.
34. Doorenbos, R.B., Etzioni, O., Weld, D.S.: A scalable comparison-shopping agent for the world-wide web. In: AGENTS '97: Proceedings of the first international conference on Autonomous agents, New York, NY, USA, ACM Press (1997) 39–48
35. Ehrig, M., Maedche, A.: Ontology-focused crawling of web documents. In: SAC '03: Proceedings of the 2003 ACM symposium on Applied computing, New York, NY, USA, ACM Press (2003) 1174–1178
36. Fetterly, D., Manasse, M., Najork, M., Wiener, J.: A large-scale study of the evolution of web pages. In: WWW '03: Proceedings of the 12th international conference on World Wide Web, New York, NY, USA, ACM Press (2003) 669–678 http://www2003.org/cdrom/papers/refereed/p097/P97sources/p97-fetterly.html.
37. Furnas, G.W.: Effective view navigation. In: CHI '97: Proceedings of the SIGCHI conference on Human factors in computing systems, New York, NY, USA, ACM Press (1997) 367–374
38. Furnas, G.W., Landauer, T.K., Gomez, L.M., Dumais, S.T.: The vocabulary problem in human-system communication. Communications of the ACM **30**(11) (1987) 964–971

39. Gasparetti, F., Micarelli, A.: Adaptive web search based on a colony of cooperative distributed agents. In Klusch, M., Ossowski, S., Omicini, A., Laamanen, H., eds.: Cooperative Information Agents. Volume 2782., Springer-Verlag (2003) 168–183

40. Gasparetti, F., Micarelli, A.: Swarm intelligence: Agents for adaptive web search. In: Proceedings of the 16th European Conference on Artificial Intelligence (ECAI 2004). (2004) 1019–1020 http://citeseer.ist.psu.edu/738711.html.

41. Gasparetti, F., Micarelli, A.:   User profile generation based on a memory retrieval theory.   In: Proc. 1st International Workshop on Web Personalization, Recommender Systems and Intelligent User Interfaces (WPRSIUI'05). (2005) 59–68 http://citeseer.ist.psu.edu/gasparetti05user.html.

42. Gauch, S., Speretta, M., Chandramouli, A., Micarelli, A.: User profiles for personalized information access. In Brusilovsky, P., Kobsa, A., Nejdl, W., eds.: The Adaptive Web: Methods and Strategies of Web Personalization. Volume 4321 of Lecture Notes in Computer Science. Springer-Verlag, Berlin Heidelberg New York (2007) this volume

43. Gulli, A., Signorini, A.: The indexable web is more than 11.5 billion pages. In: WWW '05: Special interest tracks and posters of the 14th international conference on World Wide Web, New York, NY, USA, ACM Press
(2005) 902–903 http://www.cs.uiowa.edu/~asignori/web-size/.

44. Hersovicia, M., Jacovia, M., Maareka, Y.S., Pellegb, D., Shtalhaima, M., Ura, S.:   The shark-search algorithm – an application: tailored web site mapping.   In: Proceedings of the 7th World Wide Web Conference (WWW7), Brisbane, Australia (1998) 317–326 http://www7.scu.edu.au/1849/com1849.htm.

45. Heydon, A., Najork, M.: Mercator: A scalable, extensible web crawler. World Wide Web 2(4) (1999) 219–229

46. Joachims, T., Freitag, D., Mitchell, T.M.: Webwatcher: A tour guide for the world wide web. In: Proceedings of the 15h International Conference on Artificial Intelligence (IJCAI1997). (1997) 770–777 http://citeseer.ist.psu.edu/16829.html.

47. Kleinberg, J.: Authoritative sources in a hyperlinked environment. In: Proceedings of the 9th annual ACM-SIAM symposium on Discrete algorithms, San Francisco, CA, USA (1998) 668–677 http://www.cs.cornell.edu/home/kleinber/auth.pdf.

48. Kruger, A., Giles, C.L., Coetzee, F.M., Glover, E., Flake, G.W., Lawrence, S., Omlin, C.: Deadliner: building a new niche search engine.   In: CIKM '00: Proceedings of the ninth international conference on Information and knowledge management, New York, NY, USA, ACM Press (2000)
272–281 http://citeseer.ist.psu.edu/kruger00deadliner.html.

49. Kumar, R., Novak, J., Raghavan, P., Tomkins, A.:   On the bursty evolution of blogspace.   In: WWW '03: Proceedings of the 12th international conference on World Wide Web, New York, NY, USA, ACM Press (2003) 568–576 http://www2003.org/cdrom/papers/refereed/p477/p477-kumar/p477-kumar.htm.

50. Lage, J.P., da Silva, A.S., Golgher, P.B., Laender, A.H.F.: Automatic generation of agents for collecting hidden web pages for data extraction. Data and Knowledge Engineering 49(2) (2004) 177–196

51. Lau, T., Horvitz, E.: Patterns of search: analyzing and modeling web query refinement. In: UM '99: Proceedings of the seventh international conference on User modeling, Secaucus, NJ, USA, Springer-Verlag New York, Inc. (1999) 119–128

52. Lawrence, S., Giles, L.C.: Searching the world wide web. Science 280 (1998) 98–100

53. Lawrence, S., Giles, L.C.: Accessibility of information on the web. Nature 400 (1999) 107–109

54. Levene, M., Poulovassilis, A.: Web dynamics. Software Focus 2(2) (2001) 60–67

55. Lu, W.H., Chien, L.F., Lee, H.J.: Translation of web queries using anchor text mining. ACM Transactions on Asian Language Information Processing (TALIP) 1(2) (2002) 159–172

56. Marchiori, M.: The quest for correct information on the web: Hyper search engines. In: Proceedings of the 6th World Wide Web Conference (WWW6), Santa Clara, CA, USA (1997) 1225–1235 http://www.w3.org/People/Massimo/papers/WWW6/.
57. Menczer, F.: Lexical and semantic clustering by web links. Journal of the American Society for Information Science and Technology 55(14) (2004) 1261–1269
58. Menczer, F., Belew, R.K.: Adaptive retrieval agents: Internalizing local context and scaling up to the web. Machine Learning 31(11–16) (2000) 1653–1665
59. Menczer, F., Pant, G., Srinivasan, P.: Topical web crawlers: Evaluating adaptive algorithms. ACM Transactions on Internet Technology 4(4) (2004) 378–419
60. Metropolis, N., Rosenbluth, A.W., Rosenbluth, M.N., Teller, A., Teller, E.: Equations of state calculations by fast computing machines. Journal of Chemical Physics 21(6) (1953)
61. Micarelli, A., Gasparetti, F., Sciarrone, F., Gauch, S.: Personalized search on the world wide web. In Brusilovsky, P., Kobsa, A., Nejdl, W., eds.: The Adaptive Web: Methods and Strategies of Web Personalization. Volume 4321 of Lecture Notes in Computer Science. Springer-Verlag, Berlin, Heidelberg, and New York (2007) this volume
62. Micarelli, A., Sciarrone, F., Marinilli, M.: Web document modeling. In Brusilovsky, P., Kobsa, A., Nejdl, W., eds.: The Adaptive Web: Methods and Strategies of Web Personalization. Volume 4321 of Lecture Notes in Computer Science. Springer-Verlag, Berlin Heidelberg New York (2007) this volume
63. Mizuuchi, Y., Tajima, K.: Finding context paths for web pages. In: Proceedings of the 10th ACM Conference on Hypertext and Hypermedia: Returning to Our Diverse Roots (HYPERTEXT99), Darmstadt, Germany (1999) 13–22
64. Najork, M., Heydon, A.: High-performance web crawling. In Abello, J., Pardalos, P.M., , Resende, M.G., eds.: Handbook of massive data sets. Kluwer Academic Publishers, Norwell, MA, USA (2002) 25–45
65. Najork, M., Wiener, J.L.: Breadth-first search crawling yields high-quality pages. In: Proceedings of the 10th World Wide Web Conference (WWW10), Hong Kong (2001) 114–118 http://www10.org/cdrom/papers/208/.
66. Ntoulas, A., Cho, J., Olston, C.: What's new on the web?: the evolution of the web from a search engine perspective. In Feldman, S.I., Uretsky, M., Najork, M., Wills, C.E., eds.: Proceedings of the 13th international conference on World Wide Web, WWW 2004, New York, NY, USA, May 17-20, 2004, ACM (2004) 1–12 http://www2004.org/proceedings/docs/1p1.pdf.
67. Ntoulas, A., Zerfos, P., Cho, J.: Downloading textual hidden web content through keyword queries. In Marlino, M., Sumner, T., III, F.M.S., eds.: ACM/IEEE Joint Conference on Digital Libraries, JCDL 2005, Denver, CA, USA, June 7-11, 2005, Proceedings, ACM (2005) 100–109
68. Page, L., Brin, S., Motwani, R., Winograd, T.: The pagerank citation ranking: Bringing order to the web. Technical report, Stanford Digital Library Technologies Project (1998) http://dbpubs.stanford.edu/pub/1999-66.
69. Pinkerton, B.: Finding what people want: Experiences with the webcrawler. In: Proceedings of the 2nd World Wide Web Conference(WWW2), Chicago, USA (1994) 821–829
70. Pirolli, P., Card, S.K.: Information foraging. Psychological Review 106 (1999) 643–675
71. Raghavan, S., Garcia-Molina, H.: Crawling the hidden web. In: VLDB '01: Proceedings of the 27th International Conference on Very Large Data Bases, San Francisco, CA, USA, Morgan Kaufmann Publishers Inc. (2001) 129–138
72. Rennie, J., McCallum, A.: Using reinforcement learning to spider the web efficiently. In: ICML '99: Proceedings of the Sixteenth International Conference on Machine Learning, San Francisco, CA, USA, Morgan Kaufmann Publishers Inc. (1999) 335–343 http://citeseer.ist.psu.edu/7537.html.

73. Rumelhart, D.E., Hinton, G.E., Williams, R.J.: Learning internal representations by error propagation. In Rumelhart, D.E., McClelland, J.L., eds.: Parallel distributed processing: explorations in the microstructure of cognition, vol. 1: foundations. MIT Press, Cambridge, MA, USA (1986) 318–362

74. Rungsawang, A., Angkawattanawit, N.: Learnable topic-specific web crawler. Journal of Network and Computer Applications **28**(2) (2005) 97–114

75. Salton, G., McGill, M.J.: Introduction to Modern Information Retrieval. McGraw-Hill, New York (1983)

76. Shkapenyuk, V., Suel, T.: Design and implementation of a high-performance distributed web crawler. In: Proceedings of the 18th International Conference on Data Engineering (ICDE'02), Washington, DC, USA, IEEE Computer Society (2002) 357

77. Spink, A., Jansen, B.J.: A study of web search trends. Webology **1**(2) (December 2004) http://www.webology.ir/2004/v1n2/a4.html.

78. Spink, A., Wolfram, D., Jansen, M.B.J., Saracevic, T.: Searching the web: the public and their queries. Journal of the American Society for Information Science **52**(3) (2001) 226–234

79. Srinivasan, P., Menczer, F., Pant, G.: A general evaluation framework for topical crawlers. Information Retrieval **8**(3) (2005) 417–447

80. Steele, R.: Techniques for specialized search engines. In: Proc. Internet Computing 2001, Las Vegas (June 25–28 2001) http://citeseer.ist.psu.edu/steele01techniques.html.

81. Sutton, R.S., Barto, A.G.: Introduction to Reinforcement Learning. MIT Press, Cambridge, MA, USA (1998)

82. Yang, C.C., Yen, J., Chen, H.: Intelligent internet searching agent based on hybrid simulated annealing. Decision Support Systems **28**(2) (2000) 269–277

83. Yuwono, B., Lam, S.L.Y., Ying, J.H., Lee, D.L.: A World Wide Web resource discovery system. In: Proceedings of the 4th World Wide Web Conference (WWW4), Boston, Massachusetts, USA (1995) 145–158 http://www.w3.org/Conferences/WWW4/Papers/66/.

8

Adaptive Navigation Support

Peter Brusilovsky

School of Information Sciences
University of Pittsburgh, Pittsburgh PA 15260
peterb@mail.sis.pitt.edu

Abstract. Adaptive navigation support is a specific group of technologies that support user navigation in hyperspace, by adapting to the goals, preferences and knowledge of the individual user. These technologies, originally developed in the field of adaptive hypermedia, are becoming increasingly important in several adaptive Web applications, ranging from Web-based adaptive hypermedia to adaptive virtual reality. This chapter provides a brief introduction to adaptive navigation support, reviews major adaptive navigation support technologies and mechanisms, and illustrates these with a range of examples.

8.1 Introduction

Adaptive hypermedia [9] is a research area at the crossroads of hypermedia and user modeling. Adaptive hypermedia systems (AHS) offer an alternative to the traditional "one-size-fits-all" hypermedia and Web systems by adapting to the goals, interests, and knowledge of individual users as they are represented in the individual *user models*. This chapter is focused on *adaptive navigation support* technologies originally developed in the field of adaptive hypermedia. By adaptively altering the appearance of links on every browsed page, using such methods as *direct guidance, adaptive ordering, link hiding and removal*, and *adaptive link annotation*, these technologies support personalized access to information. Over the last 10 years, adaptive navigation support technologies have been used in many adaptive Web systems in a range of application areas from e-learning to e-commerce. The evaluation of these technologies has demonstrated their ability to allow users to achieve their goals faster, reduce navigational overhead, and increase satisfaction [7; 18; 50; 52; 71].

After a brief introduction to the history of adaptive navigation support, this chapter offers a state-of-the-art overview of adaptive navigation support. The overview is divided into two parts. The first part focuses on adaptation technologies and attempts to answer the question: *What kind of adaptation effects may be useful to provide guidance to the users of Web hypermedia systems?* The second part focuses on adaptation mechanisms and attempts to answer the question: *How can these adaptation effects be produced?* Both parts are illustrated with a range of examples. The last section discusses the prospects of extending adaptive navigation support beyond Web hypermedia.

P. Brusilovsky, A. Kobsa, and W. Nejdl (Eds.): The Adaptive Web, LNCS 4321, pp. 263–290, 2007.
© Springer-Verlag Berlin Heidelberg 2007

8.2   Adaptive Navigation Support: From Adaptive Hypermedia to the Adaptive Web

Research on adaptive navigation support in hypermedia can be traced back to the early 1990's. By that time, several research teams had recognized standard problems found in static hypertext within different application areas, and had begun to explore various ways to adapt the behavior of hypertext and hypermedia systems to individual users. A number of teams addressed problems related to navigation in hypermedia—such as the problem of inefficient navigation or the problem of being lost—which had been discovered when the field of hypertext reached relative maturity at the end of the 1980's [46]. Within a few years, a number of navigation support technologies were proposed [4; 19; 33; 52]. While the proposed technologies were relatively different, they shared the same core idea: within a hypertext page (node), adapt the presentation of links to the goals, knowledge, and preferences of the individual user. The adaptive navigation support technologies introduced by early adaptive hypermedia systems were later classified as *direct guidance, sorting, hiding, annotation,* and *map adaptation* [8]. Most of these systems used adaptation mechanisms based on manual page indexing and provided navigation support within a closed corpus of documents.

The Web as "hypermedia for everyone" immediately provided an attractive platform for adaptive hypermedia applications. The problem of navigation support in Web hypermedia attracted many new researchers to the field. A good number of these researchers were motivated by pre-Web adaptive hypermedia and focused on exploring a set of known adaptive hypermedia technologies in the new Web context. Other researchers suggested new techniques such as link *disabling* and *generation* [9]. Several new adaptation mechanisms were explored including *content-based* and *social* mechanisms that allowed navigation support in an open corpus. As the Web has developed, the focus of work has also moved from exploring isolated techniques using "lab-level" systems to developing and exploring "real world" systems for different application areas such as e-learning, e-commerce, and virtual museums.

Altogether, pre-Web and Web-based AHS with adaptive navigation support explored a broad range of adaptation technologies and mechanisms in many application areas. The knowledge of these technologies and mechanisms and their effectiveness is important for the developers of future adaptive Web systems. The next two sections attempt to summarize this knowledge, presenting the most popular adaptation technologies and mechanisms, and pointing out relevant empirical studies.

8.3   Adaptive Navigation Support: Adaptation Technologies

8.3.1   Direct Guidance

Direct guidance is the simplest technology for adaptive navigation support. Direct guidance suggests the "next best" node (or sometimes, several alternative nodes) for the user to visit according to the user's goals, knowledge, or/and other parameters that have been represented in the user model. On the interface level, direct guidance can be presented to the user in two main forms. If the link to the suggested node is already present on the page, it can be outlined or emphasized in some other way. For example,

Welcome to the WebWatcher Project

....

Overview

WebWatcher is a "tour guide" agent for the world wide web. Once you tell it what kind of information you seek, it accompanies you from page to page as you browse the web, highlighting hyperlinks that it believes will be of interest. Its strategy for giving advice is learned from feedback from earlier tours.

Try it!

WebWatcher can help you search for information starting from any of the following pages. (but it has learned the most about the first of these).

- **CMU School of Computer Science Front Door** After arriving at this page, click on "The WebWatcher tour guide." under the heading **SCS Resources**
- ▓▓**Machine Learning Information Services**▓▓
- ARPA Intelligent Integration of Information Home Page
- ARPA Real Time Planning and Control Home Page

Publications

- *WebWatcher: A Learning Apprentice for the World Wide Web*, in the *1995 AAAI Spring Symposium on Information Gathering from Heterogeneous, Distributed Environments*, Stanford, March 1995. *Abstract:* A description of WebWatcher and a comparison of different machine learning approaches to suggest hyperlinks.
- *WebWatcher: Machine Learning and Hypertext*, to appear in Fachgruppentreffen Maschinelles Lernen, Dortmund, Germany, August 1995.

Fig. 8.1. Direct guidance in Personal WebWatcher. The recommended link (second from the top) is outlined by a pair of "curious eyes" icons. Used with permission from the author [62].

WebWatcher [1] and Personal WebWatcher [62] indicated the recommended link(s) by a pair of icons showing curious eyes (Fig. 8.1). Alternatively, the system can generate a dynamic "next" link which is connected to the "next best" node.

A known problem with direct guidance is that it provides no support for users who don't wish to follow the system's suggestions. Due to this problem, although direct guidance was popular in the early days of adaptive hypermedia, it is now mostly replaced by other navigation support technologies, which will be introduced below. The only group of systems where this approach remains popular are adaptive educational hypermedia systems, especially those that have roots in Intelligent Tutoring Systems such as HyperTutor [67], ELM-ART [75], or InterBook [14]. In this group, direct guidance became the hypermedia form of the traditional *curriculum sequencing* mechanisms. Several studies reviewed in [10] demonstrated that novice users with poor domain knowledge have problems in dealing with alternative navigation choices and can be best supported by direct guidance technology.

8.3.2 Link Ordering

The idea of an *adaptive sorting* or *ordering* technology is to prioritize all the links of a particular page according to the user model and some user-valuable criteria: the closer to the top, the more relevant the link is. While adaptive sorting was first introduced in 1990 in the Hypadapter system [49], the most frequently referred example of this technology is HYPERFLEX [52]. HYPERFLEX attempts to order links from the current page to related pages according to the user-perceived relevance of these pages to the current one. If the user thinks that the presented order is incorrect, the links can be manually reordered by dragging. Manual link reordering is considered by the system as a means of relevance feedback and is used to update the user model. If the user selects the current search goal from the list of existing goals (new goals can also be introduced), link ordering on every page also takes into account link relevance to the selected goal. Most important to the HYPERLEX work was not the specific adaptation technology, but rather the study of the user's link ordering, which was reported in the same paper [52]. The study demonstrated that adaptive link ordering significantly reduces navigation time and the number of steps that are required to locate the information that the user is looking for. These results helped to attract attention to link ordering and adaptive navigation support in general. It should be noted, though, that time reduction is not exclusively limited to sorting technologies. Similar time/steps reduction was later observed for other navigation support technologies, such as link hiding and annotation [18; 64] and is currently considered to be one of the most important values of adaptive navigation support in general.

Despite its demonstrated effectiveness, link sorting has not become very popular, due to its limited applicability. As shown in Table 8.1, it can be used for non-contextual links, but is difficult to use for an index page or a table of contents (which usually have a predefined order of links), and can never be used with contextual links or maps. Another problem with adaptive ordering is that this technology makes the order of links unstable: it may change each time the user enters the page. Since the first introduction of link sorting, several user studies have demonstrated that unstable order of options in menus and toolbars creates problems for at least some categories of users [34; 53]. As a result, this technology is presently used in only a few contexts where the unstable order of links creates no problem.

One such beneficial context is adaptation of link order to long-term user characteristics. In this context, different users may see a different order for links, but it is stable for each user for the whole time they are working with the system. For example, several adaptive e-learning systems order links to the different educational resources available for a topic according to the relevance of these resources to the user's learning style [55].

Another appropriate context includes several kinds of system where all or some pages have an unstable set of links. Since the set of links on a page is not fixed, a stable order does not exist anyway. In this situation the "conceptually stable" ordering offered by link sorting can become an attractive solution. Good examples of this may be found among adaptive news systems reviewed in Chapter 18 of this book [3] and collaborative resource gathering systems such as CoFIND [39] or COMTELLA [26].

Adaptive news systems typically present links to recommended news articles in a single list or on several pages by category. This list is unstable because new articles are constantly added and old articles removed. In this context, it is very natural to sort the links according to the modeled interests of the user. This ordering is typically performed by content-based mechanisms.

In collaborative resource gathering systems, users collect useful Web resources by adding interesting links to topics. Each topic may have a short introduction and a collection of links that is unstable by its nature (since resources are constantly added and even sometimes removed). To present these links, the cited systems use social mechanisms to sort topic links according to the perceived community interests. For similar reasons, link sorting is frequently used in combination with link generation (see section 8.3.6 and Fig. 8.10 for examples of this combination).

8.3.3 Link Hiding

The purpose of navigation support by *hiding* is to restrict the navigation space by hiding, removing, or disabling links to irrelevant pages. A page can be considered irrelevant for several reasons: for example, if it is not related to the user's current learning goal or if it presents materials which the user is not yet prepared to understand. Hiding protects users from the complexity of the whole hyperspace and reduces their cognitive overload. Educational hypermedia systems have been the main application area where adaptive hiding techniques have been suggested and explored. Indeed, beginning with just a part of the whole picture then introducing other components step by step as the student progresses through the course is a popular educational approach and adaptive hiding offers a simple way to implement this. Early adaptive hypermedia systems used a very simple method of hiding links—essentially removing the link as well as the anchor from a page. A good example is the ISIS-Tutor educational hypermedia system [18], which shows very few links when the student begins to work with the system but gradually makes more and more links visible, reacting to the growth of the student's knowledge of the subject. De Bra and Calvi [29] later called the ISIS-Tutor approach *link removal* and have suggested several other variants for link hiding based on the separation of three features of a link: the anchor, the visible indication, and the functionality. For example, link *hiding* preserves the link anchor (hot word), but removes all visual indications that it is a link (i.e., blue color and underline). Link *disabling* removes the functionality, i.e., the ability of the link to take the user to the related page. Both technologies (as well as their combination) extend the applicability of link hiding to contextual links where the anchor simply can't be removed. An example of link hiding in De Bra's AHA! framework is shown on Fig. 8.2. This example is taken from their adaptive paper, which presented their framework [31]. A number of studies of link hiding revealed that it is best used as a "unidirectional" technology. While gradual link enabling as used in ISIS-Tutor has been acceptable and effective, the reverse approach has been found questionable: users become very unhappy when previously available links become invisible or disabled.

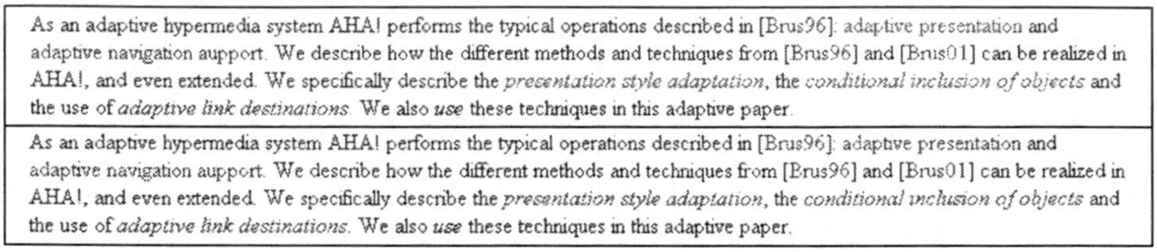

As an adaptive hypermedia system AHA! performs the typical operations described in [Brus96]: adaptive presentation and adaptive navigation support. We describe how the different methods and techniques from [Brus96] and [Brus01] can be realized in AHA!, and even extended. We specifically describe the *presentation style adaptation*, the *conditional inclusion of objects* and the use of *adaptive link destinations*. We also *use* these techniques in this adaptive paper.

As an adaptive hypermedia system AHA! performs the typical operations described in [Brus96]: adaptive presentation and adaptive navigation support. We describe how the different methods and techniques from [Brus96] and [Brus01] can be realized in AHA!, and even extended. We specifically describe the *presentation style adaptation*, the *conditional inclusion of objects* and the use of *adaptive link destinations*. We also *use* these techniques in this adaptive paper.

Fig. 8.2. Link hiding in AHA! framework taken from the adaptive paper [31]. The upper fragment shows several links leading to other sections of the paper. On the lower fragment these links are hidden—the purple color indicating the presence of a link is replaced by the black color of the surrounding text.

8.3.4 Link Annotation

The idea of *adaptive annotation* technology is to augment the links with some form of annotation, which lets the user know more about the current state of the nodes behind the annotated links. These annotations are most often provided in the form of visual cues. Manuel Excel [33] introduced link annotation with different icons, ISIS-Tutor [17] changed the color and intensity of the anchors, and Hypadapter [49] explored altering anchor font sizes. The Web generation of adaptive hypermedia systems introduced several kinds of verbal annotations that could be shown next to the anchor [45], on the browser's status bar [14], or as a *gloss* that popped up when the user moused over a link [79]. All of these approaches to link annotation are now in use, but the most popular are probably icon-based annotation and mouseovers. Naturally, annotation can be used with all possible forms of links. This technology preserves a stable order to the links, thus avoiding problems with incorrect mental maps. Annotation is generally a more powerful technology than hiding: hiding can distinguish only two states for related nodes—relevant and non-relevant—while the currently existing annotation applications can distinguish up to six states. For all the above reasons, adaptive annotation has grown into the most frequently used adaptive navigation support technology.

Some of the benefits of adaptive link annotation have been explored in several studies. For example, an early study of the ISIS-Tutor system [18] compared three versions of the ISIS-Tutor: non-adaptive, adaptive annotation, and a combination of both adaptive hiding and annotation. The results of the study demonstrated that the same educational goal is achieved with either of the adaptive versions with much less navigational overhead than with the non-adaptive version. The overall number of navigation steps, the number of unforced repetitions of previously studied concepts, and the number of task repetitions (i.e., trials to solve a previously visited task) were significantly smaller for both adaptive versions.

A popular example of adaptive annotation in Web hypermedia is ELM-ART [20], which was one of the first Web-based systems with adaptive navigation support. ELM-ART introduced the traffic light metaphor for adaptive navigation support in educational hypermedia. In this metaphor, a green bullet in front of a link indicates recommended readings, while a red bullet indicates that the student may not be able to

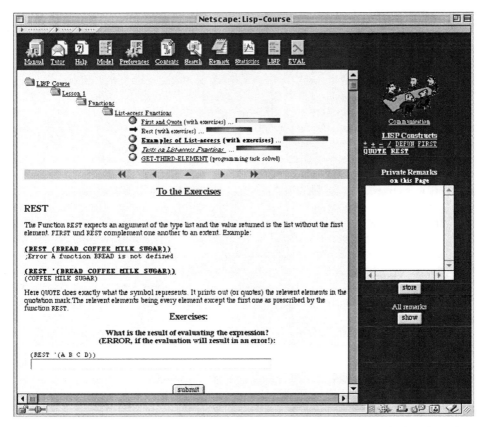

Fig. 8.3. Adaptive navigation support in ELM-ART, an electronic textbook for learning LISP. Adaptive annotation in the form of colored bullets (a traffic light metaphor) shows the educational state of pages behind the links. Adaptive annotation in the form of progress bars visualizes the student's demonstrated level of knowledge of related concepts.

understand the information behind the link yet. Other colors, like yellow or white, indicate more educational states such as the lack of new knowledge behind the link. This kind of annotation is produced by an indexing-based mechanism and will be explained in more details in section 8.4.4. In addition to link annotation, ELM-ART also supports direct guidance. Fig. 8.3 shows adaptive annotation in the most recent version of ELM-ART [75]. This version augments the traffic-light annotation (which indicates the educational status of a page) with progress-based annotation (which indicates the level of user knowledge for a LISP concept associated with this page). A combination of these two kinds of annotations is currently very popular in adaptive educational hypermedia. A study of ELM-ART [75] demonstrated that casual users stay longer within a system when adaptive navigation support is provided. The study also provided evidence that direct guidance works best for users with little previous knowledge while adaptive annotation is most helpful for users with a reasonable amount of subject knowledge.

8.3.5 Link Generation

Link generation is the "newest" adaptive navigation support technology. There has been little need to introduce link generation in the context of pre-Web adaptive hypermedia with its small, well-linked, closed corpus document collections. This technology was introduced in several early adaptive Web systems in 1996 [14; 75; 78] and became very popular in Web hypermedia with its abundance of resources. Unlike classic annotation, sorting or hiding technologies that adapt the presentation of *pre-authored* links, link generation actually creates new, non-authored links on a page. There are three known kinds of link generation: (1) discovering new, useful links between the documents and adding them permanently to the set of existing links; (2) generating links for similarity-based navigation between items; and (3) the dynamic recommendation of links that are useful within the current context to the current user (i.e., the current goal, knowledge, or interests, as reflected in the user model). The first two kinds of link generation are typically non-adaptive. We should mention, however, several known projects that explored creating new links for a group of users as a result of an analysis of group navigation patterns [5; 59; 78] and a few attempts to develop adaptive similarity-based navigation [23]. The third technology is naturally adaptive, since link generation is driven by the user's profile and context.

Since link generation is now very popular in several kinds of adaptive Web-based systems, this section is a good place to comment on the similarities and differences between using this technology in adaptive navigation support systems and the various Web recommender systems that are presented in chapters 9, 10, 11, and 12 of this book [24; 66; 69; 70]. Recommender systems attempt to suggest a list of items that are relevant to the user's short- or long-term interests. These items may or may not be part of a hyperspace. If they are, the recommendation can be presented as a set of generated links. Even those systems that attempt to recommend items in hyperspace typically do not take the current user location in hyperspace (context) into account, and instead offer links that should be of interest in general. On the other hand, navigation support systems focus on helping users to find their way through hyperspace by adapting links on a page. Link adaptation can take into account various features of the user and may take many forms as well, including, as a specific case, link generation adapted to the user's interests. In all cases, navigation support techniques provide guidance that takes into account the user's current location in hyperspace. So, when guidance is provided by link generation, a navigation support system attempts to introduce additional links that may be useful in the current context. Since navigation support systems focus on the interface and recommender systems focus on the underlying technology, the difference between these two groups is not clear-cut. Evidently, a small class of systems that generates links according to the user's interests and takes into account the user's current location can be classified as both a Web recommender system and an adaptive navigation support system. A well-known example of this is Amazon.com (http://amazon.com). This system recommends links to products that were considered or purchased by other users who viewed the current product.

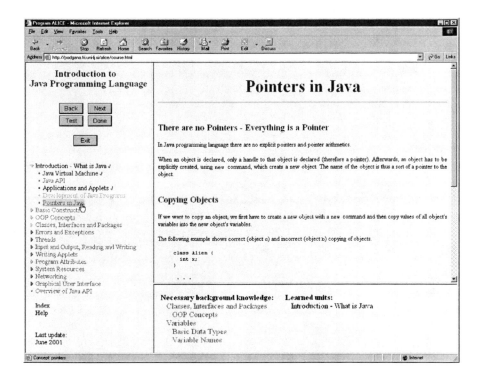

Fig. 8.4. Link generation and link annotation in ALICE. Follow-up links are generated in the bottom right frame in three groups - next possible units, necessary background units, and all learned units. The example on the figure doesn't suggest next possible units since the current unit "Pointers in Java" is not yet ready to be learned (note that it is annotated with red color on the table of contents in the left frame).

A good example of link generation adapted to user knowledge is ALICE [54], an electronic textbook about the Java programming language. ALICE includes 13 chapters and 97 sections devoted to different Java concepts and uses link generation as the main navigation support approach. There are no stable links between sections; instead, the links are generated dynamically according to the current user level of knowledge. These dynamically-generated links are added to the end of the viewed section in three groups—next possible units, necessary background units, and all learned units (Fig 8.4). The system uses a sophisticated approach to model the user's knowledge of Java, which is reviewed in more detail in Chapter 1 of this book [16]. The evaluation of navigation support in ALICE revealed that students who follow the generated navigation suggestions score better on tests.

8.3.6 Comparing and Combining the Technologies

The link adaptation technologies reviewed above have a lot in common, since they are motivated by the same need, guiding the user in hyperspace. At the same time, these technologies are quite different in their applicability. Part of this is due to the technical applicability of each specific technology for adapting different kinds of links.

Hypertext links (i.e., visible and "clickable" representations of the related pages to which the user can navigate) can be classified in several groups (Table 8.1):

Contextual links or "real hypertext" links. This type comprises "hotwords" in texts, "hot spots" in pictures, and other kinds of links, which are embedded in the context of the page content and cannot be removed from it. These links and the corresponding anchors, can be annotated or disabled, but cannot be sorted or completely hidden.

Local non-contextual links. This type includes all kinds of links on regular hyper-media pages, which are not embedded in the context of the page. They can appear as a set of buttons, a list, or a pop-up menu. These links are easy to manipulate—they can be sorted, removed, generated, or annotated, although disabling or hiding this kind of links (with the anchor preserved) makes little sense.

Links from index and table of contents. An index or a table of contents page can be considered to be a special kind of page, which contains only links that are organized in a specific order (content order for content pages and alphabetic order for index pages). As a rule, links from index and content pages are non-contextual, yet these links can't be sorted and application of all hiding technologies in this context has questionable usability.

Links on local maps and links on global hyperspace maps. Maps usually graphi-cally represent a hyperspace or a local area of hyperspace as a network of nodes con-nected by arrows. Using maps, the user can directly navigate to all nodes visible on the map by merely clicking on a representation of the desired node. From a navigation point of view, these clickable representations of nodes are navigational links, while paradoxically, the arrows serving as a representation of links are not used for direct navigation.

In brief, the analysis of technical applicability demonstrates that some technologies have much wider applicability than others. It is not surprising that the most universal technologies—annotation and generation—are also currently the most popular. How-ever, there is also another aspect to the applicability: A range of studies of adaptive navigation support systems indicates that the effect of a specific technology may be different for different classes of users. For example, a number of studies provide evi-dence that direct guidance is beneficial to users with a low level of domain knowl-edge, while link annotation works best with users who are already above the starting level of knowledge [10]. The applicability of different technologies is important to consider when developing adaptive navigation support systems.

In addition to the applicability limits, different technologies may be best suited for the different needs of an adaptive system. As a result, we see fewer and fewer "purist" systems that use exactly one of the technologies. The majority of practical systems use different technologies in parallel or in different parts of the system. For example, among the systems already mentioned above, ISIS-Tutor uses direct guidance, hiding, and annotation; Hypadapter uses sorting, hiding, and annotation; AHA! uses hiding and annotation; ALICE uses generation and annotation, and both InterBook and ELM-ART use direct guidance, annotation, and generation. Sometimes different technologies used in the same system are based on different mechanisms, but more frequently the same mechanism powers all adaptation technologies in a system. An example of using an index–based mechanism to produce direct guidance, annotation, and generation in InterBook is reviewed in section 8.4.4.

Table 8.1. Adaptive navigatßion support technologies and their applicability.

	Direct guidance	Sorting	Hiding	Annotation	Generation
Contextual links	OK		Disabling	OK	
Non-contextual links	OK	OK	OK	OK	OK
Table of contents	OK			OK	
Index	OK			OK	
Hyperspace maps	OK		OK	OK	

8.4 Adaptation Mechanisms for Adaptive Navigation Support

8.4.1 Simple Adaptation Mechanisms

To make the presentation complete, we must start with simple adaptation mechanisms that do not require advanced adaptation algorithms and yet can be of real use in a range of contexts. The most popular examples are history-based and trigger-based mechanisms

History-Based Mechanisms. History-based mechanisms simply count how many times each node in the hyperspace is accessed and attempt to represent this information visually. The oldest example is the rendering of visited links in an alternative color—a feature of every Web browser since Mosaic times (and actually inherited from hypertext research). Early research on adaptive navigation support attempted to extract more value from the stored history. For example, the MANUEL EXCEL system [33] dynamically annotated hypertext links with three different icons (a clear, gray, or black magnifying lens) to express the extent to which the area of hyperspace behind each link had previously been visited by the user (Fig. 8.5). Experiments with the system provided early evidence in favor of adaptive link annotation.

Trigger-Based Mechanisms. A trigger-based mechanism can be considered as an extension of a simple history-based adaptation. The idea of trigger-based adaptation is to connect a link with some simple event. Once this event has happened, the state of a binary trigger associated with a link is changed, resulting in a changed link appearance. A number of Learning Management Systems such as TopClass [74] use the simple trigger-based mechanism to control student access to learning content. A link to a section with learning content can be disabled or enabled at a specified time or after a specific quiz is completed by the user with a score under or above a threshold. A combination of these triggers allows teachers to provide some amount of class-level and individual personalization.

Progress-Based Mechanisms. The power of simple history-based mechanisms can be expanded if the adaptive system is able to track user visit to a page on a deeper level. For example, an information system may track time spent reading a page [63] or amount of page exploration (using eye-tracking or mouse tracking). Educational systems can measure the success of user work, e.g., a quiz that a link leads to can be solved partially, completely, or not yet attempted. The progress can be shown graphically next to each link to pages with educational activities helping the user to decide

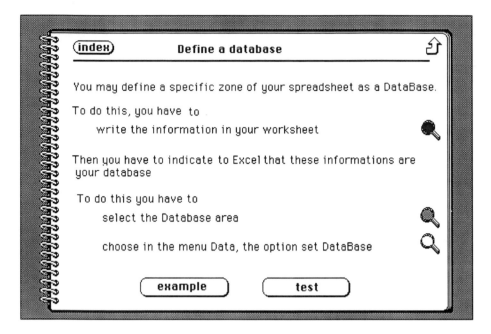

Fig. 8.5. Annotations for topic states in MANUEL EXCEL: not seen (clear magnifying lens), partially seen (grey lens), and completed (black lens).

whether to visit these pages or not. The use of information about the hypertext structure can further expand the power of progress-based adaptation. For example, in a hierarchically organized hyperspace, progress can be propagated up the hierarchy. Visual presentation of user progress for the top-level hyperspace topics provides an easy-to-grasp overview of the current state of work.

An example of using a progress-based mechanism with propagation in an educational context is provided by QuizGuide [21]. This system attempts to guide students to the most relevant self-assessment quizzes. Quizzes are grouped into topics. Once a topic link is "expanded," the links to all topic quizzes become available. Adaptive navigation support is provided on the topic level. The system traces correct and incorrect answers for all questions, calculating mastery levels for each quiz. These levels are propagated to the topic level, forming the mastery view of the whole topic. The icon annotating the link to the topic expresses this mastery in a target-arrow metaphor: the more arrows, the higher the level of mastery achieved for the topic (Fig. 8.6). These annotations allow students to see which topics are sufficiently mastered and which require additional work. The color of the target in QuizGuide attempts to express how important it is to attend to the topic, from the perspective of the class schedule. Current topics are marked by bright blue targets, their prerequisites by light blue targets, and other past topics by gray targets. Topics that are not yet introduced in class are crossed out, suggesting that the student is not ready to attempt them. This kind of annotation is supported by a trigger-based mechanism controlled by the teacher through the class schedule. The evaluation of progress-based navigation support in QuizGuide demonstrated that this technology

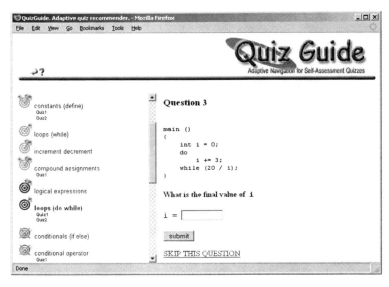

Fig. 8.6. Progress-based adaptive navigation support in QuizGuide. Depending on the percentage of correct answers to questions belonging to a topic, the icon annotating the link to the topic shows from zero to three arrows.

has succeeded in guiding the user to the most appropriate quizzes (as demonstrated by an increased rate of correct answers). In addition, the provision of adaptive visual cues significantly increased user motivation to work with the system, more than doubling the amount of non-mandatory work with the self-assessment quizzes that the students were willing to do [22].

8.4.2 Content-Based Mechanisms

Content-based adaptive navigation support mechanisms make a decision whether to suggest the user a path to a specific page by analyzing page content. Most of these mechanisms process pages to obtain keyword vectors and compare them with the profile of user interests. Link following is treated as an expression of user interests and is used for updating the user profile. More information on user profiles and document modeling can be found in Chapters 2 and 5 of this book [44; 60].

Content-based approaches were rarely used in pre-Web hypermedia. Interest in this area was attracted by the development of several pioneer systems in 1995-1996, such as WebWatcher [1], Letizia [58], Syskill & Webert [65], and Personal WebWatcher [62]. These systems influenced a number of more recent projects on both Web recommenders and content-based navigation support. While some of the pioneer systems with content-based navigation support were applied in the closed corpus context (i.e., a single Web site), others clearly demonstrated the most important innovation of content-based approaches: the ability to work with the open corpus Web. This idea was most clearly spelled out in the Letizia system, which was designed as an agent assisting user browsing by "running ahead" of the user, checking the content of pages behind the links, and suggesting the most relevant links to follow.

Fig. 8.7. Content-based navigation support in Syskill & Webert. Thumb icons identify pages that were previously rated, smiley icons point to a potentially interesting but not yet visited page. Used from [65] with author's permission.

It is probably due to this common root that a number of systems with content-based navigation use essentially the same decision-making mechanisms as content-based recommender systems. A review of these mechanisms is provided in Chapter 10 of this book [66]. A good example of the application of content-based recommender approaches in the context of navigation support is provided by the Syskill & Webert system [65]. Syskill & Webert attempts to learn user interests related to several topics while assisting user browsing. To provide relevance feedback to the system, the user explicitly rates encountered pages as hot, cold, or lukewarm. User ratings along with page representations as a bag-of-words are used to build a profile of user interests on different topics. As soon as the topic profile is discovered, the system starts suggesting interesting links on the current page by pre-fetching pages behind the links and classifying them according to the profile. Navigation support is provided by link annotation, i.e., links annotated with icons. Several different icons allow the user to differentiate previously rated pages and new, potentially interesting pages (Fig. 8.7). The prefetching-classification-annotation approach suggested in Syskill & Webert is straightforward, powerful, and universal. It could be used to recommend links on any page, whether a regular hypertext page with embedded links, a generated page with recommended links, or a page with links returned by a search engine in response to a query. Syskill & Webert demonstrated this flexibility by providing link annotation on a page generated by the Lycos search engine.

An example of content-based navigation support that differs from Syskill & Webert in several aspects is ScentTrails [64]. To start with, this system adapts to the user's search goal (formulated as a query), not to user interests. While this system also applies link annotations, it uses font size, not icons, in order to express more levels, when judging the relevancy of the path started by this link to the user goal (Fig. 8.8). However, most importantly, ScentTrails demonstrates the ability to look more than one step ahead when guiding users in hyperspace. The size of the link font shows the cumulative relevance of a whole region of hyperspace behind this link, i.e., a larger link font may indicate that the relevant page is not directly behind

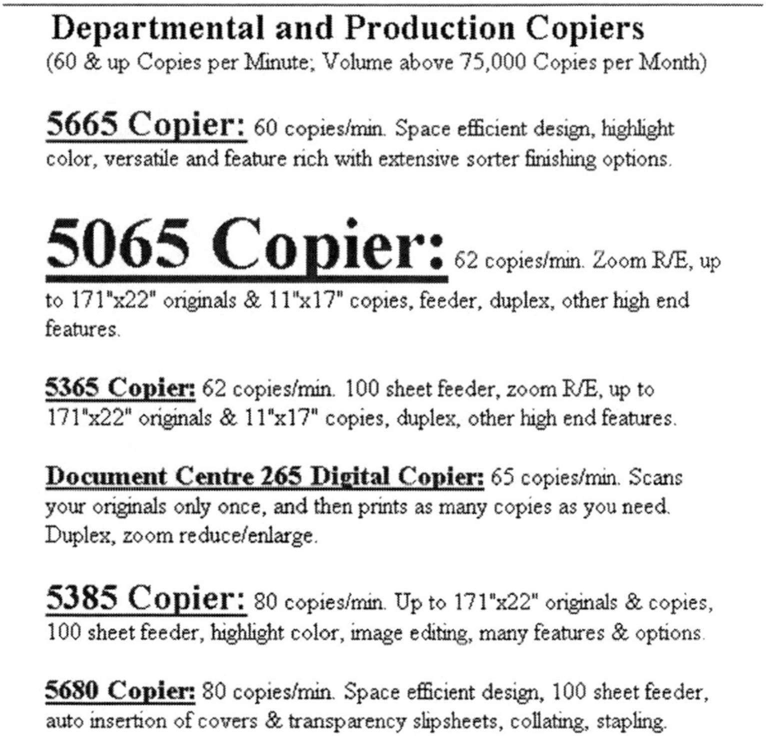

Departmental and Production Copiers
(60 & up Copies per Minute; Volume above 75,000 Copies per Month)

5665 Copier: 60 copies/min. Space efficient design, highlight color, versatile and feature rich with extensive sorter finishing options.

5065 Copier: 62 copies/min. Zoom R/E, up
to 171"x22" originals & 11"x17" copies, feeder, duplex, other high end features.

5365 Copier: 62 copies/min. 100 sheet feeder, zoom R/E, up to 171"x22" originals & 11"x17" copies, duplex, other high end features.

Document Centre 265 Digital Copier: 65 copies/min. Scans your originals only once, and then prints as many copies as you need. Duplex, zoom reduce/enlarge.

5385 Copier: 80 copies/min. Up to 171"x22" originals & copies, 100 sheet feeder, highlight color, image editing, many features & options.

5680 Copier: 80 copies/min. Space efficient design, 100 sheet feeder, auto insertion of covers & transparency slipsheets, collating, stapling.

Fig. 8.8. Content-based navigation support in ScentTrails. The size of the link font indicates how relevant the region of hyperspace behind this link is to the user's search goal. Used from [64] with the author's permission.

the link, but several steps ahead, on the path started by this link. The system is able to generate this advanced level of guidance by taking into account not only page content but also links between pages. The mechanism used in ScentTrails is based on the idea of an *information scent*. The simple scent of a page is its relevance to the user goal (a query) that is calculated using traditional information retrieval techniques. The full scent of pages in a connected set is calculated by propagating simple scents along the links. The assumption is that scent emanates equally from a page along each of its links, but decreases on each iteration. Potentially, this approach can work in the open corpus context by calculating information scent on the fly, but it is very time consuming due to the large number of pages that have to be processed. To keep response time small, the authors suggested a scent-calculating approach that is based on a relevance matrix recursively computed in advance. This effectively restricted the scope of this approach to a closed corpus context, such as a Web site. The evaluation of ScentTrails demonstrated that a full-scent version of the system allowed the user to achieve their goal significantly faster and with a much higher rate of success.

8.4.3 Social Mechanisms

Social mechanisms are based on the idea of *social navigation,* which capitalizes on the natural tendency of people to follow the direct and indirect cues of the activities of others, e.g., going to a restaurant that seems to draw many customers, or asking others what movies to watch. Social navigation in information space was originally introduced by Dourish and Chalmers as "moving towards clusters of people" or "selecting subjects because others have examined them" [38]. Social navigation support can be offered in a *direct* or *indirect* form. Direct social navigation means the direct interaction of users with each other in an information space. Indirect social navigation traces the activities of the community of users in the information space to guide new users in the system.

A typical approach to implement direct social navigation in hyperspace is to annotate links to pages that are currently being visited by other users with special icons. Several projects suggested technical solutions on how to augment links with this information [2]. Once visiting the same page, the users can typically communicate with each other. An elaborate implementation of this approach, using link annotation on a document map, was implemented in the EDUCO system [56].

Systems with indirect social navigation are typically classified into two groups: *history-enriched environments* and *collaborative filtering systems* [36]. History-enriched environments provide support for navigating through an information space by making the aggregated or individual action of others visible. This form is predominantly used by social navigation support mechanisms. The term history-rich information space was introduced by Wexelblat and Maes who implement this concept in their Footprints system [77], which visualizes usage paths throughout a web site. With the Footprints system, new users can see the popularity of each link on the current page and make navigation decisions. This approach is based on counting user passage through a link or user visits to a page and is known as a *footprint-based approach.* It was later implemented in several other systems such as CoWeb [37] and the first version of Knowledge Sea II [11]. A more recent version of the Knowledge Sea II system [40; 41] extended the footprint-based approach and explored annotation-based social navigation support. The extended version of the footprint-based approach takes into account time spent reading each page in order to scale footprints left by incomplete and accidental page visits and to obtain more reliable evidence of this page's relevance to the community of users [40]. Annotation-based social navigation support creates a history-rich environment by visualizing page annotations made by a community of users. This system is presented in more detail in Chapter 22 of this book [15].

Collaborative filtering is a technique for providing recommendation based on earlier expressed preferences or the interests of similar users. Collaborative filtering mechanisms are frequently powered by explicit user ratings, although recent systems have explored the use of implicit interest indicators [28]. While collaborative filtering mechanisms are mostly used in collaborative Web recommender systems reviewed in Chapter 9 of this book [69], a few of systems used it for providing social navigation support. A straightforward example of navigation support based on community ratings is provided by collaborative resource gathering systems such as CoFIND [39] or COMTELLA [26], which were reviewed in section 8.3.2 above. A more elaborate example is shown by the CourseAgent system [42].

CRN	Course No	Title	Day	Time	Instructor	Workload	Relevance	Action
2692	TELCOM 2940	PRACTICUM	apt					Plan It
16084	INFSCI 2120	INFORMATION AND CODING THEORY	tue	6:00-8:50 P		⚖	👍👍👍	Plan It
16077	INFSCI 2130	DECISION ANALYSIS AND DECISION SUPPORT SYSTEMS	wed	6:00-8:50		⚖	👍👍👍	Plan It
16086	LIS 2194	ETHICS IN THE INFORMATION SOCIETY	mon	3:00-5:50 P				Plan It
16099	INFSCI 2350	HUMAN FACTORS IN SYSTEMS	thu	6:00-8:50 P		⚖	👍	Register It
16056	INFSCI 2470	INTERACTIVE SYSTEM DESIGN	wed	6:00-8:50 P	Peter Brusilovsky	⚖	👍👍	Evaluate It

Fig. 8.9. Social navigation support in the schedule view of the CourseAgent system. Thumbs-up icons express the predicted usefulness of the course for the student. Darker background colors (blue and gold) indicate previously taken or planned courses.

CourseAgent attempts to recommend relevant courses to graduate students in Information Science taking into account the ratings of users who already took these courses. To make recommendations more reliable, the system uses a taxonomy of career goals. Every user is expected to select several career goals. Every course is rated independently in regard to each career goal of the rater. To predict the usefulness of a course for a student with a specific set of career goals, the system integrates existing ratings of this course in regard to these career goals. Course ratings are presented to students through link annotation. Wherever a link to a useful course is shown in the system (i.e., in a course schedule for the current semester or in a course catalog), it is augmented with thumb-up icons. The number of icons (one to three) expresses the predicted usefulness of the course (Fig. 8.9). The system also applies simple history-based navigation support, using special background colors to mark previously taken (gold) or planned (blue) courses.

8.4.4 Indexing-Based Mechanisms

Indexing-based mechanisms are the most popular and powerful mechanisms for providing adaptive navigation support in adaptive hypermedia. The idea of the indexing-based approach is similar to that of the content-based approach: represent some information about each page that can be matched to the user model and used to make a decision about whether and how to provide guidance. The difference between these two appoaches come from the representation. Content-based mechanisms use automatically-produced word-level document representations (presented in Chapter 5 of this book [60]) and similar user profiles (presented in Chapter 2 of this book [44]). Indexing-based mechanisms use manually-produced concept-level document representation and concept-level overlay models (presented in Chapter 1 of this book [16]). Concept-level representation is more powerful and precise, but due to involved manual processing it is rather expensive, which limits the application of indexing-based mechanisms to the closed corpus context.

The concept-level page representation is produced by expressing the content of each page in terms of external concept-level models. It means that each page is connected (associated) to one or more concepts that describe some aspect of this page. This process is known as indexing, because specifying a set of underlying concepts for every page is similar to indexing a page with a set of keywords. To provide a match between page indexing and user models, the same external model must be used for both building an overlay user model and page indexing. In the majority of adaptive hypermedia systems, the external model used for indexing is simply a concept-level *domain model* introduced in Chapter 1 of this book [16]. However, a number of systems use different kinds of models for indexing, such as a hierarchy of tasks, a taxonomy of learning styles, etc. These models are reviewed as *generalized models* in Chapter 1 of this book [16]. Since the aspects of page representation by indexing are not covered anywhere else in this book, the following subsections provide a brief review of these indexing approaches. For simplicity, this section refers to the elements of the external models as *concepts* regardless of their nature. Following this review, we present an example of using the indexing-based approach in the InterBook system.

Classification of Indexing Approaches. There are three attributes that are important to distinguish different indexing approaches, from the adaptive navigation support perspective: cardinality, expressive power, and navigation.

From the *cardinality* aspect, there are essentially only two different cases: single-concept indexing, where each page is related to one and only one external model concept; and multi-concept indexing, where each page can be related to many concepts. Single-concept indexing (categorization) is simpler and more intuitive for the authors. Multi-concept indexing is more powerful, but it makes the system more complex and requires more elaborate external models. In many cases, the choice of single or multi-concept indexing is a design decision for the authors of the system. To provide some simple navigation support functionalities the authors can use or build a coarse-grain model and use single-concept indexing. To provide more elaborate adaptations, they may need a finer-grained model and apply multi-concept indexing.

The *Navigation* aspect is important when distinguishing between cases where the link between a concept and a page exists only on a conceptual level (used only by internal adaptation mechanisms of the system) from cases where each link also defines a navigation path.

Expressive power concerns the amount of information that the authors can associate with every link between a concept and a page. Of course, the most important information is the very presence of the link. This case could be called flat indexing and is used in the majority of existing systems. Still, some systems with a large hyperspace and advanced adaptation techniques may want to associate more information with every link by using roles and/or weights. Assigning a *role* to a link helps distinguish several kinds of connections between concepts and pages. For example, some systems want to distinguish whether a page provides an introduction, a core explanation or a summary of a concept. Other systems use *prerequisite* role to mark the case when the concept is not presented on a page, but instead, the page is a required prerequisite for understanding the concept [14]. A case for a more elaborate indexing with multiple roles can be found in [12]. Another way to increase the expressive power of the indexing is to specify the *weight* of the link between a concept and a page. The weight may specify, for example, the percentage of knowledge about a concept presented on this page [30; 68].

Existing AH systems suggest various ways of indexing that differ in all the aspects listed above. However, for simplicity, all this variety can be described in terms of two basic approaches that are described in the remaining part of this section. Systems using the same indexing approach have a similar hyperspace structure and share specific adaptation techniques that are based on this structure. Thus the indexing approach selected by developers to a large extent defines the navigation support functionality of the system.

Concept-Based Hyperspace. The simplest approach to organizing connections between external models and hyperspace pages is known as *concept-based hyperspace*. This approach is naturally appearing in any system that uses single-concept indexing. It is useful to distinguish simple and enhanced concept-based hyperspace. *Simple concept-based hyperspace* is used in systems that have exactly one page for every concept. With this approach, the hyperspace is built as an exact replica of the external model. Each concept of the external model is represented by exactly one node of the hyperspace, while the semantic links between the concepts constitute main paths between hyperspace nodes [17; 19; 49]. The simple concept based approach was quite popular among early educational AH systems that have their roots in the ITS field. For these systems the concept-based hyperspace was simply the easiest and the most natural way to produce a well-structured hyperspace. Currently it is rarely used in AH systems in its pure form because it requires each page of the hyperspace to be devoted to exactly one concept. It is very appropriate for developing encyclopedically structured hyperspaces such as encyclopedias [6; 61] or glossaries [14], but too restrictive for other cases.

With an *enhanced concept-based hyperspace* design approach, each concept has a corresponding "hub" page in the hyperspace. The concept hub page is connected by links to all pages categorized with this concept. For example, news articles can be classified by category and presented on a dedicated category page; Web links can be assembled under Web directory categories. The links can be typed or weighted [68]. This approach is typical for adaptive e-learning systems with rich content. In this context, a variety of educational resources can be used to present different aspects of the same topic in different ways. Each page (resource) can be typed with the kind of material (video, audio, text, etc) and this typing is used for both presenting and adaptive ordering of links. With the enhanced concept-based approach users can navigate between concept pages along links that connect concepts in external models and from concept pages to the pages categorized under the concept. This approach was used for creating relatively large hyperspaces with quite straightforward structure and meaningful adaptation techniques.

The concept-based hyperspace design approach sets strong requirements to the external model. It always requires a model with established links between concepts (preferably, several types of links) that will be used to establish hyperlinks. Another restriction is that this approach can hardly be used "post-hoc" to turn an existing traditional hypermedia system into an AH system. It has to be used from the early steps of a hypermedia system design [76]. However, this approach is quite powerful and provides excellent opportunities for adaptation. With concept-based approach, the system knows exactly the type and content of each page and the type of each link. This knowledge can be used by various adaptive navigation support techniques. Annotation is the most popular technology here. For example, ISIS-Tutor [17], ELM-ART [75], InterBook [14] use different kinds of link annotation to show the current educational state of the concept (not known, known, well known). ISIS-Tutor, ELM-ART, and a number of other systems use annotation to show that a concept page is not ready to be learned (i.e., its prerequisite concepts are not learned yet). Hiding technology can be used to hide links to concept pages that are not relevant to the user knowledge or interests. For example, links to news categories that the user wants to ignore, links to concepts that do not belong to the current educational goal [17] or with not yet learned prerequisite concepts [17; 43].

Note that the concept-based hyperspace is just one of the possible design approaches for AHS with single concept indexing. There are a few known systems, especially among early AHS [67] with single concept indexing but without concept-based navigation. The concept-based hyperspace in these systems is not formed since concepts have no external hyperspace representation and/or links between concepts and pages are purely conceptual and not used for hyperspace navigation. However, once discovered, the concept-based hyperspace approach became most popular in systems with single-concept indexing.

Page Indexing. Page indexing is the standard design approach for systems with multi-concept indexing. With this approach, the hypermedia page is indexed with several external model concepts. In other words, links are created between a page and each concept that describes the page. The simplest indexing approach is flat, content-based indexing when a concept is included in a page index if it expresses some aspect of page content. For example, the content is relevant to a specific task (a concept in a taxonomy of tasks) [73] or it presents knowledge designated by a specific domain concept [17; 47; 57]. A more general but less often used way to index the pages is to add the role for each concept in the page index (role-based indexing) as was discussed above.

Page indexing can be applied even to vector external models that have no links between concepts [32; 57]. At the same time, indexing is a very powerful mechanism, because it provides the system with knowledge about the content of its pages. With content-based indexing, the system knows quite reliably what each page is about. This knowledge can be used in multiple ways by various navigation support techniques.

Concept-Based Navigation. An interesting combination of concept-based hyperspace and page indexing known as *concept-based navigation* was introduced in Inter-Book [14]. This approach merges a hyperspace of multi-concept-indexed pages and a hyperspace of concepts. Each concept used to index hyperspace pages becomes a node in the hyperspace and a navigation hub. Every link between a page and the concept established during indexing becomes visible as a two-way navigational link between this page and the hub page of the concept. Thus, from any content page, users can navigate to hubs of all concepts used to index this page. Vice versa, the concept hub page provides links to all content pages indexed with this concept. This approach creates rich navigation opportunities. A user can start from a content page, move to one of the related concepts and then move to another page connected to the same concept. Concept hubs are used here as bridges for navigation to concept-related pages that have no direct hypertext links. A similar tag-based navigation approach is now popular in collaborative tagging systems, in order to navigate from one resource to another resource through *tags*.

The Indexing-Based Mechanism in InterBook. The InterBook system [18], the first authoring platform for Web-based adaptive hypermedia, refined the ideas of the adaptive electronic textbook introduced by ELM-ART (see section 8.3.4). A document collection in InterBook was formed by grouping several hierarchically structured textbooks into bookshelves. The books on the same shelf shared the same domain model. This domain model was used to create an overlay model of user knowledge (see Chapter 1 of this book [16])and to index each section of each book on the shelf. Connections between pages and concepts were typed: a concept served either as a *prerequisite* of a page or as an *outcome*. Following the concept-based navigation approach, each domain model concept was represented in the hyperspace as a *glossary page* that contained a brief description of the concept and links to all pages indexed by this concept. To complete concept-based navigation, every book page included a sidebar with links to all concepts used to index this page. In both contexts, the links were grouped by type, i.e., prerequisite and outcome links were not intermixed (Fig. 8.10).

InterBook offered several kinds of navigation support. The most important was link annotation, using the traffic light metaphor for adaptive navigation support in educational hypermedia (Fig. 8.10). Propagated by ELM-ART and InterBook, this metaphor has later been used in numerous adaptive educational hypermedia systems, including AST [72], KBS-HyperBook [48], and SIGUE [25]. The traffic-light annotation was produced taking into account the current model of user knowledge and the type of links between pages and concepts. A page with all outcome concepts already learned was marked with a white bullet. A page with at least some outcome concepts not learned, but with all prerequisite concepts learned was marked with a green "go" bullet. A page with at least one prerequisite concept not yet learned was marked with a red "stop" bullet. Regardless of the type of annotation, all links were functional; there was no hiding, removing or disabling. Surprisingly, the study recorded that some percentage of users most frequently chose the red link. However, it harmed their performance on tests [13]. In addition, concepts links on the concept bar were annotated with checkmarks of several difference sizes, where each size corresponded to a specific knowledge level. This feature allowed the users to see immediately which new concepts are introduced on a page and which unknown prerequisite concepts made this page hard to understand.

For users who have troubles selecting a link, the system offered direct guidance using a "teach me" button. The sequencing algorithm was simple: the system selected the most sequentially close green link. Finally, the system included *link generation* to answer help requests. The idea of providing help was to assemble a list of links to pages that could be useful for understanding the current not-ready-to-be-learned page. To assemble this list, the system collected all pages that might be useful for teaching the missing prerequisite concepts of the current page and ordered them adaptively, according to a polynomial "usefulness" measure. The measure took into account how many goal concepts were introduced on a page (the more, the better), how many non-goal concepts (the fewer, the better), and what the page's current state was (green is better than red). In addition to adaptive ordering, all links were also annotated (Fig. 8.10).

Altogether, InterBook produced several kinds of navigation support using the same concept-level models for the user and the documents. A study of InterBook demonstrated that adaptive navigation support encourages non-sequential navigation and helps users who follow the system's guidance achieve a better level of knowledge [13].

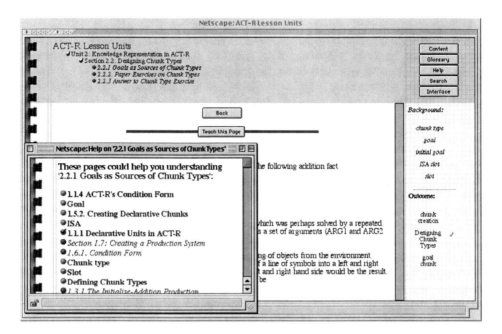

Fig. 8.10. Adaptive Navigation support in InterBook. Icons using the traffic light metaphor annotate links to book pages. Checkmarks annotate links to glossary items. By user "help" request, links to pages that can help the student to understand the current page were adaptively generated, ordered, and annotated (lower left window).

8.5 Beyond Hypermedia: Adaptive Navigation Support for Virtual Environments

Adaptive navigation support techniques have demonstrated their ability to help individual users of hypermedia and Web systems. A review of adaptation techniques and mechanisms provided in this chapter could possibly serve as a collection of useful recipes for future developers of Web hypermedia systems who are interested in providing personalized assistance to users. However, a hyperspace of connected pages—which is the context of existing AH technologies—is not the only kind of "virtual space" that is available for Internet users. Even in the early days of the Internet, a lot of people were navigating in text-based virtual environments, now called MUDs and MOOs (http://www.moo.mud.org/moo-faq/) that are currently still accessible over the Web. More recently, Web-based virtual reality has become an alternative type of virtual environment for browsing and exploration on the Web. While MUD/MOO, hypermedia, and the 3D virtual environments are quite different in their nature, all these environments are targeted for user-driven navigation and exploration. As a result, in all these contexts, users can benefit from the navigation support provided by an adaptive system. We believe that the theories behind adaptive navigation support go beyond the scope of hypermedia, although a different set of technologies may be required to provide support in these different contexts.

A pioneer attempt to develop navigation support in the MOO context, using social navigation mechanisms was done by Dieberger [35] in his system Juggler. While Juggler's concept of employing history-rich environments has been explored before, Juggler suggested a unique implementation of this idea adapted to the narrative, text-based information presentation context of MOO. A number of more recent projects explored the use of navigation support in the context of Web virtual reality. For example, [51] attempted to develop virtual reality analogs to direct guidance and link annotation. A review of work in this direction is presented in Chapter 14 of this book [27].

Acknowledgments. This material is partially based upon work supported by the National Science Foundation under Grants No. 0310576 and 0447083. The author also thanks Science Foundation Ireland for its support through the E.T.S. Walton Award.

References

1. Armstrong, R., Freitag, D., Joachims, T., Mitchell, T.: WebWatcher: A learning apprentice for the World Wide Web. In: Knoblock, C., Levy, A. (eds.) Proc. of AAAI Spring Symposium on Information Gathering from Distributed, Heterogeneous Environments. AAAI Press (1995) 6-12
2. Barrett, R., Maglio, P.P.: Intermediaries: New places for producing and manipulating Web content. In: Ashman, H., Thistewaite, P. (eds.) Proc. of Seventh International World Wide Web Conference. Computer Networks and ISDN Systems, Vol. 30. Elsevier Science B. V. (1998) 509 - 518
3. Billsus, D., Pazzani, M.: Adaptive news access. In: Brusilovsky, P., Kobsa, A., Nejdl, W. (eds.): The Adaptive Web: Methods and Strategies of Web Personalization. Lecture Notes in Computer Science, Vol. 4321. Springer-Verlag, Berlin Heidelberg New York (2007) this volume
4. Böcker, H.D., Hohl, H., Schwab, T.: ΥπAdaptερ - Individualizing Hypertext. In: Diaper, D. (ed.) Proc. of IFIP TC13 Third International Conference on Human-Computer Interaction. North-Holland (1990) 931-936
5. Bollen, I., Heylighen, F.: A system to restructure hypertext networks into valid user models. The New Review of Multimedia and Hypermedia 4 (1998) 189-213
6. Bontcheva, K., Wilks, Y.: Tailoring automatically generated hypertext. User Modeling and User Modeling and User Adapted Interaction 15, 1-2 (2005) 135-168
7. Boyle, C., Encarnacion, A.O.: MetaDoc: an adaptive hypertext reading system. User Modeling and User-Adapted Interaction 4, 1 (1994) 1-19
8. Brusilovsky, P.: Methods and techniques of adaptive hypermedia. User Modeling and User-Adapted Interaction 6, 2-3 (1996) 87-129
9. Brusilovsky, P.: Adaptive hypermedia. User Modeling and User Adapted Interaction 11, 1/2 (2001) 87-110
10. Brusilovsky, P.: Adaptive navigation support in educational hypermedia: the role of student knowledge level and the case for meta-adaptation. British Journal of Educational Technology 34, 4 (2003) 487-497
11. Brusilovsky, P., Chavan, G., Farzan, R.: Social adaptive navigation support for open corpus electronic textbooks. In: De Bra, P., Nejdl, W. (eds.) Proc. of Third International Conference on Adaptive Hypermedia and Adaptive Web-Based Systems (AH'2004). Lecture Notes in Computer Science, Vol. 3137. Springer-Verlag (2004) 24-33
12. Brusilovsky, P., Cooper, D.W.: Domain, Task, and User Models for an Adaptive Hypermedia Performance Support System. In: Gil, Y., Leake, D.B. (eds.) Proc. of 2002 International Conference on Intelligent User Interfaces. ACM Press (2002) 23-30

13. Brusilovsky, P., Eklund, J.: A study of user-model based link annotation in educational hypermedia. Journal of Universal Computer Science 4, 4 (1998) 429-448

14. Brusilovsky, P., Eklund, J., Schwarz, E.: Web-based education for all: A tool for developing adaptive courseware. In: Ashman, H., Thistewaite, P. (eds.) Proc. of Seventh International World Wide Web Conference. Vol. 30. Elsevier Science B. V. (1998) 291-300

15. Brusilovsky, P., Henze, N.: Open corpus adaptive educational hypermedia. In: Brusilovsky, P., Kobsa, A., Neidl, W. (eds.): The Adaptive Web: Methods and Strategies of Web Personalization. Lecture Notes in Computer Science, Vol. 4321. Springer-Verlag, Berlin Heidelberg New York (2007) this volume

16. Brusilovsky, P., Millán, E.: User models for adaptive hypermedia and adaptive educational systems. In: Brusilovsky, P., Kobsa, A., Neidl, W. (eds.): The Adaptive Web: Methods and Strategies of Web Personalization. Lecture Notes in Computer Science, Vol. 4321. Springer-Verlag, Berlin Heidelberg New York (2007) this volume

17. Brusilovsky, P., Pesin, L.: An intelligent learning environment for CDS/ISIS users. In: Levonen, J.J., Tukianinen, M.T. (eds.) Proc. of The interdisciplinary workshop on complex learning in computer environments (CLCE94). EIC (1994) 29-33, also available at http://cs.joensuu.fi/~mtuki/www_clce.270296/Brusilov.html

18. Brusilovsky, P., Pesin, L.: Adaptive navigation support in educational hypermedia: An evaluation of the ISIS-Tutor. Journal of Computing and Information Technology 6, 1 (1998) 27-38

19. Brusilovsky, P., Pesin, L., Zyryanov, M.: Towards an adaptive hypermedia component for an intelligent learning environment. In: Bass, L.J., Gornostaev, J., Unger, C. (eds.) Proc. of 3rd International Conference on Human-Computer Interaction, EWHCI'93. Lecture Notes in Computer Science, Vol. 753. Springer-Verlag (1993) 348-358

20. Brusilovsky, P., Schwarz, E., Weber, G.: ELM-ART: An intelligent tutoring system on World Wide Web. In: Frasson, C., Gauthier, G., Lesgold, A. (eds.) Proc. of Third International Conference on Intelligent Tutoring Systems, ITS-96. Lecture Notes in Computer Science, Vol. 1086. Springer Verlag (1996) 261-269

21. Brusilovsky, P., Sosnovsky, S., Shcherbinina, O.: QuizGuide: Increasing the Educational Value of Individualized Self-Assessment Quizzes with Adaptive Navigation Support. In: Nall, J., Robson, R. (eds.) Proc. of World Conference on E-Learning, E-Learn 2004. AACE (2004) 1806-1813

22. Brusilovsky, P., Sosnovsky, S., Yudelson, M.: Addictive links: The motivational value of adaptive link annotation in educational hypermedia. In: Wade, V., Ashman, H., Smyth, B. (eds.) Proc. of 4th International Conference on Adaptive Hypermedia and Adaptive Web-Based Systems (AH'2006). Lecture Notes in Computer Science, Vol. 4018. Springer Verlag (2006) 51-60

23. Brusilovsky, P., Weber, G.: Collaborative example selection in an intelligent example-based programming environment. In: Edelson, D.C., Domeshek, E.A. (eds.) Proc. of International Conference on Learning Sciences, ICLS'96. AACE (1996) 357-362

24. Burke, R.: Hybrid Web recommender systems. In: Brusilovsky, P., Kobsa, A., Neidl, W. (eds.): The Adaptive Web: Methods and Strategies of Web Personalization. Lecture Notes in Computer Science, Vol. 4321. Springer-Verlag, Berlin Heidelberg New York (2007) this volume

25. Carmona, C., Bueno, D., Guzmán, E., Conejo, R.: SIGUE: Making Web Courses Adaptive. In: De Bra, P., Brusilovsky, P., Conejo, R. (eds.) Proc. of Second International Conference on Adaptive Hypermedia and Adaptive Web-Based Systems (AH'2002). Lecture Notes in Computer Science, Vol. 2347. Springer-Verlag (2002) 376-379

26. Cheng, R., Vassileva, J.: Adaptive Reward Mechanism for Sustainable Online Learning Community. In: Looi, C.-K., McCalla, G., Bredeweg, B., Breuker, J. (eds.) Proc. of 12th International Conference on Artificial Intelligence in Education, AIED'2005. IOS Press (2005) 152-159

27. Chittaro, L., Ranon, R.: Adaptive 3D Web sites. In: Brusilovsky, P., Kobsa, A., Neidl, W. (eds.): The Adaptive Web: Methods and Strategies of Web Personalization. Lecture Notes in Computer Science, Vol. 4321. Springer-Verlag, Berlin Heidelberg New York (2007) this volume

28. Claypool, M., Le, P., Wased, M., Brown, D.: Implicit interest indicators. In: Proc. of 6th International Conference on Intelligent User Interfaces. ACM Press (2001) 33-40

29. De Bra, P., Calvi, L.: AHA! An open Adaptive Hypermedia Architecture. The New Review of Hypermedia and Multimedia 4 (1998) 115-139

30. De Bra, P., Ruiter, J.-P.: AHA! Adaptive hypermedia for all. In: Fowler, W., Hasebrook, J. (eds.) Proc. of WebNet'2001, World Conference of the WWW and Internet. AACE (2001) 262-268

31. De Bra, P., Smits, D., Stash, N.: The design of AHA! In: Wiil, U.K., Nürnberg, P.J., Rubart, J. (eds.) Proc. of Seventeenth ACM Conference on Hypertext and Hypermedia (Hypertext 2006). ACM Press (2006) 133

32. De Bra, P.M.E.: Teaching Hypertext and Hypermedia through the Web. Journal of Universal Computer Science 2, 12 (1996) 797-804

33. de La Passardiere, B., Dufresne, A.: Adaptive navigational tools for educational hypermedia. In: Tomek, I. (ed.) Proc. of ICCAL'92, 4-th International Conference on Computers and Learning. Lecture Notes in Computer Science, Vol. 602. Springer-Verlag (1992) 555-567

34. Debevc, M., Meyer, B., Donlagic, D., Rajko, S.: Design and Evaluation of an Adaptive Icon Toolbar. User Modeling and User-Adapted Interaction 6, 1 (1996) 1-21

35. Dieberger, A.: Supporting social navigation on the World Wide Web. International Journal of Human-Computer Interaction 46 (1997) 805-825

36. Dieberger, A., Dourish, P., Höök, K., Resnick, P., Wexelblat, A.: Social navigation: Techniques for building more usable systems. interactions 7, 6 (2000) 36-45

37. Dieberger, A., Guzdial, M.: CoWeb - experiences with collaborative Web spaces. In: Lueg, C., Fisher, D. (eds.): From Usenet to CoWebs: Interacting with Social Information Spaces. Springer-Verlag, New York (2003) 155-166

38. Dourish, P., Chalmers, M.: Running out of space: Models of information navigation. In: Cockton, G., Draper, S.W., Weir, G.R.S. (eds.) Proc. of HCI'94. (1994)

39. Dron, J., Boyne, C., Mitchell, R., Siviter, P.: CoFIND: steps towards a self-organising learning environment. In: Davies, G., Owen, C. (eds.) Proc. of WebNet'2000, World Conference of the WWW and Internet. AACE (2000) 75-80

40. Farzan, R., Brusilovsky, P.: Social navigation support in E-Learning: What are real footprints. In: Anand, S.S., Mobasher, B. (eds.) Proc. of IJCAI'05 Workshop on Intelligent Techniques for Web Personalization. (2005) 49-56, also available at http://maya.cs.depaul.edu/~mobasher/itwp05/final/Paper7Farzan.pdf

41. Farzan, R., Brusilovsky, P.: Social navigation support through annotation-based group modeling. In: Ardissono, L., Brna, P., Mitrovic, A. (eds.) Proc. of 10th International User Modeling Conference. Lecture Notes in Artificial Intelligence, Vol. 3538. Springer Verlag (2005) 463-472

42. Farzan, R., Brusilovsky, P.: Social navigation support in a course recommendation system. In: Wade, V., Ashman, H., Smyth, B. (eds.) Proc. of 4th International Conference on Adaptive Hypermedia and Adaptive Web-Based Systems (AH'2006). Lecture Notes in Computer Science, Vol. 4018. Springer Verlag (2006) 91-100

43. Forcheri, P., Molfino, M.T., Moretti, S., Quarati, A.: An Approach to the Development of Re-Usable and Adaptive Web Based Courses. In: Fowler, W., Hasebrook, J. (eds.) Proc. of WebNet'2001, World Conference of the WWW and Internet. AACE (2001) 1043-1048

44. Gauch, S., Speretta, M., Chandramouli, A., Micarelli, A.: User profiles for personalized information access. In: Brusilovsky, P., Kobsa, A., Neidl, W. (eds.): The Adaptive Web: Methods and Strategies of Web Personalization. Lecture Notes in Computer Science, Vol. 4321. Springer-Verlag, Berlin Heidelberg New York (2007) this volume

45. Geldof, S.: Con-textual navigation support. New Review of Multimedia and Hypermedia 4 (1998) 47-66
46. Hammond, N.: Hypermedia and learning: Who guides whom? In: Maurer, H. (ed.) Proc. of 2-nd International Conference on Computer Assisted Learning, ICCAL'89. Lecture Notes in Computer Science, Vol. 360. Springer-Verlag (1989) 167-181
47. Henze, N., Naceur, K., Nejdl, W., Wolpers, M.: Adaptive hyperbooks for constructivist teaching. Künstliche Intelligenz, 4 (1999) 26-31
48. Henze, N., Nejdl, W.: Adaptation in open corpus hypermedia. International Journal of Artificial Intelligence in Education 12, 4 (2001) 325-350
49. Hohl, H., Böcker, H.-D., Gunzenhäuser, R.: Hypadapter: An adaptive hypertext system for exploratory learning and programming. User Modeling and User-Adapted Interaction 6, 2-3 (1996) 131-156
50. Höök, K.: Evaluating the utility and usability of an adaptive hypermedia system. In: Moore, J., Edmonds, E., Puerta, A. (eds.) Proc. of 1997 International Conference on Intelligent User Interfaces. ACM (1997) 179-186
51. Hughes, S., Brusilovsky, P., Lewis, M.: Adaptive navigation support in 3D e-commerce activities. In: Ricci, F., Smyth, B. (eds.) Proc. of Workshop on Recommendation and Personalization in eCommerce at the 2nd International Conference on Adaptive Hypermedia and Adaptive Web-Based Systems (AH'2002). (2002) 132-139, also available at http://ectrl.itc.it/home/rpec/RPEC-Papers/15-hughes.pdf
52. Kaplan, C., Fenwick, J., Chen, J.: Adaptive hypertext navigation based on user goals and context. User Modeling and User-Adapted Interaction 3, 3 (1993) 193-220
53. Kaptelinin, V.: Item recognition in menu selection: The effect of practice. In: Proc. of INTERCHI'93. (1993) 183-184
54. Kavcic, A.: Fuzzy User Modeling for Adaptation in Educational Hypermedia. IEEE Transactions on Systems, Man, and Cybernetics 34, 4 (2004) 439-449
55. Kelly, D., Tangney, B.: Matching and Mismatching Learning Characteristics with Multiple Intelligence Based Content. In: Looi, C.-K., McCalla, G., Bredeweg, B., Breuker, J. (eds.) Proc. of 12th International Conference on Artificial Intelligence in Education, AIED'2005. IOS Press (2005) 354-361
56. Kurhila, J., Miettinen, M., Nokelainen, P., Tirri, H.: EDUCO - A collaborative learning environment based on social navigation. In: De Bra, P., Brusilovsky, P., Conejo, R. (eds.) Proc. of Second International Conference on Adaptive Hypermedia and Adaptive Web-Based Systems (AH'2002). Lecture Notes in Computer Science, Vol. 2347. (2002) 242-252
57. Laroussi, M., Benahmed, M.: Providing an adaptive learning through the Web case of CAMELEON: Computer Aided MEdium for LEarning on Networks. In: Alvegård, C. (ed.) Proc. of CALISCE'98, 4th International conference on Computer Aided Learning and Instruction in Science and Engineering. (1998) 411-416
58. Lieberman, H.: Letizia: An agent that assists Web browsing. In: Proc. of the Fourteenth International Joint Conference on Artificial Intelligence. (1995) 924-929
59. Lutkenhouse, T., Nelson, M.L., Bollen, J.: Distributed, real-time computation of community preferences. In: Proc. of Proceedings of the sixteenth ACM conference on Hypertext and hypermedia. ACM Press (2005) 88-97
60. Micarelli, A., Sciarrone, F., Marinilli, M.: Web document modeling. In: Brusilovsky, P., Kobsa, A., Neidl, W. (eds.): The Adaptive Web: Methods and Strategies of Web Personalization. Lecture Notes in Computer Science, Vol. 4321. Springer-Verlag, Berlin Heidelberg New York (2007) this volume
61. Milosavljevic, M.: Augmenting the user's knowledge via comparison. In: Jameson, A., Paris, C., Tasso, C. (eds.) Proc. of 6th International Conference on User Modeling, UM97. SpringerWienNewYork (1997) 119-130

62. Mladenic, D.: Personal WebWatcher: Implementation and Design, Technical Report No. IJS-DP-7472, Department of Intelligent Systems, J. Stefan Institute (1996)
63. Ng, M.H., Hall, W., Maier, P., Armstrong, R.: The Application and Evaluation of Adaptive Hypermedia Techniques in Web-based Medical Education. Association for Learning Technology Journal 10, 3 (2002) 19-40
64. Olston, C., Chi, E.H.: ScentTrails: Integrating browsing and searching on the Web. ACM Transactions on Computer-Human Interaction 10, 3 (2003) 177-197
65. Pazzani, M., Muramatsu, J., Billsus, D.: Syskill & Webert: Identifying interesting Web sites. In: Proc. of the Thirteen National Conference on Artificial Intelligence, AAAI'96. AAAI Press /MIT Press (1996) 54-61
66. Pazzani, M.J., Billsus, D.: Content-based recommendation systems. In: Brusilovsky, P., Kobsa, A., Neidl, W. (eds.): The Adaptive Web: Methods and Strategies of Web Personalization. Lecture Notes in Computer Science, Vol. 4321. Springer-Verlag, Berlin Heidelberg New York (2007) this volume
67. Pérez, T., Gutiérrez, J., Lopistéguy, P.: An adaptive hypermedia system. In: Greer, J. (ed.) Proc. of AI-ED'95, 7th World Conference on Artificial Intelligence in Education. AACE (1995) 351-358
68. Pilar da Silva, D., Durm, R.V., Duval, E., Olivié, H.: Concepts and documents for adaptive educational hypermedia: a model and a prototype. In: Brusilovsky, P., De Bra, P. (eds.) Proc. of Second Adaptive Hypertext and Hypermedia Workshop at the Ninth ACM International Hypertext Conference Hypertext'98. Eindhoven University of Technology (1998) 35-43, also available as http://wwwis.win.tue.nl/ah98/Pilar/Pilar.html
69. Schafer, J.B., Frankowski, D., Herlocker, J., Sen, S.: Collaborative filtering recommender systems. In: Brusilovsky, P., Kobsa, A., Neidl, W. (eds.): The Adaptive Web: Methods and Strategies of Web Personalization. Lecture Notes in Computer Science, Vol. 4321. Springer-Verlag, Berlin Heidelberg New York (2007) this volume
70. Smyth, B.: Case-base recommendation. In: Brusilovsky, P., Kobsa, A., Neidl, W. (eds.): The Adaptive Web: Methods and Strategies of Web Personalization. Lecture Notes in Computer Science, Vol. 4321. Springer-Verlag, Berlin Heidelberg New York (2007) this volume
71. Specht, M.: Empirical evaluation of adaptive annotation in hypermedia. In: Ottmann, T., Tomek, I. (eds.) Proc. of ED-MEDIA/ED-TELECOM'98 - 10th World Conference on Educational Multimedia and Hypermedia and World Conference on Educational Telecommunications. AACE (1998) 1327-1332
72. Specht, M., Weber, G., Heitmeyer, S., Schöch, V.: AST: Adaptive WWW-Courseware for Statistics. In: Brusilovsky, P., Fink, J., Kay, J. (eds.) Proc. of Workshop "Adaptive Systems and User Modeling on the World Wide Web" at 6th International Conference on User Modeling, UM97. Carnegie Mellon Online (1997) 91-95, also available at http://www.contrib.andrew.cmu.edu/~plb/UM97_workshop/Specht.html
73. Vassileva, J.: A task-centered approach for user modeling in a hypermedia office documentation system. User Modeling and User-Adapted Interaction 6, 2-3 (1996) 185-224
74. WBT Systems: TopClass, ver. 3.0, Dublin, Ireland, WBT Systems (1999) http://www.wbtsystems.com/
75. Weber, G., Brusilovsky, P.: ELM-ART: An adaptive versatile system for Web-based instruction. International Journal of Artificial Intelligence in Education 12, 4 (2001) 351-384
76. Weber, G., Kuhl, H.-C., Weibelzahl, S.: Developing adaptive internet based courses with the authoring system NetCoach. In: Bra, P.D., Brusilovsky, P., Kobsa, A. (eds.) Proc. of Third workshop on Adaptive Hypertext and Hypermedia. Technical University Eindhoven (2001) 35-48, also available at http://wwwis.win.tue.nl/ah2001/papers/GWeber-UM01.pdf
77. Wexelblat, A., Mayes, P.: Footprints: History-rich tools for information foraging. In: Proc. of ACM Conference on Human-Computer Interaction (CHI'99). (1999) 270-277

78. Yan, T.W., Jacobsen, M., Garcia-Molina, H., Dayal, U.: From user access patterns to dynamic hypertext linking. Computer Networks and ISDN Systems. (1996) 1007-1014
79. Zellweger, P.T., Chang, B.-W., Mackinlay, J.D.: Fluid links for informed and incremental link transitions. In: Grønbæk, K., Mylonas, E., Shipman III, F.M. (eds.) Proc. of Ninth ACM International Hypertext Conference (Hypertext'98). ACM Press (1998) 50-57

Collaborative Filtering Recommender Systems

J. Ben Schafer[1], Dan Frankowski[2], Jon Herlocker[3], and Shilad Sen[2]

[1] Department of Computer Science
University of Northern Iowa
Cedar Falls, IA 50614-0507
schafer@cs.uni.edu

[2] Department of Computer Science
University of Minnesota
4-192 EE/CS Building
200 Union St. SE
Minneapolis, MN 55455
{dfrankow , ssen}@cs.umn.edu

[3] School of Electrical Engineering and Computer Science
Oregon State University
102 Dearborn Hall
Corvallis, OR 97331
herlock@eecs.oregonstate.edu

Abstract. One of the potent personalization technologies powering the adaptive web is collaborative filtering. Collaborative filtering (CF) is the process of filtering or evaluating items through the opinions of other people. CF technology brings together the opinions of large interconnected communities on the web, supporting filtering of substantial quantities of data. In this chapter we introduce the core concepts of collaborative filtering, its primary uses for users of the adaptive web, the theory and practice of CF algorithms, and design decisions regarding rating systems and acquisition of ratings. We also discuss how to evaluate CF systems, and the evolution of rich interaction interfaces. We close the chapter with discussions of the challenges of privacy particular to a CF recommendation service and important open research questions in the field.

9.1 Introduction

Collaborative Filtering is the process of filtering or evaluating items using the opinions of other people. While the term collaborative filtering (CF) has only been around for a little more than a decade, CF takes its roots from something humans have been doing for centuries - sharing opinions with others.

For years, people have stood over the back fence or in the office break room and discussed books they have read, restaurants they have tried, and movies they have seen – then used these discussions to form opinions. For example, when enough of

P. Brusilovsky, A. Kobsa, and W. Nejdl (Eds.): The Adaptive Web, LNCS 4321, pp. 291–324, 2007.

Amy's colleagues say they liked the latest release from Hollywood, she might decide that she also should see it. Similarly, if many of them found it a disaster, she might decide to spend her money elsewhere. Better yet, Amy might observe that Matt recommends the types of films that she finds enjoyable, Paul has a history of recommending films that she despises, and Margaret just seems to recommend everything. Over time, she learns whose opinions she should listen to and how these opinions can be applied to help her determine the quality of an item.

Computers and the web allow us to advance beyond simple word-of-mouth. Instead of limiting ourselves to tens or hundreds of individuals the Internet allows us to consider the opinions of thousands. The speed of computers allows us to process these opinions in real time and determine not only what a much larger community thinks of an item, but also develop a truly personalized view of that item using the opinions most appropriate for a given user or group of users.

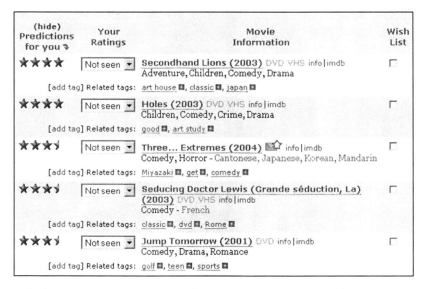

Fig. 1. MovieLens uses collaborative filtering to predict that this user is likely to rate the movie "Holes" 4 out of 5 stars.

9.1.1 Core Concepts

While this chapter considers a variety of CF systems, we introduce the topic through MovieLens[1]. MovieLens is a collaborative filtering system for movies. A user of MovieLens rates movies using 1 to 5 stars, where 1 is "Awful" and 5 is "Must See". MovieLens then uses the ratings of the community to recommend other movies that user might be interested in (Fig. 1), predict what that user might rate a movie, or perform other tasks.

To be more formal, a *rating* consists of the association of two things – user and item – often by means of some value. One way to visualize ratings is as a matrix

---

[1]   http://www.movielens.org/

(Table 1). Without loss of generality, a ratings matrix consists of a table where each row represents a user, each column represents a specific movie, and the number at the intersection of a row and a column represents the user's rating value. The absence of a rating score at this intersection indicates that user has not yet rated the item.

Table 1. A MovieLens ratings matrix. Amy rated the movie Sideways a 5. Matt has not seen The Matrix

	The Matrix	Speed	Sideways	Brokeback Mountain
Amy	1	2	5	
Matt		3	5	4
Paul	5	5	2	1
Cliff	5	5	5	5

The term *user* refers to any individual who provides ratings to a system. Most often, we use this term to refer to the people using a system to receive information (e.g., recommendations) although it also refers to those who provided the data (ratings) used in generating this information.

Collaborative filtering systems produce predictions or recommendations for a given user and one or more *items*. Items can consist of anything for which a human can provide a rating, such as art, books, CDs, journal articles, or vacation destinations.

Ratings in a collaborative filtering system can take on a variety of forms.

- Scalar ratings can consist of either numerical ratings, such as the 1-5 stars provided in MovieLens or ordinal ratings such as strongly agree, agree, neutral, disagree, strongly disagree.
- Binary ratings model choices between agree/disagree or good/bad.
- Unary ratings can indicate that a user has observed or purchased an item, or other wise rated the item positively. The absence of a rating indicates that we have no information relating the user to the item (perhaps they purchased the item somewhere else).

Ratings may be gathered through explicit means, implicit means, or both. *Explicit ratings* are those where a user is asked to provide an opinion on an item. *Implicit ratings* are those inferred from a user's actions. For example, a user who visits a product page perhaps has some interest in that product while a user who subsequently purchases the product may have a much stronger interest in that product. The issues of design decisions and tradeoffs regarding collection of different types of ratings are discussed in section 9.4.

9.1.2 The Beginning of Collaborative Filtering

As a formal area of research, collaborative filtering got its start as a means to handle the shifting nature of text repositories. As content bases grew from mostly "official" content, such as libraries and corporate document sets, to "informal" content such as discussion lists and e-mail archives, the challenge of finding quality items shifted as well. Pure content-based techniques were often inadequate at helping users find the documents they wanted. Keyword-based representations could do an adequate job of describing the content of documents, but could do little to help users understand the application of the keywords or the quality of those documents. Hence, a keyword search for "Chicago Rocks" might yield not only scholarly articles by the Chicago Rocks and Minerals Society but also the "shallower" posting to a music bulletin board regarding one visitor's opinion of the 1970s rock band.

In the early 1990s there seemed to be two possible solutions to this new challenge:

1. wait for improvements in artificial intelligence that would allow better automated classification of documents, or
2. bring human judgment into the loop.

While the challenges of automated classification have yet to be overcome, human judgment has proved valuable and relatively easy to incorporate into semi-automated systems[2].

The Tapestry system, developed at Xerox PARC, took the first step in this direction by incorporating user actions and opinions into a message database and search system [19]. Tapestry stored the contents of messages, along with metadata about authors, readers, and responders. It also allowed any user to store annotations about messages, such as "useful survey" or "Phil should see this!" Tapestry users could form queries that combined basic textual information (e.g. contains the phrase "recommender systems") with semantic metadata queries (e.g. written by John OR replied to by Joe) and annotation queries (e.g. marked as "excellent" by Chris). This model has become known as pull-active collaborative filtering, because it is the responsibility of the user who desires recommendations to actively pull the recommendations out of the database.

Soon after the emergence of Tapestry, other researchers began to recognize the potential for exploiting the human "information hubs" that seem to naturally occur within organizations. Maltz and Ehrlich [42] developed a push-active collaborative filtering recommender system that made it easy for a person reading a document to push that document on to others in the organization who should see it. This type of push-recommender role has become popular, with many people today serving as "joke hubs" who receive jokes from all over and forward them to those they believe would appreciate them (though often with far less discriminating thought than was envisioned).

A limitation of active collaborative filtering systems is that they require a community of people who know each other. Pull-active systems require that the user know whose opinions to trust; push-active systems require that the user know to whom particular content may be interesting. Automated collaborative filtering (ACF) systems relieve users of this burden by using a database of historical user opinions to automatically match each individual to others with similar opinions.

The early ACF systems included GroupLens [51,34] in the domain of Usenet newsgroup articles, Ringo [57] in the domain of music and musical artists, and Bellcore's Video Recommender [27] in the domain of movies. While a more formal dis-

---

[2] For a slightly more broad discussion on the differences between collaborative filtering and content filtering, see Section 9.2.4 of this chapter.

cussion of recommendation algorithms follows in section 9.3, each of these systems follow a process of gathering ratings from users, computing the correlations between pairs of users to identify a user's "neighbors" in taste space, and combining the ratings of those neighbors to make recommendations. GroupLens used a very explicit interface where ratings of Usenet newsgroup articles were entered manually by keystroke or button, and ratings were displayed numerically or graphically (Fig. 2). Taking this a step further, both Ringo and Video Recommender were accessible through the web and email and provided simple features for community interaction.

Fig. 2. A modified Xrn news reader. The GroupLens project added article predictions (lines of 0-9 #s on the top right) and article rating buttons (bottom)

9.1.3 Collaborative Filtering and the Adaptive Web

These early collaborative filtering systems were designed to explicitly provide users with information about items. That is, users visited a website for the purpose of receiving recommendations from the CF system. Later, websites began to use CF systems behind the scenes to adapt their content to users, such as choosing which news articles a website should be presenting prominently to a user.

Providers of information on the web must deal with limited user attention and limited screen space. Collaborative filtering can predict what information users are likely to want to see, enabling providers to select subsets of information to display in the limited screen space. By placing that information prominently, it enables the user to maximize their limited attention. In this way, collaborative filtering enables the web to adapt to each individual user's needs.

The remainder of this chapter will discuss collaborative filtering in more depth by considering:

- The tasks for which users might use a CF system, things a CF system is good at, and the kinds of domains for which CF is appropriate (Section 9.2)
- Algorithms that CF systems employ (Section 9.3)
- How types of ratings in a CF system affect design choices (Section 9.4)
- How to evaluate and compare recommenders (Section 9.5)
- Trends in the development of more interactive and explicitly social interfaces (Section 9.6)
- The challenges to privacy and trust within CF systems (Section 9.7)
- Open questions in the continuing development of CF systems (Section 9.8)

9.2 Uses for Collaborative Filtering

Thus far, we have only briefly introduced collaborative filtering systems. However, we may have still left readers asking the question "for what purposes is CF appropriate?" In this section we consider this question by exploring user tasks that CF supports, then the services that CF systems provide, and finally, contrasting CF with content filtering, a technique that supports many of the same tasks, but using different technology. Throughout, we explore both well-understood technologies, and thought-provoking proposals that are not as well understood.

9.2.1 User Tasks

Designers of web services should carefully identify the possible tasks users may wish to accomplish with their site as different tasks may require different design decisions. From a marketing perspective, this is the value added by the CF system. In this section, we consider user tasks for which collaborative filtering is useful.

Tasks for which people use collaborative filtering that have been studied include:

1. *Help me find new items I might like.* In a world of information overload, I cannot evaluate all things. Present a few for me to choose from. This has been applied most commonly to consumer items (music, books, movies), but may also be applied to research papers, web pages, or other ratable items.
2. *Advise me on a particular item.* I have a particular item in mind; does the community know whether it is good or bad?
3. *Help me find a user (or some users) I might like.* Sometimes, knowing who to focus on is as important as knowing what to focus on. This might help with forming discussion groups [39], matchmaking, or connecting users so that they can exchange recommendations socially.

4. *Help our group find something new that we might like.* CF can help groups of people find items that maximize value to group as a whole [46]. For example, a couple that wishes to see a movie together or a research group that wishes to read an appropriate paper.
5. *Help me find a mixture of "new" and "old" items.* I might wish a "balanced diet" of restaurants, including ones I have eaten in previously; or, I might wish to go to a restaurant with a group of people, even if some have already been there; or, I might wish to purchase some groceries that are appropriate for my shopping cart, even if I have already bought them before.
6. *Help me with tasks that are specific to this domain.* For example, a research paper recommender [60] might also wish to support tasks such as "recommend papers that my paper should cite" and "recommend papers that should cite my paper." Similarly, a recommender for a movie and a restaurant might be designed to distinguish between recommendations for a first date versus a guys' night out. Recommenders for some domain-specific tasks have been explored; many have not. To date, much research has focused on more abstract tasks (like "find new items") while not probing deeply into the underlying user goals (like "find a movie for a first date").

9.2.2 Collaborative Filtering System Functionality

There are also broad abstract families of tasks that CF systems support. It is no accident that this system functionality is related to the user tasks of the previous section. Ideally, the system would support all user tasks, although mapping a real application to the functionality of an actual CF system can be challenging. In any case, here are the broad families of common CF system functionality:

1. *Recommend items.* Show a list of items to a user, in order of how useful they might be. Often this is described as predicting what the user would rate the item, then ranking the items by this predicted rating. However, some successful recommendation algorithms do not compute predicted rating values at all. For example, Amazon's recommendation algorithm aggregates items similar to a user's purchases and ratings without ever computing a predicted rating [38]. Instead of displaying a personalized predicted rating, their user interface displays the average customer rating. As a result, the recommendation list may appear out of order with respect to the displayed average rating value. In many applications, picking the top few items well is crucial; producing predicted values is secondary.
2. *Predict for a given item.* Given a particular item, calculate its predicted rating. Note that prediction can be more demanding than recommendation. To recommend items, a system only needs to be prepared to offer a few alternatives, but not all. Some algorithms take advantage of this to be more scalable by saving memory and computation time [38, 52]. To provide predictions for a particular item, a system must be prepared to say something about any requested item, even rarely rated ones. How does a system decide how a particular user would rate a requested item if very few users – let alone users similar to the particular user – have rated the item? Personalized predictions may be challenging, if not impossible.

3. *Constrained recommendations: Recommend from a set of items.* Given a particular set or a constraint that gives a set of items, recommend from within that set. For example:

> "Consider the following scenario. Mary's 8-year-old nephew is visiting for the weekend, and she would like to take him to the movies. She would like a comedy or family movie rated no "higher" than PG-13. She would prefer that the movie contain no sex, violence or offensive language, last less than two hours and, if possible, show at a theater in her neighborhood. Finally, she would like to select a movie that she herself might enjoy." [55]

Schafer *et al.* [55] propose a "meta-recommendation system" that generates recommendations from a blending of multiple recommendation sources. Users define preferences and requirements through a web form that restricts the set of potential candidate items. Recommendations are based on a ranking of how well the items within this set match the provided preferences. Adomavicius *et al.* [1] call this "flexibility," and propose a SQL-like language as a desired extension in a "next-generation" recommendation system. Such a system might accept queries such as "RECOMMEND Movie TO User BASED ON Rating FROM MovieRecommender WHERE Movie.Length < 120 AND Movie.Rating < 3 AND User.City = Movie.Location." Similar techniques are discussed in Chapter 11 of this book [58].

9.2.3 Properties of Domains Suitable for Collaborative Filtering

One might simply take a user application, implement it with a CF system, and hope it will work. However, CF is better known to be effective in domains with certain properties. It seems useful to acquaint ourselves with them, and consider whether the user application is a good fit. We group these properties below into data distribution, underlying meaning, and data persistence.

Note that with special consideration, CF can be successfully applied in domains that do not have some of the properties below. We simply list them to provoke thought and discussion about what domains are easy or hard with collaborative filtering.

Data Distribution. These properties are about the numbers and shape of the data:
1. *There are many items.* If there are few items to choose from, the user can learn about them all without need for computer support.
2. *There are many ratings per item.* If there are few ratings per item, there may not be enough information to provide useful predictions or recommendations.
3. *There are more users rating than items to be recommended.* A corollary of the previous paragraph is that often you will need more users than the number of items that you want to be able to capably recommend. More precisely, if there are few ratings per user, you will need many users. Lots of systems are like this. For example, this makes web pages a challenging domain, especially if the system requires explicit ratings. Google[3], a popular search engine, claims to index 8 billion web pages at present, which is more than the number of people in the world, not to

---

[3]   http://www.google.com/

mention the number who have access to computers. As another example, with one million users, a CF system might be able to make recommendations for a hundred thousand items, but may only be able to make confident predictions for ten thousand or fewer, depending on the distribution of ratings across items. The ratings distribution is almost always very skewed: a few items get most of the ratings, a long tail of items that get few ratings. Items in this long tail will not be confidently predictable.

4. *Users rate multiple items.* If a user rates only a single item, this provides some information for summary statistics, but no information for relating the items to each other.

Underlying Meaning. These properties are of the underlying meaning of the data:

1. *For each user of the community, there are other users with common needs or tastes.* CF works because people have needs or tastes in common. If a person has tastes so unique that they are not shared by anybody else, then CF cannot provide any value. More generally, CF works better when each user can find many other users who share their tastes in some fashion.

2. *Item evaluation requires personal taste.* In cases where there are objective criteria for goodness that can be automatically computed, those criteria may be better applied by means other than collaborative filtering, e.g., search algorithms. Collaborative filtering allows users with similar tastes to inform each other. CF adds substantial value when evaluation of items is largely subjective (e.g., music), or when those items have many different objective criteria that need to be subjectively weighed against each other (e.g., cars). Sometimes there are objective criteria that can help (e.g., only recommend books written in English), but if recommendation can be performed using only objective criteria, then CF is not useful.

3. *Items are homogenous.* That is to say, by all objective consumption criteria they are similar, and they differ only in subjective criteria. Music albums are like this. Most are similarly priced, similar to buy, of a similar length. Books or research papers are also like this. Items sold at a department store are not like this: some are cheap, some very expensive. For example, if you buy a hammer, perhaps you should not be recommended a refrigerator.

Data Persistence. These are properties of how long the data is relevant:

1. *Items persist.* Not only does a CF system need a single item to be rated by many people, but also requires that people share multiple rated items – that there is overlap in the items they rate. Consider the domain of news stories. Many appear per day, and many probably are only interesting for a few days. In order for a CF system to generate a prediction for me regarding a recently appeared news story, a typical CF algorithm requires that a) one or more users have rated the story and b) these users have also rated some other stories that I have also rated. In a domain like news stories, stories are most interesting when they are new, fresh, and unfortunately, not as likely to have been rated by a large number of people. All of this means that if items are only important for a short time, these requirements are hard to meet.

2. *Taste persists.* CF has been most successful in domains where users' tastes don't change rapidly: e.g., movies, books, and consumer electronics. If tastes change frequently or rapidly, then older ratings may be less useful. An example might be clothing, where someone's taste from five years ago may not be relevant.

The properties of the preceding sections represent simplifications of the world where CF is most easily applied. In fact, applying CF in domains where these properties do not hold can provide both interesting applications and interesting research areas. For example, one might try to apply CF to non-homogenous items by using constrained recommendations, or applying external constraints (called *business rules* in the business world). Likewise, in order to perform system tasks for non-persistent items, one might try to apply content filtering, which is explored in the next section.

9.2.4 Comparing Collaborative Filtering to Content-Based Filtering

Collaborative filtering uses the assumption that people with similar tastes will rate things similarly. *Content-based filtering* uses the assumption that items with similar objective features will be rated similarly. For example, if you liked a web page with the words "tomato sauce," you will like another web page with the words "tomato sauce." The challenge is to cleanly extract the features of items that are most predictive. One then builds a user profile of features from the items a user has rated, and then compares that user profile to item profiles of new items whose features are extracted [4]. Content-based recommendations are discussed in Chapter 10 of this book [48].

Content-based filtering and collaborative filtering have long been viewed as complementary [1]. Content-based filtering can predict relevance for items without ratings (e.g., new items, high-turnover items like news articles, huge item spaces like web pages); collaborative filtering needs ratings for an item in order to predict for it. On the other hand, content-based filtering needs content to analyze. For many domains content is either scarce (e.g., restaurants and books without text reviews available) or it is difficult to obtain and represent that content (e.g., movies and music). Collaborative filtering does not require content. A content filtering model can only be as complex as the content to which it has access. For instance, if the system only has genre metadata for movies, the model can only incorporate this one extremely coarse dimension. Furthermore, if there is no easy way to automatically extract a feature, then content-based filtering cannot consider that feature. For example, while people find the quality of multimedia data (e.g., images, video, or audio) for web pages important, it is difficult to automatically extract this information [4]. Collaborative filtering allows evaluation of such features, because people are doing the evaluation.

Content-based filtering may over-specialize. Items are recommended that match the content features in the user's interest profile or query. Items that do not contain the exact features specified in the interest profile may not get recommended even if they are similar (e.g., due to synonymy in keyword terms). Researchers generally believe collaborative filtering leads to more unexpected or different items that are equally valuable. Some people call this property of recommendations *novelty* or *serendipity* [24]. (See 9.5.2 for a more complete discussion.) However, collaborative filtering has also been shown to over-specialize in some cases [62].

Content-based filtering (CBF) and collaborative filtering may be manually combined by the end-user specifying particular features, essentially constraining recommendations to have certain content features [55]. More often they are automatically combined, sometimes called a *hybrid* approach. There are many ways to combine them, and no consensus exists among researchers [5, 12, 13, 21, 49]. However, such systems generally use the content analysis to identify items that meet the immediate need of the user, and use CF to try and capture features like quality that are hard to automatically analyze. For a more detailed look at these techniques, refer to Chapter 12 of this book [10].

9.3 Collaborative Filtering Algorithms: Theory and Practice

Over the past decade, collaborative filtering algorithms have evolved from research algorithms intuitively capturing users' preferences to algorithms that meet the performance demands of large commercial applications. In this section we explore some of the most widely known collaborative filtering algorithms. Although a good deal of theoretical literature describes CF algorithms, little information is available to assist practitioners in building CF systems. We highlight not only the theoretical definition of these algorithms but their practical challenges and, where applicable, suggest techniques to address these challenges.

Breese *et al.* [9] describes CF algorithms as separable into two classes: *memory-based* algorithms that require all ratings, items, and users be stored in memory and *model-based* algorithms that periodically create a summary of ratings patterns offline. Pure memory-based models do not scale well for real-world application. Thus, almost all practical algorithms use some form of pre-computation to reduce run-time complexity. As a result, current practical algorithms are either pure model based algorithms or a hybrid of some pre-computation combined with some ratings data in memory.

Here, we explore a different organization of collaborative filtering algorithms: *non-probabilistic* algorithms and *probabilistic* algorithms. We consider algorithms to be probabilistic if they are based on an underlying probabilistic model. That is, they represent probability distributions when computing predicted ratings or ranked recommendation lists. In general, non-probabilistic models are widely used by practitioners. Probabilistic models have been gaining favor, however, particular in the machine learning community.

9.3.1 Non-probabilistic Algorithms

The most well-known CF algorithms are nearest neighbor algorithms. We introduce the two different classes of nearest neighbor CF algorithms: user-based nearest neighbor and item-based nearest neighbor. We also explore more briefly non-probabilistic algorithms that transform or cluster the ratings space to reduce the ratings space dimensionality. Other commonly cited algorithms not discussed here include graph-based algorithms [2], neural networks [8], and rule-mining algorithms [23].

User-Based Nearest Neighbor Algorithms

Early algorithms generated predictions for users based on ratings from similar users. We call these similar users *neighbors*. If a user n is similar to a user u, we say that n is a *neighbor* of u. User-based algorithms generate a prediction for an item i by analyzing ratings for i from users in u's neighborhood. Naively, we could average all neighbors' ratings for item i. Equation 1 gives this average-user formulation, where r_{ni} is neighbor n's rating for item i.

$$pred(u,i) = \frac{\sum_{n \subset neighbors(u)} r_{ni}}{number\ of\ neighbors} \tag{1}$$

Equation 1 is considered naïve because it fails to account for the fact that some members of u's neighborhood have a higher level of similarity to u than others. We should be able to generate more accurate predictions by weighting ratings from users who are similar to u more heavily. Thus, if $userSim(u,n)$ is a measure of the similarity between a target user u and a neighbor n, a prediction can be given by equation 2.

$$pred(u,i) = \sum_{n \subset neighbors(u)} userSim(u,n) \cdot r_{ni} \tag{2}$$

Unfortunately, if the similarities of the neighbors do not add up to one, this prediction will be incorrectly scaled. Accordingly equation 3, normalizes the prediction by dividing by the sum of the neighbors' similarities.

$$pred(u,i) = \frac{\sum_{n \subset neighbors(u)} userSim(u,n) \cdot r_{ni}}{\sum_{n \subset neighbors(u)} userSim(u,n)} \tag{3}$$

Finally, users vary in their use of rating scales. That is, one optimistic happy user may consistently rate things 4 of 5 stars that a pessimistic sad user rates 3 of 5 stars. They mean the same thing ("one of my favorite moves"), but use the numbers differently.

To compensate for ratings scale variations, equation 4 *average adjusts* for users' mean ratings.

$$pred(u,i) = \bar{r}_u + \frac{\sum_{n \subset neighbors(u)} userSim(u,n) \cdot (r_{ni} - \bar{r}_n)}{\sum_{n \subset neighbors(u)} userSim(u,n)} \tag{4}$$

The GroupLens system for Usenet newsgroups, one of the first CF systems, defined *userSim()* in equation 4 using the Pearson correlation [51]. The Pearson correlation coefficient is calculated by comparing ratings for all items rated by both the target user and the neighbor (e.g. *corated* items). Equation 5 gives the formula for Pearson correlation between user u and neighbor n, where $CR_{u,n}$ denotes the set of corated items between u and n.

$$userSim(u,n) = \frac{\sum_{i \subset CR_{u,n}} (r_{ui} - \bar{r}_u)(r_{ni} - \bar{r}_n)}{\sqrt{\sum_{i \subset CR_{u,n}} (r_{ui} - \bar{r}_u)^2} \sqrt{\sum_{i \subset CR_{u,n}} (r_{ni} - \bar{r}_n)^2}} \tag{5}$$

Pearson correlation ranges from 1.0 for users with perfect agreement to -1.0 for perfect disagreement users. Negative correlations are generally believed to not be valuable in increasing prediction accuracy [25] and one may choose to not use negative correlations.

Practical Challenges of User-Based Algorithms

The user-based nearest neighbor algorithm captures how word-of-mouth recommendation sharing works and it can detect complex patterns given enough users; however it has practical challenges.

Ratings data is often sparse, and pairs of users with few coratings are prone to skewed correlations. For example, if users share only three corated items, it is not uncommon for the ratings to match almost exactly (a similarity score of 1). If such similarities are not adjusted, these skewed neighbors can dominate a user's neighborhood.

Another problem with Pearson correlation is that it fails to incorporate agreement about a movie in the population as a whole. For instance, two users agreement about a universally loved movie is much less important than agreement for a controversial movie. Pearson correlation does not capture this distinction. Some user-based algorithms account for global item agreement by including weights inversely proportional to an item's popularity when calculating user correlations [9].

The original user-based algorithm as implemented in GroupLens included all users in a CF system in a prediction neighborhood [50]. Later algorithms improved accuracy and efficiency by limiting the prediction calculation to a user's closest k neighbors [25].

Most importantly, calculating a user's perfect neighborhood is expensive - requiring comparison against all other users. Thus, in a naïve implementation, the time and memory requirements of user-based algorithms scale linearly with the number of users and ratings. Amazon.com has tens of millions of customers and probably wishes recommendations to take no more than a small fraction of a second. It would be immensely resource intensive to scan the ratings of millions of customers to return a recommendation under this time constraint.

Researches have tried many techniques to reduce processing time and memory consumption:

- *Subsampling* - In sampling, a subset of users is selected prior to prediction computation. Neighborhood computation time remains fixed, and schemes have been proposed to intelligently choose neighbors in order to achieve virtually identical accuracy.
- *Clustering* - Clustering algorithms have been used to quickly locate a user's neighbors [38]. In these schemes, a user is compared to groups of users, rather than individual users. Clusters of users similar to the target are quickly discovered, and nearest neighbors can be selected from the most similar clusters. Both k-means clustering [40], and hierarchical divisive [31] and agglomerative clustering [35] can segment users into clusters. One challenge in using clustering is that clustering schemes use distance functions, such as Pearson correlation, to both form the clusters and measure distance from a cluster. However, due to missing data, distance functions generally do not obey the triangle equality and are not true mathematical metrics[4]. This can lead to unintuitive and unstable clustering.

---

[4] A distance metric has four properties: it is non-negative, the identity distance is 0, it is reflexive, and the triangle equality holds. The triangle equality is generally most difficult requirement to meet.

Item-Based Nearest Neighbor Algorithms

Item-based nearest neighbor algorithms are the transpose of the user-based algorithms. While user-based algorithms generate predictions based on similarities between users, item-based algorithms generate predictions based on similarities between items [52]. The prediction for an item should be based on a user's ratings for similar items. Consider the ratings matrix shown in **Table 2**. Assume we are trying to predict a rating for "Speed" for user #3 (marked by the X). First, we observe that the ratings for "Speed" are very similar to the ratings for "Sideways", but not as similar to the ratings for "The Matrix." We now try to predict the rating "X" by building a weighted average of user #3's other ratings (3 for "The Matrix" and 4 for "Sideways"). Since "Speed" is similar to "Sideways," we might guess that the rating for "Sideways" is more important. We conclude that a good guess is *0.25*3 + 0.75*4 = 3.75.*

Table 2. An item-based nearest-neighbor algorithm generates predictions based on similarities between items. Observe that "Speed" is fairly similar to "Sideways" and moderately similar to "The Matrix."

	The Matrix	**Speed**	**Sideways**	**Brokeback Mountain**
User 1	5	4	3	
User 2	4	5	5	3
User 3	3	X	4	
User 4	5	3	3	4

We have just outlined the item-based prediction algorithm, which we formalize in equation 6. A prediction for a user u and item i is composed of a weighted sum of the user u's ratings for items most similar to i.

$$pred(u,i) = \frac{\sum_{j \in ratedItems(u)} itemSim(i,j) \cdot r_{ui}}{\sum_{j \in ratedItems(u)} itemSim(i,j)} \tag{6}$$

Note that in equation 6, *itemSim()* is a measure of item similarity, not user similarity. Average correcting is not needed when generating the weighted sum because the component ratings are all from the same target user.

Several variations exist for calculating the similarity for a pair of items (i, j). Adjusted-cosine similarity, the most popular (and believed to be most accurate) similarity metric, is computed using all users who have rated both item i and j. Equation 7 gives the formula for adjusted-cosine similarity, where $RB_{i,j}$ denotes the set of users who have rated both item i and item j.

$$itemSim(i,j) = \frac{\sum_{u \subset RB_{i,j}} (r_{ui} - \bar{r}_u)(r_{uj} - \bar{r}_u)}{\sqrt{\sum_{u \subset RB_{i,j}} (r_{ui} - \bar{r}_u)^2} \sqrt{\sum_{u \subset RB_{i,j}} (r_{uj} - \bar{r}_u)^2}} \tag{7}$$

The only difference from Pearson correlation is that average adjusting is performed with respect to the user, not the item. As in the user Pearson correlation, the correlation value ranges from −1.0 to 1.0.

There is evidence that item-based nearest neighbor algorithms are more accurate in predicting ratings than their user-based counterparts [52].

Practical Challenges in Item-Based Algorithms
Theoretically, the size of the model could be as large as the square of the number of items. In practice, we can substantially reduce this size by only storing correlations for item pairs with more than k coratings. Sarwar *et al.* prune the model even further by only retaining the top n correlations for each item. Such modifications yield item-based algorithms that are relatively efficient in both memory usage and CPU performance. Note that pruning many of the correlations means that it may be more difficult to make a prediction for a given target item and user, since the items correlated with the user's ratings may not contain the target item.

As in the user algorithm, item pairs with few coratings can lead to skewed correlations and care must be exercised to not let skewed correlations dominate a prediction.

Non-probabilistic Dimensionality Reduction Algorithms
Large CF applications may support millions of users and items [38]. Other domains may have such a sparsity of ratings that there are few coratings. Several algorithms reduce domain complexity by mapping the item space to a smaller number of underlying "dimensions." Intuitively, these dimensions might represent the latent topics or tastes present in those items. The smaller "latent" dimensions reduce run-time performance needs and lead to larger numbers of co-rated dimensions. These techniques define a mapping between a user's ratings and their underlying tastes. An item's prediction can then be generated based on a user's underlying tastes. Mapping functions generally consist of simple vector operations, and predictions for an item can be calculated in constant time. Vector-based techniques for extracting underlying dimensions include support vector decomposition [53], principal component analysis [20], and factor analysis [11].

Practical Challenges in Dimensionality Reduction Algorithms
Mathematical dimensionality reduction techniques such as singular value decomposition [53] and principal component analysis [20] require an extremely expensive offline computation step to generate the latent dimensional space. Practical implementation of these techniques generally requires the use of heuristic methods for incrementally updating the latent dimensional space without having to entirely recompute – such as the folding-in technique for singular value decomposition [7, 16]. However, the primary challenge to utilizing such techniques is the mathematical complexity – which can lead to challenges debugging and maintaining software utilizing those techniques. While there is some evidence that these techniques can improve accuracy in predicting ratings [54], for the most part, the improvement has not been substantial enough to overcome the practical challenges of complexity.

Association Rule Mining

Association mining techniques build models based on commonly occurring patterns in the ratings matrix [23, 37]. For example, we may observe that users who rated item 1 highly often rate item 2 highly. A particular rule is represented by an input condition (e.g. item 1 rated highly) and a result condition (e.g. item 2 rated highly). The *support* of a rule represents the fraction of users who have rated both the input and result conditions, and the *confidence* of a rule is the fraction of users with the input condition that exhibit the result condition.

In order to generate a predicted rating for a user u and item i, we first select the rules with a result condition of item i that only include items rated by user u. We then use a heuristic to translate the support, accuracy, and ratings for input conditions into a predicted rating.

For more information, refer to Chapter 3 of this book [44].

Practical Challenges in Association Rule Mining

Naïve association rules can treat each rating value as independent. For example, a rating of 1 for a particular item is different than a rating of 2, even though both may be interpreted as the user indicating dissatisfaction with the item. This independence can dramatically increase the sparsity of an already sparse space. To overcome this, implementers generally place "similar" ratings into bins using one of several strategies:

- *High and low ratings bins* – Divide ratings into two bins; those above and those below a user's average rating.
- *High ratings* – Only consider ratings above a user's average when building rules.
- *All ratings* – Treat all ratings as identical when building rules.

A general drawback in association mining is that, since rating bins are treated discretely, we lose any notion of the numeric relationship among ratings. Although this relationship is theoretically meaningful, in practice it seems to have little impact.

Association rule mining in non-CF domains often looks for input patterns consisting of multiple items (e.g. if the user rated items 1 and 2 highly, they will rate item 3 highly). While these patterns may be useful, mining the patterns is too slow in CF domains due to the extremely high dimensionality.

9.3.2 Probabilistic Algorithms

Probabilistic CF algorithms explicitly represent probability distributions when computing predicted ratings or ranked recommendation lists. In general, probabilistic algorithms try to leverage well-understood formalisms of probability.

Most probabilistic CF algorithms calculate the probability that, given a user u and a rated item i, the user assigned the item a rating of r: $p(r|u,i)$. We calculate a predicted rating based on either the most probable rating value or the expected value of r. Equation 8 gives the formula for user u's expected rating for item i.

$$E(r \mid u,i) = \sum_r r \cdot p(r \mid u,i) \tag{8}$$

The most popular probabilistic framework involves Bayesian-network models that derive probabilistic dependencies among users or items. Some of the earliest probabilistic CF algorithms were proposed by Breese *et al.*, who describe a method for deriving and applying Bayesian networks using decision trees to compactly represent probability tables [9]. For example, Fig. 3 shows that users who do not watch "Beverly Hills, 90210" are very likely to *not* watch Melrose Place.   A separate tree is constructed for every recommendable item. The branch chosen at a node in the tree is dependent on the user's rating (or lack of rating) for a particular item. Nodes in the tree store a probability vector for user's ratings of the predicted item. In theory, non-naïve Bayesian networks improve upon standard item-based algorithms by modeling dependencies between input items used to calculate a prediction. However for multi-valued ratings, there has been no published evidence of Bayesian networks consistently outperforming item-based nearest neighbor algorithms.

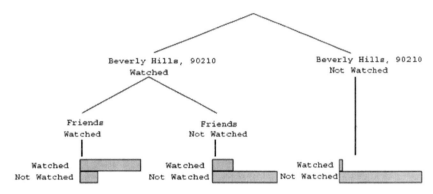

Fig. 3. A decision tree regarding whether a user watches "Melrose Place" based on whether or not they watch "Friends" and/or "Beverly Hills, 90210." Probabilities for Watched vs. Not Watched are displayed at the leaves of the tree and are dependent on the condition of viewing the programs at the parent nodes. (From [9]. Used with permission.)

There has also been a good amount of work on developing probabilistic clustering/dimensionality reduction techniques. Probabilistic dimension reduction techniques introduce a hidden variable $p(z|u)$ that represents the probability a user belongs to the hidden class z. Equation 9 gives the formula for calculating the probability of user u rating item i value r.

$$p(r \mid u, i) = \sum_z p(r \mid i, z) p(z \mid u) \tag{9}$$

The corresponding prediction is the expectation of the rating value (equation 10).

$$E(r \mid u, i) = \sum_r \left(r \cdot \sum_z p(r \mid z, i) p(z \mid u) \right) \tag{10}$$

Hoffman presents an expectation maximization (EM) algorithm for CF that estimates latent classes z with Gaussian probability distributions [29]. Clustering algorithms also have been used to estimate latent classes [61].

One advantage of probabilistic algorithms is that they can produce a probability distribution across possible rating values – information that captures the likelihood of each possible rating value. From this information, not only can you compute the most probable rating, you can also compute a likelihood of that rating being correct – thus capturing the algorithm's confidence. There has been a recent attempt to create a hybrid approach that utilizes the nearest neighbor algorithm, but represents ratings as discretized probability distributions rather than a point rating [41].

9.3.3 Over-Arching Practical Concerns

Regardless of choice of algorithm, real-world CF systems need to address several problems that are generally not covered in research literature.

Adjust for Few Ratings
Items and users with few ratings can inappropriately bias CF results. Algorithms may take steps to adjust for users, items, and user and item pairs with few co-ratings (we will generally call these *rarely-rated entities*). We will compare techniques for adjusting for rarely-rated entities, using a user-based algorithm as an example:

1. *Discard rarely-rated entities* – Algorithms often only incorporate data with greater than k ratings. In a user-based algorithm, for example, we would discard neighbors with fewer than k co-ratings with the target user. Although this is a simple and clean approach it can decrease the coverage of the CF system.
2. *Adjust calculations for rarely-rated entities*– This technique adjusts calculations for rarely-rated entities by pulling them closer to an expected mean. For instance, Pearson similarities for users with few co-ratings may be adjusted closer to 0. CF systems often make the adjustment amount inversely proportional to the number of ratings. Although adjustment can be effective, tuning adjustment parameters can be difficult and unstable.
3. *Incorporate a prior belief* – We can avoid skew by incorporating artificial data points that match an expected distribution. For example, we may believe that user's ratings will generally match a probability distribution p. We can incorporate this prior belief into user correlation calculation by including k artificial co-rated items whose ratings are independently drawn from p.

Prediction Versus Recommendation
Prediction and Recommendation tasks place different requirements on a CF system. To recommend items, a system must be prepared to know about a subset of items, but perhaps not all. Some algorithms save memory and computation time by taking advantage of this [38, 52]. To provide predictions for a particular item, a system must store information about every item, even rarely rated ones. Algorithms that are required to present personalized predictions for many items often have larger memory requirements.

On the other hand, recommendation tasks require calculation of predictions or some scoring function for many (if not all) items. A single prediction request can therefore afford a more expensive prediction calculation than a recommendation request.

Confidence Metrics

CF systems can supply a confidence metric that indicates the support for a particular prediction. Applications may choose to not display predictions with confidence measures below a certain threshold.

Confidence measures can also be used when selecting items for recommendation. CF algorithms generally choose to recommend those items with highest predicted ratings. Some CF systems may choose to tradeoff items with high predictions and low confidence for items with less-high predictions and high confidence.

Confidence measures are specific to each CF algorithm. Probabilistic algorithms may be able to use their computed probability distributions to estimate confidence. User-based algorithms often use confidence measures that incorporate the agreement for an item in a user's neighborhood, and the number of corated items between neighbors and the user. Item-based algorithms may measure the number of ratings for correlated pairs of items contributing to a prediction.

9.4 Acquiring Ratings: Design Tradeoffs

Ratings data from users on items are what enable collaborative filtering. In this section we will discuss in more depth the different kinds of ratings data that can be used and key concepts and decisions involved with acquiring ratings for collaborative filtering systems.

9.4.1 Explicit Versus Implicit Ratings: Tradeoff

Explicit ratings provided by users offer the most accurate description of a user's preference for an item with the least amount of data. However, because explicit ratings require additional work from the user, it can be challenging to collect ratings – particularly when creating a new CF service. On the other hand, implicit ratings – observations of user behavior from which preference can be inferred – are collected with little or no cost to the user, but ratings inference may be imprecise. As an example, consider using "time spent reading information about a product" as an implicit rating for that product. Intuitively, if a user spends a lot of time reading about a product, we might conclude that they would be interested in purchasing that product. However, there are reasons that this inference could be inaccurate – the user may have taken a coffee break just after opening the product info page, or the user may have concluded that the product was inappropriate after spending time to read about it. Thus if implicit ratings are used, there is more uncertainty in the computation. Other examples of implicit ratings are discussed in Oard and Kim [45].

The more ratings you have, the more uncertainty in the ratings you can handle. Uncertainty in rating values, including implicit ones is handled by aggregating ratings – collecting multiple observations of variables that are predictive of a rating and combining them into a single estimated rating – either by voting [17] or averaging [51, 57]. Thus if you are able to collect large numbers of ratings, then the errors introduced by uncertainty of implicit ratings can be canceled out by aggregation. In such a situation, you may be able to build a very successful CF system without explicit ratings.

Examples from the music domain are AudioScrobbler[5] and MusicStrands[6], which track every single song you play. With music, after enough ratings (plays) have been accumulated, these implicit ratings may represent user taste much better than small explicit ratings scales. A five point rating scale only allows you to group a user's rated items into five ranks – the CF system cannot distinguish difference in taste between items with the same rating value. When using the implicit play count, user may play individual songs thousands of times, and since each song is likely to be played a different number of times, a more complete ranking of items a user likes can be created. If you cannot capture large numbers of implicit ratings, then you will most likely need some form of explicit rating.

9.4.2 The Challenge of Collecting Explicit Ratings

Explicit ratings require dedicated attention of the user. Early researchers believed that users would not invest the time rating items required for CF systems. From an economic perspective it would appear that if incremental recommendations are free, then everybody would wait for others to identify what was good and there would be insufficient ratings [3]. However, during the past decade, experience has demonstrated that collecting explicit ratings is not as challenging as previously thought.

The first reason is that – in order to succeed – a CF system doesn't need lots of ratings from all people. Instead you just need a relatively small number of "early adopters" who rate frequently and continuously. These early adopters provide sufficient information to generate recommendations for the remaining users of the system. The remaining users must each then just provide a limited number of ratings in order for the system to learn their preferences.

The second reason that collecting explicit ratings is easier than previously expected is that users appear to gain many benefits from rating other than higher quality recommendations. Researchers and practitioners have proposed that users gain the following rewards from rating [22]:

- An increased feeling of having contributed to advancing a community
- Gratification from having one's opinion's voiced and valued
- An ability to use the CF system as an extension of their memory of what they like and dislike.

Maintainers of CF systems sometimes use incentives to encourage users to provide more explicit ratings. For example, sites may exchange user ratings for "site points." These site points can be exchanged for rewards (e.g. t-shirts and hats) or privileges (e.g. the right to view privileged content). While incentives may increase the number of ratings provided by users we are unaware of any studies that confirm this correlation.

---

[5] AudioScrobbler is owned by Last.fm which can be found at http://www.last.fm/index.php
[6] http://www.musicstrands.com/

9.4.3 Rating Scales

Another significant design decision involves choosing the explicit rating scale. The finer grained the scale, the more information you will have regarding each user's preference. Finer grained scales require more complex user interfaces. The most common types of ratings are shown in Table 3.

At some point, increasing the precision of the rating scale further may fail to add value. If a very precise scale is selected, such as 1-100, you are unlikely to get a user to give the same rating for an item if you ask them at different points in time – thus you increase the uncertainty in the rating. Perhaps the most important consideration is the desires of the user population. It may also be that people desire a fine-grained ratings scale in order to rank different movies (I liked 'The Matrix' a little better than 'Sideways' so I'll rate it 4.5 instead of 4), even if that added granularity does not help predictive accuracy. Users may feel that they cannot fully describe their tastes with too few possible rating values. In MovieLens, users were frustrated that they were not able to give ratings as precise as the systems predictions of their ratings – predicted ratings were to the closest half point while user ratings were integers [14].

Table 3. Most common explicit rating scales

Rating Scale	Description
Unary	Good or "don't know"
Binary	Good or Bad
Integer "Likert"-like	Integers: 1-5, 1-7, or 1-10

9.4.4 Cold Start Issues

The "cold-start" problem describes situations in which a recommender is unable to make meaningful recommendations due to an initial lack of ratings. This problem can significantly degrade CF performance. It can occur under three scenarios.

New User. When a user first registers with a CF service, they have no ratings on record. Thus no personalized predictions can be given. For example, a new user to MovieLens has no ratings in the system, so a neighborhood of similar users can not be calculated. This may be solved in several ways. For example, by a) having the user rate some initial items before they can use the service; b) displaying non-personalized recommendations (population averages) until the user has rated enough; c) asking the user to describe their taste in aggregate, e.g., "I like science fiction movies"; d) asking the user for demographic information, or e) using ratings of other users with similar demographics as recommendations.

New Item. When a new item is added to a CF system, it has no ratings, so it will not be recommended. For example, MovieLens is unable to recommend new Hollywood releases until someone has entered an initial rating. Unfortunately, in many domains, users are less likely to rate items that are not recommended to them. In many domains this is not a show-stopper because most good items can be discovered through means other than the CF system and will get eventually rated. However, in domains with high item turnover (such as news articles) the cold-start problem can be particularly troublesome.

Users also tend to be forgiving of systems that don't recommend obscure items. However, in domains where there may be many "sleepers" (unrated items that are very good) several techniques can be used, including: a) recommending items through non-CF techniques such as content analysis or metadata, and b) randomly selecting items with few or no ratings and asking users to rate those items.

New Community. The biggest cold-start problem is bootstrapping a new community. If a new service's value is in its personalized CF recommendations, then without ratings it may not have sufficient differentiating value – thus not retain users long enough to build up ratings. The most common solution is to provide rating incentives to a small "bootstrap" subset of the community, before inviting the entire community to use the service. Other approaches are to maintain users' interest through alternate services, initially generate recommendations using non-CF approaches, or to start with a set of ratings from another source outside the community.

9.5 Evaluation

Evaluation measures how well a collaborative filtering system is meeting its goals, either in absolute terms or in relation to alternative CF systems. Unfortunately, there is no well-accepted metric that can evaluate all-important criteria related to the performance of a CF system. The appropriate metric to choose may depend on the type of items being recommended, the user tasks supported by the CF system, and any external goals that the service providers may have (e.g., promotional or inventory depletion). An in-depth discussion of evaluation considerations of collaboration filtering systems can be found in Herlocker *et al.* [24]. In this section, we first discuss accuracy, which is generally considered the most important criteria to evaluate, and then briefly deal with some of the other criteria that may be important to evaluate and their associated metrics.

9.5.1 Accuracy

The most prominent evaluation metrics in the research literature measure the *accuracy* of the system's predictions. Accuracy can either be measured as the magnitude of error between the predicted rating and the true rating, or the magnitude of error between the predicted ranking and the "true" ranking. *Predictive accuracy* is the ability of a collaborative filtering system to predict a user's rating for an item. The standard method for computing predictive accuracy is *mean absolute error (MAE)* – the average absolute difference between the predicted rating and the actual rating given by a user. The advantage of MAE is that it is simple, well understood, and traditional significance tests can be applied to it. Furthermore, MAE seems to intuitively capture the quality of a CF system – system builders want predictions to be as close as possible to the true ratings. However, MAE has proven to be an unreliable measure of a ranked recommendation list [41]. Users perceive errors at the top of a recommendation list as much more costly than similar errors at the bottom of lists. MAE does not differentiate between errors at the top and errors at the bottom of lists.

Rank accuracy metrics attempt to compute the utility of a recommendation list to a user. Common rank accuracy metrics include precision [41, 52] and half-life utility

[9]. Precision is the percentage of items in a recommendation list that the user would rate as useful. In CF, it is often computed at varying lengths of recommendation list (1, 3, 5, etc). The half-life utility metric computes a value for a ranked list that is intended to capture percentage of the maximum utility achieved by the ranked list in question. The maximum utility is achieved if all of the items rated as useful appear above all the items rated as not useful. In the half-life utility metric, mistakes at the top of the ranked list are weighted exponentially greater than mistakes further down the list.

If the user interface of the collaborative filtering system primary provides ranked lists of "best-bet" recommendations, then the accuracy of the system should be evaluated with a rank accuracy metric. If the system displays predictions of ratings directly to the user, then it is important to evaluate the system with a predictive accuracy metric. In many cases, it may make sense to use both.

9.5.2 Beyond Accuracy

While many of the published evaluations of CF systems measure accuracy, researchers and practitioners have come to learn that accuracy is not the only criteria of interest, and in some cases, may not even be the most important. Several other evaluation criteria have been explored.

- *Novelty* is the ability of a CF system to recommend items that the user was not already aware of. While non-novel recommendations can still be valuable, for many applications novelty is one of the most valued characteristics of the CF system's recommendations. Even stronger than novelty is the idea of *serendipity*, where users are given recommendations for items that they would not have seen given their existing channels of discovery. To illustrate the distinction, consider a news article recommender. A traditional content-based personalization system may generate recommendations that are not novel, because if I say I like a particular news article, then it will recommend other news articles with similar text, including stories about the exact same news event. A system tuned for novelty will actively avoid recommending news stories of which I am already aware. A serendipitous system would recommend to me news articles about topics that I have never read about before. Researchers have studied how to adjust algorithms to promote serendipity and novelty [32], but measuring novelty is challenging because it requires live user studies where participants indicate if a recommendation was novel.
- *Coverage* is the percentage of the items known to the CF system for which the CF system can generate predictions. It is also possible to compute variants such as the percentage of items that have the potential of being recommended to users, as performance optimizations in recommendations may prevent certain items from ever being recommended [54].
- *Learning Rate* measures how quickly the CF system becomes an effective predictor of taste as data begins to arrive. Generally these are computed per-user, measuring the number of ratings that a user has to provide before they are getting high quality personalized predictions [56].

- *Confidence* describes a CF system's ability to evaluate the likely quality of its predictions. Most CF systems generate rankings based on the most probable predicted rating. A CF system that can accurately compute its confidence in a prediction has the ability to limit recommendations to high confidence ones, leading to a tradeoff of fewer false positives in return for decreased coverage and possibly decreased novelty. If confidence in predictions can be computed, it can be displayed to users to help them decide if the risk-return ratio is appropriate [26].
- *User satisfaction metrics*. The metrics described above are only a sample of possible evaluation metrics. In particular there are many more metrics that can be applied if researchers have the ability to present a system to users, and measure how users perceive the system. This can be accomplished either by surveying the users or measuring retention and use statistics. Good examples include Swearingen and Sinha [59] and Dahlen *et al.* [15].
- *Site performance metrics*. In addition to the more mathematical and often "offline" metrics described above, websites may choose to use fairly simple site analysis metrics when adding a recommender to a site or modifying the design of an existing recommender. Such metrics might include tracking an increase in items purchased or downloaded, an increase in overall user revenue, or an increase in overall user retention. While such trends are easy to track and measure, they may be difficult to correlate to specific changes to an active website.

In conclusion, it is best to select a suite of metrics that will evaluate the criteria that are most important for the successful operation of a particular CF system. For example, if you are using CF to generate a top-5 recommendations list for your website, then you might compute precision at top-5, top-3, and top-1. Furthermore, if the goal of your website recommendations is to introduce your users to new things, then you might also do some user studies where you shown recommendation lists to users and ask them to rate the novelty of those recommendations. Predictive accuracy metrics like MAE may not be so useful if you are not displaying predicted rating values to users.

9.6 Rich Interfaces and Social Navigation

Early user interfaces for CF systems simply provided ranked list of recommendations, potentially with predicted ratings. The recommendation engine was a "black-box" – there was no transparency into how a prediction was computed [15, 19, 27, 51]. A critical trend in recent years is the exploration of user interfaces that enable more rich interaction with the underlying data of a collaborative filtering system, and CF systems that expose more information about the users from whom recommendations are (or can be) generated. In this section, we describe explanation and social navigation – two of these trends, and why they are so important.

One limitation of the black box approach was that the user interfaces to the CF systems were unable to communicate to the user when predictions were more or less risky than normal. Yet the need for this was common – when users were new or when items were new, predictions are more risky because there is less data on which to base inferences. More generally, the black box approach does not expose the reasoning or the data used in a recommendation. As a result, the user has little data on which to base decisions such as a) should they trust the recommendation process, b) is the current recommendation highly confident – either through trusted sources or overwhelming evidence, or c) is this recommendation appropriate for the user's immediate context or need.

9.6.1 Explanation

Initial work on the use of explanation in CF recommendations was promising [26], and has more recently been adopted commercially by Amazon.com, which has a link "why was I recommended this item" – the link will list previous ratings or purchases that you made that strongly influenced the recommendation at hand (Fig. 4). Explanations of CF recommender systems are challenging because the underlying predictive models are complex aggregations of large quantities of data, often with significant probabilistic reasoning. Yet initial research suggests that users are overwhelmed if they are presented with too much data within an explanation [26]. While the current work on recommendations is far from conclusive, promising approaches that have been explored include: showing histograms of a user's neighbors' ratings for the recommended item and showing key items that the user rated that influenced the recommendation.

There is also a correlation between persuading a user that the recommendation is correct and explaining the recommendation to them. For many contexts, it may be sufficient to supply data from other sources not used in the recommendation that confirms the recommendations – such as reviews from critics. This may help persuade the user that the recommendation is good, yet reveals nothing about the reasoning behind the recommendations.

9.6.2 Social Navigation

Most of the CF systems we have discussed so far have been systems that use the group as a whole to help each individual user. Such systems tend to ignore the importance of the groups themselves. *Social navigation* systems encompass a variety of techniques that help people work together to help each other by making the aggregate behavior of the community visible. Users can employ this behavior to find their way through often crowded web spaces.

Höök *et al.* consider one type of social navigation system in which each visitor to a website leaves "footprints" – telltale signs regarding what information the visitor considered and how frequently or in-depth. These footprints help other users find their way more readily through that same space [30]. This type of visualization has been called "read-wear" or "edit-wear" [28]. Early users leave footprints that help later users make sense of the wealth of alternatives available to them. Later users benefit from the footprint, because they are able to direct their attention to the parts of the site that are most valuable to them. As information spaces become more crowded with users it may become important to have systems that show us only those footprints that are most useful to us.

Fig. 4. Amazon.com provides customers with list of previous purchases and ratings that strongly influenced a particular recommendation

While these early CF and social navigation systems were clearly "collaborative," they almost always have provided "implicit" collaboration. Users benefited from the ratings and footprints left by other users in an anonymous and virtually untraceable manner. Some second generation collaborative filtering services have begun to experiment with allowing more "explicit" collaboration by exposing more of the identity of the other members of the community whose ratings are being used to generate a user's recommendations.

One example is epinions.com, which is a site designed to help users make purchasing decisions. On epinions.com, users rate and review products that they have purchased and these reviews are made available as recommendations to others. When a user views a recommendation/review, she can also look at the profile of the user who made the review, seeing information such as what other reviews they have written and how other people have responded to those reviews. She can explicitly state that she "trusts" a user as a reviewer. She can also "block" a reviewer, so that user's ratings/reviews are not shown.

Interfaces like epinions.com attempt to mimic more accurately the social process of word-of-mouth recommendations. A user could choose those people whose tastes he agreed with to provide recommendations, yet could choose different people to trust for different contexts. He could base his trust of another user on his observations of their activity within the community (their ratings) or on other's expressed opinions of their value. As users began to rate each other, explicit social networks could be expressed – "webs of trust." Users could then navigate these social networks in their search for items or products that would meet their needs.

CF web services that offered this social navigation often evolved to be much more than recommendation sites. Particularly interesting was that the CF aspects of the system would bring together communities of common interest that would then engage in direct social interaction through discussion groups, chat rooms, or email. In theory, this direct social connection is the ultimate rich interface for recommendation. The CF software enables a user to navigate a potentially immense social network and find exactly those people who most closely share their tastes.

9.7 Ongoing Challenges to Collaborative Filtering

9.7.1 Privacy and Security

In order to provide personalized information to users, CF systems need to know things about those users. In fact, the more the system knows about a user, the better predictions it can provide to that user. With this increased information stored by a system often comes an increased concern on the part of the user regarding what information is collected, where and how it is stored, and how it is used. In centralized CF architectures, a single repository stores all user ratings. If the central server becomes compromised or corrupt, a user's anonymity can be destroyed. Users must trust that the CF provider will not use their preferences except for providing ratings and recommendations.

Distributed architectures may deploy ratings or models to each user, risking exposure of information to every peer [51]. To protect against this, researchers have developed security techniques building on encryption and shared keys [11]. In these schemes, a user can encrypt their ratings, and peers can tally encrypted ratings. Once ratings are totaled, distributed agents use shared keys to decrypt the rating tallies, without being able to see the original ratings.

Even systems that maintain the security of their users' ratings can be exploited to reveal personal information, particularly for users with unusual tastes. Ramakrishnan et al [50] use a graph-theoretic framework to explore these concerns. They found that "weak ties" (users who connect clusters of different tastes) are most susceptible to exploitation. Unfortunately, it is often these esoteric users that are most valuable to recommender systems, because they can provide users with unexpectedly novel recommendations. For more on the issue of privacy, see Chapter 21 of this book [33].

9.7.2 Trust

Recommender systems may break trust when malicious users give ratings that are not representative of their true preferences. What happens to a CF system if one or more users decide to "attack" an item by purposefully lowering their rating(s) of the item? What happens if a company bombards a recommender with inflated ratings of its own products (e.g. Sony using quotes from made-up critics to promote its films [6])? There have been many examples of these "shilling" attacks. O'Mahoney et al [47] showed that users could, in fact, artificially raise and lower predicting ratings. User-based algorithms are more susceptible to shilling than item-algorithms, as are new or rarely rated items. Unfortunately, this vulnerability remains a significant challenge to collaborative filtering systems. Methods researchers use to detect attacks are not even sensitive enough to detect harsh attacks [32].

While these shilling attacks may seem slightly benign on the surface, further research has suggested that their effect may be more influential than originally feared. Cosley *et al.* demonstrated that users may not only perceive biases in ratings, but also adjust their own ratings to match recommenders' biases [14]. This observation indicates that shilling effects may be compounded as having viewed predictions based on the biased ratings potentially skews later users' ratings. More research is needed to understand how to identify attacks and protect systems from them.

For more on the issues of Privacy and Trust, see [18][36].

9.8 Open Questions

This section discusses some open questions in the field of collaborative filtering. They are grouped into algorithmic questions (with an emphasis on temporal questions), and questions of broader access to collaborative filtering systems.

9.8.1 Algorithmic Questions

Evaluation metrics. There have been many metrics of recommendation quality proposed [24]. Which ones capture what people perceive as good quality? Which ones are important?

Predicting well and recommending well at the same time. As we discussed in Section 2.2, efficient algorithms for recommendation may choose not to produce predicted values at all, or may choose to only store a small amount of information necessary to recommend some items. However, predicting a rating for a given user and item is an appealing application. Are there efficient, scalable algorithms that both recommend and predict well?

Tagging. Social systems such as flickr and del.icio.us, which allow users to tag things (photos and websites, respectively) with keywords, are increasing in popularity and have captured the imagination of many people. These are collaborative filtering systems surely, though without much automation as yet. Other tagging systems have been around for years (e.g., IMDB's movie "keywords"). There are many interesting research questions. How can collaborative filtering algorithms be applied to tags? Can tags be used in conjunction naturally with ratings?

Tags without ratings are missing information. Tagging a movie "high-speed car chase" does not indicate whether that was a good thing or not. Is there a hybrid solution, where tags have associated explicit or inferred ratings?

9.8.2 Temporal Questions

These questions are about the behavior of a collaborative filtering system over time.
1. *Item lifecycle*
    a) When does an item have enough ratings to be accurately recommendable?
    b) When is an item a rising trend, falling trend, or a fad? Are many items like that?

2. *User lifecycle*
    a)   When does a user have enough ratings to get good recommendations?
    b)   Can one identify the items for which it is possible to give good recommendations for a given user?
    c)   At what point do additional user ratings fail to improve his recommendations because the system has built a sufficiently accurate model (diminishing returns)? Can users detect this point and do they change the way they use the system?
    d)   Are more ratings useful again as items are added?
    e)   How do old ratings affect a user's recommendations, versus new ratings? Do user tastes shift over time? Can we detect it?
3. *Ratings database lifecycle*
    a)   When is a rating "stale" (i.e., no longer reflective of the opinion of the rater)?
    b)   When does a database have enough ratings to give good recommendations?
    c)   Can one identify which items are likely recommendable?
    d)   How does the transition from not enough ratings to enough ratings look? Is there a critical threshold?
    e)   Is it useful to expire (not use) ratings for the purposes of recommendations?

9.8.3 Broader Access

Collaborative filtering systems have been around for at least a decade. However, for the most part only large companies or research labs actually run them, because they require unusual expertise, considerable resources, or both. Many more people might be interested in giving opinions to each other in an automated system if appropriate infrastructure were present, and the range of items, domains, and opinions might be far more diverse. What are other effective ways to access or deliver the power of collaborative filtering?

User Interfaces

The most well-known collaborative filtering systems are centralized web-based applications with explicit ratings. Other interfaces are emerging that bring the technology closer to users, who are more likely to use it if it is easy. Wikipedia and SourceForge list several applications with embedded collaborative filtering. For example, Audioscrobbler offers a plug-in to several music players (Winamp, Windows Media Player, iTunes, and several others) that collects data about which songs are played, sends it to a central website, and produces music recommendations (Fig. 5.)

Other systems have been proposed, but are not yet well studied. Miller investigates algorithms for portable, user-controlled, accurate recommendations on palmtop-sized devices [43]. These allow the users to remain anonymous and autonomous.

Fig. 5. After installing an Audioscrobbler plugin for your media player (eg: Winamp,) information about every song you listen to on your computer is sent to Last.fm to update your profile

Libraries or Toolkits
Once well understood, a technology can be bundled into a library or toolkit[7] available for embedding into an application. In addition to companies that do this commercially, there are several free, open-source alternatives[8]. However, there are no CF toolkits or libraries that have wide usage, as Apache and Internet Information Server (IIS) do in the web server space. Even though many designers see clear value in recommenders, and there seems to be increasing numbers of them on the web, few toolkits or libraries are gaining wide use. Why is this? What is the right functionality and interface for a toolkit suitable for a wide audience?

Data
Increased public availability of ratings datasets will enable more effective research into collaborative filtering, will allow practitioners to prototype CF system, as well as solve the "cold-start" problem for communities. Organizations often keep that data private, whether for competitive advantage or privacy concerns. Some are starting to open up their data. The EachMovie movie rating dataset was the most popular CF dataset until it was retired in October 2004. Remaining freely available datasets include MovieLens, Jester, and Book Crossing[9].

---

[7] For more information on collaborative filtering toolkits, consult http://en.wikipedia.org/wiki/Collaborative_filtering#Software_libraries.

[8] Open source toolkits include CoFe (http://eecs.oregonstate.edu/iis/CoFE/), MultiLens (http://www.cs.luther.edu/~bmiller/dynahome.php?page=multilens), and Taste (http://taste.sourceforge.net/).

[9] All three data sets are available from http://www.grouplens.org/

9.9 Summary

Collaborative filtering is one of the core technologies that will power the adaptive web. Content-based personalization can be effective in limited circumstances, but for the most part, it will likely be decades or longer before our hardware and software technology can begin to automatically recognize the subtleties of information that are important to people – particularly aspects of aesthetic taste. Until then, in order to filter information based on such complex dimensions, we need to include people in the loop, who analyze the information and condense their opinions into data that can be easily processed by software – ratings. In this chapter, we have attempted to provide a snapshot of the current understanding of collaborative filtering systems and methods. By necessity, as masses of information become ubiquitously available, collaborative filtering will also become ubiquitous. In the process, we will continue to gain a deeper understanding of the dynamics of collaborative filtering.

References

1. Adomavicius, G., Tuzhilin A.: Toward the Next Generation of Recommender Systems: A Survey of the State-of-the-Art and Possible Extensions. IEEE Transactions on Knowledge and Data Engineering, (2005) 17(6): p. 734-749
2. Aggarwal, C.C., Wolf J., Wu K.L., Yu P.S.: Horting Hatches an Egg: A New Graph-Theoretic Approach to Collaborative Filtering. In Proceedings of the Fifth ACM SIGKDD International Conference on Knowledge discovery and data mining. (1999). San Diego, California. ACM Press p. 201-212
3. Avery, C., Resnick P., Zeckhauser, R.: The Market for Evaluations. American Economic Review, (1999) 89(3): p. 564-584
4. Balabanovíc, M., Shoham, Y.: Fab: Content-Based, Collaborative Recommendation. Communications of the ACM, (1997) 40(3): p. 66-72
5. Basu, C., Hirsh, H., Cohen, W.W.: Recommendation as Classification: Using Social and Content-Based Information in Recommendation. In Proceedings of the Fifteenth National Conference on Artificial Intelligence. (1998) Madison, Wisconsin. AAAI Press p. 714-720
6. BBC News Online, "Sony Admits Using Fake Reviewer." June 4, 2001 http://news.bbc.co.uk/1/hi/entertainment/film/1368666.stm
7. Berry, M.W., Dumais, S.T., O'Brian, G.W.: Using Linear Algebra for Intelligent Information Retrieval. Siam Review, (1995) 37(4) p. 573-595
8. Billsus, D., Pazzani. M.J.: Learning Collaborative Information Filters. In Proceedings of the Fifteenth National Conference on Artificial Intelligence (AAAI-98). (1998) Menlo Park, CA. Morgan Kaufmann Publishers Inc. p 46-94
9. Breese, J.S., Heckerman, D., Kadie, C.: Empirical Analysis of Predictive Algorithms for Collaborative Filtering. In Proceeding of the Fourteenth Conference on Uncertainty in Artificial Intelligence (UAI). (1998) Madison, Wisconsin. Morgan Kaufmann p. 43-52
10. Burke, R.: Hybrid Web Recomender Systems. In: Brusilovsky, P., Kobsa, A., Nejdl, W. (eds.): The Adaptive Web: Methods and Strategies of Web Personalization, Lecture Notes in Computer Science, Vol. 4321. Springer-Verlag, Berlin Heidelberg New York (2007) this volume
11. Canny, J.: Collaborative Filtering with Privacy via Factor Analysis. In Proceedings of the 25th annual international ACM SIGIR conference on Research and development in information retrieval. (2002) Tampere, Finland. ACM Press p. 238-245

12. Claypool, M., Gokhale, A., Miranda, T., Murnikov, P, Netes, D., Sartin, M.: Combining Content-Based and Collaborative Filters in an Online Newspaper. In Proceedings of the ACM SIGIR '99 Workshop on Recommender Systems: Algorithms and Evaluation. (1999) Berkeley, California
13. Condliff, M.K., Lewis, D., Madigan, D., Posse, C.: Bayesian Mixed-Effect Models for Recommender Systems. In Proceedings of the SIGIR-99 Workshop on Recommender Systems: Algorithms and Evaluation. (1999). Berkeley, California
14. Cosley, D., Lam, S.K., Albert, I., Konstan, J.A., Riedl, J.: Is Seeing Believing?: How Recommender System Interfaces Affect Users' Opinions. In: Proceedings of the SIGCHI conference on Human factors in computing systems. (2003) ACM Press: Ft. Lauderdale, Florida, USA. p. 585-592
15. Dahlen, B.J., Konstan, J.A., Herlocker, J., Riedl, J.: Jump-starting Movielens: User Benefits Of Starting A Collaborative Filtering System With "Dead Data". TR 98-017, University of Minnesota
16. Deerwester, S., Dumais, S.T., Furnas, G.W., Landauer, T.K., Harshman, R.: Indexing by Latent Semantic Analysis. Journal of the American Society for Information Science. (1998) 41(6): p. 159-168
17. Delgado, J. Ishii, N.: Memory-Based Weighted Majority Prediction for Recommender Systems. In 1999 SIGIR Workshop on Recommender Systems. (1999) University of California, Berkeley p. 1-5
18. Frankowski, D., Cosley, D., Sen, S., Terveen, L., Riedl, J.: You Are What You Say: Privacy Risks Of Public Mentions. In Proceedings of SIGIR 2006 (2006 p. 562-572
19. Goldberg D, Nichols, D., Oki, B.M., Terry, D.: Using Collaborative Filtering To Weave An Information Tapestry. Communications of the ACM, 35(12): pp. 61–70
20. Goldberg, K., Roeder, T., Gupta, D., Perkins, C.: Eigentaste: A Constant-Time Collaborative Filtering Algorithm. Information Retrieval, (2001) 4(2): p. 133-151
21. Good, N., Schafer, J.B., Konstan, J.A., Borchers, A., Sarwar, B., Herlocker, J., Riedl, J.: Combining Collaborative Filtering With Personal Agents For Better Recommendations. In Proceedings of the Sixteenth National Conference on Artificial Intelligence (AAAI-99). (1999) Orlando, Florida. AAAI Press p. 439-446
22. Harper, F., Li, X., Chen, Y., Konstan, J.: An Economic Model Of User Rating In An Online Recommender System. In Proceedings of the 10th International Conference on User Modeling, (2005) Edinburgh, UK p. 307-216
23. Heckerman, D., Chickering, D.M., Meek, C., Rounthwaite, R., Kadie, C.: Dependency Networks for Inference, Collaborative Filtering, and Data Visualization. Journal of Machine Learning Research, (2001) p. 49-75
24. Herlocker, J., Konstan, J.A., Terveen, L.G., Reidl, J.: Evaluating Collaborative Filtering Recommender Systems. ACM Transactions on Information Systems, (2004) 22(1): p. 5-53
25. Herlocker, J.L., Konstan, J.A., Borchers, A., Riedl, J.: An Algorithmic Framework For Performing Collaborative Filtering. In Proceedings of the 22nd International Conference on Research and Development in Information Retrieval (SIGIR '99). (1999) Berkeley, California. ACM Press p. 230-237
26. Herlocker, J.L., Konstan, J.A., Riedl, J.: Explaining Collaborative Filtering Recommendations. In Proceedings of the 2000 ACM conference on Computer supported cooperative work. (2000) Philadelphia, Pennsylvania. ACM Press p. 241-250
27. Hill, W., Stead, L., Rosenstein, M., Furnas, G.: Recommending and Evaluating Choices in a Virtual Community of Use. In Proceedings of ACM CHI'95 Conference on Human Factors in Computing Systems. (1995) Denver, Colorado. ACM Press p. 194-201
28. Hill, W.C., Hollan, J.D., Wroblewski, D., McCandless, T.: Edit Wear and Read Wear. In Proceedings of the SIGCHI conference on Human factors in Computing Systems. (1992) Monterey, California. ACM Press p. 3-9

29. Hofmann, T.: Latent Semantic Models For Collaborative Filtering. ACM Transactions on Information Systems (TOIS) (2004) 22(1): p. 89-115
30. Höök, K., Benyon, D., Munro, A.: Footprints in the snow. In: Höök, K., Benyon, D., Munro, A.(eds): Social Navigation of Information Space. (2003) Springer-Verlag: London.
31. Johnson, S.C.: Hierarchical Clustering Schemes. Psychometrika, (1967) 32(3): p. 241-254
32. Karypis, G.: Evaluation of Item-Based Top-N Recommendation Algorithms. 10th Conference of Information and Knowledge Management (CIKM). (2001) pp. 247—254
33. Kobsa, A. Privacy-Enhanced Web Personalization. In: Brusilovsky, P., Kobsa, A., Nejdl, W. (eds.): The Adaptive Web: Methods and Strategies of Web Personalization, Lecture Notes in Computer Science, Vol. 4321. Springer-Verlag, Berlin Heidelberg New York (2007) this volume
34. Konstan, J.A., Miller, B., Maltz, D., Herlocker, J., Gordon, L., Riedl, J.: GroupLens: Applying Collaborative Filtering To Usenet News. Communications of the ACM, 40(3), pp. 77—87
35. Lam, S.K. Riedl, J.: Shilling Recommender Systems For Fun And Profit. Proceedings of the 13th international conference on World Wide Web. (2004) ACM Press: New York, NY, USA. p. 393-402
36. Lam, S.K., Frankowski, D., Riedl, J.: Do You Trust Your Recommendations? An Exploration Of Security And Privacy Issues In Recommender Systems. In Proceedings of the 2006 International Conference on Emerging Trends in Information and Communication Security (ETRICS), (2006) Freiburg, Germany p. 14-29
37. Lin, W.: Association Rule Mining for Collaborative Recommender Systems. Master's Thesis, Worcester Polytechnic Institute, May 2000.
38. Linden, G., Smith, B., York, J.: Amazon.Com Recommendations: Item-To-Item Collaborative Filtering. Internet Computing, IEEE, 2003. 7(1): p. 76-80.
39. Ludford, P.J., Cosley, D., Frankowski, D., Terveen, L.: Think Different: Increasing Online Community Participation Using Uniqueness And Group Dissimilarity. Proceedings of the SIGCHI conference on Human factors in computing systems (2004) ACM Press: Vienna, Austria p. 631-638
40. MacQueen, J.: Some Methods for Classification and Analysis of Multivariate Observations. In Proceedings of the Fifth Berkeley Symposium on Mathematical Statistics and Probability. (1967) p. 281-297
41. McLaughlin, M., Herlocker, J.: A Collaborative Filtering Algorithm and Evaluation Metric that Accurately Model the User Experience. In Proceedings of the SIGIR Conference on Research and Development in Information Retrieval. (2004) p. 329-336
42. Maltz D, Ehrlich, E.: Pointing The Way: Active Collaborative Filtering. In Proceedings of ACM CHI'95 Conference on Human Factors in Computing Systems, ACM, pp. 202—209
43. Miller, B.N., Konstan, J.A., Riedl, J.: Pocketlens: Toward A Personal Recommender System. ACM Trans. Inf. Syst., 2004. 22(3): p. 437-476
44. Mobasher, B.: Data Mining for Web Personalization. In: Brusilovsky, P., Kobsa, A., Nejdl, W. (eds.): The Adaptive Web: Methods and Strategies of Web Personalization, Lecture Notes in Computer Science, Vol. 4321. Springer-Verlag, Berlin Heidelberg New York (2007) this volume
45. Oard, D.W., Kim, J.: Implicit Feedback for Recommender Systems. In Proceedings of the AAAI Workshop on Recommender Systems. (1998) Madison, Wisconsin
46. O'Connor, M., Cosley, D., Konstan, J.A., Riedl, J.: PolyLens: A Recommender System for Groups of Users. In Proceedings of ECSCW 2001 (2001) Bonn, Germany p. 199-218
47. O'Mahoney, M.P., Hurley, N., Kushmerick, N., Silvestre, G.: Collaborative Recommendation: A Robustness Analysis. ACM Transactions on Internet Technology, (2003) 4(3): p. 344-377
48. Pazzani, M., Billsus, D.: Content-based Recommendation Systems. In: Brusilovsky, P., Kobsa, A., Nejdl, W. (eds.): The Adaptive Web: Methods and Strategies of Web Personal-

ization, Lecture Notes in Computer Science, Vol. 4321. Springer-Verlag, Berlin Heidelberg New York (2007) this volume

49. Popescul, A., Ungar, L.H., Pennock, D.M., Lawrence, S.: Probabilistic Models for Unified Collaborative and Content-Based Recommendation in Sparse-Data Environments. (2001): p. 437-444

50. Ramakrishnan, N., Keller, B.K., Mirza, B.J.: Privacy Risks in Recommender Systems. IEEE Internet Computing. 2001. p. 54-62

51. Resnick, P., Iacovou, N., Suchak, M., Bergstrom, P., Riedl, J.: Grouplens: An Open Architecture For Collaborative Filtering Of Netnews. In Proceedings of the 1994 ACM conference on Computer supported cooperative work. (1994) Chapel Hill, North Carolina. ACM Press p. 175-186

52. Sarwar, B., Karypis, G., Konstan, J.A., Riedl, J.: Item-Based Collaborative Filtering Recommendation Algorithms. Proceedings of the 10th international conference on World Wide Web. (2001) Hong Kong. ACM Press p. 285-295

53. Sarwar, B., Karypis, G., Konstan, J.A., Riedl, J.: Incremental SVD-Based Algorithms for Highly Scaleable Recommender Systems. Proceedings of the Fifth International Conference on Computer and Information Technology (2002)

54. Sarwar, B., Karypis, G., Konstan, J.A., Riedl, J.: Application of Dimensionality Reduction in Recommender System--A Case Study. ACM WebKDD 2000 Web Mining for E-Commerce Workshop. 2000. Boston, Massachusetts

55. Schafer, J.B., Konstan, J.A., Riedl, J.: Meta-Recommendation Systems: User-Controlled Integration Of Diverse Recommendations. Proceedings of the Eleventh International Conference on Information And Knowledge Management (2002) ACM Press: McLean, Virginia, USA p. 43-51

56. Schein, A.I., Popescul, A., Ungar, L.H.: Generative Models for Cold-Start Recommendations. Proceedings of the Twenty-third Annual International ACM SIGIR Workshop on Recommender Systems. (2001) New Orleans, Louisiana

57. Shardanand, U., Maes, P.: Social Information Filtering: Algorithms for Automating "Word of Mouth". (1995) New York. ACM p. 210-217

58. Smyth, B.: Case-based Recommendation. In: Brusilovsky, P., Kobsa, A., Nejdl, W. (eds.): The Adaptive Web: Methods and Strategies of Web Personalization, Lecture Notes in Computer Science, Vol. 4321. Springer-Verlag, Berlin Heidelberg New York (2007) this volume

59. Swearingen, K., Sinha, R.: Beyond Algorithms, An HCI perspective on Recommender Systems. In 2001 SIGIR Workshop on Recommender Systems. (2001) New Orleans, LA

60. Torres, R., McNee, S.M., Abel, M., Konstan, J.A., Riedl, J.: Enhancing Digital Libraries With Techlens+. Proceedings of the 4th ACM/IEEE-CS joint conference on Digital Libraries (2004) ACM Press: Tuscon, AZ, USA p. 228-236

61. Ungar, L.H., Foster, D.P.: Clustering Methods for Collaborative Filtering. Proceedings of the 1998 Workshop on Recommender Systems. (1998) Menlo Park, California. AAAI Press

62. Ziegler, C.N., McNee, S.M., Konstan, J.A., Lausen, G.: Improving Recommendation Lists Through Topic Diversification. Proceedings of the Fourteenth International World Wide Web Conference (WWW2005). (2005) p. 22-32

10

Content-Based Recommendation Systems

Michael J. Pazzani[1] and Daniel Billsus[2]

[1] Rutgers University, ASBIII, 3 Rutgers Plaza
New Brunswick, NJ 08901
pazzani@rutgers.edu
[2] FX Palo Alto Laboratory, Inc., 3400 Hillview Ave, Bldg. 4
Palo Alto, CA 94304
billsus@fxpal.com

Abstract. This chapter discusses content-based recommendation systems, i.e., systems that recommend an item to a user based upon a description of the item and a profile of the user's interests. Content-based recommendation systems may be used in a variety of domains ranging from recommending web pages, news articles, restaurants, television programs, and items for sale. Although the details of various systems differ, content-based recommendation systems share in common a means for describing the items that may be recommended, a means for creating a profile of the user that describes the types of items the user likes, and a means of comparing items to the user profile to determine what to recommend. The profile is often created and updated automatically in response to feedback on the desirability of items that have been presented to the user.

10.1 Introduction

A common scenario for modern recommendation systems is a Web application with which a user interacts. Typically, a system presents a summary list of items to a user, and the user selects among the items to receive more details on an item or to interact with the item in some way. For example, online news sites present web pages with headlines (and occasionally story summaries) and allow the user to select a headline to read a story. E-commerce sites often present a page with a list of individual products and then allow the user to see more details about a selected product and purchase the product. Although the web server transmits HTML and the user sees a web page, the web server typically has a database of items and dynamically constructs web pages with a list of items. Because there are often many more items available in a database than would easily fit on a web page, it is necessary to select a subset of items to display to the user or to determine an order in which to display the items.

Content-based recommendation systems analyze item descriptions to identify items that are of particular interest to the user. Because the details of recommendation systems differ based on the representation of items, this chapter first discusses alternative item representations. Next, recommendation algorithms suited for each representation are discussed. The chapter concludes with a discussion of variants of the approaches,

P. Brusilovsky, A. Kobsa, and W. Nejdl (Eds.): The Adaptive Web, LNCS 4321, pp. 325–341, 2007.
© Springer-Verlag Berlin Heidelberg 2007

the strengths and weaknesses of content-based recommendation systems, and directions for future research and development.

10.1.1 Item Representation

Items that can be recommended to the user are often stored in a database table. Table 10.1 shows a simple database with records (i.e., "rows") that describe three restaurants. The column names (e.g., Cuisine or Service) are properties of restaurants. These properties are also called "attributes," "characteristics," "fields," or "variables" in different publications. Each record contains a value for each attribute. A unique identifier, ID in Table 10.1, allows items with the same name to be distinguished and serves as a key to retrieve the other attributes of the record.

Table 10.1. A restaurant database

ID	Name	Cuisine	Service	Cost
10001	Mike's Pizza	Italian	Counter	Low
10002	Chris's Cafe	French	Table	Medium
10003	Jacques Bistro	French	Table	High

The database depicted in Table 10.1 could be used to drive a web site that lists and recommends restaurants. This is an example of structured data in which there is a small number of attributes, each item is described by the same set of attributes, and there is a known set of values that the attributes may have. In this case, many machine learning algorithms may be used to learn a user profile, or a menu interface can easily be created to allow a user to create a profile. The next section of this chapter discusses several approaches to creating a user profile from structured data.

Of course, a web page typically has more information than is shown in Table 10.1, such as a text description of the restaurant, a restaurant review, or even a menu. These may easily be stored as additional fields in the database and a web page can be created with templates to display the text fields (as well as the structured data). However, free text data creates a number of complications when learning a user profile. For example, a profile might indicate that there is an 80% probability that a particular user would like a French restaurant. This might be added to the profile because a user gave a positive review of four out of five French restaurants. However, unrestricted text fields are typically unique and there would be no opportunity to provide feedback on five restaurants described as "A charming café with attentive staff overlooking the river."

An extreme example of unstructured data may occur in news articles. Table 10.2 shows an example of a part of a news article. The entire article can be treated as a large unrestricted text field.

Table 10.2. Part of a newspaper article

Lawmakers Fine-Tuning Energy Plan
SACRAMENTO, Calif. -- With California's energy reserves remaining all but depleted, lawmakers prepared to work through the weekend fine-tuning a plan Gov. Gray Davis says will put the state in the power business for "a long time to come." The proposal involves partially taking over California's two largest utilities and signing long-term contracts of up to 10 years to buy electricity from wholesalers.

Unrestricted texts such as news articles are examples of unstructured data. Unlike structured data, there are no attribute names with well-defined values. Furthermore, the full complexity of natural language may be present in the text field including polysemous words (the same word may have several meanings) and synonyms (different words may have the same meaning). For example, in the article in Table 10.2, "Gray" is a name rather than a color, and "power" and "electricity" refer to the same underlying concept.

Many domains are best represented by semi-structured data in which there are some attributes with a set of restricted values and some free-text fields. A common approach to dealing with free text fields is to convert the free text to a structured representation. For example, each word may be viewed as an attribute, with a Boolean value indicating whether the word is in the article or with an integer value indicating the number of times the word appears in the article.

Many personalization systems that deal with unrestricted text use a technique to create a structured representation that originated with text search systems [34]. In this formalism, rather than using words, the root forms of words are typically created through a process called stemming [30]. The goal of stemming is to create a term that reflects the common meaning behind words such as "compute," "computation," "computer" "computes" and "computers." The value of a variable associated with a term is a real number that represents the importance or relevance. This value is called the $tf*idf$ weight (term-frequency times inverse document frequency). The $tf*idf$ weight, $w(t,d)$, of a term t in a document d is a function of the frequency of t in the document (tf_t, d), the number of documents that contain the term (df_t) and the number of documents in the collection (N).[1]

$$w(t,d) = \frac{tf_{t,d} \log\left(\frac{N}{df_t}\right)}{\sqrt{\sum_i (tf_{t_i,d})^2 \log\left(\frac{N}{df_{t_i}}\right)^2}} \qquad (10.1)$$

Table 10.3 shows the $tf*idf$ representation (also called the vector space representation) of the complete article excerpted in Table 10.2. The terms are ordered by the $tf*idf$ weight. The intuition behind the weight is that the terms with the highest weight occur more often in that document than in the other documents, and therefore are more central to the topic of the document. Note that terms such as "util" (a stem of "utility"), "power," "megawatt," are among the highest weighted terms capturing the meaning.

---

[1] Note that in the description of $tf*idf$ weights, the word "document" is traditionally used since the original motivation was to retrieve documents. While the chapter will stick with the original terminology, in a recommendation system, the documents correspond to a text description of an item to be recommended. Note that the equations here are representative of the class of formulae called $tf*idf$. In general, $tf*idf$ systems have weights that increase monotonically with term frequency and decrease monotonically with document frequency.

Table 10.3. *tf*idf* representation of the article in Table 10.2

util-0.339 power-0.329 megawatt-0.309 electr-0.217 energi-0.206 california-0.181
debt-0.128 lawmak-0.128 state-0.122 wholesal-0.119 partial-0.106 consum-0.105
alert-0.103 scroung-0.096 advoc-0.09 testi-0.088 bail-out-0.088 crisi-0.085 amid-
0.084 price-0.083 long-0.082 bond-0.081 plan-0.081 term-0.08 grid-0.078 reserv-
0.077 blackout-0.076 bid-0.076 market-0.074 fine-0.073 deregul-0.07 spiral-0.068
deplet-0.068 liar-0.066.

Of course, this representation does not capture the context in which a word is used. It
loses the relationships between words in the description. For example, a description of
a steak house might contain the sentence, "there is nothing on the menu that a vegetarian would like" while the description of a vegetarian restaurant might mention
"vegan" rather than vegetarian. In a manually created structured database, the cuisine
attribute having a value of "vegetarian" would indicate that the restaurant is indeed a
vegetarian one. In contrast, when converting an unstructured text description to structured data, the presence of the word vegetarian does not always indicate that a restaurant is vegetarian and the absence of the word vegetarian does not always indicate that
the restaurant is not a vegetarian restaurant. As a consequence, techniques for creating
user profiles that deal with structured data need to differ somewhat from those techniques that deal with unstructured data or unstructured data automatically and impre-
cisely converted to structured data.

One variant on using words as terms is to use sets of contiguous words as terms.
For example, in the article in Table 10.2, terms such as "energy reserves" and "power
business" might be more descriptive of the content than these words treated as individual terms. Of course, terms such as "all but" would also be included, but one would
expect that these have very low weights, in the same way that "all" and "but" individually have low weights and are not among the most important terms in Table 10.3.

10.2   User Profiles

A profile of the user's interests is used by most recommendation systems. This profile
may consist of a number of different types of information. Here, we concentrate on
two types of information:

1. A model of the user's preferences, i.e., a description of the types of items that
   interest the user. There are many possible alternative representations of this description, but one common representation is a function that for any item predicts
   the likelihood that the user is interested in that item. For efficiency purposes, this
   function may be used to retrieve the n items most likely to be of interest to the user.
2. A history of the user's interactions with the recommendation system. This may
   include storing the items that a user has viewed together with other information
   about the user's interaction, (e.g., whether the user has purchased the item or a rating that the user has given the item). Other types of history include saving queries
   typed by the user (e.g., that a user searched for an Italian restaurant in the 90210
   zip code).

There are several uses of the history of user interactions. First, the system can simply display recently visited items to facilitate the user returning to these items. Second, the system can filter out from a recommendation system an item that the user has already purchased or read.[2] Another important use of the history in content-based recommendation systems is to serve as training data for a machine learning algorithm that creates a user model. The next section will discuss several different approaches to learning a user model. Here, we briefly describe approaches of manually providing the information used by recommendation systems: user customization and rule-based recommendation systems.

In user customization, a recommendation system provides an interface that allows users to construct a representation of their own interests. Often check boxes are used to allow a user to select from the known values of attributes, e.g., the cuisine of restaurants, the names of favorite sports teams, the favorite sections of a news site, or the genre of favorite movies. In other cases, a form allows a user to type words that occur in the free text descriptions of items, e.g., the name of a musician or author that interests the user. Once the user has entered this information, a simple database matching process is used to find items that meet the specified criteria and display them to the user.

There are several limitations of user customization systems. First, they require effort from the user and it is difficult to get many users to make this effort. This is particularly true when the user's interests change, e.g., a user may not follow football during the season but then become interested in the Superbowl. Second, customization systems do not provide a way to determine the order in which to present items and can find either too few or too many matching items to display.

Figure 10.1 shows book recommendations at Amazon.com. Although Amazon.com is usually thought of as a good example of collaborative recommendation (see Chapter 9 of this book [35]), parts of the user's profile can be viewed as a content-based profile. For example, Amazon contains a feature called "favorites" that represents the categories of items preferred by users. These favorites are either calculated by keeping track of the categories of items purchased by users or may be set manually by the user. Figure 10.2 shows an example of a user customization interface in which a user can select the categories.

In rule-based recommendation systems, the recommendation system has rules to recommend other products based on the user history. For example, a system may contain a rule that recommends the sequel to a book or movie to people who have purchased the early item in the series. Another rule might recommend a new CD by an artist to users that purchased earlier CDs by that artist. Rule-based systems may capture several common reasons for making recommendations, but they do not offer the same detailed personalized recommendations that are available with other recommendation systems.

---

[2] Of course, in some situations it is appropriate to recommend an item the user has purchased and in other situations it is not. For example, a system should continue to recommend an item that wears out or is expended, such as a razor blade or print cartridge, while there is little value in recommending a CD or DVD a user owns.

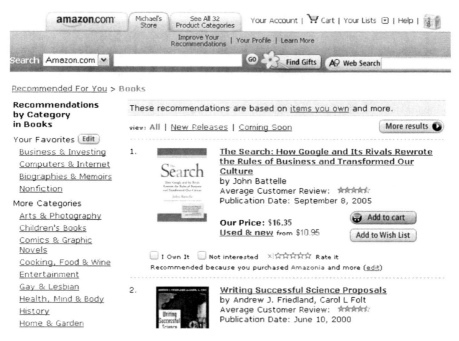

Fig. 10.1. Book recommendations by Amazon.com.

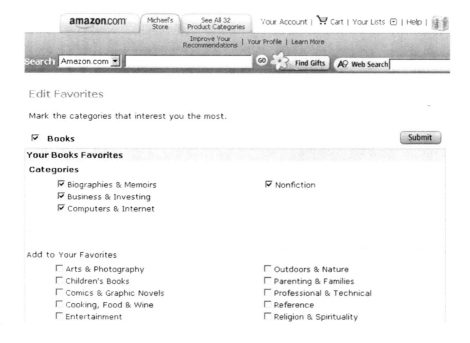

Fig. 10.2. User customization in Amazon.com

10.3   Learning a User Model

Creating a model of the user's preference from the user history is a form of classification learning. The training data of a classification learner is divided into categories, e.g., the binary categories "items the user likes" and "items the user doesn't like." This is accomplished either through explicit feedback in which the user rates items via some interface for collecting feedback or implicitly by observing the user's interactions with items. For example, if a user purchases an item, that is a sign that the user likes the item, while if the user purchases and returns the item that is a sign that the user doesn't like the item. In general, there is a tradeoff since implicit methods can collect a large amount of data with some uncertainty as to whether the user actually likes the item. In contrast, when the user explicitly rates items, there is little or no noise in the training data, but users tend to provide explicit feedback on only a small percentage of the items they interact with.

Figure 10.3 shows an example of a recommendation system with explicit user feedback. The recommender "MyBestBets" by ChoiceStream is a web based interface to a television recommendation system. Users can click on the thumbs up or thumbs down buttons to indicate whether they like the program that is recommended. By necessity, this system requires explicit feedback because it is not integrated with a television [1] and cannot infer the user's interests by observing the user's behavior.

Fig. 10.3. A recommendation system using explicit feedback

The next section reviews a number of classification learning algorithms. Such algorithms are the key component of content-based recommendation systems, because they learn a function that models each user's interests. Given a new item and the user model, the function predicts whether the user would be interested in the item. Many of the classification learning algorithms create a function that will provide an estimate of the probability that a user will like an unseen item. This probability may be used to sort a list of recommendations. Alternatively, an algorithm may create a function that directly predicts a numeric value such as the degree of interest.

Some of the algorithms below are traditional machine learning algorithms designed to work on structured data. When they operate on free text, the free text is first converted to structured data by selecting a small subset of the terms as attributes. In contrast, other algorithms are designed to work in high dimensional spaces and do not require a preprocessing step of feature selection.

10.4   Decision Trees and Rule Induction

Decision tree learners such as ID3 [31] build a decision tree by recursively partitioning training data, in this case text documents, into subgroups until those subgroups contain only instances of a single class. A partition is formed by a test on some feature -- in the context of text classification typically the presence or absence of an individual word or phrase. Expected information gain is a commonly used criterion to select the most informative features for the partition tests [38].

Decision trees have been studied extensively in use with structured data such as that shown in Table 10.1. Given feedback on the restaurants, a decision tree can easily represent and learn a profile of someone who prefers to eat in expensive French restaurants or inexpensive Mexican restaurants. Arguably, the decision tree bias is not ideal for unstructured text classification tasks [29]. As a consequence of the information-theoretic splitting criteria used by decision tree learners, the inductive bias of decision trees is a preference for small trees with few tests. However, it can be shown experimentally that text classification tasks frequently involve a large number of relevant features [17]. Therefore, a decision tree's tendency to base classifications on as few tests as possible can lead to poor performance on text classification. However, when there are a small number of structured attributes, the performance, simplicity and understandability of decision trees for content-based models are all advantages. Kim et al. [18] describe an application of decision trees for personalizing advertisements on web pages.

RIPPER [9] is a rule induction algorithm closely related to decision trees that operates in a similar fashion to the recursive data partitioning approach described above. Despite the problematic inductive bias, however, RIPPER performs competitively with other state-of-the-art text classification algorithms. In part, the performance can be attributed to a sophisticated post-pruning algorithm that optimizes the fit of the induced rule set with respect to the training data as a whole. Furthermore, RIPPER supports multi-valued attributes, which leads to a natural representation for text classification tasks, i.e., the individual words of a text document can be represented as multiple feature values for a single feature. While this is essentially a representational convenience if rules are to be learned from unstructured text documents, the approach can lead to more powerful classifiers for semi-structured text documents. For example, the text contained in separate fields of an email message, such as sender, subject, and body text, can be represented as separate multi-valued features, which allows the algorithm to take advantage of the document's structure in a natural fashion. Cohen [10] shows how RIPPER can classify e-mail messages into user defined categories.

10.5 Nearest Neighbor Methods

The nearest neighbor algorithm simply stores all of its training data, here textual descriptions of implicitly or explicitly labeled items, in memory. In order to classify a new, unlabeled item, the algorithm compares it to all stored items using a similarity function and determines the "nearest neighbor" or the k nearest neighbors. The class label or numeric score for a previously unseen item can then be derived from the class labels of the nearest neighbors.

The similarity function used by the nearest neighbor algorithm depends on the type of data. For structured data, a Euclidean distance metric is often used. When using the vector space model, the cosine similarity measure is often used [34]. In the Euclidean distance function, the same feature having a small value in two examples is treated the same as that feature having a large value in both examples. In contrast, the cosine similarity function will not have a large value if corresponding features of two examples have small values. As a consequence, it is appropriate for text when we want two documents to be similar when they are about the same topic, but not when they are both not about a topic.

Fig. 10.4. Gixo presents personalized news based on similarity to articles that have previously been read

The vector space approach and the cosine similarity function have been applied to several text classification applications ([11], [39], [2]) and, despite the algorithm's unquestionable simplicity, it performs competitively with more complex algorithms. The Daily Learner system uses the nearest neighbor algorithm to create a model of the user's short term interests [7]. Gixo, a personalized news system, also uses text similarity as a basis for recommendation (Figure 10.4). The headlines are preceded by an icon that indicates how popular the item is (the first bar) and how similar the story is to stories that have been read by the user before (the second bar). The fact that these bars differ shows the value of personalizing to the individual.

10.6   Relevance Feedback and Rocchio's Algorithm

Since the success of document retrieval in the vector space model depends on the user's ability to construct queries by selecting a set of representative keywords [34], methods that help users to incrementally refine queries based on previous search results have been the focus of much research. These methods are commonly referred to as relevance feedback. The general principle is to allow users to rate documents returned by the retrieval system with respect to their information need. This form of feedback can subsequently be used to incrementally refine the initial query. In a manner analogous to rating items, there are explicit and implicit means of collecting relevance feedback data.

Rocchio's algorithm [33] is a widely used relevance feedback algorithm that operates in the vector space model. The algorithm is based on the modification of an initial query through differently weighted prototypes of relevant and non-relevant documents. The approach forms two document prototypes by taking the vector sum over all relevant and non-relevant documents. The following formula summarizes the algorithm formally:

$$Q_{i+1} = \alpha\, Q_i + \beta \sum_{rel} \frac{D_i}{|D_i|} - \gamma \sum_{nonrel} \frac{D_i}{|D_i|} \tag{10.2}$$

Here, Q_i is the user's query at iteration i, and α, β, and γ are parameters that control the influence of the original query and the two prototypes on the resulting modified query. The underlying intuition of the above formula is to incrementally move the query vector towards clusters of relevant documents and away from irrelevant documents. While this goal forms an intuitive justification for Rocchio's algorithm, there is no theoretically motivated basis for the above formula, i.e., neither performance nor convergence can be guaranteed. However, empirical experiments have demonstrated that the approach leads to significant improvements in retrieval performance [33].

In more recent work, researchers have used a variation of Rocchio's algorithm in a machine learning context, i.e., for learning a user profile from unstructured text ([15], [3], [29]). The goal in these applications is to automatically induce a text classifier that can distinguish between classes of documents. In this context, it is assumed that no initial query exists, and the algorithm forms prototypes for classes analogously to Rocchio's approach as vector sums over documents belonging to the same class. The result of the algorithm is a set of weight vectors, whose proximity to unlabeled documents can be used to assign class membership. Similar to the relevance feedback version of Rocchio's algorithm, the Rocchio-based classification approach does not have any theoretic underpinnings and there are no performance or convergence guarantees.

10.7   Linear Classifiers

Algorithms that learn linear decision boundaries, i.e., hyperplanes separating instances in a multi-dimensional space, are referred to as linear classifiers. There are a large number of algorithms that fall into this category, and many of them have been successfully applied to text classification tasks [20]. All linear classifiers can be described in a common representational framework. In general, the outcome of the learning process is an n-dimensional weight vector w, whose dot product with an n-dimensional instance, e.g., a text document represented in the vector space model, results in a numeric score prediction. Retaining the numeric prediction leads to a linear regression approach. However, a threshold can be used to convert continuous predictions to discrete class labels. While this general framework holds for all linear classifiers, the algorithms differ in the training methods used to derive the weight vector w. For example, the equation below is known as the Widrow-Hoff rule, delta rule or gradient descent rule and derives the weight vector w by incremental vector movements in the direction of the negative gradient of the example's squared error [37]. This is the direction in which the error falls most rapidly.

$$w_{i+1,j} = w_{i,j} - 2\eta(w_i \cdot x_i - y_i)x_{i,j} \qquad (10.3)$$

The equation shows how the weight vector w can be derived incrementally. The inner product of instance x_i and weight vector w_i is the algorithm's numeric prediction for instance x_i. The prediction error is determined by subtracting the instance's known score, y_i, from the predicted score. The resulting error is then multiplied by the original instance vector x_i and the learning rate η to form a vector that, when subtracted from the weight vector w, moves w towards the correct prediction for instance x_i. The learning rate η controls the degree to which every additional instance affects the previous weight vector.

An alternative algorithm that has experimentally been shown to outperform the approach above on text classification tasks with many features is the exponentiated gradient (EG) algorithm. Kivinen and Warmuth [19] prove a bound for EG's error, which depends only logarithmically on the number of features. This result offers a theoretic argument for EG's performance on text classification problems, which are typically high-dimensional.

An important advantage of the above learning schemes for linear algorithms is that they can be performed on-line, i.e., the current weight vector can be modified incrementally as new instances become available. This is a crucial advantage for applications that operate under real-time constraints.

Finally, it is important to note that while the above approaches tend to converge on hyperplanes that separate the training data accurately, the hyperplane's generalization performance might not be optimal. A related approach aimed at improving generalization performance is known as support vector machines [36]. The central idea underlying support vector machines is to maximize the classification margin, i.e., the distance between the decision boundary and the closest training instances, the so-called support vectors. A series of empirical experiments on a variety of benchmark data sets indicated that linear support vector machines perform particularly well on text classification tasks [17]. The main reason for this is that the margin maximization is an inherently built-in overfitting protection mechanism. A reduced tendency to overfit training data is particularly useful for text classification algorithms, because in this domain high dimensional concepts must often be learned from limited training data, which is a scenario prone to overfitting.

10.8   Probabilistic Methods and Naïve Bayes

In contrast to the lack of theoretical justifications for the vector space model, there has been much work on probabilistic text classification approaches. This section describes one such example, the naïve Bayesian classifier. Early work on a probabilistic classifier and its text classification performance was reported by Maron [24]. Today, this algorithm is commonly referred to as a naïve Bayesian Classifier [13]. Researchers have recognized Naïve Bayes as an exceptionally well-performing text classification algorithm and have frequently adopted the algorithm in recent work ([27], [28], [25]).

The algorithm's popularity and performance for text classification applications have prompted researchers to empirically evaluate and compare different variations of naïve Bayes that have appeared in the literature (e.g. [26], [21]). In summary, McCallum and Nigam [26] note that there are two frequently used formulations of naïve Bayes, the multivariate Bernoulli and the multinomial model. Both models share the following principles. It is assumed that text documents are generated by an underlying generative model, specifically a parameterized mixture model:

$$P(d_i \mid \theta) = \sum_{j=1}^{|C|} P(c_j \mid \theta) P(d_i \mid c_j; \theta) \tag{10.4}$$

Here, each class c corresponds to a mixture component that is parameterized by a disjoint subset of θ, and the sum of total probability over all mixture components determines the likelihood of a document. Once the parameters θ have been learned from training data, the posterior probability of class membership given the evidence of a test document can be determined according to Bayes' rule:

$$P(c_j \mid d_i; \hat{\theta}) = \frac{P(c_j \mid \hat{\theta}) P(d_i \mid c_j; \hat{\theta})}{P(d_i \mid \hat{\theta})} \tag{10.5}$$

While the above principles hold for naïve Bayes classification in general, the multivariate Bernoulli and multinomial models differ in the way $p(d_i|c_j; \theta)$ is estimated from training data.

The multivariate Bernoulli formulation was derived with structured data in mind. For text classification tasks, it assumes that each document is represented as a binary vector over the space of all words from a vocabulary V. Each element B_{it} in this vector indicates whether a word appears at least once in the document. Under the naïve Bayes assumption that the probability of each word occurring in a document is independent of other words given the class label, $p(d_i|c_j; \theta)$ can be expressed as a simple product:

$$P(d_i \mid c_j; \theta) = \prod_{t=1}^{|V|} (B_{it} P(w_t \mid c_j; \theta) + (1 - B_{it})(1 - P(w_t \mid c_j; \theta))) \quad (\textbf{10.6})$$

Bayes-optimal optimal estimates for $p(w_t|c_j; \theta)$ can be determined by word occurrence counting over the data:

$$P(w_t \mid c_j; \theta) = \frac{1 + \sum_{i=1}^{|D|} B_{it} P(c_j \mid d_i)}{2 + \sum_{i=1}^{|D|} P(c_j \mid d_i)} \quad (\textbf{10.7})$$

In contrast to the binary document representation of the multivariate Bernoulli model, the multinomial formulation captures word frequency information. This model assumes that documents are generated by a sequence of independent trials drawn from a multinomial probability distribution. Again, the naïve Bayes independence assumption allows $p(d_i|c_j; \theta)$ to be determined based on individual word probabilities.

$$P(d_i \mid c_j; \theta) = P(|d_i|) \prod_{t-1}^{|d_i|} P(w_t \mid c_j; \theta)^{N_{it}} \quad (\textbf{10.8})$$

Here, N_{it} is the number of occurrences of word w_t in document d_i. Taking word frequencies into account, maximum likelihood estimates for $p(w_t|c_j; \theta)$ can be derived from training data:

$$P(w_t \mid c_j; \theta) = \frac{1 + \sum_{i=1}^{|D|} N_{it} P(c_j \mid d_i)}{|V| + \sum_{s=1}^{|V|} \sum_{i=1}^{|D|} N_{is} P(c_j \mid d_i)} \quad (\textbf{10.9})$$

Empirically, the multinomial naïve Bayes formulation was shown to outperform the multivariate Bernoulli model. This effect is particularly noticeable for large vocabularies (McCallum and Nigam, 1998).

Even though the naïve Bayes assumption of class-conditional attribute independence is clearly violated in the context of text classification, naïve Bayes per-forms very well. Domingos and Pazzani [12] offer a possible explanation for this paradox by showing that class-conditional feature independence is not a necessary condition for the optimality of naïve Bayes. The naïve Bayes classifier has been used in several content-based recommendation systems including Syskill & Webert [29].

10.9   Trends in Content-Based Filtering

Belkin & Croft [5] surveyed some of the first content-based recommendation systems and noted that they made use of technology related to information retrieval such as tf*idf and Rocchio's method. Indeed, some of the early work on content-based recommendation used the term "query" to refer to user models. In this view, a user model is a saved query (or a set of saved queries) that can retrieve additional or new information of interest to the user. Some representative early systems include a system at Bellcore [14] that found new technical reports related to previously read reports and LyricTime [22] that recommended songs in a multimedia player based on a profile learned from the user's feedback on prior songs played.

The creation and rapid growth of the World Wide Web in the mid 1990s made access to vast amounts of information possible and created problems of locating and

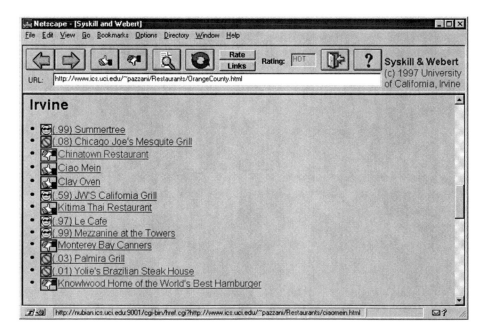

Fig. 10.5. The Syskill & Webert system learns a model of the user's preference for web pages

identifying personally relevant information. Some in the Machine Learning community applied traditional machine learning methods to user modeling of document interests. These methods reduced the text training data to a few hundred highly relevant words using techniques such as information theory or tf*idf. Some representative systems included WebWatcher [16] and Syskill & Webert [29].

10.10 Limitations and Extensions

Although there are different approaches to learning a model of the user's interest with content-based recommendation, no content-based recommendation system can give good recommendations if the content does not contain enough information to distinguish items the user likes from items the user doesn't like. In recommending some items, e.g., jokes or poems, there often isn't enough information in the word frequency to model the user's interests. While it would be possible to tell a lawyer joke from a chicken joke based upon word frequencies, it would be difficult to distinguish a funny lawyer joke from other lawyer jokes. As a consequence, other recommendation technologies, such as collaborative recommenders [35], should be used in such situations.

In some situations, e.g., recommending movies, restaurants, or television programs, there is some structured information (e.g., the genre of the movie as well as actors and directors) that can be used by a content-based system. However, this information might be supplemented by the opinions of other users. One way to include the opinions of other users in the frameworks discussed in Section 10.2 is to add additional data associated to the representation of the examples. For example, Basu et al. [4] add features to examples that indicate the identifiers of other users who like an item. Ripper was applied to the resulting data that could learn profiles with both collaborative and content-based features (e.g., a user might like a science fiction movie if USER-109 likes it). Although not strictly a content-based system, the same technology as content-based recommenders is used to learn a user model. Indeed, Billsus and Pazzani [6] have shown that any machine learning algorithm may be used as the basis for collaborative filtering by transforming user ratings to attributes. Chapter 12 of this book [8] discusses a variety of other approaches to combining content and collaborative information in recommendation systems.

A final usage of content in recommendations is worth noting. Simple content-based rules may be used to filter the results of other methods such as collaborative filtering. For example, even if it is the case that people who buy dolls also buy adult videos, it might be important not to recommend adult items in a particular application. Similarly, although not strictly content-based, some systems might not recommend items that are out of stock.

10.11 Summary

Content-based recommendation systems recommend an item to a user based upon a description of the item and a profile of the user's interests. While a user profile may be entered by the user, it is commonly learned from feedback the user provides on items. A variety of learning algorithms have been adapted to learning user profiles, and the choice of learning algorithm depends upon the representation of content.

References

1. Ali, K., van Stam, W.: TiVo: Making Show Recommendations Using a Distributed Collaborative Filtering Architecture. In: Proceedings of the Tenth ACM SIGKDD International Conference on Knowledge Discovery and Data Mining. Seattle, WA. (2004) 394-401
2. Allan, J., Carbonell, J., Doddington, G., Yamron, J., Yang, Y.: Topic Detection and Tracking Pilot Study Final Report. In: Proceedings of the DARPA Broadcast News Transcription and Understanding Workshop. Lansdowne, VA (1998) 194-218
3. Balabanovic, M., Shoham Y.: FAB: Content-based, Collaborative Recommendation. Communications of the Association for Computing Machinery 40(3) (1997) 66-72
4. Basu, C., Hirsh, H., Cohen W.: Recommendation as Classification: Using Social and Content-Based Information in Recommendation. In: Proceedings of the 15th National Conference on Artificial Intelligence, Madison, WI (1998) 714-720
5. Belkin, N., Croft, B.: Information Filtering and Information Retrieval: Two Sides of the Same Coin? Communications of the ACM 35(12) (1992) 29-38
6. Billsus, D., Pazzani, M.: Learning Collaborative Information Filters. In: Proceedings of the International Conference on Machine Learning. Morgan Kaufmann Publishers. Madison, WI (1998) 46-54
7. Billsus, D., Pazzani, M., Chen, J.: A Learning Agent for Wireless News Access. In: Proceedings of the International Conference on Intelligent User Interfaces (2002) 33-36
8. Burke, R.: Hybrid Web Recommender Systems. In: Brusilovsky, P., Kobsa, A., Nejdl, W. (eds.): The Adaptive Web: Methods and Strategies of Web Personalization. Lecture Notes in Computer Science, Vol. 4321. Springer-Verlag, Berlin Heidelberg New York (2007) this volume
9. Cohen, W.: Fast Effective Rule Induction. In: Proceedings of the Twelfth International Conference on Machine Learning, Tahoe City, CA. (1995) 115-123
10. Cohen, W.: Learning Rules that Classify E-mail. In: Papers from the AAAI Spring Symposium on Machine Learning in Information Access (1996) 18-25
11. Cohen, W., Hirsh, H. Joins that Generalize: Text Classification Using WHIRL. In: Proceedings of the Fourth International Conference on Knowledge Discovery & Data Mining, New York, NY (1998) 169-173
12. Domingos, P., Pazzani, M. Beyond Independence: Conditions for the Optimality of the Simple Bayesian Classifier. Machine Learning 29 (1997) 103-130.
13. Duda, R., Hart, P.: Pattern Classification and Scene Analysis. New York, NY: Wiley and Sons (1973)
14. Foltz, P., Dumais, S.: Personalized Information Delivery: An Analysis of Information Filtering Methods. Communications of the ACM 35(12) (1992) 51-60
15. Ittner, D., Lewis, D., Ahn, D.: Text Categorization of Low Quality Images. In: Symposium on Document Analysis and Information Retrieval, Las Vegas, NV (1995) 301-315
16. Joachims, T., Freitag, D., Mitchell, T.: WebWatcher: A Tour Guide for the World Wide Web. In: Proceedings of the 15th International Joint Conference on Artificial Intelligence. Nagoya, Japan (1997) 770 -775
17. Joachims, T.: Text Categorization With Support Vector Machines: Learning with Many Relevant Features. In: European Conference on Machine Learning, Chemnitz, Germany (1998) 137-142
18. Kim, J., Lee, B., Shaw, M., Chang, H., Nelson, W.: Application of Decision-Tree Induction Techniques to Personalized Advertisements on Internet Storefronts. International Journal of Electronic Commerce 5(3) (2001) 45-62
19. Kivinen, J., Warmuth, M.: Exponentiated Gradient versus Gradient Descent for Linear Predictors. Information and Computation 132(1) (1997) 1-63

20. Lewis, D., Schapire, R., Callan, J., Papka, R.: Training Algorithms for Linear Text Classifiers. In: Proceedings of the 19th Annual International ACM SIGIR Conference on Research and Development in Information Retrieval, Konstanz, Germany (1996) 298-306
21. Lewis, D.: Naïve (Bayes) at Forty: The Independence Assumption in Information Retrieval. In: European Conference on Machine Learning, Chemnitz, Germany (1998) 4-15
22. Loeb, S.: Architecting Personal Delivery of Multimedia Information. Communications of the ACM 35(12) (1992) 39-48
23. Mandel, M., Poliner, G., Ellis, D.: Support Vector Machine Active Learning for Music Retrieval. ACM Multimedia Systems Journal 12(1) (2006) 3-13
24. Maron, M.: Automatic Indexing: An Experimental Inquiry. Journal of the Association for Computing Machinery 8(3) (1961) 404-417
25. McCallum, A., Rosenfeld, R., Mitchell T., Ng, A.: Improving Text Classification by Shrinkage in a Hierarchy of Classes. In: Proceedings of the International Conference on Machine Learning. Morgan Kaufmann Publishers. Madison, WI (1998) 359-367
26. McCallum, A., Nigam, K.: A Comparison of Event Models for Naive Bayes Text Classification. In: AAAI/ICML-98 Workshop on Learning for Text Categorization, Technical Report WS-98-05, AAAI Press (1998) 41-48
27. Mitchell, T.: Machine Learning. McGraw-Hill (1997)
28. Nigam, K., McCallum, A., Thrun, S., Mitchell, T.: Learning to Classify Text from Labeled and Unlabeled Documents. In: Proceedings of the 15th International Conference on Artificial Intelligence, Madison, WI (1998) 792-799
29. Pazzani M., Billsus, D.: Learning and Revising User Profiles: The Identification of Interesting Web Sites. Machine Learning 27(3) (1997) 313-331
30. Porter, M.: An Algorithm for Suffix Stripping. Program 14(3) (1980) 130-137
31. Quinlan, J.: Induction of Decision Trees. Machine Learning 1(1986) 81-106
32. Quinlan, J.: C4.5: Programs for Machine Learning. Morgan Kauffman (1993)
33. Rocchio, J.: Relevance Feedback in Information Retrieval. In: G. Salton (ed.). The SMART System: Experiments in Automatic Document Processing. NJ: Prentice Hall (1971) 313-323
34. Salton, G. Automatic Text Processing. Addison-Wesley (1989)
35. Schafer, B., Frankowski, D., Herlocker, J., Sen, S.: Collaborative Filtering Recommender Systems. In: Brusilovsky, P., Kobsa, A., Nejdl, W. (eds.): The Adaptive Web: Methods and Strategies of Web Personalization. Lecture Notes in Computer Science, Vol. 4321. Springer-Verlag, Berlin Heidelberg New York (2007) this volume
36. Vapnik, V.: The Nature of Statistical Learning Theory. Springer: New York (1995)
37. Widrow, A., Hoff, M.: Adaptive Switching Circuits. WESCON Convention Record 4 (1960) 96-104
38. Yang, Y., Pedersen J.: A Comparative Study on Feature Selection in Text Categorization. In: Proceedings of the Fourteenth International Conference on Machine Learning, Nashville, TN (1997) 412-420
39. Yang, Y.: An Evaluation of Statistical Approaches to Text Categorization. Information Retrieval 1(1) (1999) 67-88

11

Case-Based Recommendation

Barry Smyth[1,2]

[1] The School of Computer Science and Informatics,
University College Dublin, Belfield, Dublin 4, Ireland
[2] ChangingWorlds Ltd.
South County Business Park, Leopardstown,
Dublin 18, Ireland.
Barry.Smyth@ucd.ie

Abstract. Recommender systems try to help users access complex information spaces. A good example is when they are used to help users to access online product catalogs, where recommender systems have proven to be especially useful for making product suggestions in response to evolving user needs and preferences. Case-based recommendation is a form of content-based recommendation that is well suited to many product recommendation domains where individual products are described in terms of a well defined set of features (e.g., *price, colour, make,* etc.). These representations allow case-based recommenders to make judgments about product similarities in order to improve the quality of their recommendations and as a result this type of approach has proven to be very successful in many e-commerce settings, especially when the needs and preferences of users are ill-defined, as they often are. In this chapter we will describe the basic approach to case-based recommendation, highlighting how it differs from other recommendation technologies, and introducing some recent advances that have led to more powerful and flexible recommender systems.

11.1 Introduction

Recently I wanted to buy a new digital camera. I had a vague idea of what I wanted—a 6 mega-pixel digital SLR from a good manufacturer—but it proved difficult and time consuming to locate a product online that suited my needs, especially as these needs evolved during my investigations. Many online stores allowed me to *browse* or *navigate* through their product catalog by choosing from a series of static features (e.g., *manufacturer, camera type, resolution, level of zoom* etc.). Each time I selected a feature I was presented with the set of cameras with this feature and I could then go on to choose another feature to further refine the presented products. Other stores allowed me to *search* for my ideal camera by entering a query (e.g. *"digital slr, 6 mega-pixels"*) and presented me with a list of results which I could then browse at my leisure.

Both of these access options were helpful in different ways—in the beginning I preferred to browse through catalogs but, after getting a feel for the various features and compromises, I tended to use search-based interfaces—however neither provided

P. Brusilovsky, A. Kobsa, and W. Nejdl (Eds.): The Adaptive Web, LNCS 4321, pp. 342–376, 2007.
© Springer-Verlag Berlin Heidelberg 2007

me with the flexibility I really sought. For a start, all of the stores I tried tended to slavishly respect my queries. This was especially noticeable when no results could be returned to satisfy my stated needs; this is often referred to as *stonewalling* [17]. For instance, looking for a 6 mega-pixel digital SLR for under $200 proved fruitless—unsurprising perhaps to those 'in the know'—and left me with no choice but to start my search again. This was especially frustrating when there were many cameras that were similar enough to my query to merit suggestion. Moreover, stonewalling is further compounded by a *diversity problem*: I was frequently presented with sets of products that were all very similar to each other thus failing to offer me a good set of alternatives. At other times I would notice a camera that was almost perfect, aside from perhaps one or two features, but it was usually difficult to provide this form of feedback directly. This *feedback problem* prevented me from requesting *"another camera like this one but with more optical zoom and/or a lower price"*, for instance.

In all, perhaps one of the most frustrating aspects of my search was the apparent inability of most online stores to learn anything about my preferences over time. In my opinion shopping for an expensive item such as a digital camera is an exercise in patience and deliberation, and one that is likely to involve many return visits to particular online stores. Unfortunately, despite the fact that I had spent a significant time and effort searching and browsing for cameras during previous visits none of the stores I visited had any facility to remember my previous interactions or preferences. For instance, my reluctance to purchase a very expensive camera—I never accepted recommendations for cameras above $1000—should have been recognised and factored into the store's recommendations, but it was not. As a result many of my interactions turned out to be requests for less expensive suggestions. This *preference problem* meant that starting my searches from scratch became a regular feature of these visits.

Recommender systems are designed to address many of the problems mentioned above, and more besides, by offering users a more intelligent approach to navigating and searching complex information spaces. They have been especially useful in many e-commerce domains with many stores using recommendation technologies to help convert browsers into buyers by providing intelligent and timely sales support and product suggestions; see for example Chapter 16 of this book [38] for a survey of recommendation techniques in an e-commerce setting. One of the key features of many recommendation technologies is the ability to consider the needs and preferences of the individual when it comes to generating *personalized* recommendations or suggestions. We will return to this issue later in this chapter but also refer the interested reader to related work on the development of personalization technologies. For example, Chapters 2 [35] and 4 [44] of this book consider different approaches to learning and modeling the preferences of users while Chapters 3 [62], 6 [61], and 18 [8] of this book consider different ways in which user models may be harnessed to provide users with more personalized access to online information and services. Indeed, while many recommendation and personalization technologies focus on the needs of the individual, some researchers have begun to consider group recommendation scenarios where the potentially competing preferences of a number of individuals need to be considered; see for example Chapter 20 [41] of this book.

Recommendation techniques come in two basic flavours. *Collaborative filtering* approaches rely on the availability of user ratings information (e.g. *"John likes items A, B and C but dislikes items E and F"* and make suggestions for a target user based on the items that similar users have liked in the past, without relying on any information about the items themselves other than their ratings; see Chapter 9 [83] of this book for a more detailed account of collaborative filtering approaches. In contrast *content-based* techniques rely on item descriptions and generate recommendations from items that are similar to those the target user has liked in the past, without directly relying on the preferences of other users; see Chapter 10 [69] of this book for a detailed account of pure content-based approaches.

Case-based recommenders implement a particular style of content-based recommendation that is very well suited to many product recommendation scenarios; see also [16]. They rely on items or products being represented in a structured way using a well defined set of features and feature values; for instance, in a travel recommender a particular vacation might be presented in terms of its *price, duration, accommodation, location, mode of transport, etc.* In turn the availability of similarity knowledge makes it possible for case-based recommenders to make fine-grained judgments about the similarities between items and queries for informing high-quality suggestions to the user. Case-based recommender systems are the subject of this chapter, where we will draw on a range of examples from a variety of recommender systems, both research prototypes and deployed applications. We will explain their origins in case-based reasoning research [1, 31, 46, 101] and their basic mode of operation as recommender systems. In particular, we will look at how case-based recommenders deal with the issues highlighted above in terms of their approach to selection similarity, recommendation diversity, and the provision of flexible feedback options. In addition we will consider the use of case-based recommendation techniques to produce suggestions that are personalized for the needs of the individual user and in this way present case-based approaches as one important solution for Web personalization problems; see also Chapters 2 [35], 3 [62], and 16 [38] in this book for related work in the area of Web personalization.

11.2 Towards Case-Based Recommendation

Case-based recommender systems have their origins in *case-based reasoning* (CBR) techniques [1, 46, 101, 48, 99]. Early case-based reasoning systems were used in a variety of problem solving and classification tasks and can be distinguished from more traditional problem solving techniques by their reliance on concrete experiences instead of problem solving knowledge in the form of codified rules and strong domain models. Case-based reasoning systems rely on a database (or case base) of past problem solving experiences as their primary source of problem-solving expertise. Each case is typically made up of a *specification* part, which describes the problem at hand, and a *solution* part, which describes the solution used to solve this problem. New problems are solved by retrieving a case whose specification is similar to the current target problem and then adapting its solution to fit the target situation. For example, CLAVIER [39] is a case-based reasoning system used by Lockheed to assist in determining the layout of materials to be cured in an autoclave (i.e., a large convection oven used, in this

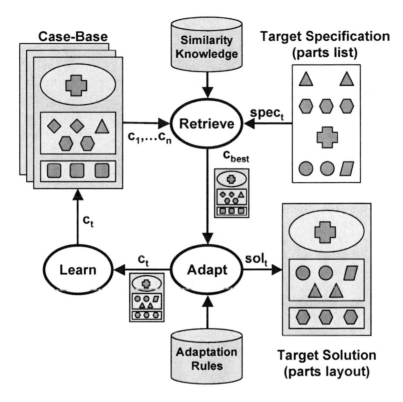

Fig. 11.1. CLAVIER uses CBR to design layout configurations for a set of parts to be cured in an autoclave. This is a complex layout task that does not lend itself to a traditional knowledge-based approach. However a case base of high-quality past layouts can be readily assembled. New layouts for a target parts-list can then be produced by retrieving a case with a similar parts-list and adapting its layout. If successful this new layout can then be learned by storing it in the case base as a new case.

case, for the curing of composite materials for aerospace applications). CLAVIER has the job of designing a good layout—one that will maximise autoclave throughput—for a new parts-list. The rules for determining a good layout are not well understood but previous layouts that have proved to be successful are readily available. CLAVIER uses these previous layout examples as the cases in its case base. Each case is made up of a parts-list (its specification) and the particular layout used (its solution). New layouts for a new parts-list are determined by matching the new parts-list against these cases and adapting the layout solution used by the most similar case; see Figure 11.1. CLAVIER has been a huge practical success and has been in use for a number of years by Lockheed, virtually eliminating the production of low-quality parts that must be scrapped, and saving thousands of dollars each month.

Case-based recommenders borrow heavily from the core concepts of retrieval and similarity in case-based reasoning. Items or products are represented as cases and recommendations are generated by retrieving those cases that are most similar to a user's query or profile. The simplest form of case-based recommendation is presented in Figure 11.2. In this figure we use the example of a digital camera recommender system, with the product case base made up of detailed descriptions of individual digital cameras. When the user submits a target query—in this instance providing a relatively vague description of their requirements in relation to *camera price* and *pixel resolution*—they are presented with a ranked list of *k* recommendations which represent the top *k* most similar cases that match the target query. As a form of content-based recommendation

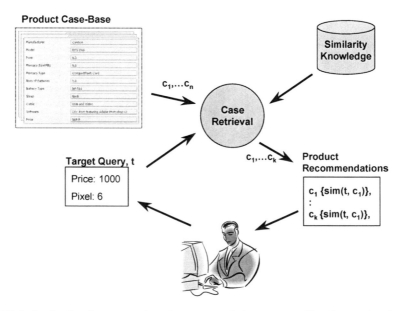

Fig. 11.2. In its simplest form a case-based recommendation system will retrieve and rank product suggestions by comparing the user's target query to the descriptions of products stored in its case base using similarity knowledge to identify products that are close matches to the target query.

(see, for example, [5, 26, 63, 78, 94] and also Chapter 10 [69] of this book) case-based recommenders generate their recommendations by looking to the item descriptions, with items suggested because they have similar descriptions to the user's query. There are two important ways in which case-based recommender systems can be distinguished from other types of content-based systems: (1) the manner in which products are represented; and (2) the way in which product similarity is assessed. Both of these will be discussed in detail in the following sections.

11.2.1 Case Representation

Normally content-based recommender systems operate in situations where content items are represented in an unstructured or semi-structured manner. For example, the NewsDude content-recommender, which recommends news articles to users, assumes text-based news stories and leverages a range of keyword-based content analysis techniques during recommendation; see for example, [9] and Chapter 18 [8] in this book. In contrast, case-based recommender systems rely on more structured representations of item content. These representations are similar to those used to represent case-knowledge in case-based reasoners. For example, they often use a set of well-defined features and feature values to describe items, rather than free-form text. This reliance on structured content means that case-based recommenders are particularly well adapted to many consumer recommendation domains, particularly e-commerce domains, where detailed feature-based product descriptions are often readily available.

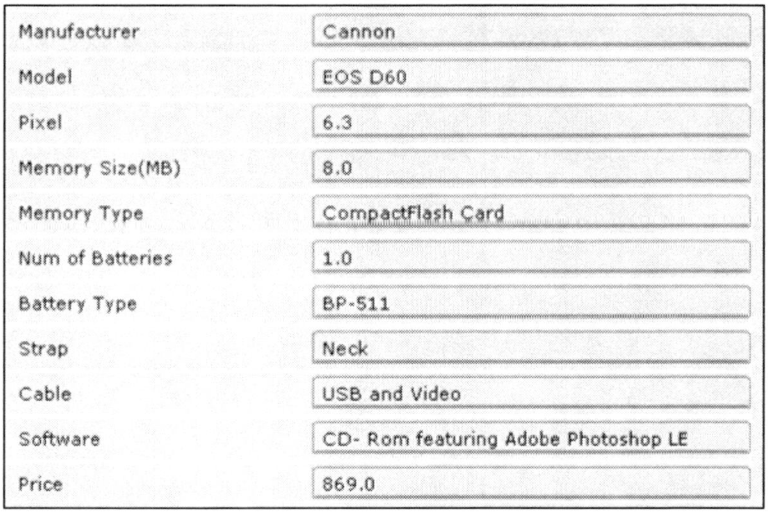

Manufacturer	Cannon
Model	EOS D60
Pixel	6.3
Memory Size(MB)	8.0
Memory Type	CompactFlash Card
Num of Batteries	1.0
Battery Type	BP-511
Strap	Neck
Cable	USB and Video
Software	CD- Rom featuring Adobe Photoshop LE
Price	869.0

Fig. 11.3. An example product case from a digital camera product catalog.

Figure 11.3 shows one such example product case from a catalog of cameras. The case is for a *Canon* digital camera and, as can be seen, the product details are captured using 11 different features (e.g., *manufacturer, model, memory type, price, etc.*) with each feature associated with one of a well-defined space of possible feature values (e.g., the *manufacturer* feature values are drawn from a well-defined set of possible manufacturers such as *Canon, Nikon, Sony etc.*). The example also highlights how different types of features can be used within a product description. In this case, a mixture of *numeric* and *nominal* features are used. For instance, *price* is an example of a numeric feature, which obviously represents the cost of the camera, and can take on values anywhere in a range of possible prices, from about $100 to upwards of $3000. Alternatively, *memory type* is a nominal feature, whose values come from a well-defined set of alternatives corresponding to the 4 or 5 different memory card options that are commonly used by digital cameras. The Entree recommender is another good example of a case-based recommender system. This system will be explored in more detail in Section 11.4 but suffice it to say that Entree is designed to make restaurant suggestions; see Figure 11.4. In terms of its core representation, Entree also uses a structured case format—although

Fig. 11.4. Entree [21, 22] recommends restaurants to users based on a variety of features such as *price, cuisine type, atmosphere,* etc..

the presentation in Figure 11.12 is largely textual the basic case representation is fundamentally feature-based—using features such as *price, cuisine type, atmosphere,* etc. to represent each restaurant case.

11.2.2 Similarity Assessment

The second important distinguishing feature of case-based recommender systems relates to their use of various sophisticated approaches to similarity assessment when it comes to judging which cases to retrieve in response to some user query. Because case-based recommenders rely on structured case representations they can take advantage of more structured approaches to similarity assessment than their content-based cousins. For example, traditional content-based techniques tend to use keyword-based similarity metrics, measuring the similarity between a user query and a product in terms of the frequency of occurrence of overlapping query terms within the product text. If the user is looking for a *"$1000 6 mega-pixel DSLR"* then cameras with all of these terms will be rated highly, and depending on the strength of the similarity criterion used, if no cameras exist with all of these terms then none may be retrieved. We have already highlighted this type of retrieval inflexibility (stonewalling) as a critical problem in the introduction to this chapter. Any reasonable person would be happy to receive a recommendation for a *"$900 6.2 mega-pixel digital SLR"*, for the above query, even though strictly speaking there is no overlap between the terms used in this description and the query.

Case-based recommenders can avail of more sophisticated similarity metrics that are based on an explicit mapping of case features and the availability of specialised feature level similarity knowledge. An online property recommender might use case-based techniques to make suggestions that are similar to a target query even when exact matches are not available. For example, a user who looking for a *"2 bedroom apartment in Dublin with a rent of 1150 euro"* might receive recommendations for properties that match the *bedroom* feature and that are *similar* to the target query in terms of *price* and *location*; the recommendations might offer slightly higher or lower priced properties in a nearby location when no exact matches are available.

$$Similarity(t,c) = \frac{\sum_{i=1..n} w_i * sim_i(t_i, c_i)}{\sum_{i=1..n} w_i} \tag{11.1}$$

Assessing similarity at the case level (or between the target query and a candidate case) obviously involves combining the individual feature level similarities for the relevant features. The usual approach is to use a weighted sum metric such as that shown in Equation 11.1. In brief, the similarity between some target query, t and some candidate case (or item), c, is the weighted sum of the individual similarities between the corresponding features of t and c, namely t_i and c_i. Each weight encodes the relative importance of a particular feature in the similarity assessment process and each individual feature similarity is calculated according to a similarity function that is defined for that feature, $sim_i(t_i, c_i)$. For instance, looking to the property recommender example above, if *rent* is very important to the user then the weight associated with this feature will be higher than the weights associated with less important features. In turn, when it comes to comparing the query and a case in terms of their *rent* the recommender system may draw on a specialised similarity metric designed for comparing monthly rents. A different metric might be used for comparing the *number of bedrooms* or the *property type*.

We must also consider the source of the individual feature level similarities and how they can be calculated. For example, returning to our camera recommender system, consider a numeric feature such as *pixel resolution*. The target query and a candidate case might be compared in terms of this feature using a similarity metric with the sort of similarity profile shown in Figure 11.5(a); maximum similarity is achieved when the *pixel resolution* of a candidate case matches that of the target query, and for cases with higher or lower *pixel resolution* there is a corresponding decline in similarity. This is an example of a *symmetric* similarity metric because there is no bias in favour of either higher or lower resolution cases.

$$sim_{price}(p_t, p_c) = 1 - \frac{|p_t - p_c|}{max(p_t, p_c)} \tag{11.2}$$

Sometimes symmetric similarity metrics are not appropriate. For instance, consider the *price* feature: it is reasonable to expect that a user will view cameras with prices (p_c) that are lower than their target price (p_t) to be preferable to cameras with higher prices, all other things being equal. The similarity metric in Equation 11.2 is used in many recommender systems (e.g., see [52, 75]) as one way to capture this notion and the metric displays a similarity profile similar to that shown in Figure 11.5(b). For instance, consider a $1000 target price and two candidate cases, one with a price of $500 and one for $1500. In terms of the symmetric similarity metric represented by Figure 11.5(a), the latter candidate corresponds to the point x in Figure 11.5(a) and the former to point y.

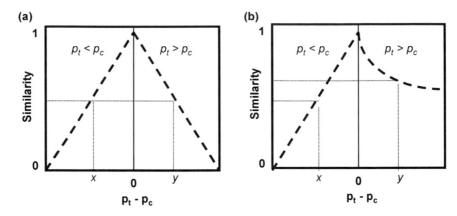

Fig. 11.5. Two example similarity profiles for numeric similarity metrics: (a) corresponds to a standard symmetric similarity metric; (b) corresponds to an asymmetric metric that gives preference to features values that are lower than the target's value.

For both cases the similarity assessment is the same, reflecting that both differ from the target price by the same amount, $500, with no preference given to whether a case is less or more expensive. These cases are plotted in the same way in Figure 11.5(b) but now we see that the more expensive case (point x) has a lower similarity than the cheaper camera (point y). Even though both candidates differ by $500, preference is given to the cheaper case.

To evaluate the similarity of non-numeric features in a meaningful way requires additional domain knowledge. For example, in a vacation recommender it might be important to be able to judge the similarities of cases of different *vacation types*. Is a *skiing* holiday more similar to a *walking* holiday than it is to a *city break* or a *beach* holiday? One way to make such judgments is by referring to suitable domain knowledge such as an ontology of vacation types. In Figure 11.6 we present part of what such an ontology might look like with different feature values represented as nodes and similar feature values grouped near to each other. In this way, the similarity between two arbitrary nodes can be evaluated as an inverse function of the distance between them or the distance to their nearest common ancestor. Accordingly, a *skiing* holiday is more similar to a *walking* holiday (they share a direct ancestor, *activity* holidays) than it is to a *beach* holiday, where the closest common ancestor is the ontology root node.

11.2.3 Acquiring Similarity Knowledge

Similarity assessment is obviously a key issue for case-based reasoning and case-based recommender systems. Of course the availability and use of similarity knowledge (feature-based similarity measures and weighting functions) is an important distinguishing feature of case-based recommendation. Although further detailed discussion of this particular issue is beyond the scope of this chapter, it is nonetheless worth considering the origin of this knowledge in many systems. For the most part this knowledge is hand-coded: similarity tables and trees, such as the vacation ontology above, are made

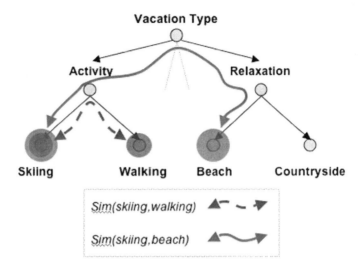

Fig. 11.6. A partial ontology of vacation types can be used as the basis for similarity judgments for non-numeric features.

available and codified by a domain knowledge expert. Similarly, importance weights might be assigned by the user at retrieval time or by the domain expert during system design. Hand-coding this knowledge is, of course, expensive and so increasingly researchers have begun to explore how machine learning techniques can be used to relieve these knowledge acquisition costs.

A number of researchers have looked at the issue of automatically learning the feature weights that are be used to influence the level of importance of different features during the case similarity calculations. Wettschereck and Aha [102], for example, describe an evaluation of a number of weight-learning algorithms that are *knowledge-poor*, in the sense that they avoid the need for detailed domain knowledge to drive the learning process. They show how even these knowledge-poor techniques can result in significant improvements in case-based classification tasks and present a general framework for understanding and evaluating different weight-learning approaches; see also the work of [42, 81] for approaches to local weight-learning in CBR based on reinforcement learning.

Stahl [96] also looks at feature-weight learning but describes an alternative approach in which feedback is provided by a "similarity teacher" whose job it is to evaluate the ordering a given retrieval set. For example, in a recommender context a user may play the role of the similarity teacher because her selections can be interpreted as retrieval feedback; if our user selects product number 3 in the list of recommendations first then we can conclude that the correct ordering should have placed this product at the top of the list. Stahl's learning algorithm attempts to minimise the average ordering error in retrieval sets. The work of Cohen et al. [25] looks at the related issue of learning to order items given feedback in the form of preference judgements. They describe a technique for automatically learning a preference function to judge how advisable it is to rank some item i ahead of item j, and go on to show how a set of items can then be ranked by attempting to maximise agreements with this learned preference function; see also the work of [13].

Finally, it is worth highlighting recent similarity learning work by O'Sullivan et al. [66, 67, 68]. This work has not been used directly by case-based recommenders but instead has been used to improve the quality of collaborative filtering recommenders (see Chapter 9 [83] in this book) by using case-based style similarity metrics when evaluating profile similarities. Normally a collaborative filtering recommender system can only evaluate the similarity between two profiles if they share ratings. For example in a TV recommender two users that have both rated *ER* and *Frasier* can be compared. But if one user has only rated *ER* and the other has only rated *Frasier* then they cannot be compared. O'Sullivan et al. point out that the ratings patterns within a collaborative filtering database can be analysed to estimate the similarity between programmes like *ER* and *Frasier*. They show that by using data-mining techniques it is possible to discover that, for example, 60% of the people who have liked *ER* have also liked *Fraiser*, and use this as a proxy for the similarity between these two programmes. They demonstrate how significant improvements in recommendation accuracy can be obtained by using these similarity estimates with more sophisticated case-based profile similarity metrics.

11.2.4 Single-Shot Recommendation

Many case-based recommenders operate in a *reactive* and *single-shot* fashion, presenting users with a single set of recommendations based on some initial query; thus the user is engaged in a single (short-lived) interaction with the system. For example, the Analog Devices OpAmp recommender presents a user with a set of available OpAmps that closely match the user's query [100, 103]. The online property recommender referred to earlier operate similarly, responding with a selection of suitable apartments in response to a user's rental constraints; see also the *DubLet* system by [40].

The point to make here is that single-shot recommendation has its shortcomings. In particular, if users do not find what they are looking for among the initial recommendations—as is frequently the case—then their only option is to revise their query and start again. Indeed the pure similarity-based nature of most case-based recommender systems increases the chances of this happening in certain situations because, as we discussed earlier, the top ranked recommendations may differ from the target query in more or less the same ways. As a result they will be very similar to each other—they will lack diversity—and if the user doesn't like the first recommendation she is unlikely to be satisfied with the similar alternatives either. In the remaining sections of this chapter we will explore how this simple model of case-based recommendation has been extended to provide a more sophisticated recommendation framework, one that provides for more sophisticated interaction between recommender and user, generating personalized recommendations that are more diverse, through an extended dialog with the user.

11.3 Similarity and Beyond

Let us look at a concrete example of the diversity problem referred to above. Consider a vacation recommender where a user submits a query for a 2-week vacation for two in the sun, costing less than $750, within 3 hours flying time of Ireland, and with good night-life and recreation facilities on-site. The top recommendation returned is for an apartment in the Hercules complex in the Costa Del Sol, Spain, for the first two weeks in July. A good recommendation by all accounts, but what if the second, third, and fourth recommendations are from the same apartment block, albeit perhaps for different two-week periods during the summer, or perhaps for different styles of apartments? While the k ($k = 4$ in this case) best recommendations are all very similar to the target query, they are also very similar to each other. The user has not received a useful set of alternatives if the first recommendation is unsuitable. This scenario is not uncommon

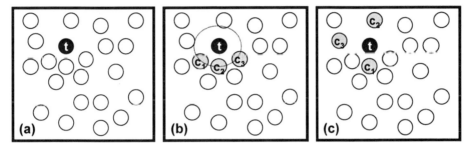

Fig. 11.7. Similarity vs diversity during case retrieval: (a) a case base with highlighted target query, t; (b) a conventional similarity-based retrieval strategy returns the cases that are individually closest to the target query, thus limiting their potential diversity; (c) an alternative retrieval strategy that balances similarity to the target and the relative diversity of the selected cases produces a more diverse set of recommendations.

in recommender systems that employ similarity-based retrieval strategies: they often produce recommendation sets that lack diversity and thus limit user options (see Figure 11.7(a&b)). These observations have led a number of researchers to explore alternatives to similarity-based retrieval, alternatives that attempt to explicitly improve recommendation diversity while at the same time maintaining query similarity. [1]

11.3.1 Similarity vs. Diversity

How then can we improve the diversity of a set of recommended cases, especially since many of the more obvious approaches are likely to reduce the similarity of the selected

---

[1] Incidentally, related concerns regarding the primacy of similarity in other forms of case-based reasoning have also come to light, inspiring many researchers to look for alternative ways to judge the *utility* of a case in a given problem solving context (e.g. [7, 19, 34, 45, 49, 91]). For example, researchers have looked at the importance of adaptability alongside similarity, arguing that while a case may appear to be similar to a target problem, this does not mean it can be successfully adapted for this target (see [49, 91]).

cases compared to the target query? In case-based recommenders, which implement a similarity-based retrieval strategy, the trade-off between similarity and diversity is often straightforward when we look at the similarity and diversity characteristics for the top k items. For low values of k, while similarity to the target query tends to be high, the diversity between the top k recommendations tends to be very low. In other words, the top ranking cases are often similar to the target query in more or less the same ways; of course what we really need is a set of recommendations that are equally similar to the target query but in different ways. As we move through the top ranking recommendations we tend to find cases that are similar to the target query but increasingly different from those that have gone before. These are the interesting cases to consider from a recommendation diversity perspective.

We attempt to capture this visually in Figure 11.8(a) by depicting a list of the top 9 recommendations in decreasing order of their similarity to the target query. Each recommendation is shaded to reflect its diversity relative to the others. For example, the top 3 recommendations are all shaded to the same degree, indicating that the are all very similar to each other. Hence there is little variation among the top 3 results; perhaps these are all examples of vacation suggestions for the same Spanish apartment complex for the first few weeks of July. It should be clear in this example how more diverse recommendations only begin to appear for higher values of k. Recommendations 4, 6 and 9, for example, are more diverse alternatives; perhaps these suggestions correspond to vacations in Tuscany or on the Costa Brava. One solution then is to look for ways of identifying and promoting these more diverse recommendations so that the user is presented with a more diverse list of suggestions, which still remain true to their target query. Figure 11.8(b) illustrates this: recommendations from positions 4,6 and 9 in Figure 11.8(a) are promoted to positions 2, 3 and 4 (or perhaps the less diverse suggestions of 2, 3, and 5 are simply removed from the suggestion list) thus providing the user with variation in the top recommendations.

One way to improve diversity is to simply select k random cases from the top bk most similar cases to the target query. This so-called *bounded random selection* strategy was proposed by [92] but it was shown to result in an unacceptable drop in query similarity, and so is hardly practical in many recommendation scenarios. However, more principled approaches are available, which rely on an explicit model of diversity. We can define the diversity of a set of retrieved cases, $c_1, ...c_k$, to be the average *dissimilarity* between all pairs of these cases (Equation 11.3). Then the *bounded greedy selection* strategy proposed by [92] offers a way to improve diversity, while at the same time maintaining target query similarity; see also [11]. This strategy incrementally builds a diverse retrieval set of k cases, R, by starting from a set of the bk most similar cases to the target query. During each step the remaining cases are ordered according to their *quality* with the highest quality case added to R. The key to this algorithm is a quality metric that combines diversity and similarity (Equation 11.4). The quality of a case c is proportional to the similarity between c and the current target t, and to the diversity of c *relative* to those cases so far selected, $R = \{r_1, ..., r_m\}$; see Equation 11.5. The first case to be selected is always the one with the highest similarity to the target. However during subsequent iterations, the case selected is the one with the highest combination of similarity to the target and diversity with respect to the set of cases selected so far.

(a) Similarity-Based   **(b) Diversity-Based**

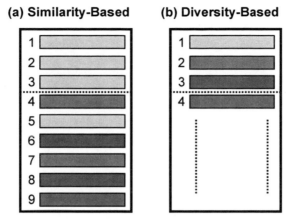

Fig. 11.8. Typical approaches to similarity-based recommendation tend to produce recommendation lists with limited diversity characteristics, such as the list shown in (a). Individual items are shaded to reflect their diversity characteristics so that in (a) items 1,2,3 and 5 are very similar to each other as are items 4 and 7 and items 6 and 8. In (b) a different ordering of items is presented, one that maximises the diversity of the top items.

$$Diversity(c_1,...c_n) = \frac{\sum_{i=1..n} \sum_{j=i..n}(1 - Similarity(c_i, c_j))}{\frac{n}{2} * (n-1)} \qquad (11.3)$$

```
1.    define BoundedGreedySelection (t, C, k, b)
2.    begin
3.       C' := bk cases in C that are most similar to t
4.       R := {}
5.       For i := 1 to k
6.          Sort C' by Quality(t,c,R) for each c in C'
7.          R  := R + First(C')
8.          C' := C' - First(C')
9.       EndFor
10.   return R
11.   end
```

Fig. 11.9. The Bounded Greedy Selection strategy for producing a diverse set of k: t refers to the current target query; C refers to the case base; k is the size of the desired retrieval/recommendation set; b refers to the bound used for the initial similarity-based retrieval.

$$Quality(t,c,R) = Similarity(t,c) * RelDiversity(c,R) \qquad (11.4)$$

$$RelDiversity(c,R) = 1 \ if \ R = \{\};$$
$$= \frac{\sum_{i=1..m}(1 - Similarity(c,r_i))}{m}, otherwise \qquad (11.5)$$

Empirical studies presented in [11, 92] demonstrate the diversity benefits of the above approach. In particular, the bounded greedy algorithm is found to provide a cost effective diversity enhancing solution, resulting in significant improvements in recommendation diversity against relatively minor reductions in target query similarity. For example, [11, 92] applied the technique in a number of different recommender systems including a vacation recommender and a recruitment advisor. In the vacation recommender, examining the similarity and diversity characteristics of the top 3 recommendations reveals that the bounded greedy technique manages to achieve a 50% improvement in relative diversity when compared to the standard similarity-based recommendation approach but suffers a minor loss of less than 10% in similarity to the target query. Similar results are found in the recruitment domain and in practice users are seen to benefit from a much more varied selection of alternatives that remain similar to their stated needs.

11.3.2 Alternative Diversity-Preserving Approaches

The *bounded greedy* technique discussed above was among the first practical attempts to explicitly enhance the diversity of a set of recommendations without significantly compromising their query similarity characteristics; although it is worth noting that some loss of similarity is experienced with this approach. In parallel Shimazu [86, 87] introduced an alternative method for enhancing the diversity of a set of recommendations. In brief, a set of 3 recommendations, c_1, c_2 and c_3, are chosen relative to some query q such that c_1 is maximally similar to q, c_2 is maximally dissimilar to c_1 and then c_3 is maximally dissimilar to c_1 and c_2. In this way, the triple of cases are chosen to be maximally diverse but, unlike the bounded greedy technique above, the similarity of c_2 and c_3 to the query is likely to be compromised. As such the value of this approach is limited to situations where the set of recommended cases is drawn from a set of cases that are all sufficiently similar to the user query to begin with.

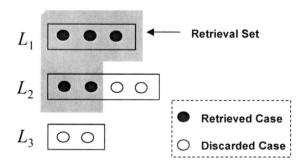

Fig. 11.10. The approach described in [55, 56] partitions the case base into similarity layers—groups of cases with equivalent similarity to the target query—and the retrieved cases are chosen starting with the highest similarity layer and until k cases have been selected. The final cases selected from the lowest necessary similarity layer are chosen based on an optimal diversity maximizing technique.

Recently a number of alternative diversity enhancing selection techniques have been proposed. For example, [55] shows that it is sometimes possible to enhance diversity without loss of query similarity and a related approach based on the idea of *similarity layers* is described [56]. Very briefly, a set of cases, ranked by their similarity to the target query are partitioned into similarity layers, such that all cases in a given layer have the same similarity value to the query. To select a set of k diverse cases, the lowest similarity layer that contributes cases to the recommendation set is identified and a subset of cases from this layer are selected for inclusion in the final recommended set with all cases in higher similarity layers automatically included; see Figure 11.10. Cases are selected from this lowest similarity layer using an optimal diversity maximizing algorithm. This approach has the ability to improve diversity while at the same time fully preserving the similarity of cases to the user query. However, the diversity improvements obtained are typically less than those achieved by the bounded greedy algorithm, because all cases from higher similarity layers are always included without any diversity enhancement. An alternative, and more flexible, diversity enhancing approach is also introduced based on the analogous notion of *similarity intervals*; see also [56]. The advantage of this approach is that it can achieve greater diversity improvements by relaxing the constraint that query similarity must be preserved. Query similarity is reduced but within a tolerance level defined by the width of the similarity intervals.

It is also worth noting that a retrieval technique may not be designed to explicitly enhance diversity but may nonetheless have a beneficial effect by its very nature. *Order-based retrieval* is a good example of such a technique [14, 17]. It is based on the idea that the relative similarities of cases to a query of *ideal* feature values is one way of ordering a set of cases for recommendation. Order-based retrieval constructs an ordering relation from the query provided by the user and applies this relation to the case base of products returning the k items at the top of the ordering. The order relation is constructed from the composition of a set of canonical operators for constructing partial orders based on the feature types that make up the user query. While the technical details of order-based retrieval are beyond the scope of this chapter the essential point to note is that an empirical evaluation of order-based retrieval demonstrates that it has an inherent ability to enhance the diversity of a set of retrieval results; that is, the cases at the top of the ordering tend to be more diverse than an equivalent set of cases ranked based on their pure similarity to the user query.

In [58] McSherry proposes a *compromise-driven* approach to retrieval in recommender systems. This approach is inspired by the observation that the most similar cases to the user's query are often not representative of compromises that the user may be prepared to accept. Compromise-driven retrieval is based on a variation of the usual similarity assumption: that a given case is more acceptable than another if it is more similar to the user's query *and* it involves a subset of the compromises that the other case involves. As well as being less likely to be contradicted by user behaviour, this assumption serves as the basis for a more principled approach to deciding which cases are included in the retrieval set than setting an arbitrary similarity threshold over the candidate cases. For example, no case is included in the retrieval set if there is a more similar case that involves a subset of the compromises it involves. Though not relying explicitly on diversity as an additional measure of recommendation quality, compromise-driven retrieval does offer users a better (usually more diverse) set of recommendation alternatives. Moreover, the recommendation set is guaranteed to provide full coverage of the available cases in the sense that for any case that is not included in the retrieval set, one of the recommended cases is at least as good in terms of its similarity to the user's query and the compromises it involves. While the size of the retrieval set required to provide full coverage cannot be predicted in advance, experimental results suggest that retrieval-set sizes tend to remain within reasonable limits even for queries of realistic complexity; see [58].

In summary then, we have seen how recent developments in case-based recommendation have relaxed the conventional wisdom of the similarity assumption, in favour of retrieval strategies that are more likely to deliver a recommendation set that offers users a more diverse set of alternatives. In the next section will will revisit the diversity issue in a slightly different context. While accepting the value of diversity during recommendation, we will question whether it should always be used as a retrieval constraint or whether there are occasions when it is more useful to focus on similarity or diversity.

11.4 The Power of Conversation

As mentioned earlier, the single-shot model of recommendation, whether similarity-based or diversity-enhanced, is limited to a single interaction between the user and the recommender system. If the user is not satisfied with the recommendations they receive then their only option is to modify their query and try again. Indeed the single-shot approach also makes the assumption that the user is in a position to provide a detailed query from the start, an assumption that does not often hold in practice. For example, in many product recommendation scenarios users may start with an initial query, which they will ultimately come to adapt and refine as they learn more about a particular product-space or the compromises that might be possible in relation to certain product features. Sometimes a user will come to disregard features that, initially at least, were important, as they recognise the value of other features, for example. These observations have motivated the development of conversational recommender systems which engage the user in an extended, interactive recommendation dialog during which time they attempt to elicit additional query information in order to refine recommendations[2]. Today most case-based recommenders employ conversational techniques, engaging users in an extended dialog with the system in order to help them navigate through a complex product space by eliminating items from consideration as a result of user feedback; note that these dialogs are normally restricted in the form of feedback solicited from the user, rather than offering free form natural language style dialogs.

Two different forms of conversational recommender systems can be distinguished according to the type of feedback that they solicit from users. In the nomenclature of Shimazu [86, 87], conversational recommenders can adopt a *navigation by asking* or a *navigation by proposing* style approach. In the case of the former, recommenders ask

---

[2] Conversational recommender systems have their origins in *conversational case-based reasoning (CCBR)* [3, 4, 12, 64], which apply similar techniques to elicit query information in problem solving domains and diagnostic tasks.

their users a series of questions regarding their requirements; this form of feedback is sometimes termed *value elicitation*. For example, a digital camera recommender might ask *"What style of camera do you want? Compact or SLR?"* or *"How much optical zoom do you need?"*. Alternatively, systems that employ navigation by proposing avoid posing direct questions in favour of presenting users with interim recommendations and asking for their feedback, usually in the form of a simple preference or a rating. Both styles of conversation have their pros and cons when it comes to user costs and recommendation benefits as we shall discuss in the following sections.

11.4.1 Navigation by Asking

Navigation by asking is undoubtedly the most direct way to elaborate a user's requirements and can lead to very efficient conversational dialogs in many situations. The Adaptive Place Advisor, which helps users to choose destinations such as restaurants, is good example of a system that employs navigation by asking [37, 98]. For instance, Figure 11.11 shows a sample Adaptive Place Advisor conversation between the user (the *inquirer*) and the recommender system (the *advisor*). Clearly, the Adaptive Place Advisor supports a sophisticated form of conversational dialog, employing natural language processing techniques in order to respond to freeform user answers. In addition, it is worth highlighting that its conversational form allows the user to ask questions of the recommender system too (see Line 3 of Figure 11.11), something that we will return to in Section 11.4.3, under the heading of mixed-initiative systems. The essential point is that each question that the Adaptive Place Advisor asks is designed to narrow down the possible recommendations that might be suggested to the user. Indeed in this system recommendations are only finally made when a few items remain.

One of the key issues that must be addressed with this style of recommendation concerns the important issue of what sets of questions should be asked during each session. For instance the order and number of questions can have a significant impact on usability and user effort. Doyle & Cunningham were among the first to report experimental results on this topic within CBR [32] by evaluating different question-selection criteria, including an entropy-based method which evaluates the informativeness or *information gain* of given features and is inspired by the decision-tree building work of [71, 73]; see also [86, 87]. Schmitt and his colleagues [84] propose an alternative approach, called *simVar*, that is especially tailored to e-commerce recommendation scenarios. Instead of using entropy-based information gain measures, this approach considers the influence that a given feature can have on the similarity of cases in the case base. Thus questions are asked of features that are deemed to have the highest influence on the similarity of cases. In addition, an estimate of the user's likely ability to answer such questions is also considered. Evaluations show that SimVar is capable of producing question dialogs that successfully prioritise questions that are both important and easy for the user to answer and that these can be more effective in product recommendation scenarios than those produced by entropy-based approaches; see also [6] for other useful dialog criteria. In Section 11.5 we will discuss how personalization techniques are used by the Adaptive Place Advisor to improve its conversational efficiency by eliminating the need to pose certain questions in situations where a learned model of the particular user can provide the answers.

1.	**Inquirer:**	*Where do you think I should eat tonight?*
2.	**Advisor:**	*What type of food would you like?*
3.	**Inquirer:**	*What types are there?*
4.	**Advisor:**	*You can say things like Chinese, Indian, and Mediterranean.*
5.	**Inquirer:**	*Oh, maybe a cheap Indian place.*
6.	**Advisor:**	*What kind of parking would you like?*
7.	**Inquirer:**	*I don't care, as long as it's in Palo Alto.*
8.	**Advisor:**	*I'm sorry, I don't know of any restaurants like that, would you like to search for any price?*
9.	**Inquirer:**	*No, I think I'd like Chinese instead.*
10.	**Advisor:**	*How does Mandarin Gourmet on 420 Ramona sound?*₁
11.	**Inquirer:**	*No, what else do you have?*
12.	**Advisor:**	*How about Jing-Jing Szechwan Hunan Gourmet on 443 Emerson Street?*
13.	**Inquirer:**	*Sure, that sounds fine.*

Fig. 11.11. An example of the type of conversational dialog between a user (the *inquirer*) and the Adaptive Place Advisor recommender system (the *advisor*) in which the user is trying to decide on a restaurant for dinner.

The value elicitation approach adopted in navigation by asking has its origins in the classification and diagnosis tasks that are the focus of many conversational case-based reasoning systems (see for e.g., [3, 4, 12, 64]), but its use in recommendation tasks is not always appropriate. For example, users are often less inclined to tolerate protracted lists of direct questions. Oftentimes they will not know the answers to questions that demand a high-level of domain knowledge or they may reject questions that ask for sensitive or personal information. Moreover, providing answers to direct questions can impose a significant interfacing burden. For example, expecting users to respond with textual answers is not appropriate in the context of recommender systems that operate over mobile devices such as PDAs and mobile phones.

11.4.2 Navigation by Proposing

The above value elicitation issues have led to an increased interest in other forms of user feedback that are amenable to the navigation by proposing style of conversational recommendation. The key feature of navigation by proposing is that the user is presented with one of more recommendation alternatives, rather than a question, during each recommendation cycle, and they are invited to offer feedback in relation to these alternatives. In general terms there are 3 important types of feedback. *Ratings-based feedback* (see [43, 89, 90]) involves the user providing an explicit rating of a recommendation, but this form of feedback is more commonly found in collaborative filtering style recommender systems (see Chapter 9 [83] in this book) and shall not be discussed further here. Alternatively, feedback might be expressed in the form of a constraint over certain features of one of the recommendations (*critique-based feedback*)

Fig. 11.12. Entree [21, 22] recommends restaurants to users and solicits feedback in the form of feature *critiques*. The screenshot shows a recommendation for *Planet Hollywood* along with a variety of fixed critiques (*less$$, nicer, cuisine* etc.) over features such as the *price, ambiance* and *cuisine* of the restaurant.

[21, 22, 33, 50, 51, 70, 75]. Even simpler again is *preference-based feedback* in which the user expresses a preference for one alternative over the others [52].

Critique-Based Feedback. Critiquing-based recommenders allow users to provide feedback in the form of a directional feature constraint. For example, a digital camera shopper might ask for a camera that is *cheaper* than the current recommendation, *cheaper* being a critique over the *price* feature. The FindMe systems [21, 22] were among the first recommenders to champion critiquing as an effective form of feedback in conversational recommender systems. For instance, the Entree recommender suggests restaurants in Chicago and each recommendation allows the user to select from seven different critiques; see Figure 11.12. When a user selects a critique such as *cheaper*, Entree eliminates cases (restaurants) that do not satisfy the critique from consideration in the next cycle, and selects that case which is most similar to the current recommendation from those remaining; thus each critique acts as a filter over the cases. As a form of feedback, critiquing has much to recommend it. It proves to be an effective way to guide the recommendation process and yet provides users with a straightforward mechanism to provide feedback, one that requires limited domain knowledge on the part of the user and it is easy to implement with even the simplest of interfaces. The FindMe systems evaluate this form of conversational recommendation and feedback in a variety of contexts including movie, car, and accommodation recommendation [22]. In the Car Navigator recommender system, for example, individual critiques were also designed to cover multiple features, so that, for instance, a user might request a *sportier* car than the current recommendation, simultaneously constraining features such as *engine size* and *acceleration*. These *compound critiques* obviously allow the

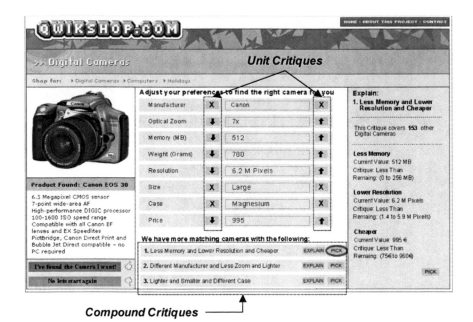

Fig. 11.13. A screenshot of the digital camera recommender system evaluated in [75, 50, 51] which solicits feedback in the form of fixed unit critiques and a set of dynamically generated compound critiques. The former are indicated on either side of the individual camera features, while the latter are presented beneath the main product recommendation as a set of 3 alternatives.

recommender to take larger steps through the product-space, eliminating many more cases than would be possible with a single-feature, *unit critique*, in a single recommendation cycle. Indeed recently the work of [50, 51, 75] has investigated the possibility of automatically generating dynamic compound critiques based on the remaining cases and the user's progress so far; see for example Figure 11.13. In short, this so-called *dynamic critiquing* approach uses data mining techniques to identify groups of unit critiques that reflect common difference patterns between the remaining cases. Evaluation results suggest that using these groups of critiques as compound critiques has the potential to offer significant improvements in recommendation performance by allowing users to navigate more efficiently through complex product-spaces.

Preference-Based Feedback. Perhaps the simplest form of feedback involves the user indicating a simple preference for one recommendation over another. It is also particularly well suited to domains where users have very little domain knowledge, but where they can readily express a preference for what they like when it is recommended. For example, Figure 11.14 shows a set of recommendations from a prototype recommendation service that uses preference-based feedback to help brides-to-be to chose a wedding dress. During each cycle 3 different suggestions are presented along with a set of technical features, and the user can select one recommendation as their preference as a lead into the next cycle. This is a good example of a domain where the average shopper is likely to have limited domain knowledge, at least in terms of the type of technical

Fig. 11.14. A set of wedding dress suggestions from a prototype recommender system. During each recommendation cycle 3 suggestions are made and the user can indicate which they prefer as a way to initiate the next recommendation cycle.

features that are presented alongside the recommendations. However most brides-to-be will be willing and able to select a preference for one dress over another.

Unfortunately, while this approach carries very little feedback overhead, from a user's perspective, it is ultimately limited in its ability to guide the recommendation process. For example, it will not always be clear why a user has selected one recommendation over another, in terms of the features behind these recommendation. They both may have many features in common and many features that distinguish them. To address this issue the *comparison-based recommendation* work of [52] proposes a variety of query revision strategies that are designed to update the current query as a result of preference-based feedback. For example the most straightforward strategy (*more like this*) simply adopts the preferred case as the new query and proceeds to retrieve the k most similar cases to it for the next cycle. This approach is not very efficient however as it does little to infer the user's true preferences at a feature level. An alternative approach only transfers features from the preferred case if these features are absent from all of the rejected cases, thus allowing the recommender to focus on those aspects of the preferred cases that are unique in the current cycle. Yet another strategy attempts to weight features in the updated query according to how confident the recommender can be that these features are responsible for the user's preference. One particular weighting strategy proposed by [52] depends on the number of alternatives for a given feature within the current recommendation set. For example, in a digital camera recommender system, if the preferred camera is a *Nikon*, and if there are many alternative manufacturers listed amongst the other $k - 1$ recommendations in the current cycle, then the recommender can be more confident that the user is genuinely interested in *Nikons* than if say one or two of the rejected cases were also *Nikons*. Both of these alternative strategies allow for more efficient recommendation sessions than the default *more like this* strategy.

11.4.3 Conversational Management and Mixed-Initiative Systems

Of course while there is a clear distinction between the navigation by asking and navigation by proposing styles of conversation, and between the different forms of feedback that they adopt, this does not limit recommender systems from implementing strategies that combine different conversational styles and methods of feedback. For example, ExpertClerk [86, 87] is designed to mimic the type of interactions that are commonly observed between shoppers and sales clerks in real-life shopping scenarios, and implements a mixture of navigation by asking and navigation by proposing. During the early stages of the recommendation session the user is asked a number of questions in order to identify their broad requirements. Questions are selected using an information theoretic approach based on ID3's information gain heuristic [72, 73]. This is followed by a navigation by proposing stage that includes critiquing-based feedback. Experiments indicate that this combination of navigational styles provides for more efficient recommendation sessions than either approach on its own and, it is argued, constitutes a more natural form of recommendation dialog.

The work of [53, 54, 93] proposes a different departure from the conventional approaches to conversational recommendation. Instead of combining different forms of feedback and conversation style, the *adaptive selection* technique proposes a diversity-enhanced approach to retrieval that has its origins in the work described in [11, 92]; see also Section 11.3 of this chapter. However, rather than simply combining similarity and diversity during each recommendation cycle, diversity is selectively introduced depending on whether or not the current recommendation cycle is judged to be on target with respect to the user's preferences. If the current set of recommendations are not judged by the user to be an improvement on previous recommendations then diversity is increased during the next recommendation cycle in order to provide a broader set of recommendations. The goal is to try to refocus the recommender system on a more satisfactory region of the product-space. If, however, the latest recommendations do appear to be on target then a similarity-based approach to retrieval is used to provide for a more fine-grained exploration of the current region of the product-space. Judging whether or not the current recommendations are on target is achieved using a technique called *preference carrying*. Specifically, the previously preferred product case is carried to the next cycle and if it is reselected then the system assumes that an improvement has not been achieved by the other cases recommended as part of this new cycle. The approach has been shown to have the potential to improve recommendation performance by offering significant reductions in recommendation session lengths when used with either preference-based or critiquing forms of feedback.

In an attempt to further improve the interactive nature of conversational recommender systems researchers have begun to look at how these systems might accommodate a wider variety of *conversational moves* [15], above and beyond the provision of simple forms of user feedback. For example, the Adaptive Place Advisor [37, 98] contemplates a number of different conversational moves in the form of dialog operators. One such operator, *ASK-CONSTRAIN* allows the recommender to ask the user for a particular feature value (standard value elicitation). A different operator (the *QUERY-VALUES* operator) allows the user to ask the recommender for the possible values of a product feature. For example, the recommendation session presented in Figure 11.11 shows examples of the recommender requesting certain feature values and also the user asking for information about the permissible values. Within recommendation sessions the availability of these operators facilitates more flexible interaction models which allow either user or system to seize and cede initiative in a more flexible manner than conventional conversational recommender systems.

This more flexible approach to the development of conversational recommender systems is closely related to recent developments in the area of *mixed-initiative intelligent systems* (see, for example, [2, 97]), which attempt to integrate human and automated reasoning to take advantage of their complementary reasoning styles and different computational strengths. Important issues in mixed-initiative systems include: the division of task responsibility between the human and the intelligent agent; how control and initiative might shift between human and agent, including the support of interrupt-driven or proactive behaviour. Mixed-initiative systems need to maintain a more sophisticated model of the current state of the human and agent(s) involved and require a well-defined set of protocols to facilitate the exchange of information between human and agent(s). For example, the mixed-initiative recommender system, Sermo [15], draws on ideas from conversational analysis techniques in the service of a recommendation framework that combines a conversational recommendation strategy with grammar-based dialog management functions; see also [12]. Sermo's dialog grammar captures the set of legal dialogs between user and recommender system, and the use of a conversational policy facilitates a reasonable balance between over-constrained fixed-role recommendation dialogs and interrupt-driven dialogs that are not sufficiently constrained to deliver coherent recommendation sessions.

11.5 Getting Personal

Personalization technologies promise to offer users a more engaging online experience, one that is tailored to their long-term preferences as well as their short-term needs, both in terms of the information that is presented and the manner in which it is presented. By reading this volume you will gain a detailed understanding of how personalisation techniques have been explored in a wide variety of task contexts, from searching and navigating the Web (see Chapter 6 [61] in this book) to e-commerce (see Chapter 16 [38] in this book), healthcare (see Chapter 15 [23]) and e-learning (Chapter 1 [18] in this book) applications.

Personalization and recommendation technologies are also intimately connected. The ultimate promise of recommendation technology goes beyond the provision of flexible, conversational information access. Indeed, arguably, the provision of a truly effective recommendation service demands an ability to respond to more than just the short-term needs of the individual. A user's personal preferences should also be considered as an important source of context with which to guide the recommendation process. Ultimately the benefits of personalization in a recommendation context come down to the potential for an improved experience for the user. In theory, personalized recommender systems should be able to respond effectively to users with less feedback; in this sense a user's learned preferences can *"fill in the gaps"*, or at least some of them, during a recommendation dialog, leading to shorter recommendation sessions. For example, recognising that a user is a fan of water sports will help a travel recommender to make more targeted recommendations without soliciting feedback on preferred vacation activities. Moreover, the quality of recommendations should improve because the recommender has access to long-term preference information that may never normally be disclosed during any individual session. For example, as a member of the One World Alliance air-miles programme our holiday-maker may prefer to fly with certain carriers to avail of air-miles. This information may be missing from their query but because it is reflected in their long-term profile these carriers can be promoted during the recommendation process.

Personalized recommender systems are distinguished by their ability to react to the preferences of the individual. Obviously there is a sense in which every recommender system reacts to the preferences of the individual, but does this make it personalized? For example, all of the conversational recommender systems discussed above react to feedback provided by users within each session. As such they react to the personal needs of the user, captured from their direct feedback, and so deliver some degree of personalization in their recommendations. However this is what might be termed *weak personalization* because the absence of a persistent user profile outside of the scope of an individual session precludes the recommender from adapting to preferences that may not be disclosed in that session. Such recommender systems can provide *in-session* personalization only and two users who respond in the same way within a session will receive the same recommendations even if they differ greatly in terms of their long-term preferences. Recommender systems that adopt a model of *strong personalization* must have access to persistent user profiles, which can be used to complement any in-session feedback to guide the recommendation process.

The key then to building a personalized recommender system relies on an ability to learn and maintain a long-term model of a user's recommendation preferences. A wide variety of approaches are available in this regard. For example, user models can represent stereotypical users [24], or they can reflect the preferences of actual users by learning from questionnaires, user ratings or usage traces (e.g., [47, 90]). Moreover, user models can be learned by soliciting long-term preference information from the user directly; for example, by asking the user to weight a variety of areas of interest. Alternatively, models can be learned in a less obtrusive manner by mining their normal online behaviour patterns (e.g., [65, 85]). Further discussion of the user modeling research literature is beyond the scope of this chapter but the interested reader is directed to Chapters 2 [35] and 4 [44] of this book for further material on this important topic.

11.5.1 Case-Based Profiling

Case-based recommendation techniques facilitate a form of *case-based* user profiling that leverages available content descriptions as an intrinsic part of the user profiles. Accordingly a user profile is made up of a set of cases plus some indication of whether the user has liked or disliked these cases. These profiles can then be used in different ways to help influence subsequent recommendations as we will see. First, it is worth highlighting how this approach to profiling stands in contrast to the type of ratings-based profiles that are exploited by collaborative filtering systems (see, for example [5, 30, 47, 74, 77, 85, 89] and also Chapter 9 [83] of this book). Ratings-based profiles are *content-free*, in the sense that they are devoid of any item content; ratings may be associated with a particular item identifier but the information about the item itself (other than the ratings information) is generally not available, at least in pure collaborative filtering systems. In contrast, case-based profiles do contain information about the items themselves because this information is stored within the item cases.

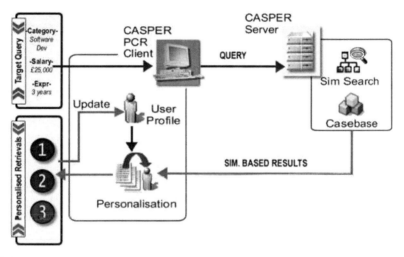

Fig. 11.15. The CASPER case-based recommender system combines case-based recommendation and user profiling in an online recruitment application.

CASPER is online recruitment system (see Figure 11.15) that uses single shot case-based recommendation to suggest jobs to users based on some initial query. It monitors the responses of users to these recommendations in order to construct case-based profiles that can be used to personalise future recommendation sessions; see [10, 74, 88]. Upon receiving a job recommendation a user has various action choices. For example, they might choose to save the advert, or email it to themselves, or submit their resume online. These actions are interpreted by CASPER's profiling mechanism as examples of positive feedback and stored in a persistent profile as rating-case pairs. Each action is translated into a rating according to how reliable an indicator of user interest it is; for example, submitting a resume as part of an online application is treated as the highest level of interest. In addition, negative ratings can be provided by rating adverts directly. Thus each user profile is made up of a set of job cases and their ratings for a given user.

CASPER's recommendation process is a two-step one. As indicated in Figure 11.15, CASPER first generates a set of recommendations based on similarity to the user query. Ordinarily these similarity-based results would be suggested to the user. However, the availability of a user profile allows CASPER to further process these recommendations in order to re-rank them according to the preferences encoded in the user's profile. To do this CASPER scores each new recommendation according to how similar it is to the cases stored in the user's profile (see Figure 11.16). This score measures the relevance of a new recommendation to the user in question. Recommendations that are similar to cases that the user has liked are preferred over recommendations that are similar to cases that the user has disliked, for example. Thus, if, over time the user has responded positively to jobs in a particular region, or with a particular salary range then, in the future, recommendations that include similar features will tend to be prioritised even if these features are absent from the current query.

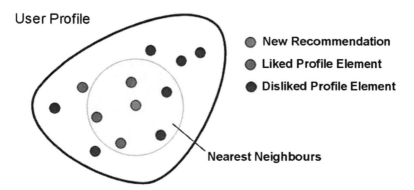

Fig. 11.16. To judge the relevance of a new recommendation, which has been suggested because of its similarity to the target query, CASPER compares the recommendation to the other cases that are similar in the user's profile and score the recommendation according to the relative distribution of positive and negative ratings within this set of similar cases.

The final ordering of recommendations can then be influenced by a combination of their similarity to the user's target query and their relevance to the user's past preferences. Live and artificial-user studies have demonstrated that this form of personalization has the potential to significantly improve the quality of CASPER's recommendations. As an aside, it is worth highlighting that in CASPER profiles are stored and updated on the client-side and the personalized re-ranking is performed as a client-side process. This enhances the system's privacy characteristics because personal preferences do not have to be revealed to the recommendation server.

A similar form of case-based profiling is used in the Personal Travel Assistant (PTA) [27, 28]. Like CASPER, the Personal Travel Assistant uses its profiles to re-order an initial set of recommendations based on the preferences that are encoded within a profile. A complimentary approach to recommendation reordering is proposed by [79, 80], but instead of relying on the target user's profile, reordering is performed with reference to a set of similar recommendation sessions that have occurred in the past with similar users.

11.5.2 Profiling Feature Preferences

The methods discussed in the previous section refer to the uses of cases themselves as part of the user profile; the user profile is made up of a set of positive (and possibly negative) cases. Collectively these rated cases capture a user's preferences. An alternative, and somewhat more direct approach to profiling in case-based recommender systems, sees the use of user profiles that attempt to capture feature level preferences. For example, the user models employed in the Adaptive Place Advisor [37, 98] capture the preferences that a user may have in relation to items, attributes and their values including the relative importance of a particular attribute, the preferred values for a particular attribute, and specific preferred items encoded as frequency distributions. For instance, for a restaurant, the user model might indicate that the type of *cuisine* has an importance weight of 0.4, compared to say the *parking* facilities, which might have a preference weight of only 0.1. Moreover a user may have a 0.35 preference weight for Italian cuisine but only a 0.1 weighting for German food. Individual item preferences are represented as the proportion of times that an item has been accepted in past recommendations. These preferences then guide the recommendation engine during item selection. Item, attribute and value preferences are all expressed as part of an item selection similarity metric so that items which have been preferred in the past, or that have important attributes and/or preferred values for these attributes, are prioritised over items that have weaker preferences.

When it comes to updating the user model, the Adaptive Place Advisor interprets various user actions as preferences for or against certain items, attributes or values. For example, when a user accepts a recommendation, this is interpreted as a preference, not just for this item, but also for its particular attributes and values. However, when a user rejects an item, this is interpreted as a rejection of the item itself, rather than a rejection of its attribute-value combinations. Early evaluations indicate that the combination of Adaptive Place Advisor's conversational recommendation technique and its personalization features has the potential to improve recommendation performance, with efficiency improvements noted as subjects learned how to interact with the system and as their user models were updated as a result of these interactions.

11.6 Conclusions

Case-based recommendation is a form of content-based recommendation that emphasises the use of structured representations and similarity-based retrieval during recommendation. In this chapter we have attempted to cover a number of important lines of research in case-based recommendation, from traditional models of single-shot, similarity-based recommendation to more sophisticated conversational recommendation models and personalization techniques. This research has helped to distinguish case-based recommendation techniques from their content-based and collaborative filtering cousins.

Of course it has only been possible to scratch the surface of this vibrant area of research and many emerging topics have been omitted. For example, there is now an appreciation that no one recommendation technology or strategy is likely to be optimal in any given recommendation scenario. As a result considerable attention has been paid to the prospect of developing hybrid recommendation strategies that combine individual approaches, such as content-based and collaborative filtering techniques; see for example Chapter 12 [20] in this book.

Another area of recent focus concerns the ability of case-based recommender systems (and indeed recommender systems in general) to *explain* the results of their reasoning. Certainly, when it comes to buying expensive goods, people expect to be skillfully steered through the options by well-informed sales assistants that are capable of balancing the user's many and varied requirements. But in addition users often need to be educated about the product space, especially if they are to come to understand what is available and why certain options are being recommended by the sales-assistant. Thus recommender systems also need to educate users about the product space: to *justify* their recommendations and *explain* the reasoning behind their suggestions; see, for example, [29, 36, 57, 59, 60, 76, 82, 95]

In summary then, case-based recommendation provides for a powerful and effective form of recommendation that is well suited to many product recommendation scenarios. As a style of recommendation, its use of case knowledge and product similarity, makes particular sense in the context of interactive recommendation scenarios where recommender system and user must collaborative in a flexible and transparent manner. Moreover, the case-based approach enjoys a level of transparency and flexibility that is not always possible with other forms of recommendation.

Acknowledgements. This work was supported by Science Foundation Ireland under Grant No. 03/IN.3/I361.

References

1. Aamodt, A., Plaza, E.: Case-Based Reasoning: Foundational Issues, Methodological Variations, and System Approaches. AI Communications **7(1)** (1994) 39–52
2. Aha, D.W.: Proceedings of the Workshop in Mixed-Initiative Case-Based Reasoning, Workshop Programme at the 6th European Conference in Case-Based Reasoning. http://home.earthlink.net/ dwaha/research/meetings/eccbr02-micbrw/ (2002)
3. Aha, D.W., Maney, T., Breslow, L.A.: Supporting Dialogue Inferencing in Conversational Case-Based Reasoning. In Smyth, B., Cunningham, P., eds.: Proceedings of the 4th European Workshop on Case-Based Reasoning, Springer-Verlag (1998) 262–273
4. Aha, D., Breslow, L.: Refining Conversational Case Libraries. In Leake, D., Plaza, E., eds.: Proceedings of the 2nd International Conference on Case-Based Reasoning, Springer-Verlag (1997) 267–278
5. Balabanovic, M., Shoham, Y.: FAB: Content-Based Collaborative Recommender. Communications of the ACM **40(3)** (1997) 66–72

6. Bergmann, R., Cunningham, P.: Acquiring customer's requirements in electronic commerce. Artificial Intelligence Review **18**(3–4) (2002) 163–193
7. Bergmann, R., Richter, M., Schmitt, S., Stahl, A., Vollrath, I.: Utility-Oriented Matching: A New Research Direction for Case-Based Reasoning. In: Proceedings of the German Workshop on Case-Based Reasoning. (2001) 264–274
8. Billsus, D., Pazzani, M.J.: Adaptive news access. In Brusilovsky, P., Kobsa, A., Nejdl, W., eds.: The Adaptive Web: Methods and Strategies of Web Personalization, Lecture Notes in Computer Science, Vol. 4321. Springer-Verlag (2007) this volume
9. Billsus, D., Pazzani, M.J., Chen, J.: A learning agent for wireless news access. In: IUI '00: Proceedings of the 5th international conference on Intelligent user interfaces, ACM Press (2000) 33–36
10. Bradley, K., Rafter, R., Smyth, B.: Case-Based User Profiling for Content Personalization. In Brusilovsky, P., Stock, O., Strapparava, C., eds.: Proceedings of the 1st International Conference on Adaptive Hypermedia and Adaptive Web-Based Systems (AH2000), Springer-Verlag (2000) 62–72
11. Bradley, K., Smyth, B.: Improving Recommendation Diversity. In O'Donoghue, D., ed.: Proceedings of the 12th National Conference in Artificial Intelligence and Cognitive Science, Maynooth, Ireland (2001) 75–84
12. Branting, K., Lester, J., Mott, B.: Dialog Management for Conversational Case-Based Reasoning. In Funk, P., Calero, P.A.G., eds.: Proceedings of the 7th European Conference on Case-Based Reasoning, Springer-Verlag (2004) 77–90
13. Branting, K.: Acquiring customer preferences from return-set selections. In Aha, D.W., Watson, I., eds.: Proceedings of the 4th International Conference on Case-Based Reasoning, Springer-Verlag (2001) 59–73
14. Bridge, D.: Diverse Product Recommendations Using an Expressive Language for Case-Retrieval. In Craw, S., Preece, A., eds.: Proceedings of the 6th European Conference on Case-Based Reasoning, Springer-Verlag (2002) 43–57
15. Bridge, D.: Towards conversational recommender systems: A dialogue grammar approach. In D.W.Aha, ed.: Proceedings of the Workshop in Mixed-Initiative Case-Based Reasoning, Workshop Programme at the 6th European Conference in Case-Based Reasoning. (2002) 9–22
16. Bridge, D., Goker, M., McGinty, L., Smyth, B.: Case-based recommender systems. Knowledge Engineering Review **20**(3) (2006) 315–320
17. Bridge, D.: Product recommendation systems: A new direction. In R.Weber, Wangenheim, C., eds.: Procs. of the Workshop Programme at the Fourth International Conference on Case-Based Reasoning. (2001) 79–86
18. Brusilovsky, P., Millan, E.: User models for adaptive hypermedia and adaptive educational systems. In Brusilovsky, P., Kobsa, A., Nejdl, W., eds.: The Adaptive Web: Methods and Strategies of Web Personalization, Lecture Notes in Computer Science, Vol. 4321. Springer-Verlag (2007) this volume
19. Burke, R.: Conceptual Indexing and Active Retrieval of Video for Interactive Learning Environments. Knowledge-Based Systems **9**(8) (1996) 491–499
20. Burke, R.: Hybrid web recommender systems. In Brusilovsky, P., Kobsa, A., Nejdl, W., eds.: The Adaptive Web: Methods and Strategies of Web Personalization, Lecture Notes in Computer Science, Vol. 4321. Springer-Verlag (2007) this volume
21. Burke, R., Hammond, K., Young, B.: Knowledge-based navigation of complex information spaces. In: Proceedings of the Thirteenth National Conference on Artificial Intelligence, AAAI Press/MIT Press (1996) 462–468 Portland, OR.
22. Burke, R., Hammond, K., Young, B.: The FindMe Approach to Assisted Browsing. Journal of IEEE Expert **12**(4) (1997) 32–40

23. Cawsey, A., Grasso, F., Paris, C.: Adaptive information for consumers of healthcare. In Brusilovsky, P., Kobsa, A., Nejdl, W., eds.: The Adaptive Web: Methods and Strategies of Web Personalization, Lecture Notes in Computer Science, Vol. 4321. Springer-Verlag (2007) this volume
24. Chin, D.N.: Acquiring user models. Artificial Intelligence Review 7(3-4) (1989) 185–197
25. Cohen, W.W., Schapire, R.E., Singer, Y.: Learning to order things. In: NIPS '97: Proceedings of the 1997 conference on Advances in neural information processing systems 10, MIT Press (1998) 451–457
26. Cotter, P., Smyth, B.: Waping the web: Content personalisation for wap-enabled devices. In: Proceedings of the International Conference on Adaptive Hypermedia and Adaptive Web-Based Systems, Trento, Italy., Springer-Verlag (2000) 98–108
27. Coyle, L., Cunningham, P.: Improving Recommendation Ranking by Learning Personal Feature Weights. In Funk, P., Calero, P.A.G., eds.: Proceedings of the 7th European Conference on Case-Based Reasoning, Springer-Verlag (2004) 560–572
28. Coyle, L., Cunningham, P., Hayes, C.: A Case-Based Personal Travel Assistant for Elaborating User Requirements and Assessing Offers. In Craw, S., Preece, A., eds.: Proceedings of the 6th European Conference on Case-Based Reasoning, Springer-Verlag (2002) 505–518
29. Cunningham, P., Doyle, D., Loughrey, J.: An Evaluation of the Usefulness of Case-Based Explanation. . In Ashley, K., Bridge, D., eds.: Case-Based Reasoning Research and Development. LNAI, Vol. 2689., Springer-Verlag (2003) 191–199
30. Dahlen, B., Konstan, J., Herlocker, J., Good, N., Borchers, A., Riedl, J.: Jump-starting movieLens: User benefits of starting a collaborative filtering system with "dead-data". In: , University of Minnesota TR 98-017 (1998)
31. de Mantaras, R.L., McSherry, D., Bridge, D., Leake, D., Smyth, B., Craw, S., Faltings, B., Maher, M.L., Cox, M.T., Forbus, K., Keane, M., Aamody, A., Watson, I.: Retrieval, reuse, revision, and retention in case-based reasoning. Knowledge Engineering Review **20**(3) (2006) 215–240
32. Doyle, M., Cunningham, P.: A dynamic approach to reducing dialog in on-line decision guides. In Blanzieri, E., Portinale, L., eds.: Proceedings of the 5th European Workshop on Case-Based Reasoning, Springer-Verlag (2000) 49–60
33. Faltings, B., Pu, P., Torrens, M., Viappiani, P.: Design Example-Critiquing Interaction. In: Proceedings of the International Conference on Intelligent User Interface(IUI-2004), ACM Press (2004) 22–29 Funchal, Madeira, Portugal.
34. Fox, S., Leake, D.B.: Using Introspective Reasoning to Refine Indexing. In: Proceedings of the 14th International Joint Conference on Artificial Intelligence, Morgan Kaufmann (1995) 391 – 397
35. Gauch, S., Speretta, M., Chandramouli, A., Micarelli, A.: User profiles for personalized information access. In Brusilovsky, P., Kobsa, A., Nejdl, W., eds.: The Adaptive Web: Methods and Strategies of Web Personalization, Lecture Notes in Computer Science, Vol. 4321. Springer-Verlag (2007) this volume
36. Gervas, P., Gupta, K.: Proceedings European Conference on Case-Based Reasoning (ECCBR-04) Explanation Workshop. (2004) Madrid, Spain.
37. Goker, M., Thompson, C.A.: Personalized Conversational Case-Based Recommendation. In Blanzieri, E., Portinale, L., eds.: Proceedings of the 5th European Workshop on Case-Based Reasoning, Springer-Verlag (2000) 99–111
38. Goy, A., Ardissono, L., Petrone, G.: Personalization in e-commerce applications. In Brusilovsky, P., Kobsa, A., Nejdl, W., eds.: The Adaptive Web: Methods and Strategies of Web Personalization, Lecture Notes in Computer Science, Vol. 4321. Springer-Verlag (2007) this volume

39. Hinkle, D., Toomey, C.: Applying case-based reasoning to manufacturing. Artificial Intelligence Magazine **16**(1) (1995) 65–73
40. Hurley, G., Wilson, D.C.: DubLet: An online CBR system for rental property accommodation. In Aha, D.W., Watson, I., eds.: Proceedings of the 4th International Conference on Case-Based Reasoning, Springer-Verlag (2001) 660–674
41. Jameson, A., Smyth, B.: Recommending to groups. In Brusilovsky, P., Kobsa, A., Nejdl, W., eds.: The Adaptive Web: Methods and Strategies of Web Personalization, Lecture Notes in Computer Science, Vol. 4321. Springer-Verlag (2007) this volume
42. Juell, P., Paulson, P.: Using reinforcement learning for similarity assessment in case-based systems. IEEE Intelligent Systems **18**(4) (2003) 60–67
43. Kim, Y., Ok, S., Woo, Y.: A Case-Based Recommender using Implicit Rating Techniques. In DeBra, P., Brusilovsky, P., Conejo, R., eds.: Proceedings of the 2nd International Conference on Adaptive Hypermedia and Adaptive Web-Based Systems (AH2002) , Springer-Verlag (2002) 62–72
44. Kobsa, A.: Generic user modeling systems. In Brusilovsky, P., Kobsa, A., Nejdl, W., eds.: The Adaptive Web: Methods and Strategies of Web Personalization, Lecture Notes in Computer Science, Vol. 4321. Springer-Verlag (2007) this volume
45. Kolodner, J.: Judging which is the "best" case for a case-based reasoner. In: Proceedings of the Second Workshop on Case-Based Reasoning, Morgan Kaufmann (1989) 77–81
46. Kolodner, J.: Case-Based Reasoning. Morgan Kaufmann (1993)
47. Konstan, J., Miller, B., Maltz, D., Herlocker, J., Gorgan, L., Riedl, J.: GroupLens: Applying collaborative filtering to Usenet news. Communications of the ACM **40(3)** (1997) 77–87
48. Leake, D.: Case-Based Reasoning: Experiences,Lessons and Future Directions. AAAI/MIT Press (1996)
49. Leake, D.B.: Constructive Similarity Assessment: Using Stored Cases to Define New Situations. In: Proceedings of the 14th Annual Conference of the Cognitive Science Society, Lawrence Earlbaum Associates (1992) 313–318
50. McCarthy, K., Reilly, J., McGinty, L., Smyth, B.: On the Dynamic Generation of Compound Critiques in Conversational Recommender Systems. In DeBra, P., ed.: Proceedings of the 3rd International Conference on Adaptive Hypermedia and Web-Based Systems (AH04), Springer-Verlag (2004) 176–184 Eindhoven, The Netherlands.
51. McCarthy, K., Reilly, J., McGinty, L., Smyth, B.: Experiments in dynamic critiquing. In: IUI '05: Proceedings of the 10th international conference on Intelligent user interfaces, ACM Press (2005) 175–182
52. McGinty, L., Smyth, B.: Comparison-Based Recommendation. In Craw, S., Preece, A., eds.: Proceedings of the 6th European Conference on Case-Based Reasoning, Springer-Verlag (2002) 575–589
53. McGinty, L., Smyth, B.: On The Role of Diversity in Conversational Recommender Systems. In Ashley, K.D., Bridge, D.G., eds.: Proceedings of the 5th International Conference on Case-Based Reasoning, Springer-Verlag (2003) 276–290
54. McGinty, L., Smyth, B.: Tweaking Critiquing. In: Proceedings of the Workshop on Personalization and Web Techniques at the International Joint Conference on Artificial Intelligence (IJCAI-03), Morgan-Kaufmann (2003) Acapulco, Mexico.
55. McSherry, D.: Increasing Recommendation Diversity Without Loss of Similarity. In: Proceedings of the Sixth UK Workshop on Case-Based Reasoning. (2001) 23–31 Cambridge, UK.
56. McSherry, D.: Diversity-Conscious Retrieval. In Craw, S., Preece, A., eds.: Proceedings of the 6th European Conference on Case-Based Reasoning, Springer-Verlag (2002) 219–233
57. McSherry, D.: Explanation in Case-Based Reasoning: An Evidential Approach. . In Lees, B., ed.: Proceedings of the 8th UK Workshop on Case-Based Reasoning. (2003) Cambridge, UK.

58. McSherry, D.: Similarity and Compromise. In Ashley, K.D., Bridge, D.G., eds.: Proceedings of the 5th International Conference on Case-Based Reasoning, Springer-Verlag (2003) 291–305

59. McSherry, D.: Explaining the Pros and Cons of Conclusions in CBR. . In Calero, P.A.G., Funk, P., eds.: Proceedings of the 6th European Conference on Case-Based Reasoning (ECCBR-04), Springer-Verlag (2004) 317–330 Madrid, Spain.

60. McSherry, D.: Incremental Relaxation of Unsuccessful Queries. . In Calero, P.A.G., Funk, P., eds.: Proceedings of the 6th European Conference on Case-Based Reasoning (ECCBR-04), Springer-Verlag (2004) 331–345 Madrid, Spain.

61. Micarelli, A., Gasparetti, F., Sciarrone, F., gauch, S.: Personalized search on the worldwide web. In Brusilovsky, P., Kobsa, A., Nejdl, W., eds.: The Adaptive Web: Methods and Strategies of Web Personalization, Lecture Notes in Computer Science, Vol. 4321. Springer-Verlag (2007) this volume

62. Mobasher, B.: Data mining for web personalization. In Brusilovsky, P., Kobsa, A., Nejdl, W., eds.: The Adaptive Web: Methods and Strategies of Web Personalization, Lecture Notes in Computer Science, Vol. 4321. Springer-Verlag (2007) this volume

63. Mooney, R.J., Roy, L.: Content-based book recommending using learning for text categorization. In: DL '00: Proceedings of the fifth ACM conference on Digital libraries, ACM Press (2000) 195–204

64. Munoz-Avila, H., Aha, D., Breslow, L.: Integrating Conversational Case Retrieval with Generative Planning. In Blanzieri, E., Portinale, L., eds.: Proceedings of the 5th European Workshop on Case-based Reasoning, Springer-Verlag (2000) 210–221

65. Nichols, D.: Implicit rating and filtering. In: Proceedings of 5th DELOS Workshop on Filtering and Collaborative Filtering. (November 1997)

66. O'Sullivan, D., Smyth, B., Wilson, D.: In-Depth Analysis of Similarity Knowledge and Metric Contributions to Recommender Performance. In: Proceedings of the 17th International FLAIRS Conference, May 17-19, Miami Beach, Florida, FLAIRS 2004 CD, AAAI Press. (2004)

67. O'Sullivan, D., Wilson, D., Smyth, B.: Improving Case-Based Recommendation: A Collaborative Filtering Approach. In Craw, S., Preece, A.D., eds.: 6th European Conference on Case-Based Reasoning. Volume 2416., Springer-Verlag (2002) 278 – 291

68. O'Sullivan, D., Wilson, D., Smyth, B.: Preserving Recommender Accuracy and Diversity in Sparse Datasets. In Russell, I., Haller, S., eds.: Proceedings of the 16th International FLAIRS Conference, AAAI Press (2003) 139 – 144

69. Pazzani, M.J., Billsus, D.: Content-based recommendation systems. In Brusilovsky, P., Kobsa, A., Nejdl, W., eds.: The Adaptive Web: Methods and Strategies of Web Personalization, Lecture Notes in Computer Science, Vol. 4321. Springer-Verlag (2007) this volume

70. Pu, P., Faltings, B.: Decision Tradeoff Using Example Critiquing and Constraint Programming. Special Issue on User-Interaction in Constraint Satisfaction. CONSTRAINTS: an International Journal. 9(4) (2004)

71. Quinlan, J.R.: Induction of decision trees. Machine Learning 1 (1986) 81–106

72. Quinlan, J.R.: Learning decision tree classifiers. ACM Comput. Surv. 28(1) (1996) 71–72

73. Quinlan, J.R.: C4.5: programs for machine learning. Morgan Kaufmann Publishers Inc. (1993)

74. Rafter, R., Bradley, K., Smyth, B.: Automatic Collaborative Filtering Applications for Online Recruitment Services. In P. Brusilovsky, O.S., Strapparava, C., eds.: Proceedings of the International Conference on Adaptive Hypermedia and Adaptive Web-based Systems, Springer-Verlag (2000) 363–368

75. Reilly, J., McCarthy, K., McGinty, L., Smyth, B.: Dynamic Critiquing. In Funk, P., Calero, P.A.G., eds.: Proceedings of the 7th European Conference on Case-Based Reasoning, Springer-Verlag (2004) 763–777

76. Reilly, J., McCarthy, K., McGinty, L., Smyth, B.: Explaining compound critiques. Artificial Intelligence Review (In Press)

77. Resnick, P., Iacovou, N., Suchak, M., Bergstrom, P., Riedl, J.: GroupLens: An open architecture for collaborative filtering of netnews. In: Proceedings of the 1994 Conference on Computer Supported Collaborative Work. (1994) 175–186

78. Resnick, P., Varian, H.R.: Recommender systems. CACM **40**(3) (1997) 56–58

79. Ricci, F., Venturini, A., Cavada, D., Mirzadeh, N., Blaas, D., Nones, M.: Product Recommendation with Interactive Query Management and Twofold Similarity. In Ashley, K.D., Bridge, D.G., eds.: Proceedings of the 5th International Conference on Case-Based Reasoning, Springer-Verlag (2003) 479–493

80. Ricci, F., Arslan, B., Mirzadeh, N., Venturini, A.: Itr: A case-based travel advisory system. In Craw, S., Preece, A., eds.: Proceedings of the 6th European Conference on Case-Based Reasoning, Springer-Verlag (2002) 613–627

81. Ricci, F., Avesani, P.: Learning a local similarity metric for case-based reasoning. In: ICCBR '95: Proceedings of the First International Conference on Case-Based Reasoning Research and Development, Springer-Verlag (1995) 301–312

82. Roth-Berghofer, T.R.: Explanations and Case-Based Reasoning: Foundational Issues. In Funk, P., Calero, P.A.G., eds.: Proceedings of the 7th European Conference on Case-Based Reasoning, Springer-Verlag (2004) 389–403

83. Schafer, J.B., Frankowski, D., Herlocker, J., Sen, S.: Collaborative filtering recommender systems. In Brusilovsky, P., Kobsa, A., Nejdl, W., eds.: The Adaptive Web: Methods and Strategies of Web Personalization, Lecture Notes in Computer Science, Vol. 4321. Springer-Verlag (2007) this volume

84. Schmitt, S.: *simVar*; a similarity-influenced question selection criterion for e-sales dialogs. Artificial Intelligence Review **18**(3–4) (2002) 195–221

85. Shardanand, U., Maes, P.: Social Information Filtering: Algorithms for Automating "Word of Mouth". In: Proceedings of the Denver ACM CHI 1995. (1995) 210–217

86. Shimazu, H.: ExpertClerk : Navigating Shoppers' Buying Process with the Combination of Asking and Proposing. In Nebel, B., ed.: Proceedings of the Seventeenth International Joint Conference on Artificial Intelligence (IJCAI-01), Morgan Kaufmann (2001) 1443–1448 Seattle, Washington, USA.

87. Shimazu, H., Shibata, A., Nihei, K.: ExpertGuide: A conversational case-based reasoning tool for developing mentors in knowledge spaces. Applied Intelligence **14**(1) (2002) 33–48

88. Smyth, B., Bradley, K., Rafter, R.: Personalization techniques for online recruitment services. Commun. ACM **45**(5) (2002) 39–40

89. Smyth, B., Cotter, P.: Surfing the Digital Wave: Generating Personalized Television Guides Using Collaborative, Case-based Recommendation. In: Proceedings of the Third International Conference on Case-based Reasoning. (1999)

90. Smyth, B., Cotter, P.: A Personalized TV Listings Service for the Digital TV Age. Journal of Knowledge-Based Systems **13**(2-3) (2000) 53–59

91. Smyth, B., Keane, M.: Adaptation-Guided Retrieval: Questioning the Similarity Assumption in Reasoning. Artificial Intelligence **102** (1998) 249–293

92. Smyth, B., McClave, P.: Similarity v's Diversity. In Aha, D., Watson, I., eds.: Proceedings of the 3rd International Conference on Case-Based Reasoning, Springer-Verlag (2001) 347–361

93. Smyth, B., McGinty, L.: The Power of Suggestion. In: Proceedings of the International Joint Conference on Artificial Intelligence (IJCAI-03), Morgan-Kaufmann (2003) 127–132 Acapulco, Mexico.

94. Smyth, B., Cotter, P.: A personalized television listings service. Communications of the ACM **43**(8) (2000) 107–111

95. Sørmo, F., Cassens, J.: Explanation goals in case-based reasoning. In Gervás, P., Gupta, K.M., eds.: Proceedings of the ECCBR 2004 Workshops, Departamento de Sistemas Informáticos y Programación, Universidad Complutense Madrid (2004) 165–174
96. Stahl, A.: Learning feature weights from case order feedback. In Aha, D.W., Watson, I., eds.: Proceedings of the 4th International Conference on Case-Based Reasoning. (2001) 502–516
97. Tecuci, G., Aha, D.W., Boicu, M., Cox, M., Ferguson, G., Tate, A.: Proceedings of the Workshop on Mixed-Initiative Intelligent Systems, Workshop Programme at the 18th International Joint Conference on Artificial Intelligence. http://lalab.gmu.edu/miis/ (2003)
98. Thompson, C.A., Goker, M.H., Langley, P.: A Personalized System for Conversational recommendation. Journal of Artificial Intelligence Research **21** (2004) 1–36
99. Veloso, M., Munoz-Avila, H., Bergmann, R.: Case-Based Planning: Methods and Systems. AI Communications **9(3)** (1996) 128–137
100. Vollrath, I., Wilke, W., Bergmann, R.: Case-based reasoning support for online catalog sales. IEEE Internet Computing **2**(4) (1998) 45–54
101. Watson, I., ed.: Applying Case-Based Reasoning: Techniques for Enterprise Systems. Morgan Kaufmann (1997)
102. Wettschereck, D., Aha, D.W.: Weighting features. In Veloso, M.M., Aamodt, A., eds.: Proceedings of the 1st International Conference on Case-Based Reasoning, Springer-Verlag (1995) 347–358
103. Wilke, W., Lenz, M., Wess, S.: Intelligent sales support with CBR. In et al., M.L., ed.: Case-Based Reasoning Technology: From Foundations to Applications. Springer-Verlag (1998) 91–113

12

Hybrid Web Recommender Systems

Robin Burke

School of Computer Science, Telecommunications and Information Systems
DePaul University, 243 S. Wabash Ave.
Chicago, Illinois, USA
rburke@cs.depaul.edu

Abstract. Adaptive web sites may offer automated recommendations generated through any number of well-studied techniques including collaborative, content-based and knowledge-based recommendation. Each of these techniques has its own strengths and weaknesses. In search of better performance, researchers have combined recommendation techniques to build hybrid recommender systems. This chapter surveys the space of two-part hybrid recommender systems, comparing four different recommendation techniques and seven different hybridization strategies. Implementations of 41 hybrids including some novel combinations are examined and compared. The study finds that cascade and augmented hybrids work well, especially when combining two components of differing strengths.

12.1 Introduction

Recommender systems are personalized information agents that provide recommendations: suggestions for items likely to be of use to a user [18, 41, 42]. In an e-commerce context, these might be items to purchase; in a digital library context, they might be texts or other media relevant to the user's interests.[1] A recommender system can be distinguished from an information retrieval system by the semantics of its user interaction. A result from a recommender system is understood as a recommendation, an option worthy of consideration; a result from an information retrieval system is interpreted as a match to the user's query. Recommender systems are also distinguished in terms of personalization and agency. A recommender system customizes its responses to a particular user. Rather than simply responding to queries, a recommender system is intended to serve as an information agent.[2]

---

[1] In this chapter, I use the e-commerce term "products" to refer to the items being recommended, with the understanding that other information-seeking contexts are also pertinent.

[2] Techniques such as relevance feedback enable an information retrieval engine to refine its representation of the user's query, and therefore can be seen as a simple form of recommendation. The search engine Google (http://www.google.com) blurs this distinction further, using "authoritativeness" criteria in addition to strict matching [6].

P. Brusilovsky, A. Kobsa, and W. Nejdl (Eds.): The Adaptive Web, LNCS 4321, pp. 377–408, 2007.

A variety of techniques have been proposed as the basis for recommender systems: collaborative, content-based, knowledge-based, and demographic techniques are surveyed below. Each of these techniques has known shortcomings, such as the well-known cold-start problem for collaborative and content-based systems (what to do with new users with few ratings) and the knowledge engineering bottleneck in knowledge-based approaches. A hybrid recommender system is one that combines multiple techniques together to achieve some synergy between them. For example, a collaborative system and a knowledge-based system might be combined so that the knowledge-based component can compensate for the cold-start problem, providing recommendations to new users whose profiles are too small to give the collaborative technique any traction, and the collaborative component can work its statistical magic by finding peer users who share unexpected niches in the preference space that no knowledge engineer could have predicted. This chapter examines the landscape of possible recommender system hybrids, investigating a range of possible hybridization methods, and demonstrating quantitative results by which they can be compared.

Recommendation techniques can be distinguished on the basis of their knowledge sources: where does the knowledge needed to make recommendations come from? In some systems, this knowledge is the knowledge of other users' preferences. In others, it is ontological or inferential knowledge about the domain, added by a human knowledge engineer.

Previous work [10] distinguished four different classes of recommendation techniques based on knowledge source[3], as shown in Figure 12.1:

- Collaborative: The system generates recommendations using only information about rating profiles for different users. Collaborative systems locate peer users with a rating history similar to the current user and generate recommendations using this neighborhood. Examples include [17, 21, 41, 46].
- Content-based: The system generates recommendations from two sources: the features associated with products and the ratings that a user has given them. Content-based recommenders treat recommendation as a user-specific classification problem and learn a classifier for the user's likes and dislikes based on product features [14, 22, 25, 38].
- Demographic: A demographic recommender provides recommendations based on a demographic profile of the user. Recommended products can be produced for different demographic niches, by combining the ratings of users in those niches [24, 36]
- Knowledge-based: A knowledge-based recommender suggests products based on inferences about a user's needs and preferences. This knowledge will sometimes contain explicit functional knowledge about how certain product features meet user needs. [8, 9, 44].

Each of these recommendation techniques has been the subject of active exploration since the mid-1990's, when the first recommender systems were pioneered, and their capabilities and limitations are fairly well known.

---

[3] It should be noted that there is another knowledge source: context, which has not yet become widely used in web-based recommendation, but promises to become important particularly for mobile applications. See, for example, [7].

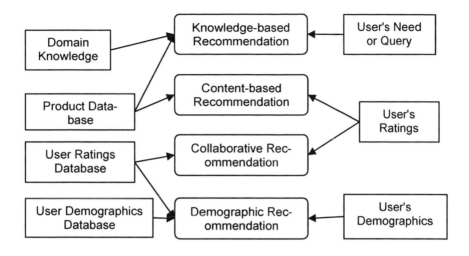

Fig. 12.1. Recommendation techniques and their knowledge sources

All of the learning-based techniques (collaborative, content-based and demographic) suffer from the *cold-start* problem in one form or another. This is the well-known problem of handling new items or new users. In a collaborative system, for example, new items cannot be recommended to any user until they have been rated by some one.. Recommendations for items that are new to the catalog are therefore considerably weaker than more widely rated products, and there is a similar failing for users who are new to the system.

The converse of this problem is the stability vs. plasticity problem. Once a user's profile has been established in the system, it is difficult to change one's preferences. A steak-eater who becomes a vegetarian will continue to get steakhouse recommendations from a content-based or collaborative recommender for some time, until newer ratings have the chance to tip the scales. Many adaptive systems include some sort of temporal discount to cause older ratings to have less influence [4, 45], but they do so at the risk of losing information about interests that are long-term but sporadically exercised. For example, a user might like to read about major earthquakes when they happen, but such occurrences are sufficiently rare that the ratings associated with last year's earthquake might no longer be considered by the time the next big one hits. Knowledge-based recommenders respond to the user's immediate need and do not need any kind of retraining when preferences change.

Researchers have found that collaborative and demographic techniques have the unique capacity to identify cross-genre niches and can entice users to jump outside of the familiar. Knowledge-based techniques can do the same but only if such associations have been identified ahead of time by the knowledge engineer. However, the cold-start problem has the side-effect of excluding casual users from receiving the full benefits of collaborative and content-based recommendation. It is possible to do simple market-basket recommendation with minimal user input: Amazon.com's "people who bought X also bought Y" but this mechanism has few of the advantages commonly associated with the collaborative filtering concept. The learning-based technologies work best for dedicated users who are willing to invest some time making their preferences known to the system. Knowledge-based systems have fewer problems in this regard because they do not rely on having historical data about a user's preferences.

Hybrid recommender systems are those that combine two or more of the techniques described above to improve recommendation performance, usually to deal with the cold-start problem.[4] This chapter will examine seven different hybridization techniques in detail and evaluate their performance. From a large body of successful research in the area, we know that hybrid recommenders can be quite successful. The question of interest is to understand what types of hybrids are likely to be successful in general or failing such a general result, to determine under what domain and data characteristics we might expect different hybrids to work well. While this chapter does by necessity fall short of providing a definitive answer to such questions, the experiments described below do point the way towards answering this important question for recommender system design.

12.2 Strategies for Hybrid Recommendation

The term *hybrid recommender system* is used here to describe any recommender system that combines multiple recommendation techniques together to produce its output. There is no reason why several different techniques of the same type could not be hybridized, for example, two different content-based recommenders could work together, and a number of projects have investigated this type of hybrid: NewsDude, which uses both naive Bayes and kNN classifiers in its news recommendations is just one example [4]. However, we are particularly focused on recommenders that combine information across different sources, since these are the most commonly implemented ones and those that hold the most promise for resolving the cold-start problem.

The earlier survey of hybrids [10] identified seven different types:
- Weighted: The score of different recommendation components are combined numerically.
- Switching: The system chooses among recommendation components and applies the selected one.
- Mixed: Recommendations from different recommenders are presented together.
- Feature Combination: Features derived from different knowledge sources are combined together and given to a single recommendation algorithm.
- Feature Augmentation: One recommendation technique is used to compute a feature or set of features, which is then part of the input to the next technique.
- Cascade: Recommenders are given strict priority, with the lower priority ones breaking ties in the scoring of the higher ones.
- Meta-level: One recommendation technique is applied and produces some sort of model, which is then the input used by the next technique.

---

[4] Some hybrids combine different implementations of the same class of technique – for example, switching between two different content-based recommenders. The present study only examines hybrids that combine different types of recommenders.

Table 12.1. The space of possible hybrid recommender systems (adapted from [10])

	Weight.	Mixed	Switch.	FC	Cascade	FA	Meta
CF/CN	▨	▨	▨	▨	▨	▨	
CF/DM	▨						■
CF/KB	▨		▨	■			
CN/CF	▨	▨	▨	▨			▨
CN/DM	░			■			■
CN/KB				■			
DM/CF	▨	▨	▨	▨			■
DM/CN	▨	▨	▨	▨			
DM/KB				■			
KB/CF	▨	▨	▨	▨	▨		
KB/CN	▨	▨	▨	▨			
KB/DM	▨	▨	▨	▨			■

FC = Feature Combination, FA = Feature Augmentation
CF = collaborative, CN = content-based, DM = demographic, KB = knowledge-based

▨	Redundant
■	Not possible
░	Existing implementation

The previous study showed that the combination of the five recommendation approaches and the seven hybridization techniques yields 53 possible two-part hybrids, as shown in Table 12.1. This number is greater than 5x7=35 because some of the techniques are order-sensitive. For example, a content-based/collaborative feature augmentation hybrid is different from one that applies the collaborative part first and uses its features in a content-based recommender. The complexity of the taxonomy is increased by the fact that some hybrids are not logically distinguishable from others and other combinations are infeasible. See [10] for details.

The remainder of this section will consider each of the hybrid types in detail before we turn our attention to the question of comparative evaluation.

12.2.1 Weighted

The movie recommender system in [32] has two components: one, using collaborative techniques, identifies similarities between rating profiles and makes predictions based on this information. The second component uses simple semantic knowledge about the features of movies, compressed dimensionally via latent semantic analysis, and recommends movies that are semantically similar to those the user likes. The output of the two components is combined using a linear weighting scheme.

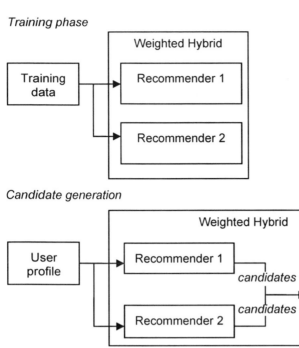

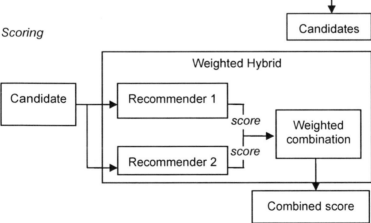

Fig. 12.2. Weighted hybrid

Perhaps the simplest design for a hybrid system is a weighted one. Each component of the hybrid scores a given item and the scores are combined using a linear formula. Examples of weighted hybrid recommenders include [15] as well as the example above. This type of hybrid combines evidence from both recommenders in a static manner, and would therefore seem to be appropriate when the component recommenders have consistent relative power or accuracy across the product space..

We can think of a weighted algorithm as operating in the manner shown in Figure 12.2. There is a training phase in which each individual recommender processes the training data. (This phase is the same in most hybrid scenarios and will be omitted in subsequent diagrams.) Then when a prediction is being generated for a test user, the recommenders jointly propose candidates. Some recommendation techniques, such as content-based classification algorithms, are able to make predictions on any item, but others are limited. For example, a collaborative recommender cannot make predictions about the ratings of a product if there are no peer users who have rated it. Candidate generation is necessary to identify those items that will be considered.

The sets of candidates must then be rated jointly. Hybrids differ in how candidate sets are handled. Typically, either the intersection or the union of the sets is used. If an intersection is performed, there is the possibility that only a small number of candidates will be shared between the candidate sets. When union is performed, the system must decide how to handle cases in which it is not possible for a recommender to rate a given candidate. One possibility is to give such a candidate a neutral (neither liked nor disliked) score. Each candidate is then rated by the two recommendation components and a linear combination of the two scores computed, which becomes the item's predicted rating. Candidates are then sorted by the combined score and the top items shown to the user.

Usually empirical means are used to determine the best weights for each component. For example, Mobasher and his colleagues found that weighting 60/40 semantic/collaborative produced the greatest accuracy in their system [32]. Note that there is an implicit assumption that each recommendation component will have uniform performance across the product and user space. Each component makes a fixed contribution to the score, but it is possible that recommenders will have different strengths in different parts of the product space. This suggests the application of the next type of hybrid, one in which the hybrid switches between its components depending on the context.

12.2.2 Mixed

> *PTV recommends television shows [48]. It has both content-based and collaborative components, but because of the sparsity of the ratings and the content space, it is difficult to get both recommenders to produce a rating for any given show. Instead the components each produce their own set of recommendations that are combined before being shown to the user.*

A mixed hybrid presents recommendations of its different components side-by-side in a combined list. There is no attempt to combine evidence between recommenders. The challenge in this type of recommender is one of presentation: if lists are to be combined, how are rankings to be integrated? Typical techniques include merging based on predicted rating or on recommender confidence. Figure 12.3 shows the mixed hybrid design.

It is difficult to evaluate a mixed recommender using retrospective data. With other types of hybrids, we can use user's actual ratings to determine if the right items are being ranked highly. With a mixed strategy, especially one that presents results side-by-side, it is difficult to say how the hybrid improves over its constituent components

Candidate generation

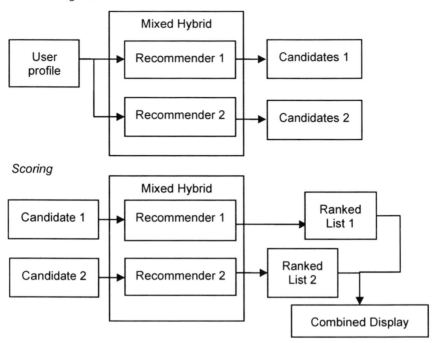

Fig. 12.3. Mixed hybrid (Training phase omitted, same as Weighted Hybrid)

without doing an on-line user study, as was performed for PTV. The mixed hybrid is therefore omitted from the experiments described below, which use exclusively retrospective data.

12.2.3 Switching

NewsDude [4] recommends news stories. It has three recommendation components: a content-based nearest-neighbor recommender, a collaborative recommender and a second content-based algorithm using a naive Bayes classifier. The recommenders are ordered. The nearest neighbor technique is used first. If it cannot produce a recommendation with high confidence, then the collaborative recommender is tried, and so on, with the naive Bayes recommender at the end of line.

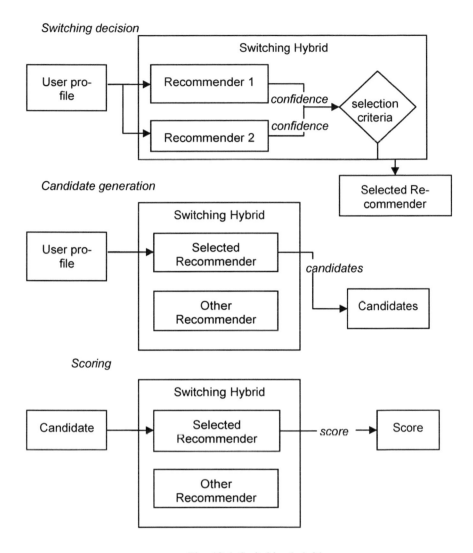

Fig. 12.4. Switching hybrid

A switching hybrid is one that selects a single recommender from among its constituents based on the recommendation situation. For a different profile, a different recommender might be chosen. This approach takes into account the problem that components may not have consistent performance for all types of users. However, it assumes that some reliable criterion is available on which to base the switching decision. The choice of this switching criterion is important. Some researchers have used confidence values inherent in the recommendation components themselves as was the case with NewsDude and van Setten's Duine system [49]; others have used external criteria [33]. The question of how to determine an appropriate confidence value for a recommendation is an area of active research. See [13] for recent work on the assessing the confidence of a case-based recommender system.

As shown in Figure 12.4 the switching hybrid begins the recommendation process by selecting one of its components as appropriate in the current situation, based on its switching criteria. Once that choice is made, the component that is not chosen has no role in the remaining recommendation process.

A switching recommender requires a reliable switching criteria, either a measure of the algorithm's individual confidence levels (that can be compared) or some alternative measure and the criterion must be well-tuned to the strengths of the individual components.

12.2.4 Feature Combination

> *Basu, Hirsh and Cohen [3] used the inductive rule learner Ripper [16] to learn content-based rules about user's likes and dislikes. They were able to improve the system's performance by adding collaborative features, thereby treating a fact like "User1 and User2 liked Movie X" in the same way that the algorithm treated features like "Actor1 and Actor2 starred in Movie X".*

The idea of feature combination is to inject features of one source (such as collaborative recommendation) into an algorithm designed to process data with a different source (such a content-based recommendation). This idea is shown schematically in Figure 12.5. Here we see that in addition to a component that actually makes the recommendation, there is also a virtual "contributing recommender". The features which would ordinarily be processed by this recommender are instead used as part of the input to the actual recommender. This is a way to expand the capabilities of a well-understood and well-tuned system, by adding new kinds of features into the mix [3, 34].

The feature combination hybrid is not a hybrid in the sense that we have seen before, that of combining components, because there is only one recommendation component. What makes it a hybrid is the knowledge sources involved: a feature combination hybrid borrows the recommendation logic from another technique rather employing a separate component that implements it. In the example above from Basu, Hirsh and Cohen, the content-based recommender works in the typical way by building a learned model for each user, but user rating data is combined with the product features. The system has only one recommendation component and it works in a content-based way, but the content draws from a knowledge source associated with collaborative recommendation.

12.2.5 Feature Augmentation

> *Melville, Mooney and Nagarajan [30] coin the term "content-boosted collaborative filtering." This algorithm learns a content-based model over the training data and then uses this model to generate ratings for unrated items. This makes for a set of profiles that is denser and more useful to the collaborative stage of recommendation that does the actual recommending.*

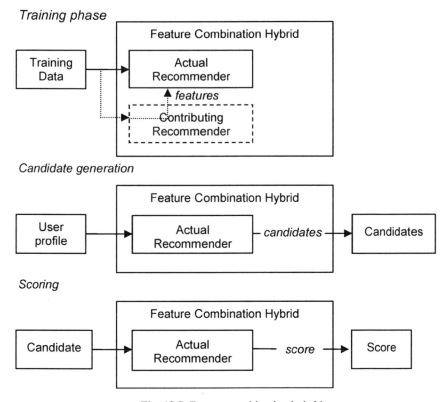

Fig. 12.5. Feature combination hybrid

Feature augmentation is a strategy for hybrid recommendation that is similar in some ways to feature combination. Instead of using features drawn from the contributing recommender's domain, a feature augmentation hybrid generates a new feature for each item by using the recommendation logic of the contributing domain. In case-based recommendation, for example, Smyth and his colleagues [35, 50] use association rule mining over the collaborative data to derive new content features for content-based recommendation.

This difference can be seen in the schematic diagram (Figure 12.6). At each step, the contributing recommender intercepts the data headed for the actual recommender and augments it with its own contribution, not raw features as in the case of feature combination, but the result of some computation. A feature augmentation recommender would be employed when there is a well-developed strong primary recommendation component, and a desire to add additional knowledge sources. As a practical matter, the augmentation can usually be done off-line, making this approach attractive, as in the case of feature combination, when trying to strengthen an existing recommendation algorithm by adjusting its input.

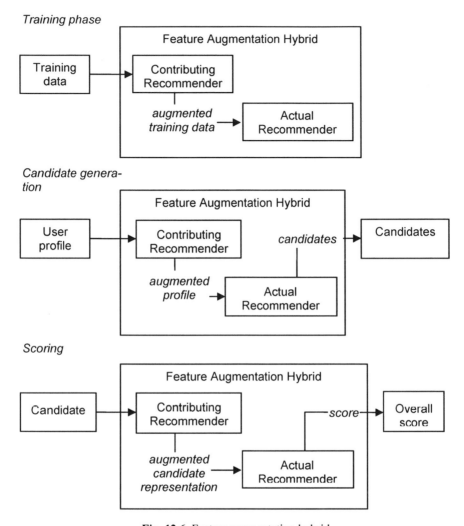

Fig. 12.6. Feature augmentation hybrid

There are a number of reasons why a feature augmentation hybrid might be preferred to a feature combination one. It is not always easy or even possible to create a feature combination hybrid for all possible hybrid combinations: the feature augmentation approach is more flexible. Also, the primary recommender in a feature combination hybrid must confront the added dimensionality of the larger training data, particularly in the case of collaborative ratings data. An augmentation hybrid adds a smaller number of features to the primary recommender's input.

Still, it is not always immediately obvious how to create a feature augmentation recommender for any two recommendation components. A recommendation component, after all, is intended to produce a predicted score or a ranking of items, not a feature for consumption by another process. What is required, as in feature combination, is attention to the knowledge sources and recommendation logic. In the example above, a content-based recommender uses the features of the items in a profile to induce a classifier that fits a particular user. The classifier can then be used to rate additional items on the user's behalf, making for a denser and more fleshed-out set of ratings, which then become input for a collaborative algorithm – we can more precisely describe this as a content-based / collaborative feature augmentation hybrid.

12.2.6 Cascade

The knowledge-based Entree restaurant recommender [10] was found to return too many equally-scored items, which could not be ranked relative to each other. Rather than additional labor-intensive knowledge engineering (to produce finer discriminations), the hybrid EntreeC was created by adding a collaborative re-ranking of only those items with equal scores.

The idea of a cascade hybrid is to create a strictly hierarchical hybrid, one in which a weak recommender cannot overturn decisions made by a stronger one, but can merely refine them. In its order-dependence, it is similar to the feature augmentation hybrid, but it is an approach that retains the function of the recommendation component as providing predicted ratings. A cascade recommender uses a secondary recommender only to break ties in the scoring of the primary one. Figure 12.7 shows a schematic depiction of this style of hybrid.

Many recommendation techniques have real-valued outputs and so the probability of actual numeric ties is small. This would give the secondary recommender in a cascade little to do. In fact, the literature did not reveal any other instances of the cascade type at the time that the original hybrid recommendation survey was completed in 2002. In the case of EntreeC, the knowledge-based / collaborative cascade hybrid described above, the knowledge-based component was already producing an integer-valued score, and ties were observed in every retrieval set, so the cascade design was a natural one.

The cascade hybrid raises the question of the uncertainty that should be associated with the real-valued outputs of a recommendation algorithm. It is certainly not the case that our confidence in the algorithms should extend to the full 32 bit precision of double floating point values. And, if the scoring of our algorithms is somewhat less precise, then there may be ties in ranks to which the cascade design can be applied. As we shall see below, recommenders operating at reduced numeric precision do not suffer greatly in accuracy and so the cascade hybrid is a reasonable option. McSherry [29] uses a similar idea in creating regions of similarity in which scores vary no more than a given ε to satisfy the goal of increasing recommendation diversity.

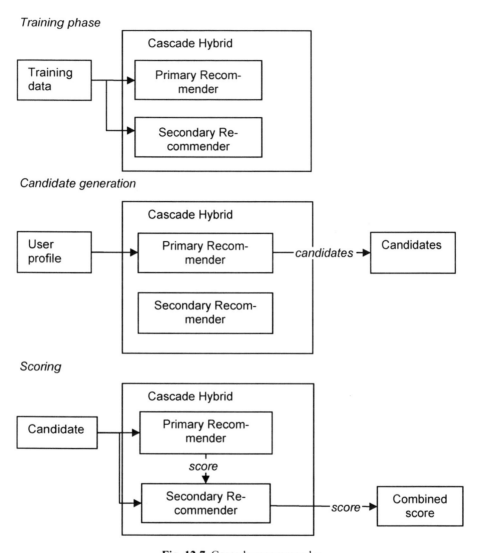

Training phase

Candidate generation

Scoring

Fig. 12.7. Cascade recommender

12.2.7   Meta-level

Pazzani [36] used the term "collaboration through content" to refer to his restaurant recommender that used the naive Bayes technique to build models of user preferences in a content-based way. With each user so represented, a collaborative step was then be performed in which the vectors were compared and peer users identified.

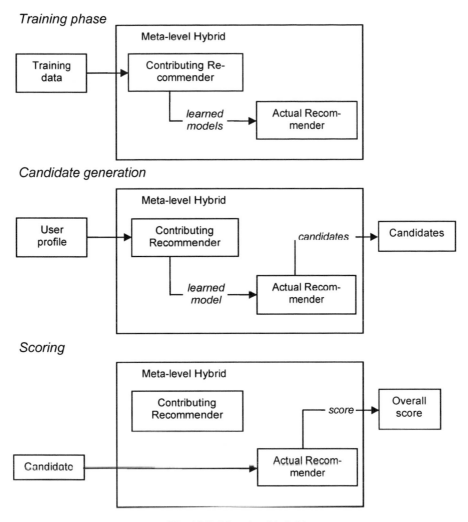

Fig. 12.8. Meta-level hybrid

A meta-level hybrid is one that uses a model learned by one recommender as input for another. Another classic example is Fab [1], a document recommender that used the same "collaboration through content" structure. Figure 12.8 shows the general schematic for this type of recommender. Note that this type is similar to the feature augmentation hybrid in that the contributing recommender is providing input to the actual recommender, but the difference is that in a meta-level hybrid, the contributing recommender completely replaces the original knowledge source with a learned model that the actual recommender uses in its computation. The actual recommender does not work with any raw profile data. We can think of this as a kind of "change of basis" in the recommendation space.

It is not always straightforward (or necessarily feasible) to derive a meta-level hybrid from any given pair of recommenders. The contributing recommender has to produce some kind of model that can be used as input by the actual recommender and not all recommendation logics can do so.

12.3   Comparing Hybrids

There have been a few studies that compared different hybrids using the same data. Pazzani's study is notable for comparing both a meta-level and a weighted scheme for hybrid recommenders using content, collaborative and demographic data. He found a significant improvement in precision for both hybrid techniques. Good and colleagues [18] examined an assortment of hybrids involving collaborative, content-based and very simple knowledge-based techniques in the movie recommendation domain. The study did find that a hybridized recommender system was better than any single algorithm and that multi-part hybrids could be successful.

To compare the full scope of the hybrid design space from Table 12.1 would require recommendation components of each of the four types: collaborative, content-based, knowledge-based and demographic. Given appropriate rating and product data, collaborative and content-based components can easily be constructed and most studies of hybrid recommendation have looked at just these components. Constructing a demographic recommendation component is more difficult as it requires access to users' personal demographic data, which is not found in the commonly-used ratings data sets used for evaluating recommender systems, such as MovieLens[5]. Constructing a knowledge-based recommendation component is a matter of knowledge engineering, and while there are a number of extant examples, there is only one that is associated with publicly-available user profile data, namely the Entree restaurant recommender system [8, 9].[6]

The benefit of using the Entree data is that it allows us to examine some of the particularly under-explored portions of the hybrid design space – those with knowledge-based components. The tradeoff is that this data set has some peculiarities (discussed in detail below), which may limit the applicability of the results. However, the experiments do allow us to examine some of the interactions between recommendation approaches and hybridization techniques, and hopefully to provide some guidance to researchers and implementers seeking to build hybrid systems.

12.3.1   The Entree Restaurant Recommender

To understand the evaluation methodology employed in this study and the operation of the knowledge-based recommendation component, we will need to examine the characteristics of the Entree restaurant recommender and the Entree data set. Entree is a restaurant recommendation system that uses case-based reasoning [23] techniques to select and rank restaurants. It operated as a web utility for approximately three years starting in 1996. The system is interactive, using a critiquing dialog [11, 47] in which

---

[5]   The MovieLens data sets are at http://www.cs.umn.edu/research/GroupLens/index.html.
[6]   The Entree data set is available from the UC Irvine KDD archive at
      http://kdd.ics.uci.edu/databases/entree/entree.html

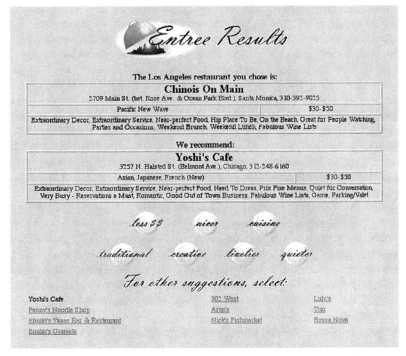

Fig. 12.9. Results of a query to the Entree restaurant recommender

users' preferences are elicited through their reactions to examples that they are shown. Recent user studies [39] have shown this technique to be an effective one for product catalog navigation, and the refinement of this model is an area of active research. See, for example, [28, 40].

Consider a user who starts browsing by entering a query in the form of a known restaurant, Wolfgang Puck's "Chinois on Main" in Los Angeles. As shown in Figure 12.9, the system finds a similar Chicago restaurant that combines Asian and French influences, "Yoshi's Cafe," as well as other similar restaurants that are ranked by their similarity. Note that the connection between "Pacific New Wave" cuisine and its Asian and French culinary components is part of the system's knowledge base of cuisines. The user might however be interested in a cheaper meal, selecting the "Less $$" button. The result would be a creative Asian restaurant in a cheaper price bracket, if one could be found. Note that the critiques are not "narrowing" the search in the sense of adding constraints, but rather changing the focus to a different point in the feature space. The user can continue browsing and critiquing until an acceptable restaurant has been located.

12.3.2 The Entree Data Set

Each user session in the Entree data set therefore consists of an entry point, which may be a restaurant or a query, a series of critiques, and finally an end point. For example, the session that began in Figure 12.9 might consist of three actions

(<"Chinois on Main", entry>, <"Yoshi's", too expensive>, <"Lulu's", end>). To turn this action sequence into a rating profile, we make the simplifying assumption that the entry and ending points are "positive" ratings and the critiques are "negative" ones. (Earlier research showed that a more nuanced interpretation of the critiques was not helpful [10].) If we look at a session consisting of ten interactions, we would have eight or nine negative ratings and one or two positive ratings. This is quite different than the typical recommender system that has a more even mix of ratings and usually more positive than negative ratings [45]. The Entree data set is also much smaller than some other data sets used for collaborative filtering research, containing about 50,000 sessions/users and a total of just under 280,000 ratings. The small number of ratings per user (average 5.6) means that collaborative and especially content-based algorithms cannot achieve the same level of performance as is possible when there is more training data.

Another way to look at the data set however is that it foregrounds the most vexing problems for recommender systems, the twin "cold start" problems of new users (short profiles) and new items (sparse ratings). Since the major motivation for using recommendation hybrids is to improve performance in these cold start cases, the Entree data set is a good trial for the effectiveness of hybrids in precisely these conditions. It is also the case that users are often reluctant to allow lengthy personal profiles to be maintained by e-commerce sites, so good performance with single session profiles is important.

The assumption that the end point is a positive rating is a rather strong assumption. Effectively, this assumption amounts to the proposition that most users are satisfied with the recommendations that they receive. It is of course possible that users are abandoning their searches in frustration. To examine the validity of this assumption, we experimented with a subset of the data that contains entry point ratings. Entry points can be confidently labeled as implicit positive ratings – users would not ask for restaurants similar to those they did not like. Experiments found extremely strong correlation (0.92) between the two conditions, demonstrating that the behavior of the algorithms does not differ markedly when exit points are treated as positive ratings. Therefore, in the experiments below, we will use the full data set and assume that both entry and exit points are positive ratings, with the understanding that there is some noise associated with this assumption.

12.3.3 Evaluation

[20] is a recent survey that compares a variety of evaluation techniques for collaborative filtering systems, and although this article looks at a larger class of recommendation systems, these results are still informative. Herlocker and colleagues identify three basic classes of evaluation measures: discriminability measures, precision measures and holistic measures. In each group, many different metrics were found to be highly correlated, effectively measuring the same property. For restaurant recommendation, we are interested in a precision-type measure, and Herlocker's results tell us that we need not be extremely picky about how such a measure is calculated.

With short sessions and a dearth of positive ratings, there are some obvious constraints on how the Entree sessions can be employed and recommendations evaluated. An evaluation technique that requires making many predictions for a given user will not be applicable, because if many ratings are held out for testing, there would not be enough of a profile left on which a recommender could base its prediction. This rules out such standard metrics as precision/recall and mean absolute error. Ultimately, in order to find good recommendations, the system must be able to prefer an item that the user rated highly. How well the system can do this is a good indicator of its success in prediction. We would like to measure how well each system is able to give a good item as a recommendation. So, the method used here is to record the rank of a positively-rated test item in a recommendation set. Averaging over many trials we can compute the "average rank of the correct recommendation" or ARC. The ARC measure provides a single value for comparing the performance of the hybrids, focusing on how well each can discriminate an item known to be liked by the user from the others.[7]

To calculate this value for each recommender system design, the set of sessions is divided randomly into training and test parts of approximately equal size. This partition was performed five times and results from each test/training split averaged. Each algorithm is given the training part of the data as its input and handles it in its own way. Evaluation is performed on each session of the test data. From the session, a single item with a positive rating is chosen to be held back.[8] This item will be the test item on which the recommender's performance will be evaluated. All of the other ratings are considered part of the test profile.

The recommendation algorithm is then given the test profile without the positively-rated item, and must make its recommendations. The result of the recommendation process is a ranked subset of the product database containing those items possibly of interest to the user. From this set, we record the rank of the positively-rated test item. Ideally, that rank would be a low as possible – the closer to the front the preferred item is placed, the more precisely the recommender is reflecting the user's preferences.

12.3.4 Sessions and Profiles

The Entree data contains approximately 50,000 sessions of widely differing lengths. Some sessions consist of only an entry and exit point, others contain dozens of critiques. To examine differences in recommender performance due to profile size, we fix the session size for each evaluation test set, discarding sessions shorter than this size and randomly discarding negative ratings from longer sessions.

Longer profiles are available if we examine user behavior over multiple visits. There are approximately 20,000 multi-session profiles. These longer multiple-visit profiles are somewhat less reliable as user profiles because they are collated using IP address alone [31]. So, we understand that they will be noisier than the ones derived from single visits.

The evaluation examined six different session sizes: three from single visits and three from multi-visit profiles. We used 5, 10 and 15 rating sessions from single vis-

---

[7] The significance of ARC results is computed with paired ANOVA analysis using the Bonferroni t test for rank with $\alpha = 0.01$. The significance calculations were performed in SAS 8.0 using the Generalized Linear Model procedure. (http://www.sas.com/)

[8] If there are no positive ratings, the session is discarded. We cannot evaluate a recommendation if we have no information about what the user prefers.

its; and 10, 20 and 30 rating sessions from multi-visit profiles. In the figures below, the single-visit profiles will be marked with a capital "S" and the multi-visit profiles with a capital "M". In the case of 5-rating sessions, we used a 50% sample of the data for testing due to the large number of profiles of this size.

12.3.5 Baseline Algorithms

Four basic algorithms were used in the study.

Collaborative Pearson – CFP. This algorithm recommends restaurants based on a collaborative filtering algorithm using Pearson's correlation coefficient to compute the similarity between users [18]. A threshold is used to select similar users and the top 50 are retained as the user's peer group. The restaurants rated by this peer group and not rated by the user are considered the candidate set. These candidates are scored using the average rating from the peer group.[9]

Collaborative Heuristic – CFH. This recommender uses a collaborative variant that computes the similarity between users, taking into account the semantics of the Entree ratings. This algorithm is described more fully in [9]. Rather than treating all of the critiques in each user session as negative ratings (as is done in the CFP algorithm), the heuristic algorithm has a distance matrix for comparing critiques directly. For example, a "nicer" critique and a "cheaper" critique are considered dissimilar, while a "nicer" and "quieter" critique are considered similar. Earlier experiments suggested that this variant was more effective than methods that treat the ratings as binary-valued.

Content-Based – CN. This technique uses the naive Bayes algorithm to compute the probability that a restaurant will be liked by the user. The training data is used to compute prior probabilities and the test session data is used to build a user-specific profile. In most recommender systems, the profile is then used to classify products into liked and disliked categories and the liked category becomes the candidate set, with the classification score becoming the rating. Because of the skewed distribution of ratings, however, this approach was not found to be effective – too few restaurants are rated as "liked". In these experiments, I instituted a candidate generation phase that retrieves all those restaurants with some features in common with the "liked" vector of the naive Bayes profile. Some of these restaurants would not be rated as "liked", but restaurants that do not have at least one such feature cannot be assigned to the "liked" category. The ranking of candidates is then determined by the prediction of the "liked" classifier.

Knowledge-Based (KB). The knowledge-based recommender recommends restaurants using Entree's knowledge-based retrieval. Entree has a set of metrics for knowledge-based comparison of restaurants. It knows, for example, that Thai and Vietnamese food are more similar to each other than Thai and German food would be. Other

---

[9] Weighting user's rating by the proximity to the test user as some authors suggest [5] was not found to be effective.

Table 12.2. Average rank of correct recommendation (ARC) for basic recommendation algorithms at each session size.

	5S	10S	15S	10M	20M	30M
CFH	80	83	124	229	231	230
CFP	113	99	158	183	213	240
KB	207	220	154	298	305	311
CN	296	276	273	313	310	336
Ave	294	304	307	316	317	317

knowledge enables it to reason about price, atmosphere and other characteristics of restaurants. In order to evaluate this component from historical user sessions, the system reissues the last query or critique present in the session and returns the candidate set and its scores. Because longer sessions are truncated, the query will rarely correspond to the one immediately prior to the exit point (which may or may not be the test item) but it will be the available rating chronologically closest to the exit point.

12.3.6 Baseline Evaluation

A starting point for analysis of the hybrids is the evaluation of the four basic algorithms, and for a baseline, we can also examine the performance of the "average" recommender, which recommends restaurants based on their average rating from all users, and does not take individual user profiles into account.

Table 12.2 shows the average rank of the correct recommendation (ARC) for each of the basic algorithms over the six different session size conditions. Figure 12.10 shows the same data in graphical form. (Brackets above the bars indicate places where differences between algorithm performance are not significant.) There are several points to make about these results. First, we should note that this recommendation task is, as expected, rather difficult. The best any of these basic algorithms can manage is average rank of 80 for the correct answer. The primary reason is the paucity of data. With only a small number of ratings to work from, collaborative algorithms cannot narrow their matching neighborhoods to precise niches, and the content-based algorithm has fewer patterns from which to learn. It is not surprising that the results are not exactly inspiring in an e-commerce context where the user might be expected only to look at the first dozen results or so. The top result for single-visit profiles is obtained by the heuristic collaborative algorithm. However, when we look at multiple visit profiles, the standard collaborative algorithm is preferred. In three of the six cases, however, the differences are not significant.

This data also demonstrates something of the task-focused nature of the Entree data, a characteristic that it shares with other consumer-focused recommendation domains. Users coming to the Entree system are planning for a particular dining occasion and their preferences undoubtedly reflect many factors in addition to their own particular tastes. (Since restaurant meals are often taken in groups, the task is effectively one of group recommendation [27].) These extra-individual factors may change radically from session to session and therefore add to the difficulty of extracting a

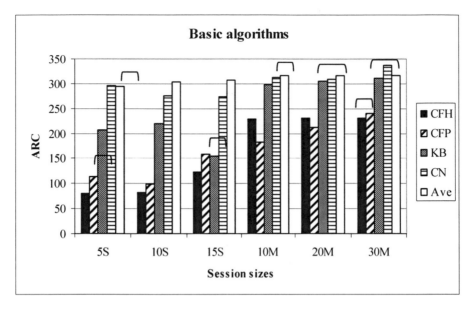

Fig. 12.10. Average rank of correct answer, basic algorithms. Non-significant differences between results are indicated with brackets

consistent multi-session profile. We can see this in the performance of the naïve Bayes (CN) recommender across the different session sizes from 5S to 15S, where it improves steadily, but as we step up to the multi-session profile of size of 30, we see that the performance of this recommender actually goes down, not statistically better than the simple average. Also, the performance of the knowledge-based recommender is weaker in the multi-session profiles, due most likely to the same lack of user consistency across visits: the constraints that the recommender can use to find restaurants in one session may not be valid in a later one.

There are two conclusions to be drawn from the performance of the basic algorithms. One is that the techniques vary widely in their performance on the Entree data. The content-based technique is generally weak. The knowledge-based technique is much better on single-session profiles than on multi-session ones. The heuristic collaborative technique may have a relative advantage over the correlation-based one for short profiles but does not have it for multi-visit ones. The second point is that there is much room for improvement in the results shown by these algorithms acting alone. This is particularly the case for the multi-session profiles.

12.4 Results

This section describes the general findings of a comparative study, examining 41 different types of hybrid recommenders using the four basic components evaluated above. Full details are omitted for resasons of space, but can be found in [11]. There are no results for "Mixed" hybrids as it is not possible to evaluate retrospectively its effectiveness from usage logs. There is no demographic data in the Entree data set, and so no demographic components were examined. The study did not examine hybrids combining recommenders of the same type, such as CFP/CFH, and some designs that are theoretically possible were not implemented due to constraints of the existing algorithms and knowledge-bases. In the interest of a level playing field, no enhancements were added that would specifically benefit only certain hybrid types.

Table 12.3 shows the two best results from each hybrid type. Grey cells indicate conditions in which no synergy was found: the hybrid was no better than one of its components taken individually. Figure 12.11 shows the ARC results for the top hybrid of each type.

12.4.1 Weak Performers

The weighted, switching, feature combination and meta-level hybrids were not particularly effective designs for this data set. They showed only scant and spotty improvement over the unhybridized algorithms, and in the case of meta-level designs, no synergy whatsoever.

Weighted. The weighted hybrid was created by setting the component weights empirically, determining which weighting yielded the best ARC value over the training data. The results were rather surprising, although [49] found a similar effect. In only 10 of the 30 conditions was the performance of the combined recommenders better than the best component working alone. The CN/CFP hybrid does show consistent synergy (5 of 6 conditions), as [36] also found. The most likely explanation is that the recommenders, especially KB and CFH, do not have uniform performance across the product and user space.

Switching. A switching hybrid is one that selects a single recommender from among its constituents and uses it exclusively for a given recommendation situation. For a different profile, a different recommender might be chosen. This approach takes into account the problem that components may not have consistent performance for all types of users. However, it assumes that some reliable criterion is available on which to base the switching decision: a confidence value. Each of the recommenders in the experimental set required a different confidence calculation. (See [11] for additional details.) However, none of the switching hybrids were particularly effective, perhaps due to the lack of reliable switching criteria: only the KB/CFP hybrid showed overall synergy.

Feature Combination. Feature combination requires that we alter the input of a recommendation component to use data from another knowledge source. Such an arrangement will, by necessity, be different for each pair of techniques being combined. The content-based recommender with a contributing collaborative part (CF/CN) was built by augmenting the representation of each restaurant with new features corresponding to the reaction of each profile in the training data to that restaurant. For example, if profiles A and B had negative ratings for restaurant X and profile C had a positive rating, the representation of restaurant X would be augmented with three new features, which can be thought of as A^-, B^- and C^+. Now an ordinary content-based

Table 12.3. Top two best results for each hybrid type. (Meta-level omitted.) Grey cells indicate conditions in which synergy was not achieved.

		5S	10S	15S	10M	20M	30M
Basic	**CFH**	80	83	124	229	231	230
	CFP	113	99	158	183	213	240
Weighted	**CN/CFP**	102		142	168	202	224
	CFP/KB		90	92			238
Switching	**CFH/KB**	65		65	205	203	211
	KB/CFP	65	93	79			239
Feature Comb.	**CN/CFP**	111	98	143	184	215	228
	CN/CFH	80	83	117	233	224	228
Feature Aug.	**CN/CFP**	23	23	24	31	33	40
	KB/CFP	18	19	20	30	31	37
Casacade	**CFP/CN**	20	16	23	29	31	38
	CFP/KB	19	16	22	30	32	38

algorithm can be employed using the test user's profile to learn a model of user interests, but this model will now take into account similarities between restaurants that have a collaborative origin. The CF/CN feature combination hybrid showed modestimprovement over the hybridized CN algorithm but it still falls far short of the basic collaborative algorithms. [10]

A collaborative recommender with a contributing content-based component turns this process around and creates artificial profiles corresponding to particular content features; these are sometimes called "pseudo-users" [43] or "genreBots" (Good et al. 1999). For example, all of the restaurants with Tex-Mex cuisine would be brought together and a profile created in which the pseudo-user likes all of the Tex-Mex restaurants in the database. Similar profiles are generated for all the other content features.[11] The CN/CFH and CN/CFP results are nearly identical to the non-hybrid results as one might expect given that the pseudo-users add only about 1% more data to the training set. Even when the training data was downsampled, the contribution of the psuedo-users was minimal.

---

[10] The naive Bayes implementation performed poorly with this augmented model, including over 5,000 new collaborative features in addition to the 256 content ones. So, for this hybrid only, the Winnow algorithm was used [26], because of its ability to handle large numbers of features. Winnow was found to be inferior to naive Bayes for the other content-based recommendation tasks.

[11] Other possibilities for feature combination hybrids turn out to be either illogical or infeasible.

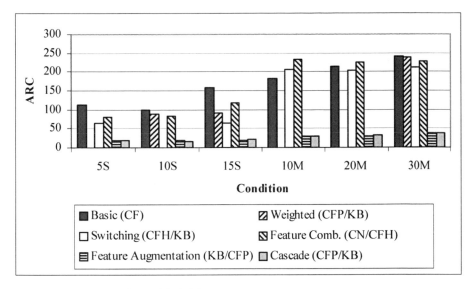

Fig. 12.11. Best results for each hybrid type.

Meta-level. The construction of a meta-level hybrid is highly dependent on the characteristics of the recommendation components being combined. It is not always feasible to derive a meta-level hybrid from any given pair of recommenders. Because none of the basic recommenders are particularly good on their own, we might expect low reliability in any learned model they might produce. This expectation is borne out by experiment: none of the six meta-level hybrids examined in the study achieved synergy in any condition, and these results are omitted from Table 12.3 and Figure 12.11. It is evident that to build a working meta-level hybrid, both recommendation components must be strong individual performers.

12.4.2 Cascade

Our recommendation components all produce real-valued outputs, making them unsuitable at first glance for the use of the cascade, which uses a secondary recommender to break ties. However, reduced-precision versions of these algorithms were implemented in which predicted ratings were limited to two decimal digits. The reduction in precision was found to have minimal impact on accuracy, and these versions of the algorithms were used to produce output that could contribute to a cascade.

The cascade hybrid was designed for a strong recommender to get a boost from a weaker one. It assumes that the primary recommender is the more reliable of its components. Therefore, it is not surprising that the collaborative-primary cascade hybrids work well. These implementations, especially the CFP/KB and CFP/CN hybrids, show great improvement over the other hybrids seen so far and over the basic recommenders.

12.4.3 Feature Augmentation

In feature augmentation, we seek to have the contributing recommender produce some feature or features that augment the knowledge source used by the primary recommender. To preserve the recommendation logic of our recommenders in this study required some ingenuity to create the eight hybrids studied. Several different methods are used as described in detail below. In each, the goal was to preserve the comparative nature of the study, to avoid adding new recommendation logic to the hybrid – where it was unavoidable, the simplest possible technique was employed, which in some cases was unsupervised clustering.

Content-Based Contributing / Collaborative Actual – CN/CF. A content-based recommender uses the features of the items in a profile to induce a classifier that fits a particular user. The features that such a classifier can produce are classifications of items into liked / disliked categories. This capability is used as follows:

1. The content-based algorithm is trained on the user profile.
1. The collaborative algorithm retrieves candidate restaurants from users with similar profiles.
2. These candidates are rated by the content-based classifier and those ratings are used to augment the profile thus filling it out with more ratings.
3. Then the collaborative recommendation process is performed again with a new augmented profile. This is Melville's "content-boosted collaborative filtering" [30].

Collaborative Contributing – CF/CN and CF/KB. A collaborative recommender deals with similarities between users. The other recommenders are interested in comparing restaurants, so the problem for a collaborative recommender contributing to a knowledge-based or content-based hybrid is how to turn user data into features associated with restaurants. One way to do this is to cluster restaurants into groups based on user preferences about them. The cluster to which a given restaurant belongs can be considered a new feature that augments the restaurant representation. To incorporate these new features into the knowledge-based recommendation, the recommender's domain knowledge was augmented with a simple metric that prefers restaurants that share the same cluster id.

Knowledge-Based Contributing – KB/CF and KB/CN. A knowledge-based recommender can be used like the content-based one to classify restaurants into liked / disliked categories by assuming that the restaurants retrieved by the recommender are in the "liked" category and all others are disliked. The algorithm given above for the CN/CF hybrid can then be employed.

This is not an adequate solution for a KB / CN feature augmentation hybrid where the knowledge-based recommender needs to augment the representation of restaurants rather than user profiles. In this case, however, we treat the knowledge-based system as a source of "profiles". For each restaurant, we retrieve a set of similar restaurants known to the system. Each such set is treated like a user profile, and this profile matrix can be transposed and clustered as in the collaborative case.

Results. The feature augmentation hybrids show the best performance seen so far, particularly where the content-oriented recommenders are contributing to the collaborative ones. The KB and CN recommenders did not make good primary components, as might be expected from the performance of the basic algorithms and none of these hybrids showed synergy. Both strong components were greatly enhanced by the addition of content-derived features. Performance is particularly good for the multi-session profiles for which none of the previous hybrids were adequate.

12.5 Discussion

Given this survey of 41 different hybrid recommenders, we can return to our initial purpose in this survey, to determine the best hybrids for the Entree data and to determine if any lessons can be learned that apply to the construction of hybrids in the more general case.

It is quite clear, as others have also shown, that there is an unqualified benefit to hybrid recommendation, particularly in the case of sparse data. This can be seen in Figure 12.11. Nowhere was this effect more striking than in the noisy multi-session profiles, which proved so much more difficult for even the stronger basic algorithms. Where the best results obtained on the 30-rating sessions by a basic algorithm was only an ARC of 227, the top hybrids all have ARC scores under 40. Note that this synergy is found under the twin difficulties of smaller profile size and sparse recommendation density, showing that hybridization does help conquer the cold start problem.

Of course, not all hybrid designs were successful, leading to a second question: What is the best hybrid type? This answer can be found by examining the relative performance over all the hybrids on the different conditions. If we rank the hybrids by their ARC performance and look at the top hybrids in each condition, feature augmentation and cascade recommenders dominate. None of the other hybrid types achieve a rank higher than 9th best for any condition, and the only non-FA or cascade hybrids that appears twice in the top ten are two switching recommenders: CFH/KB and KB/CFP. Table 12.4 shows the top ten hybrids ranked by their average ARC over all conditions. Beyond the top four (two feature-augmentation and two cascade), performance drops off markedly.

In retrospect, given the performance of the basic algorithms, the performance of the cascade recommenders is fairly predictable. The KB and CN algorithms are relatively weak, but do take into account different knowledge sources than the collaborative algorithms. A cascade design allows these recommenders to have a positive impact on the recommendation process with little risk of negative impact – since they are only fine-tuning the judgments made by stronger recommenders. What is particularly interesting is that this performance was achieved by explicitly sacrificing numeric precision in the scoring of the primary recommender. The other top performing hybrids were the feature augmentation hybrids. Again, we see that the feature augmentation design allows a contributing recommender to make a modest positive impact without the danger of interfering with the performance of the better algorithm.

Generalizing from these results is by necessity speculative, since all we have are results in a particular product domain with a somewhat sparse and unorthodox data set. These experiments show that standard recommenders with widely varying per-

Table 12.4. Top ten hybrids by ARC

Type	Recommenders used	Average ARC
FA	KB/CFP	25.8
Cascade	CN/CFP	26.2
Cascade	KB/CFP	26.3
FA	CN/CFP	29.0
FA	KB/CFH	79.9
Cascade	CN/CFH	91.1
Cascade	KB/CFH	92.2
FA	CN/CFH	95.1
Switching	CFH/KB	139.1
Switching	KB/CFP	155.6

formance can be combined to achieve strong synergies on a fairly difficult recommendation task with limited data. In particular, it is clear that even recommendation algorithms with weak knowledge sources can have a strong positive impact on performance if they are combined in an appropriate hybrid.

No hybrids were tested in which both components could be considered strong, and while it seems likely that the feature augmentation and cascade designs would work well in this best case strong-strong scenario, it is seems likely that other techniques such as the meta-level hybrid would also succeed. Clearly, other researchers have had success with meta-level designs [1, 36, 45].

We see significant differences between the hybridization techniques, particularly their sensitivity to the relative strength and consistency of each component part. Some hybrids can make the most of a weak-strong combination; others cannot. Some hybrids work under the assumption that their components have uniform performance across the recommendation space (weighted, augmentation, meta-level); others are effective even if this is not true. In choosing a hybrid recommendation approach, therefore it seems particularly important to examine the design goals for a hybridized system (overall accuracy, cold-start performance, etc.) and evaluate the relative performance of each component of the hybrid under those conditions. For example, consider an implementer interested in improving collaborative recommendation results for cold-start users by building a hybrid that adds a content-based technique. We know that new users would have small usage profiles and the content-based recommender would be weak in these cases. This situation would suggest a cascade or feature augmentation approach.

Another consideration in the choice of hybridization techniques for recommendation is efficiency, particularly run-time efficiency, since recommendations are typically made on the fly to users expecting a quick interactive response. Of the basic algorithms, the collaborative algorithms are the slowest since they must compare the user's profile against the database of other users. A number of approaches have been developed to improve the efficiency of collaborative algorithms, for example clustering and indexing [31] and these would be of interest in any hybrid scheme as well. Of the hybrid designs, the weighted approach is the least efficient since it requires that both recommenders process every request; depending on the implementation, a meta-level hybrid may have the same drawback. Among the strong performers, the cascade hybrid also requires computation from both recommenders, but since the secondary recommender is only breaking ties, it is not required to retrieve any candidates and need only rate those items that need to be further discriminated. This can be done on demand as the user requests portions of the retrieval set. On the other hand, the other top performing hybrid, the feature augmentation hybrid, the contributing recommender operates by adding features to the underlying representation. This step can be performed entirely off-line. So, the feature augmentation hybrid offers accuracy on par with the cascade hybrid with virtually no additional on-line computation.

12.6 Conclusion

This chapter has more fully characterized each of 53 hybrid types shown in Table 12.1 and described experiments that compare the performance of a subset of the design space. The experiments cover the space of possible hybrid recommender systems available with four basic recommendation algorithms: content-based, standard collaborative, heuristic collaborative and knowledge-based. Six types of combinations were explored: weighted, switching, feature combination, feature augmentation, cascade and meta-level, for a total of 41 different systems. Due to data and methodological limitations, demographic recommendation and mixed hybrids were not explored. Because two different collaborative algorithms were explored, the 41 systems evaluated represent 24 of the 53 spaces in this table, including 12 recommenders with no previous known examples.

Of course, any such study is by its nature limited by the peculiarities of the data and the recommendation domain. The Entree data set is relatively small (just over ¼ million ratings), the profiles are short and the ratings are implicit and heavily skewed to the negative. It would be valuable to repeat this study in a different recommendation domain with different products and a set of user profiles with different characteristics. In particular, it is unfortunate that the circumstances of this study allow only very limited findings with respect to meta-level recommendation.

Three general results, however, can be seen. First, the utility of a knowledge-based recommendation engine is not limited strictly to its ability to retrieve appropriate products in response to user queries. Such a component can be combined in numerous ways to build hybrids and in fact, some of the best performing recommenders seen in these experiments were created by using the knowledge-based component as a secondary or contributing component rather than as the main retrieval component. Second, cascade recommendation, although rare in the hybrid recommendation literature, turns out to be a very effective means of combining recommenders of differing strengths. Adopting this approach requires treating the scores from a primary recommender as rough approximations, and allowing a secondary recommender to fine-tune the results. None of the weak/strong cascade hybrids that were explored ranked less than eighth in any condition, and in the average results, they rank in four of the top seven positions. This is despite the fact that the primary recommender was operating in a state of reduced precision. Finally, the six hybridization techniques examined have very different performance characteristics. An implementer should evaluate the relative accuracy and consistency of each component of the hybrid to determine its best role in a hybrid system.

Acknowledgements. Entree was developed at the University of Chicago in collaboration with Kristian Hammond, with the support of the Office of Naval Research under grant F49620-88-D-0058. Parts of the experimental design were done with the assistance of Dan Billsus at the University of California, Irvine. Thanks to Bomshad Mobasher and Alfred Kobsa for helpful comments on early drafts of this chapter.

References

1. Balabanovic, M.: An Adaptive Web Page Recommendation Service. In: Agents 97: Proceedings of the First International Conference on Autonomous Agents, Marina Del Rey, CA, pp. 378-385. (1997)
2. Balabanovic, M.: Exploring versus Exploiting when Learning User Models for Text Representation. UMUAU 8(1-2), 71-102. (1998)
3. Basu, C., Hirsh, H. and Cohen W.: Recommendation as Classification: Using Social and Content-Based Information in Recommendation. In: ProC. of the 15th Natl. Conf. on AI, Madison, WI, 714-720. (1998)
4. Billsus, D. and Pazzani, M.: User Modeling for Adaptive News Access. UMUAI 10(2-3), 147-180. (2000)
5. Breese, J. S., Heckerman, D. and Kadie, C.: Empirical analysis of predictive algorithms for collaborative filtering. In: Proc. of the 14th Annual Conf. on Uncertainty in AI, 43-52. (1998)
6. Brin, S. and Page, L The anatomy of a large-scale hypertextual web search engine. Comp. Networks and ISDN Systems, 30 (1–7):107–117. (1998)
7. Brunato, M. and Battiti, R.: 2003. A Location-Dependent Recommender System for the Web. In: Proceedings of the MobEA Workshop, Budapest, May 20, 2003. Accessed at http://www.science.unitn.it/~brunato/pubblicazioni/MobEA.pdf. (2003)
8. Burke, R.: The Wasabi Personal Shopper: A Case-Based Recommender System. In: Proc. of the 11th Nat. Conf. on Innovative Appl. of AI, 844-849. (1999)
9. Burke, R.: Knowledge-based Recommender Systems. In: A. Kent (ed.): Encyclopedia of Library and Information Systems, 69, Sup. 32. (2000)
10. Burke, R.: Hybrid Recommender Systems: Survey and Experiments. UMUAI 12 (4), 331-370. (2002)
11. Burke, R. Hybrid Recommender Systems: A Comparative Study. CTI Technical Report 06-012. 2006. (Available at http://www.cs.depaul.edu/research/technical.asp.)
12. Burke, R., Hammond, K., and Young, B.: The FindMe Approach to Assisted Browsing. IEEE Expert, 12 (4), 32-40. (1997)
13. Cheetham, W., and Price J.: Measures of Solution Accuracy in Case-Based Reasoning Systems. In Proc. of the 6th Eur. Conf. on CBR. Lecture Notes in Computer Science 3155. Springer-Verlag, London, pp. 106 – 118. (2004)
14. Chen, L. and Sycara, K.: WebMate: A personal agent for browsing and searching. In Proc. of the 2nd Intl Conf. on Autonomous Agents (Agents'98), 132-139, New York. ACM Press. (1998)
15. Claypool, M., Gokhale, A., Miranda, T., Murnikov, P., Netes, D. and Sartin, M.: Combining Content-Based and Collaborative Filters in an Online Newspaper. SIGIR '99 Workshop on Recommender Systems: Algorithms and Evaluation. Berkeley, CA. Accessed at http://www.cs.umbc.edu/~ian/sigir99-rec/papers/claypool_m.ps.gz (1999)
16. Cohen, W. W.: Fast effective rule induction'. In Machine Learning: Proc. of the 12th Intl Conf., Lake Tahoe, CA, 115-123. (1995)

17. Goldberg, D., Nichols, D., Oki, B, & Terry, D. Using collaborative filtering to weave an information tapestry. 35(12):61–70. CACM. (1992)
18. Good, N., Schafer, B., Konstan, J., Borchers, A., Sarwar, B., Herlocker, J., and Riedl, J.: Combining Collaborative Filtering With Personal Agents for Better Recommendations. In: Proceedings of the AAAI'99 Conf., 439-446. (1999)
19. Herlocker, J., Konstan, J., Borchers, A., Riedl, J.: An Algorithmic Framework for Performing Collaborative Filtering. Proc. of the 1999 Conf. on Research and Development in Information Retrieval. Berkeley (SIGIR '99), 230-237. (1999)
20. Herlocker, J. L., Konstan, J. A., Terveen, L. G., and Riedl, J. T.: Evaluating Collaborative Filtering Recommender Systems. ACM Trans. on Inf. Sys. 22 (1). (2004)
21. Hill, W., Stead, L., Rosenstein, M. and Furnas, G.: Recommending and evaluating choices in a virtual community of use. In: CHI '95: Conf. Proc. on Human Factors in Computing Sys., Denver, CO, 194-201. (1995)
22. Jennings, A. and Higuchi, H.: A User Model Neural Network for a Personal News Service. UMUAI 3, 1-25. (1993)
23. Kolodner, J.: Case-Based Reasoning. San Mateo, CA: Morgan Kaufmann. (1993)
24. Krulwich, B.: Lifestyle Finder: Intelligent User Profiling Using Large-Scale Demographic Data. AI Magazine 18 (2), 37-45. (1997)
25. Lang, K.: Newsweeder: Learning to filter news. In: Proc. of the 12th Intl Conf. on Machine Learning, Lake Tahoe, CA, 331-339. (1995)
26. Littlestone, N. and Warmuth, M.: The Weighted Majority Algorithm. Information and Computation 108 (2), 212-261. (1004)
27. McCarthy, J. F. and Anagnost, T. D.: MUSICFX: An Arbiter of Group Preferences for Computer Supported Collaborative Workouts. In Proc. of the ACM 1998 Conf. on Comp. Support. Coop. Work (CSCW 98), 363-372. (1998)
28. McCarthy, K., McGinty, L. Smyth, B. and Reilly, J.: A Live-User Evaluation of Incremental Dynamic Critiquing'. In H. Muñoz-Avila and F. Ricci (eds.): CBR Research and Develop. (6th Intl Conf., ICCBR 2005). Chicago, IL, 339-352. (2005)
29. McSherry, D.: Diversity-Conscious Retrieval. In S. Craw & A. Preece (eds.): Advances in CBR (6th Eur. Conf., ECCBR 2002). Aberdeen, Scotland, 219-233. (2002)
30. Melville, P., Mooney, R. J. and Nagarajan, R.: Content-Boosted Collaborative Filtering for Improved Recommendations. In: Proc. of the 18th Natl. Conf. on AI (AAAI-2002), pp. 187-192. (2002)
31. Mobasher, B., Dai, H., Luo, T. and Nakagawa, M.: Discovery and evaluation of aggregate usage profiles for web personalization. Data Mining and Knowledge Discovery, 6:61-82. (2002)
32. Mobasher, B., Jin, X., and Zhou, Y.: Semantically Enhanced Collaborative Filtering on the Web'. In B. Berendt, et al. (eds.): Web Mining: From Web to Semantic Web. LNAI Volume 3209, Springer. (2004)
33. Mobasher, B. and Nakagawa, M.: A Hybrid Web Personalization Model Based on Site Connectivity. In Proc. of the WebKDD Workshop at the ACM SIGKDD Intl Conf. on Knowledge Discovery and Data Mining, Washington, DC. (2003)
34. Mooney, R. J. and Roy, L.: Content-Based Book Recommending Using Learning for Text Categorization. SIGIR '99 Workshop on Recommender Systems: Algorithms and Evaluation. Berkeley, CA. Accessed at http://www.cs.umbc.edu/~ian/sigir99-rec/papers/mooney_r.ps.gz (1999)
35. O'Sullivan, D., Smyth, B., Wilson, D. C.: Preserving recommender accuracy and diversity in sparse datasets. Intl J on AI Tools 13(1): 219-235. (2004)
36. Pazzani, M. J.: A Framework for Collaborative, Content-Based and Demographic Filtering. AI Review 13 (5/6), 393-408. (1999)
37. Pazzani, M. and Billsus, D.: Learning and Revising User Profiles: The Identification of Interesting Web Sites. Machine Learning 27, 313-331. (1997)

38. Pazzani, M., Muramatsu, J., and Billsus, D.: Syskill & Webert: Identifying Interesting Web Sites. In Proc. of the 13th Natl Conf. on AI, 54-61. (1996)
39. Pu, P. H. Z. and Kumar, P.: Evaluating Example-based Search Tools. In EC'04; Proc. of the 5th ACM Conf. on Electronic Commerce. New York, 208-215. (2004)
40. Reilly, J., McCarthy, K, McGinty, L. and Smyth, B.: Incremental Critiquing. In M. Bramer, et al. (eds.): Research and Devel. in Int. Sys. XXI. AI-2004, Cambridge, UK. (2004)
41. Resnick, P., Iacovou, N., Suchak, M., Bergstrom, P. and Riedl, J.: GroupLens: An Open Architecture for Collaborative Filtering of Netnews. In: Proc. of the Conf. on Comp. Supp. Coop. Work, Chapel Hill, NC, 175-186. (1994)
42. Resnick, P. and Varian, H. R.: Recommender Systems. Comm. of the ACM 40 (3), 56-58. (1997)
43. Sarwar, B. M., Konstan, J. A., Borchers, A., Herlocker, J. Miller, B. and Riedl, J.: Using Filtering Agents to Improve Prediction Quality in the GroupLens Research Collaborative Filtering System. In: Proc. of the ACM 1998 Conf. on Comp. Supp. Coop. Work, Seattle, WA, 345-354. (1998)
44. Schmitt, S. and Bergmann, R.: Applying case-based reasoning technology for product selection and customization in electronic commerce environments. In: 12th Bled Electronic Commerce Conf. Bled, Slovenia, June 7-9, (1999)
45. Schwab, I. and Kobsa, A.: Adaptivity through Unobstrusive Learning. Künstliche Intelligenz 16(3): 5-9. (2002)
46. Shardanand, U. and Maes, P.: Social Information Filtering: Algorithms for Automating "Word of Mouth". In: CHI '95: Conf. Proc. on Human Factors in Comp. Sys. Denver, CO, 210-217. (1995)
47. Shimazu, H.: ExpertClerk: Navigating Shoppers' Buying Process with the Combination of Asking and Proposing. In B. Nebel, (ed.): Proceedings of the 17th Intl J Conf on AI, 1443-1448. (2001)
48. Smyth, B. and Cotter, P.: A Personalized TV Listings Service for the Digital TV Age. Knowledge-Based Systems 13: 53-59. (2000)
49. Van Setten, M.: Supporting People in Finding Information: Hybrid Recommender Systems and Goal-Based Structuring. Report No. 016 (TI/FRS/016). Enschede, the Netherlands: Telematica Institut. (2005)
50. Wilson, D. C., Smyth, B., O'Sullivan, D.: Sparsity Reduction in Collaborative Recommendation: A Case-Based Approach. Intl J of Pattern Recognition and AI 17 (5): 863-884. (2003)

13

Adaptive Content Presentation for the Web

Andrea Bunt, Giuseppe Carenini, and Cristina Conati

Department of Computer Science
University of British Columbia
{bunt, carenini, conati}@cs.ubc.ca

Abstract. In this chapter we describe techniques for adaptive presentation of content on the Web. We first describe techniques to select and structure the content deemed to be most relevant for the current user in the current interaction context. We then illustrate approaches that deal with the problem of how to adaptively deliver this content.

13.1 Introduction

Previous chapters in this book have described types of adaptation for Web-based systems that include adaptive navigation support (see Chapter 8 of this book [8]), adaptive search (see Chapter 6 of this book [39]) and personalized recommendation of items of interest (see Chapters 9 [47], 10 [42], 11 [49], and 12 [9] of this book). In this chapter, we will focus on an additional type of adaptation widely known as *adaptive presentation of content*: how to present Web-based content in a manner that best suits individual users' needs. This type of adaptation involves determining, based on the user and context, what information the system should present and how the information should be organized and displayed. While adaptive presentation of content can serve many purposes, as we will demonstrate throughout the chapter, it can also complement several of the adaptation types discussed in previous chapters. For instance, the content of Web pages pointed to by a tailored link in a system that provides adaptive navigation support (Chapter 8 of this book [8]), or returned by adaptive search (Chapter 6 of this book [39]), can be modified to highlight the parts that are more interesting for the current user. Similarly, the description of the items returned by a recommender system (see Chapters 9 [47], 10 [42], 11 [49], and 12 [9] of this book) and can be adapted to play up the items' features that are more relevant to the user's needs, or changed to be more suitable to the user's level of familiarity with the items.

The focus of this chapter will be on computational techniques necessary to provide the user with a tailored presentation of content, rather than implementation details and technologies. Also, the chapter is not limited to techniques currently used in adaptive Web-based applications. It aims to suggest areas of future research by discussing alternative approaches that have a strong potential to augment the set of existing techniques for adaptive presentation on the Web.

P. Brusilovsky, A. Kobsa, and W. Nejdl (Eds.): The Adaptive Web, LNCS 4321, pp. 409–432, 2007.

The process of adapting content to specific user needs comprises two sub processes: content *adaptation* and *presentation*. Content adaptation involves deciding what content is most relevant to the current user and how to structure this content in a coherent way, *before* presenting it to the user. The second sub process of *content presentation* involves deciding how to most effectively adapt the presentation of the selected content to the user.

The chapter is structured as follows. In section 13.2, we address techniques for content adaptation. Although traditionally these techniques required the existence of pre-crafted versions of the relevant content, new techniques are emerging which can automatically adapt content from abstract knowledge sources. Given that the latter lead to greater flexibility and robustness, our discussion focuses on these. In section 13.3, we discuss techniques for content presentation. We first introduce techniques that deal with the problem of how to present this content so that user focus/attention is drawn to the most relevant information (possibly defined by using any of the techniques described in section 13.2) while still preserving the contextual information that can often be provided by content of secondary importance. We then discuss techniques to decide which media/modality to use to best convey the selected content.

13.2 Techniques for Content Adaptation

Content adaptation involves identifying the content most relevant to a given user and context (jointly referred to as the interaction context), as well as how this content should be organized. Relevant properties of the interaction context can include the user's preferences, interests, and expertise, as well as the presentation goals. Content adaptation of Web pages can be characterized along the following key dimension: the nature of the content provided as input. Along this dimension, we first briefly describe two rather simple approaches in which adaptation is achieved by selecting appropriate canned pages or page fragments. These approaches are referred to in the literature as *page* and *fragment variants* respectively, and they have been extensively discussed in previous surveys (e.g., [32]). After a brief description of page and fragment variants, we provide an in-depth discussion of more sophisticated approaches to content adaptation in which the input is abstract information, since to the best of our knowledge, these approaches have never been covered in detail in any previous survey on adaptive hypermedia.

13.2.1 Approaches Based on Page and Fragment Variants

The simplest form of content adaptation is the page-variant approach [32]. Here, the input of the adaptation process consists of different versions of each page that is to be adapted along with a model of the interaction context. These versions have to be written in advance. At runtime, the adaptation mechanism selects and presents the page version that is most appropriate to the current interaction context. Clearly, this approach does not scale up to complex adaptation. If several aspects of the page must be adapted in many different ways, an unmanageably large number of variants need to be written. Nevertheless, in some domains, where only high-level adaptation is needed, this approach has been effectively applied. For instance, in the ORIMUHS system

[14] page variants are applied to support user interaction in two complex software systems: a CAD modeler and a medical application. Page variants are also applied in the KBS Hyperbook system [24] to develop educational courseware on Java programming.

Moving up in the ladder of adaptation complexity, we have the fragment-variant approach. In this approach, the adaptation is performed at a finer level of granularity. More specifically, the page presented to the user is not selected from a pool of fixed pages. Rather, it is constructed by selecting and combining an appropriate set of fragments, where each fragment typically corresponds to a self-contained information element, such as a text paragraph or a picture. As with the page-variant approach, these fragments are written in advance. Two common strategies for fragment variants are: *optional fragments* and *altering fragments*. In optional fragments, a page is specified as a set of fragments, where each fragment is associated with a set of applicability conditions. At runtime, the page is generated by selecting only those fragments whose conditions are satisfied in the current interaction context. For instance in [16], different optional fragments are selected depending on the user's knowledge, interests and abilities. Altering fragments are rather different from optional fragments. In altering fragments, a page is specified as a set of constituents, and for each constituent there is a corresponding set of fragments. At runtime, the page is created by selecting for each constituent the fragment that is most appropriate in the current interaction context. Altering fragments are applied, for instance, in the AHA system [13], in which different presentations of the same entity can be selected depending on whether the target user has the necessary background knowledge.

In general, a noticeable disadvantage of fragment variants compared to page variants is that the selection and assembly of a suitable set of fragments may involve a substantial overhead at runtime. Furthermore, it may sometimes be difficult to combine the set of independently selected fragments into a coherent whole. On the other hand, the key advantage of this approach is that, once a set of fragments and their applicability conditions have been written, a large number of pages can be automatically generated to cover a corresponding large number of interaction contexts. For pointers to specific techniques to implement the fragment-variant approach the reader should refer to [32].

Note that because in the two approaches above the units of content adaptation are either whole pages or predefined page components, the two sub processes of content adaptation and presentation actually coincide. That is, the decision of what content is most relevant to the user (i.e. the page to be displayed) uniquely identifies what will be presented to the user. On the one hand, this simplifies and speeds up the complete adaptation process. On the other hand, it reduces flexibility because it eliminates the possibility to further tailor the information through adaptive presentation techniques once the first level of adaptive content presentation, content selection, has been achieved, as we will see in section 13.3.

13.2.2 Approaches Based on Abstract Information

Although many adaptive Web systems have been designed in recent years by relying only on page or fragment variants, in this section we describe more sophisticated adaptation techniques that allow a system to reason about the input content and the interaction context, both of which are expressed in more abstract terms. These techniques permit the adaptation to be more flexible, robust and scalable. Notice that part of the research on sophisticated content adaptation has been developed in the field of Natural Language Generation (NLG) [43], which investigates how natural language text can be generated from abstract non-linguistic information.

Sophisticated content adaptation, also called tailoring in NLG, requires an abstract representation of the domain from which the content is selected, as well as the features of the interaction context to which the content is tailored. Several formalisms have been used in the literature, including:

- *Traditional Knowledge Bases* [46] expressing domain entities and relationships between them. For instance, one application of the ILEX system [40] generates tailored jewel labels by relying on a large object-centered knowledge base about jewelers, materials, designers, etc. This knowledge base includes both abstract propositions, such as the fact that a necklace is a jewel, and specific propositions, such as the fact that a particular jewel was made in Birmingham in 1905.
- *Bayesian Networks* [46] expressing probabilistic relationships between random variables representing the domain. For instance, one application of the NAG system [33],[55] generates arguments about the expected rate of a researcher's future publications by relying on a Bayesian Network. This network specifies probabilistic relationships between the publication rate of a researcher and the factors that influence it, such as the strength of the institution from which the researcher graduated (e.g., the stronger the institution, the higher the likelihood of a high publication rate).
- *Preference Models* [46] expressing the user's preferences about different aspects of the domain. For instance, one application of the GEA system [10] generates user-tailored arguments on whether the user will like/dislike a given house by relying on a model specifying what aspects of a house the user cares most about (e.g., location, amenities). The PRACMA system [29] also employs a model of user preferences to tailor its description of an individual recommended item (e.g., a car) by focusing on the aspect (e.g., price) that will have the largest impact on the user's overall evaluation of that item.

Depending on the application, the same or different formalisms can be used to represent the domain model and the interaction context. For instance, in NAG [55], a system for generating factual arguments (claiming that something is or is not the case), both the domain and user model are represented as Bayesian Networks. Similarly, in HYLITE+ [5], a system for generating adaptive hypertext encyclopedia-style explanations, both the domain and the user models are expressed as traditional knowledge bases, more precisely as conceptual graphs [50]. In contrast, in GEA, a system for generating evaluative arguments (claiming that something is good vs. bad), the domain model is represented as a traditional knowledge base while the user model is expressed as a value tree [46], which is a preference model commonly used in decision theory.

The process of sophisticated content adaptation involves the two conceptually distinct phases of *content selection/determination* and *content structuring*, also jointly referred to as content planning. Although we will describe them separately to simplify the presentation, it should be noted that content selection and structuring are often implemented as one single process that simultaneously performs both phases [43].

Content Selection. During content selection, a subset of the domain knowledge is identified as relevant for the current user and situation. Strategies for content selection rely on domain-specific knowledge to different degrees. For instance, the content selection strategy used in STOP, a system for generating smoking cessation letters, is quite domain specific as it refers to psychological knowledge about addictive behavior and smoking [44]. In contrast, the content selection strategy used by the GEA system does not rely on any domain-specific knowledge (as we will see later in this section) and can be therefore applied in any domain [10]. Because of their generality, in this section we focus on strategies that are primarily domain-independent. For a discussion of more domain-specific strategies and in particular of how they can be acquired, the reader should refer to [43].

In practice, most domain-independent strategies for content selection compute a measure of relevance for each content element (i.e., fact) and then use this measure to select an appropriate subset of the available content. Content adaptation is achieved by having this measure of relevance take into account features of the current user and context. For illustration, let's consider three systems that provide a representative overview of how the measure of relevance can be computed and how it can be used for content selection.

The Intelligent Labeling Explorer (ILEX). We start with ILEX [40], a system for generating contextually-relevant hypertext descriptions of objects (e.g., museum artifacts, computer components). In ILEX, the measure of relevance for content selection combines a measure of structural relevance of a knowledge element/fact with its intrinsic score. Structural relevance takes into account the structure of the domain knowledge base - a semantic net. More specifically, structural relevance is computed starting from the focal entity (i.e., the entity being described) by considering two basic heuristics: (i) information becomes less relevant the more distant it is from the focal object, in terms of semantic links; (ii) different semantic link types (e.g., GENERALISE) maintain relevance to different degrees. The intrinsic score of a knowledge element combines numerical estimates of three factors: (i) the potential interest of the information to the current user, (ii) the importance of the information to the system's informational goals and (iii) to what degree the user may already know this information. Once the two measures of structural relevance and intrinsic score have been computed, they are combined in a single measure of relevance by straight multiplication.

In ILEX, the content selection strategy is then to return the n most relevant knowledge elements. However, if the selection process based on relevance cannot find a sufficient number of knowledge elements, additional content selection routines are activated. For instance, one technique applied by ILEX is to identify an entity which is sufficiently similar to the focal entity, so that an interesting comparison between the two can be also selected for presentation. In general, when the goal of a content selection component is to return a fixed amount of content, it may be necessary to supplement the main selection strategy with a set of ancillary strategies.

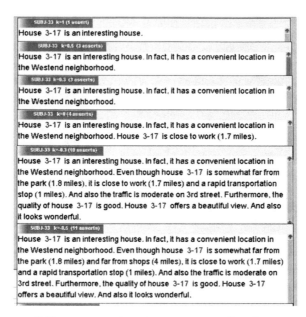

Fig. 13.1. GEA system [10]: arguments about the same house tailored to the same user containing an increasing (or equal) amount of content.

The Generator of Evaluative Arguments (GEA). As mentioned before, GEA is a system for generating evaluative arguments (claiming that something is good vs. bad). Here the measure of relevance is computed by applying a quantitative model of the user's preferences to the entity being evaluated. Generally speaking, GEA's user model relies on the notion that if something is valued, it is valued for multiple reasons. More specifically, for each user, the model specifies a decomposition of the user's overall assessment of entities in a given class (e.g., houses) into a hierarchy of aspects of the entities (e.g., location, number-of-bedrooms). The model hierarchy is annotated with numerical weights and functions that specify the relative importance/preferability of each attribute and domain-value (e.g., two bedrooms) for the particular user. Once the model is applied to an entity, it is possible to compute for each attribute how much its evaluation contributes to the overall evaluation of that entity for the current user. Based on this, a measure of relevance is defined by assuming that an attribute is relevant either because of its strength or because of its weakness in contributing to the value of the entity. For instance, if distance-from-work is an important attribute for the current user, this attribute will have high relevance with respect to the evaluation of a house that is very close to the user's workplace (because of its strong contribution), as well as to the evaluation of a house that is very far from it (because of its weak contribution).

Once the relevance of all attributes is assessed, the content selection strategy in GEA is to return for each level in the user model hierarchy, only those attributes whose relevance is greater than a (customizable) threshold. By setting this threshold to different values it is possible to generate, in a principled way, arguments that con-

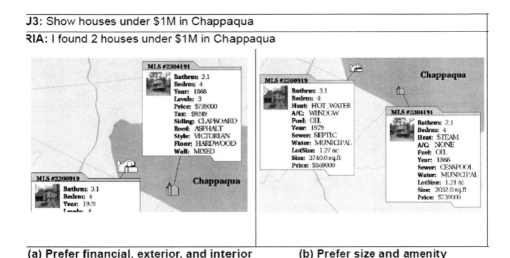

J3: Show houses under $1M in Chappaqua

RIA: I found 2 houses under $1M in Chappaqua

(a) Prefer financial, exterior, and interior   **(b) Prefer size and amenity**

Fig. 13.2. RIA system [53]: examples of queries and user-tailored responses for two users with different preferences.

tain different amounts of user-tailored relevant content, as shown in Fig. 13.1 (see [10] for details).

The Responsive Information Architect (RIA). In RIA [53], a multimedia conversation system to support information-seeking tasks, content selection is formalized as an optimization problem. The goal is to identify the most desirable subset of data dimensions (e.g., price and style in the real-estate domain) in the current interaction context. The desirability of each data dimension is computed as the linear combination of a large set of feature-based metrics that characterize how important the dimension is with respect to the interaction context. Most of these features are labeled as content-relevance features and include features of the data (e.g., inherent importance of the dimension in the domain, such as the importance of price in the real-estate domain), and features of the user (e.g., relation of the dimension to the user's interests, such as how much the user cares about price). There are also features relating the dimension to the user request (e.g., the user may be requesting information on a particular dimension, such as explicitly requesting information on price) and to the interaction history, which tries to maintain a coherent presentation of dimensions across multiple queries. So, in RIA, desirability is fundamentally what we have been referring to as a measure of content relevance. Once all the available data dimensions have been assigned their desirability, RIA's content-selection strategy returns the set of data dimensions such that their overall desirability is maximized and their cost is within the given space and time allocated for the target presentation. Special techniques are introduced to take care of dependencies between data dimensions. For instance, some groups of dimensions are considered as a single bundle (e.g., number of bedrooms and number of bathrooms) and are either all included or all excluded from the final presentation.

Fig. 13.2 shows how RIA, given the same query, selects different data dimensions for two users with different preferences in the real estate domain.

To summarize, we have seen three ways in which a measure of content relevance can be computed and three prototypical ways in which such measures can be used for content selection. In general, there is no accepted set of guidelines in the field to choose the most appropriate measure of relevance and selection strategy given a target application. For any new application, designers should first consider solutions presented in the literature (of which the systems we have just described provide a representative sample), and devise alternative solutions only if no existing one is satisfactory.

Content Structuring. Once the most relevant content elements are selected they must be organized in order to be effectively communicated/presented. This involves not only ordering and grouping them, but also specifying what discourse relations (e.g., contrast, evidence) [31] must hold between the resulting groups. Schemas [43] are the method of choice to accomplish all these tasks and are commonly implemented with task-decomposition planners (technically referred to as HTN planners [46]).

With respect to content selection, we provide a much more limited treatment of content structuring because adaptation of the latter is rather less common than adaptation of the former. One form of structure adaptation is to rely on the measure of relevance used in content selection for ordering the selected content elements. For instance, in GEA the selected elements are ordered according to the measure of relevance by following principles from argumentation theory [10]. Another form of content structuring adaptation is the selection of the discourse relations. For instance, in GEA a given fact can be selected as supporting or contrasting evidence depending on the user's preference for that fact, as determined by the evaluation of the user model.

Note that any structuring information derived from this phase of content adaptation can serve as a guide to decide how to actually present the selected content to the user. The various techniques for adaptive content presentation that we will overview in the next section can be used to adaptively render the information defined by content structuring. For instance, techniques for content emphasis can be used to express relevance information implicit in content ordering. Discourse structure can be used to identify portions of texts that can be made available on demand instead of being displayed up-front [41]. Finally, relationship information can be used to define which medium to use to display the related content (e.g., graphics to highlight quantitative relationships such as a comparison between two sets of numbers [30]). In practice, most previous and current work on adaptive content presentation relies on simpler text-based fragments, which are selected according to a simple user model. However, we believe that work on integrating adaptive content structuring with adaptive content presentation promises to be of great value for the advancement of adaptive content presentation on the Web.

We conclude this section on content adaptation from abstract information with a brief discussion of issues related to knowledge acquisition and evaluation. While all of the techniques for content adaptation from abstract information discussed so far have been engineered by researchers, recent and promising work is exploring how content adaptation strategies can be learned from user feedback on sample (multimedia) presentations [20]. This work relies on machine learning techniques that have been already successfully used in other NLG adaptation tasks [51]. As for evaluation, recent years have witnessed a surge of interest in empirically testing techniques for content adaptation from abstract information. Human judges [11], human designers [53] and task efficacy (e.g., [5], [10]) are the three basic methods that have been applied to evaluate such techniques. These methods are described under the category "controlled experiments" in Chapter 24 of this book [18], which provides a comprehensive overview of all empirical and non-empirical evaluation methods for adaptive Web systems.

13.3 Techniques for Content Presentation

In the previous section we discussed techniques to identify and structure the content most relevant to the interaction context. Here, we focus on techniques to effectively present content once its degree of relevance and its structure have been determined by content adaptation. In particular, we present techniques that decide how to present content based on its relevance, and techniques to select the type of media most appropriate to deliver the content, given the interaction context.

13.3.1 Relevance-Based Techniques

Most of the techniques that we categorize here as relevance-based were introduced in [7] and [32] as ways to manage canned content fragments. Following [52], in this section we will discuss them along two general dimensions, which we see as critical to both canned and generated content:

- Maintaining *focus*, i.e., how these techniques emphasize the content that has been classified as most relevant for the current user.
- Maintaining *context*, i.e., if and how they allow for access of the less relevant content so as to preserve the contextual information that it may provide.

There is an obvious tradeoff between these two dimensions: context is more easily maintained if much of the original content is visible to the user. However, the more content is shown, the higher the chance of generating information overload and reducing attention to the most relevant information, defeating one of the very reasons for having adaptive hypermedia in the first place.

The techniques we present here can be grouped in two main categories, depending upon how they address the context-focus tradeoff: Priority on Focus and Priority on Context.

Priority on Focus. All of the techniques in the Priority on Focus category choose to maximize focus by (a) showing the user only the content that is deemed most relevant, and (b) precluding access to the rest of the content. They include not only the fragment-variant techniques that we discussed in the previous section (i.e., optional fragments and altering fragments), but also any strategy for sophisticated content selection in which only the most relevant knowledge elements are presented to the user (see for instance Fig. 13.1 and Fig. 13.2 in section 13.2).

In addition to potentially loosing contextual information when limiting the content the user can see, Priority on Focus techniques suffer from two main drawbacks:

- They are highly impacted by the validity of the adaptation mechanism, as the user has no way to recover from bad adaptation.
- They do not allow for user control, one of the dimensions that defines usability in human-computer interaction and that significantly influences acceptance of adaptive interfaces ([27] and [28]).

Priority on Focus techniques seem to be mostly used in Adaptive Educational Hypermedia (see [13] and [38]), possibly because these systems are less subject to the above drawbacks. The pedagogical nature of the interaction makes it both easier for the system to create an accurate user model of relevant user traits, and possibly more acceptable for the user to have limited control over the computer tutor's adaptation decisions.

Priority on Context. The Priority on Context category includes the techniques known as (1) *stretchtext*, (2) *dimming fragments 3) colouring fragments,* 4) *sorting fragments* and 5) *scaling fragments.* The first four techniques have been around for a while and were previously addressed in adaptive hypermedia reviews (e.g.,[7]). The scaling fragments technique is a more recent attempt to adapt the well known *fisheye* visualization technique [17] to content organization in adaptive hypermedia. For this reason, we will describe it in slightly more detail than the other four. In general, all techniques presented in this section try to preserve the context around the most relevant content by providing different ways to make the less relevant information visible without distracting the user from the primary content. They differ, however, along the following dimensions: 1) whether or not the surrounding context is visible, 2) whether or not they permit structural information to be maintained, and 3) whether or not they can convey different levels of relevance and/or priority information. We first describe the techniques and then discuss how they differ along these key dimensions.

Stretchtext, arguably the most well established of the techniques, relies on placeholders to signal the presence of and allow access to information of secondary importance (see for instance [6] and [26]). Usually the place holders are short headers summarizing the hidden information, as shown in the screen shot of the Push system [26] in Fig. 13.3 (see items labeled as IE – Information Entities – and Hotlist). While this technique has been mostly used with text (hence the name), researchers have started generalizing it to the adaptive presentation of multimedia content, for instance in Interactive TV applications [36].

Stretchtext preserves focus by hiding the less relevant content when a page is first presented.   In contrast, the remaining four techniques *deemphasize* rather then hide the less relevant content. Dimming deemphasizes the less relevant content by fading its color [6].   Colouring, which has traditionally been used for adaptive navigation support (see Chapter 8 of this book [8]), highlights the more relevant content using one or more colours. Sorting, also most frequently used for adaptive navigation support, deemphasizes less relevant content through fragment ordering.   [25]. Scaling deemphasizes by reducing size. In scaling, size increases as a function of the Degree of Interest (DOI) of each content fragment for the user [52], as assessed by a similarity measure between vectors representing the user's focus of interest and the content of each available fragment. In this respect, scaling is a variation of the fisheye visualization technique, with the following differences:

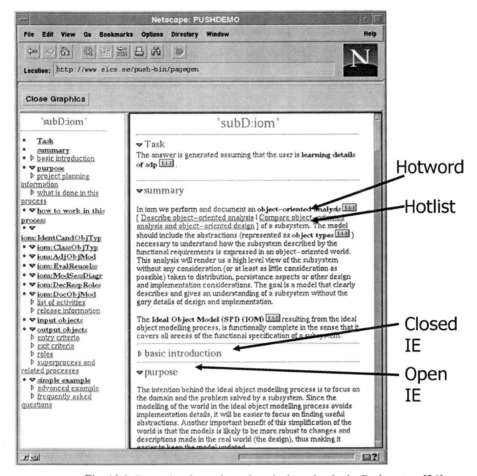

Fig. 13.3. Example of stretchtext-based adaptation in the Push system [26].

- In a traditional fisheye visualization, there is a unique focal point, based on the user's current focus of attention. In scaling, there can be multiple focal points based on the user's focus of interest.
- In traditional fisheye views, DOI (and thus content size) decreases with geometrical distance from the focal point, while in scaling, DOI decreases with *semantic* distance.

Fig. 13.4 shows an example of scaling from [52]. Here the user's focus of interest is assessed to be theater, thus paragraphs with different degree of relation with this topic are presented in different font sizes.

Table 13.1 compares the five techniques discussed in this section. Of the five techniques, stretchtext is the only technique that does not make the surrounding context visible, while sorting is the only one that doesn't maintain structural information. As a result, sorting is suitable only for fragments that are not structurally

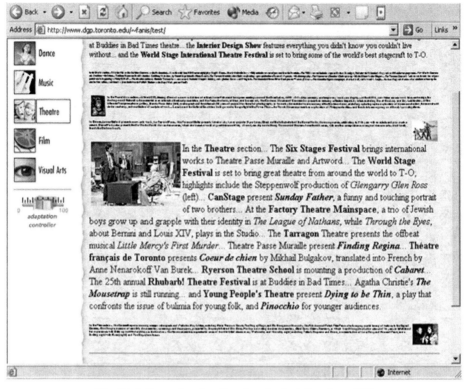

Fig. 13.4. Example of scaling-based adaptation from [52]

Table 13.1. Comparision of *Priority on Context* techniques

Technique	Context Visible	Structural Information Preserved	Priority Conveyed
Strechtext	No	Yes	No
Dimming	Yes	Yes	No
Colouring	Yes	Yes	To a limited degree
Sorting	Yes	No	Yes
Scaling	Yes	Yes	Yes

related, such as bulleted lists or self-contained sub topics. In terms of conveying priority information, the existing incarnations of dimming equally deemphasize all of the less relevant content; the same is true of stretchtext. In comparison, scaling, sorting and colouring do allow priority information to be displayed. In scaling, size conveys different degrees of relevance. With sorting, the order of appearance on the page conveys the relative relevance. With colouring, more than one colour can be used; however, there is a limit to the number of colours to which the user will be able to attribute meaning.

Although all five techniques in this category do provide more contextual information than those in the Priority on Focus category, their relative effectiveness in different applications is still an open question.

The scaling technique preserves distinctive structural elements of the deemphasized information (such as pictures, layout, number and relative length of paragraphs), and thus has better potential to provide contextual cues than stretchtext. Tsandilas and Shraefel [52] conducted a preliminary study to compare stretchtext and scaling, but the small number of subjects does not allow drawing general conclusions. They found no difference in terms of task completion time, but identified a potential interaction between technique effectiveness and page size. Because stretchtext generally presents less content, it performed better on large pages, where scaling required more scrolling to access relevant content. However, 4 out of the 6 subjects in the study gave a higher overall score to scaling because they felt it provided better information on the content of the deemphasized paragraphs.

Like scaling, dimming and colouring also preserve distinctive structural elements of the available content, but do not reduce page size (like stretchtext and scaling do). Thus, they may require additional scrolling to access the content of interest, potentially reducing focus. Finally, sorting does not preserve structure, however, placing the more relevant material at the top of the page has the advantage that scrolling is necessary for only the less relevant content. Further evaluations are needed to better understand the pros and cons of the above techniques.

Even as more empirical results become available, practitioners interested in adopting these techniques should be aware that their effectiveness depends on a number of design elements that should be carefully tested before drawing general conclusions on each technique's overall effectiveness. These elements include:

1. Quality of the headers used to indicate the presence of stretchtext ([26] and [48])
2. Cost associated with reading the deemphasized content (e.g., fatigue generated by reading small or faded text)
3. Presence and quality of ways to summarize deemphasized content so that a user can get contextual information without reading it (see for instance the mouse-over glosses described in [52])
4. Presence and effectiveness of mechanisms provided to the user to change content emphasis (e.g., double clicking on deemphasized content to bring it in focus [52])

In relation to the first element above, we describe a technique that, although currently not used in adaptive hypermedia, has an interesting potential for enriching the Priority on Context category. This technique, known as *summary thumbnail* [35] has been devised to address the problem of how to display Web pages designed for desktop-sized monitors on small screen devices. One common way to address this problem is to rescale the page to fit the width of the small screen, thus creating what is known as a *thumbnail view*. The idea is that the user should use this view to rapidly identify the content of interest and then use provided zooming mechanisms to view it. The problem is that often text in the thumbnail view becomes unreadable, forcing the user to resort to zooming to browse the content. Summary thumbnail addresses this problem by adding fragments of readable text to the thumbnails (as opposed to shrinking the text of the original page). The text is automatically generated from the original Web page, by either removing common words (as defined in a standard word frequency list) or by cropping paragraphs, so as to maintain the total number of lines and overall page layout.

Lam and Baudisch [35] showed that this technique generated better user performance and satisfaction in a browsing task, compared to two other standard techniques for page reduction. This indicates that having readable, although incomplete and possibly not fully coherent text is an effective place holder for hidden content. Thus, this technique could be adapted as a variation of stretchtext for adaptive content presentation, where place holders for hidden text are summary thumbnails. However, it should be noted that in Lam and Baudisch's experiment [35], all content was equally reduced. In order to verify the applicability of this technique to adaptive approaches, it will be necessary to test whether it remains effective when summary thumbnails are used in combination with fully-displayed content during adaptive presentation.

13.3.2   Techniques for Media Adaptation

The previous sections discussed techniques to adaptively select, structure, and present relevant information. Presentation has been addressed, however, with respect only to the problem of how to highlight relevant information and how to allow the user access to relevant context. Here we will address a different form of presentation tailoring: adapting the medium (e.g., text, graphics, spoken language) through which the selected information is conveyed to the user. We begin by discussing factors that can influence a system's choice of media. We then provide illustrative examples of adaptive hypermedia systems that adapt the medium through which information is presented. Finally, we discuss at a conceptual level two common approaches to media adaptation: the rule-based approach and the optimization approach.

Factors Relevant for Media Adaptation. The following is a description of the types of factors that a system may want to consider when deciding how to adapt the media:

- *User-Specific Features*: Relevant user features include preferences, abilities and accessibility issues. Users may have preferences for receiving information in different modalities, for example, a user may explicitly request the information to be presented in a graphical way [54]. In terms of abilities, the user may be better able to reason about information presented using a given medium. For example, if the user has poor language abilities, the material would be better presented visually [21]. In contrast, a visually-impaired user should not be presented with information in this manner, but rather through speech (e.g., [16]).
- *Information Features:* Given the presentation goals, certain types of information are better presented using specific media. For example, graphics should be used to highlight quantitative relationships [30], while text should be used if the system wants to convey a precise value, such as the name of a city [54].
- *Contextual Information:* Under certain circumstances it may be appropriate for the system to consider properties of the user's environment when deciding how to best present information. As an example, for systems designed for use in vehicles, variables such as the current weather conditions and other relevant operating conditions

Fig. 13.5. The AVANTI system [16] with information presented graphically.

Fig. 13.6. The AVANTI system [16] with information presenting using text.

(e.g., speed and traffic) should influence both the appropriate media type and the quantity of information presented [12]. For instance, in situations where the weather is poor, or the user is driving in a high speed zone, designers are exploring the use of media such as haptics and speech to avoid placing additional burden on the already loaded visual channel (e.g., [4] and [12]).

- *Media Constraints:* When using multiple media, at times these media should cooperate to ensure the best overall presentation. For example, a mix of compatible media (e.g., text and speech) can increase the recallability of the information presented [54], and contrasting pieces of information should be presented using the same medium (e.g., [1] and [54]).
- *Limitations of Technical Resources:* This factor relates to the device on which the information will eventually be displayed and the possible media limitations of this device. Examples of resource limitations include the available bandwidth (e.g., large images should not be displayed if there is low bandwidth [16, 34]), the general availability of different media (e.g., whether or not speech is available [45]) or the available screen real estate.

Example Applications. We now describe some examples of systems that aim to perform some sort of media adaptation in an adaptive hypermedia system or adaptive Web site. The first two examples are forms of tourist information systems. The remaining examples are: a mobile navigation system, a learning environment and a pointer to a more general framework for generating multimedia presentations on the Web.

The AVANTI system [16] adapts the media through which the relevant information is presented to the user according to 1) accessibility issues, for example, using spoken language for visually-impaired users, and 2) resources issues, such as not presenting too many graphics in low-bandwidth situations. Fig. 13.5 and Fig. 13.6 illustrate two example versions of a page containing information on a city map: one for sighted users and another for visually-impaired users. For visually-impaired users, the map is described using text (which could presumably be read by a speech synthesizer) (Fig. 13.6). Alternatively, a regular graphical map is displayed for sighted users (Fig. 13.5).

The MASTROCARONTE system [12], an in-car tourist information system, adapts the medium through which recommendations are made to the user by considering 1) the user-specific factors, such as preferences and the user's current level of fatigue (estimated based on the time of day and the length of the current trip), and 2) contextual factors, such as speed and traffic volume. For example, if the user is requesting restaurant recommendations, the system determines the current level of risk as indicated by the contextual factors (e.g., traffic or visibility) and the user's current level of fatigue to decide a) whether to present the recommendations visually or using speech, and b) how many recommendations to present at one time. If the contextual factors indicate a high-level of risk (e.g., high traffic or low visibility) *or* if the user is fatigued, the system will elect to present the recommendations visually, and will present only one recommendation at a time (requiring the user to press a button to retrieve additional recommendations). If there is both a high-level of risk *and* the user is fatigued, the system will present the recommendations using speech. If neither condition holds, the system will present several recommendations at once and will present them visually.

In [34], the authors describe guidelines to be used by a mobile navigation system to decide which medium to use when presenting route information. These guidelines take into account technical resource limitations (the last factor described in the previous section), as well as the user's context and cognitive resources. Route descriptions using text and speech are appropriate only if the system knows the user's location and are useful when the amount of available bandwidth is small (since no images have to be sent to the mobile device). Speech is preferred to text in this scenario if the user is also experiencing a high level of cognitive load since it does not require visual attention. Graphical route instructions do not require location information (in their system), but require more cognitive and technical resources.

The CUMAPH adaptive hypermedia environment [21] adapts hypermedia documents according to a user profile that describes the user's cognitive abilities, such as the user's ability to explore visual and spatial information, and the user's visual and auditory memory. A description of how the profile is acquired can be found in [22]. The environment has so far been applied to an instructional domain involving a course about the human brain. As an example, when adapting a Web page on the topic of "memory", the system uses the optimization approach described in the next section to decide whether to display the content of two relevant sub items ("definition" and "mechanism") using text, graphics, sound or a combination of two of the three media.

Finally, the hypermedia document formatter described in [45] includes capabilities to adapt media presentation to device characteristics and user preferences. The paper does not illustrate an application of the framework. Instead, the authors focus primarily on the architecture necessary to realize multimedia presentations on the Web.

General Approaches. The most common techniques for media adaptation fall into 2 general categories: 1) *rule-based* or *planning* approaches and 2) *optimization* approaches. We will now describe each type of technique. Once again, we extend our discussion to include systems that are not adaptive Web-based systems since we feel that the techniques could certainly be generalized to adaptive Web applications.

Rule-Based Approach. The vast majority of systems that perform media adaptation do so using rules that describe how to best convey the target information given subsets of the factors described in earlier in the section. Examples of such systems include [2], [3], [15], [16], [19], [23], [37] and [45].

Andre [1] discusses key differences among several of the above systems. A primary difference is how integrated the content selection process (as discussed in section 13.2.2) is with the media allocation process (i.e., determining which medium to use for a given information element). The systems also differ in the type and amount of communication among the components in charge of realizing the media-specific information (to ensure a coherent and feasible overall result). To illustrate these differences and to provide a more concrete understanding of how media allocation using rules can work, we will now elaborate on two examples: the work by Arens *et al.* [3] and the WIP system [2].

Arens *et al.* [3] describe a system that can adapt the media based on characteristics of the information to be conveyed, media constraints, the user's interests and abilities, and the overall goals of the information presentation (referred to as the "presenter's goals"). The system begins by selecting and structuring the content (using NLG techniques similar those discussed in section 13.2.2), which produces a tree that represents the discourse structure. The system then applies media allocation rules to the discourse structure to obtain a presentation structure. It applies these rules by traversing the tree in a bottom-up manner. As the rule-application process moves up the tree, earlier media decisions can be reconsidered based on information about the more global structure of the presentation, which is only visible at the higher-level nodes. Such a situation could occur if, for example, content item A is assigned to be presented using text and content item B graphics, but the higher-level discourse goal is to compare the two items. With this higher-level goal, both items should be presented using the same medium. Once the presentation structure is complete, each element is then sent to the appropriate generator, and all of the results are sent to a final layout specialist, which decides how to arrange them on the target display.

The WIP system, described in [2], interleaves content planning and media assignment. WIP focuses primarily on adapting the media based on characteristics of the information to be displayed, but the authors indicate that their technique could be extended to take additional factors into consideration. Generating the presentation involves a task-decomposition schema-based approach similar to what is often used for content structuring in NLG (see section 13.2.2). To permit media allocation, WIP's schemas (referred to as "presentation strategies") have an additional slot for the medium through which the information should be conveyed. Some schemas specify the appropriate medium, other schemas leave this slot open, to be filled later in the planning process. Using the schemas, the system engages in a top-down planning process, starting from the high-level presentation goal. When more than one schema is applicable, but they have different media assignments, the system selects which schema to use based on meta-rules relating to how well the media accomplish the current presentation goal. As soon as a media slot is filled, the given presentation goal is sent to the generator in charge of realizing information through that medium. This has the advantage that a given presentation goal can be refined by a generator, if it is not able to fulfill the request. Such a situation may occur if, for example, a generator is asked to display a piece of information using an image, but doing so would require an image that is larger than the amount of remaining screen real estate.

Andre [1] argues for content selection/structuring to be interleaved with media allocation because it allows the media allocation process to inform the discourse structure. However, if one does not want media selection to influence content selection/structuring, or wants to use an existing content selection/structuring process as is, it makes sense to employ the sequential approach. Arens *et al.* [3] also argue for the simplicity of the sequential approach.

Optimization Approach. An alternative to a rule-based approach is to formulate the media adaptation process as an optimization problem. That is, given information on the relevant factors, the goal is to find the media combination that produces the best overall result. Examples of systems that follow this approach are the RIA system [54] and the CUMAPH system [21]. As discussed in section 13.2.2, RIA optimizes a set of feature-based metrics to perform tailored content selection. RIA follows a similar optimization approach to adapt the media once the relevant information elements have been selected. The authors present a comprehensive list of feature-based metrics that

U2 *Speech*: How much are those?	**U3** *Speech*: How much is this one? *Gesture*: Point to a house on the screen
RIA The prices are shown on the screen.	**RIA** The asking price is $499,000.

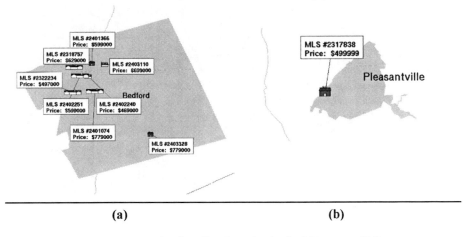

(a)	(b)

Fig. 13.7. Example of media adaptation in the RIA system [54]

they use to find the optimal media allocation. These metrics fall into two categories: 1) media selection metrics that indicate which media best suits a particular content item (based on user preferences, properties of the user's task, and properties of the content items themselves) and 2) presentation coordination metrics that describe the level of coordination among the different media. These metrics are combined in an overall objective function and the media allocation process then becomes the task of maximizing this objective function.

Fig. 13.7 shows the RIA system responding to two similar user requests (U2 and U3) with different media allocation decisions. In (a), the user (U2) is requesting the prices of a number of houses, whereas in (b) the user (U3) is requesting the price of a particular house. We use this example to show how the system's optimization procedure is able to deal with three metrics: the suitability of the information to the media, the desire to increase recallability, and the desire to maintain presentation consistency (the first is a media selection metric and the latter are two presentation coordination metrics). In (a), because of the high-volume of data, the price information should be displayed visually and speech can be used only to focus the user's attention to the screen. In contrast, the low data volume in (b) allows the system to display the price information visually and also to mention the price using speech, which increases recallability. In both examples, because house location, city name and city boundary are considered to be inter-related, all three are presented visually.

The CUMAPH system [21] follows a similar, but scaled-down approach. They have 2 metrics: one assigns the highest value to the media combination that best fits the user profile; the other favors combining multiple media (e.g., using both graphics and text as opposed to just graphics). The system generates all possible combinations of media assignments to information items and picks the one whose sum of the two metrics is the highest. This type of exhaustive generation and testing is likely only possible when the number of combinations is small.

Both of these systems perform media allocation after the content has been selected. Future work in the area could involve investigating ways to use the optimization approach to interleave the two processes.

Comparison of Approaches. According to [53] and [54], the advantages of the optimization approach are: 1) it does not require a large set of rules and/or plans to be authored, 2) it allows the system to handle issues with conflicting or interdependent factors without a large amount of communication among different system components, 3) it is more easily extended, and 4) it is more easily transferred to different domains. A downside of the optimization approach is that there do appear to be cases where the system is required to repair the media allocation after the optimization process is complete [54]. In the case of [54], these situations are detected and repaired by relying on rules.

We are not aware of any evaluations that have directly compared the rule-based and the optimization approaches. In fact, there has been very little evaluation of systems that perform media adaptation. One exception can be found in [54], where RIA's multimedia presentations were evaluated using the human-designer approach (described in Chapter CH6 [18]). The results of the experiment were promising. Out of 50 test cases, RIA was rated (or co-rated) highest in 17 cases. In the 28 of the remaining cases, the difference between the winner and RIA was minor. For the other 5 cases, the authors felt that the problem was due to content selection rather than media selection. Thus, in the majority of cases (45 out of 50), RIA was able to perform as well, almost as well, or even better than a human interface designer.

13.4 Conclusions

In this chapter we discussed techniques for providing adaptive content presentation for the Web. We structured the discussion by first introducing techniques to select and coherently structure the content deemed to be most relevant for the current interaction context, and then by illustrating approaches to further adapt the selected content by tailoring the way in which it is actually presented to the user. In particular, we have described techniques to adapt the presentation based on content relevance, as well as techniques to adapt the type of media for optimal content delivery.

Adaptive presentation of Web content can complement several of the other types of adaptations presented in this book. For instance, it can be integrated with both adaptive navigation support and adaptive search to further adapt the content of the Web pages returned by these two techniques to a user's needs; or it can be coupled with recommender systems to provide tailored descriptions of the recommended items. Although there has been some work on exploiting the integration of adaptive presentation with other types of adaptations (e.g., [10]) we see this as an area that still yields a high potential for innovative research in the adaptive Web. Another area that calls for strong research efforts is the validation of techniques for adaptive presentation. In this chapter we have discussed the results of some evaluations; however, the actual effectiveness of the many of the techniques is still largely open to investigation.

Acknowledgements. We would like to thank Kasia Muldner for her comments on multiple versions of the chapter. The anonymous cross reviewers and the editors also provided us with numerous valuable suggestions.

References

1. Andre, E.: The Generation of Multimedia Documents. In: R. Dale, H. Moisl, Somers, H. (eds.): A Handbook of Natural Language Processing: Techniques and Applications for the Processing of Language as Text, Marcel Dekker Inc,(2000) 305-327
2. Andre, E., Rist, T.: Generating Coherent Presentations Employing Textual and Visual Material. AI Review 9 (1995) 147-165
3. Arens, Y., Hovy, E. H., van Mulken, H.: Structure and Rules in Automated Multimedia Presentation Planning. In: Proc. of the 13th. International Joint Conference on Artificial Intelligence (1993) 1253-1259
4. Bellotti, F., Gloria, A. D., Montanari, R., Dosio, N., Morreale, D.: Comunicar: Designing a Multimedia, Context-Aware Human-Machine Interface for Cars. Cognition, Technology & Work 7 (2005) 36-45
5. Bontcheva, K., Wilks, Y.: Tailoring Automatically Generated Hypertext. User Modeling and User-Adapted Interaction 15 (2005) 135-168
6. Boyle, C., Encarnacion, A. O.: Metadoc: An Adaptive Hypertext Reading System. User Modeling and User-Adapted Interaction 4(1) (1994) 1-19
7. Brusilovsky, P.: Methods and Techniques of Adaptive Hypermedia. User Modeling and User-Adapted Interaction 6(2-3) (1996) 87-129.
8. Brusilovsky, P.: Adaptive Navigation Support. In Brusilovsky, P., Kobsa, A., Nejdl, W. (eds.): The Adaptive Web: Methods and Strategies of Web Personalization, Lecture Notes in Computer Science, Vol. 4321. Springer-Verlag, Berlin Heidelberg New York (2007) this volume
9. Burke, R.: Hybrid Web Recommender Systems. In: Brusilovsky, P., Kobsa, A., Nejdl, W. (eds.): The Adaptive Web: Methods and Strategies of Web Personalization, Lecture Notes in Computer Science, Vol. 4312. Springer-Verlag, Berlin Heidelberg New York (2007) this volume
10. Carenini, G., Moore, J. D.: Generating and Evaluation Evaluative Arguments. Artificial Intelligence 170(11) (2006) 925-952
11. Chu-Carroll, J., Carberry, S.: Collaborative Response Generation in Planning Dialogues. Computational Linguistics 24(2) (1998) 355-400
12. Console, L., Torre, I., Lombardi, I., Gioria, S., Surano, V.: Personalized and Adaptive Services on Board a Car: An Application for Tourist Information. Journal of Intelligent Information Systems 21(3) (2003) 249-285
13. De Bra, P., Aerts, A., Berden, B., Lange, B. d., Rousseau, B., Santic, T., Smits, D., Stash, N.: Aha! The Adaptive Hypermedia Architecture. In: *Proc. of the Fourteenth ACM Conference on Hypertext and Hypermedia* (2003) 81-84
14. Encarnacao, L. M., Stoev, S. L.: An Application-Independent Intelligent User Support System Exploiting Action-Sequence Based User Modelling. In: *Proc. of UM'99, International Conference on User Modeling* (1999)
15. Feiner, S., McKeown, K.: Automating the Generation of Coordinated Multimedia. IEEE Computer 24(10) (1994) 33-41
16. Fink, J., Kobsa, A., Nill, A.: Adaptable and Adaptive Information Provision for All Users, Including Disabled and Elderly People. The New Review of Hypermedia and Multimedia 4 (1998) 163-188
17. Furnas, G. W.: Generalized Fisheye Views. In: Proc. of CHI '86, Conference on Human Factors in Computing Systems (1986) 16-23

18. Gena, C., Weibelzahl, S.: Usability Engineering for the Adaptve Web. In: Brusilovsky, P., Kobsa, A., Nejdl, W. (eds.): The Adaptive Web: Methods and Strategies of Web Personalization, Lecture Notes in Computer Science, Vol. 4321. Springer-Verlag, Berlin Heidelberg New York (2007) this volume
19. Green, N. L., Carenini, G., Kerpedjiev, S., Mattis, J., Moore, J. D., Roth, S. F.: Autobrief: An Experimental System for the Automatic Generation of Briefings in Integrated Text and Information Graphics. International Journal on Human-Computer Studies 61(1) (2004) 32-70
20. Guo, H., Stent, A.: Trainable Adaptable Multimedia Presentation Generation. In: *Proc. of the 7th International Conference on Multimodal Interfaces* (2005)
21. Habieb-Mammar, H., Tarpin-Bernard, F.: Cumaph: Cognitive User Modeling for Adaptive Presentation of Hyper-Documents. An Experimental Study. In: *Proc. of AH 2004, Third International Conference on Adaptive Hypermedia and Adaptive Web-Based Systems* (2004) 136-145
22. Habied-Mammar, H., Tarpin-Bernard, F., Prevot, P.: Adaptive Presentation of Multimedia Interface Case Study: "Brain Story" Course. In: *Proc. of UM'03, International Conference on User Modelling* (2003) 15-24
23. Han, Y., Zukerman, I.: Using Cooperative Agents to Plan Multimodal Presentations. In: *Proc. of Multimodal Human-Computer Communication* (1995) 122-157
24. Henze, N., Nejdl, W.: Extendible Adaptive Hypermedia Courseware: Integrating Different Courses and Web Material. In: *Proc. of the International Conference on Adaptive Hypermedia and Adaptive Web-Based Systems* (2000) 109-120
25. Hohl, H., Bocker, H.-D., Gunzenhauser, R.: Hypadapter: An Adaptive Hypertext System for Exploratory Learning and Programming. User Modeling and User Adapted Interaction 6(2-3) (1996) 131-156
26. Hook, K.: Evaluating the Utility and Usability of an Adaptive Hypermedia System. In: *Proc. of IUI'97, International Conference on Intelligent User Interfaces* (1997) 179-186
27. Hook, K.: Steps to Take before Intelligent User Interfaces Become Real. Interacting with Computers 12 (2000) 409-426
28. Jameson, A.: Adaptive Interfaces and Agents. In: Jacko, J., Sears, A. (eds.): Human-Computer Interaction Handbook, Vol. Erlbaum, Mahwah, NJ (2003) 305-330
29. Jameson, A., Schafer, R., Simons, J., Weis, T.: Adaptive Provision of Evaluation-Oriented Information: Tasks and Techniques. In: Proc. of International Joint Conference on Artificial Intelligence (1995) 1886-1893
30. Kerpedjiev, S., Carenini, G., Green, N., Moore, J., Roth, S.: Saying It in Graphics: From Intentions to Visualizations. In: Proc. of the IEEE Symposium on Information Visualization (1998) 97-101
31. Knott, A., Dale, R.: Choosing a Set of Coherence Relations for Text Generation: A Data-Driven Approach. In: Adorni, G., Zock, M. (eds.): Trends in Natural Language Generation: An Artificial Intelligence Perspective, Springer-Verlag, Berlin (1996) 47-67
32. Kobsa, A., Koenemann, J., Pohl, W.: Personalized Hypermedia Presentation Techniques for Improving Online Customer Relationships. The Knowledge Engineering Review 16(2) (2001) 111-155
33. Korb, K., McConachy, R., Zukerman, I.: A Cognitive Model of Argumentation. In: Proc. of Proc. of the Nineteenth Annual Conference of the Cognitive Science Society (1997) 400-405
34. Kray, C., Laakso, K., Elting, C., Coors, V.: Presenting Route Instructions on Mobile Devices. In: Proc. of IUI'03, International Conference on Intelligent User Interfaces (2003) 117-124
35. Lam, H., Baudisch, P.: Summary Thumbnails: Readable Overviews for Small Screen Web Browsers. In: Proc. of CHI 2005, Conference on Human Factors in Computing Systems (2005) 681-690

36. Masthoff, J., Pemberton, L.: Adaptive Hypermedia for Personalized Tv. In: Chen, S., Magoulas, G. (eds.): Adaptable and Adaptive Hypermedia Systems, IDEA group publishing,(2005)
37. Maybury, M.: Planning Multimedia Explanations Using Communicative Acts. In: Maybury, M. (ed.) Intelligent Multimedia Interfaces, AAAI Press ' The MIT Press,(1993) 60-74
38. Melis, E., Andres, E., Franke, A., Frischauf, A., Goguadse, G., Libbrecht, P., Pollet, M., Ullrich, C.: Activemath: A Web-Based Learning Environment. International Journal of AI in Education 12 (2001) 385-407.
39. Micarelli, A., Gasparetti, F., Sciarrone, F., Gauch, S.: Personalized Search on the World Wide Web. In: Brusilovsky, P., Kobsa, A., Nejdl, W. (eds.): The Adaptive Web: Methods and Strategies of Web Personalization, Lecture Notes in Computer Science, Vol. 4321. Springer-Verlag, Berlin Heidelberg New York (2007) this volume
40. O'Donnell, M., Mellish, C., Oberlander, J., Knott, A.: ILEX: An Architecture for a Dynamic Hypertext Generation System. Journal of Natural Language Engineering 7(3) (2003) 225-250
41. Paris, C., Wan, S., Wilkinson, R., Wu, M.: Generating Personal Travel Guides - and Who Wants Them? In: *Proc. of UM'01, International Conference on User Modeling* (2001) 251-253
42. Pazzani, M. J., Billsus, D.: Content-Based Recommendation Systems. In: Brusilovsky, P., Kobsa, A., Nejdl, W. (eds.): The Adaptive Web: Methods and Strategies of Web Personalization, Lecture Notes in Computer Science, Vol. 4321. Springer-Verlag, Berlin Heidelberg New York (2007) this volume
43. Reiter, E., Dale, R.: Building Natural Language Generation Systems. Cambridge University Press (2000)
44. Reiter, E., Robertson, R., Osman, L.: Types of Knowledge Required to Personalise Smoking Cessation Letters. In: *Proc. of the Joint European Conference on Artificial Intelligence in Medicine and Medical Decision Making (AIMDM'99)* (1999) 389-399
45. Rodrigues, R. F., Rodrigues, P. S. L., Feijo, B., Velho, L., Soares, L. F. G.: Cross-Media and Elastic Time Adaptive Presentations: The Integration of a Talking Head Tool into a Hypermedia Formatter. In: *Proc. of AH 2004, Third International Conference on Adaptive Hypermedia and Adaptive Web-Based Systems* (2004) 215-224
46. Russell, S., Norvig, P.: Artificial Intelligence: A Modern Approach. second edn. Morgan-Kaufman, Los Altos, CA (2003)
47. Schafer, J. B., Frankowski, D., Herlocker, J. L., S.Sen: Collaborative Filtering Recommender Systems. In: Brusilovsky, P., Kobsa, A., Nejdl, W. (eds.): The Adaptive Web: Methods and Strategies of Web Personalization, Lecture Notes in Computer Science, Vol. 4321. Springer-Verlag, Berlin Heidelberg New York (2007) this volume
48. Schraefel, M. C.: Contexts: Adaptable Hypermedia. In: *Proc. of AH 2000, International Conference on Adaptive Hypermedia and Adaptive Web-based Systems* (2000) 369-374
49. Smyth, B.: Case-Base Recommendation. In: Brusilovsky, P., Kobsa, A., Nejdl, W. (eds.): The Adaptive Web: Methods and Strategies of Web Personalization, Lecture Notes in Computer Science, Vol. 4321. Springer-Verlag, Berlin Heidelberg New York (2007) this volume
50. Sowa, J.: Conceptual Structures: Information Processing in Mind and Machine. Addison Wesley (1984)
51. Stent, A., Prasad, R., Walker, M.: Trainable Sentence Planning for Complex Information Presentation in Spoken Dialog Systems. In: *Proc. of 42nd Annual Meeting of the Association for Computational Linguistics* (2004)
52. Tsandilas, T., schraefel, m.: Usable Adaptive Hypermedia. The New Review of Hypermedia and Multimedia 61(6) (2004) 5-29

53. Zhou, M. X., Aggarwal, V.: An Optimization-Based Approach to Dynamic Data Content Selection in Intelligent Multimedia Interfaces. In: *Proc. of UIST'04, the 17th Annual ACM Symposium on User Interface Software* (2004) 227-236
54. Zhou, M. X., Wen, Z., Aggarwal, V.: A Graph-Matching Approach to Dynamic Media Allocation in Intelligent Multimedia Interfaces. In: *Proc. of IUI'05, International Conference on Intelligent User Interfaces* (2005) 114-121
55. Zukerman, I., McConachy, R., Korb, K.: Using Argumentation Strategies in Automated Argument Generation. In: *Proc. of the First International Natural Language Generation Conference* (2000) pp. 55-62

14

Adaptive 3D Web Sites

Luca Chittaro and Roberto Ranon

HCI Lab, Dept. of Math. and Computer Science, University of Udine,
via delle Scienze 206, 33100 Udine, Italy
{chittaro, ranon}@dimi.uniud.it

Abstract. In recent years, technological developments have made it possible to build interactive 3D models of objects and 3D Virtual Environments that can be experienced through the Web, using common, low-cost personal computers. As in the case of Web-based hypermedia, adaptivity can play an important role in increasing the usefulness, effectiveness and usability of 3D Web sites, i.e., Web sites distributing 3D content. This paper introduces the reader to the concepts, issues and techniques of adaptive 3D Web sites.

14.1 Introduction

In recent years, technological developments have made it possible to build interactive 3D models of objects and 3D Virtual Environments (hereinafter, 3D VEs) that can be experienced through the Web, using common, low-cost personal computers. As a result, 3D content is increasingly employed in different Web application areas, such as education and training [18, 30, 40], e-commerce [26, 36], architecture and tourism [42, 44], virtual communities [2,45] and virtual museums [4].

Web sites distributing 3D content (hereinafter, we call them *3D Web sites* for simplicity) can be divided into two broad categories:

- sites that display interactive 3D models of objects embedded into Web pages, such as e-commerce sites allowing customers to examine 3D models of products [26], and
- sites that are mainly based on a 3D VE which is displayed inside the Web browser, such as tourism sites allowing users to navigate inside a 3D virtual city [44].

In the first case, the primary information structure and user's interaction methods are still based on the hypermedia model, with the additional possibility of inspecting 3D objects. In the second case, the primary information structure is a 3D space, within which users move and perform various actions. For example, a furniture e-commerce site might be based on a 3D virtual house where users can walk, choose furniture from a catalogue, and place it in the various rooms [36].

3D Web sites are not meant to substitute the hypermedia model which is the mainstream in today's Web, but they can be more effective when there is added value in

P. Brusilovsky, A. Kobsa, and W. Nejdl (Eds.): The Adaptive Web, LNCS 4321, pp. 433–462, 2007.

interacting with a 3D visualization, or in providing a first-person virtual experience close to a real-world one. For example, in the case of e-commerce, 3D models give customers the ability to visually inspect, manipulate, try and customize products before purchasing as they are accustomed to do in the real world [27]. In the case of cultural heritage, a Web museum implemented as a 3D VE allows one not only to display the museum items, but also to convey their "cultural setting" by placing them in a proper environment.

As in the case of Web-based hypermedia, adaptivity can play an important role in increasing the usefulness, effectiveness and usability of 3D Web sites. For example, an intelligent adaptive navigation support system could help users with different navigation abilities in finding targets, orienting themselves, and gaining spatial knowledge of the environment. Unfortunately, there are currently no well-established techniques or commercial tools to build adaptive 3D Web sites. Moreover, because of conceptual and technical peculiarities of 3D Web sites, most approaches, techniques and software tools developed for the Adaptive Web cannot be straightforwardly applied to personalize 3D Web content, navigation and presentation. However, some research projects have addressed the issue of adaptivity for 3D Web sites. For example, a first software architecture [17] for dynamic construction of personalized 3D Web content has been proposed and applied to e-commerce [14,16] and virtual museums [13]. Some researchers have developed methods for personalized navigation support [12,27], adaptive interaction [11] and content presentation [24] in 3D VEs. Recently, there have been some attempts at experimenting with general-purpose frameworks for Web adaptivity to deliver personalized 3D content [15,21].

This Chapter will introduce the reader to the concepts, issues and techniques of adaptive 3D Web sites. We will mainly focus on 3D Web sites based on 3D VEs, since this category is the most general and complex one (but most of the techniques we will present can be applied also to Web sites with interactive 3D objects). The Chapter is structured as follows. Section 14.2 provides an introduction to 3D Web sites for the novice reader, overviewing the major application areas, and mentioning the main technologies, with a focus on standards. Section 14.3 discusses adaptivity in the context of 3D Web sites and with respect to Web-based hypermedia, separating the problems of modeling and adaptation. Section 14.4 describes an example of a full generic architecture for adapting 3D Web content, which is instantiated in Section 14.5 considering a detailed example in the domain of e-commerce. Finally, Section 14.6 concludes the Chapter.

14.2 3D Web Basics

The languages, protocols and software tools that make it possible to build 3D models and 3D VEs that can be experienced through the Web are collectively identified with the term *Web3D technologies*. Nowadays, thanks to the increase in network bandwidth and processing power (especially 3D graphics capabilities), Web3D technologies allow a large number of users worldwide to experience complex 3D Web content, such as virtual cities, visualizations of scientific data, or virtual museums.

Web3D technologies are based on the basic technical and architectural choices typical of Web technologies: content, represented in a proper (and typically textual) format, is stored on a server, requested by a client, typically through HTTP, and displayed by a browser, or, more often, by a plug-in for a Web browser. As a result, 3D content can be strongly integrated with other kinds of Web content, by augmenting Web sites with 3D interactive objects (a 3D model can appear into a Web page together with HTML content) as well as by displaying most types of Web content (such as images, sounds, videos) inside a 3D VE accessible through the Web. This is the main distinctive features of Web3D technologies with respect to other kinds of interactive 3D graphics-related technologies, such as those historically employed in Virtual Reality. Moreover, while Virtual Reality typically focuses on immersive 3D experiences, for example employing head-mounted displays and data gloves, 3D Web content is typically experienced with the input/output devices of today's common personal computers (CRT or LCD monitor, keyboard and mouse).

14.2.1   Applications and Motivations

In the following, we overview the main application domains for 3D Web sites, present possible advantages for using 3D content on the Web, and cite some available systems.

14.2.1.1   Learning and Training

3D VEs offer the possibility to reproduce the real world or to create imaginary worlds, providing experiences that can help people in understanding concepts as well as learning to perform specific tasks in a safe environment. The possibility of delivering educational 3D VEs through the Web allows one to reach potentially large numbers of learners worldwide, at any time (see [18] for a thorough discussion of 3D Web applications in education, learning and training). Employing 3D graphics allows for more realistic representations of subjects or phenomena, offering the possibility of analyzing the same subject from different points of view. Examples in medical education [30] include 3D reconstructions of parts of the human body [47] and 3D simulators [39], like the one shown in Figure 1. Other applications have been developed for foreign language education [40], maintenance training [19, 37], special needs education [31] and optics teaching [50].

14.2.1.2   E-commerce and Product Visualization

Although almost all e-commerce Web sites use hypermedia-based interfaces, a few sites have attempted to provide users also with 3D interfaces [1], allowing them to explore a 3D VE representing a store, as in Figure 2. A 3D Web store can have some advantages, if properly implemented:

- it is closer to the real-world shopping experience, and thus more familiar to the customer,
- it supports customer's natural shopping actions (such as walking, looking around the store, picking up products,...),

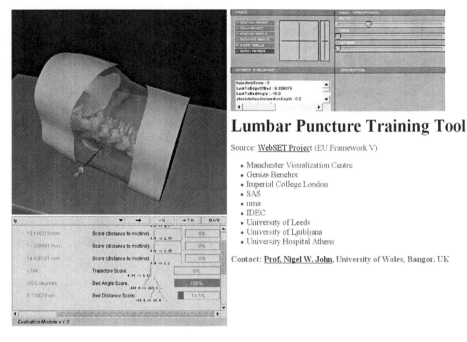

Fig. 1. 3D Web medical training simulator. Image from the WebSET project, reproduced with permission of Nigel W. John.

Fig. 2. @Mart 3D Shopping Mall [1].

- it can satisfy emotional needs of customers, by providing a more immersive and visually attractive experience,
- it can satisfy social needs of customers, by allowing them to meet and interact with people (e.g., other customers or salespeople).

On today's e-commerce sites, the simple integration of interactive 3D objects into Web pages, rather than full 3D store environments, is more common, for example in the automotive market [26].

14.2.1.3 Virtual Museums

Online collections of cultural information are useful if the digital representations of physical items contain enough detail to support the needs of visitors, e.g. researchers. Collections such as photographs or manuscripts can often be effectively acquired and displayed with 2D digital images. However, images are less effective as surrogates for three-dimensional items, such as sculptures, since much spatial information is lost, as the 3D shape of an object has to be flattened onto a two-dimensional view from a single perspective. In these cases, using 3D models can better support the needs of virtual museums visitors.

One can also build 3D VEs that contain representations of cultural objects as well as their contextual environment, e.g. to:

- provide a situated representation of objects;
- virtually reconstruct objects, structures and environments that have been damaged in the past, or do not exist anymore;
- build environments that never existed physically, but represent an appropriate conceptual or architectural environment, such as the virtual reconstruction of Leonardo's ideal city [4] shown in Figure 3.

Fig. 3. The virtual reconstruction of Leonardo's ideal city [4]. Image reproduced with permission of Thimoty Barbieri.

14.2.1.4 Architecture and Virtual Cities

Many 3D Web sites allow users to move inside 3D models of buildings and virtual cities [42,44], sometimes providing the capability of seeing each other and chatting. Although most of these sites, such as the one shown in Figure 4, focus on simply reproducing real-world places, there are many possible applications for virtual cities, such as:

- improving the planning, design and management of real cities (e.g., developers looking for sites for new buildings, local authorities managing urban infrastructure),
- providing tourists with detailed guides
- providing community resources for residents.

Fig. 4. Virtual Ljubljana [44].

14.2.1.5 Virtual Communities

3D virtual communities on the Web allow a large number of users to build and inter-act among each other inside a visual 3D space. In the last years, the number of these 3D VEs and their users has grown steadily: for example, Alphaworld [2], one of the oldest multi-user 3D VEs on the Internet, has hundreds of thousands of users, is roughly as large as the state of California, and contains more than 60 million virtual objects. The main distinctive feature of this kind of 3D Web sites is that users are allowed to build and inhabit visual community spaces, collaboratively engaging in the construction of large scale spaces (including artwork, buildings and full towns) and other social activities, like the virtual ceremony illustrated in Figure 5.

Fig. 5. Social activity in a multi-user 3D VE: wedding in Alphaworld [2].

14.2.2   Available Web3D Technologies

The history of 3D Web sites begins in 1995 with the birth of *VRML* (*Virtual Reality Modeling Language*), which is still the most known and used technology for building and delivering 3D Web content. More specifically, VRML is an open ISO standard [46] for a file format and corresponding run-time behavior to describe interactive 3D objects and 3D VEs delivered through the Web.

Recently, a new ISO standard, called *eXtensible 3D Graphics* (*X3D*) [49], has been proposed as a successor of VRML. Both VRML and X3D are managed by the Web3D Consortium [41], and result from the effort of several organizations, researchers and developers worldwide. Parts of VRML and X3D have been also integrated into the MPEG-4 standard [34], which adopts most of their concepts and instructions to describe interactive multimedia content that includes 3D objects and 3D VEs.

Access to VRML/X3D Web content is possible through one of the available Web browser plug-ins, such as (at the time of writing this Chapter) Parallelgraphics Cortona [37], Bitmanagement Contact [3], Octaga Player [35], and Mediamachines Flux [33].

Besides open ISO standards, there are many other (non-standardized) technologies for 3D on the Web. The best known examples are probably Java3D [29], an extension of the Java language for building 3D applications and applets and Shockwave 3D [32] from Macromedia. Although most of these technologies can be effectively used to build 3D Web content, in this paper we will focus on open Web3D standards. In general, open standards allow for lower costs, easier reusability of content, and easier integration with existing and future content and applications. In the following, we briefly describe the main technical features of VRML and X3D, referring the reader to published books and manuals for complete and detailed explanations.

14.2.2.1   The Virtual Reality Modeling Language (VRML)
The idea of a language for building 3D content for the Web originated back in 1994, when Mark Pesce and Tony Parisi built an early prototype of a 3D browser for the Web, called Labyrinth. Later that year, at the Second International Conference on the World Wide Web, the first specification of VRML was published. In the following years, the language underwent a series of improvements, leading to version 2.0, which was published as an ISO standard in 1997 with the name *VRML97* [46].

VRML is a language that integrates 3D graphics, 2D graphics, text, and multimedia into a coherent model, and combines them with scripting and network capabilities [10]. The language includes most of the common primitives used in 3D applications, such as geometry, light sources, viewpoints, animation, material properties, and texture mapping.

From a more technical point of view, VRML documents are text files that describe 3D objects and 3D VEs using a *hierarchical scene graph* (i.e., a directed acyclic graph). Entities in the scene graph are called *nodes*. VRML defines 54 different node types, including geometry primitives, appearance properties, sound and video, and nodes for animation and interactivity. For example, hyperlinks are implemented in VRML using the Anchor node, through which clicking on a 3D object has the effect of retrieving the resource at a specific URL.

Nodes store their properties in *fields*; the language defines 20 different types of fields that can be used to store different types of data, from single integers to arrays of 3D rotations. It is also possible for the programmer to define new nodes (i.e., extend the language) using a mechanism called *prototyping* through a statement called Proto. For example, this mechanism has been used to extend VRML with nodes to represent and animate 3D humanoids [28] and to implement distributed simulations in multi-user, networked 3D VEs [8].

VRML defines a message-passing mechanism that allows nodes in the scene graph to communicate with each other by sending events. This mechanism, together with special types of nodes, called *sensors* and *interpolators*, enables user interaction and animation. For example, the TimeSensor node generates temporal events as time passes and is the basis for all animated behaviors. Interpolators nodes are then able to continuously translate temporal events into data needed for animation. For example, the PositionInterpolator node is able to translate temporal events into 3D coordinates, allowing one to move objects in space. Other sensors are useful in managing user interaction, by generating events as the user moves through the 3D VE or when the user interacts with some input device (e.g. mouse pointing or clicking). For example, the ProximitySensor node is able to detect the user's position in the 3D VE, while the TouchSensor node is able to detect mouse clicks on 3D objects.

More complex behaviors (such as realistic physics simulation) can be implemented by using Script nodes, that allow one to manage VRML nodes with programs written in Java or JavaScript.

14.2.2.2 eXtensible 3D (X3D)

The eXtensible 3D (X3D) language for defining interactive 3D Web content was recently released as the successor of VRML, and was approved in 2004 as an ISO standard [49]. X3D inherits most of the design choices and technical features of VRML described in the previous section. As a result, it is mostly backward-compatible, that is, many VRML files require only minimal changes for translation to X3D.

X3D improves upon VRML mainly in three areas. First, it adds new nodes and capabilities, mostly to support advances in 3D graphics techniques and hardware, such as programmable shaders and multi-texturing. Second, it introduces additional data encoding formats. More specifically, it is possible to represent, store and transmit X3D content using a VRML-like textual encoding, an XML-based textual encoding, and a binary encoding, that enables better data compression and thus faster downloads. Third, similarly to XHTML, it divides the language into functional areas called *components*, which can be combined to form different *profiles* (i.e., subsets of the entire language) that are suited to specific classes of applications or devices. For example, this feature would enable one to create a specific profile to take into account the limited capabilities of mobile devices.

14.3   Adaptivity for 3D Web Sites

In Web-based hypermedia, which is the mainstream model in today's Web, information is organized and presented into (a graph of connected) pages using various media, with text being the main form of content/medium. Users interact with information mainly by reading, filling forms (e.g., using search engines), and navigating from one page to another by selecting the desired link from those contained in the current page. Many approaches to Web adaptivity presented in this book are targeted towards this model. For example, most techniques for adaptive content presentation discussed in Chapter 7 of this book [9] work with pages and textual content.

The 3D Web model is more complex than Web-based hypermedia, as Table 1 shows. In general, multimedia information, which can include 3D models, images, text, and audio, is organized and presented into a 3D space (or even in multiple 3D spaces connected by hyperlinks), following an arbitrarily complex spatial arrangement, such as a building or an entire city. Users navigate 3D space by controlling the position of their viewpoint through mouse, keyboard, or, more rarely, 3D pointing devices, and sometimes have the ability to teleport from place to place or to other 3D Web sites. As in Web pages, users can exploit hyperlinks to reach other Web resources. Besides navigation, additional interaction possibilities include the manipulation of 3D objects (e.g., clicking them to perform an action, moving them in space) and even building new objects.

Given these conceptual differences, it is not surprising that the techniques and tools for adaptivity in Web-based hypermedia cannot be straightforwardly applied to personalize 3D Web content, navigation and presentation. As mentioned above, most adaptive hypermedia techniques have been developed for content organized in pages (and not in a 3D space) and mainly made up of text (which is not the prevalent medium in 3D Web sites). With respect to adaptive navigation support, for example, link manipulation as presented in Chapter 8 of this book [5] could accommodate only navigation through hyperlinks in 3D VEs. Moreover, there are also technical differences to be taken into account, namely different file formats. Therefore, alternative techniques, or modifications of existing ones, needs to be developed for adaptivity in 3D Web sites.

Table 1. Analogies and differences between Web-based hypermedia and 3D Web sites

	Web-based hypermedia	3D Web sites
presentation container	page	3D space
content media	mainly text, but also images, videos, ...	mainly 3D models, but also text, images, videos, ...
structural organization	graph of pages	3D space or graph of 3D spaces
navigation	through hyperlinks	by moving in 3D space (e.g., walking, flying) and teleporting; also through hyperlinks
other common users' activities	reading pages, filling forms	3D object manipulation (clicking, moving, ...),

In the following, we will describe these techniques, highlighting the main differences with respect to their Web-based hypermedia counterparts. To make practical comparisons, we will use *AHA!* [22] (also discussed in Chapter 1 [7] and Chapter 13 [9] of this book) as a representative example of Web-based hypermedia adaptive systems. First, we will discuss how to build and update the user model, i.e., the *modeling* task, and then how to deliver personalized 3D content, i.e., the *adaptation* task.

14.3.1 Modeling

The approaches to adaptive 3D Web content developed so far have reused standard user model representation and reasoning techniques, such as stereotypes, graphs of concepts, and inference rules. Those techniques indeed are not specific to the hypermedia model. However, the task of user model acquisition (building and updating the model) requires a different approach in 3D Web sites.

With adaptive Web-based hypermedia, user model updates are typically triggered each time the browser requests a page. For example, in *AHA!* the adaptation engine starts by executing the rules associated with the attribute *access* of the requested page. Then, the user model is updated assuming that the requested page will be read, for example increasing the user's knowledge level about the concepts described in the page. This technique is effective under the assumption that *the user will fully read the page*, or, in other words, that all content accessed from the server will be read by the user. This is a strong assumption, since the user might skip parts of the page and thus cause inappropriate updates to the user model, but there are no easy methods to track which parts of a page have been actually read. Although there are available techniques for this purpose, such as eye tracking, they are costly or unpractical to adopt for Web sites and their visitors, except in special situations such as marketing research.

With 3D Web sites, assuming that all content accessed from the server is going to be seen or properly employed by the user is even more likely to cause erroneous user model updates. In many cases, users see only a part of the downloaded content (3D models, images, ...) that constitutes the 3D VE, for example because exploring a large or complex environment can require hours. Even in a smaller 3D VE, users might not see some objects because they are occluded by other objects (from the user's path during the visit) or simply do not notice them while navigating. Moreover, when some object manipulation is possible, users might not perform it or do it in unexpected ways. For example, in a medical training application where the trainee is required to virtually perform a certain sequence of actions with virtual medical tools, one would like to update the user model according to how actions were actually performed.

A solution proposed [16, 17] consists in closely monitoring users' behavior in the 3D VE, and send relevant time-stamped users' actions (e.g., movements, objects clicked) to the server, where they can trigger user model updates. In this way, we can update the user model not when content is accessed from the server, but only when we are confident the user has actually seen it or interacted with it. For example, by recording user's position in the 3D VE every few seconds and sending it to the server, it is possible to know which parts of the environment were actually visited and update the user model accordingly.

This approach does not require much implementation effort or special hardware because most Web3D technologies include mechanisms (called *sensors* in VRML and X3D, see Section 14.2.2) to monitor low-level events, such as mouse movements, as they are necessary for interactivity. Relevant interaction data gathered through sensors can be collected and sent to the server through programs (e.g., VRML Script nodes). For example, such technique has been used:

-   to monitor user's position in 3D space, and determine which parts of the 3D VE have been actually visited,
-   to check whether the virtual head of the user is oriented towards a certain 3D object, and determine whether the object might have been actually seen by the user (e.g., considering distance),
-   to check whether and how a certain 3D object has been clicked or dragged by the user, and determine whether a certain action has been properly performed.

A more detailed technical explanation of the proposed solution, in the case of VRML-based 3D Web sites, is presented in Sections 14.4.1 and 14.5.2.

14.3.2   Adaptation

In this section, we discuss techniques for adaptive navigation support and adaptive presentation of content in 3D Web sites. A general issue concerns how frequently adaptation can and should be made. With adaptive Web-based hypermedia, adaptation is normally performed on each requested page, although it might be desirable, for some content, to reduce the frequency of the adaptation process, for example once per session [22]. However, since users typically read one page at a time, adapting each requested page enables them to see the effects of adaptation during a browsing session and at the right time.

So far, the approaches to adaptive 3D Web sites have adopted a similar solution, i.e., adaptation is performed when 3D content is requested from the server [17,21]. However, in the typical situation where the full 3D VE is downloaded at the beginning of the user's visit, with this solution only adaptations between visits are possible. For example, an adaptive 3D virtual store where all content (store building, 3D models of products, advertisement banners) is downloaded at the beginning of the user's visit does not allow the user to see adaptations taking into account which products have been more examined since the beginning of the visit.

With most Web3D technologies, one can however download or update parts of the 3D VE during the user's visit. For example, both VRML and X3D provide this possibility, but developers are required to write ad-hoc scripts. Alternatively, there are extensions to VRML, such as *X-VRML* [48], that provide easier mechanisms to implement updates or downloads of content during visits, and thus carry out adaptations during visits. A simple but effective example of this strategy has been used in a 3D virtual museum [13]. The museum features a virtual human acting as a guide, leading the user around and describing museum items using speech synthesis. Each time an item needs to be presented, the text to be spoken is requested to the server, where it is tailored according to the user model, and then downloaded and fed to the speech synthesizer.

In general, which kinds of adaptations are best suited during visits, and their optimal frequency, are open issues. Typically, user's experience of 3D VEs should be as continuous as possible to maintain user engagement, while in Web-based hypermedia adaptive changes among pages are not (or much less) perceived as annoying break-downs since the experience is already 'divided into pages'. For example, modifying the position, appearance or behavior of visible objects while the user is visiting the 3D VE, even if the user model would suggest to do so, should be carefully performed, otherwise it will likely turn out as annoying or counter-productive for the user's experience. In the following, we first discuss how to adaptively support navigation and interaction, and then how to adaptively present 3D content. Finally, in Section 14.3.2.3, we consider adaptivity in the context of multi-user 3D VEs.

14.3.2.1 Adaptive Navigation and Interaction Support

Although Web-based hypermedia and 3D VEs are different, they are both targeted for user-driven navigation and exploration [27]. Like in the case of Web-based hyperme-dia, it seems thus interesting to develop adaptive navigation and interaction support techniques that can help users in finding and using information more efficiently, and prevent navigation and interaction problems. Moreover, navigation is a very relevant usability issue in the context of 3D VEs. In current 3D VEs, people often become disoriented and tend to get lost, and these problems are exacerbated by difficulties such as controlling movements in a 3D space, and limited field of view compared to the real-world experience. Inadequate navigation support is likely to result in users taking wrong directions, leaving the 3D VE before reaching their targets of interest, or with the feeling of not having adequately explored the visited 3D VE. These problems become even more critical in the case of novice users, who might become easily frus-trated in learning how to navigate.

Although many techniques (called *electronic navigation aids*), such as electronic 2D and 3D maps, have been developed to help users in navigating 3D VEs, they are not able to adapt to users with different navigation and interaction abilities. For this reason, some researchers [6] have proposed to develop adaptive navigation support techniques, mostly by deriving them from established methods in adaptive Hypermedia.

Fig. 6. Annotation by means of flashlight (left) and arrows (right) [27]. Image courtesy Stephen Hughes.

Hughes et al. [27] propose a number of adaptive navigation support techniques based on computing *ideal viewpoints* in the 3D VE on the basis of the user model, and then use them to prevent erroneous directions, disorientation or missed parts. The ideal viewpoints correspond to locations in the 3D VE (more specifically, positions and corresponding orientations in 3D space) from which objects or parts of the 3D VE that are interesting for the user are well visible. The idea is to constrain navigation or draw additional information to help the user in reaching the ideal viewpoints. The proposed techniques are derived from the link manipulation techniques discussed in Chapter 8 of this book [5]:

- *direct guidance* (a strict linear order through the navigation space) computes a path through the 3D VE that encompasses all ideal viewpoints, and then automatically moves the user's viewpoint along this path;
- *hiding* (restricting the number of navigation options to a limited subset) hides all irrelevant orientations by letting the user move her position freely, but having the system dictate the orientation of the users' virtual head to force it to fixate on certain objects while moving;
- *sorting* (altering the order in which navigation decision are presented to the user) orders ideal viewpoints and let the user move freely, but, as with hiding, the system dictates the orientation of the users' virtual head to force it to fixate on certain objects in the computed order. In this case, the user still has the possibility to override system decisions and orient the virtual head to explore other objects;
- *annotation* (displaying additional information on navigation options) displays attention-drawing signs, such as the arrows in the right part of Figure 6, to indicate interesting objects, or highlights them using a flashlight while unimportant features are left in the dark as in the left part of Figure 6.

An alternative approach [12,16] to implement sorting and annotation-like adaptive support exploits virtual characters, such as the ones in Figure 7 and Figure 14, that act as navigation guides to:

- show users the path to an object, or the path through a sorted list of objects of interest, i.e. implementing sorting-like navigation support;
- provide annotations in the form of additional information on navigation and interaction possibilities; for example, the virtual character in Figure 7 is showing a new user that an object can be opened to see its interior.

This style of adaptive support has been employed in two different contexts. In a 3D virtual museum [12], the virtual character acts as the museum guide, leading the user around, giving information on museum items and showing possible interactions. The first time the user visits the museum, a sequence of museum items (i.e., a museum tour) is generated on the basis of the user profile, and the virtual character guides the user through them. In successive visits, only those items that have not been seen are included in the tour (this has similarities with the hiding technique explained above). In a 3D virtual store [14,16], multiple animated characters are employed to guide the user to different products (this technique is described in more detail in Section 14.5). The animated characters look like products (see Figure 14), and their actual appearance (i.e., the specific product they represent) is adapted to take into account user's potential buying interests.

While using 3D virtual characters does not directly help the user in controlling navigation as direct guidance, hiding, and sorting, it has the following distinctive features:

- it can draw the user's attention with natural and familiar methods. For example, the humanoid character in Figure 7 uses gaze, pointing gestures, body orientation, and provides textual information through voice;
- it may have an emotional impact on the user, and increase motivation and engagement: users tend to experience presentations given by animated characters as lively and engaging [43]. Moreover, it can make the virtual place more lively, attractive, and less intimidating to the user;
- it does not restrict the navigation possibilities, since the user can choose whether to employ adaptive support or not by not following the virtual character and explore the 3D VE on its own.

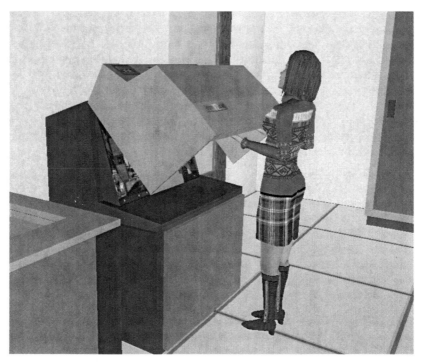

Fig. 7. A humanoid character shows the user how an object can be opened [12]

Another kind of adaptive navigation and interaction support has been proposed by Celentano and Pittarello [11]. Their idea is to monitor user's behavior and to exploit the acquired knowledge for anticipating user's needs in forthcoming interactions. More specifically, the approach is based on using sensors (as described in Section 14.3.1) to collect usage data, and compare them with previous patterns of interaction stored in the user profile. The patterns of interaction are sequences of activities which the user performs in some specific situation during the interactive execution of a task, and are encoded as Finite State Machines (FSM). Whenever the system detects that the user is entering a recurrent pattern of interaction, it may perform some activities of that pattern on behalf of the user. For example, figure 8 shows an example of interaction adaptation in a virtual fair application. The FSM on the top of the Figure shows the sequence of actions that must be performed to interact with an object inside a showcase. The FSM on the bottom of the Figure is computed by the interaction support system after the first FSM has been detected as recurring. In the FSM in Figure 8, the dotted arrow represents an automatic execution of actions performed by the system. More specifically, if the user is closer than 3 meters from the showcase, the open button, even if it is not visible, is automatically pressed to open the showcase on behalf of the user.

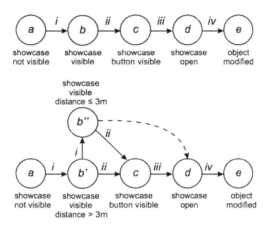

Fig. 8. Interaction adaptation in a virtual fair application [11]. Image courtesy Fabio Pittarello.

14.3.2.2 Adaptive Presentation of Content

Adaptive presentation of content concerns deciding what content is most relevant to the user, how to structure it in a coherent way, and how to present it in the best way. For the first two tasks, the most widely used techniques in Web-based hypermedia are *optional fragments* and *altering fragments*. As mentioned in Chapter 7 of this book [9], those techniques build adaptive pages by selecting and combining an appropriate set of fragments, where each fragment typically corresponds to a self-contained information element, such as text paragraphs or pictures.

The techniques for adaptive presentation of 3D content developed so far follow the same fragment-based approach, and can therefore be thought as variations of the above mentioned adaptation approaches.

The approach proposed in [17] uses the VRML PROTO construct to define each kind of self-contained adaptive fragment. In general, PROTO defines a new VRML node by specifying its *interface*, i.e. fields and events the node receives and sends, and its *body*, i.e. how the node is implemented in terms of existing or previously defined VRML nodes. As with any other VRML node, each time the new node is inserted, or *instantiated*, in the 3D VE, one can change the values of the fields declared in the interface to customize the features of the node. For example, the following code

defines a very simple node for a box-shaped product in a 3D store, where the size of
the box and the image printed on its sides are encoded as fields:

```
PROTO BoxProduct
[ field SFVec3f bsize 0 0 0      // size of the box in x, y, z
  field MFString imageURL []  // url of image that will appear on the box
]
{
    Shape {                           // node to define a 3D object
    appearance Appearance {      // appearance of the 3D object
            texture ImageTexture {
                url IS imageURL }   // applies the image to the box
    }
    geometry Box {        // the geometry of the 3D object is defined by a box
            size is bsize } // size of the box
    }
}
```

The idea is that fields in the interface define the adaptive features of the node, ab-
stracting from other non-adaptive details. In the product example, therefore, the adaptive features are the size of the box, and the image printed on its sides. With this approach, 3D adaptive content is defined by a set of BoxProduct node instantiations,
such as in the following code fragment, which includes a milk box in a 3D VE:

```
BoxProduct {
    bsize 1 2 1
    imageURL "milkBox.jpg"
}
```

The idea is that field values (such as "milkBox.jpg") are chosen among a set of
alternatives (that have to be stored separately) or computed by the adaptive engine
when content is requested.

The alternative technique proposed in [15] for the X3D language does not uses a
prototyping mechanism (which is available also with X3D), but requires an additional
file, called *Content Personalization Specification (CPS)*, for each X3D document with
adaptive content. The CPS file defines adaptive features and may also specify possible variants. With this technique, the milk box example above would be implemented
by the following X3D code fragment:

```
<Shape>
  <Box DEF="size1" />
  <Appearance>
    <ImageTexture DEF="imgUrl1" />
  </Appearance>
</Shape>
```

and a separate CPS file specifying that the size of the box and the image on its side
are adaptive features. The following CPS does that, also defining two possible actual
adaptations for the product image:

```
<CPS>
   <adaptiveContent DEF="imgUrl1" attribute="url">
      <value>"milkBox.jpg"</value>
      <value>"cerealBox.jpg"</value>
   <adaptiveContent DEF="size1" attribute="size"/>
</CPS>
```

One of the advantages of using XML-encoded content (such as X3D) is the possibility of using adaptation techniques developed for other kinds of XML-based content. For example, the approach proposed by Dachselt et al. [21] uses the Amacont general-purpose architecture [25] with X3D content or more high-level formats [20]. For example, the fact that the image printed on the sides of the box-shaped product is an adaptive parameter would be expressed in the approach of Dachselt et al. by the following code fragment, which, contrary to the techniques above, includes also the logic of adaptation:

```
<Parameter name="url" dataType="CoAnyURI" ... >
 <Variants>
  <Logic>
   <If>
    <Expr>
     <Term type="=">
      <UserParam>Favorite Product</UserParam>
      <Const>Milk</Const>
     </Term>
    </Expr>
    <Then>
      <ChooseVariant>milk</ChooseVariant>
    </Then>
    <Else>
     <ChooseVariant>cereals</ChooseVariant>
    </Else>
   </If>
  </Logic>
  <Variant name="milk">
    <CoAnyURI>"milkBox.jpg"</CoAnyURIs>
  <Variant>
  <Variant name="cereals">
   <CoAnyURIs>"cerealsBox.jpg"</CoAnyURIs>
  </Variant>
 </Variants>
</Parameter>
```

The Parameter element encodes an adaptive feature (in this case, an image depicting a product). The enclosed Variants element define possible variants for the feature. Inside the Variants element, a Logic element defines the logic of adaptation (if the user's favorite product is milk, we will use the milk variant, else we will use the cereals variant. Then, a list of Variant elements defines the possible variants as URLs of the images.

While these approaches provide fragment-based techniques to perform adaptation of content, using them is not as easy as in Web-based hypermedia. Text fragments or images can be simply juxtaposed in a page, with the only possible drawback of not preserving a good graphic layout. On the contrary, special care has to be taken in the case of 3D content to preserve a meaningful and understandable 3D space. Once relevant fragments have been chosen, one needs to properly arrange them in 3D space and time (if there are animations) such as, for example, included objects do not intersect each other, are adequately visible from the positions the user will take in space, and free space is enough for the user to move. Unfortunately, it is very difficult to develop general algorithms for this purpose. This forces one to limit the space of possible adaptations to a few variants that are guaranteed to be safe with respect to the above mentioned constraints, or to implement adaptation strategies that might work only in a specific 3D VE.

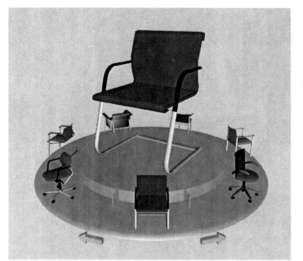 

Fig. 9. On the left, a ring menu for choosing a chair; on the right, the same menu adapted for smaller displays, such as PDAs [21] Image courtesy Raimund Dachselt.

Even if one could easily implement any kind of adaptation, there are presently no studies that investigate the effect on users of content adaptation in 3D VEs. Therefore, we can only try to hypothesize which adaptations might be useful and which might be counterproductive. For example, it is likely that changes in the navigational structure of a 3D VE will disorient the user and will make it much harder to learn how to navigate the environment. Therefore, structural changes need to be chosen carefully and be limited in scope and frequency. In the following, we mention some examples of adaptations of 3D content that have been proposed in the literature.

In the adaptive 3D e-commerce example we will discuss in Section 14.5, the number of instances of products in shelves can vary in a given range (one to four) to adaptively increase or decrease the visibility of the product itself (see Figure 15). The limited number of variants guarantees that each product will not take the space reserved to other products.

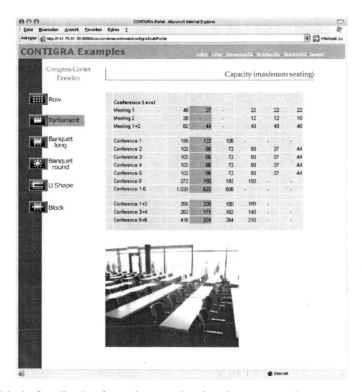

Fig. 10. Web site for adjusting the seating capacity of conference rooms Image courtesy Raimund Dachselt.

A 3D adaptive e-learning system [24] organizes learning content into a building made of rooms, and the adaptation engine places rooms (by just exchanging content among equally-sized rooms) that correspond to the areas of higher user's interest before rooms whose contents are less interesting for the user.

The 3D menus shown in Figure 9 [21] are examples of adaptation to the user's device. The idea is to provide different alternatives, with respect to screen space usage, for the same 3D interface element and information presented. In particular, the screenshot on the right shows a smaller-sized version of the ring menu on the left, and is better suited to small displays, such as PDAs.

Finally, 3D content could also be considered in media adaptation. Figures 10 and 11 shows two different versions of the same Web site, whose purpose is to adjust the seating capacity of conference rooms [21]. Figure 10 shows an HTML-based version, which might be more suited to low-bandwidth connections or users that are not familiar with 3D. Figure 11 shows a 3D-based version, where the conference room is represented by a 3D VE to better visualize the final result.

14.3.2.3 Multi-user 3D VEs
No examples of adaptivity in multi-user 3D VEs have been reported in the literature. This might be due to the fact that multi-user 3D VEs can conflict with personalization

aspects, making some of the adaptations presented in the previous section trouble

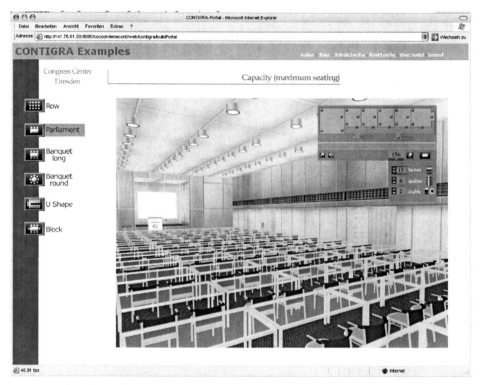

Fig. 11. Adapted version of the content in Figure 10, using a VRML 3D VE Image courtesy Raimund Dachselt.

some. In general, if multiple users navigate and interact together in the same 3D VE, adaptation of content cannot safely target the specific profile of a single user. For example, adaptations that cause one person to see the 3D VE differently from others could cause deep misunderstandings (e.g., a reference to a highlighted object that is not highlighted for another person) that may hinder social activities.

There are however strategies that could be pursued to prevent this kind of problems. For adaptations that conflict with multi-user activities, one could try to find the best common adaptation which maximizes the match with the different user models. However, considering that the set of users could continuously change, this might not be easy to implement. A second possibility could be to clearly mark what is personalized in the 3D VE, and see if users are able to adopt new conventions. Another possibility would be to find useful adaptations that do not conflict with multi-user activities, or even result from them. For example, an idea that has been developed for adaptive multi-user textual environments is to change the description of objects in the environment to reflect usage [23], such as doors or books showing signs of wear. A similar idea could be used in a multi-user 3D VE to visually represent frequently accessed paths or objects.

14.4   A Generic Software Architecture for Adaptive 3D Web Sites

A few software architectures for adaptive 3D Web sites can be found in the literature. The AWe3D (Adaptive Web 3D) architecture [17] is a general purpose architecture for generating and delivering adaptive VRML content which was proposed in 2002. More recently, a few researchers [15,21] have focused on integrating 3D content into existing technologies, such as the Amacont general-purpose architecture [25], for Web adaptivity.

In the following, we describe a generic architecture (depicted in Figure 12) that generalizes the ideas of AWe3D, for delivering adaptive content in 3D Web sites. The architecture is composed by the following modules:

- a *Usage Data Sensing* module, whose purpose is to monitor user's interaction with the 3D VE, and send the relevant events through the Internet. This module is located on the client side, run by the user's browser;
- a *Usage Data Recorder* module, whose purpose is to receive, on the server side, the events sent by the Usage Data Sensing module, and record them in the *User Model Database*;
- an *Adaptivity Engine*, that: (i) performs inferences needed to update the user model on the basis of recorded usage data, and (ii) given the current user model, computes a set of adaptation choices for requested adaptive content;
- a *3D Content Creator* module, that: (i) accepts content requests from the client; (ii) when adaptive content is requested, asks the Adaptivity Engine to provide the correct adaptation choices, and uses them to build the adapted 3D content, retrieving needed files (3D models, images, sounds,...) from the *3D Content Database*; (iii) delivers the requested 3D content to the client.

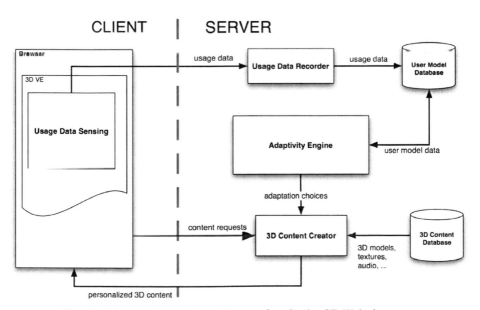

Fig. 12. Schema of a general architecture for adaptive 3D Web sites

We now describe in more detail a possible set of technical choices to implement each module in the case of VRML-based 3D Web sites.

14.4.1   Usage Data Sensing

Following the technique outlined in Section 14.3.1, this module is implemented by a set of VRML sensors whose output is routed to a Script node, which transmits relevant usage events to the Usage Data Recorder module by using a HTTP connection.

The type, number and specific settings of VRML sensors in this module depend on the type and number of usage data that needs to be collected for a specific application. In the simplest case, one would need one sensor for each event that has to be sensed. VRML sensors allow one to track the user's position, or user's collisions with an object, or mouse actions on an object, or visibility of an object. By combining the output of multiple sensors in a Script node, one can obtain higher level sensing of the user's actions: for example, a complex action that requires a sequence of clicks and drags can be monitored by using appropriate sensors to detect these low-level events, and a Script node that receives the sensors' output, recognizes the correct sequence, and send the resulting high-level event to the server.

14.4.2   Usage Data Recorder

The Usage Data Recorder is implemented by a simple server-side program that receives usage data and stores them with a DBMS. A more elaborate version could also perform calculations on the usage data before storing, for example filtering, averaging, sums, …

14.4.3   3D Content Creator Module

The *3D Content Creator* receives requests for 3D content, and returns that content to the client. Adaptive fragments are represented through VRML PROTO constructs (using the technique illustrated in Section 14.3.2.2), whose fields encode the adaptive features, such as object geometry, color, and size. The 3D Content Creator Module asks the Adaptivity Engine to compute actual values for each PROTO field (i.e., a set of *adaptation choices*), and use the result to instantiate the PROTO in the file that is returned to the client, possibly retrieving needed code (such as 3D models, animations, and images) from the 3D Content Database.

14.4.4   Adaptivity Engine

The technical choices that have to be taken in implementing the *Adaptivity Engine* depend on how complex are the inferences that have to be performed. A simple solution, using a rule-based approach, is to write a set of *User Model Update* rules to update the user model on the basis of collected usage data, and a set of *Content Adaptation* rules to compute personalized field values for adaptive content. The User Model Update rules can be activated each time usage data are received from the client, or periodically, at given intervals of time or after a certain number of user's visits to the Web site. The Content Adaptation Rules are activated each time the 3D Content creator asks for personalized versions of adaptive fragments.

14.5 An Application in E-Commerce

In the following, we describe a detailed example in the domain of e-commerce implemented using the architecture introduced in the previous section. We first describe the considered 3D store, then we discuss specific technical choices to implement an adaptive version of it. The example we propose is a simplified version of the 3D adaptive store presented in [17], to which we refer the reader for more detailed technical specifications and code examples.

14.5.1 A 3D Store VE

The 3D VE we consider is composed by a 3D model of a department store, displaying products on several shelves. The customer can wander through the store, obtain information on products by clicking on them, put them in the cart, which is also represented in 3D (see Figure 13), and go to the checkout counter to conclude her shopping session. Besides shelves, customers' attention towards products is sought by exploiting special rotating display spots in prominent places, advertisements on the walls, and audio messages. Moreover, the store is populated by *Walking Products* (*WPs*, see Figure 14), a navigation support feature to help users in finding products [14]. WPs are 3D animated representations of products that move through the store and walk to the place where the corresponding type of products is. A customer in the 3D store sees a number of WPs wandering around: if she is looking for a specific type of products, she has just to follow any WP of that type and will be quickly and easily lead to the desired destination.

Fig. 13. A 3D store with products on shelves

Fig. 14. Example of a Walking Product

14.5.2 Usage Data Sensing and Recording

Usage data we are interested in concern typical interactions with products in the store. More specifically, the data collected by the *Usage Data Sensing* module by monitoring customer's actions are:

- *Seen Products.* While the customer wanders around the store, she voluntarily or involuntarily looks at the products which fall in her field of view;
- *Clicked Products.* When the customer wants to know more about a product, she clicks on it to get the product description;
- *Cart Products.* The product description allows the customer to put the product in the shopping cart for a possible later purchase;
- *Purchased Products.* A product in the cart can be later purchased by going to the checkout counter.

Seen Products and *Clicked Products* data are acquired through Visibility and Touch sensors associated to each product. The following is a slightly simplified PROTO defining a product node with the required sensing capabilities:

```
PROTO Product
[   field MFNode product3DModel []
    eventOut SFTime productSeen
    eventOut SFTime productClicked   ]
{
    Group { children IS product3DModel }

    TouchSensor   {
                 touchTime IS productClicked }

    VisibilitySensor {
                 enterTime IS productSeen }
}
```

The interface of the node `Product` (the first three lines after the `PROTO` statement) include a field for the 3D model of the product (`product3DModel`), and two events that can be sent to other nodes, respectively indicating when the product is seen (`productSeen`) and clicked (`productClicked`). The `Product3DModel` field is an adaptive feature of the product: its 3D model can be chosen among different alternatives, for example to occupy less or more space in shelves, as shown in Figure 15. The body of the node `Product` (the code between braces) includes a reference to the 3D Model of the product, a Touch Sensor to detect click events, and a Visibility Sensor to detect when the product is visible.

In similar ways, *Cart Products* and *Purchased Products* data is acquired in the VRML code describing the cart and the checkout counter, respectively.

14.5.3 User Model

Customer models in the User Model Database of the 3D store contain the following information:

- *demographic data*, including gender, year of birth, and product categories of interest among those available in the store, which the customer can enter through an HTML form the first time she enters the store;
- *user preferences about the store*, such as presence of audio and music, and preferred music genre, which are also entered or modified by the user through the HTML form;
- *usage data*, described in the previous section, and exploited to dynamically update the user model. Usage data allows one to obtain a precise quantitative measurement of which brands, product categories, specific products, price categories, and special offers have been respectively seen, clicked, put in the shopping cart or purchased by the customer;
- *Product Interest Ranking*, which ranks products and products categories according to customer's interests.

To determine the Product Interest Ranking, an initial value is determined by using a HTML form that allows the customer to indicate her products of interests: if she chooses to fill it, the information is used to initialize the ranking. If the customer does not provide product interests in the HTML form, one can try to predict interests by using demographic profiles. Then, regardless of the quality of the initial value, product interests will be continuously updated by the Adaptivity Engine which exploits usage data: each purchase, cart insertion, and click of a product increases (with different weights) the level of interest in the corresponding product and product category or in related products.

14.5.4 3D Store Adaptivity

The adaptive features of the 3D store mainly concern where and how products are displayed:

- each product is displayed in the shelf assigned to its product category, but the amount of shelf space devoted to the product is adaptively changed to increase or decrease product visibility;
- additionally, a product may appear also in display spots, banners, or WPs to increase its exposure towards the user.

Other adaptive features concern the music that is played, and the audio messages that advertise products.

In the following, we present some examples of rules that perform adaptations in the 3D store. Simple rules are given by the direct associations between user's preferences about presence of music and preferred genres, and songs that are played during visit. More complex examples concern the exploitation of the user model to change the level of product exposure in the 3D store. The level of exposure of each product can vary the product visibility and attractiveness, for example by increasing space devoted to the product in the store or adding banners advertising the product. We call *ExposureLevel(X)* the parameter which represents the level of exposure for product X. The value of *ExposureLevel(X)* is determined by five more specific parameters:

- *ShelfSpace(X)* indicates the space assigned to product X on the shelf. It can take four different values: higher values make X more visible to the customer, increasing *ExposureLevel(X)*. The products in Figure 15 show two different allocations of shelf space;
- *DisplaySpot(X)* is false if product X is displayed only on its shelf, while it is true if product X is displayed also in a separate display spot in a prominent place (we could have also used numerical values to allow the same product to be displayed on more than one display spot);
- *Banner(X)* is true if there is a banner advertising product X in the store;
- *AudioMessage(X)* is true if audio advertisements for product X are played;
- *WP(X)* is true if there is a WP representing product X in the store.

A true value for any of the last four boolean parameters increases *ExposureLevel(X)*. Personalization rules first suggest changes to exposure level by asserting increase or decrease goals for specific products. Then, they focus on achieving those goals, by changing one or more of the above described parameters, according to the availability of store resources (e.g., if a shelf is full, shelf space for products cannot be increased on that shelf).

Fig. 15. Two different possible allocations of shelf space for the same product

We now examine some specific rules and how they relate to the information recorded in the user model. Suppose that a product X has never been seen by the customer or that changes in the Product Interest Ranking show an increasing attention towards the product. In both cases, a seller would like to increase the exposure of the product (in the first case, to give the customer the opportunity of seeing the product; in the second case, to better match customer interests). The rules that implement the two cases can be expressed as follows, where seen(X) is the recorded number of times a product has been seen, ProductInterest(X) is the rank in the product interest ranking, and NumberOfVisits is the number of times the user has visited the store:

```
IF seen(X)=0 AND NumberOfVisits>3 THEN
  goal(IncreaseExposureLevel(X))

IF increasing(ProductInterest(X)) THEN
  goal(IncreaseExposureLevel(X))
```

As another example, consider the cross-sell case where the purchase of a specific product X is an indicator of a likely future interest for related products and we want to update the user model accordingly. For example, if a customer buys a computer and has never purchased a printer, she could be soon interested in a printer. The rule can be expressed as follows, where purchased(X) is the recorded number of times a product has been purchased, lastVisit extracts the value of data considering only the last visit to the store, and RelatedProduct(X,Y) relates products by using associations provided by the seller:

```
IF lastVisit(purchased(X))>0 AND RelatedProduct(X,Y)
AND purchased(Y)=0 THEN increase(ProductInterest(Y))
```

As an effect of the increased product interest, the second rule examined above will then suggest an increase in the exposure level of related products which have not been purchased yet. Note that the RelatedProduct relation cannot be used transitively, because this could lead to counterproductive merchandising strategies. For example, an ink cartridge is obviously related to a printer, and a printer is obviously related to a computer, but it does not make sense to increase the exposure level of ink cartridges if a customer has purchased a computer but not a printer.

Finally, to prevent an excessive number of changes to the 3D store from one session to another, we impose a limit on their number for any given session. The idea is to keep the experience of returning to the 3D store consistent with the familiar experience of returning to a known real-world store: the store layout remains essentially the same, and a limited number of changes concern what products are displayed, and how the attention of the customer towards those products is sought.

14.6 Conclusions

Adaptivity of 3D content for the Web is a very recent and largely unexplored research topic. As shown in Section 14.3, there are only a few examples of adaptation of 3D content in the literature, and no thorough evaluations with users have been carried out.

To understand the true potential of adaptivity of 3D content, we need to explore in more depth the space of possible adaptations, including less obvious ones. For example, most 3D VEs (including those built with VRML or X3D) allow the use of spatial audio. An interesting possibility could be to use adaptive spatial audio to provide information to the user, for example navigation support.

It is also important to investigate users' reactions to adaptive changes in 3D content. As discussed in the Chapter, adaptivity may break or hinder important features of a user's experience in a 3D VE, such as the construction of spatial knowledge, and the continuity of the experience. Studies on users are therefore needed to establish when and how it is useful to adaptively change a 3D VE.

We hope that this Chapter has provided an easy-to-read introduction for students, as well as a stimulating starting point for researchers that aim at advancing this line of research.

References

1. @mart 3D store, www.activeworlds.com (last access on August 2006)
2. AlphaWorld virtual community, www.activeworlds.com (last access on August 2006)
3. Bitmanagement Contact, www.bitmanagement.com (last access on August 2006)
4. Barbieri, T., Paolini, P.: Reconstructing Leonardo's Ideal City - from Handwritten Codexes to Webtalk-II: a 3D Collaborative Virtual Environment System. In: Proc. of the 2001 Conference on Virtual Reality, Archeology, and Cultural Heritage (VAST 2001), Athens, Greece. ACM Press, 61-66
5. Brusilovsky, P.: Adaptive navigation support. In Brusilovsky, P., Kobsa, A., Nejdl, W. (eds.): The Adaptive Web: Methods and Strategies of Web Personalization, Lecture Notes in Computer Science, Vol. 4321. Springer-Verlag, Berlin Heidelberg New York (2006) this volume
6. Brusilovsky, P.: Adaptive Navigation Support: From Adaptive Hypermedia to the Adaptive Web and Beyond. Psychology Journal 2 1 (2004) 7-23
7. Brusilovsky, P., Millán, E.: User models for adaptive hypermedia and adaptive educational systems. In Brusilovsky, P., Kobsa, A., Nejdl, W. (eds.): The Adaptive Web: Methods and Strategies of Web Personalization, Lecture Notes in Computer Science, Vol. 4321. Springer-Verlag, Berlin Heidelberg New York (2006) this volume
8. Brutzman, D.: The Virtual Reality Modeling Language and Java. Communications of the ACM 41 6 (1998) 57-64
9. Bunt, A., Carenini, G., Conati, C.: Adaptive Content Presentation for the Web. In Brusilovsky, P., Kobsa, A., Nejdl, W. (eds.): The Adaptive Web: Methods and Strategies of Web Personalization, Lecture Notes in Computer Science, Vol. 4321. Springer-Verlag, Berlin Heidelberg New York (2006) this volume
10. Carey, R., Bell, G.: The annotated VRML97 Reference Manual. Addison Wesley (1997)
11. Celentano A., Pittarello F.: Observing and Adapting User Behavior in Navigational 3D Interfaces. In: Proc. of 7th International Conference on Advanced Visual Interfaces 2004 (AVI 2004), Gallipoli, Italy. ACM Press (2004) 275-282
12. Chittaro L., Ieronutti L., Ranon R.: Navigating 3D Virtual Environments by Following Embodied Agents: a Proposal and its Informal Evaluation on a Virtual Museum Application. Psychology Journal 2 1 (2004) 24-42
13. Chittaro L., Ranon R., Ieronutti L.: Guiding Visitors of Web3D Worlds through Automatically Generated Tours. In: Proc. of the 8th International Conference on 3D Web Technology (Web3D 2003), St. Malo, France. ACM Press (2003) 85-91

14. Chittaro L., Ranon R.: New Directions for the Design of Virtual Reality Interfaces to E-Commerce Sites. In: Proc. of the 5th International Conference on Advanced Visual Interfaces (AVI 2002), Trento, Italy. ACM Press (2002) 308-315

15. Chittaro L., Ranon R.: Using the X3D Language for Adaptive Manipulation of 3D Web Content. In: Proc. of the 3rd International Conference on Adaptive Hypermedia and Adaptive Web-based Systems (AH 2004), Eindhoven, Netherlands. Springer-Verlag, Berlin Heidelberg New York (2004) 287-290

16. Chittaro L., Ranon R.: Adding Adaptive Features to Virtual Reality Interfaces for E-Commerce. In: Proc. of the 1st International Conference on Adaptive Hypermedia and Adaptive Web-based Systems (AH 2000), Trento, Italy. Springer-Verlag, Berlin Heidelberg New York (2000) 86-97.

17. Chittaro L., Ranon R.: Dynamic Generation of Personalized VRML Content: a General Approach and its Application to 3D E-Commerce. In: Proc. of the 7th International Conference on 3D Web Technology (Web3D 2002), Tempe, Arizona. ACM Press (2002) 145-154.

18. Chittaro L., Ranon R.: Web3D Technologies in Learning, Education and Training: Motivations, Issues, Opportunities. Computers and Education, in press. doi:10.1016/j.compedu.2005.06.002, published online in 2005.

19. Corvaglia D., Virtual Training for Manufacturing and Maintenance based on Web3D Technologies. In: Proc. of the 1st International Workshop on Web3D Technologies in Learning, Education and Training (LET-WEB3D 2004), Udine, Italy, (2004) 28-33

20. Dachselt, R. , Hinz , M., Meissner , K. : Contigra: an XML-based architecture for component-oriented 3D applications. In Proc. of the 7th International Conference on 3D Web Technology (Web3D 2002), Tempe, Arizona, USA. ACM Press (2002) 155-163

21. Dachselt , R., Hinz, M., Pietschmann, S. Using the Amacont Architecture for Flexible Adaptation of 3D Web Applications. In: Proc. of the 11th International Conference on 3D Web Technology (Web3D 2006), Columbia, Maryland, USA. ACM Press (2006), 75-84.

22. De Bra, P., Aerts, A., Berden, B., De Lange, B., Rousseau, B., Santic, T., Smits, D., Stash, N.: AHA! The Adaptive Hypermedia Architecture. In: Proc. of the 14th Conference on Hypertext and Hypermedia (Hypertext 2003), Nottingham, UK. ACM Press (2003) 81-84.

23. Dieberger, A.: Browsing the WWW by interacting with a textual virtual environment - A framework for experimenting with navigational metaphors. In: Proc. of the 7th Conference on Hypertext (Hypertext 1996), Washington DC, USA. ACM Press (1996) 170-179.

24. dos Santos, C. T., Osorio, F. S.: An Intelligent and Adaptive Virtual Environment and its Application in Distance Learning. In: Proc. of the 6th International Conference on Advanced Visual Interfaces (AVI 2004), Gallipoly, Italy. ACM Press (2004) 362-365.

25. Fiala, Z., Hinz, M., Meissner, R. K., Wehner, F. A Component-based Approach for Adaptive, Dynamic Web Documents. Journal of Web Engineering 2 1-2 (2003) 58–73.

26. Fiat Ireland home page, www.fiat.ie (last access on August 2006)

27. Hughes, S., Brusilovsky, P., Lewis, M.: Adaptive navigation support in 3D e-commerce activities. In: Proc. of Workshop on Recommendation and Personalization in eCommerce at AH 2002 (2002) 132-139.

28. ISO/IEC FCD 19774 (2004) Humanoid animation (H-Anim) Available: www.web3d.org/x3d/specifications/ISO-IEC-19774-HumanoidAnimation/ (last access on August 2006)

29. Java3D media framework, java.sun.com/products/java-media/3D/ (last access on August 2006)

30. John, N. W.: The Impact of Web3D Technologies on Medical Education and Training. Computers and Education, in press. doi:10.1016/j.compedu.2005.06.003, published online in 2005.

31. Karpouzis, K., Caridakis, G., Fotinea, S. E., Efthimiou, E.: Educational Resources and Implementation of a Greek Sign Language Synthesis Architecture. Computers and Education, 2005, in press. doi:10.1016/j.compedu.2005.06.004, published online in 2005.

32. Macromedia Shockwave, www.adobe.com/products/shockwaveplayer/ (last access on August 2006)
33. Mediamachines Flux, www.mediamachines.com (last access on August 2006)
34. MPEG-4 International Standard (2002). MPEG-4 Specification. International Standard ISO/IEC JTC1/SC29/WG11 N4668.
35. Octaga Player, www.octaga.com (last access on August 2006)
36. Outline 3D Web site, www.outline3d.com (last access on August 2006)
37. ParallelGraphics Cortona, www.parallelgraphics.com /products/cortona/ (last access on August 2006)
38. ParallelGraphics Virtual Manuals, www.parallelgraphics.com/virtual-manuals (last access on August 2006)
39. Phillips, N., John, N.W.: Web-based Surgical Simulation for Ventricular Catheterisation. Neurosurgery **46** 4 (2000) 933-937.
40. Sims, E.M.: Reusable, Lifelike Virtual Humans for Mentoring and Role-Playing. Computers and Education, in press. doi:10.1016/j.compedu.2005.06.006, published online in 2005.
41. The Web3D Consortium, www.web3d.org (last accessed on August 2006).
42. Udine3D, udine3d.uniud.it (last access on August 2006)
43. van Mulken, S., André, E., Muller, J.: The Persona Effect: How Substantial is it? In: Proc. of the 1998 Human Computer Interaction Conference (HCI'98), Sheffield, UK. Springer-Verlag, Berlin Heidelberg New York (1998) 53-66.
44. Virtual Ljubljana, www.ljubljana-tourism.si/en/ljubljana/virtual_ljubljana/ (last access on August 2006)
45. VR for all community site, www.vr4all.net (last access on August 2006)
46. VRML International Standard (1997). VRML97 Functional Specification. International Standard ISO/IEC 14772-1:1997. Available at www.web3d.org/x3d/specifications/vrml/ISO-IEC-14772-VRML97/ (last access on August 2006).
47. Wakita, A., Hayashi, T., Kanai, T., Chiyokura, H.: Using Lattice for Web-based Medical Applications. In: Proc. of the 6th International Conference on 3D Web Technology (Web3D 2003), Saint Malo, France. ACM Press (2003), 29-34.
48. Walczak, K., Cellary, W.: X-VRML for Advanced Virtual Reality Applications. IEEE Computer, **36** 3 (2003) 89-92.
49. X3D International Standard (2004). X3D framework & SAI. ISO/IEC FDIS (Final Draft International Standard) 19775:200x. Available at www.web3d.org/x3d/specifications (last access on August 2006).
50. Mzoughi, T., Davis Herring, S., Foley, J. T., Morris, J. M., Gilbert, P. J: WebTOP: A 3D Interactive System for Teaching and Learning Optics. Computers and Education, in press. doi:10.1016/j.compedu.2005.06.008, published online in 2005.

Part III

Applications

15

Adaptive Information for Consumers of Healthcare

Alison Cawsey[1], Floriana Grasso[2], and Cécile Paris[3]

[1] School of Mathematical and Computer Sciences, Heriot Watt University, Edinburgh, UK,
alison@macs.hw.ac.uk
[2] Department of Computer Science, University of Liverpool, UK,
F.Grasso@csc.liv.ac.uk
[3] CSIRO ICT Centre, Sydney, Australia, Cecile.Paris@csiro.au

Abstract. This chapter discusses the application of some of the technologies of the adaptive web to the problem of providing information for healthcare consumers. The particular issues relating to this application area are discussed, including the goals of the communication, typical content of a user model, and commonly used techniques. Two case studies are presented, and evaluation approaches considered.

15.1 Introduction

So far this book has looked at some of the techniques that have been developed for the adaptive web, focusing on how we model the user, and how we use that information in adapting the user's experience. In this chapter we show how some of these ideas apply to one particular application area: the provision of information to consumers of health care.

In recent years the way in which people are involved in their own health care has changed dramatically [47]. While, in the past, the almost exclusive source of information was the medical staff directly concerned with the provision of care, nowadays the Internet and the World Wide Web have provided new opportunities for a new generation of users, the "health information consumers". These have been defined by organisations like the American Medical Informatics Association as people who seek information on various aspects related to health and well being, like health promotion, disease prevention, management of long term conditions, and so on. Health information consumers are therefore not only patients, but also their family and friends, or simply people concerned about health.

An increasing number of people are now using the Internet to support their healthcare [58], and the amount of information available on the Web continues to grow. The information needs of healthcare consumers are different from those of the members of the healthcare team (see [59, 53, 28] for some examples of research in health information systems aimed at health care providers). For example, patient-oriented health information systems may include providing information to promote patient choice, informed consent, self-care and shared patient-doctor decision-making (e.g., [46]). Providing such health information via adaptive web-based systems offers new possibilities

P. Brusilovsky, A. Kobsa, and W. Nejdl (Eds.): The Adaptive Web, LNCS 4321, pp. 465–484, 2007.

for pursuing public health objectives like providing knowledge and inducing behaviour change. Furthermore, recent studies have shown that web-based interventions (to provide knowledge and induce behavior change) can have more impact than non web-based interventions [73]. This includes increased knowledge about conditions and treatment, increased participation in health and more uptake of behaviour changes. In addition, sites that pointed readers to relevant, individually tailored material reported longer session times per web-visits and more visits.

There is also evidence that decontextualised, impersonal and generic health information, as typically found on the Internet, has less impact than health information tailored to the individual, at least in some situations (e.g., [3, 49, 18, 69, 70, 68]).

There has therefore been much interest in how we can design systems capable of tailoring information to the health care consumer, and exploiting the great potential to enhance health information and education through web delivery – applying ideas from adaptive web-based presentations and adaptive hypermedia to the problem of providing users with relevant, appropriate, understandable, and potentially persuasive information relating to their needs. There are particular issues in this area to be aware of, focusing now on patients as our main healthcare consumer.

First, we need to consider some of the goals of patient information and education. Patient information may be intended to *inform*, to *enable decision-making* or to *persuade*. We may, for example, want to: inform the user about their condition or about the side-effects of their treatment; give them enough information to enable them to take an active role in the decision-making concerning whether or not to have surgery; or persuade the patient to improve their diet. Persuading the user of a course of action may be part of encouraging patient *compliance* (or *adherence*) – we may want to encourage and motivate them to go along with the treatment regime proposed and take the necessary actions.

Whatever the objective of a healthcare communication, different patients have different individual needs. A good healthcare professional will recognise this and adjust the content and level of verbal information to the patient's perceived needs (both informational and emotional) and their level of understanding. He or she may also ensure that the language employed is both understandable and appropriate for a specific patient, remembering that, first, most patients are not medical experts, and, second, they might already be under considerable cognitive load and stress due to the situation.

This contrasts with current written sources of information (e.g., leaflets and websites) which are normally targeted at the typical patient, not at the individual. Yet written information is also of vital importance in healthcare communication. Verbal messages are often forgotten, while written information is there for reference, and potentially provides a shared information source for patient, family and friends. Recognising this, for example, a genetic counsclor will always provide patients or carers a one- to two-page letter summarising the information that was given to them verbally during the consultation [4].

Given the need for personalised or tailored information and the benefits of written sources, many researchers have explored how we can automatically adapt the content of healthcare messages to the patient (or more generally, to the user). Information may be delivered through printed leaflets, online via adaptive websites, or through phone/text messages. Similar methods of content adaptation can often be applied whatever the means of delivery. Conversely, while there are some peculiarities in each application area, general techniques for adaptive content presentation on the web apply whatever the domain. These general techniques are well described in, for example, Chapter 13 of this book [17]). In this chapter we will therefore concentrate on the issues that specifically arise in healthcare information.

First, we must take seriously issues of privacy, security and trust. Patients are unlikely to use a system where their personal medical details are potentially accessible by others. Furthermore, they need to trust the source of information. Second, in healthcare information, we are not just concerned with informing and educating, but also with the patient's emotional state and attitude (e.g., [3, 43, 35]). We have to take account of the patient's emotional needs and their willingness to accept and commit to change. An effective communication is not the one that is merely learned and remembered, but the one that enables the patient to talk about their problems and come to decisions or acceptance concerning their medical problems [55]. Finally, there may be an issue of control: i.e., patients may want to be able to control what information a system has about them and know how it is being used.

The rest of this chapter will look first at what we may be trying to achieve in personalised health communication, then at the user model (e.g., the attributes of the patient to whom the system is adapting the information) and at techniques that can be used to produce personalised healthcare information. Two case studies will be given, illustrating the range of applications and techniques. We will then look in detail at how personalised health care communication systems can be evaluated, and in particular whether evaluation methodologies from the medical domain can be usefully applied.

15.2 Health Education Goals

Before looking at *how* we personalise health materials, it is worth considering *why* in more detail. While different health professionals have different perspectives on this, two objectives are frequently discussed. The first objective is to support the patient in making decisions about their treatment (shared of course with the health professional team):

> *The overriding goal of patient education should be to support the patient's autonomous decision-making, not (as it has been conceptualized) to get patients to follow doctor's orders.*   [62]

The second objective often discussed is compliance (i.e., following the prescribed treatment and care plan). Compliance is a very important problem in health care, with many implications, both medical and socio-economical. It is estimated, for example, that in the European Union between 2% and 20% of the medical prescriptions never get to the pharmacy, and that about 125,000 deaths and 5-10% hospitalisations per year can be attributed to lack of compliance. Compliance might be achieved by a number of ways (e.g., [40]), and, conversely, non-compliance might be explained by a number of factors. For example, compliance has been shown to be correlated to the patient's understanding of their condition and prescribed treatment (e.g., [30, 42]). Indeed, some patients need to understand the rationale for their treatment, and why it will work. For example, they may need to understand what a specific drug does. They might also need to understand why and how their own actions (e.g., exercise, taking medication) are necessary for success. With this understanding, they are more likely to follow the treatment regime recommended. But understanding alone is not enough. Patients also need to be committed to the treatment, and this may require convincing them of its necessity, by ensuring they both understand and truly believe the consequences of failing to follow a specific treatment (which might include a change of lifestyle). Finally, patients are more likely to follow a treatment or advice if they trust it and its prescriber.

Compliance with the doctor's treatment plan and autonomous decision making by the patient are sometimes presented as opposing points of view. However it seems more likely that both perspectives should be supported. It is not always appropriate to leave the patient to make the decisions, and they will often not want that role, while they may want to participate in the decision-making. Note also that patients might be more likely to comply to a treatment if they were involved in its choice.

Depending on the objective, different types of information might be provided to a patient. Where treatment choice is an issue, patients may receive background information about their conditions (e.g., what causes it if it is known, its symptoms, its consequences, what can be done about it), and specific information about the alternative treatments and why a particular one is more appropriate for them. For patients with chronic disorders (e.g., asthma, diabetes), appropriate information might include information that helps them manage their own care effectively, and that provides advice as to when to call out a health professional. In addition, there are today broader time-independent health promotion objectives, addressed to groups or the population at large, as opposed to an individual at a particular point in time. For a healthier society we want to promote a good diet, exercise, stopping smoking, avoiding direct exposure to the sun, and so on. While these are almost universally recognised goals, they may be more effectively achieved by addressing the individual – by personalising the advice and the information, e.g. [18, 70].

While supporting choice and promoting a particular course of action are perhaps the easiest health education goals to characterise, much of the information giving in healthcare has a less explicit objective. With more appropriate and understandable information, patients will usually feel more in control. If they know what will happen next, which health professionals will be managing their care, and how they should prepare for any treatment, then their anxiety is likely to be reduced. Anxiety and stress reduction is therefore another important objective in health education, but a difficult one to get right. Where patients have a poor prognosis, it is particularly difficult to get a balance between sensitivity and openness, and one that a machine is unlikely to achieve.

Currently most patient information is provided verbally or through leaflets, with an increasing number of patients turning to the web for further information [58] and an increasing numbers of health information websites (e.g., [13, 14, 12]). A typical website or leaflet will focus on a particular condition, and give general information, information about diagnosis, and information about treatment, including any options and alternatives and any actions that the patient can take to help themselves. These existing resources are very much disease centred. They are not tailored to the patient's specific needs and knowledge. As a result they can sometimes be confusing or overwhelming to a patient. Adaptive and personalisation techniques open the way to more patient-centred sources of information and potentially more effective means of achieving the health education goals described above.

Effective health education is not just about making life better for the healthcare consumer, but it is also about making the process more efficient, using the available money and resources as effectively as possible and potentially saving our governments' money (e.g., [45]). By providing means for the patient (and their carers) to obtain information outside a doctor or hospital visit, there is a possibility to move some of the health care services to the home or the community. If care is to be shared between health professionals, community and patients, then each must have an appropriate level of understanding of the medical issues, as well as who to call when. If this is done effectively there is the potential to make better use of specialist expertise, and save on unnecessary hospital visits. Being able to automatically create personalised communications appropriate to context and need may prove to be a vital part of this process.

15.3 The User Model

Having briefly reviewed some of the objectives of patient education, we can turn back to how we can adapt health information to a patient, taking into account the particular goals that health professionals recognise as being important for that patient. In this section, we briefly discuss what needs to be captured in a user model in order to provide tailored information that achieves the objectives discussed above.[1]

First, it will usually be necessary to acquire and capture factual information about the patient, their condition, current treatments, and so on. This information may be available in the patient record (e.g., [20]). It is thus possible (and relatively easy) to produce patient-centred information by starting with the information available from their record. Just this amount of tailoring is likely to be an improvement over a general health education leaflet that is typically disease-centric and does not take into account a patient's particular characteristics. For example, instead of including information related to all possible treatments for a condition, a patient-centric information system (or leaflet) may only contain information about treatments relevant to the patient. Similarly, if we know that the patient is being treated by a particular consultant in a specific hospital, the information might include whom to contact where, how to get to the hospital, where to park, information about visitor's hours, etc. In other words, it is possible to produce one coherent, concise and practical information source containing all the information that is important and relevant to the patient, and that he or she is likely to seek.

It is worth mentioning practical issues in using the patient record. There are of course major security and privacy issues when accessing this confidential information,

---

[1] We address in this chapter issues specific to patient oriented health information systems. For general overviews of user models for educational systems and personalised information access, see Chapters 1 [15] and 2 [31] of this book.

and using the patient record for web-based systems is still problematic. Typically web-based systems use more limited information on the patient's health obtained through an online questionnaire. Or, when they use more extensive data, they rely on a password based authentication. This is however likely to be insufficient in many cases, and it is expected that smart cards, private keys, or encryption will be increasingly used [6]. (See Chapter 21 of this book [44] for a discussion of issues related to privacy and security.)

Health education also shares the characteristics of traditional education in that it must be delivered at a level that will be understood by the individual concerned, taking into account at least the patient's literacy, medical and otherwise. (See Chapter 1 of this book [15]).

Most crucially, however, as well as information about the user's medical conditions and treatment, health education systems may need to take into account more complex factors, such as the patient's current mental and emotional state, their ability to make decisions and perform complex actions, or their acceptance of their disease. This is the case, for example, if the patient has just received news about a life-threatening disease, and his or her ability to absorb information may be impaired. In other cases, health education may be about changing attitudes and behavioural change – for example, a reason for providing information may be to convince patients to change their diet, to stop smoking or to start exercising. In these cases, then, the patients' motivation level, their willingness to accept treatment or make changes, as well as their desires and intentions all become important. It thus seems at least plausible that adapting materials to some of these factors will make written and online materials more effective. This in fact has already been shown, as discussed earlier, e.g., [18, 69, 70, 68].

So, the user model for a health education or health promotion system will very often include the information obtainable from the patient record, but may also include a whole range of cognitive factors, such as the ones mentioned above (e.g., current understanding, motivation and anxiety). The user model may capture factors related to different personality types (which might provide insight, for example, as to how a patient is likely to deal with change or bad news in different ways – e.g., [35]). This aspect is what makes the provision of healthcare information a challenge. In tackling this task, it is sensible to ground the user model and the information adaptation on well established behavioural theories. One example of such theories is the Stages of Change Model, or Trans-Theoretical Model [61]. The model assumes that people progress through very distinct stages of change on their way to improve health:

1. *precontemplation*: people at this stage see no problem with their behavior and have no intention of changing it. They mainly lack information, in the sense that they have not been presented yet with any convincing reason to change their behaviour. Often people are not very open to receiving advice.
2. *contemplation*: in this stage, people come to understand their problem, its causes, and start to think about taking action to solve it, but have no immediate plans. This is a delicate stage, as there is always the risk to miss the opportunity, and go back to precontemplation, because of laziness or old influences.
3. *preparation*: people are planning to take an action, and are putting together a plan, but have not taken any step yet. This is a sort of transition stage between the decision to act and the action itself. Often one of the causes of going back to a pre-

vious stage is that the plan is too ambitious, and the life style change planned is too drastic.

4. *action*: people are actually in the process of actively making behaviour changes. The concern here is to pay attention to negative emotions: anger, depression, anxiety, apathy, insecurity, etc., in order to prevent relapse.

5. *maintenance*: health behaviour continued on a regular basis. The state is more stable than the action one, but there is always the possibility of relapse.

6. *(termination)* at this stage, the former problem no longer presents any temptation or threat. Many people never reach this stage.

In addition to providing a classification of the user, the model suggests strategies for recognising and dealing with each stage of change, in terms of the information that should be presented at each stage.

For example, the precontemplator needs to identify the problem in the first place, so one may provide information on related problems. It is also likely that precontemplators have misconceptions about the consequences of their actions, so one should assess prior knowledge and clarify misunderstandings. On the other hand, those in the "action" stage mainly need to get things going, by means of tips and strategies to maintain and enhance their commitment. They need reinforcement too, and encouragement.

User models containing this kind of information have indeed already been used. For example, systems generating patient education with the goal of achieving behaviour change (e.g., diet, smoking) have captured patients' attitude towards a specific change, exploiting the stages of change model [70, 64].

A user model in a health application can thus be quite complex. This leads to the question of how we obtain and update such a user model. Using the medical record is easy, and changes in the patient's treatments is generally reflected in changes in their record. However, the patients' record may not always be available to the health information system. In such cases, we need other ways to obtain the appropriate attributes of the patient. This of course provides a number of challenges. For some attributes it might be possible to let the patient fill in a simple questionnaire, but more thought is needed when considering how to capture some of the more subtle aspects that might support effective information provision (e.g., patients' personality, mental and emotional state).

There are however various instruments that can be used here (mostly standardised questionnaires) which can be applied to ascertain personality type, stage of change, anxiety level, and so on. While these instruments may be seen as moderately intrusive and as potentially not always leading to accurate results, they are already used successfully in on-line health diagnostic and intervention applications, for example to treat depression using Cognitive Behavior Therapy (e.g., [21, 23]). One approach then is to use these existing tools to populate the user model of an accompanying health information system, which can then exploit this information to provide the patient with relevant information about their condition.

Increasingly, researchers are also investigating new, less intrusive, methods for capturing some characteristics of the user, in particular emotional state and stress/anxiety levels, such as by the use of physiological sensors (e.g., [60, 71, 19]). However, these are still at early stages of research, mostly applied to the domains other than healthcare, and it is as yet unclear whether or how we can use these measures reliably outside of the experimental situation.

Finally, obtaining the attributes of the patient at one point in time is not enough. Having acquired details of the patient's current state and stored them in a user model, a system needs to be able to monitor the patient's state and update these details as their state changes. As health education is often about changing the patient's mental state (e.g., their beliefs, attitudes, anxiety, etc.), a system needs to be able to monitor these as well as the attributes related to their health problem. By monitoring the user and updating the model the system can both provide more appropriate and timely information, and also assess the effectiveness of its past interventions. For certain physiological attributes it is now possible to use small wearable monitoring devices to achieve this, providing a constantly updated model. However, in general these issues are still research challenges.

15.4 Techniques for Adaptation

As discussed previously, the goals of healthcare communication can be quite varied, including persuading the patient to take an action, enabling them to manage their care, supporting informed shared decision-making between patients and health professionals, and reducing stress and anxiety. From a broader perspective, the goal is to provide the patient with information that is *relevant* to their condition and to their situation, which enables them to understand and take control of their condition at a level appropriate to them. The specific goals will then depend on the patient and their situation, while their mental and medical state and the practical situation will influence how information can be best selected and presented to be most effective.

Having reviewed the goals, we also need to consider the nature of the communication itself. Many projects have simply generated personalised materials (leaflets or simple websites, e.g., [38, 64, 20]), where the content and style of the material is adapted to the user, but the interaction style is fixed and simple. Personalised email or text messages have also been used [48], but again with little dynamic interaction with the user. While recognising that the interaction or dialogue style may be important, in particular to acquire and maintain the user model, we focus here mostly on how content is selected, adapted and presented to the patient, given a user model. Indeed, techniques for adapting the information that is generated are similar whether a system is interactive or not. Dialogue issues are taken up in more detail when we discuss the HOMEY project [39] as a case study in the next section.

The most common techniques employed to produce tailored text-based material are based on Natural Language Generation (NLG)[2]. In the health domain, several projects have used these techniques to generate adapted primarily text-based material (e.g., [54, 65, 20]). NLG techniques are concerned with the automatic production of coherent and appropriate textual documents from structured data [50, 51, 63]. They have also been applied in recent years to the generation of appropriate and coherent multimodal

---

[2] The reader is also referred to Chapter 13 of this book [17] for some techniques for adaptive content presentation on the web.

documents (e.g., [2, 29, 24] and hypertext presentations (e.g.,[26, 56, 25]). Broadly, these techniques are divided into planning what to say (content) and deciding how to express it once there is a message to express. Planning what to say usually starts with communicative goals (e.g., persuade the hearer/reader to take some action, inform the hearer/reader of a fact or situation). The content planning process typically uses domain information from a database or knowledge base, and information about the user (from the user model/profile). The process thus selects information to present and organises it into a coherent whole. The output of this process is a sequence of primitive messages (e.g., informing the hearer/reader of a simple fact) which, given the user model, should achieve the communicative goal. Given a sequence of primitive messages, the question then arises as to how they should be expressed. For example, the following questions must be addressed: Should each fact constitute one sentence, or should they be conjoined? Should facts be announced bluntly or in a more indirect way? Should they be presented formally or less formally? What specific words and constructions should be used? Adapting this stage to the user may be as important as adapting the content and organisation of the information in healthcare communication, where the emotional state (and cultural status) of the user are important.

While natural language techniques remain important, health education and information provision has recently become of wide interest (see, for example, a number of workshops and symposia related to this topic – e.g., [9, 34, 1]), and other techniques have been investigated in healthcare communication for both health care providers and consumers. These include speech (for example the generation of voice messages over the telephone, e.g., [33], as is briefly described below), search and summarisation (e.g., [52, 28]), hypermedia and virtual reality (c.f. [34]) and Embodied Conversational Agents (ECA), e.g., [8].

15.5 Case Studies

Having introduced some of the issues and techniques in personalised patient information, we can turn to two specific projects that have made use of these techniques in practical applications. The first project illustrates how fairly simple patient-centred materials can be generated given a patient record. The second project is more ambitious, providing adaptive advice in the context of a multi-modal dialogue.

15.5.1 Personalised Information for Patients with Cancer

The first case study (Piglit) is a project concerned with creating personalised materials (online and written) for patients undergoing treatment for diabetes [11] and cancer [20]. The main goal of the project has been to provide materials that are patient centred, and which allow the patient to quickly access additional materials of interest. The techniques used have been generally simple, but the systems and approaches produced have been thoroughly evaluated with many patients.

In this project the patient's medical record is used as the main source of information about the patient. While there are many formats for computerised records, it will hold information on, at least, the patient's medical conditions and treatments. The Piglit project provided patients with online access to this record, with hyperlinks allowing access to explanatory information about their conditions and treatment. These explanatory pages were generated dynamically from a simple knowledge base of medical information, used in combination with information from the patient record, allowing the explanations to be geared to the patient's likely information needs.

Figure 1 illustrates an example page of information for a patient with prostate cancer. Italicised terms were hyperlinks taking the patient to more general, but hopefully relevant information about their condition.

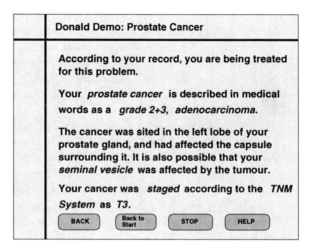

Fig. 15.1. Personalised Health Information linked to Medical Record

This system used simple planning methods from natural language generation to determine content. These techniques were used to plan what to say given a topic (e.g., prostate cancer) and a particular patient record. A simple knowledge base contained the facts about conditions, treatments, and other relevant medical concepts, and this could be exploited to generate pages of information.

The system was evaluated in a large randomised controlled trial (see section on evaluation) [41], which compared the personalised version with a general information system providing very similar information, and with standard leaflets. Patients using the computer used a touch screen system located in a room in the cancer centre where they were receiving treatment. They also received printouts of the information presented in their session. 525 patients participated in the study. Questionnaires were used to gauge the patients views of the system, and also to assess their anxiety levels (before and after the intervention). Statistical analysis of the results was done to compare personalised *vs* non personalised and computer *vs* leaflet.

The results showed that the patients receiving the personalised book were (at a statistically significant level) more likely to think they had learned something new ($p = 0.02$) and that the information was relevant ($p = 0.03$), and were more likely to show the information to family and friends ($p = 0.035$). However, this is perhaps not surprising as they received information about their own specific conditions/treatment not available to the group receiving general information. Unexpectedly, we found that patients with personalised information showed better improvement in anxiety over three months than those with more general information, despite receiving information on their condition that might be worrying. Three months after the intervention 37% of patients in the general computer information group were still anxious compared with only 19% in the personal information group, with the intervention a significant predictor of anxiety level ($p = 0.001$).

15.5.2 Personalised Home Monitoring to Support Continuity of Care

Our second case study is concerned with the needs of chronic patients, such as Hypertension or Diabetes patients. For these patients, the main objective of a health care system is to ensure compliance to the therapeutic and lifestyle regime over long term periods. In hypertensive patients, for example, it is crucial to maintain a healthy lifestyle (in terms of nutrition habits, doing exercise, stopping smoking and so on), and also to carefully monitor blood pressure, heart rate and weight. Frequent visits to the cardiovascular unit would be highly beneficial, both to keep the doctors updated with the patient's situation and to reinforce the health promotion message to the patient. However, this is difficult to implement, both because of the lack of resources from the hospital, and because of the tendency for patients, especially from long term patients, to relax their attendance to the meetings.

We describe one solution to this problem, as proposed by the HOMEY project [7, 33, 39][3], that developed a system able to efficiently communicate with the patient and to improve the information flow between patients and medical staff. The aim is to allow patients to use the system to communicate as frequently as possible their test results to their care team, while the system has the opportunity to enquire about lifestyle changes, and update the patient's record. The medical staff read the updates, and possibly make new recommendations, which in turn are stored in the system and passed to the patient at the next contact. In order to achieve this, the system makes use of natural language dialogue technology.

Many dialogue systems are based on a "scripted conversation" approach. This means that the main structure of the dialogue is fixed once and for all by the dialogue designer, in order to have control of what can happen in the dialogue. While this is simple and effective for very focussed applications (like telephone banking), it becomes rapidly expensive and too inflexible for complex situations, where there are many objectives to take into account and a very large domain knowledge, like the medical domain. In these situations, sophisticated "intelligent" dialogue systems are more appropriate. The HOMEY system is based on intelligent dialogue technology, and is able to manage a conversation with a patient, adapted to the patient's needs, preferences, and clinical history, but also taking into account the physician's goals. The system supports multimodal input and output, combining the generation of dynamic HTML pages and

[3] This is a project funded by the European Community in the programme "Health Care for Citizens" (5th Framework, Project No. IST-2001-32434). The project started in 2001 and was completed in 2004.

VoiceXML [72] sentence fragments. This allows the user to contact the system either with a simple phone call, or with a traditional computer connection. In the latter case, the user is free to choose a speech input/output, a keyboard input with a visual output, or a combination of both.

Intelligent dialogue systems need to keep track of many of the user's characteristics, in order to handle a real time dialogue. In addition to information on the evolution of the disease and the treatment, for instance, the user's goals and beliefs about the medical treatment will be needed for the system to be able to better promote, justify or reinforce the particular piece of advice that it gives. Also, the history of the past interactions with the system will give information on what to ask and what to talk about the next time round. For example, if the user had said he would try to stop smoking, the system may want to check whether the plan had been implemented, and if not, give some more motivation for the user to start doing so.

Generally speaking, when producing intelligent dialogue systems, many phenomena have to be taken into account. These can be broadly divided into *high level* and *low level* phenomena. The former include very general notions like the goals of the dialogue participants, or the strategies for producing persuasive messages, and so on, which are assumed to be independent both of the language and on the output medium. The latter include what it takes to actually produce the single message, like the grammar of the language to use, whether to use speech or text, and so on. These low level issues may be important in adaptation too. For instance the system can try to use the user's vocabulary as much as possible, in order to be better understood.

In the HOMEY project, both levels are taken into account and are dealt with in an architecture based on the concept of *abstract task specification* (see Fig. 2). This structure gives information on two important aspects of what the system should do next: the "plan" representing the high level task to be executed (such as take patient's measurements and make a decision on referral to the clinic), and the definitions of the objects involved in this plan (such as, "heart", "blood pressure", "measurement devices", etc.), together with the relationships among them, that come from a "domain ontology", that is a conceptual representation of the domain.

The abstract task specification is then transformed into the *high level dialogue specification*. The main purpose of this specification is to give some structure to the conversation. The initial dialogue structure depends on the task specification. For instance, if the plan says that a decision cannot be taken until all the patient's measurements are in, then the first part of the dialogue will involve asking the user to report his measurements. This initial structure is however flexible, and can adapt to the way in which the dialogue evolves. For instance, consider the following dialogue:

> **System**: What is your heart rate?
> **Patient**: What do you mean?

Here the user asks for clarifications before replying to the system's question, so the system will have a new "obligation" to fulfill, in addition to those coming from the task specification, and has to take a decision about what to do next (typically, the obligations coming from the user will be dealt with first).

Also, the user may take some initiative in reacting to the system's question. For example, consider the following dialogue:

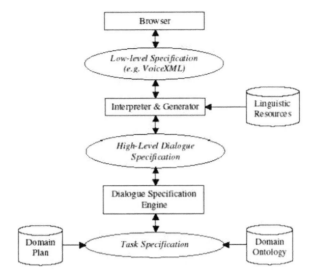

Fig. 15.2. Architecture of the HOMEY system

System: What is your heart rate?
Patient: My heart rate is 67,
                 and my blood pressure values are 90 and 120.

Here the user has anticipated a system's question by providing more information than requested, so the dialogue specification has to account for the fact that this sub-task has already been accomplished, and may move on. This knowledge comes again from the task specification, which says that two tasks (ask heart rate and ask pressure values) can be both part of a super-task (ask patient's measurements).

Another type of task that can be included in the plan concerns checking on the user's lifestyle, and perhaps reinforcing some of the recomendations coming from the physician's goals. This can lead to dialogues like the following:

System: Are you still swimming two times a week?
Patient: Yes.
System: Are you still smoking?
Patient: Yes, 5 cigarettes per day.
System: You should stop smoking.

The examples above show how dialogues might occur in the setting where the user contacts the system via telephone. In these cases the system will typically output a single new move at a time, e.g. a question. In the multi-modal context, the output could be one or more HTML forms, where several questions are presented, and where, if the user has asked for both voice and visual output, VoiceXML will utter the first question on the form, while the language model will enable the user to answer any question on the form in the preferred order.

The system has been evaluated in two Italian hospitals with a hypertension unit, in two studies. The first one involved fictitious patients, that is a number of volunteers who were assigned a disease profile. This study was mainly done to assess the usability of the system. In the second study, the aim was to assess whether using the system would actually make some difference to the health of the patients. A clinical trial was performed, where about 300 patients of the units were assigned randomly to two groups, only one of which using the (telephone based) HOMEY system. The average blood pressure of the two groups of patients was measured before and after the period under observation. While both groups had a significant decrease in blood pressure, the statistical results suggest a trend whereby the group of patients using the system had a greater systolic pressure decrease than the other group. From the point of view of the user's satisfaction, the evaluation was also successful, as it is testified by the fact that, even after the trial was completed, there is still a good number of patients that decided to continue to use the system to report their data.

An extension to the system is currently being investigated [57] in which the user model is enhanced based on three theories of behaviour change: the Social Cognitive Theory, the Health Action Process Approach [5, 67], and the already mentioned Stages of Change Model.

15.6 Evaluation and Uptake

We now return to the issue of how we evaluate adaptive or personalised healthcare information systems. Before we do so, however, it is worth mentioning that, typically, systems are first designed based on a requirements analysis: this is where the designers of the system spend time with both the expert and the intended users to elicit requirements for the system. This stage is an important aspect of usability engineering for any system. (See Chapter 24 of this book [32] for more information on Usability engineering.). For healthcare information systems researchers have used various techniques for this, including applying knowledge acquisition techniques to elicit expert knowledge of healthcare communication [66], and studying doctor-patient interaction in a natural setting [16].

Once a system designed and implemented, evaluation is crucial. We need to consider what it is that we are trying to claim for the personalised systems, and second, what techniques we can use to demonstrate that our claims or hypotheses are valid.

Some benefits of a personalised system lie in the subjective opinions of users. Perhaps a personalised system is *preferred* by users, and provides information they perceive as more *relevant*. The main way we can assess this is by questionnaire, asking users to rate or compare systems (e.g., [54]), or by simply monitoring actual uptake/use of the systems if freely available.

However, often we will be trying to influence things such as the patient's understanding of their condition, anxiety levels, level of compliance, willingness to take a test, or even their state of health (mental or physical). If a system is developed to affect these things then they naturally need to be measured in the evaluation. In the Piglit project, for example, the aim was for the patient to understand their condition better and so feel more in control. Their preferences and their state of anxiety were thus measured.

The HOMEY project was concerned more with improving the patient's ability to manage their own care, and hence the patient's blood pressure was measured, as a measure indicating good self-management for the relevant condition.

Both example projects used randomised clinical trials to measure the benefits of the system. Patients are randomly assigned to one of two (or more) interventions (e.g., personalised system versus non-personalised) and then appropriate measures are taken relating to the system goals. These measures, as we have seen, can be anything from the objective and concrete blood pressure readings to the patient's perception of information relevance. Differences in results between groups can be measured for statistical significance, and we can attempt to explain these differences as due to the different interventions.

While randomised trials are the *gold standard* for evaluation in the medical domain, we can question their utility from the perspective of the Computer Science researcher. Randomised trials are difficult to design, very expensive and time consuming to run, and, for information systems, interpreting the results is often difficult, mostly because there might be many factors outside the system itself that can affect the results. This may explain why results have often been negative or insignificant, and is not always possible to draw clear conclusions (see [65] for a discussion of negative results). Even for positive results, it may be hard to know where exactly to attribute the differences, as the differences between intervention groups is rarely reducible to one single factor. In Piglit, for example, the total *content* available to patients *and* the starting point for the navigation (medical record) was different. Yet, randomised trials are being used routinely in the medical domain, and researchers have been able to use them to draw conclusions on the benefit of various treatments, or to obtain information on the impact of various information methods (e.g., web *vs* non-web, tailored *vs* generic, etc.) [73, 68].

More such trials are in fact both planned and in progress (e.g., [27, 22]). As a community, adaptive hypermedia researchers may need to learn from medical researchers to be able to better evaluate the effectiveness of their systems. One issue will always remain, though: that of the cost involved and the need for large numbers of users. This is not always practical. At first glance it may appear that, for web based systems, the experimenter has access to a large pool of potential users at low cost – users could be recruited (e.g., by email) and allocated randomly between two or more systems. However, in most cases these users would not be truly representative of the target user group, and it may prove hard to maintain contact with the users over an extended period.

While randomised trials will always be needed in this area, and are necessary when it is the (long term) health of the user we are aiming to improve, alternative methods of evaluation should also be considered. Less expensive and time consuming methods can be used to measure usability, learning and memory, while user preferences can perhaps best be measured both through questionnaires and by looking at actual uptake and use (as in HOMEY). If we can make two different systems available to patients, the one that they choose and continue to use is clearly the preferred one.

This brings us to the question of how we get these systems accepted and used in practice. This is partly, but not just, a question of demonstrating their benefits. There is a huge range of obstacles to changing healthcare practice (see [37] for a brief discussion). Some of this relates to healthcare as a monolithic institution, somewhat resistant to change, with many stakeholders. However, other obstacles appear when we consider how to change the way information is communicated. Healthcare is an area where trust and privacy are of key importance. Healthcare information must be trusted, come from and validated by reliable sources, and automatically generated adapted information may not meet that criteria, or may not be seen as meeting that criteria. Patient information must be protected and not accessible outside the healthcare team, so even the implicit information available in a seen-over-the-shoulder personalised page of information may result in patient trust being compromised and uptake of a personalised system being reduced. For practical uptake, it is often the apparently trivial issues, like how to find a quiet and private corner of the waiting room for an information point, that prove the hardest to satisfy.

15.7 Conclusions

In this chapter, we have looked at the issue of using adaptive techniques to provide health information and education. We have argued that these techniques show great promise and may open new horizons in this domain, with potential for significant improvements over non-tailored information material. We also described some of the challenges that arise in this domain, in particular issues of trust and privacy, problems of acquiring and constantly updating the user model required to provide sophisticated tailoring, and the cost and difficulty of evaluation. Yet, it remains an exciting domain, one in which, on the one hand, even simple techniques can already bring real benefits and impact, and, on the other hand, new challenges arise. We note, in fact, that there is a growing interest in health applications, in particular for health information systems (e.g., [9, 10, 36, 34, 1]), not only to educate patients but also to assist health professionals (e.g., [28, 53], to promote better communication both amongst health professionals and between patients and their health care team, and, finally, to provide diagnostic tools and assist in health care provision [21, 23]. Further research is required to assist in these goals and provide systems capable of facilitating this communication and adapting appropriately to the context, at all the required levels.

References

1. Alm, N., Abe, S., Kuwahara, N., eds.:   The First International Workshop on Cognitive Prostheses and Assisted Communication (CPAC), Sydney, Australia (January 2006) http://www.irc.atr.jp/cpac2006/.
2. André, E., Rist, T.: Generating coherent presentations employing textual and visual material. Artificial Intelligence Review **9**(2/3) (1995) 147–165 (Special volume on the Integration of Natural Language and Vision Processing).
3. Auerbach, S., Martelli, M., Mercuri, L.: Anxiety, information, interpersonal impacts, and adjustment to a stressful health care situation. Journal of Personality And Social Psychology **44**(6) (June 1983) 1284–96
4. Baker, D., Eash, T., J., S., Uhlmann, W.: Guidelines for writing letters to patients. Journal of Genetic Counseling **11**(5) (2002) 399–418

5. Bandura, A.: Social Foundations of Thought and Action: A Social Cognitive Theory. Prentice-Hall, Englewood Cliffs, NJ (1986)
6. Bellazzi, R., Montani, S., Riva, A., Stefanelli, M.: Web-based telemedicine systems for home-care: technical issues and experiences. Computer Methods and Programs in Biomedicine **64** (2001) 175–187
7. Beveridge, M., Milward, D.: Definition of the high level task specification language. Deliverable D11, HOMEY Project, IST-2001-32434 (2003)
8. Bickmore, T.: Relational Agents: Effecting Change Through Human-Computer Relationships. PhD thesis, MIT Media Arts & Sciences (2003) Available at: http://www.ccs.neu.edu/home/bickmore/bickmore-thesis.pdf.
9. Bickmore, T., ed.: AAAI 2004 Fall Symposium on Dialogue Systems for Health Communication, Crystal City Hyatt, Washington DC (2004) http://www.aaai.org/Library/Symposia/Fall/fs04-04.php.
10. Bickmore, T., ed.: The 2005 AAAI Fall Symposium on "Caring Machines: AI in Eldercare", Arlington, Virginia (November 2005) http://www.aaai.org/Library/Symposia/Fall/fs05-02.php.
11. Binsted, K., Cawsey, A., Jones, R.: Generating personalised patient information using the medical record. In Barahona, P., Stefanelli, M., Wyatt, J., eds.: 5th Conference on Artificial Intelligence in Medicine Europe (AIME'95). Volume 934 of Lecture Notes in Artificial Intelligence. (1995) 29–41
12. http://www.bluepages.anu.edu.au BluePages, A site providing information about depression.
13. http://www.thebreastcancersite.com Breast cancer information site (1).
14. http://www.breastcancer.org/ Breast cancer information site (2).
15. Brusilovsky, P., Millan, E.: User models for adaptive hypermedia and adaptive educational systems. In Brusilovsky, P., Kobsa, A., Niejdl, W., eds.: The Adaptive Web: Methods and Strategies of Web Personalization. Volume 4321 of Lecture Notes in Computer Science. Springer-Verlag, Berlin Heidelberg New York (2007)
16. Buchanan, B., Carenini, G., Mittal, V., Moore, J.: Designing computer-based frameworks that facilitate doctor-patient collaboration. Artificial Intelligence in Medicine **12** (1995) 171–193
17. Bunt, A., Carenini, G., Conati, C.: Adaptive content presentation for the web. In Brusilovsky, P., Kobsa, A., Niejdl, W., eds.: The Adaptive Web: Methods and Strategies of Web Personalization. Volume 4321 of Lecture Notes in Computer Science. Springer-Verlag, Berlin Heidelberg New York (2007)
18. Campbell, M., DeVellis, B., Stretcher, V., Ammerman, A., DeVellis, R., Sandler, R.: Improving dietary behavior: the effectiveness of tailored messages in primary care settings. American Journal of Public Health **84**(5) (1994) 783–787
19. Carberry, S., de Rosis, F., eds.: Proceedings of the User Modelling 2005 Workshop on Adapting the Interaction Style to Affective Factors, Edinburgh, UK (2005) http://www.di.uniba.it/intint/UM05/WS-UM05.html.
20. Cawsey, A., Jones, R., Pearson, J.: The evaluation of a personalised health information system for patients with cancer. User Modeling and User-Adapted Interaction **10**(1) (2001) 47–72
21. Christensen, H., Griffiths, K.: The prevention of depression using the internet. Medical Journal of Australia **177 (suppl)** (2002) S122–5
22. Christensen, H., Griffiths, K., Jorm, A.: Delivering interventions for depression by using the internet: randomised controlled trial. British Medical Journal (23 January 2004) doi:10.1136/bmj.37945.566632.EE Downloadable from bmj.com.
23. Christensen, H., Griffiths, K., Korten, A.: A web-based cognitive behavior therapy: Analysis of site usage and changes in depression and anxiety scores. Journal of Medical Internet Research **4**(1) (2002) e3

24. Colineau, N., Paris, C.: Task-driven information presentation. In: Proceedings of OZCHI'03, Brisbane, Australia (November 2003)
25. Colineau, N., Paris, C., Wu, M.: Actionable information delivery. Revue d'Intelligence Artificielle (RSTI RIA) **18**(4) (2004) 549–576 Special Issue on Tailored Information Delivery.
26. De Carolis, B., de Rosis, F., Andreoli, C., Cavallo, V., De Cicco, M.: The dynamic generation of hypertext presentations of medical guidelines. The New Review of Hypermedia and Multimedia **4** (1998) 67–88
27. Duszynski, A., Flight, I., Wilson, C., Turnbull, D., Cole, S., Young, G.: Intersecting electronic decision support with user modelling in a web-based consumer decision aid for colorectal cancer screening. [34] http://www.csc.liv.ac.uk/ floriana/UM05-eHealth/.
28. Elhadad, N., Kan, M., Klavans, J., McKeown, K.: Customization in a unified framework for summarizing medical literature. Artificial Intelligence in Medicine **33**(2) (2005) 179–198
29. Feiner, S., McKeown, K.: Automating the generation of coordinated multimedia explanations. In Maybury, M., Wahlster, W., eds.: Readings in Intelligent Multimedia Interfaces. Morgan Kaufmann Publishers, Inc., San Francisco, Ca (1998) 89–98
30. Fisher, R.: Patient education and compliance: a pharmacist's perspective. Patient Education and Counseling **19**(3) (June 1992) 261–71
31. Gauch, S., Speretta, M., Chandramouli, A., Micarelli, A.: User profiles for personalized information access. In Brusilovsky, P., Kobsa, A., Niejdl, W., eds.: The Adaptive Web: Methods and Strategies of Web Personalization. Volume 4321 of Lecture Notes in Computer Science. Springer-Verlag, Berlin Heidelberg New York (2007)
32. Gena, C., Weibelzahl, S.: Usability engineering for the adaptive web. In Brusilovsky, P., Kobsa, A., Niejdl, W., eds.: The Adaptive Web: Methods and Strategies of Web Personalization. Volume 4321 of Lecture Notes in Computer Science. Springer-Verlag, Berlin Heidelberg New York (2007)
33. Giorgino, T., Quaglini, S., Rognoni, C., Baccheschi, J.: The homey project: a telemedicine service for hypertensive patients. [34] http://www.csc.liv.ac.uk/ floriana/UM05-eHealth/.
34. Grasso, F., Cawsey, A., Paris, C., Quaglini, S., Wilkinson, R., eds.: Working notes of the UM-2005 workshop on Personalisation for e-Health, Edinburgh, UK (July 2005) http://www.csc.liv.ac.uk/ floriana/UM05-eHealth/.
35. Green, N.: Affective factors in generation of tailored genomic information. [19] http://www.di.uniba.it/intint/UM05/WS-UM05.html.
36. Green, N., Bickmore, T., eds.: The AAAI 2006 Spring Symposium on "Argumentation for Consumers of Healthcare", Stanford University, California (March 2006) http://www.aaai.org/Library/Symposia/Spring/ss06-01.php.
37. Grol, R.: Personal paper: Beliefs and evidence in changing clinical practice. British Medical Journal **315** (1997) 418–421
38. Hirst, G., DiMarco, C., Hovy, E., Parsons, K.: Authoring and generating health-education documents that are tailored to the needs of the individual patient. In Jameson, A., Paris, C., Tasso, C., eds.: Proceedings of the Sixth International Conference on User Modeling (UM'97), Sardinia, Springer Wien New York (1997) 107–119
39. http://turing.eng.it/pls/homey/homey.home The Homey Project website.
40. Jaret, P.: 10 ways to improve patient compliance. Hippocrates **15**(2) (2001) http://www.hippocrates.com/FebruaryMarch2001/02features/02feat_compliance.html.
41. Jones, R., Pearson, J., McGregor, S., Cawsey, A., Barret, A., Craig, N., Atkinson, J., Harper Gilmour, W., McEwen, J.: Randomised trial of personalised computer-based information for cancer patients. British Medical Journal **319** (1999) 1241–47
42. Kahn, G.: Computer-based patient education: A progress report. M.D. Computing **10**(2) (1993) 93–99
43. Kessler, S., ed.: Genetic Counseling: psychological Dimensions. Academic Press, NY (1979)

44. Kobsa, A.: Privacy-enhanced web personalization. In Brusilovsky, P., Kobsa, A., Niejdl, W., eds.: The Adaptive Web: Methods and Strategies of Web Personalization. Volume 4321 of Lecture Notes in Computer Science. Springer-Verlag, Berlin Heidelberg New York (2007)

45. Kun, L.: Telehealth and the global health network in the 21st century. from homecare to public health informatics. Computer Methods and Programs in Biomedicine **64**(3) (2001) 155–167

46. Large, S., Arnold, K.:   Evaluating how users interact with nhs direct online.   [34] http://www.csc.liv.ac.uk/ floriana/UM05-eHealth/.

47. Lewis, D., Eysenbach, G., Kukafka, R., Stavri, P., Jimison, H., eds.: Consumer Health Informatics - Informing Consumers and Improving Health Care. Springer-Verlag (2005)

48. Marsden, J.:   Primary prevention of osteoporosis in young british women: a comparison of stage-based, tailored email versus non-tailored intervention.   In: Proceedings of Consumer Health Informatics Network for Scotland Conference. (October 2001)   5 http://www.phis.org.uk/pdf.pl?file=pdf/jenny%20marsden.pdf.

49. Marshall, W., Rothenberger, L., Bunnell, S.: The efficacy of personalized audiovisual patient-education materials. Journal of Family Practice **19**(5) (November 1984) 659–63

50. McKeown, K.: The TEXT system for natural language generation: An overview. In: Proceedings of the 20th Annual Meeting of the ACL (ACL'82). (1982) 113–120

51. McKeown, K.: Discourse strategies for generating natural-language text. Artificial Intelligence **27**(1) (1985) 1–42

52. McKeown, K., Chang, S., Cimino, J., Feiner, S., Friedman, C., Gravano, L., Hatzivassiloglou, V., Johnson, S., Jordan, D., Klavans, J., Kushniruk, A., Patel, V., Teufel, S.: PERSIVAL a system for personalized search and summarization over multimedia healthcare information. In: Proceedings of the 1st ACM/IEEE Joint Conference on Digital Libraries, JCDL'01, Roanoke, Virginia (June 2001) 331–340

53. McKeown, K., Pan, S., Shaw, J., Jordan, D., Allen, B.: Language generation for multimedia healthcare briefings. In: Proceedings of the fifth conference on Applied natural language processing, Washington, DC (1997) 277–282

54. Mittal, V., Carenini, G., Moore, J.: Generating patient specific explanation in migraine. In: Proceedings of the 18th Annual Symposium on Computer Applications in Medical Care, Washington DC, McGraw-Hill Inc. (1994) 5–9

55. Neumark, D.,: Providing information about advance directives to patients in ambulatory care and their families. Oncology Nursing Forum **21**(4) (1994) 771–5

56. Paris, C., Wan, S., Wilkinson, R., Wu, M.: Generating personal travel guides - and who wants them? In: Proceedings of the 8th International Conference on User Modeling (UM2001). Volume 2109 of Lecture Notes in Computer Science., Sonthofen, Germany, Springer-Verlag (July 2001) 251–253

57. Piazza, M., Giorgino, T., Azzini, I., Stefanelli, M., Luo, R.: Cognitive human factors for telemedicine systems. In Fieschi, M., ed.: MEDINFO 2004, Amsterdam, IOS Press (2004) 974–978

58. Powell, J., Clarke, A.: The www of the world wide web: Who, what and why? Journal of Medical Internet Research **4**(4) (February 2002) doi: 10.2196/jmir.4.1.e4

59. Pratt, W., Sim, I.: Physician's information customizer (PIC): using a shareable user model to filter the medical literature. In: MEDINFO 1995, Vancouver, B.C., Canada (1995) 1447–1451

60. Prendinger, H., Mori, J., Ishizuka, M.:   Recognizing, modeling and responding to users' affective states. In: Proceedings of the 10th International Conference on User Modeling (UM'05). Volume 3538 of Lecture Notes in Computer Science, Edinburgh, UK (2005) 60–69

61. Prochaska, J., Di Clemente, C.: Stages of change in the modification of problem behavior. In Hersen, M., Eisler, R., Miller, P., eds.: Progress in Behavior Modification. Volume 28. Sycamore Publishing Company, Sycamore, IL (1992) 183–218

62. Redman, B.: Advances in Patient Education. Springer (2004)

63. Reiter, E., Dale, R.: Building applied natural-language generation systems. Journal of Natural-Language Engineering **3** (1997) 57–87

64. Reiter, E., Osman, L.: Tailored patient information: some issues and questions. In: Proceedings of the ACL-1997 Workshop on From Research to Commercial Applications: Making NLP Technology Work in Practice. (1997) 29–34

65. Reiter, E., Robertson, R., Osman, L.: Lessons from a failure: Generating tailored smoking cessation letters. Artificial Intelligence **144** (2003) 41–58

66. Reiter, E., Sripada, S., Robertson, R.: Acquiring correct knowledge for natural language generation. Journal of Artificial Intelligence Research **18** (2003) 491–516

67. Schwarzer, R.: Social-cognitive factors in changing health-related behavior. Current Direction in Psychological Science **10** (2001) 47–51

68. Skinner, C., Campbell, M., Rimer, B., Curry, S., Prochaska, J.: How effective is tailored print communication? Annals of Behavioral Medicine **21** (1999) 290–298

69. Skinner, C., Strecher, V., Hospers, H.: Physicians' recommendation for mammography: do tailored messages make a difference? American Journal of Public Health **84**(1) (1994) 43–49

70. Stretcher, V., Kreuter, M., Boer, D., D., Kobrin, S., Hospers, H., Skinner, C.: The effects of computer-tailored smoking cessation messages in family practice settings. Journal of Family Practice **39**(3) (September 1994) 267–70

71. Von Wilamomitz-Moellendorff, M., Mueller, C., Jameson, A.: Recognition of time pressure via physiological sensors: is the user motion a help or a hindrance? [19] http://www.di.uniba.it/intint/UM05/WS-UM05.html.

72. W3C: Voice extensible markup language (voicexml) version 2.0. W3C Recommendation 16th March 2004 www.w3.org/TR/voicexml20/.

73. Wantland, D., Portillo, C., Holzemer, W., Slaughter, R., McGhee, E.: The effectiveness of web-based vs non-web-based interventions: A meta-analysis of behavioral change outcome. Journal of Medical Internet Research **6**(4) (2004) e40

16

Personalization in E-Commerce Applications

Anna Goy, Liliana Ardissono, and Giovanna Petrone

Dipartimento di Informatica, Università di Torino
Corso Svizzera 185, Torino, Italy
{goy,liliana,giovanna}@di.unito.it

Abstract. This chapter is about personalization and adaptation in electronic commerce (e-commerce) applications. In the first part, we briefly introduce the challenges posed by e-commerce and we discuss how personalization strategies can help companies to face such challenges. Then, we describe the aspects of personalization, taken as a general technique for the customization of services to the user, which have been successfully employed in e-commerce Web sites. To conclude, we present some emerging trends and and we discuss future perspectives.

16.1 Introduction

Electronic commerce includes different types of activities related to the online sales of goods and services. For instance, the ameris glossary defines e-commerce (EC) as follows [10]: "The conducting of business communication and transactions over networks and through computers. As most restrictively defined, electronic commerce is the buying and selling of goods and services, and the transfer of funds, through digital communications. However EC also includes all inter-company and intra-company functions (such as marketing, finance, manufacturing, selling, and negotiation) that enable commerce and use electronic mail, EDI, file transfer, fax, video conferencing, workflow, or interaction with a remote computer. Electronic commerce also includes buying and selling over the Web, electronic funds transfer, smart cards, digital cash (e.g., Mondex[1]), and all other ways of doing business over digital networks."

Initially, e-commerce mainly focused on the sales of goods; however, it has then expanded to deal with all the aspects of business interaction, at the individual and at the enterprise level. Two main areas of interest may be identified:

- *Business to Business e-commerce (B2B)* concerns the management of business interactions between enterprises.
- *Business to Consumer e-commerce (B2C)* deals with the interactions between enterprise and end customers.

The most interesting aspect which has determined the increasing success of e-commerce is the fact that geographical and time zone distance is no longer important; in fact,

---

[1] www.mondex.com

P. Brusilovsky, A. Kobsa, and W. Nejdl (Eds.): The Adaptive Web, LNCS 4321, pp. 485–520, 2007.
© Springer-Verlag Berlin Heidelberg 2007

people can be connected to one another at any time and by means of multiple interaction channels, such as e-mail, Web sites, call centers, kiosks, and similar.

- At the enterprise level (B2B), business interactions are greatly facilitated by the continuous and efficient connection between the partners. For instance, the possibility to place orders and monitor their execution and the automated management of the supply chain support an efficient management of business interactions between remote service providers; moreover, they reduce production and delivery costs.
- At the individual level (B2C), customers may purchase goods and services anywhere in the world, at any time. For instance, millions of customers use electronic marketplaces such as ebay [58] to publish products and place orders online, or they use electronic catalogs such as the online retailer Amazon.com [8] to purchase goods and services.

According to recent studies on customer behavior, the adoption of computers and internet connections in households is constantly growing; e.g., August 2004 Nielsen-NetRatings data shows that the Web penetration in the United States is 74% of the total population, up 11% more than one year before. Moreover, a relevant portion of the adult population spends some hours on the Internet every day (for leisure and/or for work) and is reducing the time devoted to TV watching and to traditional shopping. Those people are a promising target of both Internet advertising and e-shopping [65].

However, Internet-based interaction poses several challenges to B2C commerce, which traditionally established trust relationships between vendor and customers by relying on face-to-face communication. The absense of human clerks, for instance, brings up the well-known *one size fits all* issue, as in principle all customers would be offered the same solutions by electronic catalogs. Moreover, not all product types are equally suitable to e-shopping: an electronic catalog enables the customer to read textual information about products, to view pictures and play audio and video content, but the customer cannot have the goods in her[2] hands. Therefore, while several types of products, such as software, music, books and some high-tech ones, can be directly purchased in an electronic catalog, other products (e.g., garments) are suitable for electronic advertising, but they are most comfortably purchased in a physical store. Neverthless, Web sites are considered as effective tools to *advertise* most types of goods and most vendors, e.g., in the fashion industry, offer Web sites supporting the personalized search for products.

Similar considerations hold for services: information services (such as weather forecasts and traffic conditions), booking services (such as travel planning), shipping and transportation services are very good candidates for e-commerce. However, other services require a direct interaction with a human clerk assisting the customer in her decisions. For example, insurance contracts, financial investments, etc. are regulated by Service Level Agreements to be defined on a case-by-case basis with the support of an expert helping the customer to identify the most convenient solution.

A large amount of work has been devoted to address, at least partially, these challenges, with the ultimate goal of improving customer loyalty. In particular, B2C Web sites have been equipped with recommender systems supporting the personalized sug-

---

[2] In this chapter we refer to the customer by using the female gender.

gestion of goods suiting individual requirements. Moreover, dynamic configuration techniques have been employed to support the interactive customization of products and services. For instance, several online interactive services support the customization of products or the generation of draft contracts to be revised face to face with a human consultant; the Italian TeleMutuo online loan service [149] is an interesting example of the latter.

16.2 Chapter Organization

The present chapter focuses on personalization in B2C e-commerce and attempts to provide the reader with an overview of that work. In the following we will describe the main personalization techniques which have been applied to customize appearance and functionality of e-commerce Web sites to the individual customer. In order to provide the reader with a complete view on the topic, showing what could be done and what has been done in practice, we will consider both research prototypes and commercial systems. It should be noticed that:

– The former offer advanced interaction and personalization features, but they are usually developed as "closed" systems, which embed proprietary customer and product databases. Thus, they fail to support a seamless integration with the applications broadly used by vendors to manage their business activities.
– The latter are complete systems, supporting the management of product catalog, stock, orders and payments. These systems typically log the operations performed by the users on the Web sites. Moreover, they offer tools to analize the customer behavior. However, they usually support rather simple personalization features to assist the users during the navigation of the product catalog and to tailor the interaction style to individual needs.

In the remainder of this chapter, Section 16.3 describes the benefits of personalization in B2C e-commerce, as they have been perceived by retailers and consumers along the years. Section 16.4 provides some background concepts about personalization and adaptation. Section 16.5 discusses some perspectives on the management of one-to-one interaction with the customer, describing some of the most well-known e-business infrastructures and some prototype level e-commerce systems which offer advanced personalization techniques. That section ends with a brief discussion about Customer Relationship Management and mass customization. Section 16.6 describes some emerging trends in e-commerce applications and discusses which personalization perspectives are becoming important and which adaptation techniques are being developed to achieve such personalization goals. Section 16.7 closes this chapter, by presenting some future perspectives.

16.3 Benefits of Personalization in E-Commerce

Personalization strategies have attracted the attention of marketing researchers since about 1990, but their value has deeply changed along the years. Initially, online retailers aimed at extending B2C Web sites in order to show that they recognized the current user (by means of personalized greetings) and to support the personalized recommendation of *off-the-shelf* goods. This goal was based on the assumption that a Web site offering a one-to-one kind of interaction would be preferred to an anonymous one.

About 10 years later, the excitement was weaker because there was some evidence that the personalized suggestion of items, taken as a single feature, did not improve revenue enough to cover its costs, at least in small companies.[3] For instance, in 2003, Jupiter Research released a study according to which only 14% of consumers declared that a personalized Web site would lead them to buy more often from online stores; moreover, only 8% said that personalization made them more apt to visit news, entertainment and content sites more frequently. In contrast, 54% of respondents cited fast loading pages and 52% cited improved navigation as greater incentives [81, 64].

However, the interest in personalization is growing again. For instance, in 2005 ChoiceStream Inc. published an analysis according to which about 80% of customers declare to be interested in receiving personalized content, although several people are concerned with sharing personal information with vendors [49]. Moreover, according to the study "Horizons: Benchmarks for 2004, Forecasts for 2005", released in 2005 by consultants BearingPoint Inc. and the National Retail Federation, 48% of retailers placed personalization high in the list of technologies they would concentrate during 2005.

This growing interest in personalization is mainly due to the increasing demand for Customer-Centric services, which can flexibly react to dynamically changing market requirements [86]. This is a new perspective for Customer Relationship Management (CRM), which aims at improving customers' loyalty by managing, as described by Lynn Harvey (Patricia Seybold Group), "a 'ME-and-YOU' relationship building activity that's focused on companies getting to recognize, understand and ultimately serve their customers" [134].

The Customer-Centric CRM perspective includes the provision of features such as the personalized recommendation of items, but does not limit the offered services to this aspect. The idea is that the cost of a CRM solution supporting the storage and analysis of customer information can be balanced by various benefits, among which the availability of knowledge supporting the design of new products matching market trends and a substantial reduction in overhead costs.

While the former aspect is clear, the latter deserves further comments: currently, online retailers benefit from the pervasiveness of e-commerce because they can keep in contact with their customers by means of multiple channels, such as e-mail, Web sites, call centers, physical stores, and so forth. However, all those channels collect large amounts of possibly redundant information, represented in heterogeneous formats. Moreover, customer data continuously change; for instance, people change address and job, they get married, etc., and these changes may influence their needs and interests. As a result, retailers have concrete difficulties in merging the available information and in keeping it updated. Some negative effects follow:

---

[3] Neverthless, some surveys reported optimistic data about Web personalization effects; e.g., see [148].

- The lack of data integration and synchronization prevents vendors from effectively exploiting customer information to promote their products and services. For instance, electronic and paper brochures are often delivered to an undifferentiated population, instead of mailing them only to the target customer base of products. The most evident consequences of this problem are high delivery costs and possible junk mail effect on recipients [134, 65, 53].
- Retailers are forced to sustain costs caused by corrupted data; e.g., they often deliver goods to obsolete addresses and they produce wrong bills.
- The lack of tools supporting the analysis of customers' browsing behavior (e.g., shopping cart abandonment) does not enable vendors to collect feedback useful to redesign and optimize their Web sites [70].

In order to face these challenges, some Customer Data Integration (CDI) services are being developed as process neutral modules that can be embedded in complex e-commerce systems to feed operational business processes with reliable customer information. CDI services typically support the fusion of customer data collected from multiple channels and the unified analysis of information data. Besides depicting the situation of an online store in real time, CDI services support the identification of market trends, of successful and unsuccessful products, the segmentation of the customer base and also the definition of business rules supporting the provision of services targeted to specific market segments. Thus, personalization comes into play again, but this time it is supported by vital business requirements smoothing the impact of its cost.

16.4 Background

16.4.1 Adaptability and Adaptivity

The goal of this section is to clarify the main terms of our discussion; to this purpose, we first report the distinction between *adaptable* systems and *adaptive* ones [114].

- In *adaptable* systems the adaptation is decided by the user, who explicitly customizes the system to receive a personalized service.
- In *adaptive* systems, the adaptation is autonomously performed by the system, without direct user intervention.

Adaptable systems enable the user to customize several parameters by choosing the preferred values (e.g., background color and language to be applied in the User Interface). Moreover, some systems enable the user to restrict the features she is interested in. For example, as the Amazon.com retailer has now become a very large Web store, it enables the user to restrict the store catalog by selecting her "favorite stores" (e.g., Music, Electronics, Kitchen & Housewares, etc.); after the user has selected her preferred categories (stores) the system displays them at the top of the category list for easy access to the most interesting product types.

Although adaptability and adaptivity may co-exist within the same system, the former is a simpler feature and is based on standard system configuration techniques largely applied in interactive and batch software applications. In this chapter we thus address adaptivity, extensively discussed by Brusilovsky in [36, 37] and by Kobsa et al. in [91], and we focus on those aspects that are relevant to e-commerce applications.

16.4.2 Target Factors for Adaptivity

The adaptation of a system may be based on three main categories of information:

1. *Information about the user*
   The type of information to be taken into account depends on the application domain. In several domains, the following user features have been considered: socio/demographic data (e.g., age, gender, job); knowledge and skills; interests and preferences; specific needs; objectives and goals; see [91] for details. In B2C e-commerce, the user is the end customer and the following types of information may be considered:
   - The *user's knowledge* about the domain concepts and the user's skills (i.e., her "know how") can be relevant in the sales of complex products and services; e.g., computers, ADSL connections, and similar.
   - The *user's interests and preferences* usually refer to the categories of products and services sold in the online store, or to specific properties of such products/services. For instance, interest in design versus technological aspects (if the store sells, e.g., hi-tech products), in hard rock versus pop music (if the store sells MP3), in cultural-oriented travels versus sport or fitness-oriented holidays (in case of a travel agency), and so forth.
   - The *user's needs* can include different kinds of information. In particular, the information about disability is important to offer accessible services to impaired people.
   - Finally, the *user's goals* represent the information that is most closely related to the specific application domain; for this reason, goals can assume rather different meanings. For example, an e-commerce system should take into account whether the user is buying something for herself or a present for somebody else; a mobile guide might consider whether the user is traveling for business or for pleasure; an online shop assistant might help the user to find the less expensive solution, or the most reliable one, or to balance conflicting requirements.
   Although most adaptive systems only consider individual users, in some application scenarios more than one user has to be taken into account at the same time. For example, some recommender systems tailor their suggestions to possibly heterogeneous user groups, such as some tourists traveling together [20], or a family watching TV [103, 15]. For a detailed discussion about group recommendation, see Chapter 20 of this book [79].
2. *Information about the device used to interact with the system*
   The customer can access an online store by using a desktop PC, a laptop, a mobile phone, a PDA, an on-board device, or other. Every device has different characteristics, with respect to screen size, computation and memory capabilities, I/O mechanism (keyboard, touchscreen, speech, ...), type of connection, bandwidth, and so forth. These aspects have been classified as *environment data* by Kobsa et al. in [91].

3. *Information about the context of use*
   The user can interact with the online store in different situations, e.g., at home, while sitting in a train, walking or driving, or during a meeting. The *context* category, analyzed in detail by Dey, Abowd and Mynatt in [2, 56], is very broad and it is difficult to find a unique definition. Although almost everything can be considered as *context of use*, for simplicity, we split this category into two main aspects: the *physical context* and the *social context*.
   - The *physical context* includes the user location and information about the environment conditions, including light, noise, temperature, time of the connection, walking/driving speed, and other similar features.
   - The *social context* is much more difficult to define and very few applications take it into account. This element, considered in a broad sense, can include features like the social community or group the user belongs to, the task she is performing and the relation with people close to her while she is interacting with the service.

Many researchers include information about the user and the device within the notion of *context*. In this chapter, we use the term *context* in the restricted meaning of *context of use*, including the features mentioned in item 3 above, and leaving the user's characteristics and the type of device employed to access the system as separate categories.

16.4.3 Phases in the Adaptation Process

Kobsa et al. [91] identify three phases in the adaptation process: *acquisition, representation/inference, and production.*

1. The *acquisition* of data about the user, device, and context of use can be supported by the user or automatically performed by the system. In the first method, usually applied when the unobtrusive collection of data is not feasible or convenient, the user is asked to execute some action, typically to fill in a form. The second method relies on the application of intelligent inference techniques (e.g., machine learning, plan recognition, and stereotype reasoning) to acquire information about the user by analyzing her behavior. Advanced technological devices (e.g., sensors, GPS for user location, or on-board equipment for driving speed) may be employed to monitor the user's behavior and to acquire device and context variables.
2. Section 16.5 briefly presents and discusses the methodologies that can be applied in the second phase, for *representation and inference*. A detailed analysis can be found in [36, 37, 91].
3. As far as the *production* of the adaptation is concerned, it is interesting to specify *what is adapted* and *how*. With respect to this criterion, we consider three main broad adaptation features:
   - The first one is the *suggestion of the product/service*, often described as *content recommendation*. The system may play the role of a *recommender*, suggesting products and/or services tailored to the user, to the device she is using, and to context features; see Chapter 13 of this book [38]. Moreover, if the online store sells complex products or services, the system can actively guide the customer in the configuration of an item satisfying particular needs and preferences. See Chapters 9 [139], 10 [119], 11 [142] and 12 [40] of this book for details about

recommendation techniques. See also Section 16.5.7 for a discussion about product configuration.

- *The presentation of the product/service*. Orthogonally to the provision of a personalized recommendation feature, the system may tailor the presentation of items to the user, device and context. For instance, the presentation may use different media, such as written text, speech and pictures; moreover, it may adopt different presentation styles (verbose, synthetic, more or less detailed, simple or technical). The system may also personalize the kind of information about products and services it presents, as well as the presentation style, in order to satisfy different decision-making styles in the selection of products and services. For example, Popp and Lödel [125] apply stereotypical reasoning [132] about customers to recommend products satisfying typical buying preferences.
- *The user interface* (sometimes referred as *structure*) may be personalized as well. For instance, the layout, including information and navigation structures, can be modified according to various conditions, including the user's preferences, the limitations of the device and the environment.

Before addressing these adaptation features in detail, it is worth listing some additional aspects, which will not be further discussed in the present chapter. In the first place, the adaptation strategies which will be discussed in relation with e-commerce systems can be applied to a closely related field, i.e., the personalization of Web advertisements; e.g., see [95, 22, 84]. Second, three issues, strictly related to the adaptation of the user interface (and partially of the presentation), are addressed in the related work about Human Computer Interaction. Such issues concern the accessibility for users with special needs, virtual reality user interfaces, and usability and design guidelines.

- *Accessibility*. As discussed in Section 16.6, the adaptation of the presentation and of the user interface can be based not only on the usage context and on the characteristics of the device, but also on individual user needs and preferences. This kind of adaptation extends the accessibility to people with special needs and has been addressed by several researchers; among others, see [61, 102, 133, 29].
- *Virtual Reality user interfaces*. Most Web-based user interfaces for e-commerce systems are based on a traditional 2D model. However, some researchers point out that the user experience during the interaction with an online store should be as similar as possible to the real world shopping experience. In this perspective, some authors, such as Chittaro and Ranon, propose a 3D user interface that enables the user to explore a virtual reality environment representing a physical store; see [47] and Chapter 14 of this book [48]. The main goal is to satisfy the needs of those customers that have an emotional style of buying and to enable them to perform natural shopping actions.
- *Usability*. Regardless of which type of user interface (mobile/traditional, 2D/3D) is offered, an e-commerce system should be usable. Web site usability is a wide research area and a complete discussion of its main issues falls out of the scope of this chapter. We only point out that usability is a prerequisite for the success of an e-commerce site; if the user encounters difficulties in navigating the site, in finding information about products, or in managing her purchases, she might abandon the Web store. Therefore, user interfaces for e-commerce sites should be carefully designed follow-

ing usability guidelines; see, for instance, the guidelines published by Serco Ltd[4] and the article by Tilson et al. [150]. Moreover, as discussed by Benyon in [27], the adaptation of the user interface should improve the usability of the Web site in order to be useful. Therefore, adaptation strategies should be designed and implemented having usability guidelines in mind. For a thorough discussion about this topic, see Gena's survey about usability in adaptive systems [66] and Chapter 24 of this book [67]. Finally, Alpert et al. point out that usability guidelines and User-Centered Design should be seriously considered when deciding which personalization techniques have to be included in an adaptive system: according to the analysis reported in [7], it seems that the users of the ibm.com Web site do not appreciate the attempts to infer their needs and goals; indeed, those users prefer to be in full control of interaction with the system.

16.5 Perspectives on E-Commerce Personalization

16.5.1 Personalization Features in Commercial Merchant Systems

Along the last ten years, several infrastructures have been developed to facilitate the creation and management of electronic catalogs. For instance, BroadVision [35], Blue Martini Retail [33], Netscape Merchant [111], Microsoft Merchant System [13, 41] and IBM WebSphere Commerce [75] have been employed worldwide by enterprises to create e-commerce portals.

One of the main issues to be solved to enable a broad adoption of e-commerce infrastructures is the provision of transactional, secure services and the integration with legacy software that enterprises still use for doing business. All the mentioned systems are strongly focused on these aspects, at the expense of personalization, which they only partially support. However, maintaining a one-to-one relationship with the customer is recognized to be critical to the success of an e-commerce Web site (see Section 16.5.6); therefore, these systems offer some basic personalization features, mainly concerning the recommendation of products. For example, they support the long-term identification of customers, based on the assignment of identifiers, and the storage of purchase lists, as well as the possibility to track and log each customer's navigation actions. In some cases (e.g., WebSphere Commerce), they allow to define user groups, on the basis of common behavior patterns automatically detected by analysis tools. Moreover, they enable Web store administrators to define simple business rules that can be applied to promote products and propose special offers and discounts, on the basis of the products selected by the customers and/or the visited catalog pages. The discovery of behavior patterns is based on the adoption of knowledge discovery techniques developed in the data mining research area; e.g., see the book by Agrawal et al. [4] and Pierrakos et al.'s survey about Web personalization [122].

As BroadVision seems to be the most successful system, among the cited ones, we will briefly describe some of its features. BroadVision focuses the attention on two personalization strategies.

---

[4] www.usability.serco.com/research/research.htm

- In the *push* approach, the system is pro-active and it guides the user by recommending information and access to applications and features.
- In the *pull* approach, the system relies on the user who requests information and features and it handles the user's requests in a personalized way.
- Moreover, BroadVision offers the *qualifier matching* as a filtering strategy, allowing companies to target the delivery of content, the access to applications, and different navigation paths to individual visitors and groups by using qualifiers that match appropriate content and capabilities.

If the company has less complex personalization goals, the *rule-based matching* can be used. This technique consists of IF/THEN, AND/OR statements to tell the system that it should take certain actions if the Web site visitor meets certain conditions. The actions result in displaying content or granting the access to a part of the site. The business rules depend on the retailer's goals and may be very simple; for example, a typical rule triggers the generation of electronic coupons (based on previous purchases) that are sent by e-mail to each customer who has not purchased goods for a while. Of course, the underlying activity carried out by the system in order to choose the rules to be applied along time is very complex, but the Web store administrator does not need to consider it.

It should be noticed that, although the services offered by BroadVision support limited personalization, the wide adoption of this system by e-commerce Web sites and by the enterprise can be explained on the technical side. In fact, the application is built on top of standard technologies such as the Java J2EE [146] and relational databases, which run on Linux, Windows and Solaris platforms, and exist in both commercial environments (such as IBM WebSphere [76] and Oracle [115]), and Open Source distributions (e.g., JBoss Application Server [80]). Therefore, BroadVision guarantees important robustness and scalability features that cannot be neglected in a commercial Web site. Moreover, the system can interoperate in a relatively seamless way with existing databases and legacy software already in use within the enterprise that wants to offer one-to-one interaction with its own customers.

Notice also that, while most merchant systems were initially proposed as proprietary solutions, the current trend is to offer pluggable solutions, which can be seamlessly embedded in the retailer's systems. For instance, WebSphere offers both the suite WebSphere Commerce, supporting the development of a complete B2C e-commerce Web site, and a Customer Data Integration service (WebSphere Customer Center [53]), which enables online retailers to extend their own Web sites with data collection and analysis capabilities.

16.5.2 Personalized Recommendation of Products

Section 16.5.1 shows that the main adaptive feature offered by commercial merchant systems is the provision of personalized product recommendations, depending on the customer's behavior but based on relatively simple techniques, such as business rules. The question is therefore whether other techniques could be applied, within such systems, in order to enhance their recommendation capabilities, starting from a detailed analysis of the individual customer. In the following, we describe the most popular

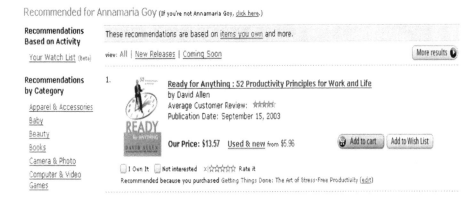

Fig. 16.1. Personalized recommendation of products in Amazon.com

techniques developed to assist the customer in the selection of items suiting her preferences. In Section 16.5.7, we will introduce the broader area of *mass customization*, which concerns the sales of customized products at mass-production costs and relies on intelligent techniques for the *design* of personalized products meeting the customer's requirements.

Generalized Recommendation Techniques. Various researchers have proposed to support the selection of products in different ways. For instance:

– In some cases, an interactive approach is adopted, enabling the customer to search for products according to her own selection criteria. For instance, Sacco [135] proposes to utilize *dynamic taxonomies*, defined by interacting with the system, to classify items and progressively reduce the search space, until the customer isolates a small set of products, which can be compared in detail. The idea is that the electronic catalog should not try to identify the customer's preferences, as she knows them in detail and she can manage the search on her own, if assisted by a search tool. In the same spirit, Pu et al. propose to enable the customer to specify hard and soft constraints to be satisfied by the solution [128]. In this case, the system proposes suboptimal solutions, which can be criticized and refined in an interactive way, until the customer is satisfied. For more details about the research on "critiquing" see also the work by Burke [39] and by McCarthy and colleagues [105].
– Several e-commerce Web sites also offer additional recommendation features based on data about general customer behavior. For instance, Amazon.com [8] offers a recommendation feature based on the items already bought by the customer (see Figure 16.1). Moreover, it offers the *customers who bought* feature (see Figure 16.2), available in the customer's shopping cart, to inform her about other items purchased by the customers who bought the same items she has selected. Other Web sites offer the *similar items* or the *correlated items* features to suggest items closely related to those in the shopping cart. For a summary about recommendation techniques in e-commerce Web sites, see the survey by Schafer et al. [140]. Moreover, see [21] for another interesting perspective on customer behavior analysis: given the importance

Fig. 16.2. *Customers who bought* feature in Amazon.com

of users evaluations in the selection of products, Avery et al. propose a model to support a market for evaluations, i.e., "a mechanism for eliciting, sharing, and paying for information", which can be a valuable source of information for personalized recommendation.

– In other cases, inference techniques have been applied to elicit information about the user's preferences or to unobtrusively acquire them on the basis of the user's behavior, in order to identify and recommend the most promising items. See below for details.

Personalized Recommendation Techniques. In the User Modeling research, the acquisition of information about user preferences has received a relevant amount of attention and various techniques have been developed to personalize the recommendation of items. Two main approaches have been introduced:[5] *content-based recommendation* suggests the most suitable items for the customer by relying on information about her preferences (interests) and on a description of the item features. *Social recommendation* techniques employ the preferences of customers similar to the current one for the suggestion; see [131, 60, 104, 39] for an overview of the topic. Specifically, the most well-known techniques that have been developed are *content-based filtering* and *collaborative filtering*:

– Content-based filtering recommends goods having properties similar to those of the products that the customer selected in the past; e.g., see [31] and Chapter 10 of this book [119].
– Collaborative filtering is a social recommendation technique; the suggestion of goods is based on the identification of customers similar to the current one and on the suggestion of the items which were appreciated by such customers. Similar customers are identified by analyzing the ranks produced by the whole customer base and the idea is that, if a customer similar to the current one liked a certain item, the current customer will probably like it as well; see Chapter 9 of this book [139].

---

[5] Burke makes a slightly different distinction in Chapter 12 [40] of this book.

The two techniques have different positive and negative aspects (see also the discussion in Chapter 12 of this book [40]). For instance:

- Content-based filtering is based on a classification of items, manually or automatically derived from a specification of their features. By applying this technique, new items can be successfully recommended, if a certain amount of information about their features is available. However, the individual customer has to be monitored for a while before successfully recommending items; moreover, content-based recommender systems tend to suggest items similar to one another, at the expense of variety. Furthermore, the accuracy in the recommendations downgrades if there are too many item features to be considered.
- Collaborative filtering does not require detailed knowledge about the items to be recommended, but it is subject to bootstrapping problems. For instance, the recommender is not able to handle items until they have been ranked by a minimum number of customers; moreover, the recommendation capabilities are poor if the matrix storing the product ranks is sparse, because in that case it is difficult to identify customers having tastes similar to those of the current one. Finally, as noticed by Linden et al. in [98], the comparison of ranks produced by customers is very heavy in the case of Web sites, such as Amazon.com, visited by millions of users every day. At the same time, any attempt to reduce the complexity of the algorithm downgrades the accuracy of the recommendations.

Various solutions have been adopted to overcome the drawbacks and limitations of these techniques:

- Hybrid recommender systems have been developed to enhance the accuracy in the recommendations. For example, in the FAB recommender system [23] and in the PTV Electronic Program Guide [116], content-based and collaborative filtering are combined to support high-quality recommendations in the presence of new products and to enrich the variability in the suggestions generated by the system. As Burke discusses in Chapter 12 of this book [40], hybrid recommender systems have also been developed to integrate much more heterogeneous recommendation techniques with the aim of combining complementary types of information about the user in the preference acquisition process. For instance, some systems rely on a combination of Naive Bayes classifiers, Bayesian networks, case-based reasoning, demographic information, and fuzzy classifiers to perform content-based product recommendation; e.g., see [161], [15] and the survey in [39].
- Lightweight recommendation algorithms have been developed to solve scalability issues in heavy-loaded recommender systems. Specifically, collaborative filtering has evolved to the *item-to-item collaborative filtering*, which computes the similarity among items (instead of customers) in order to recommend items similar to those the customer liked in the past. This algorithm operates on items; thus, its performance does not depend on the number of users accessing the Web site. See [138] for details.

The selection of the recommendation techniques to be adopted in a Web site depends on the characteristics of the application domain. For instance, item-to-item collaborative filtering is particularly suitable for e-commerce sites having relatively stable product

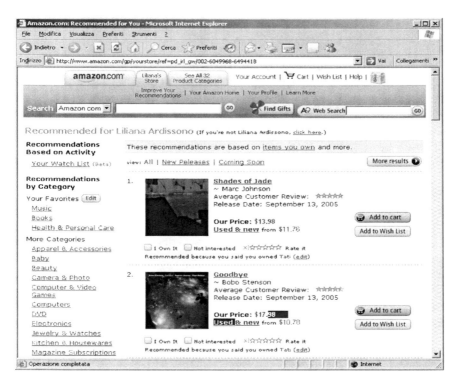

Fig. 16.3. Recommendation based on customer's owned items in Amazon.com

catalogs because the ranks may be collected along time. However, other knowledge-based techniques, such as those developed for content-based filtering, might be convenient when the pool of items to be considered changes very frequently.

Before concluding this section, it is worth mentioning the importance of transparency and explanation to enhance the customer's trust in a recommender system. In fact, having received the system's suggestions, the customer should be enabled to deeply understand the content of the electronic catalog in order to make an informed decision. As a thorough discussion about this topic would lead us far away from the focus of this chapter, we only mention that rather different perspectives on the general *usefulness* and *trust* issues have been proposed. For example, Herlocker et al. [73] propose to explain recommendations by presenting evidence about the ratings of items provided by people whose purchase histories are similar to the customer's one, or by relying on the rate of good recommendations produced by the system during previous interactions with the same user (a sort of *reputation* gained by the recommender). Similarly, Amazon.com explains its own recommendations by means of two main features: if the customer has placed some items in her shopping cart, the system may support the recommendations by means of the *customers who bought* features, discussed in Section 16.5.2 (see Figure 16.2). Moreover, if the customer has specified some of the items she already owns, the system may refer to such items in the personalized suggestions; Figure 16.3 shows a portion of the recommendation list in the music category for Liliana Ardissono, who owns Tati by Enrico Rava and has rated that disk very high.

16.5.3 Customer Information Sharing

Regardless of the technique applied within the recommender system, precise information about the individual customer's preferences can only be obtained by observing her behavior for a certain amount of time. Therefore, a delay occurs before the system is able to adapt the interaction to her. This latency issue has been recognized in various application domains and some researchers have proposed to share user information between applications in order to enhance their personalization capabilities. For instance, see the Personis user modeling server developed by Kay et al. [83], the survey about User Modeling Servers by Kobsa ([87], Chapter 4 of this book [89]), and the survey about recommender systems in e-commerce by Schafer et al. [140]. To provide a simple example, two book sellers might share the user models describing their customers in order to increase the knowledge about the common customers and to extend the set of visitors they can handle as known ones.

In commercial applications, broad initiatives have been proposed to support the identification of customers across services; for instance, the Liberty Alliance project [97] grants passports representing universal identifiers associated to users across applications. Although the passport is the enabling technology to support customer information sharing, several issues have to be addressed before customer data can be safely shared by service providers. For example:

- Information sharing has to be controlled in order to respect the customer's privacy preferences, which may be rather articulated. For details about this issue, see Chapter 21 of this book [90], [88, 92] and [158].
- The mutual trust between service providers has to be assessed in order to control the propagation of data. This is important to guarantee that data is not shared with untrusted competitors, or with service providers who would misuse it. See [14] for a preliminary proposal in this direction.

16.5.4 Personalized Presentation of Information about Products

Some efforts have been devoted to enrich electronic catalogs with the generation of personalized product descriptions. In the proposed systems, the presentations are generated "on the fly" and tailored to the individual customer's interests and preferences; see Chapter 13 of this book [38] for a detailed survey of personalized presentation techniques. The personalized generation of product descriptions is based on a declarative representation of the product features and on the adoption of customer preference acquisition techniques that can be employed, at the same time, to recommend the most promising items and to highlight their features accordingly. Different approaches have been developed. For instance:

- Jameson et al. [78, 110] propose to present evaluation-oriented information about goods in order to convince the customer to purchase a certain product. The idea is to simulate the behavior of a clerk (e.g., a car seller), who would highlight the product properties having the highest probability to impress the customer. For instance, while presenting a car, a human clerk might focus on information such as the safety

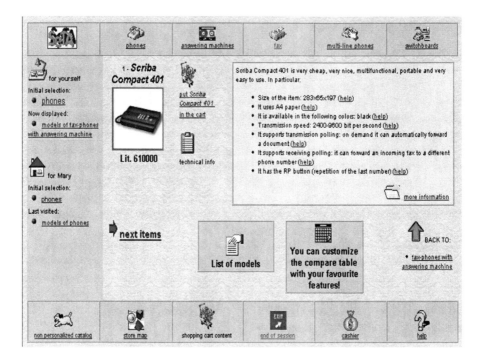

Fig. 16.4. Product presentation generated by the SeTA system

and economicity, or on different properties, such as the speed, depending on the customer's priorities. In general, the idea is to provide the customer with a high-level view of the product which satisfies her information needs and highlights the most promising product characteristics.

– Building on Jameson et al.'s interaction model, André and Rist [11] generate multimedia product presentations where animated characters, each one interested in different product properties, criticize products and discuss with one another about their properties.

– Ardissono et al. propose to distinguish the *direct user* interacting with the system from the *indirect users* on behalf of whom the user is operating. In the SeTA system [17], multiple beneficiaries are modeled in order to manage the B2C scenarios where expert Internet users may purchase goods on behalf of other people. The system can thus dynamically generate product presentations targeted to the direct user's interests and expertise; however, the properties of items that make them suitable to the beneficiary are highlighted in order to help the customer in the evaluation of products from the perspective of the person who is going to receive them [19].

In the following, we present the approach adopted in the SeTA system in some detail. See the work by Milosavljevic for another initiative concerning the dynamic generation of electronic encyclopedia entries [108]. Moreover, see the article by McKeown [107] for details about a seminal work on tailoring the presentation of information to the individual user.

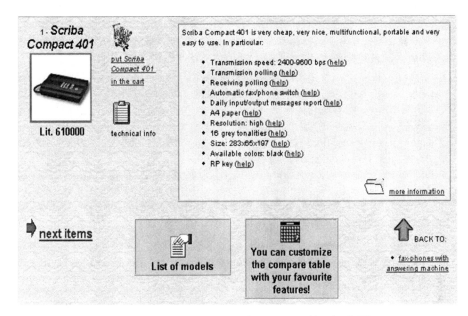

Fig. 16.5. Detail of product presentation generated by the SeTA system

16.5.5 A Case Study: the SeTA System

The SeTA system [17, 19] manages a product catalog that is dynamically generated while the customer browses it, in order to produce highly personalized product presentations. As already discussed, the system tailors the recommendation of items, which is managed by sorting items on a suitability basis, to the preferences of their beneficiary. Moreover, the system tailors the presentation of items to the characteristics of the direct user, in order to meet individual information needs. The separate management of direct user and beneficiary is achieved as follows:

- The system handles an individual user model describing the direct user. Moreover, it enables the user to describe the main characteristics of the beneficiary of the items she is looking for. The model of the direct user is acquired by eliciting some information about herself and by monitoring her navigation behavior. The models of the indirect users are initialized by applying stereotypical information about user classes; see [132] and [17].
- Each user model includes demographic data about the described person (e.g., her age and job), and information about her preferences for products (e.g., she prefers technological products). Moreover, the model of the user interacting with the system includes information about her features, such as her receptivity (i.e., the amount of information she can elaborate) and interest in different kinds of information about products (e.g., technical details or aesthetic information).

Figure 16.4 shows a typical product presentation page generated by the SeTA system. The page is structured in various portions that enable the direct user to browse the catalog, view the shopping cart, examine a particular item (the "Scriba Compact 401" fax-phone with answering machine), and so forth. Each item is presented by showing a picture, the price, and a description section which provides the user with detailed information. The description section is organized in two parts:

1. A prologue summarizes the main properties of the item in order to help the customer in the identification of the most promising products before viewing their details. For instance, in Figure 16.4, the "Scriba Compact 401" product is described by means of qualitative adjectives such as nice, portable and easy to use. These adjectives correspond to the properties of the item that make it particularly suitable for the beneficiary (in that case, the direct user).

2. A list reporting the most interesting technical and non technical features (e.g., size of the item, format of paper used to print out documents, available colors, and similar), ended by a *more information* link, that enables the user to view the complete product description.

The summaries and the detailed descriptions are tailored to the direct user: the number of features listed in the page depends on her receptivity and the descriptions are more or less technical, depending on her expertise about the products of the catalog. Figures 16.4 and 16.5 describe the same item, but they are tailored, respectively, to a non expert and to an expert user; the two pages show different lists of features of the presented item; moreover, the former page is non technical, while the latter has a more synthetic and technical style. The product descriptions are produced by applying canned text natural language generation techniques supporting a fast generation of correct sentences. See [16] for details.

The selection of the features to be shown in the main portion of the page is aimed at reducing the information overload on the user and at focusing on the most interesting aspects of the product, leaving other details available on demand. Two factors are taken into account in the presentation of features: on the one hand, the catalog should mirror the (direct) user's interests, in order to provide her with relevant information. On the other hand, the vendor might want to guarantee that certain pieces of information are always presented. In order to take these two factors into account, each product feature is given an importance value, defined at Web store configuration time, which specifies whether it is a mandatory piece of information for the product category or not (e.g., price is mandatory for all products). Moreover, each feature is classified in a typology (technical, aesthetic, etc.) corresponding to some user interest types. During the interaction, the importance of each product feature is combined with the user's interests, in order to select the set of features to be shown in the main page and to filter out those to be made available on demand. In this way, the manager of the electronic catalog can impose constraints on the presentation of strategic or mandatory data; moreover, the customer is shown the most interesting features, from her point of view. For transparency purposes, the customer is always allowed to access complete information about products, by following *more information* links.

In addition to the personalized presentation of products, SeTA supports the generation of personal views of the catalog by providing interactive functions. One of the most interesting ones is the generation of a customized compare table (see button in the lower portion of the page in Figures 16.4 and 16.5), which enables the user to select the products and the properties or features to be considered in the comparison. This table enables the customer to evaluate products on the basis of the aspects most important to her. Moreover, it is an excellent source of information for user modeling because it enables the system to unobtrusively identify the customer's priorities.

The experience gained in the development of SeTA was very interesting, especially because it highlighted positive and negative aspects which we believe are common to many Adaptive Hypermedia systems. On the one hand, the adoption of advanced user modeling and dynamic content generation techniques, together with personalized recommendation mechanisms, supported the generation of electronic catalogs meeting individual user needs with high accuracy. The experiments which were carried out in laboratory tests proved that the users were happy about the presentations and suggestions they received and some of them were particularly interested in the interactive features offered by the system, such as the personalized compare tables. The positive evaluation received by the personalization features of the SeTA system suggested that adaptivity can effectively improve the user interaction in an e-commerce Web site and encouraged us to continue the investigation of the benefits of personalization in such systems.

On the other hand, the major obstacle to a real-world exploitation of SeTA was the knowledge intensive approach supporting the system adaptation. This effort may discourage the Web store designer, who is responsible for the introduction of detailed information about the characteristics of customers and products that have to be modeled. In most cases, the problem is not related to the amount of information to be provided, but to the conceptual effort imposed on somebody who is expert in the sales domain, but not necessarily familiar with knowledge bases and ontologies. Moreover, the designer often has problems in understanding the relation between the domain knowledge introduced at Web store configuration time and the adaptation effects achieved at run time. These difficulties, related to the knowledge acquisition process, could be mitigated by approaches aimed at supporting the knowledge engineer at different levels. For instance, natural language processing and information extraction techniques may be employed to automatically extract information about products, starting from the available documentation [5, 50]. These approaches, coupled with the current work on the automatic acquisition of ontologies (see, e.g., [113, 6]), promise to address the knowledge-based design issue in an effective way. Thus, the work to be carried out by the Web store designer can be dramatically reduced.

16.5.6 Customer-Centric CRM – Advanced Personalization in B2C E-Commerce

In the last decade, Customer Relationship Management (CRM) has become a keyword for the enterprise marketing strategies. The business model underlying CRM is defined as being customer-centered, instead of product-centered, and can be viewed as a business strategy that enables a company to implement, manage and keep long-term relationships with its customers. Obviously, the ultimate goal of any company is still to increase its profits, but the new idea is that this economic advantage can be reached by increasing customer loyalty, rather than gaining market share through the acquisition of new customers. This principle could be rephrased as "They [marketers] aim not to find customers for their products and services but to find products and services for their customers" [69]. Indeed, the principle represents the core idea of the milestone book by Peppers and Rogers [120], that introduces the idea of *share of customer*, replacing the traditional *share of market*. However, customer loyalty implies long-term relationships, that are based on customer satisfaction [130, 160]; in turn, customer satisfaction can be achieved by offering products and services fitting the customer's needs and desires and by supporting an individual and personalized interaction. In this perspective, the relationship between the company and its customers becomes one-to-one; the company takes care of any single customer, and it takes her individual needs and preferences into account in order to manage personalized dialogs and to offer tailored solutions. For these reasons, different from traditional marketing strategies, CRM requires a bi-directional, interactive form of communication between company and customer.

The inclusion of adaptation and personalization techniques in an e-commerce site could support CRM in various directions. In the first place, information about the behavior of a large number of customers can be automatically collected and stored. That information can then be further analyzed by data warehousing and CRM tools [85] in order to support *cross-selling* and *up-selling* operations[6]. Moreover, the techniques supporting a personalized interaction based on the management of individual user models can play a key role, because they enable the e-commerce site to know its customers at the individual level. Of course, each individual user model should be dynamically updated, on the basis of the user's behavior, in order to support the adaptation to changing interests and needs. The idea is that the longer a customer interacts with a company's site, the more her user model is precise and accurately reflects her needs. An accurate user model can then support the proposal of personalized offers to improve the customer's loyalty and thus the company's profit, in the medium-long term.

Notice that, as Peppers and Rogers claim [121], human interaction is still critical for most people when making decisions. In this perspective, personalization features supporting one-to-one relationships can improve the feeling of a human-like interaction, and they can contribute to improving the quality of the technology-mediated interaction between customers and companies [32]. The shopping assistants proposed by Krüger et al. are examples of systems that go in the same direction; these authors describe Web agents aimed at "providing additional value to the shopping experience in the form of conversational dialogues, multimodal interaction, augmented reality, and enhanced plan recognition" (Chapter 17 of this book [94]). A related idea is the one by Pine and Gilmore [124], who claim that, in order to be more competitive, companies will move from the provision of services to the provision of *experiences*. This means providing the customer with a *sensation*, supported by her active participation in an *event* that can be personalized according to her needs and preferences, thus improving the effectiveness of the experience itself.

One of the main applications of CRM principles is *mass customization*, which aims at enhancing the flexibility of the production model in order to create products and/or services tailored to individual customer needs, while maintaining mass production ef-

---

[6] *Cross-selling* is the improvement of sales by offering a product or service complementary with respect to the one the customer is interested in; e.g., proposing a printer when the customer is looking for a PC. *Up-selling* is the operation of selling a product/service having higher quality and price with respect to the one the customer is interested in; e.g., presenting a smart phone instead of a standard mobile.

Fig. 16.6. The miAdidas Web site

ficiency and costs. The importance of personalization within mass customization processes is underlined by Gilmore and Pine [68] and is discussed in the next section.

16.5.7 Mass Customization and the Recommendation of Complex Products and Services

In the past, the mainstream of B2C e-commerce focused on the sales of *off-the-shelf* goods, such as books, and of products available in a fixed set of pre-configured alternatives, such as phones available in different colors and including or excluding a small set of optionals. However, the manufacturing industry has recently moved towards a more flexible production model, in order to meet an increasing need for individual solutions at affordable costs.

Although enabling the customer to design her own products is very challenging, this possibility has been offered by several major firms, in order to satisfy customers by providing them with very specific products. However, the availability of several configurations would impose the production of an excessively large variety of alternatives. The term *mass customization* has been introduced to describe the solution to this problem, i.e., the production of customized products at mass-production costs [152]. The idea is to produce customized items within stable processes and structures, which enable the industry to reduce fashion risks and overstock problems.

The customization of products is typically supported by the development of configuration systems (e.g., see [77]) that enable the customer to generate solutions starting from the specification of functional and aesthetic features. In some cases, e.g., as far as high-tech products are concerned, these systems are available in Web stores and they enable the customer to remotely design products. In other cases (e.g., in the sales of garments and shoes), fit requirements impose a direct and physical contact with the customer, in order to acquire detailed information about her; after the acquisition of such information, the configuration activity can be performed online at any time.

In the footwear industry, some major firms have adopted the mass customization approach for the production of their goods. For instance, the "mi Adidas" program (see Figure 16.6 and [3]) offers a foot scanning service available in selected stores and at special events which enables the firm to store the customer's measures and produce shoes fitting her feet at any time. After the customer's footprint has been acquired, she can customize as many pairs of shoes as she wishes, by choosing fabric, color, and other aesthetic features [28, 123]. The "mi Adidas" case suggests that mass customization has the potential to enhance the relationship between customer and vendor: once the customer has spent her time to specify personal preferences and create her own product, she will likely reuse her profile when customizing other products. However, as discussed by Berger and Piller in [28], "there is still only very little understanding about the perception of choice and the joy or burden of co-design or configuration experienced by customers, who often have no clear knowledge of what solution might correspond to their needs. At times these needs are not even apparent to the customers. As a result, customers may experience uncertainty or even perplexity during the design process". As a matter of fact, the kind of choice offered by online product customization services is usually limited because of two main reasons:

– On the one hand, the configuration of a product from scratch is costly for the firm, because it leads to an excessive amount of variability, and it might generate undesired outcomes spoiling the brand image. For instance, Berger and Piller [28] report that Adidas prevented customers from designing unusual color definitions, and similar considerations may hold for fit and quality aspects of products.
– On the other hand, this activity might be lengthy and even difficult for the customer, if she does not have the required technical knowledge.

While the former issue can be addressed by introducing hard configuration constraints to reduce the solution space, the latter can be addressed by employing personalization techniques aimed at improving the quality of the interaction with the customer during the configuration process. For instance, the customer could be greatly helped in the design of her products if the configuration system supported the specification of high-level features corresponding to usage requirements; e.g., the reliability, ease of use and robustness of a technological device. In this way, the configuration system would act as an intelligent assistant masking the underlying configuration settings.

The integration of personalization and configuration techniques has been experimented in the CAWICOMS EU project [43] to support the development of user-adaptive configuration systems customizing complex technological products and services. As described by Ardissono et al. in [18], in the proposed framework an Intelligent User Interface elicits the customer's functional requirements, which are translated to technical constraints to be solved by a configuration engine. The acquisition of the customer's requirements is carried out in different steps, each one leading to a restriction of the solution space, and the user is prevented from expressing inconsistent preferences because, at each step, the admissible values for the product features are suitably restricted.

Starting from the industrial partners' needs, the CAWICOMS framework was exploited to develop a prototype configuration system supporting sales engineers in the online configuration of IP/VPNs. However, the same techniques could be used to support the customization of other products and services, which challenge the customer with complex expertise requirements. For instance, in the configuration of a loan, an automated assistant might guide the customer in the selection of technical features matching family and income constraints, but also features such as the flexibility of payments, risk assessment and so forth.

16.6 Context-Aware Applications and Ubiquitous Computing

16.6.1 The New Scenario

Recently, the introduction on the market of mobile devices has lead to the idea of *ubiquitous computing*, i.e., the possibility of accessing a service anytime, anywhere and by means of different types of (mobile) devices. The services offered should be tailored to the specific context of use, that is particularly significant if the user is mobile, e.g. walking around a city or driving a car. The importance of providing ubiquitous access to integrated services, and the consequent centrality of context modeling and adaptation to context and device, are advocated by many authors, who claim that ubiquitous computing will play a major role in the future of both academia and company research labs; e.g., see [159, 34, 54] and Chapter 17 of this book [94].

This new scenario requires service providers to consider new adaptation features, such as the user's location, her traveling speed, the environment conditions (lightness, noise, etc.), and, especially, the characteristics of the increasing variety of mobile devices used to access the services; see [136]. Besides, different interaction conditions must be considered: the mobile user does not usually interact with services for exploration purposes (as sometimes happens in standard Web interaction), but she often has a precise goal, such as finding a specific piece of information. Moreover, she is likely performing other activities in parallel, such as carrying a luggage or listening to the announcement of a train arrival. These situations generate additional constraints that have to be taken into account by ubiquitous systems; see, for instance, [159, 72, 96].

The number of applications which may be classified as *context-aware* is very large and their presentation would lead our discussion out of the scope of this chapter.[7] However, we think that the most important aspect of context-aware applications is the possibility to integrate different adaptation strategies, in order to consider the user's interests and needs, as well as context conditions and device specific requirements (e.g., bandwidth and screen size). With respect to this issue, mobile guides represent an excellent example of how adaptation can support the provision of ubiquitous services; see [93] and Chapter 17 of this book [94]. Since the first example of a mobile guide, Cyberguide [1], many research prototypes (e.g., [155, 126, 9, 45, 127, 100, 26]), and commercial systems (e.g., [63, 151]) have been developed. While car manufacturers offer more or

---

[7] Chen and Kotz [44] provide an interesting survey of context-aware applications, and Dourish [57] discusses the concept of context from different perspectives.

less sophisticated on-board navigation systems,[8] research prototypes aim at offering ubiquitous services based on the integration of different adaptation strategies enabled by heterogeneous technologies.

16.6.2 M-Commerce

Ubiquitous computing poses new challenges to e-commerce, as well. Recently, the term *m-commerce* has been introduced to refer to commercial transactions performed by using wireless devices; e.g., [51, 153, 154]. However, m-commerce has evolved differently from traditional e-commerce; for instance, Stafford and Gillenson [144] point out that, while the latter is mainly focused on supporting commercial transactions, the former is more oriented towards offering an enhanced information access. The authors highlight that this change could be due to the flop of the WAP protocol, that was explicitly designed to support transactions on wireless networks; after the WAP failure, mobile devices gained a new role, by supporting product information provision instead of business transactions. For example, the authors report that in Japan there is a common practice, according to which customers retrieve information about shopping choices by using iMode phones, but they place orders within in-store self-service kiosks. In this perspective, m-commerce can be viewed as a support for actual e-commerce transactions, providing information and promotion. Other possible exploitations of mobile devices to support e-commerce activities are digital wallets, push information services, and location-based services, like automatic updates of travel reservations, and up-to-date information about on going events.

As discussed in [145], the most innovative aspect of m-commerce is the possibility to utilize information about the user's local context to tailor offers and suggestions. The information about the user's location, for instance, can be useful to provide timely, relevant, and focused services. Moreover, her location provides rough information about the physical context and the type of activity she is involved in; for example, the user might be visiting a museum, or attending a concert, or driving on a highway. Rao and Minakakis [129] list various business opportunities for location-based services, ranging from adaptive maps and driving directions to the automatic tracking of material, people, or products along the supply chain. Moreover, in Chapter 17 of this book [94], Krüger et al. describe several mobile applications supporting the user during her shopping activity. Those applications are good examples of context-aware, ubiquitous services, providing the customer with information about products, guiding her around the (physical) store, suggesting personalized shopping lists, and so forth.

Given all these opportunities, at the very beginning of the new millennium, many researchers, service providers and mobile device vendors, had foreseen a great expansion of m-commerce, thinking that it would have soon been the main business [59, 141]. However, the expansion of m-commerce has been much slower than expected, and the enthusiasm had to be put in perspective. There are various reasons for this mismatch between expectations and reality. In the first place, mobile devices are not very usable and they are still limited in processing power, bandwidth and computational efficiency, I/O

---

[8] A query on www.globalsources.com lists more than one hundred products under the category "Car Navigation System".

capabilities, etc., coupling high service costs with poor quality of service. Moreover, the lack of standards and shared protocols represents a serious obstacle to the provision of truly ubiquitous services. All these reasons together made the acceptance of mobile devices very low. Sarker and Wells [137] list several factors that influence the adoption of mobile devices, and thus the fruition of ubiquitous services, ranging from individual attitudes to cultural context, from the quality of the technological support to the user's goals. However, according to the authors, the key factor is that, taking all these aspects into account, the use process must result in a positive experience for the user.

16.6.3 Future Perspectives

From these considerations, it is clear that the ubiquitous/mobile access to services poses new challenges to the design and development of e-commerce systems, and, in particular, usability seems to be a key factor for its success. In fact, the traditional user interfaces design methods, developed for desktop computers, cannot be applied as they are to the development of user interfaces for mobile devices. New design methods are required [62, 156, 96] and adaptation techniques are widely recognized as key tools to handle the interaction in mobile user interfaces and to enhance the usability of mobile and wireless services; see, e.g., [30, 143, 37, 72]. In this perspective, new features become relevant for the usability, ad henceforth for the adaptation; for instance, the physical location of the user, but also environment conditions like noise, lightness, whether she is driving or walking, and the characteristics of the device she is using, as claimed, e.g., by Wahlster [159] and Hinz et al. [74], among the others.

An important form of adaptation is the generation of product and service presentations whose length is tailored to the screen size. Many systems adapt the quantity of information presented to the size of the screen. For instance, the UbiquiTO system [9] presents items at different levels of detail on the basis of the screen size and of the user's interests; to this purpose, the system applies adaptation rules which support the selection of one out of four pre-stored versions of the item descriptions (long-essential, long-detailed, short-essential and short-detailed).

However, the most important feature for context-aware applications is the adaptation of the layout of the user interface to the characteristics of the device used to access the service. This adaptation form can be implemented following two approaches [74]:

- The first one aims at adjusting existing (HTML) Web pages to the requirements of mobile devices. In this approach, the limitations imposed by display size and interaction capabilities are taken into account; e.g. [99].
- The second approach is based on an abstract, device-independent definition of the content of the user interface (usually based on XML), which can be used to dynamically generate different instances of the user interface, tailored to the device features and constraints.

The former approach has the advantage that it can be applied to already existing Web resources, but the latter is more flexible and extensible; for instance, new rules for generating user interfaces for new types of devices can be easily added. The automatic generation of alternative layouts starting from an abstract definition of the user interface has been addressed in several projects; e.g., [55, 118, 52, 109, 9, 71]. A standard approach is based on the application of XSL stylesheets that transform an XML object into the object representing the actual user interface (HTML, XHTML, CHTML, WML, Voice-XML, etc.), by taking the characteristics of the devices and, possibly, other context features into account. For example, in a dark place voice output is preferred to written text, but the opposite preference holds in a noisy place.

16.7 Discussion

This chapter has offered an overview of the techniques developed both at the research level and in commercial settings to support personalization in B2C e-commerce applications. In this overview, we have focused on the techniques applied in Web stores and electronic catalogs, which are nowadays used by millions of customers. Moreover, we have analyzed the provision of personalized services customizing the definition of Service Level Agreements; that application scenario poses interesting challenges because it requires the integration of personalization and configuration techniques to support a user friendly interaction with the customer during the negotiation of the service to be defined.

B2C e-commerce is the area where personalization techniques have been most frequently applied. However, before closing this chapter, we would like to shortly discuss the potential of personalization in a in B2B e-commerce context, where we believe that personalization may play a key role in dynamic supply chain management. Usually, service providers rely on alternative suppliers, which may offer similar services but at different costs and supporting different Quality of Service (QoS) levels. When a customer requests a service, the appropriate providers should be selected, in order to offer a service matching functional and QoS requirements. We identify two main issues:

- *Service Discovery:* this issue concerns the dynamic selection and composition of services satisfying the customer's needs and preferences.
- *Logic interoperability between service providers.* This second issue concerns the adaptation to diverse business strategies during the interaction with suppliers and includes adapting to the requirements of business partners, such as QoS levels, business protocols, and similar.

Within this scenario, the new paradigm of Web Services comes into play. Web Services are distributed software modules, wrapped by standard communication interfaces enabling them to interoperate with each other.

1. The first phase of Web Service interaction is the *discovery* of the service fulfilling the consumer requirements. In Enterprise Application Integration, this phase is trivial because all the available services are known. However, in an open environment, such as the Web, the discovery phase can be critical because it requires advanced match-making techniques.
2. After discovery, in many cases the services need to be integrated in order to offer a composite service, made up of several, simpler ones, possibly organized within a workflow (Web Service *composition*).

3. Finally, the last phase is the *execution* of the selected service, that implies an interaction between the Web Service provider and its consumer. This type of interaction can be as simple as a request/response exchange, but in several cases can have a complex structure, which can be modeled as a *conversation*.

In the Service Oriented Computing research [117], Web Services description languages, such as WSDL [157], enable the specification of service public interfaces. Moreover, Web Service orchestration languages, such as WS-BPEL [12, 112], support the definition of composite services based on the orchestration of multiple providers within possibly complex workflows.

Although some proposals have been presented to address the service discovery and logic interoperability issues mentioned above, it is fair to say that personalization has not attracted yet the attention of Web Service application developers, who are focused on other, primary issues, such as security and trust management, and low-level interoperability enablement. In fact, the first goal to be achieved is to abstract service management from the details of the deployment environment of the applications, and to support a seamless interaction between the possibly heterogeneous Enterprise Resource Management software employed by the various partners. For this reason, Web Services are currently mainly intended for B2B integration, but we can foresee that networked services will probably move soon into the area of B2C interactions. With the number of services and also their diversity expected to grow, adequate techniques for user-centric and preference-based service discovery and selection will be needed. Even though UDDI[9] and WSDL are standards today to implement service catalogs, they still lack concepts for service personalization; the work by McIlraith et al. [106] and by Sycara et al. [147] are examples of research in this area. Considering the interaction with a user-centered, personalized service, all of the three discovery, composition and execution phases may be enhanced by personalization, but in particular the service discovery and composition. Semantic Web techniques have been used to improve standard Web Service languages in order to add information used in personalization; see, e.g., [106, 101]. Moreover, some attempts to conform to the customer's preferences in the selection of suitable service providers have been done in the work by Balke and Wagner [24, 25]. At the same time, research in the design of composite Web Services has started. The creation of complex e-services requires innovative approaches to the design process, in order to guarantee the service quality; in particular the service must be reliable and available, and it must comply to quality dimensions, such as QoS, temporal constraints, stability, etc.; e.g., see [42].

As a last consideration, we would like to remark that, regardless of the scenario to which it may be applied (e.g., B2B or B2C), personalization is not a value per se. As discussed in the previous sections, a Web site tailoring the interaction to the individual user, or customizing products and services on an individual basis, is not necessary better than a Web site offering standard solutions. In other words, personalization should not be considered as a goal, but as a mean providing the company and the customers with some advantages. Indeed, the personalization can be considered as an added value only if it represents an advantage in terms of:

---

[9] The current basic standard for service discovery. See www.uddi.org/

- CRM: i.e., if it supports long-term relationships between the company and its customers, increasing customer loyalty;
- Quality of the offer: e.g., the products and services are tailored to individual customer needs;
- Usability of the Web site: i.e., if the personalization of the interaction makes it easier for the user to navigate the e-commerce site;
- Backoffice integration: e.g., if the personalization supports flexible interoperability between suppliers.

Many researchers who design and develop personalized e-commerce systems seem to forget that the added value of personalization is not given a-priori, but must be demonstrated within the context of any specific application. For this reason, testing personalized e-commerce systems with real users is of paramount importance; see, e.g., the discussion in [82], [46] and in Chapter 24 of this book [67].

References

1. Abowd, D.A., Atkeson, C.G., Hong, J., Long, S., Pinkerton, M.: Cyberguide: a mobile context-aware tour guide. Wireless Networks 3(5) (1996) 421–433
2. Abowd, G., Mynatt, E.: Charting past, present and future research in ubiquitous computing. ACM Transactions on Computer-Human Interaction, Special Issue on HCI in the new Millennium 7(1) (2000) 29–58
3. Adidas: mi adidas - customised shoes for you, http://www.adidas.com/products/miadidas04/content/UK/container.asp (2005)
4. Agrawal, R., Mannila, H., Srikant, R., Toivonen, H., Verkamo, A.: Fast discovery of association rules. Advances in Knowledge Discovery and Data Mining. AAAI/MIT Press, Cambridge, MA (1995)
5. Alani, H., Kim, S., Millard, D., Weal, M., Hall, W., Lewis, P., Shadbolt, N.: Automatic ontology-based knowledge extraction from Web documents. IEEE Intelligent Systems 18(1) (2003) 14–21
6. Alani, H., Kim, S., Millard, D., Weal, M., Hall, W., Lewis, P., Shadbolt, N.: Using Protégé for automatic ontology instantiation. In: Proc. of 7th Int. Protégé Conf., Bethesda, Maryland (2004)
7. Alpert, S., Karat, J., Karat, C., Brodie, C., Vergo, J.: User attitudes regarding a user-adaptive eCommerce Web Site. User Modeling and User-Adapted Interaction 13(4) (2003) 373–396
8. Amazon.com: Amazon.com: online shopping for electronics, apparel, etc., http://www.amazon.com (2006)
9. Amendola, I., Cena, F., Console, L., Crevola, A., Gena, C., Goy, A., Modeo, S., Perrero, M., Torre, I., Toso, A.: UbiquiTO: a multi-device adaptive guide. In Brewster, S., Dunlop, M., eds.: LNCS 3160, Mobile Human-Computer Interaction (Mobile HCI 2004). Springer-Verlag, Berlin Heidelberg New York (2004) 409–114
10. ameris: Glossary of IT & internet terms, http://www.ekeda.com/glossary_of_terms.cfm (2005)
11. André, E., Rist, T.: Presenting through performing: on the use of multiple lifelike characters in knowledge-based presentation systems. In: Proc. 2000 Int. Conf. on Intelligent User Interfaces (IUI'00), New Orleans, Louisiana, ACM Press (2000) 1–8

12. Andrews, T., Curbera, F., Dholakia, H., Goland, Y., Klein, J., Leymann, F., Liu, K., Roller, D., Smith, D., Trickovic, S.T.I., Weerawarana, S.: Business Process Execution Language for Web Services version 1.1, http://www-106.ibm.com/developerworks/webservices/library/ws-bpel/ (2003)

13. Arden, R.: Safe internet shopping with Microsoft Merchant System, http://www.windowsitpro.com/Windows/Article/ArticleID/2799/2799.html (2005)

14. Ardissono, L., Botta, M., Costa, L.D., Petrone, G., Bellifemine, F., Difino, A., Negro, B.: Customer information sharing between e-commerce applications. In: Proc. WOA 2004 - Sistemi Complessi e Agenti Razionali, Torino, Italy (2004)

15. Ardissono, L., Gena, C., Torasso, P., Bellifemine, F., Difino, A., Negro, B.: User modeling and recommendation techniques for personalized Electronic Program Guides. In: Personalized Digital Television. Targeting Programs to Individual Users. Kluwer Academic Publishers (2004)

16. Ardissono, L., Goy, A.: Dynamic generation of adaptive Web catalogs. In Brusilovsky, P., Stock, O., Strapparava, C., eds.: LNCS n. 1892: Adaptive Hypermedia and Adaptive Web-Based Systems, Int. Conference (AH 2000). Springer-Verlag, Berlin Heildelberg New York (2000) 5–16

17. Ardissono, L., Goy, A.: Tailoring the interaction with users in Web stores. User Modeling and User-Adapted Interaction **10**(4) (2000) 251–303

18. Ardissono, L., Goy, A., Petrone, G., Felfernig, A., Friedrich, G., Jannach, D., Zanker, M., Schaefer, R.: A framework for the development of personalized, distributed Web-based configuration systems. AI Magazine **24**(3) (2003) 93–110

19. Ardissono, L., Goy, A., Petrone, G., Segnan, M.: Personalization in Business-to-Consumer interaction. Communications of the ACM, Special Issue "The Adaptive Web" **45**(5) (2002) 52–53

20. Ardissono, L., Goy, A., Petrone, G., Segnan, M., Torasso, P.: INTRIGUE: personalized recommendation of tourist attractions for desktop and handset devices. Applied Artificial Intelligence, Special Issue on Artificial Intelligence for Cultural Heritage and Digital Libraries **17**(8-9) (2003) 687–714

21. Avery, C., Resnick, P., Zeckhauser, R.: The market for evaluations. American Economic Review **89**(3) (1999) 564–584

22. Bae, S., Park, S., Ha, S.: Fuzzy Web ad selector based on Web usage mining. IEEE Intelligent Systems **18**(6) (2003) 62–69

23. Balabanovic̀, M., Shoham, Y.: Content-based, collaborative recommendation. Communications of the ACM **40**(3) (1997)

24. Balke, W., Wagner, M.: Towards personalized selection of Web Services. In: Proc. of 12th Int. World Wide Web Conference (WWW'2003), Budapest (2003)

25. Balke, W., Wagner, M.: Through different eyes - assessing multiple conceptual views for querying Web Services. In: Proc. of 13th Int. World Wide Web Conference (WWW'2004), New York (2004)

26. Baus, J., Kray, C., Krüger, A.: Visualization of route descriptions in a resource-adaptive navigation aid. Cognitive Processing **2**(2-3) (2001) 323–345

27. Benyon, D.: Adaptive systems: a solution to usability problems. User Modeling and User-Adapted Interaction **3** (1993) 65–87

28. Berger, C., Piller, F.: Customers as co-designers. IEE Manufacturing Engineer (August/September) (2003) 42–45

29. Bertini, E., Kimani, S.: Mobile devices: opportunities for users with special needs. In Chittaro, L., ed.: LNCS 2795, Human Computer Interaction with Mobile Devices and Services, 5th Int. Symposium Mobile HCI 2003. Springer-Verlag, Berlin Heidelberg New York (2003) 486–491

30. Billsus, D., Brunk, C., Evans, C., Gladish, B., Pazzani, M.: Adaptive interfaces for ubiquitous Web access. Communications of the ACM, Special Issue on The Adaptive Web **45**(5) (2002) 34–38

31. Billsus, D., Pazzani, M.: A personal news agent that talks, learns and explains. In: Proc. 3rd Int. Conf. on Autonomous Agents (Agents '99), Seattle, WA (1999) 268–275

32. Blom, J., Monk, A.: One-to-one e-commerce: who's the one? In: Proc. of Conf. on Human Factors in Computing Systems (CHI'01), Seattle, WA, ACM Press (2001) 341–342

33. Blue Martini Retail: Sell more proactively. Blue Martini Software (2006)

34. Borriello, G., Chalmers, M., Marca, A.L., Nixon, P.: Delivering real-world ubiquitous location systems. Communications of the ACM, Special Issue on The Disappearing Computer **48**(3) (2005) 36–41

35. BroadVision: BROADVISION, http://www.broadvision.com/ (2005)

36. Brusilovsky, P.: Methods and techniques of Adaptive Hypermedia. User Modeling and User-Adapted Interaction **6**(2-3) (1996) 87–129

37. Brusilovsky, P.: Adaptive Hypermedia. User Modeling and User-Adapted Interaction **11**(1-2) (2001) 87–110

38. Bunt, A., Carenini, G., Conati, C.: Adaptive content presentation for the Web. In Brusilovsky, P., Kobsa, A., Nejdl, W., eds.: The Adaptive Web: Methods and Strategies of Web Personalization, Lecture Notes in Computer Science, Vol. 4321. Springer-Verlag, Berlin Heidelberg New York (2007) this volume

39. Burke, R.: Hybrid recommender systems: survey and experiments. User Modeling and User-Adapted Interaction **12**(4) (2002) 289–322

40. Burke, R.: Hybrid Web recommender systems. In Brusilovsky, P., Kobsa, A., Nejdl, W., eds.: The Adaptive Web: Methods and Strategies of Web Personalization, Lecture Notes in Computer Science, Vol. 4321. Springer-Verlag, Berlin Heidelberg New York (2007) this volume

41. Butler, P., Cales, R., Petersen, J., Banick, S., Denschikoff, C., McPherson, S., Melnick, D.: Using Microsoft Commercial Internet System, http://docs.rinet.ru/MCIS/ (2005)

42. Cappiello, C., Pernici, B., Plebani, P.: Quality-agnostic or quality-aware semantic service descriptions? In: Proc. W3C Workshop on Frameworks for Semantics in Web Services, Innsbruck, Austria (2005)

43. CAWICOMS: Customer-Adaptive Web Interface for the COnfiguration of products and services with Multiple Suppliers, http://www.cawicoms.org (2001)

44. Chen, G., Kotz, D.: A Survey of Context-Aware Mobile Computing Research, Technical Report: TR2000-381. PhD thesis, Dartmouth College, Hanover, NH, USA (2000)

45. Cheverst, K., Davies, N., Mitchell, K., Smith, P.: Providing tailored (context-aware) information to city visitors. In Brusilovsky, P., Stock, O., Strapparava, C., eds.: LNCS n. 1892: Adaptive Hypermedia and Adaptive Web-Based Systems, Int. Conference (AH 2000), Berlin Heidelberg New York (2000) 73–85

46. Chin, D.: Empirical evaluation of user models and user-adapted systems. User Modeling and User-Adapted Interaction **11**(1-2) (2001) 181–194

47. Chittaro, L., Ranon, R.: New directions for the design of virtual reality interfaces to e-commerce sites. In: Proc. Int. Working Conf. on Advanced Visual Interfaces (AVI 2002), Trento, Italy (2002) 308–315

48. Chittaro, L., Ranon, R.: Adaptive 3D Web sites. In Brusilovsky, P., Kobsa, A., Nejdl, W., eds.: The Adaptive Web: Methods and Strategies of Web Personalization, Lecture Notes in Computer Science, Vol. 4321. Springer-Verlag, Berlin Heidelberg New York (2007) this volume

49. ChoiceStream, Inc.: Choicestream personalization survey. Research brief (2005)

50. Ciravegna, F., Lavelli, L.: LearningPinocchio: Adaptive information extraction for real world applications. Journal of Natural Language Engineering **10**(2) (2004)

51. Clarke, I.: Emerging value propositions for m-commerce. Journal of Business Strategies **318**(2) (2001) 133–148
52. Coninx, K., Luyten, K., Vandervelpen, C., den Bergh, J.V., Creemers, B.: Dygimes: dynamically generating interfaces for mobile computing devices and embedded systems. In Chittaro, L., ed.: LNCS 2795, Human Computer Interaction with Mobile Devices and Services, 5th Int. Symposium Mobile HCI 2003. Springer-Verlag, Berlin Heidelberg New York (2003) 61–70
53. Corrigan, D.: Achieving advantage with operational customer data integration and management. Data integration solutions (2006)
54. Coutaz, J., Crowley, J., Dobson, S., Garlan, D.: Context is key. Communications of the ACM, Special Issue on The Disappearing Computer **48**(3) (2005) 49–53
55. Dees, W.: Device independent user interfaces. In: Proc. of 12th Int. World Wide Web Conference (WWW'2003), Budapest (2003) 209–249
56. Dey, A., Abowd, D.: Towards a better understanding of context and context-awareness. In: Proc. CHI2000 Workshop on the What, Who, Where, When and How of Context-Awareness, The Hague, Netherlands (2000)
57. Dourish, P.: What we talk about when we talk about context. Personal and Ubiquitous Computing **8**(1) (2004) 19–30
58. ebay: The world's online marketplace, http://www.ebay.com (2006)
59. Feldman, S.: Delivering real-world ubiquitous location systems. IEEE Internet Computing **4**(6) (2000) 74–75
60. Fink, J., Kobsa, A.: A review and analysis of commercial user modeling servers for personalization on the World Wide Web. User Modeling and User-Adapted Interaction, Special Issue on Deployed User Modeling **10**(2-3) (2000) 209–249
61. Fink, J., Kobsa, A., Nill, A.: Adaptable and adaptive information access for all users, including disabled and the elderly. In: Proc. 6th Conf. on User Modeling, Chia Laguna, Italy (1997) 171–173
62. Fithian, R., Iachello, G., Moghazy, J., Pousman, Z., Stasko, J.: The design and evaluation of mobile location-aware handheld event planner. In Chittaro, L., ed.: LNCS 2795, Human Computer Interaction with Mobile Devices and Services, 5th Int. Symposium Mobile HCI 2003, Berlin Heidelberg New York, Springer-Verlag (2003) 145–160
63. Garmin Ltd.: Streetpilot 2650, http://www.garmin.com/products/spIII (2003)
64. Gaudin, S.: Study: personalization not secret to e-commerce. DATAMATION - EARTH-WEB (2003)
65. Geller, M., Bruner, R.: Interactive marketing and consumer packaged goods. In search of CPG marketing gold. DoubleClick email solutions (2005)
66. Gena, C.: Evaluation methodologies and user involvement in user modeling and adaptive systems. PhD thesis, University of Torino (2002)
67. Gena, C., Weibelzahl, S.: Usability engineering for the Adaptive Web. In Brusilovsky, P., Kobsa, A., Nejdl, W., eds.: The Adaptive Web: Methods and Strategies of Web Personalization, Lecture Notes in Computer Science, Vol. 4321. Springer-Verlag, Berlin Heidelberg New York (2007) this volume
68. Gilmore, J., Pine, B.: The four faces of mass customization. Harvard Business Review **75**(January-February) (1997) 91–101
69. Greco, S.: The road to one-to-one marketing. In: Inc. Magazine, http://www.inc.com/magazine/19951001/2433.html (1997)
70. Hall, C.: Business intelligence - the personalization equation. Software Magazine (April) (2001)
71. Healey, J., Hosn, R., Maes, S.H.: Adaptive content for device independent multi-modal browser applications. In Brewster, S., Dunlop, M., eds.: LNCS 3160, Mobile Human-

Computer Interaction (Mobile HCI 2004). Springer-Verlag, Berlin Heidelberg New York (2004) 401–105

72. Herder, E., van Dijk, B.: Personalized adaptation to device characteristics. In De Bra, P., Brusilovsky, P., Conejo, R., eds.: LNCS 2347, Adaptive Hypermedia and Adaptive Web Based Systems, Second Int. Conference (AH 2002), Berlin Heidelberg New York, Springer-Verlag (2002) 598–602

73. Herlocker, J., Konstan, J., Riedl, J.: Explaining collaborative filtering recommendations. In: Proc. ACM 2000 Conf. on Computer Supported Cooperative Work, Philadelphia, PA (2000)

74. Hinz, M., Fiala, Z., Wehner, F.: Personalization-based optimization of Web interfaces for mobile devices. In Brewster, S., Dunlop, M., eds.: LNCS 3160, Mobile Human-Computer Interaction (Mobile HCI 2004). Springer-Verlag, Berlin Heidelberg New York (2004) 204–215

75. IBM: IBM WebSphere Commerce - xpress edition, http://www-306.ibm.com/software/websphere/sw-bycategory/subcategory/SWH00.html (2005)

76. IBM: WebSphere software - business integration solutions, http://www-306.ibm.com/ software/websphere/ (2005)

77. ILOG: ILOG JConfigurator, http://www.ilog.com/products/jconfigurator/ (2002)

78. Jameson, A., Schäfer, R., Simons, J., Weis, T.: Adaptive provision of evaluation-oriented information: tasks and techniques. In: Proc. 14th IJCAI, Montreal (1995) 1886–1893

79. Jameson, A., Smyth, B.: Recommending to groups. In Brusilovsky, P., Kobsa, A., Nejdl, W., eds.: The Adaptive Web: Methods and Strategies of Web Personalization, Lecture Notes in Computer Science, Vol. 4321. Springer-Verlag, Berlin Heidelberg New York (2007) this volume

80. JBoss, the Professional Open Source Company: JBoss Application Server, http://www.jboss.org/products/jbossas (2005)

81. Jupiter Research: Beyond the personalization myth. Site Technologies & Operations (2003)

82. Karat, C., Brodie, C., Karat, J., Vergo, J., Alpert, S.: Personalizing the user experience on ibm.com. IBM Systems Journal 42(4) (2003) 686–701

83. Kay, J., Kummerfeld, B., Lauder, P.: Personis: a server for user models. In De Bra, P., Brusilovsky, P., Conejo, R., eds.: LNCS 2347, Adaptive Hypermedia and Adaptive Web Based Systems, Second Int. Conference (AH 2002), Berlin Heidelberg New York, Springer-Verlag (2002) 203–212

84. Kazienko, P., Adamski, M.: Personalized Web advertising method. In De Bra, P., Nejdl, W., eds.: LNCS 3137, Adaptive Hypermedia and Adaptive Web-Based Systems, 3rd Int. Conference, AH2004. Springer-Verlag, Berlin Heidelberg New York (2004) 146–155

85. Kelly, S.: Data warehousing: the route to mass customization. Wiley, New York, NY (1996)

86. Kinsgstone, S.: Balancing cost reduction with value generation when delivering customer-centric services. Customer-Centric Strategies (2005)

87. Kobsa, A.: Generic user modeling systems. User Modeling and User-Adapted Interaction, Ten Year Anniversary Issue (2000)

88. Kobsa, A.: Personalized hypermedia and international privacy. Communication of the ACM 45(5) (2002) 64–67

89. Kobsa, A.: Generic user modeling systems. In Brusilovsky, P., Kobsa, A., Nejdl, W., eds.: The Adaptive Web: Methods and Strategies of Web Personalization, Lecture Notes in Computer Science, Vol. 4321. Springer-Verlag, Berlin Heidelberg New York (2007) this volume

90. Kobsa, A.: Privacy-enhanced Web personalization. In Brusilovsky, P., Kobsa, A., Nejdl, W., eds.: The Adaptive Web: Methods and Strategies of Web Personalization, Lecture Notes in Computer Science, Vol. 4321. Springer-Verlag, Berlin Heidelberg New York (2007) this volume

91. Kobsa, A., Koenemann, J., Pohl, W.: Personalized hypermedia presentation techniques for improving online customer relationships. The Knowledge Engineering Review **16**(2) (2001) 111–155
92. Kobsa, A., Schreck, J.: Privacy through pseudonymity in user-adaptive systems. ACM Transactions on Internet Technology **3**(2) (2002) 149–183
93. Kray, C., Baus, J.: A survey of mobile guides. In: Proc. HCI Workshop on mobile guides, at Mobile HCI 2004, Udine, Italy (2004)
94. Krüger, A., Baus, J., Heckmann, D., Kruppa, M., Wasinger, R.: Web-based mobile guides. In Brusilovsky, P., Kobsa, A., Nejdl, W., eds.: The Adaptive Web: Methods and Strategies of Web Personalization, Lecture Notes in Computer Science, Vol. 4321. Springer-Verlag, Berlin Heidelberg New York (2007) this volume
95. Langheinrich, M., Nakamura, A., Abe, N., Kamba, T., Koseki, Y.: Unintrusive customization techniques for Web advertising. Computer Networks **31**(11-16) (1999) 1259–1272
96. Lee, Y.E., Benbasat, I.: Interface design for mobile commerce. Communications of the ACM **46**(12) (2003) 49–52
97. Liberty Alliance Developer Forum: Liberty alliance project specifications, http://www.projectliberty.org/specs/ (2004)
98. Linden, G., Smith, B., York, J.: Amazon.com recommendations - item-to-item collaborative filtering. IEEE Internet Computing (January-February) (2003) 76–80
99. MacKay, B., Watters, C., Duffy, J.: Web page transformation when switching devices. In Brewster, S., Dunlop, M., eds.: LNCS 3160, Mobile Human-Computer Interaction (Mobile HCI 2004). Springer-Verlag, Berlin Heidelberg New York (2004) 228–239
100. Malaka, R., Zipf, A.: DEEP MAP - challenging IT research in the framework of a tourist information system. In: Proc. ENTER 2000, Information and Communication Technologies in Tourism, Barcelona, Spain, Springer-Verlag (2000) 15–27
101. Mandell, D.J., McIlraith, S.A.: Adapting BPEL4WS for the Semantic Web: The bottom-up approach to Web Service interoperation. In: LNCS 2870, Proc. 2nd International Semantic Web Conf. (ISWC 2003). Springer-Verlag, Sanibel Island, Florida (2003) 227–241
102. Marcias, M., Gonzalez, J., Sanchez, F.: On adaptability of Web sites for visually handicapped people. In De Bra, P., Brusilovsky, P., Conejo, R., eds.: LNCS 2347, Adaptive Hypermedia and Adaptive Web Based Systems, Second Int. Conference (AH 2002). Springer-Verlag, Berlin Heidelberg New York (2002) 264–273
103. Masthoff, J.: Group modeling: Selecting a sequence of television items to suit a group of viewers. User Modeling and User-Adapted Interaction **14**(1) (2004) 37–85
104. Maybury, M., Brusilovsky, P., eds.: The adaptive Web. Volume 45. Communications of the ACM (2002)
105. McCarthy, K., Reilly, J., McGinty, L., Smyth, B.: On the dynamic generation of compound critiques in conversational recommender systems. In De Bra, P., Nejdl, W., eds.: LNCS 3137, Adaptive Hypermedia and Adaptive Web-Based Systems, 3rd Int. Conference, AH2004. Springer-Verlag, Berlin Heidelberg New York (2004) 176–184
106. McIlraith, S., Son, T., Zeng, H.: Semantic Web Services. IEEE Intelligent Systems **16**(2) (2001) 46–53
107. McKeown, K.: Discourse strategies for generating natural-language text. Artificial Intelligence **27** (1985) 1–41
108. Milosavljevic, M.: The automatic generation of comparison in descriptions of entities. PhD thesis, Macquarie University, Sydney (1999)
109. Mori, G., Paternò, F., Santoro, C.: Tool support for designing nomadic applications. In: Proc. 8th Int. Conf. on Intelligent User Interfaces (IUI'03), Miami Beach, Florida, ACM Press (2003) 141–148

110. Ndiaye, A., Jameson, A.: Predictive role taking in dialog: global anticipation feedback based on transmutability. In: Proc. 5th Int. Conf. on User Modeling, Kailua-Kona, Hawaii (1996) 137–144

111. Netscape: Netscape MerchantXpert, http://wp.netscape.com/merchantxpert/index.html (2005)

112. OASIS: OASIS Web Services Business Process Execution Language, http://www.oasis-open.org/committees/documents.php?wg_abbrev=wsbpel (2005)

113. Omelayenko, B.: Learning of ontologies from the Web: the analysis of existent approaches. In: Int. ICDT'01 Workshop on Web Dynamics, London, UK (2001)

114. Opperman, R.: Adaptive user support - Ergonomic design of manually and automatically adaptable software. Lawrence Erlbaum Associates, Hillsdale, NJ (1994)

115. Oracle: Oracle Application Server, http://www.oracle.com/appserver/index.html (2005)

116. O'Sullivan, D., Smyth, B., Wilson, D., Donald, K.M., Smeaton, A.: Interactive television personalisation. From guides to programmes. In Ardissono, L., Maybury, M., Kobsa, A., eds.: Personalized Digital Television. Targeting programs to individual users. Kluwer, Dordrecht (2004) 73–92

117. Papazoglou, M., Georgakopoulos, D.: Service-Oriented Computing. Communications of the ACM **46**(10) (2003)

118. Pashtan, A., Kollipara, S., Pearce, M.: Adapting content for wireless Web Services. IEEE Internet Computing **7**(1) (2003) 74–78

119. Pazzani, M., Billsus, D.: Content-based recommendation systems. In Brusilovsky, P., Kobsa, A., Nejdl, W., eds.: The Adaptive Web: Methods and Strategies of Web Personalization, Lecture Notes in Computer Science, Vol. 4321. Springer-Verlag, Berlin Heidelberg New York (2007) this volume

120. Peppers, D., Rogers, M.: The one to one future: building relationships one customer at a time. Currency Doubleday, New York, NY (1993)

121. Peppers, D., Rogers, M.: Enterprise one to one. Currency Doubleday, New York, NY (1997)

122. Pierrakos, D., Paliouras, G., Papatheodorou, C., Spyropoulos, C.: Web usage mining as a tool for personalization: a survey. User Modeling and User-Adapted Interaction **13**(4) (2003) 311–372

123. Piller, F., Müller, M.: A new marketing approach to mass customisation. Int. Journal of Computer Integrated Manufacturing **17**(7) (2004) 583–593

124. Pine, B., Gilmore, J.: The experience economy. Harvard Business School Press, Boston, MA (1999)

125. Popp, H., Lödel, D.: Fuzzy techniques and user modeling in sales assistants. User Modeling and User-Adapted Interaction **6** (1996) 349–370

126. Poslad, S., Laamanen, H., Malaka, R., Nick, A., Buckleand, P., Zipf, A.: CRUMPET: CReation of User-friendly Mobile services PErsonalised for Tourism. In: Proc. 3G2001 Mobile Communication Technologies, London, UK (2001) 28–32

127. Pospischil, G., Umlauft, M., Michlmayr, E.: Designing lol@, a mobile tourist guide for UMTS. In Paterno, F., ed.: LNCS 2411, Human Computer Interaction with Mobile Devices and Services, 4th Int. Symposium Mobile HCI 2002. Springer-Verlag, Berlin Heidelberg New York (2002) 140–154

128. Pu, P., Faltings, B., Torrens, M.: Effective interaction principles for online product search environments. In: Proc. IEEE/WIC/ACM Int. Joint Conf. on Intelligent Agent Technology and Web Intelligence, Beijing, China (2004) 724–727

129. Rao, B., Minakakis, L.: Evolution of mobile location-based services. Communications of the ACM **46**(12) (2003) 61–65

130. Reichheld, F.: The loyalty effect. Harvard Business School Press, Boston, MA (1996)

131. Resnick, P., Varian, H., eds.: Special Issue on Recommender Systems. Volume 40. Communications of the ACM (1997)
132. Rich, E.: Stereotypes and user modeling. In Kobsa, A., Wahlster, W., eds.: User Models in Dialog Systems. Springer-Verlag, Berlin (1989) 35–51
133. Richter, K., Enge, M.: Multi-modal framework to support users with special needs in interaction with public information systems. In Chittaro, L., ed.: LNCS 2795, Human Computer Interaction with Mobile Devices and Services, 5th Int. Symposium Mobile HCI 2003. Springer-Verlag, Berlin Heidelberg New York (2003) 286–301
134. Rutledge, P.: Let's get personal: enhancing the customer experience with Web personalization. Patrice-Anne Rutledge (2003)
135. Sacco, G.: Dynamic taxonomies: a model for large information bases. IEEE Trans. on Knowledge and Data Engineering 12(3) (2000) 468–479
136. Sadeh, N.: Mobile commerce: new technologies, services and business models. Wiley & Sons, Ltd. (2002)
137. Sarker, S., Wells, J.D.: Understanding mobile handheld device use and adoption. Communications of the ACM 46(12) (2003) 35–40
138. Sarwar, B., Karypis, G., Konstan, J., Riedl, J.: Item-based collaborative filtering recommendation algorithms. In: Proc. of 10th Int. World Wide Web Conference (WWW'2001), Hong Kong (2001)
139. Schafer, J., Frankowski, D., Herlocker, J., Sen, S.: Collaborative filtering recommender systems. In Brusilovsky, P., Kobsa, A., Nejdl, W., eds.: The Adaptive Web: Methods and Strategies of Web Personalization, Lecture Notes in Computer Science, Vol. 4321. Springer-Verlag, Berlin Heidelberg New York (2007) this volume
140. Schafer, J., Konstan, J., Riedl, J.: Recommender systems in e-commerce. In: Proc. of the ACM Conference on Electronic Commerce, Denver, Colorado (1999) 158–166
141. Senn, J.A.: The emergence of m-commerce. Computer 33(12) (2000) 148–151
142. Smyth, B.: Case-base recommendation. In Brusilovsky, P., Kobsa, A., Nejdl, W., eds.: The Adaptive Web: Methods and Strategies of Web Personalization, Lecture Notes in Computer Science, Vol. 4321. Springer-Verlag, Berlin Heidelberg New York (2007) this volume
143. Smyth, B., Cotter, P.: The plight of the navigator: solving the navigation problem for wireless portals. In De Bra, P., Brusilovsky, P., Conejo, R., eds.: LNCS 2347, Adaptive Hypermedia and Adaptive Web Based Systems, Second Int. Conference (AH 2002). Springer-Verlag, Berlin Heidelberg New York (2002) 328–337
144. Stafford, T.F., Gillenson, M.L.: Mobile commerce: what it is and what it could be. Communications of the ACM 46(12) (2003) 33–34
145. Sun, J.: Information requirement elicitation in mobile commerce. Communications of the ACM 46(12) (2003) 45–47
146. Sun Microsystems, Inc.: Java 2 Platform Enterprise Edition, http://java.sun.com/j2ee/ (2002)
147. Sycara, K., Paolucci, M., Soudry, J., Srinivasan, N.: Dynamic discovery and coordination of agent-based Semantic Web services. IEEE Internet Computing 8(3) (2004) 66–73
148. Tam, K., Ho, S.: Web personalization: is it effective? Perspectives (September-October) (2003) 53–57
149. TeleMutuo: TeleMutuo, risparmiamo i tuoi soldi., http://www.telemutuo.it (2006)
150. Tilson, R., Dong, J., Martin, S., Kieke, E.: Factors and principles affecting the usability of four e-commerce sites. In: Proc. 4th Conf. on Human Factors and the Web, Basking Ridge, NJ, USA (1998)
151. TomTom: Tomtom go, http://www.tomtom.com (2004)
152. Tseng, M., Piller, T., eds.: The customer centric enterprise. Advances in Mass Customization and personalization. Springer-Verlag, New York/Berlin (2003)

153. Turban, E., King, D., Lee, J., Viehland, D.: Electronic Commerce 2004: A Managerial Perspective. Prentice Hall (2004)
154. Urbaczewski, A., Valacich, J., Jessup, L., eds.: Mobile commerce. Opportunities and challenges. Volume 46. Communications of the ACM (2003)
155. van Setten, M., Pokraev, S., Koolwaaij, J.: Context-aware recommendations in the mobile tourist application COMPASS. In De Bra, P., Nejdl, W., eds.: LNCS 3137, Adaptive Hypermedia and Adaptive Web-Based Systems, 3rd Int. Conference, AH2004. Springer-Verlag, Berlin Heidelberg New York (2004) 235–244
156. Venkatesh, V., Ramesh, V., Massey, A.P.: Understanding usability in mobile commerce. Communications of the ACM **46**(12) (2003) 53–56
157. W3C: Web Services Definition Language, http://www.w3.org/TR/wsdl (2002)
158. W3C: Platform for Privacy Preferences (P3P) Project, http://www.w3.org/P3P/ (2006)
159. Wahlster, W.: Resource-adaptive interfaces to hybrid navigation systems (keynote talk). In De Bra, P., Brusilovsky, P., Conejo, R., eds.: LNCS 2347, Adaptive Hypermedia and Adaptive Web Based Systems, Second Int. Conference (AH 2002). Springer-Verlag, Berlin Heidelberg New York (2002) 12–13
160. Witkowski, M., Pitt, J., Fehin, P., Arafa, Y.: Indicators of the effect of agent technology on consumer loyalty. In Stanford-Smith, B., Chiozza, E., eds.: E-work and E-commerce. Novel solutions and practices for a global networked economy. IOS Press (2001) 1165–1171
161. Zimmerman, J., Kurapati, K., Buczak, A., Schaffer, D., Gutta, S., Martino, J.: TV personalization system. Design of a TV show recommender engine and interface. In Ardissono, L., Maybury, M., Kobsa, A., eds.: Personalized Digital Television. Targeting programs to individual users. Kluwer, Dordrecht (2004) 27–52

17

Adaptive Mobile Guides

Antonio Krüger[1], Jörg Baus[2], Dominik Heckmann[3],
Michael Kruppa[3], and Rainer Wasinger[3]

[1] University of Münster, Germany
[2] Saarland University, Germany
[3] DFKI GmbH, Germany
antonio.krueger@uni-muenster.de
baus@uni-sb.de
\{heckmann,kruppa,wasinger\}@dfki.de

Abstract. In this chapter we discuss various aspects of adaptive mobile guide applications. After having motivated the need for web based mobile applications, we will discuss technologies that are needed to enable adaptive mobile web applications, including not only positioning technologies but also sensor technologies needed to determine additional information on the context and situation of usage. We will also address issues of modeling context and situations before giving an overview on existing systems coming from three important classes of mobile guides: museum guides, navigation systems and shopping assistants. The chapter closes with an extensive discussion of relevant attributes of web based mobile guides.

17.1 Introduction

Accessing the world wide web from mobile terminals is not difficult for people living in developed countries. Internet service providers, computer device manufacturers and telecommunication companies offer a variety of services and devices to allow for the mobile access of web-based services. For example, attendees of scientific conferences are able to access the web through their notebooks if wireless access points are provided, e.g. to retrieve background information on the current speaker. Managers read email on their blackberry devices through GPRS on the go and football fans use their mobile phones equipped with UMTS or I-mode technology to watch video scenes of the latest match of their preferred teams. Obviously, there are no major technological hurdles that prevent users from accessing and using these kinds of services.

In contrast to traditional desktop systems, mobile systems are always used in a specific context. However most of the mobile systems nowadays do not make use of the current context or situation and hence are only usable for a very specific purpose.

In this chapter we will focus on a particular subclass of mobile systems, that of *mobile guides*. Mobile guides are applications that provide assistance in a particular, sometimes narrowly defined domain. These guides usually provide assistance very similar to those of human experts of the particular domain. Suitable domains for mobile guides are

P. Brusilovsky, A. Kobsa, and W. Nejdl (Eds.): The Adaptive Web, LNCS 4321, pp. 521–549, 2007.
© Springer-Verlag Berlin Heidelberg 2007

for example tourist applications, such as museum and tour guides, pedestrian navigation systems and shopping assistants. In these domains the mobile guide acts as an expert of the domain and provides the user with information adapted to the current situation. The high degree of adaptivity is the major differences between mobile guides and the aforementioned examples of mobile computing. Adaptivity in the scope of mobile guides is often referred to as *context-based computing*, i.e. the ability to use information on the current context to adapt the user interaction and the presentation of information to the current situations of users. An important ability of this class of systems is their adaptation to limited resources, such as technical resources (e.g. screen size, bandwidth, ergonomics, and connectivity) and the cognitive resources of the users (e.g. attention span, working memory, and haptic abilitities). Mobile guides can be classified along this dimension by distinguishing three different classes of resource sensitive guides: (a) resource *adapted* guides, (b) resource *adaptive* guides and (c) resource *adapting* guides[1]. Resource adapted guides have been optimized in advance for restricted resources that are well known and follow regular patterns. The quality of their assistance remains constant for a given input. This includes for example known limitations of screen size of a mobile pedestrian navigation system. In contrast resource adaptive and resource adapting guides can handle varying resource restrictions. Therefore, their results depend on the available resources during runtime. Resource adaptive processes rely on a single strategy to react to varying resources, whereas resource adapting processes select among a number of strategies on a meta cognitive level to comply with different resource situations. For example a pedestrian navigation system that relies on positioning technologies with varying positioning quality (e.g. GPS[2]) could apply a strategy to compensate these variations by changing the level of abstraction when giving directions to users. Such a mobile pedestrian navigation system would fall into the class of resource adaptive guides. It would fall into the class of resource adapting guides if it would use several adapted positioning strategies, e.g. strategies that help to localize users indoors and outdoors and a meta strategy that selects between these strategies when appropriate.

We claim that well-designed context-based mobile guides should incorporate this notion of resource adaptivity in one of the three described ways to increase their flexibility and ability to adapt to situations and users. Beside these abilities there are technical prerequisites common to all mobile guides, such as sensing and positioning technologies and algorithms which will be discussed in the next section. Afterwards we will discuss a particular example of a representation formalism that helps to model situations and relevant knowledge on a domain based on the semantic web framework (section 17.2.2). Section 17.3 will discuss a broad range of mobile adaptive guides coming from three expert domains: museum and tourist tours, navigation advice and shopping assistance.

---

[1] This classification is based on a classification for adaptive processes, firstly presented in [53]
[2] Global Positioning System

17.2 Context-Technologies for Web-Based Adaptive Mobile Guides

Understanding the context of mobile devices is an important prerequisite for the adaption process. Following [48], context is not just the location of the mobile device but encompasses also other interesting pieces of information like the noise or lighting level, the network connectivity or bandwidth and even the social situation of the user. We see context as all pieces of information that are required to describe situations relevant for an appropriate system behaviour. Some researchers, like Dourish [16], understand context as being a notion of interaction in a broader sense. This view claims that the representation and processing of context is much more difficult, since it is not static but changing depending on the evolution of interaction. As a consequence it is necessary to model the process of interaction and not only context information alone. We share this view and make use of a practical definition of context, encompassing all entities and processes that matter for a broad range of applications, whether being mobile or not. In section 17.3 we will present several examples of applications to clarify this practical notion of context.

To better understand how these applications work, we will now focus on two specific pieces of context information, firstly the location of devices and users and then the detection of user goals and intentions. The location is needed by definition for all types of location-based services (e.g. navigation services or restaurant finders). This might be enough for a variety of simpler tasks, but for more complex tasks, systems have to use the user goals and intentions, which might be inferred from their actions or from physiological sensors and appropriate environmental sensors (e.g. light, pressure and noise sensors) that monitor the user. Afterwards we will show one particular way to represent context in a uniform manner so that mobile applications are enabled to access, process and exchange context information more easily.

17.2.1 Detecting Location

The class of available location technologies can be divided into technologies that allow for outdoor positioning and technologies that allow for indoor positioning. Outdoor positioning (such as the technology implemented in car navigation systems) mostly relies on the satellite infrastructure of the GPS, which provides an accuracy of 10-50 meters and can be considered as good enough for a variety of navigation and tourist information applications [12]. Unfortunately indoors satellite reception is limited, demanding other technologies than GPS for tracking users. Furthermore, when tracking persons in building structures (e.g. in an apartment) one is not so much interested in knowing the exact coordinate (e.g. expressed as latitude, longitude and height) of a person at a given time, but rather the symbolic location of a person (e.g. as expressed by the sentence "in the kitchen"). This requires a positioning system to respect the boundaries (i.e. the walls) of physical indoor spaces. Several methods have been proposed in the past to solve these problems, using infrared beacons [21, 10], radio signals based on the Wireless LAN standards or FM radio stations [5, 36], ultrasonic signals[45] and a combination of infrared and RFID technology [8]. Other approaches use several cameras or microphones installed in every room of interest to detect the location of users [56]. These approaches differ significantly in the amount of infrastructure that has to be installed. Camera and microphones usually need to be connected to a central server which collects and processes the video and sound signals. However, one advantage to instrumenting the environment is that users need no or only minimal instrumentation (e.g. a special name tag or badge worn by the user). In contrast, the instrumentation of the environment can be reduced if users are equipped with a personal device (such as a PDA[3] or mobile phone). In these cases beacons (either based on infrared or radio signals) are often installed in the environment. These beacons are relatively cheap and easy to deploy [21, 10]. The personal device is used to detect the presence of beacons and to calculate the position of the user from this. One drawback is that the personal device needs to always be operational and has to always stay with the user in order to provide meaningful location information.

17.2.2 Modeling and Representation of Users, Context and Situations

One prerequisite for adaptive mobile systems is the proper assessment of the user's situation. For this purpose systems need to rely on a representation of relevant situations. Depending on the supported task, situations can be characterized by many different attributes. Therefore, designers of suitable adaptation mechanisms for mobile guides need to look at a variety of spatial, temporal, physical and activity related attributes to provide effective assistance. For example, a mobile guide that assists users in a shop, needs to know about the current spatial environment of the users (e.g. which products are nearby), the temporal constraints of the user (e.g. how much time is available for shopping), the general interests of the users and their preferences (e.g. if the user prefers red or white wine with tuna), details on the shopping task itself (e.g. which items are on the shopping list and for which purpose the products are needed) and maybe even about the physiological and the emotional state of users (e.g. whether users are enjoying the shopping or not). Having the limited resources of mobile guides in mind, most of the representation and processing of relevant knowledge needs to be carried out remotely in the infrastructure. To reduce complexity and to ensure reusability of the knowledge representations and inference mechanisms, a flexible web-based approach is required that allows different types of systems to exchange and augment information on users and particular situations.

In the following, we present UBISWORLD, the user model markup language UserML and the general user model ontology Gumo for the modeling and representation of users, context and situations, uniform interpretation of decentralized user models, and the integration of ubiquitous applications with the u2m.org user model service.

UBISWORLD can be used to represent some parts of the real world like an office, a shop, a museum or an airport. It represents persons, objects, locations as well as times, events and their properties and features (see figure 17.1 for a conceptual view of the ontology). Apart from the representational function, UBISWORLD can be used for simulation, inspection and control.

The underlying ontology provides the categories as shown in figure 17.2 to represent the level of granularity of spatial elements like city, building or room. They differ for different navigational tasks like pedestrian navigation, car navigation, or even planning a trip with an airplane.

---

[3] Personal Digital Assistant

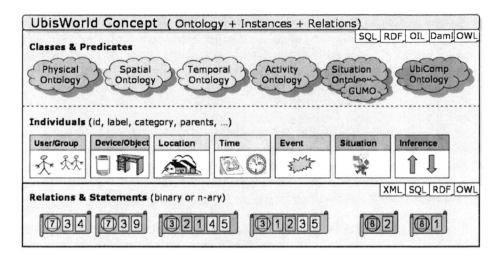

Fig. 17.1. The ontology modules of UBISWORLD

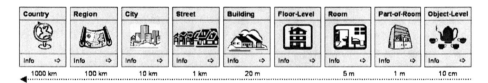

Fig. 17.2. Spatial Granulation Levels in UBISWORLD

Apart from this classification, it is also capable of expressing spatial relations like nesting or connections between elements. In particular it is possible to model various situational properties of spatial elements, such as room temperature, noise-level or humidity, which are for example especially important in museums. Additionally, the presence of persons and physical objects can also be expressed and visualized in this framework. Besides the symbolic location model, UBISWORLD is designed to model user's shopping preferences and any situational context.

UserML has been introduced in [24] as a user model exchange language. A central conceptual idea of the USERML approach is the division of user model dimensions into the three parts `auxiliary`, `predicate` and `range` as shown right below.

$$\text{subject} \{ \textit{UserModelDimension} \} \text{object}$$
$$\Downarrow$$
$$\text{subject} \{ \text{auxiliary, predicate, range} \} \text{object}$$

For example, if one wants to say *something about the user's interest in bargain-sale*, one could divide this so-called *user model dimension* into the `auxiliary` part *has interest*, the `predicate` part *bargain-sale* and the `range` part *low-medium-high*. Apart from these so called `mainpart` attributes, further important meta attributes have been identified for the user modeling domain. These are `situation` (like start, end, durability, location and position), `privacy` (like key, owner, access, purpose, retention) and `explanation` (like creator, method, evidence, confidence). `UserML` statements need not use all 25 attributes that have been arranged into groups. However each of these have a predefined meaning on which specialized meta-data inference modules work. The advantage of using `UserML` to model the user model statements is the uniform syntactical relational data structure that allows, apart from the representation in an ontology, also the storage of mass data in a database.

GUMO has been introduced in [23]. It is designed according to the approach of dividing basic user model dimensions into triples. The advantage of using GUMO in decentralized settings is the semantical uniformity. A large amount of `auxiliaries`, `predicates` and `ranges` have so far been identified and inserted into the ontology that can be inspected with a foldable tree browser at the web page www.gumo.org. However, it turned out that actually everything can be a `predicate` for the auxiliary *hasInterest* or *hasKnowledge*, what leads to a problem if the design is not modularized. The suggested solution is to identify basic user model dimensions on the one hand while leaving the more general world knowledge open for already existing other ontologies on the other hand. Candidates are the general suggested upper merged ontology SUMO, see [43], and the UBISWORLD ontology to model intelligent environments, see http://www.ubisworld.org. This insight leads to a modular approach which forms a key feature of GUMO. A commonly accepted top level ontology for user models could be of great importance for the user modeling research community. But which groups of user dimensions can be identified? In [28] and [31] rough classifications for such categories can be found. Furthermore, this ontology should be represented in a modern semantic web language like OWL and thus be available for all user-adaptive systems with web access at the same time[4]. The major advantage would be the simplification for exchanging interpretable user model information between different user-adaptive systems. and structural differences between existing user modeling systems could be overcome. We are collecting the user's dimensions that are modeled within user-adaptive systems like the *user's current position*, the *user's birthplace*, *user's ability to use stairs* or the *user's interest in modern art*.

Identified user model `auxiliaries` apart from *hasKnowledge* and *hasInterest* are for example *hasBelieve*, *hasProperty*, *hasGoal*, *hasPlan* and *hasRegularity*. User model `predicates` that fit to the auxiliary *"hasProperty"* are called *BasicUserDimensions*. Examples are `Emotional States`, `Characteristics` and `Personality`. The following listing presents the concept *PhysiologicalState* defined as `owl:Class`. It is defined as a subclass of *BasicUserDimensions*. A class defines a group of individuals that belong together because they share some properties. Classes can be organized in a specialization hierarchy using `rdfs:subClassOf`.

```
<owl:Class rdf:ID="PhysiologicalState.700016">
  <rdfs:label> Physiological State </rdfs:label>
  <rdfs:subClassOf rdf:resource="#BasicUserDimensions.700002" />
  <gumo:identifier> 700016 </gumo:identifier>
  <gumo:lexicon>state of body or bodily functions</gumo:lexicon>
  <gumo:privacy> high.640033 </gumo:privacy>
  <gumo:website rdf:resource="&GUMO;concept=700016" />
</owl:Class>
```

---

[4] Please refer for a further discussion of Semantic Web technologies to chapter 23 of this book [14]

Every concept has a unique rdf:ID, that can be resolved into a complete URI. The attribute gumo:privacy defines the default privacy status for this class of user dimensions. The attribute gumo:website points towards a web site, that has its purpose in presenting this ontology concept, to a human reader. The abbreviation &GUMO; is a shortcut for the complete URL to the GUMO ontology in the semantic web. The attribute gumo:expiry provides a default value for the average expiry which carries the qualitative time span of how long the statement is expected to be valid. In most cases when user model dimensions are measured, one has a rough idea about the expected expiry. For instance, emotional states hold normally no longer than 15 minutes, however personality traits won't change within months. Since this qualitative time span is dependent from every user model dimension, it should be defined within GUMO. Some examples of rough expiry-classifications are:

- physiologicalState.heartbeat - can change within seconds
- characteristics.inventive - can change within months
- personality.introvert - can change within years

The idea behind gumo:expiry is that if no new actual value is available on the user model server after a while, one can still work with old values, probably combined with reduced confidence values. The semantic web ontology language OWL allows to construct complex, graph-like hierarchies of user model concepts with multiple-inheritance, which is especially important for ontology integration. The GUMO vocabulary includes gumo:identifier, gumo:expiry, gumo:image, gumo:privacy, gumo:website, gumo:image and gumo:lexicon. To support the distributed construction and refinement of GUMO, we developed a specialized online editor to introduce new concepts, to add their definitions and to transform the information automatically into the required semantic web language.

The different applications or agents produce or use UserML statements to represent the user model information. UserML forms the syntactic description in the knowledge exchange process, while the each concept like the user model auxiliary hasProperty and the user model dimension timePressure points to a semantical definition of this concept which is either defined in the general user model ontology GUMO, see figure 17.3.

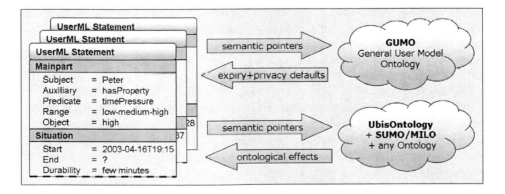

Fig. 17.3. The syntax-semantics interplay between USERML and GUMO

A user model *service* manages information about users and contributes additional benefit compared to a user model *server*. The u2m.org user model service consists of a set of application-independent servers with a distributed approach for accessing and storing user information, the possibility to exchange and understand data between different applications, as well as adding privacy and transparency to the statements about the user[5]. Applications can retrieve or add information by HTTP requests like:

```
http://www.u2m.org/UbisWorld/UserModelService.php?
subject=Peter&auxiliary=hasInterest&predicate=bargain-sale
```

We have tested the approach in a MOBILEMUSEUMSGUIDE, see [37], in a POSITIONINGSERVICE, see [8] and in an ALARMMANAGER application, see [9]. The latter one is a notification service for instrumented environments that adapts the presentation of announcements to the user's state of arousal and the user's location. Both are retrieved from the UserML and GUMO enabled user model service. The location is derived from the POSITIONINGSERVICE application. This service runs on the user's PDA and uses infrared beacons and active RFID tags that are installed in the environment to estimate the location of the user which is then send via WiFi to the user model service.

17.3 Review of mobile guide Applications

This chapter contains an overview on several existing mobile guides, including both commercial and research systems and prototypes. It shows the variety of approaches and different adaptation levels. We will discuss systems from three different, prominent application domains: museum guides, navigation systems and shopping assistants.

17.3.1 Museum Guides

In this section we will review four different museum guide research projects in order to give a brief overview on the research area. While all of the mentioned systems share a common attribute, namely the use of handheld-computers in order to allow for mobile content presentations and user interactions, each project focuses on a different aspect. While the Sotto Voce project [3] fosters communication among users while exploring the museum site and using the mobile tour guide system, the AgentSalon project [52] focuses on inter-user communication at stationary devices spread throughout the museum. Both the Hippie project [42] and the PEACH project [47] are concerned with automatic content adaptation based on technical restrictions of specific presentation devices but also influenced by user preferences and knowledge. While the Hippie museum guide system uses stationary and mobile devices in a sequential way, the PEACH museum guide combines both mobile and stationary systems in real time on site.

The basic idea behind the Sotto Voce project is to build a mobile museum guide system taking into account the special needs of groups visiting a museum. Instead of supporting only individual users, the system should be capable of not only supporting

---

[5] For more details on privacy issues in personalized web applications see chapter 21 of this book[30]

but actually encouraging communication among group members. In most cases, participants of docent-led tours are turned into a passive audience and audio tours often force users into isolation [25]. In order to engage users of the Sotto Voce system into conversation, the electronic guidebook forming the central part of the system supports technologically mediated sharing of informational audio content. This sharing mechanism is called *eavesdropping* and will be explained later.

The Sotto Voce prototype consists of several guidebook devices. These devices combine a Compaq iPAQ 3650 handheld computer featuring a high resolution colour touchscreen with a wireless local-area network (WLAN) card. In addition, headsets that do not fully occlude the ears and hence still allow users to communicate verbally with each other, are connected to the iPAQ. In order to support the audio sharing mechanism, two guidebook devices may be paired over the WLAN using standard internet protocols (UDP/IP). Since the audio content to be presented is fixed and pre-installed on each guidebook, only synchronizing messages need to be exchanged between the paired guidebooks (e.g. "stop playing clip x", "start playing clip y"). The user interface of the prototype is subdivided into two regions (see figure 17.4).

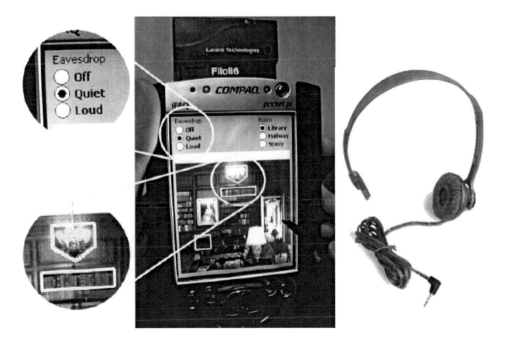

Fig. 17.4. The Sotto Voce prototype setup (source: [3])

While a small region in the upper part of the touchscreen features two different radio button groups (one for controlling the eavesdrop mechanism which we will describe below and a second one which allows users to select the room in which they are located), the major, lower part of the touchscreen is used to display a photograph of a wall of the room in which the user is standing. By pressing hardware buttons on the iPAQ, users may change the wall displayed on the touchscreen. The photographs resemble a visual interface with buttons to select audio content related to specific objects on the walls. The photographs displayed are a set of web browser imagemaps. If a user taps on the screen and hits the target region of an image map, the audio content corresponding with the tapped object will be started. If a user taps on a region which is not linked with an object, the guidebook displays transient target outlines appearing around the objects that may be tapped.

The eavesdropping mechanism allows two users of the system to share audio content. When a visitor selects an audio clip by tapping on the touchscreen, she always hears that audio clip. However, if the second person is playing an audio clip, while the first person is not currently listening to an audio clip, both users will hear the same audio clip. Audio clips are never mixed, and personally selected clips are always preferred. The playback of audio clips is synchronized, which means that, in case both users are listening to different audio clips, and one clip is shorter than the other, the corresponding user will not hear the other users audio clip from the beginning, but from the actual playback position. Users may control the eavesdrop mechanism by either turning it off or by selecting two different volume settings.

In the AgentSalon project, several mobile devices may be used in combination with a single stationary system, a so called information kiosk which is assumed to be located in a meeting place of an exhibition site. The main idea behind the system is to support face-to-face discussions and exchange of knowledge by tempting users to chat with each other. In order to achieve this goal, users are monitored while exploring the exhibition site. In this way, the system is aware of the exhibits already visited by each user. In addition, users are also asked to state personal interests prior to using the AgentSalon system and they are also allowed to rate each exhibit they visit during their tour. While exploring the exhibition site, each user is accompanied by a virtual personal agent, represented visually on the mobile device as a simple, static comic figure.

During the exploration phase, users of the system are free to visit a stationary information kiosk. Once in front of the information kiosk, the user's virtual personal agent is capable to migrate from the user's personal mobile device to the information kiosk (see figure 17.5). The migration is done in a very simple way, by turning of the characters visual representation on the mobile device and by showing the same visual character representation on the information kiosk. The information kiosk allows for direct user interaction with the system by integrating a touch-screen device. In this way, a single user can request additional information regarding the exhibition.

However, the main purpose of the information kiosk is to foster direct communication between several users of the system. The system allows up to five users to benefit of a single stationary device. Each of these user's virtual personal agent will transit to the information kiosk, as described above. Since each user chooses a different visual representation for his/her virtual personal agent, these characters may all appear on a single device and still allow users to easily distinguish the different characters. Once a virtual personal agent appears on the information kiosk, the AgentSalon system detects common as well as different parts in the different user's interests and visiting records. Based on this calculated overlap between user interests and experiences, the agents plan and begin conversations observed by the users. The dynamic script generation for

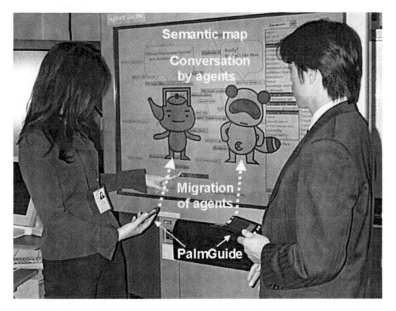

Fig. 17.5. AgentSalon: Multiuser and multidevice interaction (source: [52])

interesting inter-agent communications is based on a knowledge based system. Using strategic rules, the system combines reusable templates of conversations, resulting in a conversation script for several virtual personal agents.

The basic idea behind these inter-agent communications is to start a communication between their users (as illustrated in figure 17.5). While the personal agents talk about the interests and experiences of their users, hopefully users will be encourage to start a conversation among themselves about the same topics.

The Hippie system is an internet based museum guide which may be used in a stationary context at home and in a mobile scenario on location. The selection of content presented to the users is influenced by a number of factors like physical location, personal interests of users as well as user knowledge and preferences. While the stationary version of the Hippie systems runs on an arbitrary, internet enabled computer, the mobile version is especially designed to be used in conjunction with a handheld computer (Toshiba Libretto 100 CT). The users physical location within the museum is determined by a combination of infrared beacons and an electronic compass. While the electronic compass is used to determine the user's orientation, the infrared beacons are employed to determine the user physical location. The infrared beacons emit a unique signal which is received by an infrared receiver which is connected to the handheld-computer and is fixed to the users jacket or alternatively, to the systems earphones. The unique ID is then forwarded from the handheld-computer to a server via a wireless connection. The server selects appropriate content from a database which is then transferred back to the handheld-computer via the same communication channel.

The different configurations of mobile and stationary hardware demand different application layouts and content presentation on the different devices. The content is automatically adapted according to the capabilities and limitations of each device. Sometimes, information presented using a specific presentation modality has to be converted to another presentation modality. An example is written text, which may be presented in an appropriate fashion on the large screen of the stationary device but which is inconvenient for the small display of the mobile device. In such a case, the written text would be substituted by pre-recorded spoken utterances which would then be rendered on the mobile device. Another adaptation strategy is the alternation of quality. For example a large scale image may be shown on the stationary device, but may have to be reduced in resolution prior to its presentation on the screen of the mobile device.

Apart from the content adaptation to the different output modalities of mobile and stationary devices, the system also tries to adapt the presented content according to the knowledge and interests of each user. For this purpose, the system keeps a user model which is updated based on the user's interactions with the mobile system and the users trajectory within the museum. A rule based system is used to determine appropriate content or to choose alternative routes through the museum.

The PEACH museum guide project [51, 20] also deals with automatic content adaptation to different output devices, however the approach is quite different. While in the Hippie system, stationary and mobile devices are used in a sequential way (e.g. a user first prepares her museum visit on the personal computer at home and then uses the mobile device while actually visiting the museum), the PEACH system combines both mobile and stationary systems in parallel. While exploring the physical site of the museum, users are equipped with a handheld-computer with integrated wireless communication functionality. Location based services (i.e. information that is related to a specific object or location within the museum) are presented on the mobile device. Instead of simply resizing images that do not fit the screen of the mobile device (like in the Hippie project), in the PEACH project, large images (and corresponding spoken utterances) are automatically transformed into video clips which show particular details of the large image which correspond timely with the content of the spoken utterances.

However, in some cases the information to be presented is not available in a format which may be adapted to the mobile device. To fill in this gap, additional stationary devices are spread throughout the museum. The idea is to allow users to occasionally use one of these stationary devices to request information of such kind, that can not be presented in an appropriate way on the mobile device (for example high resolution images or video clips). While the user moves through the physical space of the museum (both user location and orientation are derived from infrared beacons very similar to the ones used in the Hippie system) and interacts with the mobile device, the system builds up and updates a user model. From time to time, the mobile guide may suggest to look for a stationary device in order to see some additional information related to recently presented information on the mobile device.

A peculiarity of the PEACH museum guide system is the use of virtual character which act as tour guides within the museum. The comic like characters may be used on both mobile- and stationary devices and may easily transit from one device to another. For both types of devices, different character layouts where developed (see figure 17.6). The virtual characters fulfill a number of purposes within the scenario.

Fig. 17.6. PEACH: Virtual characters for mobile and stationary devices

First of all, they are taking the role of a museum guide. Instead of presenting the information in a standard fashion, the virtual character helps to make the experience for users of the PEACH system more lively and engaging. While being a museum guide, a virtual character within PEACH may play different roles. The virtual character may either play the role of a presenter (i.e. using spoken utterances and gestures to present information) or it may play the role of an anchorman (similar to television, where an anchorman appears in-between different video clips to make the whole presentation a coherent experience).

A second purpose fulfilled by the virtual characters is the representation of different views on the content of the museum. By providing not a single character, but instead a number of different characters (which represent different stereotypes) to choose from, users may decide to change the perspective of their personal tour within the museum. For example, a character resembling a medieval artist is used in the same scenario together with another character, a female representing the medieval upper class. By choosing one of the characters, the tour will either focus on artistical aspects of the objects on display or it will focus on the socio- historical background of the time in which those objects were produced. Since users are free to choose a different character whenever they like, they may experience the same exhibit with different perspectives.

Finally, the virtual characters are used to guarantee for a coherent experience while using both mobile and stationary devices. The characters may transit from mobile- to stationary device and vice versa. The transition from one device to another is visualized by a disappearing character on the source device and a reappearing character on the destination device and is underlined by a sound starting on the source device and ending on the destination device. In this way, the users attentional focus is guided from one device to another. Additionally, the virtual characters offer a unified user interface on both mobile and stationary devices.

17.3.2 Navigation Systems

Mobile guides and navigational assistants have come a long way since the first research prototypes (e. g. [1]). At the moment, there are not only many different research projects focussing on the topic, some of which will be presented in the following, but there are also several commercial services available to mobile phone users and car drivers[6]. Recent developments such as the emergence of ubiquitous computing [55] and the convergence of portable computing devices (such as a PDA and a laptop computers), wireless communication (such as wireless LAN or the General Packet Radio Service (GPRS)) and localisation means (such as the Global Positioning System (GPS)) have further increased the pace of progress. The arrival of the new generation of mobile phones that provide a higher bandwidth and allow for a more precise localisation will most likely have a similar effect.

In the following we select a set of systems that offer unique features or have been influential in the development of the field.

The first project reviewed in our section about research systems is the GUIDE project [12]. The system provides information about the city of Lancaster. The mobile component of the GUIDE system was connected wirelessly to an information server. Based on the current WLAN access point the guide senses the position and provides guidance and information services from a central server through a browser-based interface. The GUIDE system was used by 'real' tourists visiting the city of Lancaster, which have been asked to rate their experience with the system. In [12] a small user study based on direct observation, audio recording and a time stamped log of user interactions is reported. In this study the majority of users appreciated the ability to use the system as a tour guide, a map or a guidebook. The main services offered by the GUIDE system consist in the provision of information on sights and guidance of the tourist to these sights. Unlike most other outdoor guidance systems, GUIDE uses the network cells defined by several strategically placed wireless access points to determine the current position of the user and does not rely on GPS. In addition, there is a simple interactive means to address the loss of network connection: the user is given a long list of thumbnails of all sights at the city and is then asked to select the one closest to her. Based on her answer, the system tries to estimate the user's current position in the city.

TellMaris was a prototype for a mobile tourist guide that was developed at Nokia Research Center [33, 38]. It was one of the first mobile systems to combine three-dimensional graphics with two-dimensional maps running on a mobile phone. TellMaris was developed for the city of Tønsberg, Norway to help boat tourists in finding locations of interest (e.g. hotels). The user of the system can dynamically navigate the rendered scene using cursor keys on the phone. The user's position is also highlighted in the map and dynamically updated to reflect the navigation of the user in the 3D scene. The prototype only provides static navigation services, which means that the user can only navigate in the 3D scene and the map in order to explore the city of Tønsberg, but so far there has been no guidance functionality implemented. The only automated assistance consists of an arrow in the 3D model that points in the direction of the previously selected target location. While in future version of the system positioning eventually

---

[6] E.g. the Garmin Streetpilot 2650

will depend on GPS, the first prototype depends entirely on the user to manually position herself on the map or in the 3D model. Within the TellMaris project, a further study was reported [6], which investigated the usefulness of combined 3D/2D presentations with a limited number of participants and a limited prototype. 3D maps were received positively, although some users complained that they had difficulties in comparing the 2D and 3D maps provided by the system, which they attributed to some lack of correspondence between them. In the 3D map the user had the possibility to choose between a walking level (pedestrian view) and a flying level (birds-eye view). In the study, the flying mode was found to be superior for navigational purposes. A more sophisticated view on 3D adaptive web sites in general is given in chapter 14 of this book [13].

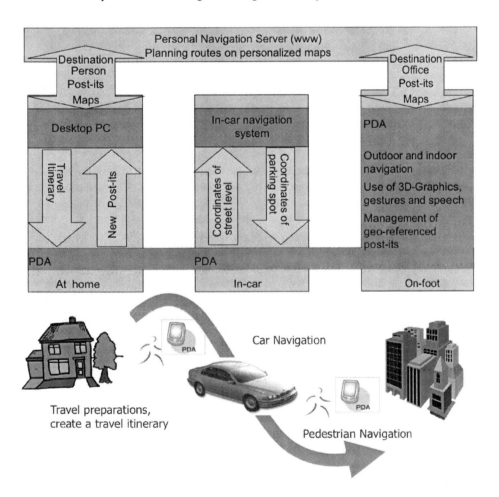

Fig. 17.7. The BPN-System combines a web-based routing services with a mobile car and a pedestrian navigation service. (source: [34])

The LoL@ (local location assistant) system [2, 44] is a mobile tourist guide for the city of Vienna designed for the next generation of mobile phone networks, the Universal Mobile Telecommunications System (UMTS). It was developed at the Forschungszentrum Telekommunikation Wien. Like the GUIDE system and TeLLMaris, LOL@ is based on a client-server paradigm: a client, in this case a mobile phone, accesses a server (i.e. a web server). While this approach allows for easy addition of multiple clients, it highly depends on a reliable connection between client and server, which is not always given. Similarly to the aforementioned GUIDE system, the main interaction metaphor in the LOL@ system is that of a web browser, which means that the system's users can click on links or buttons to access information and/or to trigger some actions. LOL@ provides three main services to its users: guidance on virtual or real tours through Vienna, provision of information on sights, and a personalized semi-automated tour diary. The guidance process requires the user to manually confirm her arrival at the end of each segment in order to trigger instructions for the next one. These instructions are either given via speech or using a map enriched with a set of predefined sights. The user can also access multi-media information on the objects that are visible on screen. Links allow for accessing structured information such as pictures, texts, or movie clips. When the user follows a tour, LoL@ automatically generates a tour diary consisting of a web page containing a chronologically ordered list of visited sights and user additions such as pictures and notes.

Based on the experiences gained in the project REAL [6] researchers of Saarland University and the car manufacturer BMW developed the BMW Personal Navigator (BPN) [34]. The system combines a desktop event and a web-based route planer, a car navigation system, and a multi-modal, in- and outdoor pedestrian navigation system for a PDA and offers a situated personalized navigation service summarized in figure 17.7. In order to prepare the route for the car, the BPN uses a route-planner freely accessible from the internet. For pedestrian navigation the systems uses externally generated navigational information by a GIS server. Information about streets, landmarks and references to maps and 3D models are stored on the mobiles device. During the trip, the information can be updated by server access via mobile phone or a BlueTooth connection to the in-car navigation system.

17.3.3 Shopping Assistants

Shopping has been identified by [18] as a "realm of social action, interaction and experience which increasingly structures the everyday practices of urban people". Sociologists have described the shopping experience as complex and ambiguous, and full of contradictions and tensions. [39] for example state that shopping is ambiguous in nature because it is essentially a private experience that occurs in a public setting. They argue that shopping is contradictory in that it is an experience that yields both pleasure and anxiety which can easily morph into a nightmare. The act of shopping can also be seen to entail tension, in the form of rationality versus impulse, and between a pleasurable social form and a necessary maintenance activity. From these perspectives and the associated intricacies, it is evident that shopping is a subject consisting of considerable depth. It plays a central role within society, and thus is a prime field of study for mobile guide applications, particularly in relation to the benefits that may result for retailers and consumers.

This section summarizes a range of shopping assistants that are currently either being developed as part of research projects, or that have already found their way into the commercial market place. Matching the diversity that entails the act of shopping itself, the described assistants cover a wide range of product domains such as everyday grocery items like bread and milk, electronic items like digital cameras, and even car sales. Some of the described implementations are location and context aware, and delve into the realms of mobile, ubiquitous and pervasive computing. Their architectures are often based on instrumented environments and shopping trolleys, and handheld devices that accompany a user around a store. Extending upon the roles of the traditional real-world sales assistant, the main areas that this type of shopping assistant focus on include guiding a user around a store, and providing users with supporting information in the form of personal shopping lists and product specifications. Being location and context aware is also a commonality of other types of mobile guides including navigational guides (see section section 17.3.2), and museum guides (see section section 17.3.1).

Some systems are now also beginning to merge these application domains together, which can be seen in [35] where a multimodal shopping assistant is tightly linked to a mobile pedestrian navigation and exploration system. A second class of shopping assistant that is briefly discussed within this section are those based on web-agents. These collate data from many different product vendors and then allow customers to access the results via the Web in the form of comparison charts. In contrast to location and context-aware shopping assistants, which need to cater for customers that are "moving" around a shop, assistants based on web agents are generally accessed through a stationary desktop computer and are closely related to the paradigm of home Internet shopping. Current research into shopping assistants tries to build on the aforementioned assistants by providing additional value to the shopping experience in the form of conversational dialogues, multimodal interaction, augmented reality, and enhanced plan recognition. They also try to cater for the personal needs of individual users and for specific user groups such as people with disabilities (e.g. sight-impaired users).

Context-aware shopping assistants have the primary goal of improving a customer's "in-store" shopping experience, while at the same time increasing the store's level of efficiency and profits. Two commercial context-aware shopping assistants include the METRO Group's Future Store[7], and IBM's Shopping Buddy [26] as can be seen in figure 17.8. The goal of the Future Store initiative was to integrate multiple emerging technologies into an existing store, and to evaluate the technologies as a preliminary step to broad integration of the technologies throughout the retail chain. Key technology components used in the store include servers, Radio Frequency Identification (RFID) readers, kiosks, desktop and mobile PCs, handheld devices, and network components. The Future Store installation can be seen to benefit both retailers and customers. From a retailer's point of view, RFID tags can be placed on pallets and individual products to allow inventory throughout the store's supply chain to be tracked. This is achieved through the use of RFID readers, which for example if attached to shelves can notify staff when products need to be replenished. The system also allows staff to access

---

[7] For more information see http://www.future-store.de

Fig. 17.8. The IBM's Shopping Buddy mounted to an ordinary shopping trolley. (source: [26])

business intelligence through mobile PDAs, via functions that for example allow stock levels to be checked, item information to be requested, and product prices to be automatically changed on electronic advertising displays. From a customer's point of view, benefits revolve around a more convenient, engaging, and customized shopping experience. A loyalty card allows the customer to begin shopping before they enter the store by selecting goods that they plan to purchase from a website and saving these to the card for later use in conjunction with an instrumented shopping trolley. Touch screen tablet PCs mounted on top of trolleys provide shopping lists, product descriptions and pictures, pricing information, and store maps, along with running totals for the products selected and placed inside the trolley. Promotional offers are also displayed on the trolley's display, based on the shopper's location in the store, and 19" displays mounted above product areas offer further promotional information using video and animation.

On a similar front, IBM's Shopping Buddy [26] has been deployed in several stores and has many of the same goals as that of the METRO Future Store. The Shopping Buddy for example displays running totals of how much consumers have spent and saved during their visit. It reminds them of past purchases, and allows them to place orders with the supermarket's deli from their trolley and pick up their requests once the system indicates they are ready. Complementing the trolley functionality, a location tracking system permits the delivery of targeted promotions, and is also capable of helping consumers navigate through the store, and helping consumers locate products. Similar to the Future Store, the system reduces checkout lines by allowing consumers to scan and bag items as they shop, and then complete their transactions using IBM's Self Checkout system.

Shopping assistants based on web agents have the goal of optimising a customer's "online" shopping experience, in the form of both time efficiency and cost savings. They generally collate product and price information from a number of different vendors, and make this information available to consumers in the form of comparison charts. They have however only had limited success in commercial markets, where many vendors do not stand to gain by competing on price. Most web-based shopping agents that exist commercially are often biased towards participating vendors, in that they only present results from companies with whom they collaborate. [40] list a number of such examples including MySimon, DealTime, PriceScan, and RoboShopper. They also outline a range of attempts at building unbiased agents, including ShopBot [15] which was later commercialized by Excite, PersonaLogic which disappeared after being bought out by AOL, and FireFly which ceased operations after it was acquired by Microsoft.

Current research into the development of shopping assistants is being conducted on a number of fronts, ranging from extensions to the general in-store scenario to cater for additional surroundings like that of the family home [32], the use of plan recognition [49] and decision theoretic planning [7] to better predict and guide a customer throughout their shop, the incorporation of augmented reality [57], conversational interfaces [11, 46], multimodal interaction [54], and the design of shopping assistant interfaces with respect for example to small displays [41] and the visually impaired [17].

Most of these research projects are extensions to location- and context-aware shopping systems rather than web-based shopping agents. The Personal Shopping Assistant [4] from AT&T Bell Laboratories is a system that closely resembles the two commercial context-aware systems described above. It is also based on a client-server architecture consisting of a shop server containing the business logic, and hand held client devices that present information to customers. In this system, the handheld devices are however seen as dumb devices and used as information portals only; aside from the communications protocol there is no computation performed by the portable units. The system aims to help a customer navigate within a store, provide them with details on products of interest via the mobile device's screen, points out items that are on sale, allows for comparative price analyses, determines price information through an attached barcode scanner, and even plays the user's favourite music over a private headset. User input primarily takes the form of spoken keywords (via a microphone), but is also possible via a bar code reader and several general purpose buttons.

The MyGROCER project [32] extends the general in-store scenario to cater for in-house and on-the-move interaction aswell. Whereas the functionality of the in-store scenario is based on an instrumented shopping trolley and includes displaying a user's shopping list as well as in-store promotions based on previous consumer buying behaviour and cross-selling product associations, the in-house scenario is based on instrumented key-storing locations that are inter-networked with RFID receivers and allow products that are removed from their original location to be added to the user's shopping list and accessed via a mobile phone connection. The on-the-move scenario incorporates notifications about products that have run out-of-stock and allows for the home delivery of such products.

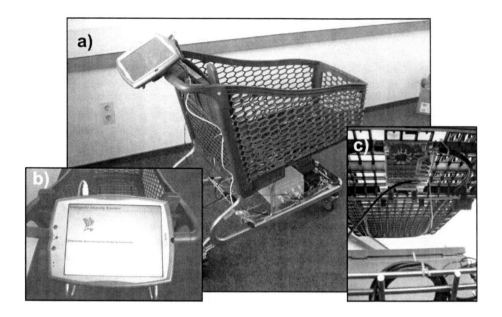

Fig. 17.9. The shopping card used by Schneider for their plan-recognition shopping guide. (source: [49])

Schneider outlines in [49] an adaptive shopping assistant that utilizes plan recognition techniques to aid the user while shopping. In particular Schneider develops a proactive user interface driven by implicit interaction in a real world shopping scenario. As a result, the system may provide detailed product information if users pick up a product that they have never handled before. Alternatively, the system may display a list of similar products, or product chart comparisons (e.g. if the user has two products in their hands). If the system infers that the user is cooking a particular dish, a list of additional products that might be required may also be displayed. Presentation output takes the form of dynamic HTML pages, displayed on the shopping trolley's display.

[7] develop a PDA-based system that gives a shopper directions through a shopping mall based on the type of products the shopper has expressed interest in, the shopper's current location, and the purchases that the shopper has made so far. The approach uses decision-theoretic planning to compute a policy that optimizes the expected utility of a shopper's walk through the shopping mall, taking into account uncertainty about whether the shopper will actually find a suitable product in a given location, and the time required for each purchase.

Another interesting design is the PromoPad [57]. This is an in-store e-commerce system that provides context-sensitive shopping assistance and personalized advertising through augmented reality techniques. Individual objects that are encountered in the real world are augmented with virtual complements so as to make the real objects more meaningful and appealing. The system is novel in that, aside from adding new imagery relative to a focal product, the system can also remove elements of the image that may

Fig. 17.10. Interacting with the ShopAssist system through multimodal input (source: [54])

distract from the focal product. This system is based on a tablet PC with a camera mounted on the back. The display on the tablet provides a modified version of the camera image which the shoppers can look at as though it were a "magic frame" (i.e. see-through).

Two shopping assistants supporting conversational interfaces include [11] and CrossTalk [46]. [11] developed an online conversational dialog system that assists users in finding notebooks by engaging them in dialog. Based on a market survey, they acquired an appropriate set of natural language user vocabulary consisting of 195 keywords and phrases, and then generated statistical n-gram models and a shallow noun phrase grammar for extracting keywords and phrases from user input. Subsequent user studies found that when compared to a menu driven system, use of the conversational interface reduced the average number of clicks by 63% and the average interaction time by 33%. In comparison to this system, which encourages direct human-computer conversation, CrossTalk is an interactive installation in which agents engage in conversational car sale dialogues. It builds on the Inhabited Market Place (IMP) [29] by adding a virtual fair hostess Cyberella to act as mediator between human visitors and the IMP application. The IMP is a virtual place (e.g. a showroom) where seller agents provide product information to potential buyer agents in the form of typical multi-party sales dialogues. This allows for human users observing the dialogue to learn about the features of a car.

The Mobile ShopAssist [54] is a shopping assistant that focuses on mobile and multimodal interaction. It provides a shopper accompanied by a mobile Pocket PC device with the ability to enquire about product attributes, and to ask for comparison information between different products like digital cameras. The input modalities provided by the assistant include speech, handwriting, gesture, and combinations thereof. An interesting feature of the system is that it caters not only for interaction with products on the mobile device's screen (e.g. intra-gestures), but also with products in the real world via extra-gestures in the form of pick-up and put-down actions. The fusion of all the inputs is performed locally on the device itself, as too the recognition and interpretation of all interactions except extra-gesture, which is recognized by RFID instrumented shelves and then sent to the mobile device via a wireless LAN connection. Studies conducted on the basis of this implementation show that modalities like intra-gesture were preferred in a public setting (e.g. in a store), while modalities like speech and extra-gesture were preferred in more private settings (e.g. at home).

In comparison to the Mobile ShopAssist, which caters for many types of users via a blanket coverage of the modalities, SAVi [17] is an assistant that specifically caters for the visually impaired, and is designed to aid the blind or sight-impaired shopper in identifying and selecting products from store shelves, by verbalizing the name, brand, and price of an item. In contrast to other systems in which product IDs are detected by readers that are built into instrumented shelves, in this implementation, the product IDs are detected via an iGlove (based on work from Intel Research) that contains the RFID reader. The proposed solution is said to have the benefit that it can also be used when putting items away at the customer's home provided the right infrastructure exists.

[41] take a different focus with their research into mobile shopping assistants and discuss on the basis of studies into people's grocery shopping habits, what an interface for mobile devices should actually look like. With this goal in mind, they design and evaluate prototypes and also perform usability tests within a true shopping environment. Based on user preferences on nine different spatial and contextual interface designs, they develop a user interface that is divided into three segments, the middle more prominent region consisting of a shopping list, the top consisting of a spatial map, and a promotional area at the bottom displaying revolving store specials.

17.4 Discussion

The previous section gives an impression of the broad range of application domains of adaptive mobile guides. Given the early state of the field and the emerging technologies that are yet to come, it seems obvious that mobile guides will have a considerable impact on our daily lives in the future. Figure 17.11 gives an overview on the most relevant features of the most prominent systems that we have discussed. The figure thus represents the current state of the art of adaptive mobile guides. When looking at the columns it seems that systems from the museum domain are the most advanced in terms of the degree of resource adaptivity (as discussed in section 17.1). Two out of four systems are resource adaptive, i.e. they are able to adapt their output to certain limited resource and one system is even resource adapting and applies different strategies to improve the quality of the adaptation process. In the class of navigation systems two approaches can be considered resource adapting (mainly because they use different strategies to determine a good fix for the location). The shopping guides are less resource adaptive. This might be due to the high amount of commercial system we have discussed. The commercial guides are more adapted towards a single dimension (e.g. the small screen size of the mobile device or a particular user preference). Two research systems of this class are however resource adaptive and demonstrate the future direction mobile shopping guides will take.

	System name	Type of Adaptivity	Positioning Technology	Knowledge Representation	# of Users	User Model / Properties	Social Context Adaptation	Presentation Metaphor	Technological Plattform
Museum Systems	SottoVoce	RAtiv	IR	DB	up to two users	Stereotypes	eavesdropping for two users	virtual museum	PDA
	AgentSalon	RAtiv	IR	DB	up to two users	none	Kiosk communication with up to two users	user centered virtual character	PDA + kiosk
	Peach	RAting	IR+WLAN+OR	GUMO	multiple users	UBISWORLD	Group presentations	role-based virtual character	PDA + kiosk
	Hippie	RAted	IR+EC	DB	one user	Stereotypes	none	textbook	Subnotebook
Navigation Systems	Guide	RAting	WLAN+SP	DB	one user	user preferences	none	guide leaflet	Notebook + PDA
	TellMaris	RAted	GPS	DB	one user	none	none	City map	Mobile phone
	LOLA	RAted	GPS+UMTS	DB	one user	user preferences	none	City map	Mobile phone
	BPN	RAting	GPS+IR	GUMO	one user	UBISWORLD	none	City map	PDA
Shopping Systems	IBM Shopping Buddy	RAted	none	DB	one user	none	none	Shopping list	Trolley + subnotebook
	Personal Shopping Assistant	RAted	none	DB	one user	user preferences	none	Shopping list	PDA
	MyGrocer	RAted	RFID	DB	one user	user preferences	none	Runs in the background	Mobile phone
	Adaptive Shopping Assistant	RAtive	RFID	GUMO	one user	UBISWORLD	none	Shopping basket and list	Trolley + subnotebook
	PromoPad	RAted	OR	DB	one user	none	none	Augmented Reality	TabletPC and camera
	Mobile Shop Assist	RAtive	RFID and IR	GUMO	one user	UBISWORLD	none	shopping shelf	PDA + headset
	SAvI	RAted	RFID	DB	one user	none	none	Audio interface	Mobile phone + rfid reader

Legend:
RAted = resource adapted
RAtiv = resource adaptive
RAting = resource adapting
IR=infrared
EC= Electronic Compass
WLAN=Wireless LAN

PDA = Personal Digital Assistant
UMTS = Universial Mobile Telecommunications Ssytems
DB = reational Database approach
OR=Optical Recognition
SP=Sell Positioning

Fig. 17.11. Comparison of the relevant features of adaptive mobile guides.

Positioning technology and algorithms and thus the adaptation to the current location is an important topic for now and for the future. Only two of the discussed systems do not look at the user's location. All the other systems rely more or less on this important piece of context information. Although orientation is an important information for most of the systems, it has to be retrieved differently in the respective domains. While museum guides adapt to users that walk rather slowly and tend to turn on the spot and therefore need explicit sensor technology to track orientation (e.g. an electronic compass or infrared beacons), outdoor navigation systems can use the GPS trajectory to infer the current user orientation. Also, these systems do not need fine grain resolution of orientation to provide proper assistance.

It is also interesting to note that most of the discussed systems use a self-designed database and only a couple of them make use of semantic web standards. However, it is very likely that we will see more and more systems using these standards. This will not only increase the degree of adaptivity of a single system, but also the degree of interoperability between several mobile guides. This will lead to mobile systems that learn across multiple domains. For example, a mobile shopping guide could take advantage of information gathered from the user in the museum domain. Such an integration of different services is an important future research topic.

Until now mobile guides are mostly targeted to a single user, although some of the discussed systems explicitly look at two or even more users at the same time. Given the high relevance of the social experience of many possible application domains we will see more and more socially aware systems that take into account other humans in the surroundings of the actual user.

It is also important to note that the technological platform of most systems is still the PDA, but more and more guides will appear for mobile phones. This will have a major impact on the distribution of commercial systems. Another trend that can be observed, is that more and more systems integrate mobile devices with devices that are installed in the infrastructure (e.g. large displays or projection devices). This is very much in line with the computing paradigm of ubiquitous computing and it seems plausible that research will continue in that direction.

17.5 Conclusion

In this chapter we have discussed a variety of aspects related to the design and implementation of mobile guides. We have highlighted that mobile guide applications differ significantly from traditional adaptive systems by their use in a specific and dynamic context. This includes not only aspects of their users that should be represented in a user model, but also characteristics of the environment and the particular situations of usage. We have briefly introduced UbisWorld, a web-based approach to formalize all the aspects that we regard as relevant for mobile applications. Finally, we have presented a review of mobile guide applications, coming from three different domains: museum guides, navigation systems and shopping assistants. This overview has shown that although commercial mobile prototypes do exist their ability to adapt to situations and the environment is rather limited.

To this extent, some of the benefits of the actual research efforts are already visible in the commercial marketplace in the form of web-based shopping agents, and in the form of instrumented shops consisting of smart trolleys and mobile hand-held devices. In the area of museum guides, rather simple commercial systems do exist that exploit the location of the visitor to provide information on exhibits nearby. In the future those systems will not only support single users, but whole groups of visitors (more details on the challenge to model groups can be found in chapter 20 of this book [50]), reflecting the fact that visiting a museum is also a social experience. This will require sophisticated group models and new forms of distributed and embodied interaction in the museum space. Commercial navigation systems are common place. However, similar to the museum guides, they usually just exploit the location of the user. In the future navigation systems will be interconnected and exchange information about the environment. Car navigation systems will broadcast traffic road information to following cars. Additional information, such as weather reports, and personal preferences will be incorporated to tailor instructions to the actual situation. Here it is apparent that the semantic integration of different services is of utmost importance.

When designing adaptive mobile systems it is important to keep in mind that these systems have to be able to process imprecise information. For example the localization of users is often only possible within certain error bounds. Inferences based on such imprecise data are therefore also subject to errors. To handle these problems, statistical methods are used to give statistical estimates of the likelihood of certain system predictions. Often used in the domain of user modelling are Bayesian Networks (BNs, an overview of systems can be found in [27]). BNs allow system designers to explicitly specify the structure of causes and effects for a given domain by the means of a network. The edges of this network represent the conditional probabilities between the observable states (e.g. the location of a user) and non-observable states (e.g. the user's goal). These probabilities can be fixed or learned over time. One advantage over a purely statistical approach to learning, such as a neural network [22], is that a BN allows incorporating expert knowledge during the design process. This in turn enables the system to provide users with explanations of the decision processes at runtime. Mobile adaptive systems that are able to process imprecise information will be much more flexible and robust. As we have seen, some approaches do exist, but they can be considered only a starting point for future research.

In general mobile applications of the future will be much more integrated into the infrastructure of the environment. Navigation and shopping guides will be able to use displays in the environment to guide or inform users. For example smart door displays will be used to navigate users through a building and projectors will be used to project information directly on products in a shop. This development will lead to a smoother interaction and will blur the difference between mobile and static parts of the application. Of course usability engineering methods have to be applied to ensure the proper ergonomics of such spatially distributed applications (more details on the topic of usability engineering for the adaptive web are provided in chapter 24 of this book [19]).

It seems feasible to assume that guide applications will run independently from the actual hardware. Instead of remaining on a mobile device, applications will be mobile themselves, migrating from device to device depending on the situation and the environment. Finally this will require a sophisticated representation and management of the user's attention. Without such a management the amount of mobile applications requesting the user's attention will be limited to a small amount. The details of such a component are still unclear and the subject of current research.

References

1. Abowd, D.A., Atkeson, C.G., Hong, J., Long, S., Kooper, R., Pinkerton, M.: Cyberguide: A Mobile Context-Aware Tour Guide. Wireless Networks **3**(5) (1996) 421–433
2. Anegg, H., Kunczier, H., Michlmayr, E., Pospischil, G., Umlauft, M.: LoL@: designing a location based UMTS application. Elektrotechnik und Informationstechnik **119**(2) (2002) 48–51
3. Aoki, P.M., Grinter, R.E., Hurst, A., Szymanski, M.H., Thornton, J.D., Woodruff, A.: Sotto voce: exploring the interplay of conversation and mobile audio spaces. In: CHI '02: Proceedings of the SIGCHI conference on Human factors in computing systems, New York, NY, USA, ACM Press (2002) 431–438
4. Asthana, A., Cravatts, M., Krzyzanowski, P.: An indoor wireless system for personalized shopping assistance. In: IEEE Workshop on Mobile Computing Systems and Applications. (1994)
5. Bahl, P., Padmanabhan, V.: Radar: An in-building rf-based location and tracking system. In: IEEE INFOCOM 2000. (2000)
6. Baus, J., Krüger, A., Wahlster, W.: A Resource-Adaptive Mobile Navigation System. In: IUI2002: International Conference on Intelligent User Interfaces, New York, ACM Press (2002) 15–22
7. Bohnenberger, T., Jameson, A., Krüger, A., Butz, A.: Location-aware shopping assistance: Evaluation of a decision-theoretic approach. In: 4th International Symposium on Mobile Human-Computer Interaction. (2002) 155–169
8. Brandherm, B., Schwartz, T.: Geo referenced dynamic bayesian networks for user positioning on mobile systems. In: Proceedings of the International Workshop on Location- and Context-Awareness (LoCA). LNCS 3479, Springer (2005)
9. Brandherm, B., Schmitz, M.: Presentation of a modular framework for interpretation of sensor data with dynamic Bayesian networks on mobile devices. In: LWA 2004, Lernen Wissensentdeckung Adaptivität, Humboldt-Universität zu Berlin, Germany (2004) 9–10
10. Butz, A., Baus, J., Krüger, A.: Augmenting buildings with infrared information. In: Proceedings of the International Symposium on Augmented Reality (ISAR), IEEE Computer Society Press (2000)
11. Chai, J., Horvath, V., Kambhatla, N., Nicolov, N., Stys-budzikowska, M.: A conversational interface for online shopping. In: first international conference on Human language technology research. (2000) 1–4
12. Cheverst, K., Davies, N., Mitchell, K., Friday, A., Efstratiou, C.: Developing a Context-aware Electronic Tourist Guide: Some Issues and Experiences. In: Proceedings of the 2000 Conference on Human Factors in Computing Systems (CHI-00), New York, ACM Press (2000) 17–24
13. Chittaro, L., Ranon, R.: Adaptive 3d web sites. In Brusilovsky, P., Kobsa, A., Nejdl, W., eds.: The Adaptive Web: Methods and Strategies of Web Personalization. Lecture Notes in Computer Science, Vol. 4321, Berlin Heidelberg New York, Springer-Verlag (2007) this volume.

14. Dolog, P., Nejdl, W.: Semantic web technologies for the adaptive web. In Brusilovsky, P., Kobsa, A., Nejdl, W., eds.: The Adaptive Web: Methods and Strategies of Web Personalization. Lecture Notes in Computer Science, Vol. 4321, Berlin Heidelberg New York, Springer-Verlag (2007) this volume.

15. Doorenbos, R., Etzioni, O., D., Weld: A scalable comparison-shopping agent for the worldwide web. In: First International Conference on Autonomous Agents. (1997) 39–48

16. Dourish, P.: What we talk about when we talk about context. Personal and Ubiquitous Computing 8(1) (2004) 19–30

17. Ebaugh, A., Chatterjee, S.: Savi: Shopping assistant for the visually impaired. Technical report, University of Washington (2004) http://www.cs.washington.edu/homes/sauravc/cs477/SAVi.pdf.

18. Falk, P., Campbell, C.: The Shopping Experience. SAGE Publications (1997)

19. Gena, C., Weinbelzahl, S.: Usability engineering for the adaptive web. In Brusilovsky, P., Kobsa, A., Nejdl, W., eds.: The Adaptive Web: Methods and Strategies of Web Personalization. Lecture Notes in Computer Science, Vol. 4321, Berlin Heidelberg New York, Springer-Verlag (2007) this volume.

20. Goren-Bar, D., Graziola, I., Pianesi, F., Zancanaro, M.: The influence of personality factors on visitor attitudes towards adaptivity dimensions for mobile museum guides. User Modeling and User-Adapted Interaction 16(1) (2006) 31–62

21. Harter, A., Hopper, A.: A distributed location system for the active office. IEEE Network 8(1) (1994) 62–70

22. Hasenjaeger, M., Ritter, H.: New learning techniques in computational intelligence paradigms. Active Learning in Neural Networks (2000)

23. Heckmann, D., Brandherm, B., Schmitz, M., Schwartz, T., von Wilamowitz-Moellendorf, B.M.: GUMO - the general user model ontology. In: Proceedings of the 10th International Conference on User Modeling, Edinburgh, Scotland, LNAI 3538: Springer, Berlin Heidelberg (Jun 2005) 428–432

24. Heckmann, D., Krüger, A.: A user modeling markup language (UserML) for ubiquitous computing. In: Proceedings of the 8th International Conference on User Modeling, Johnstown, PA, USA, LNAI 2702: Springer, Berlin Heidelberg (Jun 2003) 393–397

25. Hood, M.G.: Staying away: Why people choose not to visit museums. Museums News 61(4) (April 1983) 50–57

26. IBM: Stop and shop grocery drives sales and boosts customer loyalty with ibm personal shopping assistant. http://www.pc.ibm.com/store/products/psa/ (2004)

27. Jameson, A.: Numerical uncertainty management in user and student modeling. an overview of systems and issues. User Modeling and User-Adapted Interaction 5 (1996) 193–251

28. Jameson, A.: Systems That Adapt to Their Users: An Integrative Perspective. Department of Computer Science, Saarland University, Saarbrücken, Germany (Jun 2001)

29. Johnson, W., Ricket, J., Lester, J.: Animated pedagogical agents: Face-to-face interaction in interactive learning environments. International Journal of Artificial Intelligence in Education (2000) 47–78

30. Kobsa, A.: Privacy-enhanced web personalization. In Brusilovsky, P., Kobsa, A., Nejdl, W., eds.: The Adaptive Web: Methods and Strategies of Web Personalization. Lecture Notes in Computer Science, Vol. 4321, Berlin Heidelberg New York, Springer-Verlag (2007) this volume.

31. Kobsa, A.: Generic user modeling systems. User Modelling and User-Adapted Interaction Journal 11(1-2) (2001) 49–63

32. Kourouthanassis, P., Koukara, L., Lazaris, C., Thiveos, K.: Last-mile supply chain management: Mygrocer innovative business and technology framework. In: 17th International Logistics Congress on Logistics from A to U: Strategies and Applications. (2001) 264–273

33. Kray, C., Laakso, K., Elting, C., Coors, V.: Presenting route instructions on mobile devices. In Johnson, W.L., André, E., Domingue, J., eds.: Proceedings of IUI 03, Miami Beach, FL, ACM Press (2003) 117–124

34. Krüger, A., Butz, A., Stahl, C., Wasinger, R., Steinberg, K., Dirschl, A.: The Connected User Interface: Realizing a Personal Situated Navigation System. In: IUI2004: International Conference on Intelligent User Interfaces, New York, ACM Press (2004) 161–168

35. Krüger, A., Wasinger, W.: Multi-modal interaction with mobile navigation systems. it - Information Technology 46(6) (2004) 322–331

36. Krumm, J., Cermak, G., Horvitz, E.: Rightspot: A novel sense of location for a smart personal object. In: Proceedings of Ubicomp 2003. (2003) 36–43

37. Kruppa, M., Heckmann, D., Krüger, A.: Adaptive multimodal presentation of multimedia content in museum scenarios. KI Journal 1 (Jan 2005) 56–59

38. Laakso, K.: Evaluating the use of navigable three-dimensional maps in mobile devices. Master's thesis, Helsinki University of Technology, Helsinki (November 2002)

39. Lehtonen, T.K., Mäenpää, P.: Shopping in the east centre mall. In: The Shopping Experience. SAGE Publications (1997) 136–165

40. Menczer, F., Street, W.N., Vishwakarma, N., Monge, A., Jakobsson, M.: Intellishopper: a proactive, personal, private shopping assistant. In: Proc. of the first international joint conference on Autonomous agents and multiagent systems. (2002) 1001–1008

41. Newcomb, E., Pashley, T., Stasko, J.: Mobile computing in the retail arena. In: SIGCHI conference on Human factors in computing systems. (2003) 337–344

42. Oppermann, R., Specht, M.: A nomadic information system for adaptive exhibition guidance. In: Proceedings of ICHIM99 conference. (1999) 103–109

43. Pease, A., Niles, I., Li, J.: The suggested upper merged ontology: A large ontology for the semanticweb and its applications. In: AAAI-2002Workshop on Ontologies and the Semantic Web. Working Notes (2002) http://projects.teknowledge.com/AAAI-2002/Pease.ps.

44. Pospischil, G., Umlauft, M., Michlmayr, E.: Designing LoL@, a Mobile Tourist Guide for UMTS. In Paterno, F., ed.: Proceedings of Mobile Human-Computer Interaction 2002, Berlin, Heidelberg, New York, Springer (2002) 140–154

45. Priyantha, N., Chakraborty, A., Balakrishnan, H.: The cricket location-support system. In: In Proceedings of Proc. 6th Ann. Int'l Conf. Mobile Computing and Networking (Mobicom 00), ACM Press (2000) 32–43

46. Rist, T., Baldes, S., Gebhard, P., Kipp, M., Klesen, M., Rist, P., Schmitt, M.: Crosstalk: An interactive installation with animated presentation agents. In: 2nd Conference on Computational Semiotics for Games and New Media. (2002) 61–67

47. Rocchi, C., Stock, O., Zancanaro, M., Kruppa, M., Krüger, A.: The museum visit: generating seamless personalized presentations on multiple devices. In: IUI '04: Proceedings of the 9th international conference on Intelligent user interface, New York, NY, USA, ACM Press (2004) 316–318

48. Schilit, B., Adams, N., Want, R.: Context-aware computing applications. In: Workshop on Mobile Computing Systems and Applications. IEEE (1994)

49. Schneider, M.: Towards a transparent proactive user interface for a shopping assistant. In: Workshop on Multi-User and Ubiquitous User Interfaces (MU3I) at Intelligent User Interfaces 03. (2003) 10–15

50. Smyth, B., Jameson, A.: Recommending groups. In Brusilovsky, P., Kobsa, A., Nejdl, W., eds.: The Adaptive Web: Methods and Strategies of Web Per- sonalization. Lecture Notes in Computer Science, Vol. 4321, Berlin Heidelberg New York, Springer-Verlag (2007) this volume.

51. Stock, O., Zancanaro, M., eds.: PEACH: Intelligent Interfaces for Museum Visits. Cognitive Technologies Series. Springer (2006)

52. Sumi, Y., Mase, K. In: Interface Agents that facilitate knowledge interactions between community members. Springer-Verlag (Cognitive Technologies series) (2004) 405–427
53. Wahlster, W., Tack., W.: Sfb 378: Ressourcenadaptive kognitive prozesse. In Jarke, M., ed.: Informatik 97: Informatik als Innovationsmotor. Springer (1997) 51–57
54. Wasinger, R., Krüger, A., Jacobs, O.: Integrating intra and extra gestures into a mobile and multimodal shopping assistant. In: 3rd International Conference on Pervasive Computing. (2005) 297–314
55. Weiser, M.: The computer of the 21st century. Scientific American (1991) 94–100
56. Xuehai, B., Abowd, G., Rehg, J.: Using sound source localization in a home environment. In: Proceedings of Pervasive Computing 05, Springer (2005)
57. Zhu, W., Owen, C.B., Li, H., Lee, J.H.: Personalized in-store e-commerce with the promopad: an augmented reality shopping assistant. electronic Journal for E-Commerce Tools and Applications (eJETA) 1(3) (2004)

18

Adaptive News Access

Daniel Billsus[1] and Michael J. Pazzani[2]

[1] FX Palo Alto Laboratory, 3400 Hillview Ave., Bldg. 4,
Palo Alto, CA 94304 , USA
billsus@fxpal.com
[2] Rutgers University, ASBIII, 3 Rutgers Plaza
New Brunswick, NJ 08901
pazzani@rutgers.edu
pazzani@rutgers.edu

Abstract. This chapter describes how the adaptive web technologies discussed in this book have been applied to news access. First, we provide an overview of different types of adaptivity in the context of news access and identify corresponding algorithms. For each adaptivity type, we briefly discuss representative systems that use the described techniques. Next, we discuss an in-depth case study of a personalized news system. As part of this study, we outline a user modeling approach specifically designed for news personalization, and present results from an evaluation that attempts to quantify the effect of adaptive news access from a user perspective. We conclude by discussing recent trends and novel systems in the adaptive news space.

18.1 Introduction

The World Wide Web has had a profound impact on our everyday lives: we routinely rely on it as a ubiquitous source for timely information. In particular, the web's real-time and on-demand characteristics make it an ideal medium for news access anywhere and anytime. As a result, virtually every news organization now has a presence on the World Wide Web.

In addition to transforming how traditional news organizations distribute information, the web has enabled new forms of information dissemination. The web provides an audience and a platform for individuals to express themselves or engage in discussions. Weblogs, or *blogs*, on thousands of topics contributed by thousands of individuals have created a rich information landscape of gigantic proportions.

While the availability of continuously updated news content provides great value, it represents yet another facet of the, now omnipresent, information overload problem. How can individuals find the most interesting or relevant news content? How can they discover trusted news organizations or find relevant blog posts among thousands of choices?

Adaptive web technology provides the basic building blocks to address these challenges. We now know how to build tools that help people discover relevant content,

P. Brusilovsky, A. Kobsa, and W. Nejdl (Eds.): The Adaptive Web, LNCS 4321, pp. 550–570, 2007.

route the right information to the right people at the right time, or help aggregate content from thousands of sources.

In this section, we show how some of the adaptive web technologies discussed in previous chapters have been applied to adaptive news access. First, we provide an overview of different types of adaptivity in the context of news access and identify corresponding algorithms. For each adaptivity type, we briefly discuss representative systems that use the described techniques. Next, we discuss an in-depth case study of a personalized news system. As part of this study, we outline a user modeling approach specifically designed for news personalization, and present results from an evaluation that attempts to quantify the effect of adaptive news access from a user perspective. We conclude by discussing recent trends and systems in the adaptive news space.

18.2   Types of Adaptive News Access

The main goal of adaptive news techniques is to facilitate access to relevant news content. This goal can be achieved in several different ways. In this section, we take a closer look at the following types of adaptivity:

- *News Content Personalization.* Systems that personalize content help users find personally relevant news stories based on a model of the user's interests. These systems can recommend or automatically rank stories, so that the most relevant content is easier to find.
- *Adaptive News Navigation.* Adaptive navigation assists the user in navigating to the most frequently read sections of a news site.
- *Contextual News Access.* Contextual news access techniques provide users with news content on the basis of currently viewed information.
- *News Aggregation.* Automated aggregation and classification of news content helps users identify ongoing or emerging news topics, and assists in accessing coverage of a specific topic by multiple providers. In contrast to the adaptivity types listed above, news aggregation does not necessarily enable personalization, but automated aggregation commonly exhibits adaptive behavior: dynamically generated aggregator pages, e.g. Google News, automatically adapt to the current news landscape and provide an indication of emerging topics or trends.

While this chapter focuses on adaptive news access, it is important to point out that the analysis of news content for purposes other than adaptivity has been the focus of much research in the information retrieval and machine learning communities. For example, Topic Detection and Tracking (DTD) is a DARPA-sponsored initiative to investigate techniques for finding and following new events in streams of broadcast news stories ([35] provides an overview of the DTD sub-tasks and corpora). Clearly, research results from these areas are often directly applicable to adaptive news access techniques.

18.2.1   News Content Personalization

This section focuses on adaptive techniques that model the user's interests based on explicit or implicit feedback, and use the resulting user models to personalize news content. Collecting user feedback explicitly or implicitly has been the focus of much research. See Chapter 2 of this book [13] for an in-depth discussion of various feedback methods. In addition, the value and accuracy of implicit user feedback have been studied in the context of news personalization ([2], [21], [17], [18]).

The adaptive techniques described in this chapter are in contrast to static content customization via user-provided interest profiles. For example, many major news sites, e.g. Yahoo! News or Google News, allow users to customize the news categories to be included on the front page, or allow users to indicate interesting topics via web questionnaires. To distinguish between these two approaches, we refer to user-defined news profiles as *customization* and adaptive techniques as *personalization*.

Previous sections of this book discussed techniques that model users' individual interests to personalize content (see Chapters 2, 3, 9 and 10 of this book [13], [26], [30], [27]). Many of the described modeling and recommendation techniques apply directly to news personalization. However, news access has several characteristics that make some approaches better suited to the problem than others. In this section, we describe how news access differs from other personalization tasks and suggest a set of techniques that are particularly appropriate for this domain. The following characteristics are important factors for the design of adaptive approaches to news personalization.

Dynamic Content. News content is more dynamic than many other content types, such as movies, music or books. News stories are released and updated continuously, and many stories only remain online for a short period of time until new details emerge. This makes content-based methods better suited to news personalization than collaborative methods. As discussed in Chapter 9 of this book [30], collaborative filtering is based on using the ratings of like-minded users as predictors for relevant content. However, collaborative filtering often suffers from the "sparse matrix" problem, and this limitation is particularly noticeable for news access. For example, ratings from users who have accessed a story can often not be used as predictors for the current user because of very limited rating overlap. Related to this issue, collaborative filtering approaches suffer from a "latency" problem, i.e. depending on a site's popularity and traffic, it may take some time for news stories to receive enough user feedback to lead to accurate recommendations. In contrast, content-based methods predict the user's interest using text alone, and do not depend on the availability of ratings. In addition, the benefits of collaborative filtering, i.e. being able to take advantage of qualitative human judgments, are often not critical for services that serve news from a single provider. Once the user has selected a content provider he or she trusts and agrees with ideologically, the selection of relevant stories is primarily an issue of content and not quality or style. However, collaborative methods can certainly be very useful for services that attract a large number of users or aggregate stories from multiple providers. For example, the *GroupLens* system is well known for its application of collaborative filtering techniques to Usenet news [20].

Changing Interests. Since new news topics emerge continuously, users' interests tend to change frequently. From an algorithmic perspective, this calls for methods that can quickly adjust to changing target concepts. For example, a system that uses a machine learning algorithm to learn a model of the user's interests (as discussed in Chapter 10 of this book [27]), should be based on learning algorithms that can quickly adjust to changing interests, so that the user does not have to provide a large number of training examples until the system discovers the interest change. More generally, the notion of changing target concepts that must be tracked algorithmically is known as concept drift. The machine learning literature discusses many approaches specifically designed to address this problem. These techniques are often based on algorithms that can either explicitly detect changing concepts or limit the effects of concept drift via windowing techniques that only consider a temporally constrained subset of the available data [36], [19]. In addition, learning algorithms that require only a few training examples to approximate useful models can quickly pick up new user interests. For example, instance-based algorithms such as the k-nearest-neighbor algorithm fall into this category ([10], Chapter 10 of this book [27]).

Multiple Interests. Users are usually interested in a broad range of different news topics. This means that a user modeling approach for news must be capable of representing multiple topics of interest. If the user model is inferred by a machine learning algorithm, k-nearest-neighbor methods are a good choice to address this issue. Suppose the user model is based on the k-nearest-neighbor-based representation described in Chapter 10 of this book [27]: in this case, the user model contains labeled news stories, e.g. interesting vs not interesting, where each story is represented as a vector in the Vector Space Model [33]. A previously unseen story can now be classified using the labels of the k most similar stories in the user model. Since the classification only depends on these k nearest neighbors, stored news stories that are "further away" from the story to be classified and likely reflect the user's interests in different topics, do not influence the classification. As a result, k nearest neighbor methods lend themselves to modeling multiple disjoint areas of user interests.

Novelty. A news story is usually considered most interesting if it conveys information the user does not yet know. This has interesting algorithmic implications and further differentiates news access from other domains where finding "more of the same" can be a good thing. Algorithms that keep track of information the user has previously accessed can avoid this problem by selecting content that is similar, but not identical to, previously accessed information. For example, Newsjunkie [12] is a system that personalizes news for users by identifying the novelty of stories in the context of stories they have already reviewed. The system uses novelty-analysis algorithms that analyze inter- and intra-document dynamics by considering how information evolves over time from article to article, as well as within individual articles.

Avoiding Tunnel Vision. Clearly, personalization should not prevent the user from finding important novel information or breaking news stories. Adaptive news systems can overcome this issue by optionally integrating editorial input into the recommendation algorithm, or by explicitly boosting the diversity of stories presented to the user (see Section 18.3 for an example and more details).

Editorial Input. Adaptive news systems attempt to identify a set of stories deemed most "relevant" within a large set of potential candidate stories. Traditionally, this has been the job of a news editor, i.e. a person who decides which stories to include in a paper (or site), and how to order and categorize them. The advent of adaptive news technology does not imply that human news editors are not needed anymore. Ranking stories by their perceived importance is hard to automate and, ideally, the input from human editors should inform the automated selection of content via learned user models. In addition to obvious benefits from a user perspective, retaining editorial input is an important feature for news organizations that are interested in deploying personalization technology: loss of control over the content that users will get to see does not appeal to news organizations. Retaining editorial input can be achieved via simple methods, such as factoring the position of a news story (as a measure of its relative importance) into recommendation algorithms, or by specifying selection rules that ensure the user will always get to see the top n stories, regardless of the user's interest model [5]. This is an issue that primarily concerns news personalization for individual news organizations. For systems that aggregate news content from multiple providers this is less of an issue, and, in fact, one goal of such aggregation systems can be to overcome editorial bias.

Brittleness. A single action, such as selecting something accidentally or skipping over an article on a topic (perhaps because one heard about the details on the radio or the wireless connection is dropped) should not have a drastic or unrecoverable effect on the model of the user's interests.

Availability of Meta-Tags. News personalization algorithms can usually not rely on the availability of meta-tags. A process that requires the content provider to do extra work by hand, such as adding meta-tags or category labels, is not feasible with thousands of new items being added daily.

In Section 18.3, we outline one particular algorithm that addresses most of the characteristics identified above in more detail, and we highlight results from an experimental evaluation. In addition to the algorithm described in Section 18.3 and the examples listed above, there are many alternative ways to address the issues identified in this section, and numerous studies that discuss various facets of content-based news personalization have been published. For example, early work by Bharat et al. [2] focuses on a dynamic user interface for personalized web-based news access: according to the authors, the *Krakatoa Chronicle* was the first newspaper on the World Wide Web to provide a layout similar to that of real-world newspapers. The system is highly interactive, supports article layout customization, and provides a content-based personalization approach that allows users to control the extent to which public and personal interests affect the selection of articles. Closely related to this work, Sakagami and Kamba [29] describe the *Anatagonomy* system, a research prototype of a personalized online newspaper. This work mainly focuses on the utility of implicit vs explicit user feedback. For example, *Anatagonomy* tracks user interactions such as selecting and enlarging articles or scrolling through articles, and interprets these interactions as implicit feedback. Similarly, Balabanovic [1] describes *Slider*, a user interface specifically designed to capture users' preferences implicitly, in the context of content-based news personalization. The *Slider* interface is based on a set of on-screen panels that users can associate with topics they find interesting. At any time, users can create new panels or delete old panels in order to reflect their changing interests. When new stories arrive, users can optionally "slide" these stories onto a panel, and thereby implicitly define the topic associated with the panel. News stories are represented in the Vector Space model [33], and for each panel, the system constructs one prototype vector by averaging over all documents contained in the panel. In subsequent sessions, the system attempts to locate news stories that are similar to these prototype vectors.

In general, purely statistical Information Retrieval techniques (such as the Vector-Space-model-based approaches described in this chapter) are commonly used as the underlying foundation for content-based news personalization approaches. However, there has also been work that examines the utility of linguistic information in the context of news personalization. For example, Magnini and Strappavara [23] explore the utility of word-sense information for user profile acquisition, and report a tangible increase in recommendation accuracy, compared to a purely statistical approach. Additional content-based news personalization studies of interest include [14], [21], [16], [32] and [22].

Finally, it is important to point out that news content personalization systems have not only created academic interest, but are starting to become publicly available as part of commercial news services. An early example of a publicly available personalized news service that automatically learns from users' access patterns is Findory *(http://www.findory.com)*. For each individual user, Findory tracks accessed stories and uses this information to generate a personalized front page. For example, the page

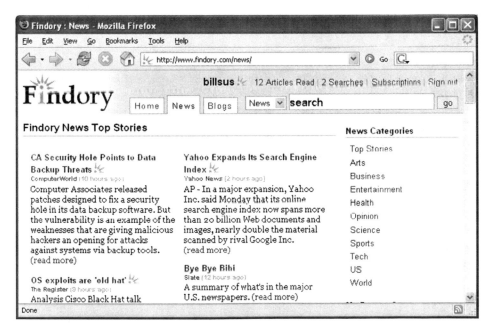

Fig. 18.1. Findory's adaptively personalized front page. Recommended stories are annotated with a "sun" icon.

shown in Figure 18.1 is the result of selecting several stories about security issues and search technology (recommended stories are annotated with a "sun" icon). While the exact details of the personalization algorithm are proprietary, the Findory web site states that the "personalization algorithm combines statistical analysis of the article's text and behavior of other users with what we know about articles you have previously viewed", which suggests that the system is based on a combination of content-based and collaborative methods.

18.2.2   Adaptive News Navigation

Similar to personalization based on content profiles, the goal of adaptive navigation is to simplify access to relevant content. However, instead of finding individual news stories that match the user's interest profile, this technique focuses on analyzing the user's access patterns to determine the position of menu items within a menu hierarchy. For example, a user who frequently accesses the technology section of a news site, but never reads sports stories, would probably prefer to see the technology category along with individual technology stories high up on the front page of a personalized news site. This personalization approach is particularly effective for mobile applications on PDAs and cell phones, because of the limited screen space of these devices. For example, Smyth and Cotter [31], describe an algorithm that personalizes the menu hierarchy of a mobile application based on menu access frequencies that are maintained for each individual user. The system estimates the probability that a user will select option o given that it is included in menu m, and uses these probabilities to construct menus that are most likely to contain options the user will select. An empirical evaluation of this approach applied to a mobile portal showed that, on average, the number of menu-select and scroll operations was reduced by over 50%, leading to a much improved user experience.

Since this approach does not use any content and is primarily geared towards adaptive menu reordering, it does not lend itself to recommendations for individual news stories. However, the simplicity of adaptive menu reordering is also its greatest strength: it does not require complex infrastructure that maintains large content-based profiles for individual users, which means that it is much easier to deploy and satisfy real-world scalability requirements than more complex techniques. In addition, the approach does not require a lengthy training period, leading to significant usability improvements even after only a single user session.

While Smyth and Cotter [31] demonstrate the approach in the context of a mobile portal, it can be applied to any application with long or complex menu hierarchies. For example, promoting a user's favorite news categories close to the top of a personalized news site is a useful application of this approach, followed by several mobile news sites. For example, at the time of this writing, the San Diego Union Tribune maintained an adaptive news site for mobile devices that reorders its news categories based on access frequencies at *http://go.sosd.com*.

Clearly, adaptive navigation and content personalization are not mutually exclusive. A combined approach could provide access to the user's favorite news categories based on access probabilities, and each accessed news category could contain a personalized set of stories based on a content-based user profile.

18.2.3 Contextual Recommendations

Contextual recommendations, sometimes referred to as just-in-time retrieval, are closely related to content-based personalization [28]. However, instead of using a model of the user's interests learned over time, the approach draws on currently displayed information, such as a web page or email message, as an expression of the user's current interests. For example, a contextual news recommender could recommend a news story about a certain company when the user visits a web page that contains information about the company. Likewise, a contextual news recommender could recommend a news story about an actor when the user receives an email message from a friend that mentions the actor's name.

Contextual recommenders typically operate as follows. First, the system extracts textual information currently displayed on the user's screen. This can be accomplished via plug-ins for commonly-used applications, such as web browsers, email clients or word processors. Alternatively, web proxies can be used to access the text currently displayed in the user's browser. The extracted text is then used to retrieve related content, such as related news stories. This step of the process is often based on statistical text processing: using statistical term-weighting techniques ([33], Chapter 5 of this book [25]) to identify informative terms, the text can be used to automatically construct a query which can subsequently be sent to a search engine to retrieve related content. In addition to statistically determined query terms, Natural Language Processing approaches can be used to assist in the query generation process. For example,

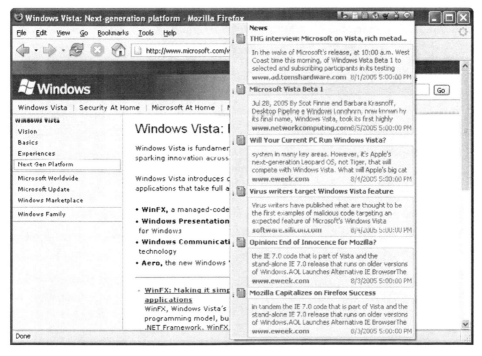

Fig. 18.2. Blinkx recommends news stories based on contextual information. In this example, Blinkx recommends news stories related to a web page the user is viewing.

a named-entity-tagger [7] can help identify names of companies, people or products for inclusion in the automatically generated query. *Watson* [8] and *FXPAL Bar* [6] are examples of this type of system. Both systems are implemented as toolbars that run within web browsers, email clients or other applications, and communicate the availability of related content via subtle icon cues, such as displaying a light bulb or changing the color of a button. Similarly, *Blinkx* is a publicly available contextual recommender (*http://www.blinkx.com*). Like *Watson* and *FXPAL Bar*, *Blinkx* is a toolbar that proactively recommends content in response to changing information on the user's screen. In addition to web pages, products and TV content, *Blinkx* recommends news stories. In the scenario shown in Figure 18.2, the user is viewing a web page about the next version of Microsoft Windows. The *Blinkx* bar in the upper right corner of the screen indicates the availability of related news stories via a small paper icon. When the user clicks on this icon, a list of closely related news stories appears (about Microsoft's new operating system, in this example).

18.2.4   News Aggregation

The adaptivity types discussed in the sections above focus on inferring users' interests based on the browsing history or currently viewed information. In contrast, news aggregators are services that automatically aggregate content from many different news sources and, as a result, adapt to the current news landscape as a whole. A technical trend that has significantly contributed to the emergence of news aggregators and other news-related services is the widespread use of RSS (Really Simple Syndication) feeds. A news or blog provider can publish RSS feeds, i.e. XML documents that provide links to currently available content, to simplify syndication. The XML excerpt below shows an example of a simple RSS feed.

```xml
<?xml version="1.0"?>
<rss version="2.0">
  <channel>
    <title>My Gaming News</title>
    <link>http://mygamingnews.com/</link>
    <description>Gaming News</description>
    <item>
      <title>600,000 Xbox 360 units sold in US</title>
      <link>http://mygamingnews.com/story1.html</link>
      <description>LOS ANGELES -- Microsoft Corp.
        has sold 600,000 of its new XBox 360
        videogame consoles …
      </description>
    </item>
    <item>
      <title>'Grand Theft Auto' slapped with lawsuit
      </title>
      <link>http://mygamingnews.com/story2.html</link>
    </item>
  </channel>
</rss>
```

For example, *Google News* (*http://news.google.com*) is a popular news aggregation service. The site currently collects stories from more than 4,500 news sources in English, automatically identifies common topics, ranks these topics by estimated "importance" (measured in terms of recency and volume), and then generates a new front page. This means that *Google News* adaptively generates a front page without explicit editorial input from a human editor, i.e. the aggregation acts as an unbiased news editor. While the exact details of the topic identification and ranking algorithm are proprietary, a news aggregation service can be implemented using statistical term weighting and text similarity techniques to automatically assess the similarity of any two stories ([33], Chapter 5 of this book [25]). In addition, text categorization approaches (see Chapter 10 of this book [27]) can be used to train classifiers that automatically categorize news stories from different providers into a set of news categories.

In addition to support for full-text search of news articles from multiple providers, news aggregation services such as *Google News* provide value by identifying topics that are generally considered important by a large number of news providers. In addition, aggregators allow users to compare coverage of a story between different providers, leading to a greater variety of perspectives than one single organization can offer.

In addition to *Google News,* there are a variety of aggregation services that support similar features. For example, *Topix* (*http://www.topix.net*) is similar to *Google News*, but emphasizes news categorization and geo-coding. The service automatically labels stories with the location of events, and also categorizes stories based on detected entities such as company names, celebrities or sports teams to enable fine-grained story categorization.

Fig. 18.3. Google News' automatically generated front page. The service aggregates news content from 4,500 sources.

18.3 Case Study

In this section, we present a case study of a personalized news service that was first released in 1999, powering the mobile version of various publicly available news services for several years [3], [4], [5]. We first describe the goals, design and user interface of the system. Next, we introduce the system's personalization algorithm – a machine learning approach specifically designed for adaptive news access. Finally, we summarize results from an empirical evaluation of the service (for additional details, see [3]).

While the study presented in this section focuses on the design and evaluation of only one individual system, numerous other studies that explore the design and utility of adaptive access to news, usenet or blog content can be found in the literature (see Section 18.2 and [14], [23], [22], [34], [11], [21], [16], [32], [2], [20]).

18.3.1 Adaptive News Personalization for Mobile Content Access

While personalization has proved to be an important supplement to web applications, the constraints of mobile information access make personalization essential to producing usable applications. Mobile devices, such as cell phones or PDAs (personal digital assistants), have much smaller screens, more limited input capabilities, slower and less reliable network connections, less memory and less processing power than desktop computers. In this section, we briefly summarize the main features of an adaptive news service for mobile content delivery that automatically infers the user's interests based on previously accessed content and personalizes content accordingly.

The news system dynamically generates a user interface that can be rendered on PDAs and cell phones. The interface displays section names (such as 'Sports'), headlines and articles. It is intended to be used by a single news site to deliver its content to readers of that site, rather than aggregating news across multiple sites. Personalization reorders sections of the news site so that the most frequently accessed sections may be reached without scrolling, reorders headlines within a section so that the most personally relevant items are displayed toward the top, and selects headlines for display on the front page.

Figure 18.18.4 illustrates how the system adapts to two different users. Both are shown the same three headlines initially. On the top row, a user reads a college football story and when the next page of headlines is requested, additional college football stories are shown. In the bottom row, a user instead reads a horse-racing story and is shown additional stories on this topic. In each case, a golf story is included on the next page, both to allow some diversity in the stories and to present the system with the opportunity to learn more about the user.

18.3.2 Learning User Models for News Access

The server that powers the described system uses a machine learning approach to automatically learn a simple model of each user's individual interests. The algorithm is a content-based approach specifically designed for news access, and addresses most of the news-specific personalization issues identified in Section 18.2.1. In short, a combination of similarity-based methods, e.g. [10], and Bayesian methods, e.g. [11], achieves the right balance of learning and adapting quickly to changing interests while avoiding brittleness. These two algorithms form a multi-strategy learning approach that learns two separate user-models: one represents the user's short-term interests, the other represents the user's long-term interests. Distinguishing between short-term and long-term models has several desirable qualities in domains with temporal characteristics [9]. Learning a short-term model from only the most recent observations may lead to user models that can adjust more rapidly to the user's changing interests. The need for two separate models can be further substantiated by the specific task at hand, i.e. classifying news stories. Users typically want to track different "threads" of ongoing recent events - a task that requires short-term information about recent events. For example, if a user has indicated interest in a story about a current Space Shuttle mission, the system should be able to identify follow-up stories and present them to the user during the following days. In addition, users have general news preferences, and modeling these general preferences may prove useful for deciding if a new story, which is not related to a recent rated event, would interest the user. With respect to the Space Shuttle example, we can identify some of the characteristic terminology used in the story and interpret it as evidence for the user's general interest in technology and science related stories.

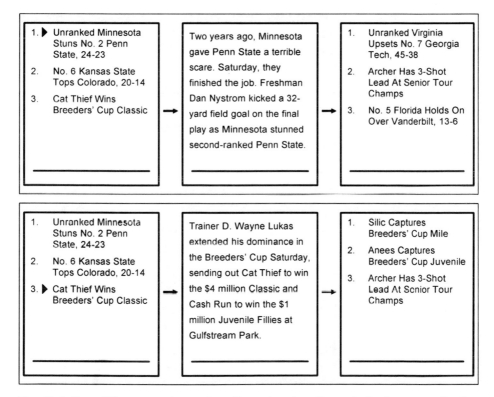

Fig. 18.4. Two different user interactions illustrating the effects of adaptive personalization. The top-row user is interested in college football, the bottom-row user is interested in horse racing. The system automatically adapts to the users' respective interests.

The purpose of the short-term model is two-fold. First, it contains recently read stories, so that other stories which belong to the same event thread can be identified. Second, it allows for identification of stories that the user already knows. A natural choice to achieve the desired functionality is the k-nearest-neighbor algorithm (Chapter 10 of this book [27]). To apply the algorithm to news stories, standard IR techniques, such as *tf-idf* term vectors and the cosine similarity measure are used ([33], Chapter 5 of this book [25]): news stories that were accessed or skipped by the user are represented as term vectors that are labeled as "interesting" or "not interesting", and the resulting set of stories is used to classify previously unseen news content. The model size is limited to the n most recent stories, so that the model remains dynamic and only reflects the user's most recent interests. To make sure that the user does not repeatedly see stories that are virtually identical to previously read content, the system artificially reduces the scores of stories that exceed a specified similarity threshold to at least one of the accessed stories in the user model. If a story to be classified does not have any near neighbors, the story cannot be classified by the short-term model at all, and is passed on to the long-term model. The nearest-neighbor-based short-term model is a reasonable choice for news recommendation, because it is able to represent a user's multiple interests, and can quickly adapt to new or changing interests. For example, a single story of a new topic is enough to allow the algorithm to identify future follow-up stories.

The system's long-term model is intended to model a user's general preferences. Since most of the words appearing in news stories are not useful for this purpose, the system periodically selects an appropriate vocabulary for each individual news category from a large sample of stories. After feature selection, the same set of features is used for all users. The goal of the feature selection process is to select informative words that recur over a long period of time. In this context, an informative word is one that distinguishes documents from one another, and can thus serve as a good topic indicator. With respect to individual documents, *tf-idf* weights can be interpreted as a measure of the amount of information that an individual word contributes to the overall content of a document. In order to determine the n most informative words for each document, the system sorts words with respect to their *tf-idf* values and selects the n highest-scoring words. A word is a useful feature for the long-term model if it frequently appears in top n lists over a large set of documents from one category (the system uses the most recent 10,000 news stories per category for feature selection). The system sorts all words that appear in the overall vocabulary with respect to the number of times they appear in top n lists. Finally, the k most frequent words are selected. This approach performs well at selecting the desired vocabulary: it selects words that occur frequently throughout one news category, but are still informative as measured by their *tf-idf* weights. For example, the following list shows the top 50 long-term features selected from a set of 10,000 science news stories:

```
drug, cancer, space, cells, patients, women, crops,
gene, launched, disease, food, virus, rocket, city,
mission, bacteria, infection, children, heart, hiv,
satellite, eclipse, blood, genetic, suns, winds,
trial, mice, orbit, antibiotics, vaccine, resistance,
russian, human, aides, storm, percent, brain, fda,
cdc, mosquitoes, energy, test, damage, hurricane,
computer, baby, government, hospital, texas.
```

The selected features are used as part of a probabilistic learning algorithm, a naïve Bayesian classifier ([11], Chapter 10 of this book [27]), to assess the probability of stories being interesting, given that they contain a specific subset of features. Each story is represented as feature-value pairs, where features are the words from the selected feature set that appear in the story, and feature values are the corresponding word frequencies. In order to take advantage of the word frequency information, the system uses a multinomial version of naïve Bayes [24].

To predict whether a user would be interested in a news story, the system applies the two models sequentially. It uses the short-term model first, because it is based on the most recent observations only, allows the user to track news threads that have previously been rated, and can label stories as already known. If a story cannot be classified with the short-term model, the long-term model is used. In addition to the user model's prediction, editorial input is incorporated by boosting the priority of lead stories. The effect of this boosting is that first-time users of the wireless news site see articles in default order (determined by an editor), and all users always see the lead story in each section. This also allows the adaptive personalization engine to learn more about each user. Finally, the similarity-based methods are also used to ensure that a variety of news articles are presented on each screen in much the same way that a newspaper does not fill the front page with articles on the same topic.

18.3.3 Evaluation

The personalization server reorders news stories with respect to users' individual interests. The main intuition is that such a modified order helps users access relevant content. However, information is rarely presented in random order. For example, editors prioritize news stories based on human judgment, which means that, in this case, users access content in an order deemed appropriate by human professionals. While such an order is static in the sense that it is the same for every user, it is possible that it is sufficient for most users to easily access relevant content. In this section, we briefly summarize the results from two experiments that compare personalized information access provided by the described personalization server to static information access. These results show that the system's user modeling algorithm generates an adaptive order that has two closely related effects: it simplifies locating relevant content and leads to an overall increase in accessed information. The main idea underlying both experiments is to present items either in static or adaptive order so that resulting differences in users' selection and browsing behavior can be quantified. The experiments described in the following sections were performed as part of a publicly deployed news service for handheld devices, such as cell phones and PDAs with wireless Internet access [3]. Since the experiments did not require any user interface changes, all collected data is based on normal system usage by regular users who did not know that their news access patterns contributed to the system's evaluation.

The system's ultimate goal is to simplify access to interesting content. A simple and informative measure that quantifies progress towards this goal is the average display rank of selected stories. If the system successfully learned to order items with respect to users' individual interests, this would, on average, result in interesting stories moving toward the top of users' personalized lists of items. Therefore, the average display rank quantifies the system's ability to recommend interesting news stories. Since this measure does not depend on a predicted numeric score or classifying news stories based on predicted interest levels, it is possible to apply it to static information access, allowing for a comparison of both strategies. Both experiments quantify the system's personalization performance using this measure.

The "Alternating Sessions" Experiment. The "alternating sessions" experiment quantifies the difference between static and adaptive information access by randomly determining whether a user receives content in static or adaptive order. During a period of two weeks, the server used its user modeling approach for approximately half of the users, while the other half received news stories in static order determined by an editor at the news source. On odd days, users with odd account registration numbers received news in personalized order and even users received a static order. On even days, this policy was reversed. To quantify the difference between the two approaches, the server recorded the mean rank of all selected stories for the personalized and static operating modes. Since a difference between static and adaptive access can only be determined for users that previously retrieved several stories, the analysis was restricted to users with a minimum of five selected stories. Comparing both access modes for this subset of users revealed a significant difference. The average display rank of selected stories was 6.7 in the static mode and 4.2 in the adaptive mode (based on 50 users that selected 340 stories out of 1882 headlines). The practical implications of this difference become apparent by analyzing the distribution of selected stories over separate headline screens (every screen contains 4 stories). Figure 18.5 illustrates these two distributions. In the static mode, 68.7% of the selected stories were on the top two headline screens, while this was true for 86.7% of the stories in the personalized mode. It is reasonable to argue that this makes a noticeable difference when working with handheld devices. In addition, this result suggests that effective personalization can be achieved without requiring any extra effort from the user.

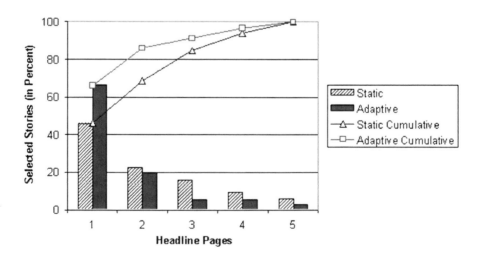

Fig. 18.5. Distribution of selected stories (alternating sessions experiment)

While the results from the first experiment look promising, the experiment has several shortcomings. First, the results are based on a small data set, consisting of only 50 users who selected 340 stories (since users can likely perceive the effect of disabling personalization, data collection was limited to two weeks). Second, due to the high cost of information access on wireless devices (some cell phone plans treat data access like regular voice minutes) users typically only select a small number of headline screens in each session. It is likely that users select from these few screens the stories that interest them most, and that this is true for both the static and adaptive access modes. Therefore, a drawback of the "alternating sessions experiment" is that users might not see stories they would have seen in the adaptive mode. Likewise, in the adaptive mode, users might not see stories they would have seen in static mode. The following experiment addresses this problem by displaying both adaptive and static stories on the same screen.

The "Alternating Stories" Experiment. The "alternating stories" experiment is similar in principle to the "alternating sessions" experiment, i.e. it is designed to quantify the difference between static and adaptive information access. However, the "alternating stories experiment" displays stories selected with respect to both the adaptive and static strategies on the same screen. During the experiment, the client was configured to display four stories on each screen, with every screen containing two adaptive stories and two static stories. The server determines randomly if the first displayed story is a static or adaptive story, and the remaining stories are selected by alternating between the two strategies. The "alternating stories" methodology has two advantages. First, the system still adapts to the users' interests, because every screen contains two stories that were selected adaptively. This results in a change of system behavior that is much more subtle from a user perspective than the resulting change of the "alternating sessions" experiment. Therefore, it is possible to run the experiment over a longer period of time, because all users still receive a useful service. Second, users see the current top-ranked adaptive and static stories on the same screen, allowing for a direct comparison between the two selection strategies. If the system learns to adjust to users' individual interests, users can be expected to select more adaptive stories when presented with a choice between adaptive and static content.

The personalization server used the "alternating stories" methodology over a period of four weeks to collect access data for 5000 adaptive stories and 5000 static stories that were shown to users who had previously selected a minimum of 5 stories. Using these criteria, data obtained from 222 different users were included in the experiment. Similar to the "alternating sessions" experiment, the average display rank can be used to quantify the difference between the two display strategies. However, using the "alternating stories" methodology, the difference between the two average display ranks was not as pronounced as in the "alternating sessions" experiment: 5.8 for the static mode vs 5.27 for the adaptive mode. Likewise, the distributions of selected stories over consecutive headline pages revealed only a small difference between the two display modes: for the static mode, 75.57% of the selected stories were on the top two headline screens, while this was true for 80.44% of the stories in the adaptive mode. The smaller difference between the two modes can be attributed to the presence of adaptive stories on every page. As a result, the user's information need might be satisfied after seeing only a small number of headline pages. If users do not have to request multiple screens to find relevant information, the observable difference in display ranks is reduced. However, this explanation only holds if users indeed select more adaptive stories than static stories. The percentage of selected stories for the two display modes clearly indicates that users are more likely to select adaptive stories than static stories. In particular, users selected 13.26% of all displayed static stories (663 stories), vs 19.02% (951 stories) of all displayed adaptive stories, which amounts to a 43% increase in selected content. Figure 18.6 shows how this difference is distributed over consecutive news headline screens. For each headline screen, this plot compares the probability that a selected story was an adaptive story to the probability that the story was presented in static order. More formally, the plot compares the conditional probabilities *p(adaptive | selected)* and *p(static | selected)* for separate headline screens.

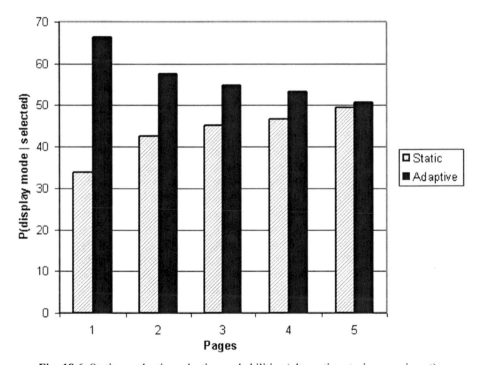

Fig. 18.6. Static vs adaptive selection probabilities (alternating stories experiment)

Figure 18.6 also shows that the difference in selection probabilities is particularly noticeable on the first headline screen and then decreases gradually from page to page. On the first headline screen, *p(static | selected)* is 0.33 vs 0.66 for *p(adaptive | selected)*. This difference indicates that the adaptive display strategy indeed helps users locate relevant content, as users prefer adaptive stories over static stories on average.

In summary, the "alternating sessions" and "alternating stories" experiments both show that adaptive information access is superior to static access. The "alternating sessions" experiment demonstrated that the adaptive order helps to move interesting items towards the beginning of personalized item lists, simplifying access to relevant content. The "alternating stories" experiment showed that the system is capable of ordering content in a way such that the top-ranked stories have a significantly higher chance of being selected than the top-ranked stories obtained from a static order.

18.4   Recent Trends and Systems

In this section, we briefly discuss recent news delivery trends, novel systems and emerging research opportunities.

18.4.1   Personalizing Audio and Video News Feeds

The Internet is rapidly turning into an on-demand delivery platform for multimedia content. For example, as part of the phenomenal success of *Apple Inc's* portable audio player, the *iPod,* online audio distribution of news content, or in short *podcasting,* is becoming increasingly popular. Thousands of regularly updated news programs can be located via countless services (including *Apple Inc*'s popular *iTunes* software) and downloaded to personal media players. Even though podcasting is still in its infancy, selecting the most informative, relevant or interesting audio content from thousands of feeds is challenging. Since information is distributed in the form of audio files, users currently cannot search for information within podcasts, and the text-based recommendation techniques discussed in this book cannot be used (because textual transcripts are usually not available for audio broadcasts). Since collaborative filtering algorithms only depend on ratings from other people and do not require any content analysis, these techniques are immediately applicable to audio feed recommendation. However, as discussed in Section 18.2.1, collaborative filtering has potential disadvantages in the context of the very dynamic nature of news content. Therefore, content-based recommendations for audio files, or fragments thereof, could significantly enhance the utility of available audio content. For example, users may not want to listen to an entire audio program if only a small segment of it is personally relevant or interesting. Techniques that automatically extract text from audio files, enable full-text search, find topic-based segments within audio or use content-based recommendation techniques to assemble personalized podcasts, is an interesting area for future work. Clearly, video feeds face similar problems: textual transcripts are often not available, and features that could be automatically extracted by image analysis techniques are not yet semantically meaningful enough to support accurate news personalization. However, since this is an area that will undoubtedly become increasingly important, adaptive access to multimedia content is an active area of research [22], [34], [15], [5].

18.4.2   Personalization and the Blogosphere

The term *blogosphere* refers to the collective set of all weblogs or *blogs.* As *blogging* becomes an increasingly popular form of self-expression, there is a need for more sophisticated tools to help navigate the blogosphere and discover relevant content. Initial systems that support personalized blog access are beginning to emerge: e.g. *Findory.com,* as described in Section 18.2.1, can adaptively personalize blogs, similar to its news personalization capabilities. *NewsGator (www.newsgator.com)* can recommend news and blog feeds, based on a collaborative approach that uses a user's subscriptions as the basis for personalized feed recommendations. In addition, blog search engines, such as *www.technorati.com* or *www.blogdigger.com* are beginning to incorporate customization features (similar to most news services) into their sites. However, these customization features are usually static and do not adapt to the user's interests.

18.4.1   News Zeitgeist

Zeitgeist is a German word that means "the spirit *(Geist)* of the time *(Zeit)*". As news and blog aggregation services are becoming more popular, many sites are incorporating *Zeitgeist* features. The goal is to automatically identify the most popular or talked-about topics, as an expression of the current blogosphere "buzz" or *Zeitgeist.* For example, *www.technorati.com* lists the most popular blog searches of the hour, the most talked about books and movies, and most-cited blogs. *Daypop (www.daypop.com)* uses blogs as a *Zeitgeist* meter for news content, by generating a list of the 40 news stories that are most frequently cited in the blogosphere. Closely related to this, *Digg (www.digg.com)* is a representative of a new breed of social content discovery sites: users submit potentially interesting news articles or blog posts to the site, and *Digg's* user community expresses interest in the submitted stories by clicking corresponding "digg it" buttons. The stories with the most "diggs" are then prominently displayed on the site, which means that, arguably, the content featured on *Digg* automatically adapts to the interests of its user base.

In summary, as news and blog services evolve, we will continue to see meta-services that not only aggregate content, but also attempt to automatically convey *Zeitgeist* information – by auto-generating pages that adapt to the spirit of the time.

18.5   Summary and Conclusions

The web is increasingly evolving into the most powerful news delivery platform of the 21st century. It provides new information dissemination channels for established news organizations, allows individuals to be heard, and enables new forms of coverage analysis. As our established reading, publishing and sense-making practices continue to evolve, we need new technology to help leverage the full potential of web-based news distribution. Continued growth of online news content is of limited use if we cannot find the personally most relevant and useful information. This section outlined early steps towards technology that is specifically designed to help us navigate the continuously evolving news landscape.

References

1. Balabanovic, M.: Learning to Surf: Multiagent Systems for Adaptive Web Page Recommendation. Ph.D. Thesis, Stanford University (1998)
2. Bharat, K., Kamba, T., Albers, M.: Personalized Interactive News on the Web. Multimedia Systems 6(5) (1998) 349-358

3. Billsus, D., Pazzani, M.: User Modeling for Adaptive News Access. User Modeling and User-Adapted Interaction 10(2/3) (2000) 147-180
4. Billsus D., Pazzani, M., Chen, J.: A learning agent for wireless news access. In: Proceedings of the 2000 International Conference on Intelligent User Interfaces (2000) 33-36
5. Billsus, D., Brunk, C., Evans, C., Gladish, B., Pazzani, M.: Adaptive Interfaces for Ubiquitous Web Access. Communications of the ACM 45(5) (2002) 34-38
6. Billsus, D., Hilbert D., Maynes-Aminzade, D.: Improving Proactive Information Systems. In: Proceedings of the 10th International Conference on Intelligent User Interfaces (2005) 159-166
7. Borthwick, A.: A Maximum Entropy Approach to Named Entity Recognition. Ph.D. Thesis. New York University (1999)
8. Budzik, J., Hammond K., Birnbaum, L.: Information Access in Context. Knowledge-Based Systems 14 (1-2) (2005) 37-53
9. Chiu, B., Webb, G.: Using decision trees for agent modeling: improving prediction performance. User Modeling and User-Adapted Interaction, 8 (2005) 131–152
10. Cover, T., Hart, P.: Nearest Neighbor pattern classification, IEEE Transactions on Information Theory, 13 (1967) 21-27
11. Duda, R., Hart, P.: Pattern Classification and Scene Analysis. New York, NY: Wiley and Sons (1973)
12. Gabrilovich, E., Dumais, S., Horvitz, E.: Newsjunkie: Providing Personalized News Feeds via Analysis of Information Novelty. In: The Thirteenth International World Wide Web Conference (2004) 482-490
13. Gauch, S., Speretta, M, Chandramouli, A., Micarelli, A.: User Profiles for Personalized Information Access. In: Brusilovsky, P., Kobsa, A., Nejdl, W. (eds.): The Adaptive Web: Methods and Strategies of Web Personalization. Lecture Notes in Computer Science, Vol. 4321. Springer-Verlag, Berlin Heidelberg New York (2007) this volume
14. Haas, N., Bolle, R., Dimitrova, N., Janevski, A., Zimmerman. J.: Personalized News through Content Augmentation and Profiling. In: IEEE 2002 International Conference on Image Processing, (ICIP 2002) (2002) 9-12
15. Haggerty, A., White, R.W. and Jose, J.M.: NewsFlash: Adaptive TV News Delivery on the Web. In: Proceedings of the 1st International Workshop on Adaptive Multimedia Retrieval (2003) 72-86
16. Jokela, S., Turpeinen, M., Kurki, T., Savia, E., Sulonen, R.: The Role of Structured Content in a Personalized News Service. In: Hawaii International Conference on System Sciences (HICSS 34) (2001) 1-10
17. Kelly, D., Belkin, N.: Reading Time, Scrolling and Interaction: Exploring Sources of User Preferences for Relevance Feedback During Interactive Information Retrieval. In: Proceedings of the 24th Annual ACM SIGIR Conference (2001) 408-409
18. Kelly, D., Teevan, J.: Implicit Feedback for Inferring User Preference: A Bibliography. SIGIR Forum 37(2) (2003) 18-28
19. Klinkenberg, R., Renz, I.: Adaptive Information Filtering: Learning Drifting Concepts. In: AAAI/ICML-98 Workshop on Learning for Text Categorization, Technical Report WS-98-05, AAAI Press (1998) 33-40
20. Konstan, J., Miller, B., Maltz, D., Herlocker, J., Gordon, L., Riedl, J.: GroupLens: Applying Collaborative Filtering to Usenet News. Communic. of the ACM 40(3) (1997) 77-87
21. Lai, H., Liang, T., Fu, Y.: Customized Internet News Services Based on Customer Profiles. In: Proceedings of the 5th International Conference on Electronic Commerce (ICEC 2003), Pittsburgh , Pennsylvania (2003) 225-229
22. Lee, H., Smeaton, F., O'Connor, N., Smyth, B.: User Evaluation of Físchlár-News: An Automatic Broadcast News Delivery System. ACM Transactions on Information Systems 24(2) (2006) 145-189

23. Magnini, B., Strapparava, C.: User Modeling for News Web Sites with Word Sense Based Techniques. User Modeling and User Adapted Interaction 14 (2004) 239-257

24. McCallum, A., Nigam, K.: A Comparison of Event Models for Naive Bayes Text Classification. In: AAAI/ICML-98 Workshop on Learning for Text Categorization, Technical Report WS-98-05, AAAI Press (1998) 41-48

25. Micarelli, A., Sciarrone, F., Marinilli, M.: Document Modeling. In: Brusilovsky, P., Kobsa, A., Nejdl, W. (eds.): The Adaptive Web: Methods and Strategies of Web Personalization. Lecture Notes in Computer Science, Vol. 4321. Springer-Verlag, Berlin Heidelberg New York (2007) this volume

26. Mobasher, B.: Data Mining for Web Personalization. In: Brusilovsky, P., Kobsa, A., Nejdl, W. (eds.): The Adaptive Web: Methods and Strategies of Web Personalization. Lecture Notes in Computer Science, Vol. 4321. Springer-Verlag, Berlin Heidelberg New York (2007) this volume

27. Pazzani, M., Billsus, D.: Content-based Recommendation Systems. In: Brusilovsky, P., Kobsa, A., Nejdl, W. (eds.): The Adaptive Web: Methods and Strategies of Web Personalization. Lecture Notes in Computer Science, Vol. 4321. Springer-Verlag, Berlin Heidelberg New York (2007) this volume

28. Rhodes, B., Maes, P.: Just-in-Time Information Retrieval Agents. IBM Systems Journal 39 (2000) 685-704.

29. Sakagami, H., Kamba, T.: Learning Personal Preferences on Online Newspaper Articles from User Behaviors. In: Proceedings of the Sixth International World Wide Web Conference (WWW6) (1997) 291-300

30. Schafer, B., Frankowski, D., Herlocker, J., Sen, S.: Collaborative Filtering Recommender Systems. In: Brusilovsky, P., Kobsa, A., Nejdl, W. (eds.): The Adaptive Web: Methods and Strategies of Web Personalization. Lecture Notes in Computer Science, Vol. 4321. Springer-Verlag, Berlin Heidelberg New York (2007) this volume

31. Smyth, B., Cotter, P.: Intelligent Navigation for Mobile Internet Portals. In: IJCAI Workshop on AI Moves to IA: Workshop on Artificial Intelligence, Information Access, and Mobile Computing. The 18th International Joint Conference on Artificial Intelligence (IJCAI-03) http://www.dimi.uniud.it/workshop/ai2ia/ (2003)

32. Sorensen, H., McElligott, M.: PSUN: A Profiling System for Usenet News. In: Proceedings of the CIKM 95 Workshop on Intelligent Information Agents. http://www.cs.umbc.edu/cikm/iia/proc.html (1995)

33. Van Rijsbergen, C. J.: Information Retrieval. London: Butterworths (1979)

34. Wang, Q., Balke, W-T., Kieling, W., Huhn, A.: P-News: Deeply Personalized News Dissemination for MPEG-7 Based Digital Libraries. In: Research and Advanced Technology for Digital Libraries: 8th European Conference (2004) 256-268

35. Wayne, C.: Multilingual Topic Detection and Tracking: Successful Research Enabled by Corpora and Evaluation. In: Proceedings of the Second International Language Resources and Evaluation Conference, Athens, Greece (2000) 1487-1493

36. Widyantoro, D., Ioerger, T., Yu, J.: An Adaptive Algorithm for Learning Changes in User Interests. In: Proceedings of the Eighth International Conference on Information and Knowledge Management (1999) 405-412

Part IV

Challenges

19

Adaptive Support for Distributed Collaboration

Amy Soller

Institute for Defense Analyses
4850 Mark Center Drive
Alexandria, Virginia, USA
asoller@ida.org

Abstract. Through interaction with others, a person develops multiple perspectives that become the basis for innovation and the construction of new knowledge. This chapter discusses the challenges facing emerging web-based technologies that enable distributed users to discover and construct new knowledge collaboratively. Examples include advanced collaborative and social information filtering technology that not only helps users discover knowledge, peers, and relevant communities, but also plays a powerful role in facilitating and mediating their interaction. As the internet extends around the world and interconnects diverse cultures, the adaptive web will be challenged to provide a personalized knowledge interface that carries new perspectives to diverse communities. It will play the role of an interface for knowledge construction, a mediator for communication and understanding, and a structured channel through which knowledge is created, interpreted, used, and recreated by other users.

19.1 Introduction

Methods for individual adaptation on the web, such as content selection and sequencing, navigation support, and presentation adaptation, focus on helping the user find and apply the knowledge he needs in the most efficient manner. These methods are effective if the knowledge is available somewhere on the web. What if it is not? Discovery, meaning-making or sensemaking, understanding, and innovation are emergent processes that develop over time through experiences and the interpretation of interaction with others [36]. This chapter discusses the challenges facing emerging web-based technologies that help users discover and construct new knowledge by facilitating the interaction between groups of internet users. Examples include integrated combinations of distributed performance support and collaborative and social information filtering technology that not only help users discover knowledge, peers, and relevant communities, but also play a powerful role in facilitating and mediating their interaction. As the internet extends around the world and interconnects diverse cultures, the adaptive web will be challenged to provide a personalized knowledge interface that helps different communities interpret and understand alternative perspectives. It will play the role of an interface for knowledge construction, a mediator for

P. Brusilovsky, A. Kobsa, and W. Nejdl (Eds.): The Adaptive Web, LNCS 4321, pp. 573–595, 2007.
© Springer-Verlag Berlin Heidelberg 2007

communication and understanding, and a structured channel through which knowledge is created, interpreted, used, and recreated by other users.

Through interaction with others, a person develops the multiple perspectives that become the basis for innovation and the construction of new knowledge. The adaptive web has the potential to facilitate this process of collaborative knowledge construction by assisting in the discovery of new business or learning partners, promoting the development of existing and new professional and social communities, and supporting and mediating the interaction between these new relationships.

The first step in developing support for distributed collaboration is enabling people to exchange the right information, at the right level of detail, using the right language, at the right time, in the right context, with the right people. Examples of tools that assist people in finding the right knowledge and expertise at the right times include collaborative filtering and social matching algorithms [31, 39] (also see Chapter 9 of this book [30]). Examples of tools that provide appropriate contexts for information sharing and learning include online communities and virtual spaces for meeting, collaborating, and constructing knowledge online [40].

The second step is effectively mediating the participants' cognitive and collaborative processes. Adaptive collaboration environments that move beyond content or social-based recommender system approaches to support the innovative processes of knowledge construction will be challenged to address the complex interplay between physical, cognitive, and social variables. These factors affect the way in which information flows between the collaborating participants, shaping their interaction. In knowledge domains such as those involving peer help or the development of trusting relationships, efforts in mediating and maintaining compatibility between collaborative processes should parallel efforts in matching static traits and attributes.

The potential for joint understanding and meaning-making is greatly affected by the degree of trust and motivation for collaboration and the policies or rules that govern these processes. People rarely follow up on face-to-face encounters unless business process, economic, political, or other factors play a role in maintaining the interaction. Throughout the examples in this chapter, incentive is provided though the intrinsic motivational characteristics of distance and organizational learning situations. The research methods and environments presented here should scale more generally to distributed collaborative environments that encourage persistent collaboration and active knowledge construction.

The next section in this chapter discusses the challenges of extending user and group modeling technology to connect people with knowledge and provide support for complex collaborative processes. The third section discusses research progress in developing, maintaining, and mediating adaptive online knowledge-sharing communities. The fourth section summarizes the adaptive collaboration support technology possibilities within the framework of a theoretical collaboration management cycle [35]. The final section discusses future trends in managing and supporting web-based collaboration.

19.2 From Social Matching to Adaptive Collaboration Support

Individual user models (also called user profiles) store information about a user's persona, behavior, and preferences. They can be used to recommend products or services that fit the user's interests or to provide help and guidance (see Chapter 1 of this book [3]). Collaborative filtering techniques traditionally compute the similarity between elements in individual user models or group models and attributes of available content to suggest appropriate information, products, services, activities, or advice [31] (also see Chapter 9 of this book [30]). Social matching systems apply similar algorithms to compute the similarity between users or groups, given their interests or information needs. These systems introduce people to each other, recommend communities or experts, and suggest opportunistic times for collaboration [39]. The next few paragraphs briefly introduce these basic concepts in more detail before moving into a discussion of more advanced methods, and can be skipped if the reader is familiar with these methods.

Collaborative *content-based filtering* methods aim to match individuals or groups to appropriate content, products, services, or activities. For example, content-based filtering can be used to recommend web sites that would be of interest to a team of students with different backgrounds and experiences collaboratively navigating the web together (see [10] for an overview of social navigation). The algorithm would attempt to find similarities within the student models and select those web sites that would be appealing and appropriate for a majority of the team members. The recommended items and the consequential student reactions are sometimes stored in group models. Group models characterize the group as a whole, including elements such as group performance and history. They may also contain individual member profiles. User and group models are examples of tools that help adaptive collaboration technology determine the best way to mediate and support online collaboration. Later in this section, we will see how user and group models can be used to model and mediate dynamic collaborative processes.

If the students in our hypothetical web site recommender example were to rate the web sites that they found most useful in their work, collaborative *social filtering* could then be used to recommend the most popular or most useful sites to other groups of students. For example, users of the Ringo system [31] rate musical artists. The system then recommends new artists to users with similar preferences, automating the "word-of-mouth" phenomenon. Users can also write reviews that might be useful to other users with similar tastes or receive lists of the "top 20" or "bottom 10" rated artists.

Social matching systems bring people together to satisfy explicit information needs, curiosity, or community-oriented or interpersonal interests. For example, the Expertise Recommender system [26] helps people in an organization locate other users who have specific expertise. The user can search and sort candidate profiles according to several criteria, including a social network that incorporates the results of personal interviews. The I2I system [4] attempts to find appropriate partners by tracking users' actions on documents. It uses this information to dynamically identify users who are working on similar documents and who might be interested in collaborating. As a user is working on a document, he is presented with visualizations depicting the other users who are working within the conceptual space defined by the document.

Users can also leave "calling cards" on documents to let other users know that they are interested in chatting about the document.

Content-based, social filtering, and social matching systems can be combined into hybrid systems (see Chapter 12 in this book [5]) and can also filter implicit, tacit knowledge. For example, OWL [24] dynamically profiles individuals as they work, attempting to capture the tacit knowledge that describes users' behavior—that about which the users themselves may not even be aware. The system observes as users apply sequences of tool functions to satisfy task-related goals. For example, using Microsoft Word, a user might select the menu items *Table → Convert → Text to Table* to convert a segment of tabbed text to a table. As the system observes groups of users over time, it identifies differences between individuals' behaviors, skills, and activities, and suggests further learning to each user based on these (knowledge gap) differences.

OWL also provides learning recommendations to users about software functions that their peers in communities of interest have found useful (e.g., support, research, or managerial staff communities). As we will see in section 19.3.2, knowledge and expertise can take on a different character when viewed through the lenses of different communities. Communities might be based on project teams, organizational roles, background, experience, community membership, or culture.

The systems described thus far identify and introduce people who may have shared interests, and recommend opportunistic times for them to collaborate within a shared context. They accomplish this through collaborative content and social filtering, matching, navigation, and visualization tools. Navigation tools follow the user as he navigates the web and inform him of other users who have navigated similar paths, while visualization tools represent the activities or characteristics of communities of users so that the user can decide for himself which communities he might like to join. The first two sections in this book discuss these systems at great length. The remainder of this chapter discusses the challenges in building upon these collaborative filtering and social matching technologies to provide adaptive support for the underlying collaborative, cognitive, and social processes involved in distributed information sharing and knowledge construction.

19.2.1 Beyond Social Matching

Once access to information or expertise on the web is obtained (e.g., via internet search, collaborative filtering, or social matching), sustained collaboration is necessary for the development of understanding, knowledge construction, and coordinated action. Supporting persistent collaboration requires attention to more than individual attributes and traits. As individuals interrelate and collaborate, levels of interdependency increase, and people begin to feel and act less like isolated individuals and more like group members [23]. Over time, it becomes more difficult to predict group performance based on individual members' characteristics. The significance of these characteristics lessens as group dynamics and process become core contributing factors in predicting group outcome. The complex ways in which information and interpretations flow between collaborating participants ultimately shapes the group's interaction and the collaboration's outcome.

Barriers to effective collaboration and knowledge sharing are pervasive. Research in social psychology has consistently shown that group members tend to discuss information that they share in common instead of discussing the knowledge they uniquely possess [37]. Hatano and Inagaki [16] showed that when knowledge is constructed during group discussions in context, individuals may have difficulty assimilating this knowledge without the support and collaboration of the other group members. This is particularly true when the information is presented by those who hold a minority opinion. Even when it is not, the way that the information is represented and the context in which it was created may prevent the receivers of the information from easily incorporating it into their own mental representations. Group productivity has been positively linked to such group processes as peer helping, hypothesis development and testing, management of competition and conflict, ability to use different viewpoints, mutual support, and ability to produce detailed, elaborated explanations [8].

Approaches to supporting and sustaining effective distributed collaboration range from systems that assist in locating experts or teammates combined with feedback and reputation updating processes [e.g., see 40] to systems that provide dynamic team facilitation and coaching. These approaches aim to promote effective collaboration in distributed knowledge environments by drawing upon user and group models in different ways.

In the first approach, user and group models are consulted, filtered, generalized, or aggregated, and a group is constructed by selecting members with the most compatible knowledge, skills, and behaviors. Because individuals may behave differently in groups, individual user models may include behaviors prevalent and productive during prior group interactions. This process of constructing the best possible group is intended to influence the team dynamics positively and increase the likelihood of group success.

In the second approach, a (human or computer) facilitator analyzes the group interaction after the users have begun to work collaboratively, and dynamically attempts to either facilitate the group interaction or modify the environment appropriately. User and group models help the facilitator determine the most effective mediation methods and record how well the users respond to the interventions. Later in this chapter, section 19.4 returns to these concepts by describing a cyclic phase-based model of collaboration management. The model starts from the user and group modeling phases and moves through the behavior analysis and knowledge visualization phases to the adaptive group facilitation phase. During the final phase, the users' responses to the environmental feedback are interpreted and used to update the group models in preparation for the next cycle. The system described in the next section introduces the notion of combining user modeling and adaptive facilitation to support online collaborative learning activities and illustrates some of the challenges in this area.

19.2.2 Strategic Pairing and Adaptive Support for Distributed Collaborative Learning

IMMEX™ (Interactive Multi-Media Exercises; http://www.immex.ucla.edu) is a web-based multimedia learning environment designed to help groups of students learn how to develop and evaluate hypotheses, and analyze laboratory tests while solving real-world problems. The single-user version has been used for over 13 years in science classes across U.S. middle and high schools, universities, and medical schools, and has logged over 250,000 student problem-solving performances [38]. The collaborative version of IMMEX includes general-purpose collaborative web navigation and synchronization facilities, and a structured chat interface [29] (see Fig. 19.1).

Fig. 19.1. The IMMEX^{TM} Collaborative problem-solving environment runs within students' web browsers. The left-hand panel enables and displays student chat communication. The bottom panel shows which student has control of the mouse. The main window is a shared, synchronized multimedia and hypertext workspace

In IMMEX, individual user profiles describe students' learning performance, progress, gender, preferred problem-solving strategies, and predicted future strategies [38]. Student ability is modeled using Item Response Theory (IRT), which estimates the likelihood that a student will correctly solve a problem given the characteristics of the problem and the characteristics of the individual. While traditional IRT has historically provided a good estimation of students' overall abilities within a domain, other approaches have been more successful in modeling the development of complex cognitive processes [see 27 for a discussion of modeling evidentiary reasoning].

Student development of problem-solving strategies in IMMEX is modeled though a self-organizing map neural network approach [20]. The neural network is designed to represent the space of student problem-solving strategies in varying stages of development. First, the 36-node network topology is developed. Then, it is iteratively trained with thousands of student performances represented by sequences of problem-solving actions. For the domain of chemistry, a student performance might include actions such as selecting a flame test, a blue litmus test, and a precipitate test, and then looking up the periodic table, and searching through the library. Each student performance is classified into one of the 36 nodes so that each node represents a different generalized subset of the population of strategies. The way in which a student's problem-solving strategy changes over time as she solves progressively more difficult problems provides an indication of her cognitive development. The neural network turns this strategy development into an observable set of variables in the student model. These student model variables can be used to provide immediate feedback to the student, input to an assessment module, or guidance for adapting content or facilitation methods to groups of collaborative learners.

Stevens, Johnson, and Soller [38] have found general qualitative strategic differences in the ways that students adopt and apply problem-solving strategies. For example, their methodology can identify students who begin problem-solving by selecting a large number of random items, and then gradually become more selective and less dependent on background information such as libraries and glossaries. Some students quickly learn specific domain-based strategies such as initially performing a flame test to segment the problem space before continuing their analysis. These strategies sometimes work for select classes of problems, but do not generally transfer to more complex classes of problems. Student problem-solving strategies vary in their degree of efficiency and transfer across problem sets. Stevens et al.'s approach assesses the degree to which students are developing efficient and transferable problem-solving strategies by linking the solution frequency to each of the neural network nodes.

The student models in IMMEX are also used to make predictions about a student's most likely future learning trajectory, given his performance history. This is done using Hidden Markov Models (HMMs) [28]. HMMs are represented by three sets of parameters: state transition probabilities, observation symbol probabilities, and prior probabilities. The state transition probability distribution describes the likelihood of a student transitioning from one general problem-solving strategy set to another (e.g., on the next problem set). The observation symbol probability distribution describes the likelihood of each student applying each problem-solving strategy at each state. The matrix of prior probabilities describes the likelihood of each state before HMM training begins. This matrix quickly becomes obsolete and is replaced by the transition probability matrix, which is iteratively trained using a series of examples that describe how students typically transit from one problem-solving strategy to another as they complete consecutive problem sets. In parallel, this process trains the observation symbol probability distribution, which describes the probabilities of each of the 36 problem-solving strategies at each state. As a student completes a series of IM-MEX problem sets, he will typically transit through several HMM states. At each state, his performance is modeled by the following characteristics:

1. The general category of problem-solving strategies the student is currently applying (given by the HMM state).

2. The student's specific problem-solving strategy (given by the most probable HMM observation symbol, which is linked directly to the 36-node neural network).

3. The next most likely strategy (and least likely strategy) that the student will apply (given by the HMM state transition matrix).

By comparing models across different classes of students, the system can strategically select collaborative learning partners who might help the user see a different point of view, thus increasing the probability that the user's future learning trajectory would follow a more productive course. Although small group research has suggested that individual characteristics are generally poor predictors of group learning performance [23, 42], the IMMEX approach is unique because it boosts the predictive capabilities of individual student models. The neural network and HMM analyses project the effects of individual tendencies into future online collaborative interactions, thus facilitating the prediction of future individual and group behavior. The challenge is determining what combinations of current and future cognitive problem-solving strategies will be the most productive. Collaborative learning studies suggest that students generally work best in heterogeneous groups with a combination of abilities, as long as the heterogeneity is not too wide-ranging [8].

One can imagine several different partnering combinations based on students' current and future strategy predictions. For example, the system might recommend that a student who is using an ineffective strategy (and whom we predict will continue to use the ineffective strategy) partner with another student who has adopted an efficient strategy. Alternatively, the system might recommend that two students work together if they are both using less effective strategies but show a high tendency to shift their strategies on the following problem set. The collaboration component of IMMEX sets up on-line collaborative sessions, introduces the team members, and helps to facilitate and guide the group learning session. Once a group is strategically constructed and begins a collaborative problem-solving session, the IMMEX neural network-based modeling software begins to predict the new group problem-solving strategy automatically. This analysis is done by examining and probabilistically modeling the sequence of group members' actions [for more detail, see 38].

Although the group strategy provides some indication of how the group problem-solving is proceeding, it may provide little information about the individuals' learning. For example, a student using an efficient strategy may solve the problem alone without explaining his actions to his partner, or he may instead give instructions to his partner about what to do, and his partner may simply follow these instructions without questioning them. In both cases, the system will recognize the overall group problem-solving strategy as efficient even though the individual learning outcomes may tell a different story. Whether or not the individual with the less efficient strategy adopts a more efficient problem-solving method depends not only on the combination of prior individual strategies, but also on the way the collaborative learning process develops over time. Another possibility is that the student with the more efficient strategy will regress. For this reason, *monitoring* and *facilitating* the collaborative interaction is important.

Monitoring and assessing collaborative interaction might be done similarly to Soller's [33] approach, in which sequences of student chat conversation (coded using sentence openers such as "I think" or "Do you know") and actions are analyzed using HMMs [also see 15]. This approach was shown to predict the effectiveness of student knowledge-sharing interaction in laboratory experiments with about 74% accuracy. Preliminary studies [14] have applied a similar approach to determine the degree to which students' conversational structures provide evidence about whether or not the group members are helping each other adopt more efficient problem-solving strategies. If the structure of students' discussions reflects the structure of their decision processes, then problem-solving strategy shifts might be recognized by modeling and characterizing interaction patterns in the context of various known strategy applications.

Work is underway to develop web-based pedagogical agents for IMMEX that use the knowledge of a group's mix of cognitive strategies to strategically take on behaviors that might nurture the development of more efficient group problem-solving strategies. Playing this complex role will require an understanding of how groups members collaborate to construct new knowledge, and an understanding of how to support this process.

Situations in which the student interaction is less likely to produce problem-solving strategy shifts might be facilitated by targeted mouse control schemes. Previous research has shown that mouse control schemes that change the way in which group members share their view of the learning environment can have significant effects on student learning [17]. For example, Chiu [6] studied the effect of four different schemes on student performance: *assign*, in which one student was assigned exclusive control of the workspace; *rotate*, in which control automatically shifted to the next student every 3 minutes; *give*, in which the student currently controlling the workspace decided when and to whom to relinquish control; and *open*, in which any member could take control at any time. The results of the study suggest that when one student is assigned control of the workspace such that the other group members cannot anticipate attaining control at some future time, the students not only perform better, but also engage in more task-oriented dialog. The inability to control the workspace directly may encourage students to express and justify their ideas in words, rather than waiting for their turn to take actions.

Modeling users and groups, and using these models to strategically construct and facilitate online groups is just one way of providing adaptive support for distributed collaborative web-based applications. The next section discusses how online communities provide adaptive virtual spaces for meeting, collaborating, learning, sharing, and constructing knowledge online.

19.3 Knowledge Sharing and Discovery in Online Communities

Professionals across distributed organizations naturally share knowledge by forming small groups based on similar interests, practices, personal affinity, and trust. These groups are termed *Communities of Practice* [2, 21] because they function as cohesive communities that share a common sense of purpose and interest. Communities of Practice facilitate the sharing and creation of new knowledge, and are therefore important to the stability and growth of organizations and the development of knowledge areas. Their members interact on an ongoing basis, sharing best practices and shaping the growth and advancement of those practices.

Communities of Interest (CoIs) [43] are less formally structured community networks linked by shared interests rather than best practices. Because of the tenuous and diverse types of links between members' peripheral relationships, processes within CoIs are difficult to identify and understand. While formal organizational learning literature has focused on topics such as understanding how peripheral community members become core members, research in CoIs recognizes the benefits of establishing peripheral community membership in many different communities and helping to bridge disparate communities into informal social networks. These loosely bridged networks can explain how innovation happens when community members interact with members of possibly far-reaching communities that may be able to offer new, different perspectives. Such chance encounters and informally planned interactions are often encouraged through referrals by peripheral community members. This section describes how adaptive collaboration support technology enables, mediates, encourages, and guides this natural process.

Effective knowledge sharing across Communities of Interest with different objectives and perspectives means sharing the right information, at the right level of detail, using the right language, at the right time, in the right context, with the right people [13]. A failure related to any one of these factors can lead to a knowledge-sharing breakdown. Some social psychology research has identified strategies that might encourage communities to share the information they uniquely possess. Such strategies include helping participants understand the nature and granularity of the knowledge held by each Community of Interest, and setting up interactive agendas specifically for information sharing so that gaps can be more readily identified. This section discusses tools and methodologies for facilitating knowledge sharing and community development

Facilitating knowledge sharing across Communities of Interest that do not yet have established processes for information sharing involves creating the infrastructure, mindset, and tools needed to support a new culture of collaboration and sharing. Several different factors influence community members' participation, involvement, and the eventual success of the collaboration. These include (1) the degree to which users are aware of the various communities, information, and knowledge available in the environment (awareness), (2) the ability of online communities to maintain knowledge and user interest, and provide access to useful information in a timely manner (maintenance), and (3) whether community members perceive an immediate benefit from collaborating with others (motivation). The next three subsections address these three processes respectively.

19.3.1 Knowledge Discovery and Awareness

The distributed and virtual nature of the adaptive web makes effective collaboration, knowledge sharing, and an understanding of collaborators' perspectives essential to creating meaningful knowledge and achieving complementary objectives. Helping communities develop their own awareness and understanding of other communities' knowledge, problems, and goals are some of the most difficult challenges.

In supporting collaborative knowledge discovery and awareness, one of the most important decisions involves the design of shared workspaces. Each Community of Interest might have a different set of complementary objectives and may still need to collaborate effectively to share the information that others need, without necessarily aiming to attain the same goals. A shared, unified workspace or common view may be helpful for providing the appropriate context for sharing knowledge but may, in some cases, also hinder collaborators' ability to engage in certain specialized activities for meeting their individual goals (e.g., exploring private databases or web portals, customizing views to perform focused analyses). The design of shared workspaces should consider the degree to which representations of shared artifacts will be viewed and interpreted differently by participants based on their backgrounds, experiences, cultures, and values [22, 25].

A simple example with which the savvy international traveler might be familiar is given by the litany of traffic signs and symbols in foreign countries. Shinar, Dewar, Summala, and Zakowska [32] asked 1000 licensed drivers from Canada, Finland, Israel, and Poland to interpret 31 traffic signs from various countries. They found highly significant differences in participants' understanding of the signs. Fig. 19.2 shows examples of two signs ("Dead End" and "No Vehicles Carrying Explosives") for which 86% and 78% of the participants answered incorrectly. Ten percent of the participants actually misinterpreted three of the signs as having the opposite of their true meaning.

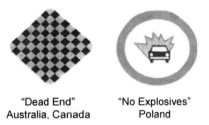

"Dead End"   "No Explosives"
Australia, Canada   Poland

Fig. 19.2. Two traffic signs for which 86% (Dead End) and 78% (No Explosives) of the participants interpreted incorrectly

Shared workspaces for supporting online communities should also take into consideration the persistence and validity of information [11]. Activities that involve transient or uncertain information may be more appropriately conducted within private workspaces or private chat rooms. Once the information reaches a level of stability appropriate for a broader audience, it should be migrated to shared community workspaces. Community members will perceive information in shared workspaces as stable and reusable because the nature and affordances of shared workspaces inherently convey information persistence. This is a common problem on the web, a medium regarded as persistent, but one in which links are often moved, updated, and deleted.

Distributed collaboration technology adds adaptivity to shared virtual workspaces by supporting awareness and tolerance, and helping users understand how their perspectives differ. Examples of technology for supporting these processes include knowledge seeking and searching tools that attempt to understand the user's core community perspective while guiding her toward the most appropriate knowledge sources tailored to her needs. Other awareness tools help communities frame their knowledge in terms and languages that are most familiar to other known communities, developing implicit links between similar concepts and programs, or suggesting meaningful analogies to facilitate this conceptual translation. Social awareness and social networking tools can be useful for connecting community members and enabling them to attach meaning to tacit knowledge that was developed in specific contexts.

Social network theory defines methods and models for analyzing and understanding the linkages between entities in social networks. Concepts such as "cliques" (defined by the interconnections between actors), "centrality and prestige," and "affiliations" (representing the links between actors and events) provide the building blocks of this applications-oriented theory [41]. Social network tools provide views of online communities and their members by drawing upon specialized user profiles that specify the communities of interest to which each user belongs. Members who enhance their profile to include more detailed information enable the system to serve them in a more meaningful way. For example, a member of a funding agency who posts detailed information about his agency's resources and funding opportunities enables the technology to assist the member in identifying potential customer communities that seek such resources. Resources might be linked to individual members' profiles and categorized in their corresponding communities (e.g., Training and Simulation community, Nanotechnology community) so that they can be retrieved either by community or individual member search criteria.

Web-based social networking technology enables community members to view visualizations of social networks and run content or member-based searches across these networks. A typical search might begin with a researcher viewing his usual community of professional colleagues and friends. Clicking on a contact in the social network might set off two different processes: (1) the system would look to see if the selected person has an existing profile, and (2) the system would use the selected participant to "grow" the social network. New contacts might be "discovered" by linking the selected person with the co-authors on their publications or the partners they list on their Curriculum Vitae. Examples of systems that apply these or similar ideas include LiveJournal (http://www.livejournal.com), iVisto [34], Referral Web [19], Friend-of-a-Friend (FOAF) (http://www.foaf-project.org), and Huminity (http://www.huminity.com). LiveJournal and IVisTo are described in more detail in the remainder of section 19.3.

Users register with LiveJournal by creating a simple profile. A user profile includes a mini-bio, a list of interests (used to find other users with similar interests), (optionally) a list of friends, and (optionally) a list of communities. Once a user has created a profile, she can create journal entries that include icons, representing her mood, and polls that request other users to vote on her ideas. She also has the option to allow peers to respond (through comments) to the ideas in her journal. Journals can be customized or embedded in web pages.

LiveJournal automatically shows the user a hypertext list of the communities related to his interests and the other members of those communities. Users can then freely navigate through communities and discover new communities by viewing the communities to which each user is a member. For example, I list "Education" as one of my interests and discover that Mary and Bob are both members of that community. They are also members of the community "Collaborative Learning Technology," through which I discover Peter, who is a member of several other communities I did not know even existed. The trail continues indefinitely.

Privacy is handled by allowing the user to control who can view his contact information and journal entries, who can send him text messages, who can leave comments (and whether or not the user wants to screen the comments posted to his journal), and who can participate in polls. The user also has full control over the communities that he moderates.

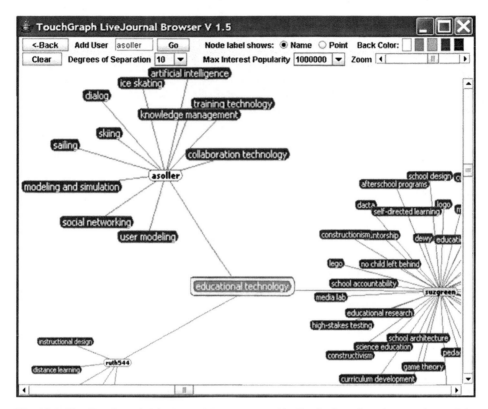

Fig. 19.3. The Touchgraph LiveJournal Browser graphically displays the other members of the user's CoI and any other related CoIs

The Touchgraph LiveJournal Browser, shown in Fig. 19.3 enables users to visualize the virtual communities within the LiveJournal environment.

In the figure, pink and white nodes represent users, and blue nodes represent communities. Pink user nodes are "expanded" to show all the user's community memberships, while white user nodes are "collapsed" to save screen real estate. The user has the option of either viewing a particular Community of Interest's members (blue node surrounded by pink and white nodes) or a particular member's Communities of Interest (pink node surrounded by blue nodes). Each node includes a link directly to the LiveJournal web page describing the user or community. For example, clicking on the green "info" box attached to the "Educational Technology" node in the center of would automatically bring up the LiveJournal web page for that community.

New nodes can also be added to the display, and the system will automatically identify the user or community links between the new node and the nodes already on the display. When the user moves the mouse over a Community of Interest, the system highlights those other users who share the same interests, thus enabling the user to identify new friends and communities.

19.3.2 Community Maintenance

Like any real-world community, online communities need support and maintenance to sustain their development and growth. Determining how to provide this support requires an understanding of what to expect over the lifetime of the online community. Communities should generally be motivated to share quality understandable information with other communities that repay the good will. The perceived and measured benefit of collaborating is predictive of the level to which community members continue to collaborate with each other over time. For example, Cho, Stefanone, and Gay [7] studied the online interaction of students using listservs and community discussion boards, and found that less information was shared and processed by the students as the term progressed. Central/prestigious actors shared more information at the beginning of the term, while less central/prestigious (more peripheral) actors were more likely to interact and share knowledge later in the term. This suggests that peripheral actors require time to enter community-based practices, providing a concrete web-based application of Lave and Wenger's [21] legitimate peripheral participation/situated learning theory.

Cho et al. [7] also found that URLs posted to the class listservs (and consequently emailed to all the participants) were visited significantly more times than those posted on the discussion boards that the students needed to access explicitly. The "push" technology was necessary to have the learners fully involved in the community-based activities. This concept may be particularly important for more established community members because their motivation for community-supported knowledge discovery may decrease over time as they reach the knowledge boundaries of the community and perceive a reduced need to use a system to discover things they think they already know.

By connecting Communities of Interest and providing more information and associations at users' fingertips, we increase the volume of data through which a user must search to find the most relevant information. Guidelines, roadmaps, metadata, structures, and tools for finding relevant information in community-based contexts are essential and must be constantly updated and maintained.

The community moderator role is also key; several moderators may be needed (e.g., perhaps one from each community). Questions should also be raised regarding the characteristics that are needed for effective moderation of community-based knowledge networks. For example, moderators may need domain knowledge or experience in professional group facilitation, or they may need time to get to know the collaborating partners personally and establish a level of trust with them.

Cross-community discussion groups that are linked to shared data sources may help to give more context and meaning to the content. For example, users and groups could collaborate in online discussion forums that are directly linked to the imagery and reports they are sharing, commenting and explicitly making linkages (e.g., arrows, highlights) to sections of the shared items being discussed. Rating or voting tools also help community members determine what information (discussion items, images, and so forth) was helpful for what purposes. The most useful information can then be maintained and enhanced as less central knowledge migrates to community peripheries. The remainder of this section discusses a social networking tool for such community management.

IVisTo (Interactive Visualization Tool) [34] is a social networking tool that operates within a peer-to-peer knowledge management environment. It enhances user and group (community) models by monitoring and analyzing users' keyword and ontology-based search behaviors. IVisTo displays a weighted combination of social networks, where each social network addresses a different user model variable, and the weights are given by the learner's social and semantic preferences. The interface contains a set of slider bars that represent the social variables in the user model (e.g., Organizational Role, Collaboration Level) (see Fig. 19.4). Using these slider bars, the user can indicate the importance, or *weight*, of each variable. Behind the scenes, the system generates a social network for each of the variables, and then computes one single network by calculating a weighted sum of the individual networks. For example, by increasing the importance of the "Organizational Role" slider bar, the tool gives more "credit" to

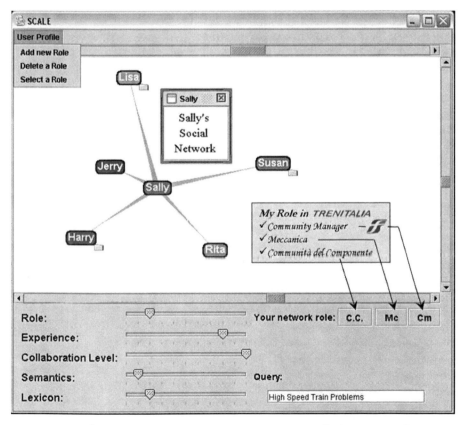

Fig. 19.4. IVisTo[1] interface showing Sally's social network. Sally is a community manager working in the Mechanical Components division, searching for knowledge about high-speed train problems

---

[1] IVisTo was developed using the Touchgraph$_{LLC}$ toolkit (see http://www.touchgraph.com).

those other Community of Interest members who hold a similar Organizational Role as the user, displaying those users as more prominent on the interface. In this way, IVisTo provides each learner across an organization access to a personalized set of visualizations from his perspective, weighted according to his interests.

As the learner carries on her day-to-day learning and collaborative work, the lengths of the links in IVisTo are recalculated as the elicited and inferred information in her user model is updated. For instance, when users interact with new colleagues from different online communities, the system updates the appropriate corresponding user model values for their level of collaboration. It also reassesses the degree of semantic and lexical similarity between users' queries and their shared resources using an ontological matching procedure [1]. These activities help the system intelligently infer and visualize different types of knowledge-sharing communities and identify potential future members and items of interest. These kinds of adaptive personalized social networks may also raise users' awareness of the social factors that define their Communities of Interest, and facilitate their access to relevant artifacts and other related communities. The next challenge is evaluating the ability of adaptive social network-based tools to perform these tasks while maintaining and serving virtual communities.

19.3.3 Motivation and Participation

Distributed communities that are actively engaged and motivated to share knowledge may experience improved learning and development, and increased productivity and growth [2]. Motivating community members to interact regularly and maintain their engagement is key to community development. Communities that experience long-term success reward members for taking actions that improve the health and progress of the community by providing positive feedback. Feedback can take the form of peer ratings, an improved reputation, a greater understanding of the domain, or privileged involvement in planning core community activities.

Online communities might encourage members to participate and interact by enabling them to rate each other and their resources via informal peer review. Ratings might be weighted and aggregated to compute values for user reputation and resource value. For example, suppose Professor Arnold searches among her Communities of Interest, finds Mr. Brown, and discovers through that link that Professor Clark might be a good scientific partner for a project proposal. She should be able to provide the system with feedback describing the crucial role Mr. Brown played in establishing this partnership, perhaps even without his knowledge. In Vassileva's [40] approach, a user's reputation is based both on feedback from other users and their level of collaboration with respect to the communities to which they belong (e.g., the number of resources contributed to each community, how many "favors" the user owes to other users, and whether or not the user is being a "free rider") An inflation rate allows "older" activity to decrease in importance and weight over time.

Reputation-based behaviors and processes in online communities mirror our behavior in face-to-face contexts. A strong link appears to exist between a person's online reputation according to his peers and his degree of perceived trustworthiness. Esfandiari and Chandrasekharan [12] explain that trust has both cognitive and mathematical foundations. From a cognitive perspective, trust is a function of one's underlying beliefs; from a mathematical perspective, trust is a metric based on variables such as competence, risk, utility, and importance. In propagating trust through structures such as social networks, Esfandiari and Chandrasekharan recommend exercising caution because different paths in the network might produce contradictory values and cycles in the graph can artificially decrease trust values (e.g., one might loop three times before reaching a neighboring agent). Using even the most stable and fair algorithm still means determining the degree to which peers are trustworthy, reliable, or knowledgeable by substituting mathematical procedures for personal judgments based on experience, culture, beliefs, and values. Depending on one's point of view, the degree of fallibility in either case can be seen as variable (poor judgment may be no better than a mediocre computer algorithm).

As discussed earlier in this section, user and group models might be updated to reflect the outcomes of positive knowledge-sharing interactions (e.g., improved reputation for knowledge sharer, improved understanding of content for knowledge receiver). They might also serve as resources by which the adaptive web provides community members with summative feedback about their participation and collaboration. Augmenting participation and activity statistics with suggestions and comments can also help community participants understand what is working and why (or why not). Evaluation and assessment should be done at each phase of development and deployment with a high level of community involvement. For example, each organization should understand what knowledge was shared and how it was used by other organizations.

19.4 Practical Collaboration Management

At the beginning of this chapter, we discussed how web-based user and group models might assist in group construction by selecting members who have the most compatible knowledge, skills, and behaviors. Strategically composing groups may provide a reasonable way to set up online collaborations, but once the collaboration begins, variables such as users' prior knowledge, motivation, roles, language, and group dynamics will interact with each other in unpredictable ways, making it difficult to measure and understand behavioral effects. In the third section of this chapter, the need for dynamic mediation and facilitation led us to a discussion of awareness, knowledge discovery, and community maintenance tools. Understanding when, how, and to what extent to employ these tools during online collaboration sessions can make a significant difference. For guidance on this, we can build upon the Collaboration Management Cycle ([35], see Fig. 19.5), a phase-based model designed to frame our understanding of how to structure and mediate distributed virtual group activity.

In the first two phases of the Collaboration Management Cycle, the online interaction is observed, recorded, formatted, and logged for later processing. Recording can happen at many different levels of granularity, from audio and video capture to embedded instrumentation of web-based software applications. Ultimately, user actions and interactions must take the form of standardized, computer analyzable log files (e.g., <time: 14:00> , <user: Tom>, <event: click-entity5>, <chat: "I'm going to paste the image of the bike now">).

The state of interaction must then be conceptualized and represented using the data gathered in the first two phases. The way that this model of interaction is conceptualized

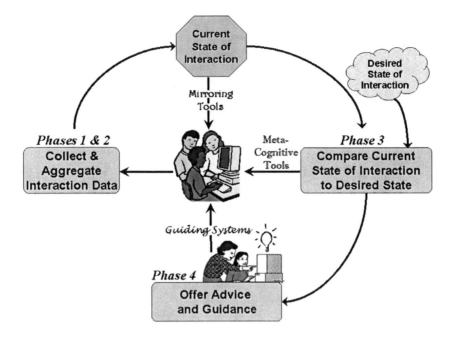

Fig. 19.5. The Collaboration Management Cycle

From Soller, A., Martínez-Monés, A., Jermann, P., Muehlenbrock, M.: From mirroring to guiding: A review of state of the art technology for supporting collaborative learning. International Journal of Artificial Intelligence in Education **15**(4) (2005) 261–290. Copyright 2005 by the International AIED Society. Reprinted with Permission.

depends on how the performance is to be measured and assessed. Typically, one or more high-level variables, such as "collaboration" or "skill competency," are selected and evaluated by algorithms that dynamically read in the log file data. Although the methods behind these algorithms vary broadly from simple statistical calculations to iterative probabilistic models and fuzzy logic, the end result should always reflect a better understanding of the collaborative process and an improvement in individual and group performance. Factors such as "group cohesion" or "shared understanding" are difficult to grasp, and even more difficult to measure quantitatively. Researchers and practitioners improve collaboration management by both theoretically grounding the selection of variables and metrics, and comprehensively evaluating the impact of those variables on human performance.

In the third phase of the Collaboration Management Cycle, the online interaction is diagnosed, and preparations are made for possible remediation. This phase requires a conception of the desired interaction formulated using the same computational representation and/or variables as the current state of interaction. The difference between the current and desired states should provide the users with an understanding of how well they are performing and how much more they could potentially achieve. This phase prepares the system for providing recommendations and advice to the users.

Remediation might be offered by intelligent interfaces, web-based computer agents, or human facilitators when discrepancies exist between the current and desired states of interaction. The labels on the arrows pointing inward in Fig. 19.5 show three different categories of adaptive collaboration support technologies that mediate online interaction: mirroring tools, metacognitive tools, and guiding systems.

Mirroring tools are termed as such because after they collect and log the interaction, they simply reflect this data back to the user. These tools are intended to provoke self-reflection and self-mediation. Users who self-reflect using mirroring tools however may have more difficulty mediating their interaction than those who self-reflect using metacognitive tools. Metacognitive tools show users representations of both their own interaction and their potential interaction, and may also hint at possible ways to improve performance. Jermann [18] found that these tools positively affect student performance online by increasing their task-related communication, and the quality and sophistication of their problem-solving plans. His system displays participation rates to pairs of collaborators as they are solving a traffic light tuning problem. The display compares the volume of messages sent by each student to the volume of problem-solving actions taken by each student. The system also displays a color-coded model of desired interaction next to the observed interaction state—the students used this standard to judge the quality of their interaction. Jermann studied the behavior of students when desirable interaction was represented as engaging in a greater proportion of talk relative to the proportion of simulation-based actions. He found that the metacognitive display positively encouraged the students to participate more through the chat interface, in particular to engage in more precise planning activities.

Guiding systems attempt to augment users' cognitive processes by assessing the collaboration activities and providing hints, guiding questions, dynamically selected and structured content, or recommendations for online partners. This guidance might be presented by a web-based animated agent serving as a coach, group facilitator, or peer. For example, the pedagogical agents in the COLER system consider the differences between students' personal, individual problem-solving workspaces and their group's shared workspace [9]. The agent provides feedback and advice to the students by using a decision tree that considers combinations of these differences and the progression of the students' collaboration. Collaboration variables include such factors as overall participation, the degree to which students have equally contributed to the shared solution, and whether or not the students would benefit from reflecting on their work (students are required to state agreement or disagreement when changes are made to the group's shared workspace).

Remediation will have an impact on students' future interactions regardless of whether or not it is offered by a system or human, and this impact must be evaluated to ensure that it produces the desired effects. The arrows in Fig. 19.5 that run from phase 4 back through the center illustration to phase 1 indicate the cyclic nature of the Collaboration Management Cycle and the importance of continual evaluation and assessment.

In less-structured environments in which goals and objectives are not as clearly defined, adaptive collaboration support technology may give the users more control over the way that their interaction is mediated. For example, the adaptive web might take the form of an interactive, personalized social network visualization that enables users to discover knowledgeable peers, online communities, and other resources in cross-cutting research areas. The user might help the system personalize the discovery process by suggesting levels of constraints on criteria such as location, expertise, organizational role, and online collaboration level. Adaptive collaboration support technology sometimes also takes the form of distributed teams of socially aware intelligent recommendation agents that might put the user in contact with an online expert or instructor from a selected Community of Practice [34, also see section 19.3.2].

19.5 Future Trends

We communicate with each other through the many flavors of voice, text, appearance, behavior, and action, the complex interplay between these forms, and even through the absence of communication itself. The outcomes of our communication are sometimes difficult to predict because they depend on the combination of forms used in a particular context and timeframe. Even the notions of "context" and "timeframe" in today's internationally networked knowledge-based society are unclear, as are the traditional characteristics that distinguish asynchronous from synchronous communication. Interaction in such a society might be supported and enhanced by harnessing the opportunities afforded by the adaptive web as a unique communication medium. This is however a unique challenge that may require researchers to design new theories of interaction, and develop new performance support tools that enable the seamless shifting between communication forms, while providing awareness and a greater understanding of the interaction as it evolves and transforms contexts over time.

In designing the next generation of collaboration tools for the adaptive web, we should continue to improve the interoperability and design of collaborative tools for voice, text, and nonverbal communication and for constructing and annotating documents, images, and videos. Awareness and support facilities should help people understand the way that the technology increases learning and work efficiency and shapes participant roles as they move between social contexts and communication tools. This suggests that we should invest more in the study of distributed collaborative work and learning processes in context and address our findings through new collaboration tool paradigms. The technology should be prepared to model and analyze unpredictable events in new contexts, learn from those events, and effectively impart its knowledge to its human collaborators.

The web indeed provides a vast knowledge resource and the opportunity to improve individual productivity through advanced filtering and adaptation algorithms, but it is also an interface for knowledge construction, a mediator for communication and learning, and a structured channel through which knowledge is created, interpreted, used, and recreated by other users.

Acknowledgments. This work was supported by the Institute for Defense Analyses Central Research Program. The opinions, assertions, and analyses in this chapter are those of the author alone. They do not necessarily reflect official positions or views of any U.S. government entity, and they should not be construed as asserting or implying U.S. government endorsement of this content.

References

1. Bouquet, P., Serafini, L., Zanobini, S.: Semantic coordination: A new approach and an application. Proceedings of the 2nd International Semantic Web Conference (2003) 130–145
2. Brown, J.S., Duguid, P.: Organizational learning and communities of practice: Toward a unified view of working, learning, and innovation. Organization Science 2(1) (1991) 40–57
3. Brusilovsky, P., Millán, E.: User models for adaptive hypermedia and adaptive educational systems. In: Brusilovsky, P., Kobsa, A., Nejdl, W. (eds.): The Adaptive Web: Methods and Strategies of Web Personalization, Lecture Notes in Computer Science, Vol. 4321. Springer-Verlag, Berlin Heidelberg New York (2007) this volume
4. Budzik, J., Bradshaw, S., Fu, X., Hammond, K.: Supporting online resource discovery in the context of ongoing tasks with proactive software assistants. International Journal of Human-Computer Studies 56(1) (2002) 47–74
5. Burke, R.: Hybrid Web Recommender Systems. In: Brusilovsky, P., Kobsa, A., Nejdl, W. (eds.): The Adaptive Web: Methods and Strategies of Web Personalization, Lecture Notes in Computer Science, Vol. 4321. Springer-Verlag, Berlin Heidelberg New York (2007) this volume
6. Chiu, C.H.: Evaluating system-based strategies for managing conflict in collaborative concept mapping. Journal of Computer Assisted Learning 20 (2004) 124–132
7. Cho, H., Stefanone, M., Gay, G.: Social information sharing in a CSCL community. In: Stahl, G. (ed.): Proceedings of the 2002 ACM CSCL Conference Lawrence Elbaum Associates, Hillsdale, NJ (2002) 43–53
8. Cohen, E.: Restructuring the classroom: Conditions for productive small groups. Review of Educational Research 64(1) (1994) 1–35
9. Constantino-González, M.A., Suthers, D., Escamilla de los Santos, J.: Coaching web-based collaborative learning based on problem solution differences and participation. International Journal of Artificial Intelligence in Education 13 (2002) 263–299
10. Dieberger, A., Dourish, P., Höök, K., Resnick, P., Wexelblat, A.: Social navigation: techniques for building more usable systems. interactions 7(6) (2000) 36-45
11. Dillenbourg, P., Traum, D.: Does a shared screen make a shared solution? In: Hoadley, C., Rochelle, J. (eds.): Proceedings of the Third Conference on Computer Supported Collaborative Learning. Lawrence Erlbaum Associates, Mahwah, NJ (1999) 127–135
12. Esfandiari, B., Chandrasekharan, S.: On how agents make friends: Mechanisms for trust acquisition. Proceedings of the 4th Workshop on Deception, Fraud and Trust in Agent Societies (2001) Montreal, Canada, 27-34
13. Frank, F., Soller, A.: Collaboration and knowledge sharing across the intelligence community. In: Sapp, A., Brown, B., Kirkhope, J., Tomes, R. (eds.): The Faces of Intelligence Reform: Perspectives on Direction and Form. The Council for Emerging National Security Affairs (CENSA), New York (2005) 99-102
14. Giordani, A., Gerosa, L., Soller, A., Stevens, R.: Extending an online individual scientific problem-solving environment to support and mediate collaborative learning. Proceedings of the Artificial Intelligence in Education (AI-ED 2005) Workshop on Representing and Analyzing Collaborative Interactions (2005) 12-22
15. Goodman, B., Linton, F., Gaimari, R., Hitzeman, J., Ross, H., Zarrella, G.: Using dialogue features to predict trouble during collaborative learning. User Modeling and User-Adapted Interaction 15(1) (2005) 85–134
16. Hatano, G., Inagaki, K.: Sharing cognition through collective comprehension activity. In: Resnick, L., Levine, J., Teasley, S. (eds.): Perspectives on socially shared cognition. American Psychological Society, Washington D.C. (1991) 331–348
17. Inkpen, K., McGrenere, J., Booth, K., Klawe, M.: The effect of turn-taking protocols on children's learning in mouse-driven collaborative environments. Proceedings of Graphics Interface '97, Kelowna, BC (1997) 138–145

18. Jermann, P.: Computer support for interaction regulation in collaborative problem solving. Unpublished doctoral dissertation, University of Geneva, Switzerland (2004)
19. Kautz, H., Selman, B., Shah, M.: The hidden web. AI Magazine 18(2) (1997) 27–36
20. Kohonen, T.: Self-organizing maps. Springer-Verlag, Berlin (2001)
21. Lave, J., Wenger, E.: Situated learning: Legitimate peripheral participation. Cambridge University Press, Cambridge (1991)
22. Lesgold, A.: Contextual requirements for constructivist learning. International Journal of Educational Research 41(6) (2004) 495-502
23. Levine, J.M., Moreland, R.L.: Small groups. In: Gilbert, D., Fiske, S., Lindzey, G. (eds.): The handbook of social psychology. McGraw-Hill, Boston, MA (1998) 415–469
24. Linton, F.: OWL: A system for the automated sharing of expertise. In: Ackerman, M., Pipek, V., Wulf, V. (eds.): Sharing expertise: Beyond knowledge management. MIT Press, Cambridge, MA (2002) 383–401
25. Mantovani, G.: Social context in HCI: A new framework for mental models, cooperation, and communication. Cognitive Science 20 (1996) 237–269
26. McDonald, D., Ackerman, M.: Expertise recommender: A flexible recommendation system and architecture. Proceedings of the ACM Conference on Computer Supported Cooperative Work (2000) 231–240
27. Mislevy, R., Steinberg, L., Almond, R.: On the structure of educational assessments. Measurement: Interdisciplinary Research and Perspectives 1(1) (2003) 3–62
28. Rabiner, L.: A tutorial on Hidden Markov Models and selected applications in speech recognition. Proceedings of the IEEE 77(2) (1989) 257–286
29. Ronchetti, M., Gerosa, L., Giordani, A., Soller, A., Stevens, R.: Symmetric synchronous collaborative navigation applied to e-learning. IADIS International Journal on WWW/Internet 3(3) (2005) 1-16
30. Schafer, J.B., Frankowski, D., Herlocker, J., Sen, S.: Collaborative Filtering Recommender Systems. In: Brusilovsky, P., Kobsa, A., Nejdl, W. (eds.): The Adaptive Web: Methods and Strategies of Web Personalization, Lecture Notes in Computer Science, Vol. 4321. Springer-Verlag, Berlin Heidelberg New York (2007) this volume
31. Shardanand, U., Maes, P.: Social information filtering: Algorithms for automating "Word of Mouth". CHI'95 - Human Factors in Computing Systems (1995) 210–217
32. Shinar, D., Dewar, R., Summala, H., Zakowska, L.: Traffic sign symbol comprehension: a cross-cultural study. Ergonomics 46(15) (2003) 1549–1565
33. Soller, A.: Understanding knowledge sharing breakdowns: A meeting of the quantitative and qualitative minds. Journal of Computer Assisted Learning 20 (2004) 212–223
34. Soller, A., Guizzardi, R., Molani, A., Perini, A.: SCALE: Supporting community awareness, learning, and evolvement in an organizational learning environment. Proceedings of the 6th International Conference of the Learning Sciences, Santa Monica, CA (2004) 489–496
35. Soller, A., Martínez-Monés, A., Jermann, P., Muehlenbrock, M.: From mirroring to guiding: A review of state of the art technology for supporting collaborative learning. International Journal of Artificial Intelligence in Education 15(4) (2005) 261–290
36. Stahl, G.: Group Cognition: Computer Support for Building Collaborative Knowledge. MIT Press, Cambridge, MA (2005)
37. Stasser, G.: The uncertain role of unshared information in collective choice. In: Thompson, L., Levine, J., Messick, D. (eds.): Shared knowledge in organizations Erlbaum, Hillsdale, NJ (1999) 49–69
38. Stevens, R., Johnson, D., Soller, A.: Probabilities and predictions: Modeling the development of scientific problem solving skills. Cell Biology Education 4(1) (2005) 42–57 (Available at http://www.cellbioed.org/)
39. Terveen, L.G., McDonald, D.W.: Social matching: A framework and research agenda. ACM Transactions on Computer-Human Interaction (ToCHI) 12(3) (2005) 401–434

40. Vassileva, J.: Supporting peer-to-peer user communities. In: Meersman, R., Tari, Z. (eds.): On the Move to Meaningful Internet Systems 2002: CoopIS, DOA, and ODBASE : Confederated International Conferences CoopIS, DOA, and ODBASE 2002. Springer-Verlag, Berlin-Heidelberg (2002) 230–247
41. Wasserman, S., Faust, K.: Social network analysis: Methods and applications. Cambridge University Press, Boston (1994)
42. Webb, N., Palincsar, A.: Group processes in the classroom. In: Berlmer, D., Calfee, R. (eds.): Handbook of educational psychology. Simon & Schuster Macmillan, New York (1996) 841–873
43. Wenger, E., McDermott, R., Snyder, W.: Cultivating Communities of Practice: A guide to managing knowledge. Harvard Business School Press, Boston (2002)

20

Recommendation to Groups

Anthony Jameson[1] and Barry Smyth[2]

[1] DFKI, German Research Center for Artificial Intelligence
[2] Department of Computer Science, University College Dublin

Abstract. Recommender systems have traditionally recommended items to individual users, but there has recently been a proliferation of recommenders that address their recommendations to groups of users. The shift of focus from an individual to a group makes more of a difference than one might at first expect. This chapter discusses the most important new issues that arise, organizing them in terms of four subtasks that can or must be dealt with by a group recommender: 1. acquiring information about the user's preferences; 2. generating recommendations; 3. explaining recommendations; and 4. helping users to settle on a final decision. For each issue, we discuss how it has been dealt with in existing group recommender systems and what open questions call for further research.

20.1 Introduction

Almost all of the techniques of web-based personalization discussed in the other chapters of this book are designed to allow effective adaptation to individual users. But often the users of such systems operate not individually but in groups, which may vary from formally established, long-term groups to ad hoc collections of individuals who use a system together on a particular occasion. This phenomenon can in principle occur with just about any form of web personalization. In this chapter, we will focus on the subclass of recommender systems (cf. the chapters in this volume by Schafer et al. [34]; Pazzani & Billsus [31]; Smyth [35]; Burke [5]; and Goy & Ardissono [15]), but many of the points made will be applicable by analogy to other types of adaptive web-based system (cf. Section 20.6.2).

Some types of items that a system can recommend (e.g., restaurants and museum exhibits; see Table 20.1 for additional examples) tend to be used at least as often by groups as by individuals, so addressing recommendations to individuals can actually be unnatural. Moreover, the evolution of computers away from the desktop PC makes it increasingly natural for systems to address groups as well as individuals: Wall displays, information kiosks, PDAs, and cell phones can be used easily by persons who are interacting with each other. And even with the traditional PC, users are being offered an increasing variety of ways to communicate with each other and perform tasks together. For these reasons, we can expect a continuing growth in the trend toward recommendation (and, more generally, adaptation) to groups of users.

In this chapter, we will identify the issues that should be addressed by designers of group recommender systems and the ways in which they have been dealt with in

P. Brusilovsky, A. Kobsa, and W. Nejdl (Eds.): The Adaptive Web, LNCS 4321, pp. 596–627, 2007.
© Springer-Verlag Berlin Heidelberg 2007

systems that have been developed so far. Our two goals are to allow designers and researchers (a) to make effective use of knowledge and experience that have already been accumulated and (b) to address the many open questions that still require careful consideration and research.

20.1.1 Existing Group Recommenders

Figure 20.1 lists almost all of the group recommender systems that, according to the authors' knowledge, have been described in the literature up to the time of the writing of this chapter. Although a number of these systems have been described only briefly and some were not implemented as web-based systems, we will refer to aspects of all of them, so as to convey an idea of the variety of application settings, design issues, and possible methods that designers of group recommender systems should be aware of.

20.1.2 Overview of Recommendation Subtasks and Issues

Relative to recommendation for individuals, there are a number of new issues that arise with group recommenders. Table 20.2 organizes these issues in terms of four high-level subtasks than must (or can) be performed by a group recommender. The issues corresponding to each of these subtasks will be addressed in one section of this chapter.

These subtasks differ greatly in the amount of attention they have attracted in research so far and hence in the length of the corresponding sections of this chapter. By far the most research has been done on Subtask 2, a fact that is understandable given that any group recommender must have some way of assessing the suitability of items for the group. Subtask 3 is the second most popular one, especially given the growing interest in making the reasoning underlying recommendations comprehensible to users (cf., e.g., the chapter in this volume by Schafer et al. [34]). Subtask 1 has attracted much less attention, because many methods for acquiring information about users' preferences are equally applicable to groups and to individuals; though the extension to groups does not always require a change in methods, it can create opportunities to introduce new methods. Subtask 4 has attracted the least attention of all, since it is usually assumed that the final decision about whether to accept a recommendation will be made by a single group member or in face-to-face discussion among group members.

20.2 Acquiring Information About Group Members' Preferences

Most group recommenders developed so far apply methods for acquiring information about users' preferences that are barely distinguishable from the methods applied in recommender systems for individuals. After briefly surveying some typical applications of such methods, we will look at preference acquisition methods that have been developed specifically for group recommendation settings.

Table 20.1. Overview of the group recommender systems mentioned in this chapter.

System	Reference	(Examples of) Groups of Users	Items recommended
		Web / news pages	
Let's Browse	Lieberman et al. (1999)	Persons browsing the web together	Web pages
G.A.I.N	Pizzutilo et al. (2005)	Persons viewing a wall display or information kiosk	News items
I–Spy	Smyth et al. (2005)	Employees of a company	Web pages
		Tourist attractions	
Intrigue	Ardissono et al. (2003)	Tourists	Sightseing tours
CATS	McCarthy et al. (2006)	Friends planning a vacation	Vacation packages
Travel Decision Forum	Jameson (2004)	Friends planning a vacation	Criteria for choosing a vacation package
Group Modeler	Kay and Niu (2005)	Persons visiting a museum together	Information about exhibits
Pocket RestaurantFinder	McCarthy (2002)	Colleagues going out to dine together	Restaurants
		Music tracks	
MusicFX	McCarthy and Anagnost (1998)	Persons working out in a gym	Music stations
Flytrap	Crossen et al. (2002)	Persons using a public area of a building	Music tracks to be played
In–vehicle multimedia recommender	Yu et al. (2005)	Passengers in a vehicle	Multimedia items to be played
Adaptive Radio	Chao et al. (2005)	Colleagues working together in an office	Songs to be played on the radio
		Television programs and movies	
FIT	Goren–Bar and Glinansky (2002)	Family members watching TV together	TV programs
TV program recommender	Yu et al. (2006)	TV viewers	Sequences of TV programs
PolyLens	O'Connor et al. (2001)	Persons planning to go to a movie together	Movies

Table 20.2. Overview of the issues to be addressed in this chapter, organized in terms of four subtasks of a group recommender system.

Subtask of the recommender system	Difference from recommendation to individuals	General issues raised
1. The system acquires information about the members' preferences.	If members specify their preferences explicitly, it may be desirable for them to be able to examine each other's preference specifications.	What benefits and drawbacks can such examination have, and how can it be supported by the system?
2. The system generates recommendations.	Some procedure for predicting the suitability of items for a group as a whole must be applied.	What conditions might such a procedure be required to fulfill; and what kinds of procedure tend to fulfill these conditions?
3. The system presents recommendations to the members.	The (possibly different) suitability of a solution for the individual members becomes an important aspect of a solution.	How can relevant information about suitability for individual members be presented effectively?
4. The system helps the members arrive at a consensus about which recommendation (if any) to accept.	The final decision is not necessarily made by a single person; negotiation may be required.	How can the system facilitate the necessary communication among group members?

20.2.1 Preference Acquisition Methods That Are Not Specifically Adapted to Group Recommendation

Acquisition of Preferences Without Explicit Specification. As we have seen in the chapters by Schafer et al. [34]; Pazzani & Billsus [31]; Smyth [35]; Burke [5]; and Goy & Ardissono [15], many recommender systems do not require their users to specify their preferences explicitly. With group recommenders as well, it may be possible for the system to get by with implicitly acquired information about users. A straightforward example is found in the system FLYTRAP (Crossen et al. [11]), which selects music for playing in a public room. The system learns about the music preferences of the potential users by (a) noticing what MP3 files each user plays on his or her own computer and (b) consulting available information about the music played to derive a model of each user's preferences.

Another example is found in LET'S BROWSE (Lieberman et al. [23]), which recommends web pages to a group of two or more persons who are browsing the web together. The system makes initial estimates of the interests of its users by analyzing the words that occur in each user's web homepage. During the actual group browsing, it analyzes the words that occur in the pages visited by the group.

Explicit Preference Specification. But there are some types of group recommender that do require an explicit specification of preferences. An example is the POCKET RESTAURANTFINDER (McCarthy [26]), which helps a group of people who are preparing to go out to eat together in selecting a restaurant. For each of 15 types of cuisine that might be represented at a given restaurant, each user must indicate his or her preference on a 5-point scale ranging from "Definitely don't want ..." to "Definitely want ...". Similar ratings are given for 17 possible restaurant amenities, 3 price categories, and 3 ranges of travel time from the current location. (Users presumably consider this rating effort worthwhile only if they intend to use the system repeatedly.)

Similarly, explicit preference specifications are required by the TRAVEL DECISION FORUM (Jameson [18]; Jameson et al. [19]), which helps a group of users to agree on the desired attributes of a vacation that they are planning to take together. The system needs to know how each user feels about dozens of attributes of vacation destinations, ranging from the facilities that are available in their rooms to the sightseeing attractions that are available in the surrounding area. Here again, only explicit elicitation is likely to be feasible.

A less explicit form of preference specification is found in POLYLENS, an extension of the MOVIELENS system (cf. the chapter in this volume by Schafer et al. [34]) that recommends movies to groups of users. Since POLYLENS (like MOVIELENS) is based on collaborative filtering, users do not explicitly describe their movie preferences, but they do rate individual movies on a scale from 1 to 5 stars. As we will see, this procedure raises some of the same issues as the explicit specification of general preferences.

An intermediate case between implicit and explicit preference specification is found in the system I-SPY (see, e.g., Smyth et al. [37]), a community-based web search engine that personalizes search results for a community of like-minded searchers on the basis of a model of community search preferences.[1] The I-SPY user indicates interest in a given search result by selecting the result in question from a query result list, and I-SPY interprets each result selection as an indication of relevance with respect to the current search query. The specification is implicit in that the user's primary intent in selecting a result is not in general to indicate his or her preferences to the system; but it has some elements of explicitness in that users are aware of the fact that their selections are being interpreted as reflecting their preferences and can, if they like, choose results that they would not otherwise have chosen, in order to influence the system's preference model (cf. the discussion of manipulation in Section 20.3.2).

20.2.2 Adapting Preference Specification to the Requirements of Group Recommendation

Focus on Negative Preferences. In the context of the system ADAPTIVE RADIO, Chao et al. [7] argue that the method used to elicit preferences from users should take into account the way in which these preference specifications will subsequently be used

---

[1] I-SPY is not a group recommender system in the most commonly assumed sense, because its recommendations are made use of by individuals, not by members of a collaborating group. But as we will see, the system addresses some of the same issues as systems that make recommendations to groups.

for the generation of recommendations. Specifically, they argue that for a system that chooses music to be played to a group, it makes more sense to elicit *negative preferences* (e.g., expressions of dissatisfaction with particular music tracks) than to elicit more detailed types of rating such as the ones mentioned in the previous subsection. If the procedure that will be used for generating recommendations is designed mainly to avoid the playing of music that is disliked by any member (cf. Section 20.3.2), effort expended by one group member in discriminating among different degrees of liking may in fact be wasted, since many of the songs that that member likes may in effect be vetoed by another member anyway. An informal evaluation of ADAPTIVE RADIO suggested that the focus on negative preferences was appropriate in the particular application setting studied.

Sharing Information About Specified Preferences. In a recommender system for individuals, there is in general no person besides the user who has an immediate interest in seeing explicitly specified preferences with a view to improving the current recommendation process. In a group recommender, each member may have some interest in knowing the other members' preferences, for several possible reasons:

1. *Saving of effort.* Specifying preferences is usually seen by users as a tedious process. (The avoidance of tedium is claimed by Chao et al. [7], as a supplementary advantage of their focus on negative preferences.) If a group member m_1 knows that another member m_2 with generally similar preferences has already specified their preferences, m_1 may be able to save time and effort by copying at least some of m_2's entries and then perhaps making some changes—especially if the system makes it easy to do such copying and postediting.
2. *Learning from other members.* Another member's preferences may be based in part on knowledge or experience (e.g., concerning a particular vacation destination) that the current member lacks.

An attempt to exploit both of these potential benefits is found in the TRAVEL DECISION FORUM: A simple extension of a typical rating-scale dialog box allows the current member optionally to view (and perhaps copy) the preferences already specified by other members (see Figure 20.1). An additional feature that makes sense mainly if other persons will be viewing the specifications is the option to add brief verbal explanations or *arguments* for specific ratings.[2] Arguments can have various forms and functions in group decision contexts (cf., e.g., Jennings et al. [21]). In a group recommendation context, two typical functions are (a) to persuade other members to specify a similar preference, perhaps by giving them information that they previously lacked; and (b) to explain and justify a member's preference even if the argument is not generalizable to other members (e.g., "I can't go hiking, because of an injury").

Experience with this method of *collaborative preference specification* has revealed further benefits beyond the two already mentioned:

---

[2] These arguments can be entered and viewed in pop-up windows that are not visible in Figure 20.1.

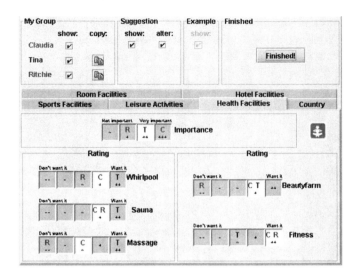

Fig. 20.1. Dialog box for the collaborative specification of preferences in the TRAVEL DECISION FORUM. (The currently active group member is Claudia, the other two are Ritchie and Tina. The preferences of each member are represented by the first letter of his or her name. Each scale refers to a single attribute and ranges from $--$ for "Don't want it" to $++$ for "Want it". The highlighting of one cell for each attribute is added only when a compromise proposal has been suggested, as explained in Section 20.3.1. Figure 1 of Jameson [18], reproduced with permission.)

1. *Taking into account attitudes and anticipated behavior of other members.* Sometimes the preference of the current member depends in part on the preferences and/or the anticipated behavior of one or more other members. For example, if m_1 sees that m_2 has specified a strong preference for tennis facilities, m_1 may want to specify a similar preference, reasoning that if a hotel is found that offers tennis, m_1 and m_2 will be able to play together. Otherwise, m_1 may genuinely not want to emphasize tennis facilities, on the grounds that she would probably have no one to play with anyway.
2. *Encouraging assimilation to facilitate the reaching of agreement.* A different reason why m_1 may assimilate her preferences to those of m_2 is simply a desire to minimize conflicts that may make it more difficult for the group to find a solution. This pattern is especially likely in cases where m_1 was originally more or less indifferent between two possible preference specifications, before seeing that m_2 has chosen the other one of them. The difference between this case and the previous one is that here, m_1's true preference has not changed, but she has strategically changed her specification of it.

In a similar vein, the more recently developed vacation recommender system CATS (McCarthy et al. [28]; McCarthy et al. [29]) allows group members to achieve some awareness of each other's activities as they explore vacation options, working simultaneously around a DIAMONDTOUCH table, an environment that facilitates synchronous work of group members on a common project (cf. Dietz and Leigh [12]). In the

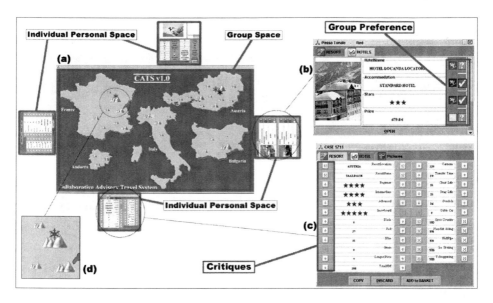

Fig. 20.2. Illustration of several ways in which the CATS system enhances mutual awareness among group members as they plan a skiing vacation. (Explanation in text. Figure 1 of McCarthy et al. [28], reproduced with permission.)

overview in Figure 20.2, several examples can be seen: Each mountain icon in the large map (a) represents one of the resorts that is being considered, the size of the icon reflecting the currently estimated overall preference of the group for that particular resort. In the description of a particular hotel (b) the check mark or question mark next to the color-coded icon for each group member indicates that group member's estimated interest in the hotel. The color-coded snowflakes on the map (a) indicate what resort each member is investigating at the moment. Finally, each member can send a "critique" that he or she is working on to the other members, thereby sharing his or her thoughts about a particular option.

Although I-SPY's preference specification is largely implicit, there are some phenomena involved in the use of I-SPY that are similar to those that arise with collaborative preference specification of the type we have seen with the TRAVEL DECISION FORUM. These are related to the fact that each user sees the effects of the choices made by other users, even if he does not recognize these effects as such. I SPY exposes the learned preferences of its community to searchers, in part by highlighting *promoted* results in a search result list (see Figure 20.3 for examples and Smyth et al. [36], and Smyth et al. [37] for further details). Thus, just as in the TRAVEL DECISION FORUM a user can intentionally copy the preferences specified by another group member, in I-SPY, the choices (and thus the implicit preference specifications) of each community member will tend to be affected by the choices of previous searchers. In recent versions of I-SPY, the current user can even see which individual users were responsible for the promotion of a particular link (see Section 20.4.2). The purpose of the provision of this type of information is to provide a sort of explanation of the recommendation of a given

Fig. 20.3. Screen shot of I-SPY illustrating several ways in which the current user is helped with information derived from the behavior of other community members. (The "eyes" icon indicates that a result has been promoted to a higher position in the result list than it would have had otherwise.)

link; this function will be discussed in more detail in Section 20.4; but this property is relevant here in that it can lead to (a) a user thinking twice about whether to choose a particular link because of knowing that others will see that he or she chose that link; and (b) the tendency of users to be influenced by the fact that particular other users have already expressed a degree of interest in a given link.

The idea of collaborative preference specification could be extended to some systems that do not yet to make use of it. For example, suppose that m_1 and m_2 jointly make use of MOVIELENS's *buddy* feature, which is the more recent implementation of the ideas introduced with POLYLENS (cf. http://movielens.umn.edu/). It is likely that the movie tastes of m_1 and m_2 overlap to a greater extent than the interests of an arbitrary pair of movie-goers. How might m_1 benefit from being able to use m_2's ratings as a starting point for her own ratings? The most obvious case would be the one in which many of m_1's ratings coincide exactly with those of m_2; but even simply knowing which movies m_2 has rated at all might be helpful: Perhaps the most tedious thing about using MOVIELENS is the job of finding movies that you have seen and so are able to rate—a process that may require scrolling through a list of movies that extends over many web pages. A list containing the set of movies that m_2 has rated but m_1 has not rated might contain a higher proportion of movies that m_1 will be able to rate.

20.3 Generating Recommendations

No matter how a group recommender acquires the members' preferences, the recommendation for the group will in general be based on information about the preferences of the individual group members. Therefore, some type of *aggregation* method is required, by which information about individual preferences is combined in such a way that the system can assess the suitability of particular items for a group as a whole.[3] The need to choose an aggregation method is the most obvious and intensively studied difference between group recommendation and recommendation for individuals. The topic of preference aggregation is a multifaceted and complex one that has been addressed in various scientific fields (see, e.g., Arrow [2], for a seminal contribution and Masthoff [24] for a summary of this literature from the perspective of group recommendation). The ideas to be discussed in this section overlap to some extent with ideas from these other fields, but a number of new elements are introduced by the technical and practical context of interactive adaptive systems. For example, much of the literature on preference aggregation considers large communities, whose members mostly do not know each other (e.g., citizens of a country who are voting in an election); and the practical contexts require aggregation methods that are technically simpler than many of those that are feasible in adaptive web-based systems.

In this section, we will first discuss ways in which the aggregation problem has been handled in group recommender systems to date. We will then discuss various complications that have not yet received as much attention as they will ultimately require.

We assume for now that the system is supposed to make recommendations concerning just one decision (e.g., one film that is to be watched by the group). The case where a sequence of decisions is to be made (e.g., concerning several TV programs that will be watched on a given evening) will be considered in Section 20.3.3.

20.3.1 Approaches to Preference Aggregation

Although the various approaches differ in the ways in which they gather and represent users' preferences, almost all approaches make use of one of three schemas: (a) merging of sets of recommendations, (b) aggregation of individuals' ratings for particular items, or (c) construction of group preference models.[4]

Merging of Recommendations Made for Individuals. In cases where what is to be presented is a *set* of candidate solutions, among which the group is to select one for

---

[3] It is not actually logically necessary for group recommendations to be based on information about the preferences of individual members. For example, a group movie recommender might somehow acquire the knowledge that Walt Disney movies tend to be suitable when parents are taking their small children to a movie, even if the recommender system has no information about how parents and children, respectively, tend to evaluate Walt Disney movies. But since in practice almost all group recommenders start with information about preferences of individual members, we will view the problem of making recommendations for a group as involving preference aggregation.

[4] Somewhat similar distinctions among aggregation approaches have been made by, among others, O'Connor et al. [30]; Kay and Niu [22]; and Yu et al. [39].

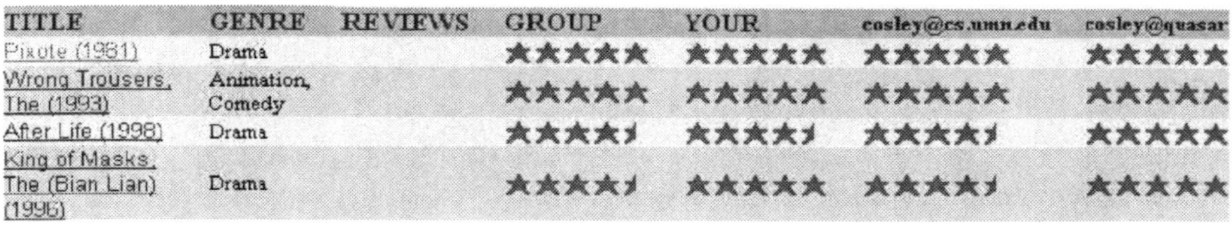

TITLE	GENRE	REVIEWS	GROUP	YOUR	cosley@cs.umn.edu	cosley@quasar
Pixote (1981)	Drama	★★★★★	★★★★★	★★★★★	★★★★★
Wrong Trousers, The (1993)	Animation, Comedy	★★★★★	★★★★★	★★★★★	★★★★★
After Life (1998)	Drama	★★★★✦	★★★★✦	★★★★✦	★★★★★
King of Masks, The (Bian Lian) (1996)	Drama	★★★★✦	★★★★★	★★★★✦	★★★★★

Fig. 20.4. Example of a display of group recommendations in POLYLENS. (Adapted with permission from an image supplied by John Riedl. Explanation in text.)

adoption, a simple aggregation method is that of generating a small number of recommended solutions for each member and then merging them into a single list:

1. For each member m_j:
   - For each candidate c_i, predict the rating r_{ij} of c_i by m_j.
   - Select the set of candidates C_j with the highest predicted ratings r_{ij} for m_j.
2. Recommend $\bigcup_j C_j$, the union of the set of candidates with the highest predicted ratings for each member.

This method was one of those considered for the POLYLENS system (O'Connor et al. [30]). Because of the simple relationship to the generation of recommendations for individuals, this method can be implemented easily as an extension of a recommender for individuals, making use, for example, of any explanation facilities of the individual recommender. In particular, if each recommendation is accompanied by a display of the (predicted) ratings of each member, the members may have quite a good basis for choosing a truly acceptable solution. On the other hand, this approach does presuppose that the members will play an important role in the final decision making, since the list of recommendations does not in itself indicate which solutions are best for the group as a whole. In fact, in the worst case each proposed solution might be excellent for one member but terrible for all of the others. More generally, this method ignores the set of solutions that are not expected to be especially appealing to any member but which might represent the best solution for the group as a whole.

Since this method is rarely even considered for use in group recommenders, we will not discuss it further.[5]

Aggregation of Ratings for Individuals. This commonly applied approach starts with the assumption that, for each candidate item c_i and each member m_j, the system can predict how m_j would evaluate (or *rate*) c_j if he or she were familiar with it:

1. For each candidate c_i:
   - For each member m_j predict the rating r_{ij} of c_i by m_j.
   - Compute an aggregate rating R_i from the set $\{r_{ij}\}$.
2. Recommend the set of candidates with the highest predicted ratings R_i.

---

[5] A closely related method is considered by Yu et al. [39] for the recommendation of *sequences* of TV programs; essentially the same drawbacks apply in that context as well.

This approach is illustrated by POLYLENS, as can be seen in Figure 20.4. The three right-hand columns in the screen shot display ratings that have been predicted for individual users via the same collaborative filtering method that MOVIELENS uses for individual users. The column labeled "GROUP" shows the aggregated rating. In POLYLENS, the aggregation method is very simple:

$$R_i = \min_j r_{ij}. \tag{20.1}$$

That is, instead of looking for the movie with the highest average rating, POLYLENS applies the strategy of "least misery": It bases its recommendation on the lowest predicted rating for each candidate, preferring candidates for which the lowest predicted rating for any group member is relatively high. Other plausible aggregation methods will be mentioned in Section 20.3.2.

Construction of Group Preference Models. The second widely applied approach to aggregation does not involve any predictions of ratings of individual users. Instead, the system somehow uses its information about the preferences of individual group members to arrive at a model of the preferences of the group as a whole:

1. Construct a preference model M that represents the preferences of the group as a whole.
2. For each candidate c_i, use M to predict the rating R_i for the group as a whole.
3. Recommend the set of candidates with the highest predicted ratings R_i.

With regard to Step 1: There are even more possible methods for the construction of group preference models than for the aggregation of individual ratings, since group preference models can take many different forms.

In some cases, the group preference model can be seen as an aggregation of individual preference models. An example is given by LET's BROWSE: Each individual user's profile is a set of keyword/weight pairs that reflects the typical content of the pages that this user likes to view. The system computes a model of the group by forming a linear combination of these individual models. From then on, it no longer has to consult the individual models when making recommendations. Similarly, in the context of an in-vehicle multimedia recommender (Yu et al. [40]) and a TV program recommender (Yu et al. [39]), Yu and colleagues introduced and evaluated a method for constructing a group preference model on the basis of individual preference models, using a notion of *distance* between preference models.

In other cases, a group preference model may represent an aggregate of preference models for subgroups, rather than for individual members. This approach is exemplified by the system INTRIGUE (Ardissono et al. [1]), which is designed to help tour guides who need to design tours for heterogeneous groups of tourists that include relatively homogeneous subgroups (e.g., "children"). The tourist group leader divides the tour group into several categories of homogeneous users and specifies a preference model for each such subgroup. The group model is then a weighted average of the subgroup models, with the weights reflecting the importance of the subgroups (e.g., the subgroup of disabled persons is considered especially important because of the special requirements of its members).

The TRAVEL DECISION FORUM takes the focus on group models one step further: In fact, the main function of the system is to help the group members, for each aspect of the vacation that the group members are planning, to arrive at a group preference model that all members have agreed to—that is, at a way of filling out each preference specification form (such as the one shown in Figure 20.1) in such a way that it reflects the preferences of the group as a whole. If we look at the system in this way, the system can be seen as one that recommends specific preferences for the group model (e.g., a rating of ++ for the attribute "Sauna" in Figure 20.1).

I-SPY creates a group (or community) preference model directly on the basis of data concerning the behavior of individual group members, bypassing the level of individual preference models—partly because of privacy considerations, as will be discussed shortly. I-SPY's basic community preference model consists of a record of queries that have been submitted (by the community of searchers) and the result pages that have been selected for these queries, along with frequency information for these selections. When deciding to what extent to promote a particular search result for a particular community, I-SPY bases its decision on an estimate of how relevant this result page is likely to be for the current query. This estimate is based on the frequency with which this page has been previously selected by community members for the current query and for similar queries.

Choosing Between Rating Aggregation and Group Preference Models. Constructing a preference model for the group has the clearest advantages when the group members will have an opportunity to examine and/or negotiate about the group's model before or after it is actually applied. In this case, for example, the users of INTRIGUE or the TRAVEL DECISION FORUM could settle among themselves once and for all the relative priorities of historical interest and entertainment, instead of debating this issue with respect to each individual attraction. This type of process will be discussed further in Sections 20.4 and 20.5.

If, on the other hand, the group model will be created and applied in the background, without inspection by the group members, the question of whether a group model is better is a more technical one that involves considerations such as efficiency and the quality of recommendations. For example, O'Connor et al. [30] discuss various ways in which POLYLENS could have been designed to create a model of each group (e.g., a "pseudo-user" who represents the interests of the group as a whole) before any recommendations were generated—and some typical consequences of such group models. For instance, a group model might (accurately or not) recommend a movie for which the predicted rating of each individual member was low—something that cannot happen with recommendation-level aggregation.

Another advantage of a group preference model concerns its potential privacy benefits. Recording and maintaining individual user profiles will typically raise privacy concerns, especially if these profiles are owned by some third-party system on the server side. In contrast, the use of a group preference model may go a considerable way toward alleviating these privacy concerns. I-SPY is a case in point. Our web search behavior can be surprisingly revealing when it comes to understanding the likes and dislikes of an individual—far more revealing and valuable to an eavesdropper than movie or music preferences, for example. I-SPY's use of a community-based profile, in which the search behavior of individual searchers is merged, means that the search preferences of any individual searcher can no longer be reconstructed.

20.3.2 Alternative Goals and Procedures for Aggregation

Even once a general approach has been chosen, the question arises of what particular computational procedure (or *mechanism*) should be used for the aggregation. This is the single question in this area that has received the most attention. The problem is that there are a number of goals that may be desirable in any given situation (e.g., total satisfaction, fairness, and comprehensibility), and conflicts between them can easily arise. In this section, we give several examples of such goals.

Whereas many treatments of these issues (see, e.g., Masthoff [24]; Yu et al. [39]) devote considerable attention to mathematical formulas, quantitative examples, and technical concepts, we will focus on the basic underlying issues and concepts and how they relate to realistic application scenarios.

Maximizing Average Satisfaction. Suppose at first that we are taking the approach of aggregating individual ratings. In this case, the goal of maximizing average satisfaction can be achieved by an aggregation function that computes some sort of average of the predicted satisfaction of each member for use as a basis for the selection of candidates (see Equation 20.2). The POCKET RESTAURANTFINDER (McCarthy [26]; cf. Section 20.2.1) applies a variant of this formula to the predicted ratings of restaurants by members of a group who are preparing to go out to dine together. The G.A.I.N. system of Pizzutilo et al. [32], which presents news items on a wall display or an information kiosk, uses a more complex variant of this formula that takes into account uncertainty about which users will be viewing the display at any given time; a similar procedure is applied in FIT (Goren-Bar and Glinansky [14]), which recommends TV shows for members of a family.

$$R_i = \text{average}(\{r_{ij}\}) = 1/n \cdot \sum_{j=1}^{n} r_{ij}. \tag{20.2}$$

If the predicted ratings are not thought to represent satisfaction accurately, some transformation of them can be used, such as the square of the rating; some results concerning transformations of this sort are given by Masthoff [24].

Minimizing Misery. Even if the average satisfaction is high, a solution that leaves one or more members very dissatisfied is likely to be considered undesirable. Even the most ego-centered group member may not want to have to interact with another member who is thoroughly dissatisfied; and such a member may refuse to go along with the solution in any case. In POLYLENS, the minimization of misery is the only criterion applied (see Equation 20.1 above). It is also possible to take this factor into account as a constraint that must be fulfilled by a solution: The lowest predicted rating must not fall below a given threshold.

Ensuring Some Degree of Fairness For similar reasons, a solution that satisfies everyone just about equally well is in general preferred to one that satisfies some at the expense of others—all other things being equal. Even more than in the case of minimizing misery, the goal of ensuring fairness is in general combined with some other goal. After all, no-one wants a perfectly fair solution that makes everyone equally miserable. For example (again assuming the approach of aggregating individuals' ratings), the aggregation of predicted individual ratings might include a penalty term that reflects the amount of variation among the predicted ratings, as in Equation 20.3:

$$R_i = \text{average}(\{r_{ij}\}) - w \cdot \text{standard-deviation}(\{r_{ij}\}), \tag{20.3}$$

where w is a weight that reflects the relative importance of fairness.

Treating Group Members Differently Where Appropriate. In some situations, it is generally agreed that the preferences of some group members need to be treated differently than those of others. If two hosts are planning a visit to a restaurant with a visitor from out of town, they are likely to give high priority to the visitor's preferences, requiring only that the solution is not entirely unsatisfactory for themselves (cf. e.g., Kay and Niu [22]). In INTRIGUE, the tourist guide is able to assign higher weights to subgroups such as those of disabled persons or children, on the assumption that these group members are less able to put up with solutions that are even partly unsatisfactory for them.

Discouraging Manipulation of the Recommendation Mechanism. The problem of *manipulation* is illustrated by experience with an early version of the system MUSICFX (McCarthy and Anagnost [27]), one of the earliest group recommender systems, which automatically selected music genres for the music to be played in a fitness studio: Although the system essentially applied an averaging procedure to construct a group model from individual preference models, the system also enforced a constraint of the type mentioned above in connection with the "least misery" criterion: Any music genre that was "hated" by any member currently in the gym was removed from the list of possible genres to play. Some users were observed to force an immediate change of genre by adapting their specifications to indicate that they "hated" the genre currently being played—even if they really didn't mind it but simply liked it less well than some other genres.

The potential for manipulation is even more obvious in the TRAVEL DECISION FORUM, in which one group member can often see the preferences specified by the other members. For example, suppose that in Figure 20.1 Claudia's true preference regarding the presence of a sauna was $\sim$ ("Don't care"): Instead of selecting the middle box in the scale, she might be inclined to select the left-most box (indicating strong disapproval of the availability of a sauna), so as to compensate for the positive preferences specified by the other group members Ritchie and Tina, expecting that the aggregated group preference for a sauna will end up being closer to her own.

When this type of insincere specification of preferences occurs, the aggregation algorithm used will be operating on false premises, since the algorithms presuppose that a group member's expressed preferences reflect his or her true preferences.

One way of making manipulation difficult is to make it impossible for users to see each others' preferences before specifying their own: If you don't know what the other members prefer, it is hard to distort the resulting recommendation in your own direction by specifying an insincere preference. But users may be able to guess other members' preferences (at least roughly); and in any case, as we have seen, there are advantages to allowing members to see each others' preferences at an early stage.

Manipulation is most likely to be possible if the input that the system uses for making its predictions consists of explicit preference specifications; with implicit inference of preferences, users are much less likely to be able to see how they could influence a recommendation by acting in some particular way. But exceptions can occur; for example, in I-SPY, user can quickly notice that, when they choose a particular link for a given query, that link gets promoted in the search result list for that query. It is then an obvious next move to click on links that one would like to promote (e.g., pages written by the user), regardless of their actual relevance to the query. One can view this type of manipulation as an alternative form of search engine spam, because subsequent users will see potentially irrelevant results being unjustifiably promoted to positions of prominence. As a potential solution to this problem, Briggs and Smyth [3] propose the use of an explicit model of trust that provides a filtering mechanism with a view to eliminating the contributions of these manipulative selections: The selections of individual users are evaluated for their reliability. In the simplest sense, a result selection is considered to be reliable if the same link is subsequently reselected by a certain minimum number of searchers for similar queries in the future. This information is used for (among other things) the evaluation of the trustworthiness of individual users, so that recommendations that stem from the activities of users with low trust values can be eliminated or demoted. Preliminary evaluation results suggest that the technique is capable of improving recommendation accuracy.

A different approach to discouraging manipulation is to have the system use an aggregation method that is inherently *nonmanipulable*: It is never in the interest of a given user to specify any preference other than the one that he or she really has. To return to the example with the TRAVEL DECISION FORUM given above: A simple nonmanipulable aggregation mechanism uses as a preference for the group as a whole concerning a given attribute the *median* of the individual preferences for that attribute (i.e., the one that falls exactly in the middle of an ordered list of all preferences). In our example, Claudia will not be able to drag down the group preference for a sauna below $+$ by specifying a low preference herself, since the median preference will be $+$ for any preference that she specifies between $--$ and $+$. (It will be left as an exercise for the reader to verify that, with the use of the median mechanism in this setting, no group member could ever benefit by specifying a preference insincerely.)

In general, many nonmanipulable mechanisms may exist for any given preference aggregation problem. In the research area of *automated mechanism design* (see, e.g., Conitzer and Sandholm [8]; Conitzer and Sandholm [9]; Sandholm [33]), methods are developed for automatically generating aggregation methods for a particular setting that (a) satisfy the constraint of being nonmanipulable (at least in that particular setting) and (b) also respect other constraints as well (e.g., maximizing average satisfaction and/or ensuring a certain degree of fairness). The methods introduced by Conitzer and Sand-

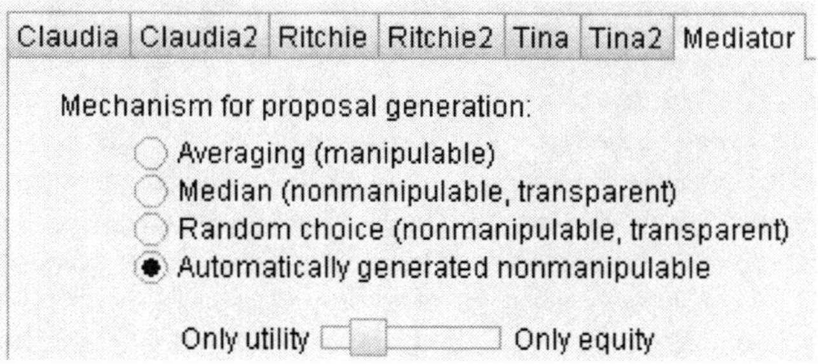

Claudia	Claudia2	Ritchie	Ritchie2	Tina	Tina2	Mediator

Mechanism for proposal generation:

○ Averaging (manipulable)
○ Median (nonmanipulable, transparent)
○ Random choice (nonmanipulable, transparent)
● Automatically generated nonmanipulable

Only utility ▭ Only equity

Fig. 20.5. Part of a dialog box in the TRAVEL DECISION FORUM via which an aggregation procedure can be selected. (The slider, which is active only when a mechanism is to be generated automatically, determines the relative weight of average satisfaction and fairness—cf. Equation 20.3. Adapted with permission from Figure 2 of Jameson [18].)

holm were implemented in the TRAVEL DECISION FORUM, along with the median mechanism just mentioned and two other mechanisms. The system administrator or an individual group member can request that an automatically designed mechanism be used for the generation of recommendations and can specify the properties that this mechanism should have (see Figure 20.5); the system then generates on the fly a mechanism that fulfills the specified constraints. The main issues that arose concerning the appropriateness of such automatically designed mechanisms were whether they were sufficiently comprehensible (or "transparent", in the terms of Figure 20.5) and acceptable to users. These issues will be discussed in the next subsection.

Ensuring Comprehensibility and Acceptability. As will be discussed in Section 20.4, group members sometimes like to be able to understand the rationale behind a recommendation. In particular, they may want to check to what extent acceptability criteria such as the ones discussed earlier in this section are being fulfilled. Even with ingenious visualizations such as those that will be shown in Section 20.4, it may be difficult for a system to explain a recommendation if the mechanism by which the recommendation was generated is inherently complex and/or counterintuitive. Therefore, it may be worthwhile to choose an inherently comprehensible mechanism even if it is not the best mechanism in terms of the other criteria.

An example of a comparison of mechanisms in terms of their inherent comprehensibility and acceptability is the exploration of automatically designed nonmanipulable mechanisms in the context of the TRAVEL DECISION FORUM (cf. the previous subsection and Jameson et al. [20]). One fundamental limitation of the automatically designed mechanisms is the fact that such a mechanism cannot be represented with a simple formula (such as the formula for the average or the median) but rather has to be represented by a table that specifies, for each possible combination of preferences of individual users, which item should be chosen.[6] Therefore, a group member cannot apply the mechanism mentally in order to predict or understand recommendations.

---

[6] Actually, automatically generated mechanisms are often nondeterministic: For each possible combination of preferences, the mechanism specifies a vector of probabilities associated with

Moreover, unless special acceptability constraints are applied in the mechanism generation process, the generated mechanism may give rise to recommendations that strike people as counterintuitive and inappropriate (e.g, proposing for some combinations of individual preferences an outcome that none of the group members likes, even though there exist outcomes that some members like). In sum, automated mechanism design is an approach that deserves further attention, but special attention must be paid to the goal of ensuring adequate comprehensibility and acceptability.

20.3.3 Further Complications Concerning Preference Aggregation

The often conflicting goals discussed in the previous subsection are in themselves enough to make the problem of choosing a suitable aggregation procedure a difficult one. But there are additional complications that need to be taken into account in some settings.

Generating Recommendations Concerning Multiple Decisions. So far, we have been focusing on the situation in which a recommender will make recommendations to a group concerning just one decision. But often the group members will expect a system to make recommendations concerning a larger set of decisions, either at the same time or in succession: INTRIGUE's tour guide will choose several sights to visit; the music selection systems will choose one song after the other; and a TV program recommender will recommend several programs for a group to watch in succession.

In this type of situation, the procedure for generating recommendations about the entire sequence can be related to a procedure with respect to individual decisions in any of several ways. Figure 20.3 compares three approaches, which differ in terms of several criteria.

1. The simplest approach is to treat each decision separately, ignoring the fact that there will be a sequence of decisions. Because of its simplicity, this approach tends to be computationally simple and easy to explain to users. One drawback is that a goal such as ensuring fairness can be taken into account only with respect to individual decisions, whereas it can be advantageous to consider it with regard to the entire set of decisions. For example, it may seem fair enough to recommend a single TV program that is much less attractive for one group member than for the others as long as that group member's overall satisfaction with the sequence of programs is comparable to that of the other members. Trying to ensure approximately equal satisfaction among a group members with each individual program in the sequence may rule out too many options that would be attractive for the group as a whole. An even clearer limitation of this approach arises in cases where the system can acquire feedback about the results of each decision before making recommendations concerning the next one. Suppose, for example, that one group member was especially dissatisfied with a given TV program, even though this dissatisfaction was not predicted in advance. It may be feasible and desirable to bias the recommendations concerning one or more subsequent programs in favor of that group member, so that he or she a ultimately reaches an appropriate level of satisfaction; but

---

the various possible outcomes, which is to be used for the random selection of one of the outcomes.

Table 20.3. Positive ($+$), negative ($-$), and intermediate ($+/-$) aspects of three approaches to the treatment of a sequence of decisions by a group recommender.

Criterion	Approach to treating a sequence of decisions		
	Independently	As one complex decision	Individually but with consideration of other decisions
Computational complexity	$+$	$-$	$+/-$
Comprehensibility	$+$	$-$	$+/-$
Appropriateness of evaluation criteria	$-$	$+$	$+/-$
Ability to take into account actual results of individual decisions	$-$	$-$	$+$
Applicability when decisions and members involved are not known in advance	$+$	$-$	$+/-$
Ability to take into account additional ordering constrains	$-$	$+$	$+/-$

this type of compensation is not possible if the decisions are treated completely independently. Similarly, this approach cannot deal with other constraints that concern relationships among members of the sequence (e.g., the possible undesirability of presenting two very similar items in succession, cf. the discussion of FLYTRAP below; or a possible tendency of earlier items in the sequence to have a generally larger impact on users than later items, cf. Masthoff [25]).

2. The opposite approach is to view the recommendations for the entire sequence as a single recommendation problem, much like that of recommending complex items (such as vacations) that differ with respect to a number of attributes. This approach is applicable only if it is known in advance what particular sequence of decisions is going to be made and what group members are going to be involved (e.g., how many times and on which particular occasions a group of diners is going to go out to dine together). This approach makes it straightforward to apply criteria such as fairness to entire sequences. On the other hand, it does not make it possible to take into account the results of previous decisions, since the decision making process for the entire sequence is completed before any decisions are executed. Also, the necessary computational procedures tend to be more complex; for example, the number of possible sequences may be too large for it to be feasible for the system to iterate through all possible sequences and evaluate their suitability.

3. An approach that lies between these two extremes starts with the idea of treating each decision problem separately but makes some adjustments to take into account the

fact that a sequence is being dealt with. For example, each individual decision might be approached with the goal of maximizing overall satisfaction; but if the system notices that, up to a given point in time, a given member has been less satisfied than the others, his or her satisfaction can be given greater weight in subsequent decisions, until the discrepancy has been eliminated. This approach is able to take into account the actual results of decisions (as opposed to only the predicted results). Although more complex than independent treatment of decisions, the method may be reasonably explainable if it corresponds with familiar decision making schemas from everyday life (e.g., "That last program was awful for Mary, so let's give her a break this time.") This approach does not require the set of decisions to be known in advance, but it can be applied most effectively if a good deal is known. For example, if the system has decided to grant a certain amount of extra satisfaction to a particular group member while making the subsequent recommendations, it will be helpful to know how many decisions remain to be made.

An example of this third approach is found in the FLYTRAP music selection system (cf. Section 20.2.1): The system has to choose songs one at a time, because the set of persons who are present to hear them frequently changes. But its selection procedure does take into account constraints imposed by a "DJ agent" that tries, for example, to avoid abrupt and distracting changes of genre.

20.3.4 Preference Specifications That Reflect More Than Personal Taste

In most analyses, it is assumed that the preferences specified by a group member represent simply the desires of that individual member (e.g., in a TV context, the programs that the group member would watch if he or she were watching alone). But in some settings, a group member may be taking into account the assumed interests of other members when expressing his or her own preferences. For example, we noted in connection with the TRAVEL DECISION FORUM that users sometimes expressed preferences in such a way as to minimize the likelihood of conflict. To take a more extreme example, consider a mother who is taking her two children to the movies and whose primary motivation is to find a movie that the children will like and that she herself will not hate having to sit through. In this case, her preference specification (e.g, a high rating for a particular children's movie) will reflect only to a minimal extent her own taste in movies. So it would not be appropriate, for example, for a recommender system to look straightforwardly for a "compromise" between the mother's expressed preference and the preferences expressed by the children (though some more sophisticated aggregation of the various expressed preferences might well make sense).

An additional complication arises when the group members' preferences reflect not just subjective tastes but also knowledge that may be relevant to the choices of the other members as well. For example, suppose that a member m_1 of a travel group who is especially familiar with Switzerland expresses a strong preference for a given Swiss ski resort: The other group members may be willing to give extra weight to m_1's opinion, even if their own evaluation criteria for resorts are somewhat different, on the grounds that the resort in question is especially likely to be good at least according to m_1's criteria, which overlap to some extent with the criteria of the other group members.[7] An ad hoc way of taking differences in knowledge into account is to assign greater weight to the preferences of more knowledgeable group members; but this method does not address in a principled way the relationship between knowledge and preferences.

20.3.5 At What Points Can These Complexities Be Dealt With?

After this lengthy discussion of conflicting goals for preference aggregation procedures and additional complexities, the reader may be wondering whether it will ever be possible to deal with these issues adequately in group recommender system. Fortunately, the issues can be dealt with by different people at different points in time:

1. *The system's designers and/or deployers may specify an appropriate means of handling each problem:* The persons who are designing a group recommender—or arranging the deployment of an existing recommender in a given context—can consider each of the issues discussed above with regard to their particular target group and application setting and work out some locally appropriate solution. For example, the designers of POLYLENS thought that the "least misery" aggregation function would be appropriate because they expected most groups of people who go to see a movie together to be small (i.e., 2 or 3 members); for settings involving larger groups, the same function would probably lead to too many cases in which a solution that would be liked by many members would in effect be vetoed by the one person who liked it least. If the designers of a restaurant recommender anticipate that there will often be some group members who are familiar with the restaurants in question and others who are not, they might look for a principled way of treating the two types of group member differently.
2. *The system's users may select a suitable preference aggregation method for each decision:* A system can allow the users to decide what aggregation mechanism is to be used, either before any recommendations are made or during an iterative process of requesting recommendations and adjusting the aggregation function. For example, with INTRIGUE the tour guide can specify a different set of subgroup weights for each tour group. As was mentioned above, a variety of aggregation mechanisms can be chosen in the TRAVEL DECISION FORUM. This idea could be adopted in many recommendation settings.
3. *Users can take any remaining factors into account when evaluating specific recommendations and negotiating about the final decision:* If the system presents a number of recommendations and allows users to choose which one(s) they want to adopt, it may not be necessary for the system itself to deal with all of the subtle problems that can arise. Instead, the users themselves may be able to take these issues into consideration, making use of their long experience with social interaction and relationships. After all, even the most subtle of the issues discussed in this section concern matters that people are accustomed to dealing with in everyday life. If

---

[7] Hastie and Kameda [16] show how aggregation procedures can be compared in terms of their suitability for arriving at accurate decisions even in settings where the group members are assumed to have identical preferences—that is, where knowledge aggregation rather than preference aggregation is involved.

the group recommender is designed on the assumption that there are some aspects of the decision problem that are better dealt with by the users themselves, the function that the system serves is mainly decision support rather than decision making. In these cases, special importance should be assigned to the third and fourth subtasks of a group recommender (explaining recommendations and supporting final decision making), which will be discussed in the next two sections, respectively.

20.4 Explaining Recommendations

20.4.1 Motivation

Given the many ways in which recommendations for a group can be derived—and the often conflicting goals that can be pursued—it is natural that group members should want to understand to some extent how a recommendation was arrived at—and in particular, how attractive a recommended item is likely to be to each individual group member.

Recommender systems for individuals often accompany each recommendation with some sort of analysis of its predicted acceptability; the analysis may range from a simple index of the system's confidence to a complex visualization of the pros and cons of the recommended solution (see, e.g., Herlocker et al. [17], and the chapter in this volume by Schafer et al. [34]). With group recommenders, it is in principle possible to present such an analysis for each individual member, for the group as a whole, and perhaps for subsets of members. A member m_1 may be interested in the analysis for m_2 because m_1 considers it important that m_2 be satisfied, because m_1 wants to make sure that she is getting "as good a deal" as m_2, or simply in order to understand how the recommendation was derived.

20.4.2 Treatment in Existing Group Recommenders

As can be seen in Figure 20.6, LET'S BROWSE explains each of its web page recommendations by listing the keywords in the page that it assumes to be of interest to all group members. By showing where these keywords are located in each member's profile, the system also allows each user to guess how interesting each member will find the page. In the example in the figure, it looks as if the (hypothetical) group member George Lucas will be less enthusiastic about the page than the member Bill Gates, given that Lucas is only marginally interested in technology. As some of the systems to be discussed below suggest, it might be a worthwhile further step for LET'S BROWSE to present an explicit estimate of the likely interestingness of the page for each group member and perhaps for the group as a whole. In this way, for example, Lucas might more readily accept the system's recommendation of the page involved in Figure 20.6, seeing quickly that it is more interesting to other members than it is to him; or, depending on his overall attitude, he might object to the recommendation for just that reason. On the other hand, since (as was mentioned in Section 20.3.1) LET'S BROWSE uses a group-level model to compute its recommendations, the computation of predictions for individual group

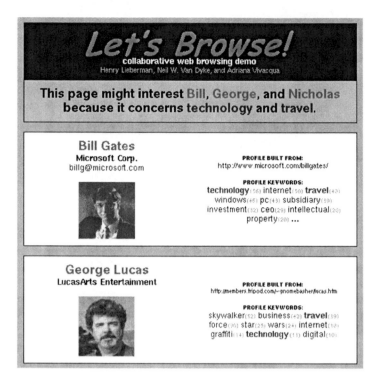

Fig. 20.6. Screen shot of LET's BROWSE that shows how the system explains a recommendation for three (hypothetical) users of a particular web page. (Adapted with permission from an image supplied by Henry Lieberman.)

members for presentational purposes would involve additional overhead, and it would not reflect the way in which the system actually arrives at its recommendations.

POLYLENS gives an idea of how a display that shows predictions for individuals might look. Since POLYLENS uses collaborative filtering, it cannot explain a movie recommendation in terms of the movie's content; but it does show the predicted rating for each group member and for the group as a whole (see Figure 20.4 above). In addition to explaining each recommendation in terms of the underlying predictions for individuals, this visualization makes it possible for the attentive user to notice how group recommendations are generated—via the "least misery" principle (Equation 20.1)—by comparing predictions for the individual members with the predictions for the group. Incidentally, more than 90% of the users surveyed stated that they had no privacy concerns about having their predicted ratings shown to other group members—a result that encourages the development of additional methods that expose individual-level predictions to all group members.

INTRIGUE offers a type of explanation (Figure 20.7) that is partly similar to that of POLYLENS: It shows the predicted attractiveness of each recommended attraction for the tourist group as a whole; but instead of simply presenting a predicted attractiveness for each subgroup, it explains verbally the aspects of the attraction that are likely to

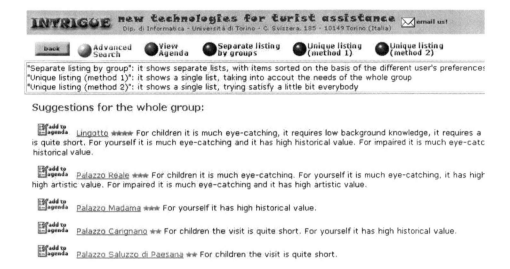

Fig. 20.7. Example of INTRIGUE's main explanation method. (Adapted with permission from Figure 3 of Ardissono et al. [1].)

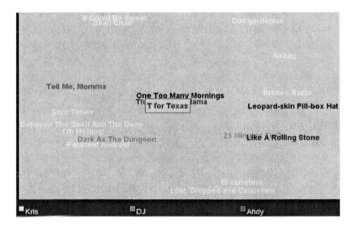

Fig. 20.8. Part of a visualization from the FLYTRAP music selection system. (The songs most likely to be played are shown in the middle of the display. The color coding reflects the estimated degree of interest of the two current listeners Kris and Andy, as well as the influence of the DJ agent. Adapted with permission from Figure 1 of Crossen et al. [11].)

appeal to that subgroup (not mentioning the less appealing aspects). This type of explanation seems helpful for the tour guide who would like all group members to accept the recommendation, but it does not convey a clear idea of how attractive the recommended tour is to each subgroup.[8]

---

[8] INTRIGUE also offers a different type of explanation that shows only predictions for individual subgroups; see Figure 8 of Ardissono et al. [1].

FLYTRAP offers a completely different type of visualization (see Figure 20.8), which indicates which users are present in the room in which music is being played, what songs are more or less likely to be selected for playing, and how these songs are expected to be evaluated by the users who are present and by the "DJ agent". Since the users of FLYTRAP do not have an opportunity to influence the choice of songs directly, the purpose of this visualization is to convey a general understanding of how the system works.

I-SPY incorporates a number of strategies for making clear the reasons for the promotion of a given link in a search result list. Since the users of I-SPY are not viewed as working together when they look for relevant links, the focus here is not on showing the desirability of an option for particular group members with a view to resolving conflicts. Nonetheless, it has proven worthwhile to provide information about how other community members have dealt with the page in question in the past, because such information helps the current user to judge its value for him- or herself. Types of information offered include (a) *related queries* for which the page in question has been selected as a promising result (see Figure 20.3); (b) quantitative and temporal information such as "10% of searchers have also selected this result for similar queries as recently as 15 minutes ago" (Coyle and Smyth [10]); and (c) the names of the users who are responsible for the promotion of the page (a by-product of the antimanipulation measures discussed in Section 20.3.2; see Briggs and Smyth [3]).

The TRAVEL DECISION FORUM introduces two novel, complementary methods that aim to provide a more detailed picture of the consequences of a given proposal for each group member:

1. The first method automatically follows from the use of the preference specification form for the presentation of proposals (see Figure 20.1). Since both the specified preferences and the recommended joint preferences are shown on the same set of scales, the user can quickly see which group members should be most / least satisfied with a given proposal (i.e., the ones whose preferences are closest to / farthest from the highlighted cells). Also, with a bit of practice the user can see more complex patterns (e.g., Claudia might notice that "Tina and Ritchie have generally similar preferences, and they usually get their way, while my preferences have little influence"). Any verbal arguments associated with the other members' stored preferences add further detail to the picture of how they would evaluate a given proposal.

2. The second method takes into account the fact that any graphical explanation of a recommendation is likely to be less interesting, vivid, and memorable than the type of feedback that group members get while they are interacting face to face: A member who is disappointed with a proposal may complain about specific aspects of it in an emotional manner, formulating (or repeating) arguments. In settings where all group members arc physically present in front of the group recommender system, this type of face-to-face discussion is likely to occur spontaneously. For settings in which no such direct communication is possible, the TRAVEL DECISION FORUM tries to recapture some of the flavor of face-to-face interaction through animated characters: It is assumed that at any given moment only one group member will be interacting with the system; each of the other members is represented by an animated character who bears that member's name. Whenever the system (represented by an animated character called

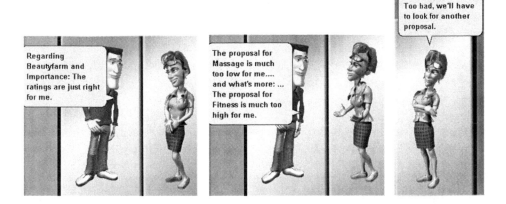

Fig. 20.9. Use of animated characters in the TRAVEL DECISION FORUM to represent the likely responses of the absent group members to a proposal. (Here, the representative of Tina is responding to the proposal shown in Figure 20.9. Adapted with permission from Figure 3 of Jameson [18].)

the *mediator*) has recommended a particular joint preference model for a given value dimension (such as "health facilities"), he asks the representatives of the absent group members to comment on it in turn. Parts of a typical performance of a representative are shown in Figure 20.9.

This type of simulated reaction can heighten the group members' awareness of the other members' points of view—including their motivational orientations—and overcome the natural tendency to focus on one's own evaluations. Also, like the explanations of INTRIGUE, these presentations are selective, focusing on the most important considerations for each group member. The user can switch attention back and forth between the animated agents and the graphical explanation, because the two types of explanation make use of largely complementary communication channels.

20.4.3 Concluding Remarks About Explanations

These examples from existing systems illustrate (a) that there are many types of information that can be conveyed by an explanation of the system's recommendations and (b) that the function of such explanations is not necessarily to convince the group members that the system's recommendations should be accepted. Instead, the system's explanations are often best seen as information that puts the group members in a better position to make a final decision, which may deviate radically from the recommendations made. The process of arriving at a final consensus is the subject of the next section.

20.5 Helping Group Members to Achieve Consensus

Even with a recommender system for individual users, no matter how appropriate and compelling the system's recommendations and explanations may be, there is usually no guarantee that any of the recommendations will be adopted. With individual recommenders, although the decision process may be complex, it typically takes place within the mind of a single person. With a group recommender, extensive debate and negotiation may be required, which may be especially problematic if the members are not able to communicate easily.

20.5.1 Treatment in Existing Systems

Group recommender systems have tended not to provide explicit support for the process of arriving at a final decision. Such support may in fact not be required if any of the following conditions hold:

1. The system simply translates the most highly rated solution into action without requesting the consent of any users.
   This method is applied by the music selection systems ADAPTIVE RADIO, FLYTRAP, and MUSICFX, which select and play music autonomously on the basis of the preferences of the group members who are present. It would in fact usually be impractical to allow the persons who happen to be present in a public space to debate about each piece of music that is to be played.
2. It is assumed that one group member is responsible for making the final decision.
   LET'S BROWSE is based largely on the assumption that one group member controls the pointing device and will therefore make most of the decisions about what pages to visit. With INTRIGUE, it appears to be assumed that the tourist guide will decide what tour should be taken.
3. It is assumed that group members will arrive at the final decision through conventional discussion (e.g., face to face or by phone).
   An especially clear example where this assumption is justified is the situation where several group members are working simultaneously with the CATS vacation recommender system (cf. Section 20.2.2) on the DIAMONDTOUCH interactive tabletop. In addition to the usual broad bandwidth of face-to-face communication, the group members can refer during their negotiations to the various information displays provided by the system.

The only other system we know of that specifically supports communication among group members for the purpose of final decision making is the TRAVEL DECISION FORUM—understandably, since this system is designed specifically for groups of users who usually cannot communicate with each other in real time. The animated representatives of absent group members (Section 20.4) do not serve only as a means of visualizing the implications of recommended solutions for the absent members. In addition, each member can grant her representative a certain amount of authority to accept proposals during interactions with another group member. For example, in Figure 20.9, Tina's representative states that she cannot accept the current proposal. If instead the representative had accepted it and the same were true of Ritchie's representative and the current user Claudia, the proposal would have been treated as finally accepted, even though Tina and Ritchie had not really seen it.

20.5.2 Possible Extensions to Existing Systems

Since in most applied scenarios the overhead of animated agents would be too great, we should consider how functions such as those of the TRAVEL DECISION FORUM can be realized in a lighter-weight manner. For example, a system like POLYLENS might allow each member to specify that they are willing to go to any movie whose predicted rating for them is above some threshold (e.g., 4 stars out of 5). In that case, the system could present not only recommendations but also a subset of the recommended movies that can be decided on without further consultation; a designated group member could then go on and buy tickets for any of these movies. If it is assumed that each group member will view the recommendations before a decision is made, a procedure could be introduced that allowed the group members to vote for movies among the recommended ones, also indicating which particular ones they are willing to accept. In this case as well, it could be agreed that a designated group member could make the final decision.

As the reader may have noticed, this type of voting mechanism can in itself be viewed as a simple recommender system that makes use of explicit preference specifications and helps the group members to choose among the recommended options. And in fact we could apply all of the concepts introduced so far to this "recommender", considering, for example, whether the group members should be allowed to see each other's votes (cf. Section 20.2), how the votes should be counted and weighted (Section 20.3), how the results of the voting should be presented (Section 20.4), and even how the really final decision ought to be made (the present section). Fortunately, we are not faced with an infinite regress, since the recommendation problem that we are dealing with now—choosing from among a small number of recommended items—is considerably simpler than the original problem, and simple solutions may be quite adequate. For example, suppose that for the original recommendations an aggregation function was used which gave especially high weight to particular group members (e.g., the visitors from out of town). It may not be necessary for the final voting mechanism to be biased in their favor as well, since the set of recommendations itself will already contain mainly films that the guests will like; and the hosts are likely in any case to vote in a way that takes into account the greater importance of the guests.

More generally, decisions are often made in several stages, and in each stage a different type of decision making procedure may be applied. Issues that require a great deal of attention in one stage may be simpler to deal with in another stage.

20.6 General Conclusions

20.6.1 Conclusions Concerning Group Recommender systems

This chapter has shown that the differences between recommending for groups and recommending for individuals are more numerous, important, and complex than one would tend to think at first glance.

1. New methods of acquiring information about users' preferences are available, especially when group members specify their preferences explicitly; but the interpretation of explicitly specified preferences can be less straightforward than in the case of individual users.

2. The process of selecting items to recommend for the group as a whole can involve much more than the application of numerical formulas, and the appropriateness of the potentially applicable methods depends on various aspects of the application setting.

3. An explanation of the considerations that underlie a recommendation can take many different forms and convey many different types of information, which can be processed further by the group members in their efforts to do justice to considerations that could not be taken into account by the system.

4. The process of arriving at a final decision can require communication and negotiation, which can be supported in various ways by the system, depending on the nature of the setting.

Because of the many specific questions raised by these differences between recommendation for groups and for individuals, it is all too natural for designers and researchers to focus on one or two of the differences, implicitly adopting with respect to other issues a default solution that may be far from optimal for their particular setting. We hope that, by accumulating and organizing ideas and experience concerning the most important differences, this chapter will enable designers and researchers to do justice to all of these differences in their own work, either by adopting ideas that have already been developed or by working out new solutions of their own.

20.6.2 Implications for Other Types of Adaptive Web-Based System

Even more generally, just about any type of system that adapts to its users can be seen in some sense as a recommender system, and it may be natural to extend it for adaptation to groups of users. For example, a system for personalized information access (cf. the chapter by Gauch, Speretta, and Micarelli [13]) can be seen as "recommending" particular documents to a user; and it is reasonable to adapt to groups of cooperating information-seeking users. Either explicitly or implicitly, we will then have to deal with issues such as the specification and aggregation of preferences of the various group members.

Similarly, a system that offers adaptive navigation support (cf. the chapter by Brusilovsky [4]) can be seen as recommending moves within a hyperspace; and it is natural to consider groups of navigating users (as is in fact done in LET's BROWSE).

With systems that aim to encourage and support collaboration (cf. the chapter by Soller [38]), one of the many functions is that of "recommending" to a set of potential collaborators courses of action that are predicted to be beneficial to them. From a group recommendation perspective, issues that arise include those of (a) how to select a form of collaboration that best takes into account the possibly diverging goals and preferences of the potential collaborators; and (b) how to convince the potential collaborators that they would in fact benefit from the proposed collaboration—or how at least to enable them to devise some form of collaboration that they consider more suitable.

Turning to the more domain-specific topic of systems for health care (cf. the chapter by Cawsey et al. [6]), consider the example of a system designed to persuade members of a family to adopt healthier eating habits: This type of intervention can be seen as involving recommendations to a group of users who have different roles and preferences. Most of the issues discussed in this chapter take a more complex form in this type of setting than in the application settings discussed so far.

In the light of these and many other possible examples, it seems natural that group recommender systems should attract increasing attention not only from designers and researchers who are interested in this particular type of system but also from those who work with other types of adaptive web-based systems.

Acknowledgments. For their support for the preparation of this chapter and of their own relevant research, the authors are grateful to the following sources: for Anthony Jameson, the German Ministry of Education and Research (BMBF) (projects MIAU and SPECTER); for Barry Smyth, Science Foundation Ireland under grant 03/IN.3/ I361 and the Informatics Research Initiative of Enterprise Ireland. The anonymous reviewers of this chapter provided valuable feedback that led to substantial improvements.

References

1. Ardissono, L., Goy, A., Petrone, G., Segnan, M., Torasso, P.: INTRIGUE: Personalized recommendation of tourist attractions for desktop and handset devices. Applied Artificial Intelligence **17**(8-9) (2003) 687–714
2. Arrow, K.: Social Choice and Individual Values. 2nd edn. Wiley, New York (1963)
3. Briggs, P., Smyth, B.: Modeling trust in collaborative web search. In: Proceedings of the Sixteenth Irish Artificial Intelligence and Cognitive Science Conference. (2005)
4. Brusilovsky, P.: Adaptive navigation support. In Brusilovsky, P., Kobsa, A., Nejdl, W., eds.: The Adaptive Web: Methods and Strategies of Web Personalization. Lecture Notes in Computer Science, Vol. 4321, Berlin Heidelberg New York, Springer-Verlag (2007) this volume.
5. Burke, R.: Hybrid web recommender systems. In Brusilovsky, P., Kobsa, A., Nejdl, W., eds.: The Adaptive Web: Methods and Strategies of Web Personalization. Lecture Notes in Computer Science, Vol. 4321, Berlin Heidelberg New York, Springer-Verlag (2007) this volume.
6. Cawsey, A., Grasso, F., Paris, C.: Adaptive information for consumers of healthcare. In Brusilovsky, P., Kobsa, A., Nejdl, W., eds.: The Adaptive Web: Methods and Strategies of Web Personalization. Lecture Notes in Computer Science, Vol. 4321, Berlin Heidelberg New York, Springer-Verlag (2007) this volume.
7. Chao, D., Balthrop, J., Forrest, S.: Adaptive Radio: Achieving consensus using negative preferences. In: Proceedings of the 2005 International ACM SIGGROUP Conference on Supporting Group Work. (2005) 120–123
8. Conitzer, V., Sandholm, T.: Complexity of mechanism design. In Darwiche, A., Friedman, N., eds.: Uncertainty in Artificial Intelligence: Proceedings of the Eighteenth Conference. Morgan Kaufmann, San Francisco (2002) 103–110
9. Conitzer, V., Sandholm, T.: Applications of automated mechanism design. Presented at the First Bayesian Modeling Applications Workshop at the Nineteenth Conference on Uncertainty in Artificial Intelligence, Acapulco, Mexico (2003)
10. Coyle, M., Smyth, B.: Explaining search results. In Kaelbling, L.P., Saffiotti, A., eds.: Proceedings of the Nineteenth International Joint Conference on Artificial Intelligence. Morgan Kaufmann, San Francisco (2005) 1553–1555
11. Crossen, A., Budzik, J., Hammond, K.: Flytrap: Intelligent group music recommendation. In Gil, Y., Leake, D., eds.: IUI 2002: International Conference on Intelligent User Interfaces. ACM, New York (2002) 184–185

12. Dietz, P., Leigh, D.: DiamondTouch: A multi-user touch technology. In: Proceedings of the 14th Annual ACM Symposium on User interface Software and Technology, Orlando, FL (2001) 219–226

13. Gauch, S., Speretta, M., Chandramouli, A., Micarelli, A.: User profiles for personalized information access. In Brusilovsky, P., Kobsa, A., Nejdl, W., eds.: The Adaptive Web: Methods and Strategies of Web Personalization. Lecture Notes in Computer Science, Vol. 4321, Berlin Heidelberg New York, Springer-Verlag (2007) this volume.

14. Goren-Bar, D., Glinansky, O.: Family stereotyping – A model to filter TV programs for multiple viewers. In: Proceedings of the Workshop on Personalization in Future TV at the Conference on Adaptive Hypermedia and Adaptive Web-Based Systems, Malaga, Spain (2002)

15. Goy, A., Ardissono, L., Petrone, G.: Personalization in e-commerce applications. In Brusilovsky, P., Kobsa, A., Nejdl, W., eds.: The Adaptive Web: Methods and Strategies of Web Personalization. Lecture Notes in Computer Science, Vol. 4321, Berlin Heidelberg New York, Springer-Verlag (2007) this volume.

16. Hastie, R., Kameda, T.: The robust beauty of majority rules in group decisions. Psychological Review **112**(2) (2005) 494–508

17. Herlocker, J., Konstan, J., Riedl, J.: Explaining collaborative filtering recommendations. In: Proceedings of the 2000 Conference on Computer-Supported Cooperative Work, Philadelphia, PA (2000) 241–250

18. Jameson, A.: More than the sum of its members: Challenges for group recommender systems. In: Proceedings of the International Working Conference on Advanced Visual Interfaces, Gallipoli, Italy (2004) 48–54

19. Jameson, A., Baldes, S., Kleinbauer, T.: Two methods for enhancing mutual awareness in a group recommender system. In: Proceedings of the International Working Conference on Advanced Visual Interfaces, Gallipoli, Italy (2004) 447–449

20. Jameson, A., Hackl, C., Kleinbauer, T.: Evaluation of automatically designed mechanisms. In: Proceedings of the First Bayesian Modeling Applications Workshop at the Nineteenth Conference on Uncertainty in Artificial Intelligence, Acapulco, Mexico (2003)

21. Jennings, N., Faratin, P., Lomuscio, A., Parsons, S., Sierra, C., Wooldridge, M.: Automated negotiation: Prospects, methods and challenges. International Journal of Group Decision and Negotiation **10**(2) (2001) 199–215

22. Kay, J., Niu, W.: Adapting information delivery to groups of people. In: Proceedings of the First International Workshop on New Technologies for Personalized Information Access at the Tenth International Conference on User Modeling, Edinburgh (2005)

23. Lieberman, H., Van Dyke, N., Vivacqua, A.: Let's Browse: A collaborative Web browsing agent. In Maybury, M., ed.: IUI99: International Conference on Intelligent User Interfaces. ACM, New York (1999) 65–68

24. Masthoff, J.: Group modeling: Selecting a sequence of television items to suit a group of viewers. User Modeling and User-Adapted Interaction **14**(1) (2004) 37–85

25. Masthoff, J.: The pursuit of satisfaction: Affective state in group recommender systems. In Ardissono, L., Brna, P., Mitrovic, A., eds.: UM2005, User Modeling: Proceedings of the Tenth International Conference. Springer, Berlin (2005) 296–306

26. McCarthy, J.: Pocket RestaurantFinder: A situated recommender system for groups. In: Proceedings of the Workshop on Mobile Ad-Hoc Communication at the 2002 ACM Conference on Human Factors in Computer Systems, Minneapolis (2002)

27. McCarthy, J., Anagnost, T.: MusicFX: An arbiter of group preferences for computer supported collaborative workouts. In: Proceedings of the 1998 Conference on Computer-Supported Cooperative Work. (1998) 363–372

28. McCarthy, K., Salamó, M., Coyle, L., McGinty, L., Smyth, B., Nixon, P.: Group recommender systems: A critiquing-based approach. In Paris, C., Sidner, C., eds.: IUI 2006: International Conference on Intelligent User Interfaces. ACM, New York (2006) 267–269

29. McCarthy, K., Salamó, M., McGinty, L., Smyth, B.: CATS: A synchronous approach to collaborative group recommendation. In: Proceedings of the Nineteenth International Florida Artificial Intelligence Research Society Conference, Melbourne Beach, FL (2006)

30. O'Connor, M., Cosley, D., Konstan, J., Riedl, J.: PolyLens: A recommender system for groups of users. In Prinz, W., Jarke, M., Rogers, Y., Schmidt, K., Wulf, V., eds.: Proceedings of the Seventh European Conference on Computer-Supported Cooperative Work. Kluwer, Dordrecht, The Netherlands (2001)

31. Pazzani, J., Billsus, D.: Content-based recommender systems. In Brusilovsky, P., Kobsa, A., Nejdl, W., eds.: The Adaptive Web: Methods and Strategies of Web Personalization. Lecture Notes in Computer Science, Vol. 4321, Berlin Heidelberg New York, Springer-Verlag (2007) this volume.

32. Pizzutilo, S., De Carolis, B., Cozzolongo, G., Ambruoso, F.: Group modeling in a public space: Methods, techniques and experiences. In: Proceedings of WSEAS AIC 05, Malta (2005) 175–180

33. Sandholm, T.: Automated mechanism design: A new application area for search algorithms. In Rossi, F., ed.: Principles and Practice of Constraint Programming — CP 2003: 9th International Conference. (2003) 19–36

34. Schafer, B., Frankowski, D., Herlocker, J., Sen, S.: Collaborative filtering recommender systems. In Brusilovsky, P., Kobsa, A., Nejdl, W., eds.: The Adaptive Web: Methods and Strategies of Web Personalization. Lecture Notes in Computer Science, Vol. 4321, Berlin Heidelberg New York, Springer-Verlag (2007) this volume.

35. Smyth, B.: Case-base recommendation. In Brusilovsky, P., Kobsa, A., Nejdl, W., eds.: The Adaptive Web: Methods and Strategies of Web Personalization. Lecture Notes in Computer Science, Vol. 4321, Berlin Heidelberg New York, Springer-Verlag (2007) this volume.

36. Smyth, B., Balfe, E., Boydell, O., Bradley, K., Briggs, P., Coyle, M., Freyne, J.: A live-user evaluation of collaborative web search. In Kaelbling, L.P., Saffiotti, A., eds.: Proceedings of the Nineteenth International Joint Conference on Artificial Intelligence. Morgan Kaufmann, San Francisco (2005) 1419–1424

37. Smyth, B., Balfe, E., Freyne, J., Briggs, P., Coyle, M., Boydell, O.: Exploiting query repetition and regularity in an adaptive community-based web search engine. User Modeling and User-Adapted Interaction 14(5) (2005) 383–423

38. Soller, A.: Adaptive support for distributed collaboration. In Brusilovsky, P., Kobsa, A., Nejdl, W., eds.: The Adaptive Web: Methods and Strategies of Web Personalization. Lecture Notes in Computer Science, Vol. 4321, Berlin Heidelberg New York, Springer-Verlag (2007) this volume.

39. Yu, Z., Zhou, X., Hao, Y., Gu, J.: TV program recommendation for multiple viewers based on user profile merging. User Modeling and User-Adapted Interaction 16(1) (2006) 63–82

40. Yu, Z., Zhou, X., Zhang, D.: An adaptive in-vehicle multimedia recommender for group users. In: Proceedings of the IEEE 62nd Semiannual Vehicular Technology Conference, Dallas (2005)

21

Privacy-Enhanced Web Personalization

Alfred Kobsa

Donald Bren School of Information and Computer Sciences
University of California, Irvine
Irvine, CA 92697-3440, U.S.A.
kobsa@uci.edu
http://www.ics.uci.edu/~kobsa

Abstract. Consumer studies demonstrate that online users value personalized content. At the same time, providing personalization on websites seems quite profitable for web vendors. This win-win situation is however marred by privacy concerns since personalizing people's interaction entails gathering considerable amounts of data about them. As numerous recent surveys have consistently demonstrated, computer users are very concerned about their privacy on the Internet. Moreover, the collection of personal data is also subject to legal regulations in many countries and states. Both user concerns and privacy regulations impact frequently used personalization methods. This article analyzes the tension between personalization and privacy, and presents approaches to reconcile the both.

21.1 Introduction

It has been tacitly acknowledged for many years that personalized interaction and user modeling have significant privacy implications, due to the fact that large amounts of personal information about users needs to be collected to perform personalization. For instance, frequent users of a search engine may appreciate that their search terms become recorded to disambiguate future queries and deliver results that are better geared towards their interests (see Chapter 6 of this book [130]). They may not appreciate though if their search history from the past few years becomes accessible to others. Secretaries may value if the help component of their text editor can give personalized advice based on a model of their individual word-processing skills that it built over time by watching how they interact with the word processor [113]. They are however likely to be concerned if the contents of their model becomes accessible to others, specifically if negative consequences may arise from a disclosure of the skills they lack. Other potential privacy concerns in the context of personalized systems include (see [37]): unsolicited marketing, computer "figuring things out" about the user [197], fear of price discrimination, information being revealed to other users of the same computer, unauthorized access to accounts, subpoenas by courts, and government surveillance.

P. Brusilovsky, A. Kobsa, and W. Nejdl (Eds.): The Adaptive Web, LNCS 4321, pp. 628–670, 2007.

Kobsa [104, 105] was arguably the first to point out the tension between personalization and privacy nearly twenty years ago, but without much impact. The only privacy solution of these days was to ascertain that users could store their models on a diskette (R. Oppermann, GMD) or a PCMCIA card (J. Orwant, MIT [143]), and carry them with. The situation changed completely in the late 1990s, for four main reasons:

Personalized Systems Moved to the Web. Web retailers quickly realized the enormous potential of personalization for customer relationship management and made their websites user-adaptive. This had significant privacy implications. While user models were previously confined to stand-alone machines or local networks, people's profiles were now collected on dozens if not hundreds of personalized websites. Widely publicized security glitches and privacy breaches as well as aggressive telemarketing led to a widespread (~60-80%) stated reluctance of Internet users to disclose personal data and being tracked online. This reluctance however endangers the basic foundations of personalization, which highly relies on such data [180].

More Sources of User Data Available. While in the 1980s the main source of user modeling was nearly exclusively textual data entered by the user, assumptions about users can nowadays be drawn from, e.g., their mouse movements, mouse clicks, eye movements, facial expression, physiological data and location data. Completely new privacy threats arose in ubiquitous computing environments where users are no longer merely IP addresses in an abstract online space, but become identified individuals who are being monitored and contacted by their physical surroundings.

More Powerful Analyses of User Data Available. More powerful computers, computer networks, sensors and algorithms have made it possible to collect, connect and analyze far more data about users than ever before. Complete digital lifetime archives replete with personal data may soon become reality.

Restrictions Imposed by Privacy Legislation. Many more countries, states and provinces have meanwhile introduced privacy laws, which severely affect not only commercial websites but also experimental research on user modeling (in many cases even when it is done "just with IP numbers", "just with our students", or "just for testing purposes") [107]. Areas that are specifically affected include data mining for personalization purposes (see Chapter 3 of this book [135]), adaptive tutoring systems that build learner models (see Chapter 1 of this book [23]), and adaptation to the needs of people with special needs [58, 174].

Consumer studies demonstrate that online users value personalized content [32, 149, 178]. At the same time, providing personalization on websites seems quite profitable for web vendors [11, 34, 77, 86]. This win-win situation is however marred by privacy concerns since personalizing people's interaction entails gathering considerable amounts of data about them. As a consequence, the topic of "Privacy and Personalization" has received considerable attention from industry and academia in the past few years. Three industry conferences with this title were held in 2000-01 (in New York, London and San Francisco) with a participation of about 150 people. The Ubiquitous Computing conferences have held privacy workshops in the past four years[1] that address, among other topics, privacy in context-aware systems. In July 2005, twenty researchers met for a first workshop on Privacy-Enhanced Personalization at the 10th International User Modeling Conference in Edinburgh, Scotland [108], and one year later for a second workshop at the CHI-2006 conference in Montréal, Canada [130].

The aim of research on privacy-enhanced personalization is to reconcile the goals and methods of user modeling and personalization with privacy considerations, and to strive for best possible personalization within the boundaries set by privacy. This research field is widely interdisciplinary, with contributions coming from information systems, marketing research, public policy, economics, computer-mediated communication, law, human-computer interaction, and the information and computer sciences. This chapter analyzes how current research results on privacy in electronic environments relate to the aims of privacy-enhanced personalization. It first discusses the impact of Internet users' privacy concerns on the disclosure of personal data. Section 21.3 then reviews the current state of research on factors that contribute to alleviating privacy concerns and to encouraging the disclosure of personal data. Section 21.4 analyzes the impact of privacy regulation on personalized systems (specifically of privacy legislation, but also industry and company self-regulation as well as principles of fair information practices). Section 21.5 finally describes privacy-enhancing technical solutions that are particularly well suited for personalized systems.

As we will see throughout these discussions, there exists no magic bullet for making personalized systems privacy-enhanced, neither technical nor legal nor social/organizational. Instead, numerous small enhancements need to be introduced, which depend on the application domain as well as the types of data, users and personalization goals involved. At the end of most sections and subsections, we will list the lessons for the privacy-minded design of personalized systems that ensue from the research results discussed in the respective section. In a concrete project, though, the applicability of these recommendations will still need to be verified as part of the normal interaction design and user evaluation process [155, 175].

21.2 Individuals' Privacy Concerns

21.2.1 Methodological Preliminaries

This section analyzes empirical results regarding people's privacy-related attitudes, and the subsequent section known motivators and deterrents for people disclosing personal information to websites. Two principal types of empirical methods are available for identifying such attitudes, motivators and deterrents:

1. *Inquiry-based methods.* In this approach, the participants of an empirical study are being asked about their privacy attitudes ("reported/perceived attitudes"), their disclosure behavior in the past ("reported/perceived behavior"), and their anticipated disclosure behavior under certain privacy-related circumstances ("stated behavioral intentions"). In the third case, these privacy-related circumstances can be merely described to subjects, or one can try to immerse subjects in them as much as possi-

---

[1] See the links at http://www.cs.berkeley.edu/~jfc/privacy/

ble (e.g. by showing them a website with characteristics that may be important in subjects' disclosure decisions).

2. *Observation-based methods.* In this approach, the privacy-related behavior of participants is being observed during the empirical study. Subjects are put into a situation that resembles the studied circumstances as much as possible (usually a lab experiment, ideally a field experiment), and they have to exhibit privacy-related behavior therein (e.g., disclose their own personal data while purchasing products) rather than merely answer questions about their likely behavior.

Both approaches have complementary strengths and weaknesses, and mixes of both approaches are therefore customary. Inquiry-based methods do not directly unveil people's actual privacy-related attitudes and disclosure behavior, but only their perception thereof, which may not be in sync with reality. Observation-based methods on the other hand often do not allow one to recognize people's higher-level behavioral patterns or rationale, which in return can be more easily accessed through inquiries. In addition, both approaches are equally subject to various potential biases that must be eliminated through careful experimental design (see e.g. [141]).

In the area of privacy, other factors also seem to come into play that may skew the results of empirical studies. Such known or suspected factors include the following:

1. *Biased self-selection.* It may be the case that predominantly those people volunteer to participate in a privacy study, or take the pains to complete it until the very end, for whom privacy is a personal concern. This may bias the responses towards higher concerns.

2. *Socially desirable responses.* It may be the case in privacy studies that subjects tend to respond and act in ways that are deemed socially desirable. For instance, in times of ever-increasing identity theft this bias may skew responses towards higher concerns since not having privacy concerns might be viewed as displaying a lack of prudence and responsibility.

3. *Discrepancies between stated attitudes and observed behavior.* In several privacy studies in e-commerce contexts, discrepancies have already been observed between users stating high privacy concerns but subsequently disclosing personal data carelessly [17, 128, 172]. Several authors therefore challenge the genuineness of such reported privacy attitudes [80, 90] and emphasize the need for experiments that allow for an observation of actual online disclosure behavior [128, 167].

It seems possible to eliminate the first two sources of bias through careful experimental design and post-hoc recalibration of socially desirable responses. The discrepancies between stated privacy attitudes and observed disclosure behavior will be discussed in more detail in Sections 21.2.5 and 21.3.5. For the time being, it seems useful though to clearly distinguish whether an experimental finding stems from the observation of actual human disclosure behavior in an experiment, or is based on subjects' reports of attitudes, past behavior or behavioral intentions. We will therefore introduce the convention of marking findings of the first kind with an asterisk (*) in the remainder of this chapter.[2]

---

[2] Note that a cited article may both describe observation-based findings (which will be marked with an asterisk) and findings that are based on subjects' reports (which will not).

21.2.2 Potential Effects of Privacy Concerns on Personalized Systems

Numerous consumer surveys and research studies have revealed that Internet users harbor considerable privacy concerns regarding the disclosure of their personal data to websites, and the monitoring of their Internet activities. These studies were primarily conducted between 1998 and 2003, mostly in the United States (see [156] for an incomplete listing). In the following, we summarize a few important findings (the percentage figures indicate the ratio of respondents who adopted the respective view). For a more detailed discussion we refer to [180].

Personal Data

1. Internet users who are concerned about the privacy or security of their personal information online: 70% [15], 83% [196], 89.5% [187], 84% [147];
2. People who have refused to give personal information to a web site at one time or another: 95% [87], 83% [82], 82% [44];
3. Internet users who would never provide personal information to a web site: 27% [63];
4. Internet users who supplied false or fictitious information to a web site when asked to register: 40% [87]; 34% [44]; 24% [63]; 15% more than half of the time [167]; 6% always, 7% often, 17% sometimes [163]; 48.9% never, 24.1% a quarter or less of the time, 18% between ¼ and over ¾ of the time [76]; 19.4% in an experiment (half of them multiple times) [127]; 39.6% in an experiment (2-3 items on average, and the likelihood of falsification was correlated with the stated sensitivity of the item).
5. People who are concerned if a business shares their data for a purpose that is different from the one for which they were originally collected: 90% [163], 89% [147];
6. Online users who believe that sites that share personal information with other sites invade privacy: 83% [45].

Significant concern about the use of personal data is visible in these results, which may cause problems for those personalized systems that depend on users disclosing data about themselves. More than a quarter of respondents stated that they would never consider providing personal information to a web site. Quite a few users indicated having supplied false or fictitious information to a web site when asked to register, which makes all personalization based on such data dubious, and may also jeopardize cross-session identification of users as well as all personalization based thereon. Furthermore, 80-90% of the respondents are concerned if a business shares their information for a different than the original purpose. This may have severe impacts on central user modeling servers that collect data from, and share them with, different user-adaptive applications (see Chapter 4 of this book [130]).

User Tracking and Cookies

1. People who are concerned about being tracked on the Internet: 54% [63], 63% [82], 62% [147];
2. People who are concerned that someone might know what web sites they visited: 31% [63];

3. Users who feel uncomfortable being tracked across multiple web sites: 91% [82];
4. Internet users who generally accept cookies: 62% [148];
5. Internet users who set their computers to reject cookies: 25% [44], 10% [63]; and
6. Internet users who delete cookies periodically: 53% [148].

These results reveal significant user concerns about tracking and cookies, which may have effects on the acceptance of personalization that is based on usage logs. Observations 3–6 directly affect machine-learning methods that operate on user log data since without cookies or registration, different sessions of the same user can no longer be linked. Observation 3 may again affect the acceptance of the above-mentioned user modeling servers which collect user information from several websites (see Chapter 4 of this book [130]).

These survey results indicate that privacy concerns may indeed severely impede the adoption of personalized web-based systems. As a consequence, personalized systems may become less used, personalization features may become switched off if this is an option, fewer personal information may become disclosed, and escape strategies may be adopted such as submitting falsified data, maintaining multiple accounts/identities, deleting cookies, etc. However, developers of personalized web-based systems should not feel completely discouraged by the abundance of stated privacy concerns in consumer surveys. As we will see, privacy concerns are only one of many factors that influence whether and to what extent people disclose data about themselves and utilize personalized systems. In Section 21.3 we will discuss numerous factors that can seemingly mitigate users' privacy concerns and prompt them to nevertheless disclose personal data about themselves. Designers of personalized systems will have to carefully analyze users' privacy concerns in their application domain and address them, but also consider those mitigating factors and ascertain that as many of them as possible are present in the design of their systems.

21.2.3 Effect of Information Type

Not surprisingly, many surveys indicate that users' willingness to disclose personal information also depends on the kind of information in question. For instance,

– Ackerman et al. [1] found that the vast majority of their respondents always or usually felt comfortable providing information about their own preferences, including favorite television show (82%) and favorite snack food (80%). In contrast, only a very small number said they would usually feel comfortable providing their credit card number (3%) or social security number (1%). The figures decreased in all categories if the data was about subjects' children and not about themselves.

– Phelps [151] found that consumers are more willing to provide marketers with demographic and lifestyle information than with financial, purchase-related, and personal identifier information. The vast majority of respondents were always or somewhat willing to share their two favorite hobbies, age, marital status, occupation or type of job, and education.

– Metzger [127] found that participants of her experiment "were most willing to provide basic demographic information (e.g., sex, age, education level, marital status), and slightly less willing to provide information about their actual online behavior (past purchases time spent online), religion, political party identification, race, hobbies/interests, and occupation. Respondents were by far most protective of their personal contact information (telephone number and email address) and financial information (credit card number, social security number, and income)."*

- In a different experiment, Metzger [126] found that participants were most likely to withhold their credit card and social security numbers. The next-most withheld items included email address, telephone number, favorite website, hobbies/ interests, and last purchase made online, income and political party affiliation. Participants were least likely to withhold general demographic information about themselves, for example, their sex, race, education, marital status, time spent online, number of people in their household, and age. Name and address were given out most frequently, but those were required for receiving a free CD.

An experiment by Huberman et al. [88] suggests that not only different data categories, but also different values within the same category may have different privacy valuations*. A group of experimental subjects participated in a reverse auction for the disclosure of certain personal information to all others (namely of individuals' age, weight, salary, spousal salary, credit rating and amount of savings). The anonymously submitted asking prices for this personal data turned out to be a (largely linear) function of the deviance of the data values from the socially desirable standard (this holds true both for individually perceived and actual deviance)*. The results seem to indicate that the more undesirable a trait is with respect to the group norm, the higher is its privacy valuation.

The lesson from these findings for the design of personalized web-based systems seems that highly sensitive data categories should never be requested without the presence of some of the mitigating factors that will be discussed later. To lower privacy concerns for data values that are possibly highly deviant, one-sided open intervals should be considered whose closed boundary does not deviate too much from the expected norm (such as "weight: 250 pounds and above" for male adults).

21.2.4 Interpersonal Differences in Privacy Attitudes

Various studies established that age [52, 132], education [151] and income [3] are positively associated with the degree of stated Internet privacy concern. Smith et al. [170] also found that people who were victims of a perceived privacy invasion or had heard of one had higher privacy concerns. Gender effects on Internet privacy concerns could not be clearly established so far.

In a broad privacy survey that was first conducted in 1991 [81] and since then repeated several times, Harris Interactive and Alan Westin clustered respondents into three groups, namely privacy fundamentalists, the privacy unconcerned, and privacy pragmatists. Privacy fundamentalists generally express extreme concern about any use of their data and unwillingness to disclose them, even when privacy protection

---

* An asterisk indicates that the data is based on an observation of human privacy-related behavior in an experiment rather than a survey of stated attitudes, reported past behavior, or stated behavioral intentions (see Section 24.2.1 for a more detailed explanation).

mechanisms would be in place. In contrast, the privacy unconcerned tend to express mild concern for privacy only, and also mild anxiety about how other people and organizations use information about them. Privacy pragmatists as the third group are generally concerned about their privacy as well. In contrast to the fundamentalists though, their privacy concerns are lower and they are far more willing to disclose personal information, e.g. when they understand the reasons for its use, when they see benefits for doing so, or when they see privacy protections in place.

In the latest edition of this survey in 2003[3], privacy fundamentalists comprise about 26% of all adults, the privacy unconcerned about 10%, and the privacy pragmatists 64% [179]. Previous editions and other studies yield slightly different figures and/or clusters. For instance, the clustering of the responses in [1] resulted in 17% privacy fundamentalists, 27% "marginally concerned", and 56% members of the "pragmatic majority". Acquisti and Grossklags [3] found four different clusters: "privacy fundamentalists with high concern toward all collection categories (26.1 percent), two medium groups with concerns either focused on the accumulation of data belonging to online or offline identity (23.5 percent and 20.2 percent, respectively), and a group with low concerns in all fields (27.7 percent)." Spiekermann et al. [172] also identified privacy fundamentalists (30%) and marginally concerned users (24%). In addition, the authors were able to split the remaining respondents into two distinct groups, namely ones who are concerned about revealing information such as their names, email or mailing addresses ("identity concerned, 30%) and others who are rather more concerned about the profiling of their interests, hobbies, health and other personal information ("profiling averse", 25%).

21.2.5 Stated Attitudes Versus Reported and Observed Behavior

What are the effects of high privacy concerns? If one looks at people's reported past behavior or intended future behavior, the effects seem straightforward:

- Sheehan and Hoy [167] found that people's stated concern for privacy correlates negatively with the reported frequency of registering with websites in the past, and positively with providing incomplete information when they do register.

- Metzger [127] more generally found that stated concern for online privacy negatively predicted reported past online information disclosure (i.e., those who expressed high privacy concerns also tended to report less information disclosure in the past, and vice versa).

- Smith et al. [170] developed and validated a survey instrument for determining individuals' level of privacy concerns, which is composed of four subscales that measure concerns about inappropriate collection, unauthorized secondary use, improper access, and errors in storing. Research by Xu et al. [195] that used this instrument indicates that if people's individual privacy "sub-concerns" are addressed, their intended data disclosure rose significantly (concerns regarding improper access and unauthorized secondary use had particularly high regression coefficients).

---

[3] See [114] for a more detailed comparison of privacy concern indicators over different years.

- Finally, Chellappa and Sin [31] found that users' stated intention to use personalization services (which necessitates their willingness to disclose information about themselves) is also negatively influenced by their individual level of privacy concern.

Other survey results however shed doubts on whether Internet users always follow through on their stated concerns. A large majority of people buy online (and thereby give out personal data) despite professing privacy concerns [16, 76, 179]. More paradoxically, Behrens [15] found that 20 percent of adults who say they have placed an order on the Internet in the past three months also say they won't put personal information such as their name and address on the Web.

If we look at observable user behavior, the discrepancy to stated privacy concerns becomes even more apparent. The experiment of Metzger [128] did not confirm the hypothesis that individuals' level of concern about online privacy and data security is negatively related to the *observed* amount of personal information they disclosed to a commercial Web site*. The experiment by Spiekermann et al. [172] showed that privacy fundamentalists in particular did not live up to their expressed attitudes*. They only answered 10 percentage points fewer questions than marginally concerned participants.

As mentioned above, a lesson from the apparent discrepancy between intended and actual disclosure behavior of highly privacy-concerned individuals is that developers in the area of personalized web-based systems should not feel completely discouraged by the abundance of stated privacy concerns in consumer surveys. User experiments and daily web practice prove that people do disclose their personal data, since other factors are in effect at the same time that override or alleviate their privacy concerns. Such factors will be discussed in the next few sections. Moreover, based on the abovementioned results of Xu et al. [195], it seems worthwhile to address people's individual privacy "sub-concerns". Section 21.5.4 will discuss methods for dealing with privacy in a more personalized manner.

21.3 Factors Fostering the Disclosure of Personal Information

This section describes factors that have been shown to influence people's willingness to disclose personal data about themselves on the Internet. Those factors include the value that people assign to personalization, their knowledge of and control over how personal information is used, users' trust in a website (and known antecedents thereof, namely positive past experience, the design, operation and reputation of a website, and the presence of privacy statements and privacy seals), as well as data disclosure benefits other than personalization. The section also discusses consequences of these findings for the design of web-based personalized systems. In Section 21.3.5 we will describe how users consider these factors in a situation-specific cost-benefit analysis when deciding on whether or not to disclose individual personal data.

21.3.1 Value of Personalization

Chellappa and Sin [31] found that the value which Internet users assign to personalization is a very important factor with regard to their stated intention to use personalized websites, and that it can "override" privacy concerns: "the consumers' value for personalization is almost two times […] more influential than the consumers' concern for privacy in determining usage of personalization services. This suggests that while vendors should not ignore privacy concerns, they are sure to reap benefits by improving the quality of personalized services that they offer" [31]. A study by White [194] also confirmed that users are more likely to provide personal information when they receive personalization benefits (the opposite seems to hold true however in the case of potentially embarrassing information in combination with a deep relationship between consumer and business, as will be explained in more detail in Section 21.3.3.1).

How much value, then, do Internet users assign to personalized services? Consumer surveys from the turn of the century (i.e. from the time when personalization features became first visible on the web) suggest that a slight majority of respondents value personalization, but that about a quarter sees no value in personalization or is not willing to disclose personal data to receive it:

1. Online users who see / do not see personalization as a good thing: 59% / 37% [82];
2. People who are willing to give information to receive a personalized online experience: 51% (15% not) [148], 43% (39% not) [163];
3. Types of information users would provide to a web site that used it to personalize/ customize their experience, compared to one that does not provide any personalization: hobbies 76% vs. 51%, address 81% vs. 60%, job title 50% vs. 32%, phone number 45% vs. 29%, income 34% vs. 19%, name 96% vs. 85%, mother's maiden name 22% vs. 14%, e-mail address 95% vs. 88%, credit card number 22% vs. 19%, social security number 6% vs. 7%.
4. Online users who find it useful if a site remembers basic information (name, address): 73% (9% not) [148];
5. Online users who find it useful if a site remembers information (preferred colors, music, delivery options etc.): 50% (20% not) [148];
6. People who are bothered if a web site asks for information one has already provided (e.g., mailing address): 62% [148].

More recent surveys found the percentage of respondents who value personalization to be significantly higher. In a 2005 study by ChoiceStream [32], 80% of respondents stated that they are interested in receiving personalized content (news, books, search results, TV/movie, music). This number is consistent with the 2004 edition of the same survey in which 81% expressed their interest in personalized content. Young people are slightly more interested in personalization than older people. No figures are available on those who are not interested. 60% indicated that they would spend at least 2 minutes answering questions about themselves and their interests in order to receive personalized content, versus 56% in 2004. 26% agreed that they would spend at least 6 minutes answering such questions, compared with 21% in 2004. Moreover, 59% (2004: 65%) of respondents indicated a willingness to provide information about their personal preferences, and 46% (2004: 57%) to provide demographics. The authors of the study attribute these decreases in people's willingness to provide personal data to a surge in societal privacy concerns during the intermittent year.

These findings suggest that developers of personalized web-based systems need to make the personalization benefits of their system very clear to users, and ascertain that those benefits are ones that people want.[4] If users perceive value in the personalization services offered, they are considerably more likely to intend to use them and provide the required information about themselves.

21.3.2 Knowledge of and Control over the Use of Personal Information

Many privacy surveys indicate that Internet users find it important to know how their personal information is being used, and to have control over this usage. In a survey of Roy Morgan Research [163], 68% of respondents indicate that it was very important (and 25% that it was important) to know how their personal info may be used. In a survey by Turow [182], 94% even agree that they should have a legal right to know everything that a web site knows about them. In a 1997 survey by Harris Interactive [119], 63% of people who had provided false information to a website or declined to provide information said they would have supplied the information if the site provided notice about how the information would be used prior to disclosure, and if they were comfortable with these uses.

As far as control is concerned, 69% of subjects said in a 2003 Harris poll [179] that "controlling what information is collected about you" is extremely important, and 24% still regarded it as somewhat important. Likewise in a direct marketing study of Phelps et al. [151], the vast majority of respondents desire more control over what companies do with their information. Sheehan and Hoy [168] even found that control (or lack of control) over the collection and usage of information is the most important factor for people's stated privacy concerns, explaining 32.8% of the variance.

Some empirical evidence also exists that people are more willing to disclose their personal data if they possess knowledge of and/or control over the use of this data. In the above-mentioned survey by Roy Morgan Research [163], 59% said they'd be more likely to trust an organization if it gave them more control over how their personal information was used (as we will see in Section 21.3.3, trust in turn is an important factor for people's willingness to disclose their personal data). In a 1998 survey [76], 73.1% indicated that they would give demographic information to a Web site if a statement was provided regarding how the information was going to be used. In a survey by Hoffman [87], 69% of Web users who do not provide data to Web sites say it is because the sites provide no information on how the data will be used. In an experiment by Kobsa and Teltzrow [111], users disclosed significantly more information about themselves when, for every requested piece of personal information, a website explained the user benefits and the site's privacy practices in connection with the requested data* (the effects of these two factors were not separated in this study).

These findings suggest that personalized systems should be able to explain to users what facts and assumptions are stored about them and how these are going to be

---

[4] Time savings and monetary savings, and to a lesser extent pleasure, received the highest approval in surveys conducted by Tan et al. [89, 177] on benefits that businesses collecting personal information should offer. In a survey by Cyber Dialogue [122], customized content provision and the remembering of preferences were quoted as the main reasons for users to personalize websites.

used.[5] Moreover, users should be given ample control over the storage and usage of this data. This is likely to increase users' data disclosure and at the same time complies with the rights of data subjects accorded by many privacy laws, industry and company privacy regulations, and Principles of Fair Information Practices (see Section 21.4).

Extensive work in this direction has been carried out by Judy Kay and her team under the notion of "scrutability" [47, 99-101]. According to Kay, "this means that the user can scrutinise the model to see what information the system holds about them. In addition, it means that the user can scrutinise the processes underlying the user modelling. These include the processes used to collect data about the user. It also includes the processes that made inferences based on that data." [102]. A qualitative evaluation was carried out which showed that "participants in the evaluation could, generally, understand how the material was adapted and how to control that adaptation" [48]. It was challenging for them though to determine what content was included/excluded on a page and what caused the adaptation, and to understand how to change their profiles to control the inclusion or exclusion of content.

21.3.3 Trust in a Website

Trust in a website is a very important motivational factor for the disclosure of personal information.[6] In a survey by Hoffman et al. [87], nearly 63% of consumers who declined to provide personal information to web sites stated as the reason that they do not trust those who are collecting the data (similar responses can be found in Milne and Boza [132]). Conversely, Schoenbachler and Gordon [165] found a positive relationship between trust in an organization and stated willingness to provide personal information. In the experiment of Metzger [127], Internet users' trust in a company's Web site positively influenced their information disclosure to the site*. Trust was also found to positively affect the intended use of an e-commerce website [68].

Several antecedents to trust have been empirically established, and for many of them effects on disclosure have also been verified. Such trust-inducing factors include[7]

- positive experiences in the past,
- the design of a website,
- the reputation of the website operator,
- the presence of a privacy seal, and
- the presence of a privacy statement (but not necessarily its content).

These factors will be discussed in the following subsections.

---

[5] In Section 24.3.3.4 we will see that "privacy statements" (aka "privacy policies"), which constitute the current best practice for privacy disclosures, are not an effective medium for providing such explanations.

[6] A number of different definitions and conceptualizations of trust in online environments have been proposed or used in the literature. For a discussion and critical analysis of those we refer to [69, 73, 124].

[7] Telling users how their personal data will be used and giving them control over this usage (see Section 24.3.2) may also increase users' trust [87]. We refrain from listing it as a factor for trust though since the empirical support for this claim seems insufficient to date.

21.3.3.1 Positive Experiences in the Past

Positive experience in the past is an established factor for trust. Almost half (47%) of the respondents in a consumer survey of the Australian privacy commissioner [163] agreed that their trust in an organization with their personal information would be based on good past experience. Pavlou [145] found a highly significant positive effect of good prior experience on trust for a number of existing websites.

The impact of positive experience in the past on the disclosure of personal information is well supported. In an open-ended questionnaire by Culnan and Boza [132], the number one reason that consumers gave for trusting organizations with personal information was past experience with the company. Culnan and Armstrong [43] found that people who agree that their personal data be used for targeted marketing purposes are more likely to have prior experience with direct marketing than people who do not agree. Metzger [127] observed what types of information subjects disclosed to an experimental website and also asked them what types of information they had disclosed in the past. The author found the total amount of past information disclosure to be a good predictor for the current amount of information disclosure*.

Of specific importance are established, long-term relationships. Sheehan and Hoy [168] prompted subjects for their privacy concerns in 15 hypothetical scenarios. They identified three factors in these scenarios that explain the stated level of privacy concerns, one of them including "items that suggest that the online user has an established relationship with the online entity, in which some level of communication and interaction has already been established between the two parties." Schoenbachler and Gordon [165] found a positive relationship between respondents' perception of a relationship with an organization and their stated willingness to provide personal information. The same was the case in a study by White [194] for two pieces of information that were determined to be specifically private (namely one's address and telephone number). Interestingly enough, the author also found that a deeper relationship with the customer lead to a *decreased* willingness to disclose two other pieces of personal information that were determined to be specifically embarrassing and might cause a loss of face when disclosed, namely one's purchase history of Playboy/ Playgirl magazine and of condoms.

The lesson for the design of personalized systems is not to regard the disclosure of personal information as a one-time matter. Users of personalized websites can be expected to become more forthcoming with personal details over time if they obtain positive experiences with the same site or comparable sites. Personalized websites should be designed in such a way that they can deliver satisfactory user experiences with any amount of personal data that users chose to disclose, and allow users to add more personal detail incrementally at later times.

21.3.3.2 Design and Operation of a Website

Various interface design elements and operational characteristics of a website have been found to increase users' trust in the website, such as

- the absence of errors, such as wrong information or incorrect processing of inputs and orders [12],
- the (professional) design of a site [59, 61],
- the usability of a site [49, 59, 162], specifically for information-rich sites such as sports, portal, and e-commerce sites [12],

- the presence of contact information, namely physical address, phone number or email address [59, 60],
- links from a believable website [59],
- links to outside sources and materials [59],
- updated since last visit [59],
- quick responses to customer service questions [59, 60],
- email confirmation for all transactions [59],
- the presence of an interactive communication channel with a site, specifically instant messaging or voice communication [13], and
- the presence of a photo of a "customer care person" (positive effect for sites with low reputation, negative effect for sites with high reputation)* [159].

While there do not seem to be studies yet that measure directly the effect of website design and operational characteristics on users' willingness to disclose personal data, the established effect of trust on user disclosure behavior (see Section 21.3.3) makes the existence of such an effect very plausible. The lesson from the above findings for the design of personalized websites is therefore to use personalization preferably in professionally designed and easy-to-use websites that also possess some other of the above-mentioned trust-increasing design elements and operational characteristics.

21.3.3.3 Reputation of the Website Operator
The reputation of the organization that operates a website is an important factor for users' trust in the website. Schoenbachler and Gordon [165] found a positive relationship between the perceived reputation of a company and stated trust in the company. Likewise, Metzger [127] established a positive correlation between individuals' subjective regard for a non-existing company whose fictitious website they saw, and their stated trust in this website. Jarvenpaa et al. [95] and Pavlou [145] also found an effect of reputation on trust for several existing websites (Jarvenpaa et al. [94, 95] moreover determined that perceived company size is positively associated with consumers' trust in these websites, though size and reputation are highly related). Metzger [128] varied the reputation of a website between subjects and found that the one with higher reputation was deemed more trustworthy than the one with lower reputation.

Not surprisingly then, reputation is positively correlated with users' willingness to disclose personal information. In a Canadian consumer survey [30], 74% indicate that a company's reputation would make them more comfortable with providing personal information. In a paper-based experiment, Andrade et al. [5] found an effect of perceived reputation on stated concern about the disclosure of personal information (this effect only approached statistical significance though). In the online survey of Earp and Baumer [53], subjects were randomly shown one of 30 web pages from higher traffic and lower traffic websites in different sectors. Subjects were significantly less willing to provide personally identifiable information (specifically their phone numbers, home and email addresses, and social security and credit card numbers) to the lower-traffic sites (which were presumably less known to them).[8]

---

[8]   Metzger [128] found that regard for the company had a stronger relationship with disclosure than did trust, which is somewhat contradictory to the current view that reputation effects disclosure indirectly via fostering trust.

The lesson for the design of personalized systems seems to be that everything else being equal, users' information disclosure at sites of well-reputed companies is likely to be higher than at sites with lower reputation. Personalization is therefore likely to be more successful at sites with higher reputation. It may of course be possible to compensate for the lack of reputation by putting more emphasis on other factors that foster the disclosure of personal data. Designers should however clearly refrain from using personalization features as a "gimmick" to increase the popularity of websites with low reputation since based on the aforesaid, it is unlikely that users will take much advantage of the personalization features if they have to disclose personal data to a low-reputation website.

21.3.3.4 Presence of a Privacy Statement

Privacy statements on websites (which are often also called "privacy policies") describe the privacy-related practices of these sites. Most countries that have privacy laws enacted require that users be informed about the data being collected and the purposes for which they are used. And even in jurisdictions where omnibus privacy legislation does not exist, special provisions at the federal or state level or simply public relation motives prompt many companies to publish privacy statements at their websites.[9] The comprehensibility of these disclosures for normal Internet users is however fairly low [96, 121].

There exists weak empirical evidence that the mere *presence* of a privacy statement at a website fosters trust.[10] For instance, 55% of the respondents in a survey of the Australian Privacy Commissioner [163] indicated that having a privacy statement would help build trust. This leads to the expectation that the presence of a privacy statement would also foster purchases and disclosure, namely via increased trust (see Section 21.3.3), which already received some empirical confirmation. In the study of Jensen et al. [97], the presence of a privacy statement proved to be one of the two best predictors for subjects' stated intent to buy from a website. In an experiment with Singaporean students, Hui et al. [90] found an effect of the presence of a privacy statement on subjects' willingness to completely fill in an online questionnaire with personal information*, but this effect only approached statistical significance $(p<0.1)$.[11] Metzger [126] however found the opposite effect. In her experiment, 43.7% of subjects who bought CDs from a fake online music store, or completed a questionnaire to receive a free CD, withheld information when a privacy policy was present. In contrast, only 15% withheld information when the privacy policy was *not* present, and the difference was statistically significant* [129].

Not too many people seem to view and read privacy policies. As far as self-reported past behavior is concerned, the percentage of respondents who indicated

---

[9] For example, as far as the U.S. is concerned, a 2004 survey of more than 1,000 websites across a spectrum of industries found that 93% of them featured a privacy statement [40].

[10] As in the case of privacy seals, Internet users seem to be confused about what protection the presence of a privacy policy affords to them. For instance, Turow [182] found that 57% of U.S. adults who use the Internet at home agree or strongly agree with the statement "When a web site has a privacy policy, I know that the site will not share my information with other websites or companies."

[11] Interestingly, the authors did not find the same effect when the tested website featured both a privacy policy and a privacy seal.

having looked at privacy policies varies between 3% ("most of the time, carefully") [83], 4.5% ("always") [133], 14.1% ("frequently") [133], 31.8% ("sometimes") [133], 33% ("sometimes, carefully") [83], 23.7% ("likely, at first visit") [97], and 43% ("likely, e-commerce site, before buying") [97]. Milne and Culnan [133] found that stated concern for privacy is positively associated with stated tendency to read online privacy notices.

Observing user behavior in experiments and real life portrays a somewhat different picture though. Jensen et al. [97] found that subjects read privacy statements in 25.9% of cases where they were available* (the authors believe though that this number is inflated since subjects knew they were being observed and what the purpose of the experiment was and therefore took more care in their decision-making than usual). In the experiment of Kobsa and Teltzrow [111], only two out of 52 subjects accessed the privacy statement*. The most reliable figures are presumably real-world server-side observations: only one percent of users or less click at links to a website's privacy statement according to Reagan [158], and less than 0.5% according to Kohavi [112]. In contrast to the above-mentioned survey results of Milne and Culnan [133], Jensen et al. [97] also found that those subjects whom they classified as privacy fundamentalists were no more likely to read privacy policies than the privacy unconcerned*.

When users do read privacy statements, the effect on users' behavior is unclear as yet. In an experiment by Metzger [126], 62.5% of the participants who clicked on the strong version of the privacy policy disclosed some information to the website and only 37.5% of those who clicked on the weak version, but the difference was not statistically significant*. Likewise in the experiment of Spiekermann et al. [172], the privacy protection that was promised in privacy statements did not have a statistically significant effect on subjects' willingness to disclose personal data* (subjects had to sign that they had read and accepted this statement prior to shopping at the experimental website). In contrast to these negative results, Andrade et al. [5] did find an effect of the length or level of detail of privacy statements: subjects who saw a 12-word statement professed considerable higher concern about the disclosure of personal information than subjects who saw a 88-word statement (a 22-word example statement was initially presented to all subjects as being "typical"). The ecological relevance of this experiment is however unclear since real-life privacy statements usually comprise several pages of text and not just a few words.

The preliminary lesson for the design of personalized systems seems to be that *traditional* privacy statements should not be posted in the expectation of increasing users' trust and/or disclosure of personal information, even when the statement describes good company privacy practices. There may of course be other reasons for posting such statements, such as legal or self-regulatory requirements (see Section 21.4), or demonstration of good will. Evidence is mounting though that privacy-minded company practices can have such a positive effect if they are communicated to web users in comprehensible forms, such as the following:

– Kobsa and Teltzrow [111] found that subjects disclose significantly more information about themselves if every website does not only display a link to a privacy policy, but if additionally every entry field for personal information is accompanied by a short summary of the website's privacy practices regarding specifically the solicited piece of information (and an explanation of why it is needed)*.

– Gideon at al. [70] asked subjects to search for vendors of a given product in a search engine and to buy the product with their own credit cards. For every site in the result list, the color of an appended "Privacy Bird" [9, 38] indicated whether the P3P [193] encoded privacy policy of the site matches typical medium-level privacy expectations, does not match them, or could not be parsed. If the website had no P3P policy posted, no bird would appear. The authors of the study found that when subjects were asked to buy a pack of condoms, they patronized websites with conforming privacy policies significantly more often than a control group that saw no privacy birds*. No such difference could be found when subjects had to buy a surge protector rather than condoms.

21.3.3.5 Presence of a Privacy Seal

Privacy seals are logos of certifying agencies such as consumer organizations, data commissioner's offices or private companies. These agencies assert to web visitors that websites that display their seals respect privacy to some extent. The amount of assured privacy protection varies from seal to seal and also over time. U.S. privacy seals originally merely asserted that a website abides to its published privacy statement, no matter how privacy-friendly this policy actually was. Meanwhile, trust organization require minimum privacy standards such as the observance of the FTC principles of notice, choice and consent [65].

A number of recent studies uncovered several problems with at least some privacy seals though:

Insufficient Scrutiny of Trust Organizations: Using webbots that analyze websites' privacy practices, Edelman [54] found that sites that used practices most Internet users would find objectionable nevertheless received a privacy seal from TRUSTe, the leading US trust mark. The percentage of untrustworthy sites certified by an TRUSTe seal even significantly increased over time (to nearly 3.5% as of January 2006). Various privacy breaches at websites that carried the TRUSTe seal, and to a much smaller extent also at sites with the BBBOnLine seal, have been reported as well [115, 137, 169].

Negative Self-Selection of Seal-Bearing Websites. Several studies came to the conclusion that websites that decide to "pay up" for certain privacy seals seem to have more questionable privacy practices than ones that don't. Larose and Rifon [115] found that sealed sites requested significantly more personal information from users than unsealed sites. Miyazaki and Krishnamurthy [134] reviewed 60 high-traffic websites and found no support for the hypothesis that participation in a seal program is an indicator of better privacy practices (Larose and Rifon [115] made similar findings). While these studies were all performed manually, Edelman [54] analyzed more than 500,000 websites with web bots. He found that the ratio of untrustworthy vs. trustworthy sites certified by TRUSTe (5.4%) is more than twice as high as for non-certified sites (2.5%). In a regression model with several site characteristics, the presence of an TRUSTe privacy seal turns out to be a statically significant negative coefficient for site trustworthiness. In contrast, the much less frequent BBBOnLine privacy seal that comes with a more cumbersome and restrictive certification process does not seem to suffer from such an adverse self-selection, and seal-bearing websites are slightly more likely to be trustworthy than a random cross-section of sites.

Seals Not Understood by Web Users: The results of a study by Portz et al. [154] on a specific privacy seal, WebTrust, "were mixed in terms of potential customers correctly understanding what WebTrust signifies." In a study by Moores [136], 42% recognized the TRUSTe logo and 29% the BBBOnline logo as a privacy seal. A whopping 15% however also mistook an officially looking fake graphic for a genuine privacy seal.

The presence of privacy seals clearly does have an effect on web users though, despite this confusion about what assurances they actually afford. Rifon et al. [160] found a positive effect on the perception of trust in a website. Miyazaki and Krishnamurthy [134] found that the presence of a privacy seal resulted in more favorable consumer perceptions regarding the privacy policies of a website. In the study of Jensen et al. [97], the presence of a privacy seal turned out to be one of the two best predictors for subjects' stated intent to buy from a website.

There is also empirical evidence for an effect of the presence of a privacy seal on users' stated willingness to disclose personal data to the website [116]. Other studies found that this effect was moderated by other factors, namely

– *Perceived self-efficacy* (i.e. confidence in one's ability to protect one's privacy): Rifon et al. [160] found that for individuals with lower self-efficacy, the presence of a privacy seal had a positive effect on anticipated disclosure of personal data. No such effect on subjects with high self-efficacy could be found.

– *Perceived online shopping risk* (when compared to transactions made at traditional brick and mortar stores): Miyazaki and Krishnamurthy [134] found a positive effect of privacy seal presence on anticipated disclosure of personal information for those subjects who experience relatively high levels of online shopping risk. No effect on subjects with low-risk experience could be found.

It remains to be seen whether these moderating factors are independent of each other, or rather correlated (which seems more likely). For designers of web-based personalized systems, the pragmatic conclusion at this point is to display privacy seals as long as web users associate trust with them since doing so is likely to foster users' disclosure behavior.

21.3.4 Benefits Other Than Personalization

Financial Rewards. In a consumer survey by Turow [182], 16% of respondents agreed or even strongly agreed with the statement "I will give out information to a website only if I am paid or compensated in some way". Hann et al. [78] found that a financial reward of 20 Singapore dollars[12] (but not of 10 or 5 dollars) had a statistically significant positive effect on intended disclosure behavior. However, this economic benefit turned out to be relatively less important by a considerable margin than three privacy concerns measured by the survey instrument of Smith [170], namely unauthorized internal/external secondary usage, unauthorized access, and errors in the data. The authors calculated that monetary compensations between about S$15.00 and S$50.00 would be needed to motivate subjects to overcome these concerns. Financial

---

[12] One Singapore dollar equaled 0.54 U.S. dollars in 2002.

rewards also had a statistically significant effect on observed disclosure behavior* in an experiment by Hui et al. [90], even though rewards only ranged from S$1 to S$9.

Social Adjustment Benefits. A study by Lu et al. [120] demonstrated that social adjustment benefits, i.e. the opportunity of establishing social identity by integrating into desired social groups [14], can also have an effect on intended disclosure behavior. The three experimental conditions were (a) no benefits (control group), (b) opportunity of face-to-face interaction with other people (namely meetings with people having similar interests, participation in focus groups, membership in downtown clubs of the Internet business), and (c) opportunity of online interaction (namely access to online chat-rooms with similar interests, exclusive membership in the online clubs of the Internet business, access to online forums featuring focus groups). For extrovert subjects, both treatment conditions had a statistically significant effect on their intended disclosure of personal data, while for introvert subjects this was only the case when online interaction was offered.

Both results seem only marginally relevant for personalized web-based systems since those normally do not offer such benefits. In special application scenarios though, the provision of personal data might open an opportunity for financial benefits (e.g., targeted advertising with special discounts) or social adjustment benefits (e.g., participation in discussion groups with people who have similar goals or interests). Designers should consider taking advantage of the increase in trust and disclosure that these benefits may entail.

21.3.5 Disclosure Behavior as the Result of a Cost-Benefit Analysis

Current privacy theory regards people's disclosure behavior to websites as the result of a situation-specific cost-benefit analysis, in which the potential risks of disclosing one's personal data are weighed against potential benefits of the data disclosure.[13] Trust thereby is an important risk-mitigating factor [31, 87, 95, 124, 128].

This cost-benefit tradeoff explains the discrepancies between stated privacy concerns and observed "inconsequent" data disclosure behavior that was discussed in Section 21.2.5. While it seems true that many Internet users are privacy-concerned, it is also a fact that most are willing to "trade off" their concerns against benefits that they value (see Sections 21.3.1 and 21.3.4) [17, 78, 173, 179], and become even more swayed to do so by the presence of trust-evoking signals such as those discussed in Section 21.3.3.

Acquisti and Grossklags [2, 3] point out however that Internet users often lack sufficient information to be able to make educated privacy-related decisions (for instance, they underestimate the probability with which they can be identified if they disclose certain data, or are unfamiliar with a site's privacy practices since they hardly ever read privacy statements (see Section 21.3.3.4). Like all complex probabilistic decisions, privacy-related decisions are moreover affected by systematic deviations from rationality [98]. For instance, Acquisti and Grossklags [3] present evidence of

---

[13] Culnan [42] coined the term "privacy calculus" to refer to this cost-benefit comparison (the term dates back to Laufer et al.'s [117, 118] notion of "calculus of behavior").

hyperbolic temporal discounting, which may lead to an overvaluation of small but immediate benefits and an undervaluation of future negative privacy impacts.

An implication of users' cost-benefit analysis for personalized systems is that developers can work in four, and possibly even five directions to encourage more liberal disclosure behavior, and thereby enhance the quality of the system's personalized services. They can

1. address the privacy concerns directly, as explained in Sections 21.2.5 and 21.5.4,
2. ensure that the user values the personalization benefits of the system (see Section 21.3.1),
3. ascertain that the user trusts the website (which mitigates privacy concerns), e.g. by establishing the trust-enhancing factors described in Sections 21.3.3.1 – 21.3.3.5,
4. ascertain that the user is made aware of, and can control, how personal information is being used (see Section 21.3.2), and
5. if meaningful, ascertain that financial rewards and social adjustment benefits are provided (see Section 21.3.4).

Interaction effects between these factors have not been established as yet. From the experiment of Chellappa et al. [31] (see Section 21.3.1) and the work of Acquisti and Grossklags [2, 3] we can conclude that *instant personalization benefits* will be a very important factor in the outcomes of users' cost-benefit analyses.

21.4 Privacy Laws, Industry and Company Regulations, and Principles of Fair Information Practices

To date, more than forty countries and numerous states have privacy laws enacted [161, 190]. Many companies and a few industry sectors additionally or alternatively adopted self-regulatory privacy guidelines. These laws and self-regulations are often based on more abstract principles of fair practices regarding the use of personal information. In this section, we will analyze the effects that these regulatory instruments have, specifically on personalization in web-based systems. We will uncover some deficits in current personalized systems, which open avenues for interesting and challenging future research. Privacy laws, industry and company regulations and Principles of Fair Information Practices may also impose requirements that are not directly related to personalization but affect any system that collects personal data. These more general implications cannot be discussed here. Readers are advised to consult their national privacy literature.

21.4.1 Privacy Laws

Since personalized systems collect personal data of individual people, they are also subject to privacy laws and regulations if the respective individuals are in principle identifiable. To date, more than forty countries and numerous states have privacy laws enacted. They lay out procedural, organizational and technical requirements for the collection, storage and processing of personal data, in order to ensure the protection of these data as well as the data subjects to whom the data apply. These requirements include disclosure duties (e.g. about the purpose of data processing), and conditions for legitimate data acquisition, data transfer (e.g., to third parties or across national borders) and the processing of personal data (e.g., their storage, modification and deletion). Other requisites include user opt-in (e.g., asking for their consent before collecting their data), opt-out (e.g., of data collection or data processing), and users' right to be informed (e.g., about what personal information has been collected and possibly how it is processed and used). Other legal stipulations establish adequate security mechanisms (e.g., access control), and the supervision and audit of personal data processing.

Some requirements imposed by privacy laws directly or indirectly affect the permissibility of personalization methods. Here are some examples:

1. *Value-added (e.g. personalized) services based on traffic or location data require the anonymization of such data or the user's consent* [56][14]. This clause requires the user's consent for any personalization based on interaction logs if the user can be identified.
2. *Users must be able to withdraw their consent to the processing of traffic and location data at any time* [56]. In a strict interpretation, this stipulation requires personalized systems to immediately honor requests for the termination of all traffic or location based personalization, i.e. even during the current session. A case can probably be made that users should not only be able to make all-or-none decisions, but also decisions with regard to individual aspects of traffic or location based personalization (such as agreeing to be informed about nearby sights but declining to receive commercial offers from nearby businesses).
3. *The personalized service provider must inform the user of the type of data which will be processed, of the purposes and duration of the processing, and whether the data will be transmitted to a third party, prior to obtaining her consent* [56]. It is sometimes fairly difficult for personalized service providers to specify beforehand the particular personalized services that an individual user would receive. The common practice is to collect as much data about the user as possible, to lay them in stock, and then to apply those personalization methods that "fire" based on the existing data (see, e.g., rule-based personalization or stereotype activation [109]). Also, internal inference mechanisms may augment the available user information by additional assumptions about the user, which in return may trigger additional personalization activities. For meeting the disclosure requirements of privacy laws in such cases of low ex-ante predictability, it should suffice to list a number of *typical* personalization examples (preferably those that entail the most severe privacy consequences) [79].
4. *Personal data that were obtained for different purposes may not be grouped* [46]. This limitation affects centralized user modeling servers (see Chapter 4 of this book [130]), which store user information from, and supply this data to, different personalized applications. Such servers must not return data to requesting personalized applications that was collected for a different purpose than the one for which the data is now being sought.

---

[14] EU directives are "Europe-wide minimum standards" in the sense that all European Union member states have to implement them in their national legislation, but are free to go beyond them.

5. *Usage data must be erased immediately after each session* (except for very limited purposes) [50]. This requirement could affect the use of machine learning methods that derive additional assumptions about users (see Chapter 3 of this book [135]), when the learning takes place over several user sessions.
6. *No fully automated individual decisions are allowed that produce legal effects concerning the data subject or significantly affect him and which are based solely on automated processing of data intended to evaluate certain personal aspects relating to him, such as his performance at work, creditworthiness, reliability, conduct, etc.* [55]. These provisions could affect, for example, personalized tutoring applications (see Chapter 22 of this book [84]), if they assign scores to users that significantly affect them.

Besides "omnibus" privacy laws at the national or state level, there also exist various sectorial laws. Examples in the U.S. include the Health Insurance Portability and Accountability Act (HIPAA [183]) for the privacy of medical data, the Gramm-Leach-Bliley Act (GLB [185]) for the privacy of financial data, and the Children's Online Privacy Protection Act (COPPA [184]) for protecting the privacy of children aged 13 and younger. The HIPPA and GLB Acts would affect personalized systems that collect or process users' medical or financial information, and COPPA those that have children among their users.

21.4.2 Industry and Company Regulations

Many companies have internal guidelines in place for dealing with personal data. Several industry associations also developed privacy standards to which their members must subject themselves (e.g., the Direct Marketing Association, the Online Privacy Alliance, and the Personalization Consortium). Both company and supra-company self-regulations may affect the aims and methods of personalized systems, as is the case for privacy legislation. For instance, the privacy principles of the members of the U.S. Network Advertising Initiative [139] prohibit the use of "personally identifiable information ("PII") [...] collected offline merged with PII collected online for online preference marketing unless the consumer has been afforded robust notice and choice about such merger before it occurs." This stipulation thus restricts the merger of clickstream data with data from legacy customer databases, which is a frequently-found functionality of commercial user modeling servers (see Chapter 4 of this book [130]).

21.4.3 Principles of Fair Information Practices

Over the past three decades, several collections of basic principles have been defined for ensuring privacy when dealing with personal information. So-called Principles of Fair Information Practices have been drafted by several countries as a foundation of their national privacy laws [10, 41], by supra-national organizations as a guidance for their member states [6, 142], and by professional societies as recommendations for policy makers and as guidance for the professional conduct of their members [186].

Developers of personalized systems should also take such privacy principles into account if those are not already indirectly considered through applicable privacy laws and industry or company guidelines. Many guidelines have direct implications on personalized systems. As an example, let us consider excerpts from the recommendations of the U.S. Public Policy Committee of the Association for Computing Machinery (ACM) [186], the largest computer science association worldwide. These recommendations have been strongly shaped by the 1980 OECD guidelines [142][15] but are more modern and concrete in their technical demands.

Minimization Principles

1. *Collect and use only the personal information that is strictly required for the purposes stated in the privacy policy.*
2. *Store information for only as long as it is needed for the stated purposes.*
3. *Implement systematic mechanisms to evaluate, reduce, and destroy unneeded and stale personal information on a regular basis, rather than retaining it indefinitely.*
4. *Before deployment of new activities and technologies that might impact personal privacy, carefully evaluate them for their necessity, effectiveness, and proportionality: the least privacy-invasive alternatives should always be sought.*

Somewhat in contradiction to these requirements, a current tacit paradigm of personalized systems seems to collect as much data as possible and lay them in stock, and to let personalization being triggered by the currently available personal data (data-driven personalization). Applications in several personalization areas[16] have now sufficiently progressed that it should be possible to determine in hindsight which of the collected data hardly ever trigger personalization, and to forego storing these less needed data in the future even when they would be readily available.

Consent Principles

5. *Unless legally exempt, require each individual's explicit, informed consent to collect or share his or her personal information (opt-in); or clearly provide a readily-accessible mechanism for individuals to cause prompt cessation [...] including when appropriate, the deletion of that information (opt-out).*

One implication of this requirement for personalized systems is that personalization based on the users' personal data must be an option that can be switched on and off at any time.

Openness Principles

8. *Whenever any personal information is collected, explicitly state the precise purpose for the collection and all the ways that the information might be used [...].*
10. *Explicitly state how long this information will be stored and used, consistent with the "Minimization" principle.*
11. *Make these privacy policy statements clear, concise, and conspicuous to those responsible for deciding whether and how to provide the data.*

---

[15] See [37] for a discussion of the effects of the OECD principles on personalized e-commerce systems.

[16] For instance, student-adaptive tutoring systems (see Chapter 22 of this book [84]), customer relationship management on the web (see [109] and Chapter 16 of this book [72]), and recommender systems (see Chapters 9-12 of this book [24, 146, 164, 171]).

The likely positive effect of such explanations on users' willingness to disclose personal data was discussed in Section 21.3.2. Some difficulties in providing a full explanation of the personalization purposes were discussed in Section 21.4.1 (3).

Access Principles

14. *Establish and support an individual's right to inspect and make corrections to her or his stored personal information, unless legally exempted from doing so.*

This principle calls for online inspection and correction mechanisms for personal data, as discussed in Section 21.3.2.

Accuracy Principles

17. *Ensure that personal information is sufficiently accurate and up-to-date for the intended purposes.*
18. *Ensure that all corrections are propagated in a timely manner to all parties that have received or supplied the inaccurate data.*

So far, allowing users to verify their data seems to be the only solution for assuring data accuracy that has been adopted in the personalization literature. Little attention has been paid to recognizing the obsoleteness of data, and to recording the provenance of data and propagating error and change notifications to the data sources.

Security Principles

19. *Use appropriate physical, administrative, and technical measures to maintain all personal information securely and protect it against unauthorized and inappropriate access or modification.*

This principle not only entails that user information must be protected when it is stored in a repository, but also while it is in transit (e.g. by only using secure channels between authenticated senders and receivers). In the case of personalized systems, the latter is currently not often considered.

21.5 Privacy-Enhancing Technology for Personalized Systems

In this section, we describe and analyze several technical approaches that may reduce privacy risks and make privacy compliance easier. They are by no means complete "technical solutions" to the privacy risks of personalized systems, and their presence is also unlikely to "charm away" users' privacy concerns. Rather, these technologies should only be employed as additional privacy protections in the context of a user-oriented system design that also takes normative aspects into account (see Section 21.4). This analysis will be restricted to technologies that are specifically intended for personalized web-based systems. For an overview of more general privacy-enhancing technologies that can be applied to wider classes of systems (including personalized web-based systems in many cases), we refer to [21, 25, 71, 188].

21.5.1 Pseudonymous Users and User Models

It is possible for users of personalized systems to enjoy anonymity and at the same time receive full personalization [110, 166]. In an anonymization infrastructure that supports personalization, users would need to have the following characteristics (using the terminology of [93, 150]):

- *Unidentifiable.* Neither the personalized system nor third parties should be able to determine the identity of pseudonymous users;
- *Linkable for the personalized system.* The personalized system can link every interaction with a specific user, even across sessions (users maintain a persistent identity);
- *Unlinkable for third parties.* Third parties cannot link two interaction steps of the same user;
- *Unobservable for third parties.* Third parties cannot recognize that a personalized system is being used by a given user.

To ensure their linkability, users would need to employ a "pseudonym" in all their transactions, i.e. a unique and persistent identifier that differentiates them from all other users. The personalized system may allow users to freely define their pseudonyms (or pick them from a list of available pseudonyms) without disclosing their true identities. Users may however also be required to reveal their identities to a registrar who assigns pseudonyms to them ("escrowed identity" [103], "initially nonpublic pseudonym" [150]). In the latter case, the pseudonym may be revoked at a later time, by an act of the registrar alone or in tandem with the website operator and/or user. This revocation of pseudonyms may be desirable in cases of misuse or when the identification of the user becomes necessary for other reasons, such as non-anonymous payment and delivery scenarios.

A number of authors proposed infrastructures for pseudonymous yet personalized user interaction with websites based on some or all of the above properties [8, 66, 85, 92, 110, 166]. Protecting the identity of users may not be enough, however. If user data is stored on a user modeling server on the Internet (see Chapter 4 of this book [130]), not only the user but also the user modeling server may need to remain anonymous. User models may reside anywhere on the network, like on the user's platform (as is envisaged, e.g., in the P3P framework [193]) or on a remote server (such as in Microsoft's Passport architecture [144]). A location close to the user (such as informatics.uci.edu or even more alfredkobsa.name) may compromise the user's anonymity. To safeguard it, Kobsa and Schreck [110, 166] extend their pseudonymity infrastructure to also protect the anonymity of user modeling servers.

Some authors expect that Internet users are more likely to provide information when they are not identified [39, 110], which may improve the quality of personalization and the benefits that users receive from it. To date, this claim has however not found much empirical substantiation. In an online survey from 1998 [76], 66.3% of respondents strongly agreed and 21.8% somewhat agreed with the statement "I value being able to visit sites on the Internet in an anonymous manner." 30.5% also strongly agreed and 22.1% somewhat agreed with the statement "I would prefer Internet payment systems that are anonymous to those that are user identified". The demographics of the survey respondents was however considerably skewed towards higher education (nearly 80% had at least some college-level education) and towards fairly advanced web skills. Ordinary consumers tend to be unfamiliar with many basic security features, and base their perception of security rather on the company's reputation, their experience with the site, and recommendations from independent third parties [181].

The implications of these limited findings for the design of personalized web-based systems seem a bit unclear. Designers should definitely allow for pseudonymous access and pseudonymous user models (and even allow for anonymization architectures with the above properties if one is readily available). This follows from the data minimization and security requirements of the Principles of Fair Information Practices that were discussed in Section 21.4.3. Some privacy laws also mandate [51] or recommend [56] the provision of pseudonymous access if it is technically possible and not unreasonable (an interesting side effect of pseudonymous access is that in most cases privacy laws do not apply any more when users cannot be identified with reasonable means).

Due to a lack of relevant studies, it is unclear though whether increased anonymity will lead to more disclosure and better personalization. Anonymity is currently also difficult and/or tedious to preserve when payments, physical goods and non-electronic services are being exchanged. It also harbors the risk of misuse and hinders vendors from cross-channel marketing (e.g. sending a products catalog to a web customer by postal mail). Finally, research shows that the anonymity of database entries [176], web trails [123], query terms [140], ratings [64] and textual data [157] can be surprisingly well defeated by a resourceful attacker who has identified data available that can be partly matched with the "anonymous" data.

21.5.2 Client-Side Personalization

A number of authors [28, 29, 36, 138] have worked on personalized systems in which users' data are located at the client rather than the server side. Likewise, all personalization processes that rely on this data are also carried out at the client side only. From a privacy perspective, this approach has two major advantages:

1. The privacy problem becomes smaller since very few, if any, personal data of users will be stored on the server. In fact, if a website with client-side personalization does not have control over any data that would allow for the identification of users with reasonable means, it will generally not be subject to privacy laws.
2. Users may possibly be more inclined to disclose their personal data if personalization is performed locally upon locally stored data rather than remotely on remotely stored data, since they may feel more in control of their local physical environment.[17]

---

[17] No empirical verification for this assumption seems to exist as yet. In times of global network connectivity, this purported feeling of local control may be illusionary though. For instance, probably not many Skype users are aware that if they are not sitting behind a firewall or broadband gateway, but have good connectivity to the network, then they are pretty likely to have other people's traffic flowing through their computers (and using their network bandwidth). The pervasiveness of malware on people's computers also does not speak for a higher safety of locally stored personal data.

Client-side personalization also poses a number of challenges though:

1. Popular user modeling and personalization methods that rely on an analysis of data from the whole user population, such as collaborative filtering and stereotype learning (see [109]), cannot be applied any more or will have to be radically redesigned (see the next section).

2. Personalization processes will also have to operate at the client side since even only a temporary or partial transmission of personal data to the server is likely to annul the abovementioned advantages of client-side personalization. However, program code that is used for personalization often incorporates confidential business rules or methods, and must be protected from disclosure through reverse engineering. Trusted computing platforms will therefore have to be developed for this purpose, similar to the one that Coroama and Langheinrich [35, 36] envisage to ensure the integrity of their client-side collection of personal data.

If these drawbacks pose no problems in a specific application domain, then developers of personalized web-based systems should definitely adopt client-side personalization as soon as suitable tools become available. Doing so would constitute a great step forward in terms of the data minimization principle (see Section 21.4.3) and is also likely to increase users' trust.

21.5.3 Distribution, Encrypted Aggregation, Perturbation and Obfuscation

A number of techniques have been proposed and partially also technically evaluated that can help protect the privacy of users of recommender systems that employ collaborative filtering (see Section 9 of this book [164]). Traditional collaborative filtering systems collect large amounts of information about their users in a central repository (e.g., users' product ratings, purchased products or visited web pages), to find regularities that allow for future recommendations. Such central repositories may not always be trustworthy though, and they are also likely to constitute an attractive target for unauthorized access. To some extent, central repositories may also be mined for individual user data by requesting recommendations using cleverly constructed profiles [27]. For instance, personal websites tend to be visited by their owners more frequently than by anyone else. In a recommender system that tracks users' website visits, websites that are highly correlated with personal websites are hence likely to have been visited by those owners as well. Requesting a recommendation for pages to visit using a profile that contains this home page only may therefore reveal frequently visited web pages of its owner. Another statistical vulnerability is that correlations between an item and others will disclose much information about the choices of its raters if this item has very few raters only.

Client-side personalization (see Section 21.5.2) alone is not a remedy against such privacy attacks in collaborative filtering systems. Even when all user profiles are stored at the clients' sides, a considerable number of them (or even all) must still be merged and compiled in order that recommendations can be generated. Below we describe several strategies that are currently investigated to thwart such risks.

21.5.3.1 Distribution

One possible strategy to better safeguard individuals' data is to abandon central repositories that contain the data of all users, in favor of distributed clusters that contain information about some users only. Distribution may also improve performance and availability of the recommender system.

For instance, in the distributed match-making system Yenta [62], agents representing a user continuously form clusters of like-minded agents by exchanging information about their users and referring agents to potentially similar other agents. While this work is not explicitly aimed at protecting privacy, it does so to some extent by virtue of the fact that at any given time, agents only maintain the data of a limited number of like-minded agents and that a pseudonymity scheme can by added to protect users' identity.

The distributed PocketLens collaborative filtering algorithm [131] goes even further in terms of data avoidance. For each user, PocketLens first searches for neighbors in a P2P network and then incrementally updates the user's individual item-item similarity model by incorporating one neighbor's ratings at a time (ratings are immediately discarded thereafter). The recommendations produced by PocketLens were shown to be as good as those of the best "centralized" collaborative filtering algorithms published to date.

21.5.3.2 Aggregation of Encrypted Data

Canny [26, 27] proposed the usage of a secure multi-party computation scheme that allows users to privately maintain their own individual ratings, and a community of such users to compute an aggregate of their private data without disclosing them by using homomorphic encryption and peer-to-peer communication. The aggregate (a single-value decomposition of a user-item matrix) then allows personalized recommendations to be generated at the client side using one's own ratings. The scheme is however still prone to the above-mentioned statistical vulnerabilities. The PocketLens system [131] was also connected to a blackboard based on the same security schemes as those used by Canny, to allow a community of users to compute a similarity model without having to reveal their individual rankings.

21.5.3.3 Perturbation

In the perturbation approach, users' ratings are submitted to a central server which performs all collaborative filtering. These ratings become systematically altered before submission though, to hide users' true values from the server. Polat and Du [152, 153] show that adding random numbers to user ratings may still yield acceptable recommendations. The quality of recommendation based on perturbed data improves when the number of items and users increases and when the standard deviation of the perturbation function decreases (the latter obviously reduces privacy). The authors conducted a series of experiments with two databases of user rankings, namely Jester [75] and MovieLens [74], using a privacy measure proposed by Agrawal and Agrawal [4] that is based on differential entropy between the unperturbed and the perturbed data. For the Jester database, the authors find that privacy levels of about 97% and 90% will introduce average errors of about 13% and 5%, respectively, compared with predictions based on unperturbed data. For MovieLens, the average relative errors due to perturbation at these privacy levels were 10% and 5%, respectively.

21.5.3.4 Obfuscation

In the obfuscation approach of Berkovsky et al. [19], a certain percentage of users' ratings become replaced by different values before the ratings are submitted to a central server for collaborative filtering. Users are supposed to be able to freely choose which of their data should be obfuscated, and to "plausibly deny" the accuracy of any of their data should they become compromised. In subsequent work, Berkovsky et al. [20] combined obfuscation with distributed recommendation generation by ad-hoc peers, which adds an additional layer of privacy protection through distribution (see Section 21.5.3.1).

The authors performed experiments on the user ratings of the Jester [75], Movie-Lens [74] and EachMovie [125] recommender systems. They varied the ratio of obfuscated data in users' submitted rankings and compared the ensuing loss of prediction accuracy. They found that obfuscation of the true rating through replacement by the following values had the smallest impact on the prediction error (in the range of 5-7% at an obfuscation rate of 90%): the means of the ratings scale, a random value from the scale, and a random value from the scale taking the means and variance of the ratings in the data set into account. In contrast, uniform replacement by the highest or lowest scale value resulted in an about 300% increased prediction error at a 90% obfuscation rate.

In all these experiments, the data to be obfuscated were randomly selected for each individual user. This strategy does not take into account that users are likely to prefer obfuscation for certain kinds of data rather than random data (see Section 21.2.3). Such a tendency is likely to further increase the prediction error. Recent experiments by the authors showed that obfuscating 10% of the ratings at the high end of the scale affected the prediction error more than obfuscating 10% of mid-scale ratings [18].

21.5.3.5 Consequences for the Design of Personalized Systems

Distribution, aggregation of encrypted user data, perturbation and obfuscation constitute promising privacy-protecting techniques. They can be supplemented by pseudonymity in applications where anonymity of users or their user models is additionally desired (see Section 21.5.1). While aggregation of encrypted user data cannot defeat attacks on statistical vulnerabilities that were discussed at the beginning of Section 21.5.3, perturbation and obfuscation may be able to thwart them (specifically if users are aware of their "weak statistical spots" and elect to obfuscate them). Experiments will need to determine the required level of perturbation or obfuscation that guarantees a high degree of protection.

While these techniques have so far only been investigated in the area of recommender systems, it is likely that distribution, perturbation and obfuscation can in principle be applied to virtually any machine learning technique that computes aggregate data based on individual user data (learning of encrypted user data will only be possible if a suitable homomorphic encryption can be found). The effects on the quality of the learning results still remain to be seen, however.

21.5.4 Personalizing Privacy

Individual privacy preferences may differ between users (see Section 21.2), and applicable privacy laws may also be different for users from different states and coun-

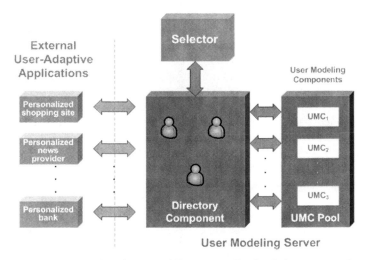

Fig. 1. Dynamic privacy-enabling personalization infrastructure (from [192])

tries (see Section 21.4). Different privacy preferences and laws impose different requirements on admissible personalization methods for each user. Personalized systems should therefore cater to the different privacy needs of individual users, i.e. they should "personalize privacy" [37, 106].

So far, there only exist two simplistic "solutions" to this problem:

1. *Largest permissible common subset approach.* In this approach, only those personalization methods are used that satisfy the privacy laws and regulations of all jurisdictions of all users of a website. The Disney website, for instance, observes both the U.S. Children's Online Privacy Protection Act [184], and the European Union Directive [51]. This solution is likely to run into problems if more than a very few jurisdictions are involved, since the largest common subset of permissible personalization methods may then become very small. The approach also does not take users' individual privacy preferences into account.

2. *Different country/region versions.* In this approach, personalized systems have different country versions, each of which uses only those personalization methods that are permitted in the respective country. If countries have similar privacy laws, these countries can be pooled using the above-described largest permissible common subset approach. For example, IBM's German-language pages comply with the privacy laws of Germany, Austria and Switzerland [91], while IBM's U.S. site meets the legal constraints of the U.S. only. As with the largest permissible common subset method, this approach also does not scale well when the number of countries/regions, and hence the number of different versions of the personalized system, increases. It also does not take users' individual privacy preferences into account.

Wang and Kobsa [191, 192] developed an architecture that allows personalized systems to provide optimal personalization benefits for each user, while at the same time satisfying the privacy constraints that apply to each individual user (e.g., their privacy preferences, and applicable laws and regulations). Figure 1 gives an overview of this architecture. The Directory Component is a repository of user models, each of which also includes the user's privacy constraints stemming from personal preferences and applicable laws and regulations. The UMC Pool contains a set of User Modeling Components, each of which encapsulates a user modeling method that operates upon the user models in the Directory Component, such as collaborative filtering (see Chapter 9 of this book [164]) or case-based recommendation (see Chapter 11 of this book [171]). On the left-hand side we see user-adaptive clients that access models of their current users in order to personalize their interaction with them.

As described so far, this architecture is similar to the one presented by Fink and Kobsa [57, 130], which was also used in a commercial user modeling server. The novel privacy enhancement consists in each user having his or her own instance of the UMC Pool, each containing only those user modeling components that meet the privacy requirements for the respective user (users with identical UMC Pool instances share the same instance). To realize this, the above architecture has been implemented as a Software Product Line (SPL) architecture [22, 33], with the UMCs as optional elements. At the beginning of the interaction with a user, a Selector verifies for every UMC whether it is allowed to operate under the privacy constraints that apply to the specific user, and creates an architectural instance with those permissible UMCs (or lets the user share this instance if one already exists). The special SPL management environment that we employ [7, 67, 189] even supports dynamic runtime (re-) configuration, which allows the Selector to react immediately should, e.g., users change their privacy preferences during the current session. The architecture therefore fully supports compliance with the consent principles discussed in Section 21.4.1 and 21.4.3, allowing a website to adjust its data practices to the user's preferences in a nuanced and highly dynamic manner.

21.6 Conclusion

A tension exists between personalization and privacy in web-based systems. On the one hand, personalization provides benefits to both users and operators of personalized websites. On the other hand, Internet users have high concerns regarding their privacy online, which may make them reluctant to disclose data about themselves to personalized systems. This poses a threat to personalization, whose quality hinges strongly on the amount of personal data supplied. The problem is exacerbated by the fact that many countries and states have privacy laws enacted that affect the permissibility of personalization methods, and that some company and industry regulations as well as principles of fair information practices have the same effect.

This chapter described a number of approaches that can be taken to render personalization more compatible with privacy. It first discussed measures that have proven to increase users' willingness to disclose data about themselves, mostly through increased trust (one of these measures, pointedly, consists in increasing the value of personalization as perceived by the user). It then analyzed how privacy legislation, self-regulation and principles of fair information practices impact the usage of personalization methods. Finally, it presented a number of technical solutions specifically intended for personalized systems that may either lessen the privacy problem in the first place (albeit no verification through user studies seems to have taken place as yet), or help developers of personalized systems adjust personalization individually to users' privacy preferences and to normative demands stemming from privacy laws, regulations and principles.

Personalization has already made some inroads into current commercial websites (see Chapter 16 of this book [72]). Given the high privacy concerns of today's Internet users, further advances are likely to only take place if privacy plays a much more important role in the future. Research on Privacy-Enhanced Personalization aims at reconciling the goals and methods of user modeling and personalization with privacy considerations, and at achieving the best possible personalization within the boundaries set by privacy. Many of the approaches described in this chapter are ready to be deployed to practical systems, and feedback from such deployments will in turn be very informative for research. Other approaches still need further technical development or evaluation in user experiments and may yield fruitful solutions in the future.

Acknowledgments
The preparation of this article has been supported by grant IIS 0308277 of the National Science Foundation, by a Trans-Coop grant, and by an Alexander von Humboldt Research Award. The author would like to thank Peter Brusilovsky, Lorrie Cranor, Miriam Metzger, Sameer Patil, Yang Wang and the anonymous reviewers for valuable comments on an earlier draft.

References

1. Ackerman, M. S., Cranor, L. F., and Reagle, J.: Privacy in E-commerce: Examining User Scenarios and Privacy Preferences. First ACM Conference on Electronic Commerce, Denver, CO (1999) 1-8, DOI 10.1145/336992.336995.
2. Acquisti, A.: Privacy in Electronic Commerce and the Economics of Immediate Gratification. EC'04 ACM Conference on Electronic Commerce, New York, NY (2004) 21-29, DOI 10.1145/988772.988777.
3. Acquisti, A. and Grossklags, J.: Privacy and Rationality in Individual Decision Making. IEEE Security & Privacy 3, (2005) 26-33, DOI 10.1109/MSP.2005.22.
4. Agrawal, D. and Aggarwal, C. C.: On the Design and Quantification of Privacy Preserving Data Mining Algorithms. 20th ACM SIGACT-SIGMOD-SIGART Symposium on Principles of Database System, Santa Barbara, CA (2001) 247–255.
5. Andrade, E. B., Kaltcheva, V., and Weitz, B.: Self-Disclosure on the Web: The Impact of Privacy Policy, Reward, and Brand Reputation. In: Advances in Consumer Research, Broniarczyk, S. M. and Nakamoto, K., Eds. Valdosta, GA: Assoc for Consumer Research (2002) 350-353, http://bear.cba.ufl.edu/weitz/papers/Andrade, Kaltcheva, Weitz.pdf.
6. APEC Privacy Framework. Asia-Pacific Economic Cooperation, Singapore, Report APEC#205-SO-01.2, (2005).
7. Archstudio: ArchStudio 3.0. (2006), http://www.isr.uci.edu/projects/archstudio/
8. Arlein, R. M., Jai, B., Jakobsson, M., Monrose, F., and Reiter, M. K.: Privacy-Preserving Global Customization. 2nd ACM Conference on Electronic Commerce, Minneapolis, MN (2000) 176-184, DOI 10.1145/352871.352891.
9. AT&T: AT&T Privacybird. (2006), http://www.privacybird.com/
10. AU-NPP: National Privacy Principles. The Office of the Privacy Commissioner, Australia (2001), http://www.privacy.gov.au/publications/npps01.html

11. Bachem, C.: Profilgestütztes Online Marketing. Personalisierung im E-Commerce, Hamburg, Germany (1999).
12. Bart, Y., Shankar, V., Sultan3, F., and Urban, G. L.: Are the Drivers and Role of Online Trust the Same for All Web Sites and Consumers? A Large-Scale Exploratory Empirical Study. Journal of Marketing 69, (2005) 133-152, DOI 10.1509/jmkg.2005.69.4.133.
13. Basso, A., Goldberg, D., Greenspan, S., and Weimer, D.: First Impressions: Emotional and Cognitive Factors Underlying Judgments of Trust E-Commerce. 3rd ACM Conference on Electronic Commerce (2001) 137-143, DOI 10.1145/501158.501173.
14. Baumeister, R. F. and Leary, M. R.: The Need to Belong: Desire for Interpersonal Attachments as a Fundamental Human Motivation. Psychological Bulletin 117, (1995) 497-529.
15. Behrens, L., Ed.: Privacy and Security: The Hidden Growth Strategy (2001), http://www.gartner.com/5_about/press_releases/2001/pr20010807d.html.
16. Bellman, S., Lohse, G. L., and Johnson, E. J.: Predictors of Online Buying Behavior. Communications of the ACM 42, (1999) 32-38, DOI 10.1145/322796.322805.
17. Berendt, B., Günther, O., and Spiekermann, S.: Privacy in E-Commerce: Stated Preferences vs. Actual Behavior. Communications of the ACM 48, (2005) 101-106, DOI 10.1145/1053291.1053295.
18. Berkovsky, S.: Personal Communication. (2006).
19. Berkovsky, S., Eytani, Y., Kuflik, T., and Ricci, F.: Privacy-Enhanced Collaborative Filtering. In: PEP05, UM05 Workshop on Privacy-Enhanced Personalization, Kobsa, A. and Cranor, L., Eds. Edinburgh, Scotland (2005) 75-83, http://www.isr.uci.edu/pep05/papers/PEPfinal.pdf.
20. Berkovsky, S., Eytani, Y., Kuflik, T., and Ricci, F.: Hierarchical Neighborhood Topology for Privacy Enhanced Collaborative Filtering. Proceedings of PEP06, CHI 2006 Workshop on Privacy-Enhanced Personalization, Montreal, Canada (2006) 6-13, http://www.isr.uci.edu/pep06/papers/PEP06_BerkovskyEtAl.pdf.
21. Borking, J. J. and Raab, C. D.: Laws, PETs and other Technologies for Privacy Protection. Journal of Information, Law and Technology, (2001), http://elj.warwick.ac.uk/jilt/01-1/borking.html.
22. Bosch, J.: Design and Use of Software Architectures: Adopting and Evolving a Product-Line Approach. New York: Addison-Wesley (2000).
23. Brusilovsky, P. and Millán, E.: User Models for Adaptive Hypermedia and Adaptive Educational Systems. In: The Adaptive Web: Methods and Strategies of Web Personalization, Brusilovsky, P., Kobsa, A., and Nejdl, W., Eds. Berlin Heidelberg New York: Springer Verlag (2007) this volume.
24. Burke, R.: Hybrid Web Recommender Systems. In: The Adaptive Web: Methods and Strategies of Web Personalization, Nejdl, W., Ed. Heidelberg, Germany: Springer Verlag (2006) In this volume.
25. Burkert, H.: Privacy-Enhancing Technologies: Typology, Critique, Vision. In: Technology and Privacy: The New Landscape, Agre, P. E. and Rotenberg, M., Eds. Boston, MA: MIT Press (1997) 126-143.
26. Canny, J.: Collaborative Filtering with Privacy. IEEE Symposium on Security and Privacy, Oakland, CA (2002) 45-57, DOI 10.1109/SECPRI.2002.1004361.
27. Canny, J.: Collaborative Filtering with Privacy via Factor Analysis. 25th Annual International ACM SIGIR Conference on Research and Development in Information Retrieval (SIGIR 2002), Tampere, Finland (2002) 238-245, DOI 10.1145/564376.564419.
28. Cassel, L. and Wolz, U.: Client Side Personalization. DELOS Workshop: Personalisation and Recommender Systems in Digital Libraries, Dublin, Ireland (2001), http://www.ercim.org/publication/ws-proceedings/DelNoe02/CasselWolz.pdf.

29. Ceri, S., Dolog, P., Matera, M., and Nejdl, W.: Model-Driven Design of Web Applications with Client-Side Adaptation. In: Web Engineering: 4th International Conference, ICWE 2004, Koch, N., Fraternali, P., and MartinWirsing, Eds. Berlin – Heidelberg: Springer Verlag (2004) 201-214, DOI 10.1007/b99180.

30. Privacy Policies Critical to Online Consumer Trust. Columbus Group and Ipsos-Reid, Canadian Inter@ctive Reid Report (2001), http://www.ipsos-na.com/news/pressrelease.cfm?id=1171.

31. Chellappa, R. K. and Sin, R.: Personalization versus Privacy: An Empirical Examination of the Online Consumer's Dilemma. Information Technology and Management 6, (2005) 181-202, DOI 10.1007/s10799-005-5879-y.

32. ChoiceStream Personalization Survey: Consumer Trends and Perceptions. (2005), http://www.choicestream.com/pdf/ChoiceStream_PersonalizationSurveyResults2005.pdf

33. Clements, P. and Northrop, L.: Software Product Lines: Practices and Patterns. New York: Addison-Wesley (2002).

34. Cooperstein, D., Delhagen, K., Aber, A., and Levin, K.: Making Net Shoppers Loyal. Forrester Research, Cambridge, MA June 1999 (1999).

35. Coroama, V.: The Smart Tachograph: Individual Accounting of Traffic Costs and Its Implications. In: Pervasive Computing: 4th International Conference, PERVASIVE 2006,, Fishkin, K. P., Schiele, B., Nixon, P., and Quigley, A., Eds. Berlin – Heidelberg: Springer Verlag (2006) 135-152, DOI 10.1007/11748625.

36. Coroama, V. and Langheinrich, M.: Personalized Vehicle Insurance Rates: A Case for Client-Side Personalization in Ubiquitous Computing. Proceedings of PEP06, CHI 2006 Workshop on Privacy-Enhanced Personalization, Montreal, Canada (2006) 56-59, http://www.isr.uci.edu/pep06/papers/PEP06_CoroamaLangheinrich.pdf.

37. Cranor, L. F.: 'I Didn't Buy it for Myself': Privacy and Ecommerce Personalization. 2003 ACM Workshop on Privacy in the Electronic Society, Washington, DC (2003), DOI 10.1145/1005140.1005158.

38. Cranor, L. F., Guduru, P., and Arjula, M.: User Interfaces for Privacy Agents. ACM Transactions on Human-Computer Interactions, (2006), DOI 10.1145/1165734.1165735.

39. Cranor, L. F., Reagle, J., and Ackerman, M. S.: Beyond Concern: Understanding Net Users' Attitudes About Online Privacy. AT&T Labs - Research, Technical Report TR 99.4.3, (1999), http://www.research.att.com/resources/trs/TRs/99/99.4/99.4.3/report.htm.

40. Largest 100 US Firms Rated on Customer Online Experience They Provide in Third Annual Customer Respect Group Study. The Customer Respect Group (2004), http://www.customerrespect.com

41. Model Code for the Protection of Personal Information. Canadian Standards Association (1996), http://www.csa.ca/standards/privacy/code/Default.asp?language=English

42. Culnan, M. J.: Managing Privacy Concerns Through Fairness and Trust: Implications for Marketing. Visions for Privacy in the 21st Century: A Search for Solutions. Conference Papers., Victoria, BC (1996), http://www.privacyexchange.org/iss/confpro/bcculnan.html.

43. Culnan, M. J. and Armstrong, P. K.: Information Privacy Concerns, Procedural Fairness, and Impersonal Trust: An Empirical Investigation. Organization Science 10, (1999) 104-115, http://links.jstor.org/sici?sici=1047-7039%28199901%2F02%2910%3A1%3C104%3AIPCPFA%3E2.0.CO%3B2-H.

44. Culnan, M. J. and Milne, G. R.: The Culnan-Milne Survey on Consumers & Online Privacy Notices: Summary of Responses. Interagency Public Workshop: Get Noticed: Effective Financial Privacy Notices, Washington, D.C. (2001), http://www.ftc.gov/bcp/workshops/glb/supporting/culnan-milne.pdf.

45.   UCO Software To Address Retailers $6.2 Billion Privacy Problem: Cyber Dialogue Survey Data Reveals Lost Revenue for Retailers Due to Widespread Consumer Privacy Concerns. Cyber Dialogue, (2001), http://www.cyberdialogue.com/news/releases/2001/ 11-07-uco-retail.pdf

46.   Act 101 of April 4, 2000 on the Protection of Personal Data and on Amendment to Some Acts. Czech Republic (2000), http://www.uoou.cz/index.php?l=en&m=left&mid=01:105

47.   Czarkowski, M. and Kay, J.: Bringing Scrutability to Adaptive Hypertext Teaching. In: Intelligent Tutoring Systems 2000, Gauthier, G., Frasson, C., and VanLehn, K., Eds. Berlin: Springer (2000) 423-433, www.springerlink.com/link.asp?id=q0qvb6qu7nbtxkxp.

48.   Czarkowski, M. and Kay, J.: How to Give the User a Sense of Control Over the Personalization of AH? AH2003: Workshop on Adaptive Hypermedia and Adaptive Web-Based Systems, Budapest, Hungary; Johnstown, PA; Nottingham, England (2003), http://wwwis.win.tue.nl/ah2003/proceedings/paper11.pdf.

49.   D'Hertefelt, S.: Trust and the Perception of Security. 3 January 2000 (2000), http://www.interactionarchitect.com/research/report20000103shd.htm.

50.   Teleservices Data Protection Act 1997, as amended on 14 Dec. 2001. (2001), http://bundesrecht.juris.de/tddsg/BJNR187100997.html

51.   Personal Communication, Chief Privacy Officer, Disney Corporation. (2002).

52.   Dommeyer, C. J. and Gross, B. L.: What Consumers Know and What They Do: An Investigation of Consumer Knowledge, Awareness, and Use of Privacy Protection Strategies. Journal of Interactive Marketing 17, (2003) 34-51, DOI 10.1002/dir.10053.

53.   Earp, J. B. and Baumer, D.: Innovative Web Use to Learn About Consumer Behavior and Online Privacy. Communications of the ACM Archive 46, (2003) 81 - 83, DOI 10.1145/641205.641209.

54.   Edelman, B.: Adverse Selection in Online "Trust" Certifications. Harvard University, Working Draft 15 Oct. 2006 (2006), http://www.benedelman.org/publications/advsel-trust-draft.pdf.

55.   Directive 95/46/EC of the European Parliament and of the Council of 24 October 1995 on the Protection of Individuals with Regard to the Processing of Personal Data and on the Free Movement of such Data. Official Journal of the European Communities, (1995) 31ff, http://158.169.50.95:10080/legal/en/dataprot/directiv/directiv.html.

56.   Directive 2002/58/EC of the European Parliament and of the Council Concerning the Processing of Personal Data and the Protection of Privacy in the Electronic Communications Sector. (2002), http://register.consilium.eu.int/pdf/en/02/st03/03636en2.pdf

57.   Fink, J.: User Modeling Servers - Requirements, Design, and Evaluation. Amsterdam, Netherlands: IOS Press (2004), http://books.google.com/books?q=isbn:1586034057.

58.   Fink, J., Kobsa, A., and Nill, A.: Adaptable and Adaptive Information Provision for All Users, Including Disabled and Elderly People. The New Review of Hypermedia and Multimedia 4, (1998) 163-188, http://www.ics.uci.edu/~kobsa/papers/1998-NRHM-kobsa.pdf.

59.   Fogg, B. J.: Persuasive Technology: Using Computers to Change What We Think and Do. San Francisco: Morgan Kaufmann Publishers (2003).

60.   Fogg, B. J., Kameda, T., Boyd, J., Marshall, J., Sethi, R., Sockol, M., and Trowbridge, T.: Stanford-Makovsky Web Credibility Study 2002: Investigating What Makes Web Sites Credible Today. A Research Report by the Stanford Persuasive Technology Lab & Makovsky & Company, Stanford, CA (2002), http://www.webcredibility.org/pdf/Stanford-MakovskyWebCredStudy2002-prelim.pdf.

61.   Fogg, B. J., Soohoo, C., Danielson, D. R., Marable, L., Stanford, J., and Tauber, E. R.: How Do Users Evaluate the Credibility of Web Sites?: a Study with over 2,500 Participants. Conference on Designing for User Experiences, San Francisco, CA (2003) 1-15, DOI 10.1145/997078.997097.

62. Foner, L. N.: Yenta: A Multi-Agent Referral-Based Matchmaking System. International Conference on Autonomous Agents, Marina del Rey, CA (1997) 301-307, DOI 10.1145/267658.267732.

63. Fox, S., Rainie, L., Horrigan, J., Lenhart, A., Spooner, T., and Carter, C.: Trust and Privacy Online: Why Americans Want to Rewrite the Rules. The Pew Internet & American Life Project, Washington, DC (2000), http://www.pewinternet.org/report_ display. asp?r=19.

64. Frankowski, D., Cosley, D., Sen, S., Terveen, L., and Riedl, J.: You Are What You Say: Privacy Risks of Public Mentions. 29th Annual International ACM SIGIR Conference on Research and Development in Information Retrieval, Seattle, WA (2006) 565-572, DOI 10.1145/1148170.1148267.

65. Privacy Online: Fair Information Practices in the Electronic Marketplace. A Report to Congress. Federal Trade Commission (2000), http://www.ftc.gov/reports/privacy2000/ privacy2000.pdf

66. Gabber, E., Gibbons, P. B., Matias, Y., and Mayer, A.: How to Make Personalized Web Browsing Simple, Secure, and Anonymous. In: Financial Cryptography'97. Berlin - Heidelberg - New York: Springer Verlag (1997), DOI 10.1007/3-540-63594-7_64.

67. Garg, A., Critchlow, M., Chen, P., van derWesthuizen, C., and van der Hoek, A.: An Environment for Managing Evolving Product Line Architectures. 19th IEEE International Conference on Software Maintenance, Amsterdam, Netherlands (2003) 358-367, http://www.ics.uci.edu/~andre/papers/C31.pdf.

68. Gefen, D., Karahanna, E., and Straub, D. W.: Trust and TAM in Online Shopping: An Integrated Model. MIS Quarterly 27, (2003) 51-90, http://search.epnet.com/login.aspx? direct=true&db=buh&an=9284295.

69. Gefen, D., Srinivasan Rao, V., and Tractinsky, N.: The Conceptualization of Trust, Risk and Their Electronic Commerce: the Need for Clarifications. 36th Annual Hawaii International Conference on Systems Sciences, Big Island, HI (2003), DOI 10.1109/ HICSS.2003.1174442.

70. Gideon, J., Cranor, L., Egelman, S., and Acquisti, A.: Power Strips, Prophylactics, and Privacy, Oh My! Second Symposium on Usable Privacy and Security, Pittsburgh, Pennsylvania (2006) 133-144, DOI 10.1145/1143120.1143137.

71. Goldberg, I. A.: Privacy-Enhancing Technologies for the Internet, II: Five Years Later. In: Privacy Enhancing Technologies – Second International Workshop, PET 2002, Dingledine, R. and Syverson, P., Eds. Berlin - Heidelberg: Springer Verlag (2003) 1-12.

72. Goy, A., Ardissono, L., and Petrone, G.: Personalization in E-Commerce Applications. In: The Adaptive Web: Methods and Strategies of Web Personalization, Brusilovsky, P., Kobsa, A., and Nejdl, W., Eds. Berlin Heidelberg New York: Springer Verlag (2007) this volume.

73. Grabner-Kräuter, S. and Kaluscha, E. A.: Empirical Research in On-line Trust: a Review and Critical Assessment. International Journal of Human-Computer Studies 58, (2003) 783-812, DOI 10.1016/S1071-5819(03)00043-0.

74. GroupLens Research: movielens: Helping You Find the Right Movies. (2007), http://movielens.umn.edu/login

75. Gupta, D., Digiovanni, M., Narita, H., and Goldberg, K.: Jester 2.0: A New Linear-Time Collaborative Filtering Algorithm Applied to Jokes. Workshop on Recommender Systems Algorithms and Evaluation, 22nd International Conference on Research and Development in Information Retrieval, Berkeley, CA (2000).

76. GVU: GVU's 10th WWW User Survey. Graphics, Visualization and Usability Lab, Georgia Tech (1998),

77. Hagen, P. R., Manning, H., and Souza, R.: Smart Personalization. Forrester Research, Cambridge, MA (1999).

78. Hann, I.-H., Hui, K.-L., Lee, T. S., and Png, I. P. L.: Online Information Privacy: Measuring the Cost-Benefit Tradeoff. Proceedings of the Twenty-Third Annual International Conference on Information Systems, Barcelona, Spain (2002) 1-10, http://aisel.isworld.org/pdf.asp?Vpath=ICIS/2002&PDFpath=02CRP01.pdf.

79. Hansen, M.: Personal communication. (2002).

80. Harper, J. and Singleton, S.: With a Grain of Salt: What Consumer Privacy Surveys Don't Tell Us. Competative Enterprises Institute (2001), http://www.cei.org/PDFs/with_a_grain_of_salt.pdf.

81. Harris, Louis and Associates and Alan F. Westin: Harris-Equifax Consumer Privacy Survey 1991. Atlanta, GA: Equifax Inc. (1991).

82. A Survey of Consumer Privacy Attitudes and Behaviors. Harris Interactive. (2000), http://www.bbbonline.org/UnderstandingPrivacy/library/harrissummary.pdf

83. Privacy Notices Research: Final Results. Harris Interactive, Inc. Study No. 15338, December 2001, http://www.bbbonline.org/UnderstandingPrivacy/library/datasum.pdf.

84. Henze, N. and Brusilivsky, P.: Open Corpus Adaptive Educational Hypermedia. In: The Adaptive Web: Methods and Strategies of Web Personalization, Brusilovsky, P., Kobsa, A., and Nejdl, W., Eds. Berlin Heidelberg New York: Springer Verlag (2007) this volume.

85. Hitchens, M., Kay, J., Kummerfeld, B., and Brar, A.: Secure Identity Management for Pseudo-Anonymous Service Access. In: Security in Pervasive Computing: Second International Conference, SPC 2005, Boppard, Germany, April 6-8, 2005. Proceedings, Hutter, D. and Ullmann, M., Eds. Berlin - Heidelberg: Springer Verlag (2005) 48-55, DOI 10.1007/b135497.

86. Hof, R., Green, H., and Himmelstein, L.: Now it's YOUR WEB. Business Week October 5, (1998) 68-75.

87. Hoffman, D. L., Novak, T. P., and Peralta, M.: Building Consumer Trust Online. Communications of the ACM 42, (1999) 80-85, DOI 10.1145/299157.299175.

88. Huberman, B. A., Adar, E., and Fine, L. R.: Valuating Privacy. Fourth Workshop on the Economics of Information Security (WEIS05), Cambridge, MA (2005), http://infosecon.net/workshop/pdf/58.pdf.

89. Hui, K.-L., Tan, B. C. Y., and Goh, C.-Y.: Online Information Disclosure: Motivators and Measurements. ACM Transactions on Internet Technology 6, (2006) 415 - 441, DOI 10.1145/1183463.1183467.

90. Hui, K.-L., Teo, H. H., and Lee, S.-Y. T.: The Value of Privacy Assurance: An Exploratory Field Experiment. MIS Quarterly 31, (2007), www.comp.nus.edu.sg/~lung/PrivacyAssurance.pdf.

91. Personal Communication, Chief Privacy Officer, IBM Zurich. (2003).

92. Ishitani, L., Almeida, V., and Wagner, M., Jr.: Masks: Bringing Anonymity and Personalization Together. IEEE Security & Privacy Magazine 1, (2003) 18-23, DOI 10.1109/MSECP.2003.1203218.

93. ISO: ISO/IEC 15408-2: Information Technology — Security Techniques — Evaluation Criteria for IT Security: Part 2: Security Functional Requirements. (1999), http://csrc.nist.gov/cc/t4/wg3/15408-2.zip

94. Jarvenpaa, S. and Tractinsky, N.: Consumer Trust in an Internet Store: A Cross-Cultural Validation. Journal of Computer Mediated Communication 5, (1999) 1-36, http://jcmc.indiana.edu/vol5/issue2/jarvenpaa.html.

95. Jarvenpaa, S. L., Tractinsky, N., and Vitale, M.: Consumer Trust in an Internet Store. Information Technology and Management 1, (2000) 45-71, DOI 10.1023/A:1019104520776.

96. Jensen, C. and Potts, C.: Privacy Policies as Decision-Making Tools: An Evaluation of Online Privacy Notices. 2004 Conference on Human Factors in Computing Systems, Vienna, Austria (2004) 471-478.

97. Jensen, C., Potts, C., and Jensen, C.: Privacy Practices of Internet Users: Self-Reports versus Observed Behavior. International Journal of Human-Computer Studies 63, (2005) 203–227, DOI 10.1016/j.ijhcs.2005.04.019.

98. Kahneman, D. and Tversky, A.: Choices, Values, and Frames. Cambridge: Cambridge Univ. Press (2000).

99. Kay, J.: A Scrutable User Modelling Shell for User-Adapted Interaction. In Basser Department of Computer Science: University of Sydney, Australia (1999), http://www.cs.usyd.edu.au/~judy/Homec/Pubs/thesis.bz2.

100. Kay, J.: Accretion Representation for Scrutable User Modeling. In: Intelligent Tutoring Systems 2000, Gauthier, G., Frasson, C., and VanLehn, K., Eds. Berlin: Springer (2000).

101. Kay, J.: Stereotypes, Student Models and Scrutability. In: Intelligent Tutoring Systems 2000, Gauthier, G., Frasson, C., and VanLehn, K., Eds. Berlin: Springer (2000).

102. Kay, J., Kummerfeld, R. J., and Lauder, P.: Foundations for Personalized Documents: a Scrutable User Model Server. Proceedings of ADCS'2001, Australian Document Computing Symposium (2001) 43-50, http://www.cs.usyd.edu.au/~judy/Homec/Pubs/2001_adcs_personis.pdf.

103. Kilian, J. and Petrank, E.: Identity Escrow. In: Advances in Cryptology — CRYPTO '98. Heidelberg - Berlin: Springer Verlag (1998) 169-185, DOI 10.1007/BFb0055715.

104. Kobsa, A.: User Modeling in Dialog Systems: Potentials and Hazards. IFIP/GI Conference on Opportunities and Risks of Artificial Intelligence Systems, Hamburg, Germany (1989) 147-165.

105. Kobsa, A.: User Modeling in Dialog Systems: Potentials and Hazards. AI & Society 4, (1990) 214-240, DOI 10.1007/BF01889941.

106. Kobsa, A.: Tailoring Privacy to Users' Needs (Invited Keynote). In: User Modeling 2001: 8th International Conference, Bauer, M., Gmytrasiewicz, P. J., and Vassileva, J., Eds. Berlin - Heidelberg: Springer Verlag (2001) 303-313, http://www.ics.uci.edu/~kobsa/papers/2001-UM01-kobsa.pdf.

107. Kobsa, A.: Personalization and International Privacy. Communications of the ACM 45, (2002) 64-67, DOI 10.1145/767193.767196.

108. Kobsa, A. and Cranor, L., Eds.: Proceedings of the UM05 Workshop 'Privacy-Enhanced Personalization'. Edinburgh, Scotland (2005), http://www.isr.uci.edu/pep05/papers/w9-proceedings.pdf.

109. Kobsa, A., Koenemann, J., and Pohl, W.: Personalized Hypermedia Presentation Techniques for Improving Customer Relationships. The Knowledge Engineering Review 16, (2001) 111-155, DOI 10.1017/S0269888901000108.

110. Kobsa, A. and Schreck, J.: Privacy through Pseudonymity in User-Adaptive Systems. ACM Transactions on Internet Technology 3, (2003) 149–183, DOI 10.1145/767193.767196.

111. Kobsa, A. and Teltzrow, M.: Contextualized Communication of Privacy Practices and Personalization Benefits: Impacts on Users' Data Sharing Behavior. In: Privacy Enhancing Technologies: Fourth International Workshop, PET 2004, Toronto, Canada, Martin, D. and Serjantov, A., Eds. Heidelberg, Germany: Springer Verlag (2005) 329-343, DOI 10.1007/11423409_21.

112. Kohavi, R.: Mining E-Commerce Data: the Good, the Bad, and the Ugly. Seventh ACM SIGKDD International Conference on Knowledge Discovery and Data Mining, San Francisco, CA (2001) 8-13, DOI 10.1145/502512.502518.

113. Krause, J., Hirschmann, A., and Mittermaier, E.: The Intelligent Help System COMFOHELP: Towards a Solution of the Practicability Problem for User Modeling and Adaptive Systems. User Modeling and User-Adapted Interaction: The Journal of Personalization Research 3, (1993) 249-282, DOI 10.1007/BF01257891.

114. Kumaraguru, P. and Cranor, L. F.: Privacy Indexes: A Survey of Westin's Studies. Institute for Software Research International, School of Computer Science, Carnegie Mellon University, Pittsburgh, PA, Technical Report CMU-ISRI-5-138, December 2005 (2005), http://reports-archive.adm.cs.cmu.edu/anon/isri2005/CMU-ISRI-05-138.pdf.

115. LaRose, R. and Rifon, N. J.: Your Privacy Is Assured—of Being Disturbed: Comparing Web Sites with and Without Privacy Seals. New Media and Society 8, (2006) 1009-1029, DOI 10.1177/1461444806069652.

116. LaRose, R., Rifon, N. J., and Lee, A.: Promoting I-Safety: Effects of Privacy Warning Boxes and Privacy Seals on Risk Assessment and Online Privacy Behaviors. Paper presented at AMA, Marketing and Public Policy Conference, Salt Lake City, UT (2004).

117. Laufer, R. S., Proshansky, H. M., and Wolfe, M.: Some Analytic Dimensions of Privacy. In: Architectural Psychology. Proceedings of the Lund Conference, Rikkard Kuller, Ed. Stroundsbourg, PA: Dowden, Hutchinson & Ross (1974) 353-372.

118. Laufer, R. S. and Wolfe, M.: Privacy as a Concept and a Social Issue: A Multidimensional Developmental Theory. Journal of Social Issues 33, (1977) 22-42.

119. Louis Harris and Associates and Westin, A. F.: Commerce, Communications, and Privacy Online: A National Survey of Computer Users. (1997), http://www.harrisinteractive.org.

120. Lu, Y., Tan, B. C. Y., and Hui, K.-L.: Inducing Customers to Disclose Personal Information to Internet Businesses with Social Adjustment Benefits. ICIS 2004: Twenty-Fifth International Conference on Information Systems, Washington, D.C. (2004) 272-281, http://aisel.isworld.org/Publications/ICIS/2004/2004RP45.pdf.

121. Lutz, W.: Statement of William Lutz. American Council of Life Insurers, et al. vs. Vermont Department of Banking, Securities, and Healthcare Administration, et al. (2004), http://www.epic.org/privacy/glba/vtlutz.pdf

122. Mabley, K.: Privacy vs. Personalization: Part III. Cyber Dialogue, Inc. (2000), http://www.cyberdialogue.com/library/pdfs/wp-cd-2000-privacy.pdf

123. Malin, B., Sweeney, L., and Newton, E.: Trail Re-Identification: Learning Who You Are From Where You Have Been. Carnegie Mellon University, Laboratory for International Data Privacy, Pittsburgh, PA, Technical Report LIDAP-WP12, March 2003 (2003), http://privacy.cs.cmu.edu/people/sweeney/trails1.pdf.

124. Mayer, R. C., Davis, J. H., and Schoorman, F. D.: An Integrative Model of Organizational Trust. Academy of Management Review 20, (1995) 709-734, links.jstor.org/sici?sici=0363-7425%281995O7%2920%3A3%3C709%3AAIMOOT%3E2.0.CO%3B2-9.

125. McJones, P.: Eachmovie Collaborative Filtering Data Set. (1997), http://research.compaq. com/SRC/eachmovie/

126. Metzger, M.: Communication Privacy Management in Electronic Commerce. Journal of Computer-Mediated Communication 12, (2007), http://jcmc.indiana.edu/vol12/issue2/metzger.html.

127. Metzger, M. J.: Privacy, Trust, and Disclosure: Exploring Barriers to Electronic Commerce. Journal of Computer-Mediated Communication 9, (2004), http://jcmc.indiana.edu/ vol9/issue4/metzger.html.

128. Metzger, M. J.: Effects of Site, Vendor, and Consumer Characteristics on Web Site Trust and Disclosure. Communication Research 33, (2006) 155-179, DOI 10.1177/0093650206287076.

129. Metzger, M. J.: Personal Communication. (2007).

130. Micarelli, A., Gasparetti, F., Sciarrone, F., and Gauch, S.: Personalized Search on the World Wide Web. In: The Adaptive Web: Methods and Strategies of Web Personalization, Brusilovsky, P., Kobsa, A., and Nejdl, W., Eds. Berlin Heidelberg New York: Springer Verlag (2007) this volume.

131. Miller, B. N., Konstan, J. A., and Riedl, J.: PocketLens: Toward a Personal Recommender System. ACM Transactions on Information Systems 22, (2004) 437-476, DOI 10.1145/1010614.1010618.
132. Milne, G. R. and Boza, M.-E.: Trust and Concern in Consumers' Perceptions of Marketing Information Management Practices. Journal of Interactive Marketing 13, (1999) 5-24, DOI 10.1002/(SICI)1520-6653(199924)13:1<5::AID-DIR2>3.0.CO;2-9.
133. Milne, G. R. and Culnan, M. J.: Strategies for Reducing Online Privacy Risks: Why Consumers Read (or Don't Read) Online Privacy Notices. Journal of Interactive Marketing 18, (2004) 15-29, DOI 10.1002/dir.20009.
134. Miyazaki, A. D. and Krishnamurthy, S.: Internet Seals of Approval: Effects on Online Privacy Policies and Consumer Perceptions. Journal of Consumer Affairs 36, (2002) 28, DOI 10.1111/j.1745-6606.2002.tb00419.x.
135. Mobasher, B.: Data Mining for Web Personalization. In: The Adaptive Web: Methods and Strategies of Web Personalization, Brusilovsky, P., Kobsa, A., and Nejdl, W., Eds. Berlin Heidelberg New York: Springer Verlag (2007) this volume.
136. Moores, T.: Do Consumers Understand the Role of Privacy Seals in E-Commerce? Communications of the ACM 48, (2005) 86-91, DOI 10.1145/1047671.1047674.
137. Moores, T. T. and Dhillon, G.: Do Privacy Seals in E-Commerce Really Work? Communications of the ACM 46, (2003) 265 - 271, DOI 10.1145/953460.953510.
138. Mulligan, D. and Schwartz, A.: Your Place or Mine?: Privacy Concerns and Solutions for Server and Client-Side Storage of Personal Information. Computers, Freedom & Privacy Conference (1999) 81-84, DOI 10.1145/332186.332255.
139. Self-Regulatory Principles for Online Preference Marketing by Network Advertisers. Network Advertising Initiative (2006), http://www.networkadvertising.org/pdfs/NAI_principles.pdf
140. Nakashima, E.: AOL Search Queries Open Window Onto Users' Worlds. washingtonpost.com (2006), http://www.washingtonpost.com/wp-dyn/content/article/2006/08/16/AR2006081601751_pf.html
141. Neale, J. M. and Liebert, R. M.: Science and Behavior: An Introduction to Methods of Research. Englewood Cliffs, NJ: Prentice-Hall (1973).
142. Recommendation of the Council Concerning Guidelines Governing the Protection of Privacy and Transborder Flows of Personal Data. OECD (1980), www1.oecd.org/ publications/e-book/9302011E.PDF
143. Orwant, J.: Heterogenous Learning in the Doppelgänger User Modeling System. User Modeling and User-Adapted Interaction: The Journal of Personalization Research 4, (1995) 107-130, DOI: 10.1007/BF01099429.
144. Microsoft Passport Network. (2006), http://www.passport.net
145. Pavlou, P. A.: Consumer Acceptance of Electronic Commerce: Integrating Trust and Risk with the Technology Acceptance Model. International Journal of Electronic Commerce 7, (2003) 101-134, http://mesharpe.metapress.com/link.asp?id=ymy1p2ngk06wt39f.
146. Pazzani, M. J. and Billsus, D.: Content-Based Recommendation Systems. In: The Adaptive Web: Methods and Strategies of Web Personalization, Brusilovsky, P., Kobsa, A., and Nejdl, W., Eds. Heidelberg, Germany: Springer Verlag (2006) In this volume.
147. Summary Report. Privacy Commissioner – Te Mana Matapono Matatapu, New Zealand (2006), http://www.privacy.org.nz/filestore/docfiles/24153322.pdf.
148. Personalization & Privacy Survey. Personalization Consortium (2000), http://www.personalization.org/SurveyResults.pdf

149. New Survey Shows Consumers Are More Likely to Purchase At Web Sites That Offer Personalization: Consumers Willing to Provide Personal Information in Exchange for Improved Service and Benefits. Personalization Consortium, Wakefield, MA, Press Release May 9, 2001, http://web.archive.org/web/20010526174824/http://www.personalization.org/pr050901.html.

150. Pfitzmann, A. and Köhntopp, M.: Anonymity, Unobservability, and Pseudonymity: A Proposal for Terminology. In: Anonymity 2000, Federrath, H., Ed. Berlin-Heidelberg, Germany: Springer-Verlag (2001) 1-9,

151. Phelps, J., Nowak, G., and Ferrell, E.: Privacy Concerns and Consumer Willingness to Provide Personal Information. Journal of Public Policy & Marketing 19, (2000) 27-41, http://search.epnet.com/login.aspx?direct=true&db=buh&an=3215141.

152. Polat, H. and Du, W.: Privacy-Preserving Collaborative Filtering. International Journal of Electronic Commerce 9, (2003) 9-35, http://ejournals.ebsco.com/direct.asp?ArticleID=1A72U87WVYJ61B9C.

153. Polat, H. and Du, W.: SVD-based Collaborative Filtering with Privacy. ACM Symposium on Applied Computing, Santa Fe, New Mexico (2005) 791-795, DOI 10.1145/1066677.1066860.

154. Portz, K., Strong, J. M., Busta, B., and Schneider, K.: Do Consumers Understand What Web-Trust Means? The CPA Journal 70, (2000) 46-52, http://www.nysscpa.org/cpajournal/2000/1000/features/f104600a.htm.

155. Preece, J., Rogers, Y., and Sharp, H.: Interaction Design: Beyond Human-Computer Interaction. New York, NY: Wiley (2002).

156. Opinion surveys. Privacy Exchange (2003), http://www.privacyexchange.org/iss/surveys/surveys.html

157. Rao, J. R. and Rohatgi, P.: Can Pseudonymity Really Guarantee Privacy? 9th USENIX Security Symposium (2000) 85–96, http://www.usenix.org/publications/library/proceedings/sec2000/full_papers/rao/rao_html.

158. Regan, K.: Does Anyone Read Online Privacy Policies? E-Commerce Times, (2001), http://www.ecommercetimes.com/story/11303.html.

159. Riegelsberger, J., Sasse, M. A., and McCarthy, J. D.: Shiny Happy People Building Trust?: Photos on E-commerce Websites and Consumer Trust. SIGCHI Conference on Human Factors in Computing Systems, Ft. Lauderdale, FL (2003) 121-128, DOI 10.1145/ 642611.642634.

160. Rifon, N. J., LaRose, R., and Choi, S. M.: Your Privacy Is Sealed: Effects of Web Privacy Seals on Trust and Personal Disclosures. Journal of Consumer Affairs 39, (2005) 339-360, DOI 10.1111/j.1745-6606.2005.00018.x.

161. Rotenberg, M.: The Privacy Law Sourcebook 2004: United States Law, International Law, and Recent Developments. Washington, DC: EPIC (2005).

162. Roy, M. C., Dewit, O., and Aubert, B. A.: The Impact of Interface Usability on Trust in Web Retailers. Internet Research 11, (2001) 388-398, DOI 10.1108/10662240110410165.

163. Roy Morgan Research: Privacy and the Community. Prepared for the Office of the Federal Privacy Commissioner, Sydney (2001), http://www.privacy.gov.au/publications/rcommunity.pdf.

164. Schafer, J. B., Frankowski, D., Herlocker, J., and Sen, S.: Collaborative Filtering Recommender Systems. In: The Adaptive Web: Methods and Strategies of Web Personalization, Brusilovsky, P., Kobsa, A., and Nejdl, W., Eds. Berlin Heidelberg New York: Springer Verlag (2007) this volume.

165. Schoenbachler, D. D. and Gordon, G. L.: Trust and Customer Willingness to Provide Information in Database-Driven Relationship Marketing. Journal of Interactive Marketing 16, (2002) 2-16, DOI 10.1002/dir.10033.

166. Schreck, J.: Security and Privacy in User Modeling. Dordrecht, Netherlands: Kluwer Academic Publishers (2003), http://www.security-and-privacy-in-user-modeling.info.

167. Sheehan, K. B. and Hoy, M. G.: Flaming, Complaining, Abstaining: How Online Users Respond to Privacy Concerns. Journal of Advertising 28, (1999) 37-51, http://search.ebscohost.com/login.aspx?direct=true&db=buh&AN=2791549&site=ehost-live.
168. Sheehan, K. B. and Hoy, M. G.: Dimensions of Privacy Concern Among Online Consumers. Journal of Public Policy & Marketing 19, (2000) 62-73, http://search.epnet.com/login.aspx?direct=true&db=buh&an=3215144.
169. Singel, R.: 'Free IPod' Takes Privacy Toll. Wired, No. Issue, March 16, 2006, http://www.wired.com/news/technology/0,70420-0.html
170. Smith, H. J., Milberg, S. J., and Burke, S. J.: Information Privacy: Measuring Individuals' Concerns about Organizational Practices. MIS Quarterly 20, (1996) 167-196, http://links.jstor.org/sici?sici=0276-7783%28199606%2920%3A2%3C167%3AIPMICA%3E2.0.CO%3B2-W.
171. Smyth, B.: Case-Based Recommendation. In: The Adaptive Web: Methods and Strategies of Web Personalization, Brusilovsky, P., Kobsa, A., and Nejdl, W., Eds. Berlin Heidelberg New York: Springer Verlag (2007) this volume.
172. Spiekermann, S., Grossklags, J., and Berendt, B.: E-privacy in 2nd Generation E-Commerce: Privacy Preferences versus Actual Behavior. EC'01: Third ACM Conference on Electronic Commerce, Tampa, FL (2001) 38-47, DOI 10.1145/501158.501163.
173. Spiekermann, S., Grossklags, J., and Berendt, B.: Stated Privacy Preferences versus Actual Behaviour in EC Environments: a Reality Check. WI-IF 2001: the 5th International Conference Wirtschaftsinformatik - 3rd Conference Information Systems in Finance, Augsburg, Germany (2001) 129-148.
174. Stephanidis, C.: Adaptive Techniques for Universal Access. User Modeling and User-Adapted Interaction: The Journal of Personalization Research 11, (2001) 159-179, DOI 10.1023/A:1011144232235.
175. Stone, D., Jarrett, C., Woodroffe, M., and Minocha, S.: User Interface Design and Evaluation. San Francisco, CA: Morgan Kaufmann (2005).
176. Sweeney, L.: k-Anonymity: A Model for Protecting Privacy. International Journal on Uncertainty, Fuzziness, and Knowledge-based Systems 10, (2002) 557-570, DOI 10.1142/S0218488502001648.
177. Tam, E.-C., Hui, K.-L., and Tan, B. C. Y.: What Do They Want? Motivating Consumers to Disclose Personal Information to Internet Businesses. Proceedings of the Twenty-Third Annual International Conference on Information Systems, Barcelona, Spain (2002) 11-21, http://aisel.isworld.org/pdf.asp?Vpath=ICIS/2002&PDFpath=02CRP02.pdf.
178. Tam, K. Y. and Ho, S. Y.: Web Personalization: is it Effective? IT Professional 5, (2003) 53-57, DOI 10.1109/MITP.2003.1235611.
179. Taylor, H.: Most People Are "Privacy Pragmatists" Who, While Concerned about Privacy, Will Sometimes Trade It Off for Other Benefits. The Harris Poll #17, March 19, 2003 (2003), http://www.harrisinteractive.com/harris_poll/index.asp?PID=365.
180. Teltzrow, M. and Kobsa, A.: Impacts of User Privacy Preferences on Personalized Systems: a Comparative Study. In: Designing Personalized User Experiences for eCommerce, Karat, C.-M., Blom, J., and Karat, J., Eds. Dordrecht, Netherlands: Kluwer Academic Publishers (2004) 315-332, DOI 10.1007/1-4020-2148-8_17.
181. Turner, C. W., Zavod, M., and Yurcik, W.: Factors that Affect the Perception of Security and Privacy of Ecommerce Web Sites. Fourth International Conference on Electronic Commerce Research, Dallas TX (2001) 628-636, http://www.sosresearch.org/publications/icecr01.pdf.
182. Turow, J.: Americans and Online Privacy: The System is Broken. Annenberg Public Policy Center, University of Pennsylvania (2003), http://www.asc.upenn.edu/usr/jturow/internet-privacy-report/36-page-turow-version-9.pdf
183. Health Insurance Portability and Accountability Act of 1996. 104th Congress Aug. 21, 1996, http://aspe.hhs.gov/admnsimp/pl104191.htm.

184. Children's Online Privacy Protection Act of 1998, http://www.ftc.gov/ogc/coppa1.htm.
185. Gramm-Leach-Bliley Act of 1999. Public Law No 106-102 (1999), http://thomas.loc.gov/ cgi-bin/bdquery/z?d106:SN00900:|.
186. USACM Policy Recommendations on Privacy. U.S. Public Policy Committee of the Association for Computing Machinery, New York, NY June 2006, http://www.acm.org/usacm/Issues/Privacy.htm.
187. Fifth Study of the Internet by the Digital Future Project Finds Major New Trends in Online Use for Political Campaigns. Center for the Digital Future, Annenberg School, University of Southern California (2005), http://www.digitalcenter.org/pdf/Center-for-the-Digital-Future-2005-Highlights.pdf
188. van Blarkom, G. W., Borking, J. J., and Olk, J. G. E., Eds.: Handbook of Privacy and Privacy-Enhancing Technologies: The Case of Intelligent Software Agents. The Hague, The Netherlands: TNO-FEL (2003), http://www.andrewpatrick.ca/pisa/handbook/Handbook_Privacy_and_PET_final.pdf.
189. van der Hoek, A., Rakic, M., Roshandel, R., and Medvidovic, N.: Taming Architectural Evolution. Sixth European Software Engineering Conference (ESEC) and the Ninth ACM SIGSOFT Symposium on the Foundations of Software Engineering (FSE-9), Irvine, CA (2001) 1-10, DOI 10.1145/503209.503211.
190. Wang, Y., Chen, Z., and Kobsa, A.: A Collection and Systematization of International Privacy Laws, with Special Consideration of Internationally Operating Personalized Websites. (2006), http://www.ics.uci.edu/~kobsa/privacy
191. Wang, Y. and Kobsa, A.: A Software Product Line Approach for Handling Privacy Constraints in Web Personalization. In: PEP05, UM05 Workshop on Privacy-Enhanced Personalization, Kobsa, A. and Cranor, L., Eds. Edinburgh, Scotland (2005) 35-45, http://www.ics.uci.edu/~kobsa/papers/2005-PEP-kobsa.pdf.
192. Wang, Y., Kobsa, A., van der Hoek, A., and White, J.: PLA-based Runtime Dynamism in Support of Privacy-Enhanced Web Personalization. 10th International Software Product Line Conference, Baltimore, MD (2006) 151-162, DOI 10.1109/SPLINE.2006.1691587.
193. Wenning, R., Ed.: The Platform for Privacy Preferences 1.1 (P3P1.1) Specification: W3C Working Draft (2006), http://www.w3.org/TR/P3P11.
194. White, T. B.: Consumer Disclosure and Disclosure Avoidance: A Motivational Framework. Journal of Consumer Psychology 14, (2004) 41-51, DOI 10.1207/s15327663jcp1401&_2_6.
195. Xu, Y., Tan, B. C. Y., Hui, K.-L., and Tang, W.-K.: Consumer Trust and Online Information Privacy. International Conference on Information Systems 2003, Seattle, WA (2003) 538-548, http://aisel.isworld.org/Publications/ICIS/2003/03CRP45.pdf.
196. Yankee: Interactive Consumers in the Twenty-First Century: Emerging Online Consumer Profiles, Access Strategies and Application Usage". Yankee Group 23 Oct. 2001 (2001), http://www.yankeegroup.com.
197. Zaslow, J.: If TiVo Thinks You Are Gay, Here's How to Set It Straight. Wall Street Journal (Eastern Edition), (2002) A.1, online.wsj.com/article_email/0,,SB1038261936872 356908,00.html.

Open Corpus Adaptive Educational Hypermedia

Peter Brusilovsky[1] and Nicola Henze[2]

[1] School of Information Sciences, University of Pittsburgh
135 North Bellefield Ave., Pittsburgh, PA 15260, USA
peterb@pitt.edu
[2] IVS – Semantic Web Group, University of Hannover, & L3S Research Center
Appelstr. 4, D-30167 Hannover, Germany
henze@l3s.de

Abstract. Despite the fact that adaptive hypermedia techniques have proven their ability to provide user guidance and orientation in hyperspace, we do not currently see the widespread adoption of these techniques. A couple of reasons may explain this phenomenon. One of them is the current lack of re-usability and interoperability between adaptive techniques/systems, which – to some degree – originates in the so-called "open corpus problem" found in adaptive hypermedia. In this article, we analyze this problem in a popular arena: adaptive hypermedia systems with an emphasis on education. The origins and effects of the open corpus problem are discussed, and recent approaches are demonstrated that have – in one way or the other – developed as strategies for solving the open corpus problem. We summarize these findings and discuss how solution strategies can be successfully employed in the future, enabling adaptive hypermedia techniques within open, dynamic information spaces, such as the Semantic Web.

22.1 Introduction

The volume of educational resources available to students is changing rapidly. A variety of educational resources such as tutorials, electronic textbooks, and topic overviews are now available on the Web for almost every domain. Dedicated repositories of educational material, such as educational digital libraries (DL), and pools of reusable learning objects are being created. Finding high quality materials is much less of a problem with the use of modern Web search engines [6] and DL search services [43]. However, the resources that one finds have different presentation styles, target audiences, and coverage. Also, many resources are highly redundant. The abundance of resources has created another problem: How to help students find, organize, and use resources that match their individual goals, interests, and current knowledge? In brief, access is not the issue; *personalized access* is.

The need to provide personalized access to information is well recognized outside of education. Numerous research projects have proposed and investigated a wide

P. Brusilovsky, A. Kobsa, and W. Nejdl (Eds.): The Adaptive Web, LNCS 4321, pp. 671–696, 2007.

range of techniques for personalized access within only the last few years. Earlier chapters of this book provided a good overview of personalization techniques for all major paradigms of information access: information retrieval in Chapters 6 [48] and 7 [47] of this book, browsing in Chapters 8 [8], and filtering/recommendation in Chapters 9 [56], 10 [52], 11 [59], 12 [16]. Successful application of personalization techniques has been achieved in such application areas as news access and e-commerce, covered in Chapters 18 [4] and 16 [28] of this book. Education, however, remains resistant to successful development while simultaneously being one of the few areas - accompanying medicine and public health - where the provision of personalized access is most important for users and society. The majority of adaptation techniques that focus on user interests and work successfully in other fields have a limited applicability in the educational context where users differ not just by their interests, but most essentially in their goals, skills, knowledge, and learning styles.

So far, the only techniques that demonstrate a good ability to provide personalized access to information in the educational context are adaptive navigation support techniques developed in the field of adaptive hypermedia (AH), presented in Chapter 8 of this book [8]. In a number of educational AH systems, adaptive navigation support techniques were able to help individual students locate, recognize, and comprehend relevant information, thus increasing learning outcomes and retention [7; 10; 13; 65].

Adaptive hypermedia techniques could provide a real difference for students who are trying to locate useful resources on the Web or in learning repositories and DL. Web resources rarely match the needs and the level of preparation of a specific class of students. Serious efforts are frequently required from students to understand which content is relevant, which is not, and how to find their way through it. Without individual guidance, students dealing with the increasing complexity of navigational-possibilities may get lost in hyperspace in a number of senses. For example, they may fail to identify learning goals and recognize coherences, relations, and causal dependencies. Even in a learning repository where resources are carefully selected and classified by subject and category, the usefulness of resources depend on the individual learner's progress: some resources may require additional knowledge that the learner does not yet have (in accordance to his/her user model), while others may teach the subject without sufficient in-depth information and are thus too easy for this learner. At this juncture, methods from adaptive hypermedia can be used to support the learner in finding the *most appropriate* learning resource; for providing awareness about the learning process (e.g., by pointing out necessary pre-knowledge that this learner might otherwise miss); for providing guidance (e.g., by providing an individually tailored sequence of learning resources—teaching the topics s/he is interested in while incorporating all required prerequisite knowledge); for providing orientation (e.g., by pointing out the next learning steps to take, or the existence of different schools-of-thought); for considering individual learning styles; and so on.

Unfortunately, traditional adaptive hypermedia, with all its power, can't be directly applied in any of these important contexts. As it may become apparent from the study of Chapter 8 of this book [8], traditional adaptive navigation support techniques are only able to work within a limited set of documents that have been manually structured and indexed with domain concepts and metadata at design time. Traditional adaptive hypermedia systems are predominantly *closed corpus* adaptive hypermedia, since the *document space* of these adaptive systems is a closed set of resources. Less than a handful of the adaptive hypermedia systems have attempted to deal with *open corpus* such as the Web's educational resources or dynamically expanding educational repositories. Closed corpus AH systems demonstrate what is possible to achieve with adaptive hypermedia technologies, but they are impractical for most real world applications because no teacher or content provider is able to invest time to structure and index thousands of documents collected from all over the Web as required by traditional adaptive hypermedia systems; worse, these systems would need constant maintenance, as new information becomes available daily.

The apparent contradiction between the potential power of adaptive hypermedia and its predominant close-corpus application content has caused a number of researchers to focus on what we call the *open corpus problem*:

Is the applicability of the adaptive hypermedia techniques restricted by nature to closed corpus of educational resources or it is possible to develop *open corpus adaptive hypermedia* that will successfully work in such contexts as the Web and educational repositories?

The goal of this paper is to convince the reader that open corpus adaptive hypermedia is feasible and to discuss possible approaches to construct it. We start with a brief review of adaptation-specific information that is used by current adaptive hypermedia applications[1] (section 22.2). In Section 22.3 we stress several problems that have made it substantially difficult to use the current techniques of adaptive hypermedia with an open corpus of documents and review a range of known approaches and systems that attempt to overcome these problems, by attacking the open corpus problem from very different angles. Based on these considerations, we re-analyze the open corpus problem, especially with respect to the functional re-usability and interoperability of adaptive hypermedia and related systems (section 22.4). Then we discuss further, emerging solutions for realizing personalized access to information in open, distributed information spaces.

22.2 Adaptation-Specific Information in Adaptive Educational Hypermedia

To understand the essential difference between open and closed corpus adaptive hypermedia we want to start with more formalized definitions:

Definition 1 (Closed Corpus Adaptive Hypermedia System)

A closed corpus adaptive hypermedia system is an adaptive hypermedia system which operates on a closed corpus of documents, where documents and relationships between the documents are known to the system at design time.

---

[1] We restrict ourselves to adaptive navigation support because we consider the first problem to solve in an open corpus setting is to adapt ------to documents to the user's needs. *Content-level adaptation* (i.e., --) is a possible future step which may later be addressed by, for example, considering information chunks or variations of documents instead of documents-as-entities, and by applying techniques from navigational-level adaptation onto these chunks / variations / etc.

Definition 2 (Open Corpus Adaptive Hypermedia System)

An open corpus adaptive hypermedia system is an adaptive hypermedia system which operates on an open corpus of documents, e.g., a set of documents that is not known at design time and, moreover, can constantly change and expand.

What makes closed corpus hypermedia special, from the adaptation point of view, is exactly the fact that all documents and relations on the documents are known to the authors of an adaptive hypermedia system at design time. It allows the authors to augment the documents and relationships with additional information that can be used later by the adaptation algorithms to deliver the adaptation effectively to every user. We refer to this kind of information as *adaptation-specific* information. This information is typically hidden from the user; however, it is the real source of power of adaptive hypermedia. The goal of this section is to reveal the kind of information that is used by adaptive educational hypermedia in order to perform the adaptation. Understanding the nature adaptation-specific information can help us to identify the information necessary for the open corpus context, and how to produce or compensate for this information.

It turns out that we can distinguish two classes of this adaptation-specific information: The first class comprises information that adds some semantics to the hypermedia, i.e., assigns specific types to the hypermedia documents and relationship and introduce additional, semantic relationships between the documents. The second class provides additional knowledge "behind" the hyperspace documents by connecting documents to external models that are separate from the hyperspace itself. The variety of these models is high: we can find conceptual models, pedagogical models, goal models, stereotype hierarchies, and more. Many approaches from artificial intelligence have been used to verify, maintain and interpret these models in order to perform the adaptation task for these adaptive educational hypermedia systems.

22.2.1 Adaptation-Specific Information: Enriching Hypertexts with the Annotation of Documents and Relationships

The first kind of annotation-specific information acts within the hyperspace itself, attempting to introduce some additional knowledge about the documents (nodes), links, and additional relationships between the documents. Additional knowledge about the documents and links is typically provided in the form of *types* assigned to documents and links. For example, in the KBS-Hyperbook [34], which is presented in more detail in section 22.3.2, documents can be marked as "problem statement," "example," "theory," etc. Links are usually typed to reflect the semantics of the structural relationship between connected documents. Some systems use an elaborate set of typed links [2]; however, other systems such as MetaLinks [51] achieve a good functionality while using just two types of links. From the modern point of view, link and document types can be seen as metadata that is added to documents. However, in adaptive educational hypermedia systems, they are normally referred to not as metadata, but as knowledge about documents and links. This knowledge about documents connected to the current document, and about the connection types, allows the adaptive hypermedia decision mechanisms to guide the user to the most appropriate documents, using such techniques as link ordering, annotation, and hiding presented in Chapter 8 of this book [8].

In adaptive educational hypermedia systems, the typing of existing links is often not sufficient, because these systems rely on knowledge dependencies or pedagogical relationships between documents that may not be directly connected by a link. To compensate, these systems introduce additional relationships between documents that are often invisible, i.e., not accessible for navigation. Most typical among these relationships is a prerequisite relationship that notes that one document should be known before another. It is used in many systems such as ELM-ART [64] or AHA! [22]. Document-to-document relationships are very powerful in adaptation. On the other hand, the drawback is that alterations to the set of documents in hyperspace normally requires a huge effort. Consider, for example, the introduction of a new document: At which points shall it be presented to learners? Which documents are prerequisites, and to which documents is this document a prerequisite? Checking all documents of the documents space one-by-one may be required to establish proper relationships.

22.2.2 Adaptation-Specific Information: Connecting Hypertexts with External Models

An alternative way to add adaptation specific information is to rely on external models that exist beyond the hyperspace, such as knowledge models, pedagogical models, usage models, etc. These models typically encapsulate some kind of knowledge. For example, conceptual domain models encapsulate knowledge about the domain while stereotype hierarchies represent knowledge about users. In this case, the necessary knowledge is added to hypertext documents by connecting them to elements of these external models. Most popular in the field of adaptive educational hypermedia are the domain concept models, presented in Chapter 1 of this book. The use of domain concept models, along with user overlay models, allows these systems to provide sophisticated adaptations to the user's level of knowledge. For example, in the InterBook system [11], the authors connect (or *index*) documents with domain concepts using two kinds of document-concept relationships – the outcome and prerequisites. These links allow the authors of the system to express what domain knowledge is presented in the page or what knowledge should have been mastered before the page is accessed. Other models, such as didactical models, provide information on a certain didactical approach, and can be seen as a new layer to both the document-to-document annotation (internal references) and the document-to-concept annotation approach (external references). Generally, storing adaptation-specific information in external models supports the application of artificial intelligence techniques for reasoning about this information.

Indexing documents in terms of external models provides for a higher level of adaptation than simply typing and connecting documents, since these models typically encapsulate additional knowledge that can be used by adaptation algorithms. However, it also adds an additional challenge to the system development since the building of sound external models is a considerable knowledge engineering effort that typically requires expert knowledge in a specific field. The initial investment into developing external models pays off, to some extent, since this allows the indexing of documents to become easier: authors can write their materials, and index it with concept models without considering the whole set of *currently* available documents. In particular, multi-author approaches are supported, where material can be designed and annotated independently. Another advantage of document-to-concept relations are achieved with respect to *maintenance*: changes in the document space affect only the altered / added documents, no further annotations of documents need to be altered.

22.3 Several Ways to Open Corpus Adaptive Educational Hypermedia

How can we achieve progress in developing open corpus adaptive educational hypermedia systems that are compatible in personalization power with existing closed corpus systems? Arguably, this goal can be achieved if we find the way to enhance open corpus resources with additional knowledge that is comparable with the knowledge behind traditional adaptive hypermedia, which was analyzed in the previous section. If comparable knowledge could be obtained in the open corpus context, existing adaptive hypermedia techniques or their modifications could be used to deliver adaptive navigation support for open corpus documents. This section attempts to analyze several known ways of developing open corpus adaptive hypermedia, i.e., several ways to collect the missing knowledge. Following the structure of the previous chapter, we separate the discussion into two subsections - one dealing with intra-hyperspace problems such as document interlinking and link typing and the other dealing with the problem of external models and the indexing of hyperspace documents.

22.3.1   How to Create a Linked Hyperspace from Open Corpus Resources

Adaptive hypermedia technologies support hypertext-browsing activities of the user, i.e., they assist the user in moving from document to document, following interdocument links. Thus, the first problem to resolve when building an open corpus adaptive hypermedia system is how to build a hyperspace from an open collection of generally independent documents. This problem could be solved in two different ways: relying on human power to create hyperspace and creating a hyperspace automatically.

As discussed below, the manual interlinking of a constantly expanding set of open corpus documents is possible in only a few contexts. A more general solution is to apply some techniques that can automatically create a linked hyperspace from a collection of independent documents. The problem of automatic hyperspace creation has been explored by researchers in the area of hypertext and information retrieval for at least 20 years. This research started originally under the term *intelligent hypertext* and focused mostly on automatic linking of documents as a help for hypertext authors who may not be able to identify all useful links. Later, the ideas of automatic linking were explored by the *open hypermedia* movement. Open hypermedia, as a research direction, specifically focused on conceptual and architectural problems of creating a hyperspace from originally independent open corpus documents [50]. A large body of literature has been produced in both areas and a range of techniques has been suggested. These techniques can be generally classified into two groups that we call keyword-based techniques and metadata-based techniques.

Manually Constructed Hyperspaces. Relying on human power to create hyperspace is a possible solution for an open corpus system, so long as the developers of this system are not involved in the hyperspace construction. In fact, the simplest way to explore open corpus techniques for educational AH is to take an existing educational hypertext-application and to add a layer of adaptive navigation support to it. This approach allows the developers of open corpus adaptive hypermedia system to simply avoid the problems of hyperspace construction and focus on navigation support techniques. A number of early explorers of open corpus educational AH have used this approach with various kinds of pre-existing hypermedia applications, such as an educational encyclopedia [36], a hypertext tutorial [32], and an educational Web site [58]. Two of these systems are presented in more detail in section 22.3.2 of this chapter.

While early projects operated in the context of pre-authored hyperspace, similar approaches could be applied to the constantly expanding yet human-linked document collections. The challenge here is to find an environment where the human-supported hypertext construction is supported naturally, i.e., where each new resource is being immediately linked to the whole collection by a human author or manager. While in most of the cases this is not feasible, there are at least two meaningful contexts that deserve further exploration. One context is organization-supported hyperspaces such as Web sites or educational portals where the integration of new resources into the previously linked hyperspace is ensured by the organization that owns or maintains the collection. Unfortunately, this context is becoming more and more rare: due to high cost of manual linking, many portals and resource collections adopt a *pool approach* where each new resource is simply added to the pool. Another context is *community-driven* hypertext creation, where linking new documents to the existing hyperspace is done by a whole community of users. Two popular examples of community created hypertexts are Wikis and blogs where the nature of these community-based systems encourages linking newly authored documents. Both kinds of expanding hypertext systems, organization-supported and community-driven, provide a really challenging but creative application area for open corpus adaptive hypermedia and we expect more work in this direction in the coming years. A pioneer example of open corpus adaptive hypermedia for Wiki is the CoWeb system [24], which uses the ideas of *social navigation* to provide annotation-based adaptive navigation support. CoWeb is briefly reviewed in sectioin 22.3.2 below.

Keyword-Based Techniques for Automatic Hyperspace Creation. This group of techniques is based on the automatic keyword-level analysis of documents. The work on keyword-based linking started at the end of 1980 with exploring similarity-based navigation. The idea of similarity-based navigation is to create links between documents that are similar on the keyword level. The techniques for calculating keyword-level similarity are well explored and covered in more detail in Chapter 5 of this book [49]. Since the pioneer work of Mayes and Kibby [40; 46], keyword level similarity-based navigation has been applied in a number of systems. This automatic linking technology is simple and straightforward and can be used in almost any context. To interlink an existing collection of documents, a similarity metrics is calculated between each pair of documents. To link a new document to a collection, the similarity is computed between the new document and all documents in the collection. After that, documents with similarity higher than a certain threshold are connected by a bi-directional hyperlink.

The negative side of this technology is that the quality of simple keyword-level similarity techniques is not perfect, so it can often link pages that are not really semantically related. In addition, a hyperspace created with classic similarity-based navigation techniques suffers from two problems – the lack of typed links and the lack of clear structure (the resulting hyperspace is rather chaotic). As we observed in section 22.2, typed links enable more advanced navigation support technologies. A clear hypertext structure helps users to find their way and position in hyperspace.

More recent research attempts have focused on overcoming these problems using more advanced keyword-based techniques. The first challenge to be addressed was link typing. By the end of 1990, a number of keyword-level techniques were suggested for generating typed links [2; 21] as well as typing existing hypertext links [1; 54]. Several researchers focused on improving the precision of keyword-level linking by replacing standard document indexing with "semantic-oriented" techniques such as latent semantic indexing [44] or lexical chaining [29].

To structure a collection of unrelated documents, several researchers applied Self-Organized Maps (SOM). The SOM technology is able to cluster documents into cells on a rectangular grid in such a way that documents allocated to the same cell are quite similar to each other and documents in the neighboring cell are also similar, although to a lesser extent [41]. This unique property of SOM allows the introduction of some reasonable level-structuring even in a large collection of Web resources [20; 42; 55]: each cell or group of cells serves as a category (section of hyperspace) with spatial proximity expressing similarities between the categories. Thus the application of SOM turns a collection of documents into a structured spatial hypertext (spatial hypertext implies implicit links between spatially co-located documents [57]). Using map-based navigation, introduced in [14], this spatial hypertext can be converted into a regular hyperspace that allows navigation from a document to the hosting map cell and then to similar documents. This technology has been applied in the Knowledge Sea II system, presented below.

Metadata-Based Techniques for Automatic Hyperspace Creation. Another branch of research that may resolve problems of simple keyword-level hypertext linking is the application of metadata-based techniques. Generally, metadata-based approaches allow the production of better quality results in the linking and structuring of hyperspace. The early focus on keyword-level techniques was justified by the lack of metadata. However, over the last several years a number of repositories have assembled a large volume of documents indexed with metadata. In addition, some progress has been achieved by extracting metadata from Web resources. As a result, metadata-based approaches now overshadow keyword-based approaches. With the exception of link typing (which is not a problem in the presence of metadata), the work on metadata-based hypertext construction has been focused on the same goals – automatic linking and automatic structuring.

The pioneer work on metadata-based linking was done in early 1990 by the team of Douglas Tudhope [60]. They explored similarity-based navigation in a richly metadata-indexed photo-archive. The core idea of similarity-based navigation is the same as for keyword-based linking: a similarity measure is computed between documents and those with similarity above a certain threshold are connected by a link. The metadata similarity is calculated as a weighted measure of similarity along each metadata facet. This process is presented in more detail in Chapter 11 of this book [59]. Since metadata expresses semantic similarity (in contrast to surface similarity expressed by keywords) this approach obtains high quality links. More recently, the focus of research on metadata-based linking moved from simple quantitative metadata (such as time, size, or difficulty) to ontology-based linking. Ontology-based linking is possible when documents are indexed (manually or automatically) with terms of ontology or a thesaurus. In this case, the process of finding similar documents is a more challenging process, since it has to take into account the position and connections of ontological tags in the ontology. A well-known approach to ontology-based linking in open corpus hypermedia is presented in [19]. Currently these ideas are being explored in the context of the Semantic Web [25].

The presence of metadata also allows some meaningful structuring of hyperspace. The complexity of possible structuring is determined by the complexity of metadata indexing.

- The simplest case of metadata indexing (single-facet non-ontological metadata) allows the *grouping* of documents that share the same metadata value and the organization of *concept-based navigation* between independent documents [15]. Concept-based navigation is based on a set of additional navigation "concept" pages – one for each value of metadata (for example, the author of a publication). Each of the concept pages provides links to all documents indexed with this value. Each document is also connected to concept pages corresponding to all concepts from its index. Concept-based navigation allows a user to navigate from a document to any of the related concept and then to any other document indexed with the selected concept.
- The presence of ontological metadata allows organizing documents into a hierarchy (which is known as the best browsing framework) along the structure of the ontology used for indexing. The user can navigate the collection of documents along the ontology tree where a visit to each node (taxon) of the tree (or, at least, of each terminal node) provides access to all documents indexed with this taxon. The user can also use an extended version concept-based navigation moving from a document to a concept related to this document, that to a concept connected to the first concept in the ontology, and then to a document connected to the second concept. This powerful navigation approach is currently used in many resource repositories and has already been considered for open corpus adaptive hypermedia [45].
- Finally, the presence of faceted ontological indexing (multi-faced indexing with ontological metadata) allows to generation an exceptionally rich lattice-based navigation structure. This case (now typical for many digital libraries) further extends ontology-based navigation with an opportunity to navigate along several taxonomies (switching them on the fly). A good example of using navigation opportunities provided by multi-faced ontological indexing is the Flamenco browser [67].

22.3.2 External Models and the Indexing of Open Corpus Resources

As explained in section 22.2, one of the keys to providing adaptive navigation support is the presence of knowledge (adaptation-specific information) behind documents. While document and link typing provides us with some knowledge, in classic adaptive hypermedia this knowledge is most frequently provided in a different way: by connecting documents to *external models* – such as domain models, pedagogical models, or stereotype hierarchies. This process is known as *indexing*. More information about it is provided in Chapter 8 of this book [8]. Both, the creation of external models and indexing are traditionally done by the authors of adaptive hypermedia. To apply comparable methods to the open corpus of documents, one needs to resolve two related problems: where to find external models and how to index open corpus documents in terms of these models.

Existing open corpus adaptive hypermedia systems have explored several ways to solve these problems. Quite similar to the case of hypertext construction and typing, we can distinguish between manual and automatic technologies, with the automatic technologies being classified as keyword-based, metadata-based, or community-based.

The Manual Indexing of Open Corpus Resources. The manual indexing approach assumes that all adaptation-specific information is added to documents by humans, although not by the system developers and possibly after the core system has been created. This approach was explored in a few classic adaptive hypermedia systems that attempted to cover open corpus documents: KBS-Hyperbook [34] and SIGUE [18]. These systems used the classic kind of external models – domain models in the form of a network of concepts (see Chapter 1 of this book [12];) however, their corpus of documents was not closed. Any Web document could be integrated into the systems as soon as it is indexed with domain concepts. This approach is limited in its applicability yet it can be used in expanding document repositories where the manual indexing of incoming documents is feasible. The positive side of this approach is the ability to use high-level models: it results in a good quality of indexing. When coupled with concept-level overlay user models, presented in Chapter 1 of this book [12], manual concept indexing could support most advanced navigation support techniques. The negative side is the high price of model development and indexing.

A representative example of the manual approach is the KBS Hyperbook system. The first prototype of the KBS Hyperbook was developed in 1998 [32]. The fundamental concept behind the adaptation component in the KBS Hyperbook system is the separation of the adaptation module from the hypermedia system itself. This is realized by a rigorous separation of the reasoning engines and the resulting adaptation functionality from the module for organizing and maintaining the hypertext structure.

The KBS Hyperbook uses an external domain model, which serves various purposes. First of all, the domain model describes the application domain by defining all concepts relevant to the domain, as well as the relationships between these concepts. The domain model is created manually, and is the only source of knowledge about the domain that the system uses. Secondly, the domain model's concepts are used to link the hyperspace to this external domain model. Thirdly, the domain model provides the main source for the creation of a Bayesian Network [53], whose main responsibility is to estimate the actual knowledge state of a user ⌐⌐at any given time.

The indexing of hypermedia documents covers two dimensions: the first dimension describes the content of the document. This is done by indexing each document against a set of concept names from the external domain model. In addition, this indexing step also provides the necessary information for linking the hyperspace to the external model. The second dimension contains attributes that state the type of document, referring to a so-called conceptual model. The conceptual model defines possible types of documents found in this domain, e.g., such categories as "problem statement," "example," "theory," etc.

Thus, if a document is added or modified, the author has to assign a set of concepts describing the type and content of the document (or modify these attributes accordingly). In this way, the system meets the requirement that the metadata annotations for the document are independent from the application domain.

KBS Hyperbook is therefore an early showcase of open corpus adaptive hypermedia, relying on external models (domain model / Bayesian Network, and conceptual model) and the metadata arising from them, which is then added to the hypermedia documents. Several adaptive e-learning systems have been realized with the KBS Hyperbook technology, the most prominent among them being the Hyperbook on Java Programming, the *Java Hyperbook*. The Java Hyperbook is an adaptive system which uses course materials from an undergraduate course on Java Programming held at the University of Hannover and guides the student through the course by showing the next reasonable learning steps, selecting projects, generating and proposing reading sequences, annotating the educational state of information, and then by selecting information that will be useful to the user, based on their actual goals and knowledge [33]. To prove the openness of the Java Hyperbook, the authors added the content of the Sun Java Tutorial [17], a freely available online tutorial, to the Java Hyperbook. The Java Hyperbook was capable of adapting to both corpora [34]. A screenshot of the Java Hyperbook is displayed in Figure 22.1.

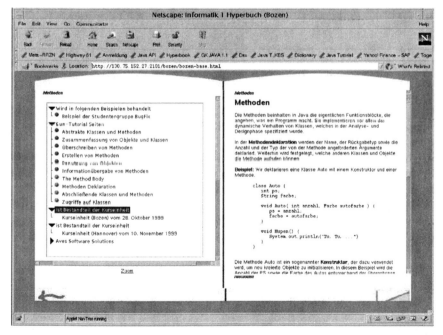

Fig. 22.1. Example of the KBS Hyperbook for Java Programming, displaying a learning unit on methods in Java (on the right hand side). The left side is composed of links to relevant learning material. Traffic light annotations of the links recommend to each learner certain navigational possibilities over others. The Sun Java pages are enriched with these recommendations, matching the previously annotated materials from the Java course corpus. Along the top, there is an array of references links, e.g., to examples, references, the Sun Java tutorial, and courses where this learning unit is used.

However, the coding of learning dependencies into the external domain model has been shown to be a drawback to the simultaneous integration of different corpora: Each collection of learning materials may follow their own learning / teaching strategy and therefore may define different learning dependencies, resulting in a different structure to the domain model. Thus, the coding of knowledge in an external domain model has the clear advantage of making the indexing and integrating of new documents is very straightforward and cost efficient. In addition, when the chosen domain model can be applied successfully to these new documents, the adaptation will work immediately. This approach is functioning very well for the Java course and Sun Tutorial, since both share the same view of the domain. In cases where this constraint is not given, the adaptation may not work correctly. Thus, the KBS Hyperbook approach is applicable to an open corpus of documents where each of the corpora shares a common domain model view.

To overcome this drawback, the KBS Hyperbook team continued with a focus on the following issue [35]: A generic knowledge modeling approach for adaptive, open corpus hypermedia systems based on ontology modeling. For each corpus of documents integrated into the open adaptive hypermedia system, a sub-graph of the ontology was calculated, with the goal of estimating the user's knowledge with respect to this corpus. This enables the KBS Hyperbook to maintain different domain models (corresponding to these sub-ontologies), which are related to each other via the common overall ontology.

Automatic Keyword-Based Indexing. In contrast to the manual approach, automatic keyword-based methods offer a low-price solution. These methods use the information retrieval approach to document modeling that is presented in Chapter 5 of this book [49]. The role of the external model is played here by a set of meaningful keywords. This is a relatively low-level model, however it can be automatically produced by document analysis and supports simple, automatic indexing of documents. Coupled with keyword-level user profiles, presented in Chapter 2 of this book [27], this approach is used in the majority of content-based recommendation systems (see Chapter 10 of this book [52] for more details). This approach has been successfully used in a number of contexts to recommend open corpus resources that are relevant to user *interests*, but its ability to adapt to other user aspects – such as knowledge or goals is very limited. Another negative side of this approach, from the educational viewpoint, is its lower precision. This reduces its applicability to educational context where adaptation to knowledge and learning goals is typically more important than adaptation to interest and where reliability of guidance is critical—because students typically aren't capable of judging how relevant an educational resource is.

An attempt to apply the keyword-level approach in adaptive educational hypermedia was done in the ML Tutor system [58]. The ML Tutor is a hypertext system that provides suggestions to the user on the basis of their recent browsing history, indicating pages that are relevant to the user's current area of interest. The ML Tutor was specifically designed to support user navigation in Web-based hypermedia, however, its internal mechanism is not essentially different from *Syskill and Webert* and other "generic" content-based recommender systems presented in Chapter 10 of this book [52]. Like many of these systems, the ML Tutor also applies machine-learning techniques to "learn" the user's profile of interests by observing browsing behavior and then recommends the most relevant pages in ML Tutor's "known" part of the Web.

The hyperspace used by ML Tutor is constructed manually, but not by the authors of the system. Instead, the authors integrated four existing independent Web sites, connecting them with additional "bookmark" links to form a joint hyperspace with 133 nodes. The role of the domain model in ML Tutor is played by a list of domain-specific keywords that were constructed manually. The indexing process is fully automatic. The result of this indexing is a binary vector for every page, where each keyword that is present in the page is indicated by 1 and each absent word by 0. The page vectors are stored in the internal database along with page IDs and URLs.

The system is implemented as an applet communicating with the server-side Machine Learning Component (MLC) of the system. At runtime, the ML Tutor applet passes the URL addresses of the last ten hypertext pages visited by a user to the MLC. Knowing their page vectors, MLC produces a list of recommended hypertext pages that focus on the same "topic" but are not yet visited and sends this information to the ML Tutor applet. The list of recommended links is displayed to the user in typical recommender-system style, with non-contextual link generation, as shown in figure 22.2.

Fig. 22.2. Adaptive navigation support using non-contextual link generation in the ML Tutor. The list of suggested pages is shown in a separate window in the upper left corner. The figure is reused from [58] with the publisher's permission.

Metadata-Based Approaches. The goal of these approaches is to improve the quality and range of supported adaptation techniques by using higher quality semantic knowledge about a document. When metadata is added to a document, it may provide important information about the document's content, intended use, primary reader group, difficulty, etc. In the area of e-learning, these metadata-based approaches benefit from the existence of standards for describing learning resources (or so-called Learning Objects). An example of an adaptive educational hypermedia system that makes use of metadata is the Personal Reader Framework (PRF) [30], which provides an environment for designing, maintaining, and running *personalization services* in the Semantic Web. The goal of the framework is to establish adaptation functionality as a Semantic Web service, which can be encapsulated and re-used.

In the run-time component of the framework, Personal Reader instances are generated by plugging together one or several of these personalization services. Each developed Reader consists of a browser for learning resources (*the reader part*) and a side-bar or remote, which displays the results of the personalization services, e.g., individual recommendations for learning resources, contextual information, pointers to further learning resources, quizzes, examples, etc. (*the personal part*). A screenshot of a Personal Reader for learning the Java programming languages is depicted in Figure 22.3.

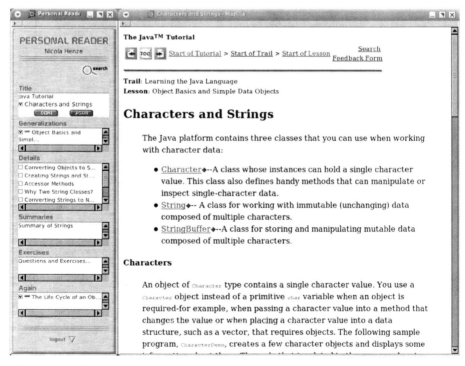

Fig. 22.3. Screenshot of the Personal Reader for Java programming The Personal Reader consists of a browser for learning resources (*the reader part*) and a side-bar or remote, which displays the results of the Personalization Services, e. g., individual recommendations for learning resources, contextual information, pointers to further learning resources, quizzes, examples, etc. (*the personal part*).

The PRF makes use of recent Semantic Web technologies for realizing the service-based environment necessary for implementing and accessing personalization services. The core component of the PRF is the so-called *connector service* whose task is to pass requests and processing results between the user interface component and available personalization services, and to supply user profile information, and available metadata descriptions on learning objects, courses, etc. In this way, the connector service is the mediator between all services in the PRF.

Two different kinds of services - apart from the connector service – are used in the PRF: personalization services and *visualization services*. Each *personalization service* offers some adaptive functionality, e.g., recommends learning objects, points to more detailed information, quizzes, exercises, etc. Personalization services are available to the PRF via a service registry using the WSDL (Web Service Description Language, [63]). Thus, service detection and invocation take place via the connector service, which asks the Web service registry for available personalization services, then selects appropriate services from this list. The task of the *visualization services* is to provide a user interface for the Personal Reader: interpret the results of the personalization services to the user, and create the actual interface, composed of reader and personalization sections.

The PRF refers—as far as possible—to standard metadata annotations: The currently implemented sample readers (for the domains "Java Programming" and "Semantic Web") make use of metadata descriptions for documents in accordance with LOM [38], while user profile information relies on the IEEE PAPI specification for describing learners [37]. Further, domain ontologies are applied: e.g., domain ontologies for Java programming or the Semantic Web. By using ontologies for describing run-time user observations and for adaptation, these models can be shared with other applications. The PRF can also implement concurrent personalization services which fulfill the same goal (e.g., provide personal recommendations for some learning object), but which consider different aspects in the metadata. For example, one personalization service can calculate recommendations based on the structure of the learning materials in some course and the user's navigation history, while another checks for keywords which describe the learning objectives of that learning object and calculates recommendations based on relationships to the corresponding domain ontology. Examples of such personalization services are described in [30].

The Community-Based Approaches. *Community-based* approaches to open corpus adaptive navigation support are based on the idea of *social navigation*. Social navigation tries to solve the navigation problem by taking advantage of the natural human tendency to follow the footsteps of other people with similar interests. Similar to collaborative filtering systems (see Chapter 9 of this book [56]), these approaches ignore the *content* of the documents, relying instead on information about the *usage* of these documents by a community of users. In a community-based approach, a document is "indexed" with all users who paid attention to this document explicitly or implicitly (i.e., rated, read carefully, bookmarked, or printed it). Thus, the community (or communities) of users of the system serves as an external model.

The CoWeb system [23; 24] mentioned in the section 22.3.1 above provides a good example of a simple social navigation system that works in manually authored hyperspace or a Wiki system CoWeb. To increase awareness of what is going on in the

CoWeb and to guide the users to most recently updated or visited pages all links inside the CoWeb were annotated with activity markers (Figure 22.4). An *access marker* showed access information using a metaphor of *footprints*. Small footprint symbols in three different colors (gray, orange, red) were placed right next to links to indicate the amount of traffic the page behind that link received in the past 24 hours. A *novelty marker* also in three different levels indicated how long ago that page was last modified.

A more sophisticated example of social navigation support is the Knowledge Sea II system [9], which attempts to automate both hypertext construction and indexing. Knowledge Sea II relies on SOM technology for the hypertext construction, which was introduced by its predecessor system Knowledge Sea [14]. Knowledge Sea applied SOM to build an 8 by 8 knowledge map from several thousands of Web pages belonging to several independent online resources for learning C programming language. As was mentioned earlier, SOM technology allowed the placement of similar pages into the same or adjacent cells on the map. Using *map-based navigation* [14], the users of the system were able to navigate from a page to the cell it belongs to, to connected cells (if necessary), and finally to pages that were similar to the page where they began their map-based navigation.

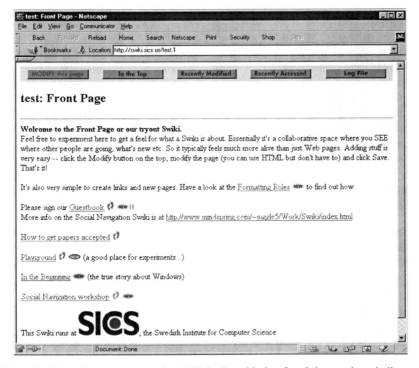

Fig. 22.4. Social navigation support in CoWeb. Two kinds of activity markers indicate when the page behind the link was last modified and also whether it was recently accessed. Used from [23] with the permission from the author.

Knowledge Sea II expanded the original Knowledge Sea with two kinds of adaptive navigation support, based on social navigation concepts: traffic- and annotation-based navigation support. Both kinds of navigation support are provided by generating visual cues that change the appearance of links on the pages and map cells presented to the user (Figure 22.5). The system generates appropriate cues individually for each user by analyzing past individual activities of the user and other users belonging to the same group.

Traffic-based navigation support attempts to express how much attention the user herself and other users from the same group paid to each of 25,000 pages that the system monitors. The level of attention for a page is computed taking into account both number of visits and time spent on the page and is displayed to the user through an icon that shows a human figure on a blue background. The color saturation of the figure expresses the level of the user's own attention while the background color expresses the average level of group attention. The higher the level of attention is, the darker the color appears to the user. The contrast between colors allows the user to compare her navigation history with the navigation of the entire group. For example, a light figure on a dark background indicates a page that is popular among group members but remains under-explored by the user. The color of the map cell and the

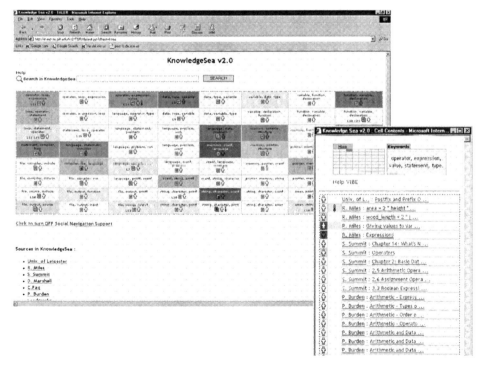

Fig. 22.5. Social navigation support in the Knowledge Sea II system. The knowledge map is shown on the left and an opened cell on the right. A darker blue background indicates documents and map cells that have received more attention from users within the same group. Human icons with darker colors indicate documents and cells that have received more attention from the user herself. Similarly, a yellow background indicates density of annotation and a thermometer icon measures how positive these annotations were.

human figure shown in the cell is computed by integrating attention parameters of all pages belonging to that cell.

Annotation-based navigation support uses a similar approach to represent the number of page annotations made by the user herself and other users from the same group. Each page in the system can be annotated by the user. The user can also indicate that this note is praise (i.e., the page is good in some aspect). While users make annotations mainly for themselves, Knowledge Sea II allows all users of the same group to benefit from collective annotation behavior. The yellow annotation icon shown next to the blue traffic icon shows the density and the "praise temperature" of annotation for each page. The more annotations a page has, the darker the yellow background color appears to the user. The temperature shown on a thermometer icon indicates the percentage of praise annotations.

Knowledge Sea II provides a good case for stressing the positive and negative sides of the use of community-based approaches with implicit feedback for adaptive navigation support. On the positive side, Knowledge Sea II requires the least effort to add a new document to the system: neither manual nor automatic page pre-processing is required for the navigation support part to work (however, note that automatic processing is required to add a page to the map since the system uses a keywords-based hyperspace creation approach). As a result, Knowledge Sea II can instantly add any new document to the system as soon as the first of its users encounters it during navigation. This gives it a ranking as the most "open" of all the approaches to open corpus adaptive navigation support. On the negative side, the navigation support provided by community-based technologies is relatively weak and is sensitive to the system's ability to identify a group of "similar" users.

22.3.3 Discussion

The previous sections demonstrate the existence of a whole range of approaches that might be able to overcome two aspects of the open corpus problem in adaptive educational hypermedia. It's interesting to note that both the existing hypertext construction and page indexing approaches can be grouped into four similar categories – manual, keyword-based, metadata-based, and community-based (Table 22.1). While community-based hypertext linking approaches have not been analyzed, they do exist [5; 66]; the authors have simply failed to find an example of these approaches, used in an appropriate context.

Table 22.1. A Summary of hyperspace construction and document indexing approaches for the open corpus, with examples of actual systems

Approaches	Hyperspace construction	Document indexing
Manual	KBS-Hyperbook[32]; MT Tutor [58]; CoWeb [24]	KBS-Hyperbook [32]; SIGUE [18]
Keyword-based	Knowledge Sea [14]	MT Tutor [58]
Metadata-based	COHSE [19]; Flamenco [67]	PRF [30]
Community-based	Bollen & Heylighen [5]	CoWeb [24]; Knowledge Sea II [9]

It is important to stress again that these approaches do not contradict but rather complement other since they have different strong and weak sides. So far, a number of existing systems have combined different approaches to hypertext construction and navigation support. For example, Knowledge Sea II uses the keyword-based approach to create the hyperspace and a community-based approach to provide navigation support. However, it is certainly wise to combine different approaches to achieve the same goal – as has already been done by the hybrid recommender systems presented in Chapter 12 of this book [16]. Moreover, an interesting challenge is to integrate approaches so that they will support each other. For example, the techniques used for analyzing social navigation patterns or for identifying Web communities may be used to help detect hidden relations between documents, where these relations might express similarity in content, as well as finding contradicting relations between documents and others. So, these techniques can be used to gather metadata—based on usage, structure, or content—for the hypermedia components, the hypertext documents, and hypertext relations. This metadata can then be used within a personalization service for recommending and visualizing information, as in Personal Reader. Vice versa, the metadata used in the Personal Reader can be used to further strengthen the pattern-detection algorithms of the social navigation process.

22.4 The Road Ahead

In this section, we analyze the effects of the open corpus problem on reusability and interoperability issues. As a conclusion to this, we will discuss the open corpus problem in relation to the Semantic Web, and give possible solutions for overcoming the open corpus problem and its implications. Although our starting point has been the open corpus problem in the field of adaptive educational hypermedia, many of the below considerations are valid for adaptive hypermedia in general.

22.4.1 Re-usability and the Open Corpus Problem

Traditional adaptive hypermedia systems operate on some fixed document space, where documents and relations between them are known at the design time and adaptation strategies are developed with respect to this specific set of documents. Especially document-to-document relations (see section 22.2.1) can only be validly assigned if the complete document space is known. Adaptation algorithms deliver faulty results if the document space is altered (e.g., if documents are modified, deleted, or new documents are introduced) as the document-to-document relations used in the algorithms become invalid. Only sophisticated re-engineering of the metadata (again on the complete document space) can recover the situation. One implication of the closed corpus in traditional adaptive hypermedia is that adaptive applications consequently fail in exchanging content with other (adaptive or non-adaptive) applications. Thus, the *re-use of content*—a very important aspect, especially when it comes to the Web—is not supported. To achieve re-usability, substantial re-engineering of particular systems is required, which cannot be realized in an on-demand basis.

In the context of e-learning, recent developments have yielded not only metadata standards for e-learning but also large collections / repositories of learning material, where both learners and teachers can store and retrieve learning objects (see section 22.1). These repositories should enable their different users to retrieve and select *appropriate* learning materials – which is the classic context for adaptation. However, successful approaches must include open corpus adaptive hypermedia, for example, becoming flexible enough to deal with varying metadata schemes and metadata details / quality.

Possible solutions discussed in section 22.3.2 of this paper include the manual indexing approach, which links hypermedia documents to external models and the automatic keyword-based indexing approach. Most promising in this context is on-the-fly metadata identification approach also presented in section 22.3.2. By analyzing usage patterns and signature structures in large hypertexts, metadata-like information can be gathered and explored to show relations between documents, rank documents or relations, and recommend relevant documents or relations for specific target groups.

22.4.2 Interoperability and the Open Corpus Problem

Apart from the re-use of content, which might be the most obvious implication of the open corpus problem, the *re-use of adaptive functionality* itself can be seen as equally important. Currently, most adaptive hypermedia systems are built from scratch, re-implementing adaptive functionality instead of re-using appropriate software modules. A first step to arrive at a re-usable adaptive functionality is to analyze and describe adaptive functionality in a system-independent manner, which, formally stated, describes the adaptation algorithms together with the required processing data. This processing data pertains to all aspects of the adaptation process: the adaptation-specific information in the adaptive hypermedia system, the user characteristics and models, as well as data that is only available at runtime (e.g., [31], which introduces a formal characterization of adaptive functionality in some of the most-cited adaptive educational hypermedia systems).

The *re-use of adaptive functionality* across applications requires interoperability solutions for adaptive systems. Interoperability is a very important aspect of today's systems, not only adaptive systems, and many issues for enabling true interoperability remain to be solved.

In section 22.3.2, we have seen an approach for solving the open corpus problem on the level of architectures. A service-oriented architecture with personalization services – each of them realizing a certain adaptive functionality – is proposed. Integration and syndication of the results of the services is realized within a dedicated reasoning component, making this reasoning a very important part of the inter-operation process.

We claim that solutions to the open corpus problem in adaptive hypermedia contribute to solving general interoperability issues, and on the other hand, interoperable adaptive hypermedia systems have—in one way or the other—have tackled and continue to contribute solutions to the open corpus problem. Furthermore, continuous efforts are required to solve re-usability of adaptive functionality and adaptive systems, and interoperability between adaptive components or systems. As of today, adaptive hypermedia systems are mainly developed at universities, with limited commercial use. While evaluations of adaptive hypermedia systems have proven their benefit, the wide use of these methods and techniques in practical systems is still pending. One of the reasons for this arises from the missing or limited re-usability. Development costs are high, since in the majority of cases the realization of adaptive hypermedia techniques starts from scratch instead of extending and re-using existing systems. Re-use can help in limiting development costs, and lower development costs will make it more attractive for developers and project managers to choose adaptive, personalized solutions.

22.4.3 Adaptive Hypermedia and the Semantic Web

Overcoming the open corpus problem in adaptive hypermedia receives special importance in the light of upcoming expectations and research on adding semantics to Web information [3]. The need for personalized, adaptive access to Web information will be high if semantic-enabled applications want to demonstrate their effectiveness: one-size-fits-all approaches will not explore the full potential that can be found with automated reasoning that is based on machine-processable semantics. On the other hand, the information space of the (Semantic) Web can be characterized as highly dynamic, open, and heterogeneous: Far from being under control of only a couple of system developers, information on the Web can emerge, be modified, altered, or disappear. User-tailored applications in the (Semantic) Web therefore require open-world solutions.

The Semantic Web (see also Chapter 23 of this book [26]) aims at machine-readable and machine-processable semantics for the Web. Metadata, together with formal ontologies providing the semantics, are a meaningful source for expressing adaptation-specific information.

As an example, the document-to-concept relations discussed in section 2.3 can be expressed in the language of the Resource Description Framework (RDF [62]), with direct references to an ontology (e.g., written in the language OWL [61]). The ontology itself can be used for expressing the required adaptation-specific information in complementary models. The crucial point in document-to-concept relations, the intrinsic dependency on specific concept models, can be tackled by using different ontologies corresponding to the different concept models, and applying techniques from ontology mapping and ontology merging to externalize this intrinsic information. Furthermore, the languages of the Semantic Web provide the required add-on to pure metadata approaches: For example, by analyzing whether adaptation-specific information can be encoded with standard metadata catalogs for learning materials, constraints on subject classifications can be identified. The embedding of the subject classifier in a concept model can be described with the languages on the Semantic Web in a machine-processable way, thus enabling adaptation algorithms to evaluate and reason about the subject classifier and its meaning with respect to a referenced ontology.

The layered architecture of the Semantic Web, accompanied by reasoning engines, rule languages, logical formalisms, and trust models, provides means for reasoning about the adaptation-specific information in a standardized but open environment. This facilitates, on the one hand, an adaptation functionality that processes this information in order to determine appropriate adaptive treatment. On the other hand, the reasoning can—at least to some degree—be proven and externalized to explain to end-users what has been done. This will improve the transparency of the whole adaptation process as users can inquire about why a certain recommendation or navigation support or whatever adaptive treatment has been determined by the adaptation module. In addition, various possibilities thus open up for extending controllability of the adaptation process, leading to *scrutable* adaptation (see the discussion on scrutable user models in [39]).

Overall, we can observe that the Semantic Web, with its languages, formalisms, and machine-processable semantics, provides excellent conditions for the use of adaptation. The Semantic Web achieves the separation of Web content from its later delivery and a certain context-of-use by enabling computer programs to reason about this Web content and its meaning. Adaptation, on the other hand, allows for the tailoring of this delivery according to the specific and individual requirements of users, within their current context. In this way, adaptation or personalization is important for optimizing the process of querying for, retrieving, selecting, and accessing information on the Web under user-specific constraints, and adaptive methods from the field of Adaptive Hypermedia should be realized very well within this Semantic Web architecture.

22.5 Conclusions

Since the mid-nineties, techniques in the field of *adaptive hypermedia* have been developed to adapt hypertexts to the needs of individual users. Success stories of adaptive hypermedia have especially been reported in the educational field, with the delivery of different individual learning paths and recommendations for learning goals or exercises, thus providing precisely attuned guidance and support during the learning process.

Despite the fact that techniques from adaptive hypermedia have proven their successfulness in providing individually optimized views on large hypertextual information spaces, wide-spread use of these techniques in e-learning is still pending. We argue that one reason for this can be identified as the *open corpus problem* in adaptive hypermedia.

This paper provides an in-depth analysis of the open corpus problem in adaptive hypermedia within the application area of educational systems. We show how document corpora and adaptation techniques are intertwined, and discuss consequences of this coupling for applying adaptive hypermedia techniques to open and dynamic information spaces. We characterize and compare the different approaches to overcome the open corpus problem and discuss their benefits and drawbacks. We reveal the relations of the open corpus problem to the re-usability and interoperability of adaptive systems, and point out the benefits of applying Semantic Web technologies to tackle and solve the open corpus problem.

Acknowledgments. This research is supported by the National Science Foundation under Grant No. 0447083 and partially supported by REWERSE - Reasoning on the Web (www.rewerse.net), Network of Excellence, 6th European Framework Program.

References

1. Allan, J.: Automatic hypertext link typing. In: Proc. of 7th ACM Conference on Hypertext. ACM Press (1996) 42-52
2. Bareiss, R., Osgood, R.: Applying AI models to the design of exploratory hypermedia systems. In: Proc. of Fifth ACM Conference on Hypertext. ACM Press (1993) 94-105
3. Berners-Lee, T., Hendler, J., Lassila, O.: The Semantic Web. Scientific American, May (2001) 35-43
4. Billsus, D., Pazzani, M.: Adaptive news access. In: Brusilovsky, P., Kobsa, A., Nejdl, W. (eds.): The Adaptive Web: Methods and Strategies of Web Personalization. Lecture Notes in Computer Science, Vol. 4321. Springer-Verlag, Berlin Heidelberg New York (2007) this volume
5. Bollen, J., Heylighen, F.: A system to restructure hypertext networks into valid user models. The New Review of Multimedia and Hypermedia 4 (1998) 189-213
6. Brin, S., Page, L.: The anatomy of a large-scale hypertextual (Web) search engine. In: Ashman, H., Thistewaite, P. (eds.) Proc. of Seventh International World Wide Web Conference. Vol. 30. Elsevier Science B. V. (1998) 107-117
7. Brusilovsky, P.: Adaptive hypermedia. User Modeling and User Adapted Interaction 11, 1/2 (2001) 87-110
8. Brusilovsky, P.: Adaptive navigation support. In: Brusilovsky, P., Kobsa, A., Neidl, W. (eds.): The Adaptive Web: Methods and Strategies of Web Personalization. Lecture Notes in Computer Science, Vol. 4321. Springer-Verlag, Berlin Heidelberg New York (2007) this volume
9. Brusilovsky, P., Chavan, G., Farzan, R.: Social adaptive navigation support for open corpus electronic textbooks. In: De Bra, P., Nejdl, W. (eds.) Proc. of Third International Conference on Adaptive Hypermedia and Adaptive Web-Based Systems (AH'2004). Lecture Notes in Computer Science, Vol. 3137. Springer-Verlag (2004) 24-33
10. Brusilovsky, P., Eklund, J.: A study of user-model based link annotation in educational hypermedia. Journal of Universal Computer Science 4, 4 (1998) 429-448
11. Brusilovsky, P., Eklund, J., Schwarz, E.: Web-based education for all: A tool for developing adaptive courseware. In: Ashman, H., Thistewaite, P. (eds.) Proc. of Seventh International World Wide Web Conference. Vol 30. Elsevier Science B. V. (1998) 291-300
12. Brusilovsky, P., Millán, E.: User models for adaptive hypermedia and adaptive educational systems. In: Brusilovsky, P., Kobsa, A., Neidl, W. (eds.): The Adaptive Web: Methods and Strategies of Web Personalization. Lecture Notes in Computer Science, Vol. 4321. Springer-Verlag, Berlin Heidelberg New York (2007) this volume
13. Brusilovsky, P., Pesin, L.: Adaptive navigation support in educational hypermedia: An evaluation of the ISIS-Tutor. Journal of Computing and Information Technology 6, 1 (1998) 27-38
14. Brusilovsky, P., Rizzo, R.: Map-based horizontal navigation in educational hypertext. Journal of Digital Information 3, 1 (2002) http://jodi.ecs.soton.ac.uk/Articles/v03/i01/Brusilovsky/
15. Brusilovsky, P., Schwarz, E.: Concept-based navigation in educational hypermedia and its implementation on WWW. In: Müldner, T., Reeves, T.C. (eds.) Proc. of ED-MEDIA/ED-TELECOM'97 - World Conference on Educational Multimedia/Hypermedia and World Conference on Educational Telecommunications. AACE (1997) 112-117
16. Burke, R.: Hybrid Web recommender systems. In: Brusilovsky, P., Kobsa, A., Neidl, W. (eds.): The Adaptive Web: Methods and Strategies of Web Personalization. Lecture Notes in Computer Science, Vol. 4321. Springer-Verlag, Berlin Heidelberg New York (2007) this volume
17. Campione, M., Walrath, K.: The Java Tutorial. Addision-Wesley, (2003)

18. Carmona, C., Bueno, D., Guzmán, E., Conejo, R.: SIGUE: Making Web Courses Adaptive. In: De Bra, P., Brusilovsky, P., Conejo, R. (eds.) Proc. of Second International Conference on Adaptive Hypermedia and Adaptive Web-Based Systems (AH'2002). Lecture Notes in Computer Science, Vol. 2347. Springer-Verlag (2002) 376-379

19. Carr, L., Hall, W., Bechhofer, S., Goble, C.: Conceptual linking: Ontology-based open hypermedia. In: Proc. of 10th International World Wide Web Conference. ACM Press (2001) 334-342

20. Chen, H., Houston, Andrea L., Sewell, Robin R., Schatz, Bruce R.: Internet browsing and searching: User evaluations of category map and concept space techniques. Journal of the American Society for Information Science 49, 7 (1998) 582-608

21. Cleary, C., Bareiss, R.: Practical methods for automatically generating typed links. In: Proc. of 7th ACM Conference on Hypertext. ACM (1996) 31-41

22. De Bra, P., Calvi, L.: AHA! An open Adaptive Hypermedia Architecture. The New Review of Hypermedia and Multimedia 4 (1998) 115-139

23. Dieberger, A.: Where did all the people go? A collaborative Web space with social navigation information (2000) available online at
http://homepage.mac.com/juggle5/WORK/publications/SwikiWriteup.html

24. Dieberger, A., Guzdial, M.: CoWeb - experiences with collaborative Web spaces. In: Lueg, C., Fisher, D. (eds.): From Usenet to CoWebs: Interacting with Social Information Spaces. Springer-Verlag, New York (2003) 155-166

25. Dolog, P., Henze, N., Neidl, W.: Logic-based open hypermedia for the semantic web. In: Proc. of 1st International Workshop on Hypermedia and the Semantic Web at Hypertext 2003. (2003) http://www.l3s.de/~dolog/pub/htsw2003.pdf

26. Dolog, P., Nejdl, W.: Semantic Web Technologies for the Adaptive Web. In: Brusilovsky, P., Kobsa, A., Neidl, W. (eds.): The Adaptive Web: Methods and Strategies of Web Personalization. Lecture Notes in Computer Science, Vol. 4321. Springer-Verlag, Berlin Heidelberg New York (2007) this volume

27. Gauch, S., Speretta, M., Chandramouli, A., Micarelli, A.: User profiles for personalized information access. In: Brusilovsky, P., Kobsa, A., Neidl, W. (eds.): The Adaptive Web: Methods and Strategies of Web Personalization. Lecture Notes in Computer Science, Vol. 4321. Springer-Verlag, Berlin Heidelberg New York (2007) this volume

28. Goy, A., Ardissono, L., Petrone, G.: Personalization in e-commerce applications. In: Brusilovsky, P., Kobsa, A., Neidl, W. (eds.): The Adaptive Web: Methods and Strategies of Web Personalization. Lecture Notes in Computer Science, Vol. 4321. Springer-Verlag, Berlin Heidelberg New York (2007) this volume

29. Green, S.: Building hypertext links by computing semantic similarity. IEEE Transactions on Knowledge and Data Engineering 11, 5 (1999) 713-730

30. Henze, N.: Personal Readers: Personalized Learning Object Readers for the Semantic Web. In: Looi, C.-K., McCalla, G., Bredeweg, B., Breuker, J. (eds.) Proc. of 12th International Conference on Artificial Intelligence in Education, AIED'2005. IOS Press (2005) 274-281

31. Henze, N., Neidl, W.: A Logical Characterization of Adaptive Educational Hypermedia. New Review of Hypermedia and Multimedia 10, 1 (2004) 77-113

32. Henze, N., Nejdl, W.: Adaptivity in the KBS Hyperbook System. In: Brusilovsky, P., Bra, P.D., Kobsa, A. (eds.) Proc. of Second Workshop on Adaptive Systems and User Modeling on the World Wide Web. Vol. 99-07. Eindhoven University of Technology (1999) 67-74, also available as http://wwwis.win.tue.nl/asum99/henze/henze.html

33. Henze, N., Nejdl, W.: Extendible adaptive hypermedia courseware: Integrating different courses and Web material. In: Brusilovsky, P., Stock, O., Strapparava, C. (eds.) Proc. of Adaptive Hypermedia and Adaptive Web-based Systems, AH'2000. Lecture Notes in Computer Science, Vol. 1892. Springer-Verlag (2000) 109-120

34. Henze, N., Nejdl, W.: Adaptation in open corpus hypermedia. International Journal of Artificial Intelligence in Education 12, 4 (2001) 325-350
35. Henze, N., Nejdl, W.: Knowledge modeling for open adaptive hypermedia. In: De Bra, P., Brusilovsky, P., Conejo, R. (eds.) Proc. of Second International Conference on Adaptive Hypermedia and Adaptive Web-Based Systems (AH'2002). Lecture Notes in Computer Science, Vol. 2347. (2002) 174-183
36. Hirashima, T., Hachiya, K., Kashihara, A., Toyoda, J.i.: Information filtering using user's context on browsing in hypertext. User Modeling and User Adapted Interaction 7, 4 (1997) 239-256
37. IEEE LTCS WG2: PAPI Learner, Draft 8 Specification. IEEE P1484.2 Learner Model Working Group (2002) available online at http://edutool.com/papi/
38. IEEE LTCS WG12: IEEE Standard for Learning Object Metadata, IEEE (2002) available online at http://ltsc.ieee.org/wg12/par1484-12-1.html
39. Kay, J., Kummerfeld, B.: Scrutability, user control and privacy for distributed personalization. In: Proc. of Workshop on Privacy-Enhanced Personalization at CHI 2006. (2006) http://www.isr.uci.edu/pep06/papers/PEP06_KayKummerfeld.pdf
40. Kibby, M.R., Hayes, J.T.: Towards intelligent hypertext. In: McAleese, R. (ed.) Hypertext: Theory into practice. Intellect, Oxford (1989) 164-171
41. Kohonen, T.: Self-organizing maps. Springer Verlag, Berlin (1995)
42. Kohonen, T.: Self-organization of very large document collections: State of the art. In: Niklasson, L., Bodén, M., Ziemke, T. (eds.) Proc. of 8th International Conference on Artificial Neural Networks, ICANN98. Vol. 1. Springer (1998) 65-74
43. Lagoze, C., al., e.: Core services in the architecture of the national science digital library (NSDL). In: Marchionini, G., Hersh, W. (eds.) Proc. of ACM/IEEE-CS Joint Conference on Digital Libraries. ACM Press (2002) 201-209
44. Macedo, A.A., Pimentel, M.d.G.C., Camacho-Guerrero, J.A.: An infrastructure for open latent semantic linking. In: Anderson, K.M., Moulthrop, S., Blustein, J. (eds.) Proc. of 13th ACM Conference on Hypertext and Hypermedia (Hypertext 2002). ACM (2002) 107-115
45. Masthoff, J.: Automatic generation of a navigation structure for adaptive Web-based instruction. In: Brusilovsky, P., Henze, N., Millán, E. (eds.) Proc. of Workshop on Adaptive Systems for Web-Based Education at the 2nd International Conference on Adaptive Hypermedia and Adaptive Web-Based Systems (AH'2002). (2002) 81-91, also available at http://www.csd.abdn.ac.uk/~jmasthof/ah02workshop.pdf
46. Mayes, J.T., Kibby, M.R., Watson, H.: StrathTutor: The development and evaluation of a learning-by-browsing on the Macintosh. Computers and Education 12, 1 (1988) 221-229
47. Micarelli, A., Gasparetti, F.: Adaptive focused crawling. In: Brusilovsky, P., Kobsa, A., Neidl, W. (eds.): The Adaptive Web: Methods and Strategies of Web Personalization. Lecture Notes in Computer Science, Vol. 4321. Springer-Verlag, Berlin Heidelberg New York (2007) this volume
48. Micarelli, A., Gasparetti, F., Sciarrone, F., Gauch, S.: Personalized search on the World Wide Web. In: Brusilovsky, P., Kobsa, A., Neidl, W. (eds.): The Adaptive Web: Methods and Strategies of Web Personalization. Lecture Notes in Computer Science, Vol. 4321. Springer-Verlag, Berlin Heidelberg New York (2007) this volume
49. Micarelli, A., Sciarrone, F., Marinilli, M.: Web document modeling. In: Brusilovsky, P., Kobsa, A., Neidl, W. (eds.): The Adaptive Web: Methods and Strategies of Web Personalization. Lecture Notes in Computer Science, Vol. 4321. Springer-Verlag, Berlin Heidelberg New York (2007) this volume
50. Millard, D.E., Moreau, L., Davis, H.C., Reich, S.: FOHM: a fundamental open hypertext model for investigating interoperability between hypertext domains. In: Proc. of Eleventh ACM on Hypertext and hypermedia. ACM Press (2000) 93-102

51. Murray, T.: MetaLinks: Authoring and affordances for conceptual and narrative flow in adaptive hyperbooks. International Journal of Artificial Intelligence in Education 13, 2-4 (2003) 199-233

52. Pazzani, M.J., Billsus, D.: Content-based recommendation systems. In: Brusilovsky, P., Kobsa, A., Neidl, W. (eds.): The Adaptive Web: Methods and Strategies of Web Personalization. Lecture Notes in Computer Science, Vol. 4321. Springer-Verlag, Berlin Heidelberg New York (2007) this volume

53. Pearl, J.: Probabilistic Reasoning in Intelligent Systems: Networks of Plausible Inference. Morgen Kaufmann Publishers, Inc., San Mateo, CA (1988)

54. Rizzo, R.: A Method of Labelling Hypertext Links in a Context - Dependent Way. In: Davies, G., Owen, C. (eds.) Proc. of WebNet'2000, World Conference of the WWW and Internet. AACE (2000) 449-453

55. Rizzo, R., Allegra, M., Fulantelli, G.: Developing hypertexts through a self-organized map. In: Maurer, H., Olson, R.G. (eds.) Proc. of WebNet'98, World Conference of the WWW, Internet, and Intranet. AACE (1998) 768-773

56. Schafer, J.B., Frankowski, D., Herlocker, J., Sen, S.: Collaborative filtering recommender systems. In: Brusilovsky, P., Kobsa, A., Neidl, W. (eds.): The Adaptive Web: Methods and Strategies of Web Personalization. Lecture Notes in Computer Science, Vol. 4321. Springer-Verlag, Berlin Heidelberg New York (2007) this volume

57. Shipman III, F.M., Marshall, C.C.: Spatial hypertext: an alternative to navigational and semantic links. ACM Computing Surveys 31, 4es (1999) Article No. 14

58. Smith, A.S.G., Blandford, A.: MLTutor: An application of machine learning algorithms for an adaptive Web-based information system. International Journal of Artificial Intelligence in Education 13, 2-4 (2003) 235-261

59. Smyth, B.: Case-base recommendation. In: Brusilovsky, P., Kobsa, A., Neidl, W. (eds.): The Adaptive Web: Methods and Strategies of Web Personalization. Lecture Notes in Computer Science, Vol. 4321. Springer-Verlag, Berlin Heidelberg New York (2007) this volume

60. Tudhope, D., Taylor, C.: Navigation via similarity: automatic linking based on semantic closeness. Information Processing & Management 33, 2 (1997) 233-242

61. W3C: OWL: Web Ontology Language Reference. World Wide Web Consortium (2004) available online at http://www.w3.org/TR/owl-ref/

62. W3C: Resource Description Framework (RDF): Concepts and Abstract Syntax. World Wide Web Consortium (2004) available online at http://www.w3.org/TR/owl-ref/

63. W3C: WSDL: Web Services Description Language, version 2.0. World Wide Web Consortium (2004) available online at http://www.w3.org/TR/2004/WD-wsdl20-20040803/

64. Weber, G., Brusilovsky, P.: ELM-ART: An adaptive versatile system for Web-based instruction. International Journal of Artificial Intelligence in Education 12, 4 (2001) 351-384

65. Weber, G., Specht, M.: User modeling and adaptive navigation support in WWW-based tutoring systems. In: Jameson, A., Paris, C., Tasso, C. (eds.) Proc. of 6th International Conference on User Modeling. SpringerWienNewYork (1997) 289-300

66. Yan, T.W., Jacobsen, M., Garcia-Molina, H., Dayal, U.: From user access patterns to dynamic hypertext linking. Computer Networks and ISDN Systems. (1996) 1007-1014

67. Yee, K.-P., Swearingen, K., Li, K., Hearst, M.: Faceted metadata for image search and browsing. In: Proc. of ACM Conference on Human Factors in Computing Systems, CHI 2003. ACM Press (2003) 401-408

23

Semantic Web Technologies
for the Adaptive Web

Peter Dolog[1] and Wolfgang Nejdl[2]

[1] Department of Computer Science, Aalborg University,
Fredrik Bajers Vej 7E, DK-9220 Aalborg, Denmark,
dolog@cs.aau.dk
http://www.cs.aau.dk/~dolog
[2] L3S Research Center, University of Hannover
Appelstrasse 9A, 30167 Hannover, Germany
nejdl@l3s.de

Abstract. Ontologies and reasoning are the key terms brought into focus by the semantic web community. Formal representation of ontologies in a common data model on the web can be taken as a foundation for adaptive web technologies as well. This chapter describes how ontologies shared on the semantic web provide conceptualization for the links which are a main vehicle to access information on the web. The subject domain ontologies serve as constraints for generating only those links which are relevant for the domain a user is currently interested in. Furthermore, user model ontologies provide additional means for deciding which links to show, annotate, hide, generate, and reorder. The semantic web technologies provide means to formalize the domain ontologies and metadata created from them. The formalization enables reasoning for personalization decisions. This chapter describes which components are crucial to be formalized by the semantic web ontologies for adaptive web. We use examples from an eLearning domain to illustrate the principles which are broadly applicable to any information domain on the web.

23.1 Introduction

Information access on the web is realized through the hypertext paradigm. Hypertext interlinks related pieces of information (pages) and allows the user to browse through the information space. The links are provided either explicitly, encoded by authors of the pages, or they are generated automatically, for example based on the results of a query.

Personalized information access in this context is concerned with user-centered bias of the hyperlinks to better support the current user context. Generating links automatically, taking user profiles into account, is a very attractive option but creates challenges as well. According to [5], adaptive web systems extend the adaptive navigation and presentation techniques from closed corpus adaptive hypermedia to the open corpus information resources available on the web and thus supporting personalized access on the web. In this chapter we discuss solutions based on semantic web techniques to

P. Brusilovsky, A. Kobsa, and W. Nejdl (Eds.): The Adaptive Web, LNCS 4321, pp. 697–719, 2007.

realize personalized link generation. Key aspects of this solution are ontologies and reasoning techniques. Ontologies represent shared and agreed upon conceptual models in a domain, which describe the main concepts of the domain and their relationships. Ontologies can thus serve as reference models for generating links in this domain, and represent hypertext, content and user information. Reasoning techniques can then work on metadata based on these ontologies, and generate links based on content, user context and user background.

As discussed in Chapter 8 [4] of this book, hypertext is a collection of text fragments interconnected by active links, used to access the information fragments addressed by them. Research in the hypertext community has concentrated on how to improve navigation in hypertext systems. The hypertext community has been concerned with several ways of browsing [18, 17]. Information retrieval concepts have been studied together with hypertext concepts [1, 35].

We can distinguish between two link concepts in hypertext: links maintained within the text (embedded links) and links maintained externally to the text as first class entities. Hypertext which utilizes the first view is often denoted as a *closed hypertext*, the latter one is denoted as an *open hypertext* [29]. Hypertext is used also in connection with hypermedia, i.e. text is augmented with other media types like pictures, video or audio.

The advantage of the embedded links is that they are bound directly to the information which utilizes the links to access related information. The advantage of the second kind of links is that we can maintain and exchange links which link information in different contexts and possibly for different users, thus providing a more flexible solution ready for personalized access. This separation of text/media items from link structures is now widely accepted in hypermedia systems [18, 17].

Information retrieval systems (especially the content-based ones) rely on index structures with terms from the documents they index. The index structures are used for making retrieval more efficient (see Chapter 10 [31] of this book for more details on content-based recommender systems). Advanced information retrieval systems maintain additional relationships between the index entries. Such structures can be seen as document models which are based on conceptual modeling approaches, semantic net approaches, Bayesian network approaches and so on (see Chapter 5 [6] of this book on document modeling). Open hypermedia research deals with links which are external to the content items. Such links can be seen as indexes of the content helping to browse and navigate the content items they index and map in an efficient way. Therefore, such conceptual structures are related to the document models and information retrieval approaches.

A notion of conceptual open hypermedia has been developed [8, 26, 33, 16]. Conceptual open hypermedia deals with knowledge representation of access structures to content items for particular context from a browsing point of view. Current semantic web technologies are very close to this notion of hypermedia, i.e. they can be used to model and represent such link structures and related objects for reasoning, querying, and processing purposes.

Though the domain ontologies are useful to generate links suited for a particular domain context, with huge corpuses it might result in too many links. Knowledge about a user might help to further constrain the links, in particular to help with some hints or annotations or simply by hiding links not suitable to his goals. Chapter 1 [6] of this book reviews several approaches to user modeling for personalized access. Semantic web technologies can be utilized for representing, sharing, processing, and reasoning on knowledge about a user in a way similar to the conceptual structures for hyperlinks.

In the following, we will start by reviewing basic hypertext concepts. We will illustrate the concepts by two examples, first with links automatically generated for a page in an eLearning application based on underlying models of content and user, and second with links generated as search results of a user query, also in an eLearning application. We will use these examples throughout our chapter to discuss how to support link generation in those applications. The examples are originally described in [12, 19]. The examples are from operational systems, the personal reader system described in [11] and personal learning assistant in [12]. We then summarize basic principles of the semantic web in terms of representation models and reasoning on the semantic web. We share this idea on reasoning with [14, 10]. Based on this background, we introduce an ontology for providing ontological hypertext links on the semantic web. The links have to be bound to specific resources either manually or as a result of reasoning process. Metadata describing instances of ontological structures are used for the binding purposes as a result of a reasoning process. To support personalized access, knowledge about a user has to be maintained and provided to the link generation systems, and an ontology for a user of an eLearning application is introduced for this purpose. The appropriateness of a resource to be bound to links provided to a user is determined according to a knowledge about the user described by instantiating the user ontology. Finally, we show how links can be generated based on these ontologies and metadata applying semantic web reasoning languages.

23.2 Hypertext and Links

Links in conceptual open hypermedia are usually described as associations between source and target information fragments. The HTML implementation of a link is a bit limited because it refers to target only; i.e. the source of the link is the fragment/page where the link is placed/anchored. The target is identified by the URL which is used to identify pages and fragments on the World Wide Web. Some more advanced applications maintain other information together with the link, for example link type. Some of them allow links to multiple sources and targets and some allow references to other links used as in sources or targets of the links.

To facilitate exchange and reuse of links across hypertext applications, an open hypermedia model has been introduced based on the paradigm of links external to fragments. The light version of the fundamental open hypermedia model [25] treats links as associations between information fragments, as sources of the link and as targets of the link. Fragments are referenced by anchors which are placeholders for information nodes.

Figure 23.1 depicts an example of a link generated (externally to the information fragments) in the *Personal Reader* framework. The link is generated in the left frame of the picture as a complex association consisting of sublinks. Sublinks are typed, providing different types of resources linked to the currently presented fragment as generalizations, details, summaries, and exercises. Furthermore, the link is annotated by a traffic light metaphor to inform the user which of the resources are ready for him to use, according to his background. The green symbol means that a link is recommended, red that it is not recommended and yellow means that user has to still acquire some prerequisite background needed to access the resource.

Such complex links in the Personal Reader system provide a user participating in a particular course with the context of currently presented information fragments relevant to his/her learning task.

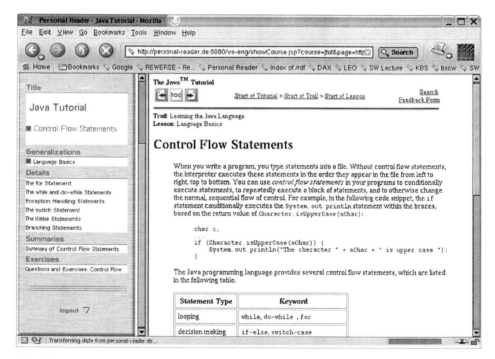

Fig. 23.1. Screenshot of the Personal Reader, showing the adaptive context of a learning resource in a course. The Personal Reader is available at www.personal-reader.de

Another example of link generation, this time in the Personal Learning Assistant (PLA), is depicted in fig. 23.2. Links are generated as search results and point to the resources relevant for a user query. These links are simpler than the one presented in fig. 23.1. Besides the identifier of a resource and its title used to generate the HTML link, it contains further information like the resource description and the concepts described by the resource. Similar to the Personal Reader links, it also provides personalization annotations as traffic light symbols. The concepts and resource descriptions are used to inform a user whether the resource really fits the user query typed at a user interface. Users formulate queries by using the concepts which annotate the resources.

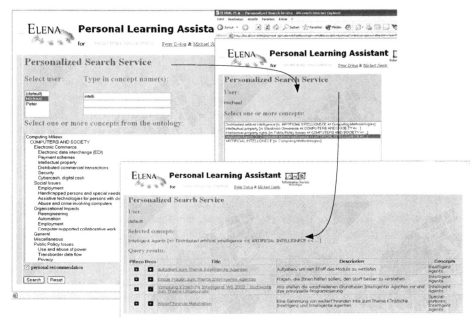

Fig. 23.2. A prototype for search user interface.

In both cases, links form more complex structures than the traditional HTML links to better support users with additional navigation information enabling them to decide which links to follow. Furthermore, links are ordered based on the knowledge about the user background. To be able to generate such links, conceptual structures for such links have to be introduced. In addition, to be able to decide on particular bindings to resources as targets of such links, information about the resources, domain of the resources and the user has to be available.

23.3 Metadata on the Semantic Web

Semantic web technologies like the Resource Description Format (RDF) [23] or RDF Schema (RDFS) [2] provide us with appropriate modeling constructs to model and represent the domain of resources, the resources themselves, as well as users and links. RDF is used to describe specific resources, RDFS serves to define domain-specific vocabularies for the metadata records represented as RDF descriptions. The following paragraphs summarize the basic principles of semantic web representation formats which we will use to describe vocabularies needed for personalized access to web resources. For more information we refer the reader to [9, 34][1].

On the Web, each resource has its own identifier provided, specified as a *Unified Resource Identifier (URI)* which is globally unique. Descriptions about resources are

---

[1] A reader who is familiar with the sematic web technologies might skip this section

represented as triples of subject, object, and predicate. For example, an assertion about the fact that the homepage of Peter Dolog was created by Peter Dolog is depicted in fig. 23.3.

Fig. 23.3. Example of an RDF graph

The subject of this triple is http://www.cs.aau.dk/~dolog, the predicate is author and the object is Peter Dolog as a literal. Predicates might be defined in different namespaces The URL prefix of author is a reference to a Dublin Core namespace in fig. 23.3. The Dublin Core is a standardization initiative for digital libraries metadata and has defined a set of predicates which are used for metadata annotations in the domain.

Object values can be resources or literals. Literals are strings of text, resources are referenced by URIs. Triples can be embedded in HTML files in an appropriate XML serialization.

Concepts and vocabularies can be provided explicitly on the semantic web and used for these RDF descriptions. The semantic web metadata model distinguishes three types of concepts: *fundamental concepts*, *schema definition concepts*, and *utility concepts*. Each concept has its own identifier in the form of an URI. The concept definitions are grouped into schemas or namespaces which are identified by URIs as well. It is possible to use abbreviated syntax for the concepts where a namespace is abbreviated into a string and separated from the concept identifier by a colon.

The *fundamental concepts* define the RDF triples, providing *rdf:Resource* as a subject, and *rdf:Property* as a predicate. A triple statement can be represented by *rdf:Statement* for reification purposes. These concepts are mandatory for all agents which claim to be developed for and operated on the semantic web.

The *schema definition concepts* are used to define custom vocabularies to be used with metadata descriptions. These concepts are usually domain specific and will be understood just by the domain specific agents, e.g. web applications for particular purposes. The new vocabulary is defined by means of classes (*rdfs:Class*). The classes can be extended with properties by defining a domain of properties (*rdfs:domain*), i.e. their inclusion in a particular class. Properties can be further restricted by defining their range of values (rdfs:Range). Classes and properties can be specialized by using subclassof and subpropertyof predicates (*rdfs:subClassOf* and *rdfs:subPropertyOf*). Any property defines a relation between resources. *rdfs:subPropertyOf* defines a subset of the property range. Similarly, *rdfs:subClassOf* relation between classes is defined as a subset inclusion. Classes define sets of resources of a certain kind. *rdf:type* is used to denote that a resource is an instance of a class or in other words that it belongs to a certain set of resources. Furthermore, typing the resources gives the resource a meaning in a certain context, defined and constrained by a schema.

The *utility concepts* are additional concepts used to define collections and for deploying RDF vocabulary on the web. Collections can be defined by one of the subclasses of the *rdfs:Container* as a bag (*rdf:Bag*), ordered sequence (*rdf:Seq*), or alternatives (*rdf:Alt*). *rdfs:seeAlso* and *rdfs:isDefinedBy* are used to point to alternative descriptions of a resource. *rdfs:label* and *rdfs:comment* are used to add human readable descriptions of a resource.

The Web Ontology Language (OWL) extends RDFS with restrictions on properties, equality between classes and properties, intersection of classes, property characteristics, 0 and 1 cardinality restrictions, and versioning in its light version. OWL Full and DL (relates to description logic) add class axioms, arbitrary cardinality, filler information, and boolean combinations of class expressions.

23.4 Reasoning on the Semantic Web

Several query and reasoning languages have been introduced to query for, and reason on, metadata on the semantic web such as QEL [27] or SPARQL [15]. The semantics of the languages are often based on Datalog, as used in the Edutella Query Language (QEL) [27, 28], and extended rule and logic programming languages.

QEL offers a full range of predicates in addition to equality, general Datalog rules, and outer join (see [28]). An example for a simple QEL query over resources is the following:

```
s(X, <dc:title>, Y),
s(X, <dc:subject>, S),
qel:equals(S, <java:OO_Class>).
```

The query tries to find resources where dc:subject equals java:OO_Class. The prefixes qel:, dc:, and java: are abbreviations for URIs of the schemas used. Variable X will be bound to URIs of resources, variable Y will be bound to titles of the resources, and variable S will be bound to subjects of the resources.

A rule language especially designed for querying and transforming RDF models is TRIPLE [32]. Rules defined in TRIPLE can reason about RDF-annotated information resources, translation tools from RDF to TRIPLE and vice versa are provided.

TRIPLE supports *namespaces* by declaring them in clause-like constructs of the form *namespaceabbrev := namespace*, resources can use these namespaces abbreviations.

```
sun_java := "http://java.sun.com/docs/books/tutorial".
```

Statements are similar to frame logic (F-Logic) [22] object syntax: An RDF statement (which is a triple) is written as *subject[predicate → object]*. Several statements with the same subject can be abbreviated in the following way:

```
sun_java:'index.html'[rdf:type->doc:Document;
   doc:hasDocumentType->doc:StudyMaterial].
```

RDF *models* are explicitly available in TRIPLE: Statements that are true in a specific model are written as "@model", for example:

```
doc:OO_Class[rdf:type->doc:Concept]@results:simple.
```

Connectives and quantifiers for building logical formulae from statements are allowed as usual, i.e. ∧, ∨, ¬, ∀, ∃, etc. For TRIPLE programs in plain ASCII syntax, the symbols AND, OR, NOT, FORALL, EXISTS, <-, ->, etc. are used. All variables must be introduced via quantifiers.

23.5 Ontologies and Metadata for Personalized Access

23.5.1 Link Structures

An ontology for link structures is used to describe structures relevant for visualization. Such an ontology adapted from FOHM [25] is depicted in fig. 23.4.

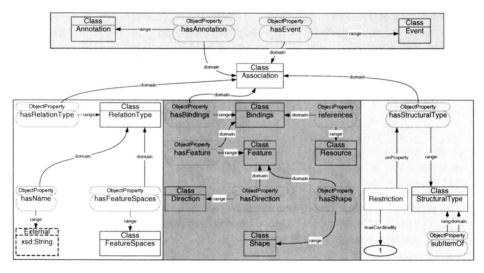

Fig. 23.4. An excerpt of the link ontology based on FOHM [25]

The main element of the ontology is the Association which links the information fragments/ pages which are relevant. Like in [25], the Association is built from three components: Bindings, RelationType, and StructuralType (in FOHM the association is a Cartesian Product of bindings, relation type and structural type). These three components (classes) are related to association through hasBindings, hasRelationType, and hasStructuralType properties.

A StructuralType is either a stack, link, bag, or sequence of resources. They are specialized forms of a general Structure. We use a subItemOf property for hierarchy specification (see fig. 23.5). The Association is restricted to have exactly one StructuralType.

Bindings references a particular Resource on the web (document, another association, etc.), and Feature-s. A Feature can be a Direction, Shape, etc. Entries for Direction are depicted in fig. 23.6b, entries for Shape are depicted in fig. 23.6c. The RelationType has a Name which is a string. The RelationType

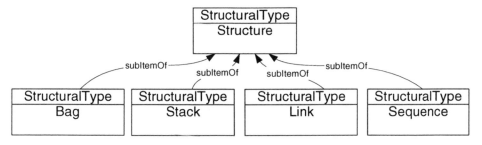

Fig. 23.5. Ontology for Structural Types.

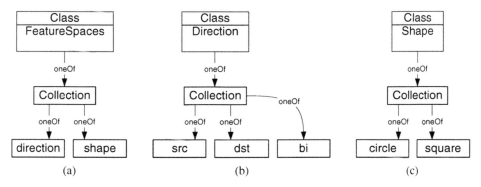

Fig. 23.6. Members of Collection of: (a) Feature Spaces, (b) Direction, (c) Shape.

also points to the `FeatureSpaces`. Entries for the `FeatureSpaces` are depicted in fig. 23.6a.

In addition, `Association` can have associated events (e.g. click events for processing user interactions) through the `hasEvent` property, and an annotation (e.g. green/red/yellow icon from traffic light metaphor technique from adaptive hypermedia [3]) through `hasAnnotation` property.

The `hasEvent` property defines an event which is provided within the document (to be able to get appropriate observation). Whenever the event is generated observation reasoning rules assigned to this type of event are triggered. The `represents` property references a resource, which is stored in observations about the learner, after an event is generated.

FOHM introduces *context* and *behavior* objects. Filtering and contextual restrictions maintained by the *context* objects in FOHM are substituted by richer reasoning language and rules in our approach. On the other hand, interactions and observations together with events substitute the notion of *behavior* objects.

Let us recall our two examples discussed in sec. 23.2, the Personal Reader and PLA. The links which are depicted there can be described using our ontology for link structures.

Figure 23.7 depicts an excerpt of a link structure visualized in fig. 23.1. The boxes represent instances (objects) and links represent specific relations between them. The box slots represent instantiations of the class attributes. The toolbar of the per-

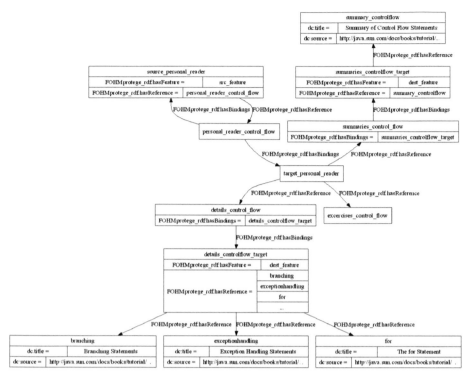

Fig. 23.7. An excerpt of a link instance for the Personal Reader and resource depicted in fig. 23.1

sonal reader is represented as a complex link (`personal_reader_control_flow`) pointing to sublinks for each part: generalizations (collapsed), details (`details_control_flow`), summaries (`summaries_control_flow`), and exercises (`exercise_control_flow`). These are treated as targets of the personal reader link and are instances of the ontology class `Association`. Furthermore, the link also contains a source (`source_personal_reader`). Each of the associations has a binding to its features which point to direct resources. For example, the `detail_control_flow` has a destination feature pointing to resources which are then used to generate click-able HTML links, i.e. URLs of web pages describing the JAVA language constructs for branching, exception handling, cycles (e.g., FOR, WHILE), and so on. The PLA links are represented similarly.

Note that the Personal Reader and PLA are just two examples of visualization agents of such links. The instantiated links can be stored for exchange and search purposes and visualized by other user interface agents in many different ways.

The resources bound to the links refer to resource metadata like title, source and others in this case from Dublin Core namespaces. They have to be selected and bound to such a link. The regeneration program is invoked whenever a user interacts with the link, i.e., the link is annotated with additional events to store user behavior and to invoke a program for regeneration.

To be able to select and bind resources to the links through its features, they have to be described in a certain way. The ontologies and metadata serve to represent knowledge about the resources and users to be used for the generation and visualization purposes.

23.5.2 Information Resources and Users

Specific domain information is usually described by concepts and their mutual relationships. The semantic web vocabularies (ontologies) in RDFS or OWL serve as domain specific models [19]. Domain ontologies consist of classes (classifying objects from a domain) and relationships between them.

The ontologies are used in annotations of specific documents/resources. The annotation metadata serves as knowledge about domain information, information fragments composition or index, and navigation which involve particular resources. In other words, the vocabulary defined by the domain specific ontologies are used to annotate/index information fragments, their compositions, and possible navigation directions in them.

The metadata can be created by the authors of information fragments or in some cases generated automatically. The ontologies described can be used to bias the descriptions of the resources and index them by the concepts from the ontology based on document analysis techniques. We have performed an experiment of automatic extraction of metadata within the framework of Personal Reader for realizing the global context. The external resources (in this case Java API) were indexed by terms from the java tutorial subject ontology (see Chapter 5 [24] and Chapter 10 [] for details on document analysis and modeling techniques which can be used to extract terms from document resources). To improve search results, a JAVA API ontology has been learned and used to cross annotate the java tutorial pages. In the following we show some examples of metadata which are created to support link generation and search.

Resource Indexing

Subject Ontologies. Subject ontologies represent organization of concepts, topics, knowledge items, or competencies in a particular domain. In the eLearning domain, the subject ontologies represent usually the domain to be taught. The concepts from ontologies are used to index information to be presented to a user and for retrieval purposes.

Figure 23.8 depicts an example of such a subject ontology, an excerpt of the java programming domain. We show a fragment of a domain knowledge base covering Java programming concepts with isa (subConceptOf) relationships between these concepts. Figure 23.8 depicts the Programming_Strategies concept with its subconcepts: Object_Oriented, Imperative, Logical, and Functional. The Object_Oriented concept is further specialized to OO_Class, OO_Method, OO_Object, OO_Inheritance, and OO_Interface. Other relations between concepts might be useful for personalization purposes as well, e.g. sequencing or dependency relations.

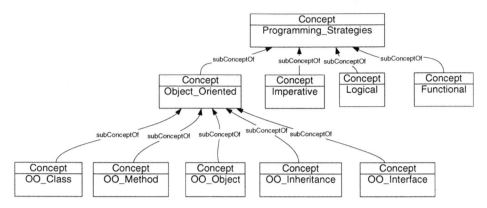

Fig. 23.8. An excerpt of application domain ontology for Java e-lecture

Resource Description Ontologies. The resource description ontologies represent the organization of metadata about resources on the web. They specify attributes which are used to describe resources and classes which categorize them.

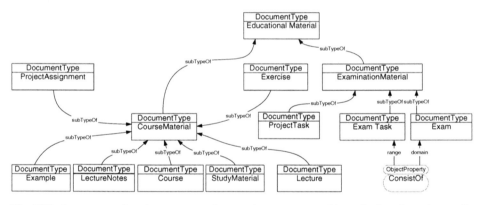

Fig. 23.9. An excerpt of environment ontology as document types hierarchy for eLearning applications

An example of a resource description ontology is depicted in fig. 23.9. The ontology depicts document types in the educational domain. The most general document type is EducationalMaterial. EducationalMaterial has two subtypes: CourseMaterial and ExaminationMaterial. ExaminationMaterial can be further specialized to ProjectTask, ExamTask, and Exam. The Exam can consist of the ExamTask-s. CourseMaterial can be further specialized into Lecture, Example, LectureNote, Course, Exercise, and Project-Assignment.

An ontology for documents and their relationships to other components is depicted in fig. 23.10. The ontology represents a context of learning material which is usually provided as a document. The class Document is used to annotate a resource which is a

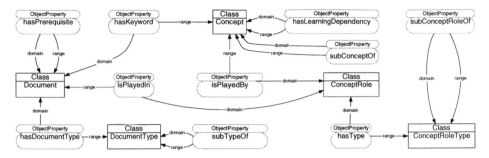

Fig. 23.10. An excerpt of environment domain ontology of documents

document. Documents describe concepts; we use class `Concept` to annotate concepts. Concepts and documents are related through the `hasKeyword` property.

Documents can be ordered by the `hasPrerequisite` property. There can be different ordering for and within applications, so multiplicity is allowed. The hasPrerequisite property is intended for navigation purposes.

Concepts play certain roles in particular document fragments. For example some concepts represent the crucial information, i.e. they are of the main information serving goal, while the others can just play a concretization role or role of comparison. In the ontology, we represent these facts by instances of `ConceptRole` class and its two properties: `isPlayedIn` and `isPlayedBy`. Document properties can be further extended by assigning a `DocumentType`. Similarly, the roles can be further extended by specifying their types. Concepts, concept role types, and document types can form hierarchies. We define `subTypeOf`, `subConceptRoleOf`, and `subConceptOf` properties for these purposes.

Information Composition and Indexing. The topics, concepts or competencies from subject ontologies represent specific content realization or composition in particular resources. An example of such a resource is a page describing sun_java: 'java/concepts/class.html'. The following example shows how such a page can be annotated based on the above mentioned resource and subject ontologies.

```
sun_java:'java/concepts/class.html'[rdf:type->doc:Document;
  hasTopic->doc:OO_Class].
doc:OO_Class[rdf:type->doc:Concept;
  doc:subConceptOf->doc:Classes_and_objects].
doc:ClassesIntroduction[rdf:type->doc:ConceptRole;
  doc:isPlayedBy->doc:OO_Class;
  doc:isPlayedIn->sun_java:'java/concepts/class.html';
  doc:hasType->doc:Introduction].
doc:Introduction[rdf:Type->doc:ConceptRoleType;
  doc:subConceptRoleOf->doc:Cover].
```

The page is a document (RDF type `Document`). The type specifies which environment the documents can be accessed through. The document describes information about classes (`OO_Class` concept). The `OO_Class` concept is annotated with type `Concept` and is a subconcept of the `Classes_and_objects` concept.

The relations and roles of concepts in particular information resources are represented by the ClassesIntroduction resource which is of type Concept-Role. The OO_Class concept plays a role of introduction (the Introduction role type) in the document which is annotated by using properties isPlayedBy and isPlayedIn respectively by references to OO_Class concept and the document. The Introduction is of type ConceptRoleType and means that the concept is covered by the content to a certain extent. Therefore, the Introduction is a subtype of Cover concept role type — a generic role type for stating that a concept is covered by document content.

Pedagogical prerequisites are encoded in the metadata to state which knowledge a user should have when accessing particular resources, (hasPrerequisite or inverse property isPrerequisiteFor of a concept or resource). In our example, the OO_Class concept is a prerequisite for the OO_Inheritance. Therefore, the above mentioned example is extended with the instance of this property.

```
sun_java:'java/concepts/class.html'[rdf:type->doc:Document;
  hasTopic->doc:OO_Class].
doc:OO_Class[rdf:type->doc:Concept;
  doc:subConceptOf->doc:Classes_and_objects;
    doc:isPrerequisiteFor->doc:OO_Inheritance].
doc:ClassesIntroduction[rdf:type->doc:ConceptRole;
  doc:isPlayedBy->doc:OO_Class;
  doc:isPlayedIn->sun_java:'java/concepts/class.html';
  doc:hasType->doc:Introduction].
doc:Introduction[rdf:Type->doc:ConceptRoleType;
  doc:subConceptRoleOf->doc:Cover].
```

User Modeling with Ontologies

User Ontologies. User modeling is used to gather knowledge about a user for personalization purposes. Personalization (or user-centered adaptation) decides about the presentation of variable resources and links on the web based on knowledge about a user. Data about a user serves to derive contextual structures. It is used to determine how to adapt the presentation of hypertext structures. In the eLearning domain, an ontology for a user profile based on IEEE [20] and IMS Global Consortium (e.g. [21]) specifications can be used. *Preference* indicates the types of devices and objects, which the user is able to recognize. Learner *Performance* and *Preference* are the main aspects relevant for personalization. Learner performance may further contain references to his *Portfolio* of projects, documents created, and experiences gained. For more discussion on learner modeling standards see for example [13].

Figure 23.11 depicts an example of an ontology for learner profiles. Learner performance is maintained according to a class Performance. Performance is based on learning experience (learningExperienceIdentifier), which is supported by particular documents. Experience implies a Concept learned from the experience, which is represented by learningCompetency property. Performance is certified by a Certificate, which is issued by a certain Institution. Performance has a certain PerformanceValue, which is in this context defined as a floating point number and restricted to the interval from 0 to 1.

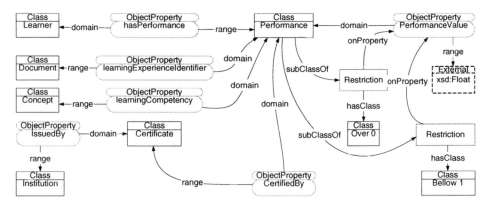

Fig. 23.11. Ontology for learner performance

Observations About a User. At run time, users interact with a web system. User interactions can be used to draw conclusions about possible user interests, user goals, tasks, knowledge, etc. These conclusions can be used for providing personalized views on hypertexts. An ontology of observations should therefore provide a structure of information about possible user observations, and - if applicable - their relations and/or dependencies.

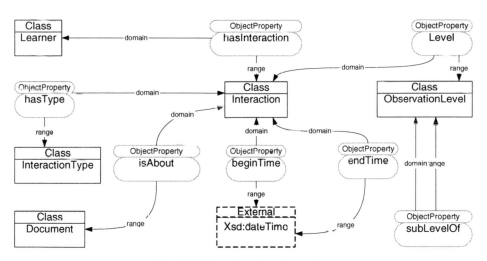

Fig. 23.12. Ontology for observations

A simple ontology for observations is depicted in fig. 23.12. The ontology allows us to state that a Learner interacted (hasInteraction property) with a particular Document (isAbout property) via an interaction of a specific type (InteractionType). Example of InteractionTypes are access or bookmark. The information that an interaction has taken place during a time interval is maintained by beginTime and endTime properties. The ObservationLevel describes particular activity types representing the purpose of the interaction. Examples for Observation-Levels are that a user has `visited` a page, has `worked` on a project, or has `solved` some exercise.

Runtime User Model. Based on these ontologies, a run-time user model can be derived, stored, maintained and used for personalization. The run-time user model is an instance of a user domain model selected for a particular application. For example, if a learner interacts with a course on JAVA, his learning performance is derived from the pages he has visited and the concepts he has worked with at particular pages. These concepts are taken from the metadata which annotate these pages. They are represented in a domain ontology for a learning outcome. Furthermore, if a page is linked to a learner assessment on particular topics, the results taken from such a learner assessment can be classified similarly as the pages and their metadata used to instantiate a learner performance record.

Let's take an example of a learner maintained as `user2` in a system. He has a performance record (maintained with `user2P` identifier in a system). Performance contains learning experience about the KBS Java objects resource. The concept covered in the resource is also stored in performance. A certificate about the performance with performance value and institution that issued the certificate is recorded in learner performance as well. Such a model in an RDF format would look like as follows:

```
user:user2[rdf:type -> learner:Learner;
   learner:hasPerformance -> user:user2P].
user:user2P[rdf:type->learner:Performance;
   learner:learningExperienceIdentifier->
      sun_java:'java/concepts/object.html';
   learner:learningCompetency->doc:OO_Object;
   learner:CertifiedBy->KBScerturi:C1X5TZ3;
   learner:PerformanceValue->0.9].
KBScerturi:C1X5TZ3[rdf:type->learner:Certificate;
   learner:IssuedBy->KBSuri:KBS].
KBSuri:KBS[rdf:type->learner:Institution].
```

23.6 Generating Links from Metadata

As discussed above, the link and hypertext paradigm can be employed in searching and in browsing. Links can be generated with the help of knowledge encoded in metadata, extracted from resources and biased by agreed upon ontologies and standards. In both cases, the process of link generation consists of several steps. Figure 23.13 describes examples of activities which support the interaction with an adaptive eLearning system. The user has the possibility of defining a learning goal, or the goal is defined implicitly by a lecture as in the Personal Reader system. In the Personal Learning Assistant, the user needs to define a learning goal by selecting some concepts from an appropriate ontology. The ontology contains competencies, skills, or concepts to be learned as described above.

If the user selects concepts from formal ontologies, a `Query` in a language appropriate for the repository can be constructed directly. If the user typed free text (into

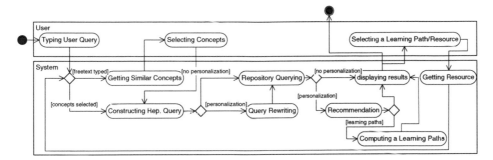

Fig. 23.13. Activities in adaptive system

fields for competencies) the system has to provide similar concepts from the ontologies for the purpose of refining his query.

After formulating such a personal learning goal, the system constructs the first version of a `Repository Query` which searches for appropriate learning resources or services. The query in Personal Reader is based on metadata on the currently presented resource. In addition, if the user requires personalization features, the query has to be rewritten taking the user profile into account (`Query Rewriting`). Such a query can then be sent to a repository.

Results returned from the repository can be either processed for display (`Displaying Results`), or the results are postprocessed by personalization algorithms (`Recommendation`) and learning path planning algorithms (`Computing Learning Path`). In both cases, the returned resources are used for generating bindings in the associations from the link ontology depicted in fig. 23.4. The sequencing information and similarities between topics in resources are used to order the bindings in associations when presented to the user. Associations are generated either based on predefined templates for user queries (specifying which information to present) or according to the local neighborhood given by several relations in the case of browsing. In both cases, annotations are used to express personalization/recommendation information. The structural types for ordering the resources bound and the visualization types represented by feature space, direction and shape are part of the specification for a particular application.

Query Rewriting. Since annotations of web resources will often vary (simpler ontologies, missing metadata, and even inconsistent metadata), we need heuristics to construct queries that cope with these difficulties. If the exact query returns no or too few results, the query needs to be relaxed by replacing some restrictions with semantically similar (usually, more general) ones, or by dropping some restrictions entirely. For this, we also need a strategy to decide which attributes to relax first (e.g., first relax dc:subject, then relax type).

The following TRIPLE predicate `similar_concept(C, CS, D)` shows how to enumerate, for a given concept `C`, similar concepts `CS` by traversing the underlying ontology and extracting superconcepts, subconcepts, and siblings with a given maximum distance `D` from `C` in the ontology. We assume that the predicate `direct_super` connects concepts with their direct superconcepts.

```
FORALL C, CS similar_concept(C, CS, 1) <- // direct super/subconcept
    direct_super(C, CS) OR direct_super(CS, C).
FORALL C, CS, D, D1  similar_concept(C, CS, D) <- // recurse
    D > 1 AND D1 is D - 1 AND similar_concept(C, CS1, D1) AND
    (direct_super(CS, CS1) OR direct_super(CS1, CS))
    AND not unify(C, CS).
```

This predicate is used iteratively to relax the query: get all similar concepts with D = 1, relax the query (by query rewriting), and send it to the remote repositories. If the returned result set is empty or too small, increment D and reiterate. The maximum number of iterations should be significantly smaller than the hierarchy depth of the ontology to avoid completely meaningless results.

Queries can also be expanded by additional restrictions from the user profile (e.g. language preferences). We have implemented a query rewriting service which adds additional constraints to a QEL query created based on the concepts selected by a user. These constraints reflect concepts and language preferences maintained in user profiles.

We illustrate query rewriting on the following simple restriction profile, implemented in TRIPLE.

```
@edu:p1 {
    edu:add1[rdf:type -> edu:AddSimpleRestriction;
        rdf:predicate -> dc:lang;
        rdf:object -> lang:de].

    edu:add2[rdf:type -> edu:AddTopicRestriction;
            edu:addTopic -> acmccs:'D.1.5'].}
```

This heuristic is used to extend a QEL query with a constraint which restricts the results to learning resources in German language (restriction edu:add1).

Another restriction derived from the user profile is a restriction on resources about *object-oriented programming* (edu:add2). The ACM Computer Classification System [30] is used to encode the subject. In that classification system, the *object-oriented programming* can be found in the category D representing *software*. The subcategory D.1 represents *programming techniques* with the fifth subcategory being *object-oriented programming*. Heuristics for query rewriting especially in case of concept or subject restrictions are usually more complex. They depend on concepts being selected or typed as a user query.

The derived restrictions profile is used in a TRIPLE view which takes as an input the profile and QEL query model. The following illustrates one of the rules for reasoning over language restrictions profiles. The view @edu:p1 encapsulates the restrictions model.

```
FORALL QUERY, VAR, PRED, OBJ, NEWLIT
  QUERY[edu:hasQueryLiteral -> edu:NEWLIT] AND
  edu:NEWLIT[rdf:type -> edu:RDFReifiedStatement;
          rdf:subject -> VAR;
          rdf:predicate -> PRED;
          rdf:object -> OBJ]
  <-
  EXISTS LITERAL, ANY (
```

```
QUERY[rdf:type -> edu:QEL3Query;
      edu:hasQueryLiteral -> LITERAL]
AND
LITERAL[rdf:type -> edu:RDFReifiedStatement;
         rdf:subject ->
             VAR[rdf:type -> edu:Variable];
         rdf:predicate -> dc:ANY])
AND
EXISTS A
  A[rdf:type -> edu:AddSimpleRestriction;
    rdf:predicate -> PRED;
    rdf:object -> OBJ]@edu:p1
AND
unify(NEWLIT, lit(VAR,PRED,OBJ)).
```

Recommendation Annotations. Recommendations can be expressed as an additional property of a resource; i.e. can annotate learning resources according to their educational state for a user. The recommendation property can take on the value of *recommend*, which specifies that a resource is recommended to a specific user, or can take a weaker value of recommendation like *might be understandable*. It can be a *not recommend* learning resource or point out that this learning resource leads to a page that the user has already visited.

To derive appropriate recommendation annotations for a particular user, prerequisite concepts for a learning resource have to be mastered by the user. The `lr:isPrerequisiteFor` relationships of concepts covered in a learning resource are analyzed for this purpose. On the other hand, a user performance profile and competencies acquired and maintained in that profile are analyzed in comparison to the prerequisites of particular learning resource.

One example of a recommendation rule is a rule which determines learning resources which are Recommended. A learning resource is recommended if *all* prerequisite concepts of all of concepts it covers have been mastered by a user:

```
FORALL LR,U learning_state(LR, U, Recommended) <-
   learning_resource(LR) AND user(U)
   AND NOT learning_state(LR, U, Already_visited)
   AND FORALL Ck ( prerequisite_concepts(LR, Ck) ->
        p_obs(Ck, U, Learned) ).
```

Predicates used in the rule derive concepts like learning resource, concepts, users, observations and learning states from metadata based on types taken from ontologies described above. We have implemented other rules to compute less strong recommendations. This includes for example a recommendation that a resource Might_be_understandable if at least one prerequisite concept has been learned.

This kind of recommendation can be used as a link annotation technique in the area of adaptive hypermedia [7], or to annotate query results with the recommendation information. On the user interface side, it is often implemented using the already mentioned traffic lights metaphor.

23.7 Discussion and Conclusions

This chapter discussed adaptive navigation support with the help of ontologies. The information access on the Web is realized through a hypertext paradigm, i.e. through provision of links. The links which are provided directly by an author usually reflect a particular context the author had when he created them. On the other hand, link generation procedures based on pure document analysis techniques may result in too many links.

The ontologies as shared conceptual models of the domain, provide context of that domain, i.e. they can be used to generate links relevant for the domain. They may serve as an input to the document analysis algorithms to take only those terms similar to the concepts from the ontology into account. Furthermore, the ontologies for a user help to further restrict the set of links which are generated only to the ones a user is interested in the most or annotate them appropriately.

The semantic web technologies in this context provide the following advantages:

- improved interoperability,
- explicit semantics,
- formal representation,
- formal reasoning.

Information resources are provided by several independent systems used in a specific context. The semantic web representation models provide uniform ways to describe, share and exchange knowledge about information resources, domains (subjects) they describe, users who use them and further knowledge needed and acquired in those systems automatically or semi-automatically. Therefore, those systems are able to *interoperate better* providing users with an extended access to information resources. In this chapter, we have shown how the ontologies represented in the semantic web format for subject, resource, user, and link can be used to realize the personalized access to information on the semantic web.

Subject ontologies which are used to index the information resources provide the *explicit semantics* about the information resource discourse which helps systems to better understand how they fit to user query, goal and background. Furthermore, user profile ontologies in semantic web representation format provide an *explicit semantics* about certain user aspects, his activities, and features what helps to improve personalization and a user satisfaction.

The semantic web technologies provide *formal representation* for knowledge on the web, thus enabling *formal reasoning* on top of them. Therefore, deduction rules can be employed for personalization. Observations about users are used to bias and reorder resources bound to the links. User models are used for personalized selection of resources. We have shown how rule-based reasoning techniques can be applied to generate such links and annotate them with recommendation information. The principles described in this chapter are general purpose but illustrated for the eLearning domain. For different domains, different ontologies have to be used, but applied similarly as we have shown in the examples from the eLearning domain.

There are three main aspects for further research challenges in this area: knowledge representation, technological, and computational. From the knowledge representation point of view, procedural knowledge in addition to the propositional knowledge about content and user is important especially in business domains and collaborative learning in workplaces. The procedures which correspond to problem solving and are related to a user's activity can be used to guide them through the problem according to real workplace settings and workflows. The connection between procedural and propositional knowledge and personalization has to be further studied.

From the technological point of view, heterogeneity of information resource is a big challenge. Information integration and approximation approaches are possibly relevant when searching large collections of heterogeneous information sources. From a practical point of view, another challenge is how to combine statistical, information retrieval models with reasoning techniques while still employing semantic web technologies. We have shown certain combinations of document analysis techniques, used to re annotate web resources of JAVA API, and formal reasoning on top of generated metadata. Further investigations are needed in this context.

From the computational point of view, performance of the reasoners is a big issue, especially when considering large semantically interconnected collections of objects on the web. For practical applications, the performance issues related with reasoning should be researched.

References

1. Agosti, M., Crestani, F., Melucci, M.: Design and implementation of a tool for the automatic construction of hypertexts for information retrieval. Inf. Process. Manage. **32**(4) (1996) 459–476
2. Brickley, D., Guha, R.V.: Resource Description Framework (RDF) Schema Specification 1.0 (2002) http://www.w3.org/TR/rdf-schema.
3. Brusilovsky, P.: Methods and techniques of adaptive hypermedia. User Modeling and User-Adapted Interaction **6**(2-3) (1996) 87 129
4. Brusilovsky, P.: Adaptive navigation support. In Brusilovsky, P., Kobsa, A., Nejdl, W., eds · The Adaptive Web: Methods and Strategies of Web Personalization. Volume this volume of Lecture Notes in Computer Science., Springer Verlag (2007)
5. Brusilovsky, P., Maybury, M.T.: From adaptive hypermedia to the adaptive web. Commun. ACM **45**(5) (2002) 30–33
6. Brusilovsky, P., Millán, E.: User models for adaptive hypermedia and adaptive educational systems In Brusilovsky, P., Kobsa, A., Nejdl, W., eds.: The Adaptive Web: Methods and Strategies of Web Personalization. Volume this volume of Lecture Notes in Computer Science., Springer Verlag (2007)
7. Brusilovsky, P., Nejdl, W.: Adaptive hypermedia and adaptive web. In Singh, M., ed.: Practical Handbook of Internet Computing. CRC Press (2004) 1–1 – 1–12
8. Bruza, P.D.: Hyperindices: a novel aid for searching in hypermedia. (1992) 109–122
9. Champin, P.A.: Rdf tutorial. http://www710.univ-lyon1.fr/ champin/rdf-tutorial/rdf-tutorial.html (April 2001)
10. Denaux, R., Dimitrova, V., Aroyo, L.: Integrating open user modeling and learning content management for the semantic web. In Ardissono, L., Brna, P., Mitrovic, A., eds.: User Modeling 2005, 10th International Conference, UM 2005. Volume 3538 of Lecture Notes in Computer Science., Edinburgh, Scotland, UK, Springer (July 2005) 9–18

11. Dolog, P., Henze, N., Nejdl, W., Sintek, M.: The personal reader: Personalizing and enriching learning resource using semantic web technologies. In Nejdl, W., Bra, P.D., eds.: Proc. of AH2004 — International Conference on Adaptive Hypermedia. Volume 3137 of LNCS., Einghoven, The Netherlands, Springer (August 2004) 85–94

12. Dolog, P., Henze, N., Nejdl, W., Sintek, M.: Personalization in distributed e-learning environments. In: Proc. of WWW2004 — The Thirteen International World Wide Web Conference, New Yourk, ACM Press (May 2004) 170–179

13. Dolog, P., Schäfer, M.: A framework for browsing, manipulating and maintaining interoperable learner profiles. In Ardissono, L., Brna, P., Mitrović, A., eds.: Proc. User Modeling 2005: 10th International Conference, UM 2005. Volume 3538 of LNAI., Edinburgh, Scotland, UK, Springer (July 2005) 397–401

14. Domingue, J., Dzbor, M.: Magpie: supporting browsing and navigation on the semantic web. In Vanderdonckt, J., Nunes, N.J., Rich, C., eds.: Proceedings of the 2004 International Conference on Intelligent User Interfaces, Funchal, Madeira, Portugal, ACM (January 2004) 191–197

15. for RDF, S.Q.L.: Sparql query language for rdf (2006) published online at http://www.w3.org/TR/rdf-sparql-query/.

16. Goble, C., Bechhofer, S., Carr, L., Roure, D.D., Hall, W.: Conceptual open hypermedia = the semantic web? In Decker, S., Fensel, D., Sheth, A., Staab, S., eds.: Proceedings of the SemWeb2001, The Second International Workshop on the Semantic Web at World Wide Web Conference — WWW10, Hong Kong (May 2001) Available at: http://CEUR-WS.org/Vol-40/Goble-et-al.pdf.

17. Grønbæk, K., Trigg, R.H.: Design issues for a dexter-based hypermedia system. Commun. ACM 37(2) (1994) 40–49

18. Hall, W.: Ending the tyranny of the button. IEEE MultiMedia 1(1) (1994) 60–68

19. Henze, N., Dolog, P., Nejdl, W.: Towards personalized e-learning in a semantic web. Educational Technology and Society Journal. Special Issue on Ontologies and the Semantic Web for E-learning 7(4) (October 2004) 82–97

20. IEEE: IEEE P1484.2/D7, 2000-11-28. draft standard for learning technology. public and private information (papi) for learners (papi learner) (2000) Available at: http://ltsc.ieee.org/archive/harvested-2003-10/working_groups/wg2.zip. Accessed on December 20, 2003.

21. IMS: IMS learner information package specification Available at: http://www.imsproject.org/profiles/index.cfm. Accessed on October 25, 2002.

22. Kifer, M., Lausen, G., Wu, J.: Logical foundations of object-oriented and frame-based languages. J. ACM 42(4) (1995) 741–843

23. Lassila, O., Swick, R.: W3C resource description framework (rdf) model and syntax specification Available at: http://www.w3.org/TR/REC-rdfsyntax/. Accessed on October 25, 2002.

24. Micarelli, A., Sciarrone, F., Marinilli, M.: Web document modeling. In Brusilovsky, P., Kobsa, A., Nejdl, W., eds.: The Adaptive Web: Methods and Strategies of Web Personalization. Volume this volume of Lecture Notes in Computer Science., Springer Verlag (2007)

25. Millard, D.E., Moreau, L., Davis, H.C., Reich, S.: FOHM: a fundamental open hypertext model for investigating interoperability between hypertext domains. In: Proceedings of the eleventh ACM on Hypertext and hypermedia, ACM Press (2000) 93–102

26. Nanard, J., Nanard, M.: Using structured types to incorporate knowledge in hypertext. In: HYPERTEXT '91: Proceedings of the third annual ACM conference on Hypertext, New York, NY, USA, ACM Press (1991) 329–343

27. Nejdl, W., Wolf, B., Qu, C., Decker, S., Sintek, M., Naeve, A., Nilsson, M., Palmér, M., Risch, T.: EDUTELLA: a P2P Networking Infrastructure based on RDF. In: In Proc. of 11th World Wide Web Conference, Hawaii, USA, ACM Press (May 2002) 604–615

28. Nilsson, M., Siberski, W.: RDF Query Exchange Language (QEL) - Concepts, Semantics and RDF Syntax. Available at: `http://edutella.jxta.org/spec/qel.html`. Accessed: 20th September 2003 (2003)

29. Nürnberg, P.J., Leggett, J.J., Wiil, U.K.: An agenda for open hypermedia research. In: HYPERTEXT '98. Proceedings of the Ninth ACM Conference on Hypertext and Hypermedia: Links, Objects, Time and Space - Structure in Hypermedia Systems, Pittsburgh, PA, USA, ACM (June 1998) 198–206

30. of Computing machinery, A.: The acm computer classification system. `http://www.acm.org/class/1998/` (2002)

31. Pazzani, M.J., Billsus, D.: Content-based recommendation systems. In Brusilovsky, P., Kobsa, A., Nejdl, W., eds.: The Adaptive Web: Methods and Strategies of Web Personalization. Volume this volume of Lecture Notes in Computer Science., Springer Verlag (2007)

32. Sintek, M., Decker, S.: TRIPLE—A query, inference, and transformation language for the semantic web. In Horrocks, I., Hendler, J.A., eds.: Proc. of ISWC 2002 — 1st International Semantic Web Conference. Volume 2342 of LNCS., Sardinia, Italy, Springer (June 2002) 364–378

33. Tudhope, D., Taylor, C.: Navigation via similarity: automatic linking based on semantic closeness. Inf. Process. Manage. **33**(2) (1997) 233–242

34. W3C: Owl web ontology language semantics and abstract syntax. Technical Report. Available at: `http://www.w3.org/TR/owl-semantics/`. Accessed: 20th September 2003 (August 2003)

35. Weiss, R., Vélez, B., Sheldon, M.A., Namprempre, C., Szilagyi, P., Duda, A., Gifford, D.K.: Hypursuit: A hierarchical network search engine that exploits content-link hypertext clustering. In: Hypertext '96, The Seventh ACM Conference on Hypertext, Washington DC, USA (March 1996) 180–193

Usability Engineering for the Adaptive Web

Cristina Gena[1] and Stephan Weibelzahl[2]

[1] Dipartimento di Informatica, Università di Torino
Corso Svizzera 185, Torino, Italy
cgena@di.unito.it
[2] School of Informatics, National College of Ireland
Mayor Street, Dublin, Ireland
sweibelzahl@ncirl.ie

Abstract. This chapter discusses a usability engineering approach for the design and the evaluation of adaptive web-based systems, focusing on practical issues. A list of methods will be presented, considering a user-centered approach. After having introduced the peculiarities that characterize the evaluation of adaptive web-based systems, the chapter describes the evaluation methodologies following the temporal phases of evaluation, according to a user-centered approach. Three phases are distinguished: requirement phase, preliminary evaluation phase, and final evaluation phase. Moreover, every technique is classified according to a set of parameters that highlight the practical exploitation of that technique. For every phase, the appropriate techniques are described by giving practical examples of their application in the adaptive web. A number of issues that arise when evaluating an adaptive system are described, and potential solutions and workarounds are sketched.

24.1 Introduction

Involving users in the design and evaluation of adaptive web-based systems has the potential to considerably improve the systems' effectiveness, efficiency and usability. Many authors have emphasized the importance of empirical studies [32, 64, 65, 96, 154], as well as the lack of suitable examples reported in the literature. Like most systems, adaptive web-based systems [100] can benefit considerably from user involvement in design and evaluation.

24.1.1 The User's Perspective for System Design

Designing adaptive web-based systems is challenging from a usability perspective [65, 73], because some of the inherent principles of these systems (e.g., automatically tailoring the interface) might violate standard usability principles such as user control and consistency (see Section 24.4.5).

Usability engineering is the systematic process of developing user interfaces that are easy to use [109, 159]. A variety of methods have been proposed to ensure that

P. Brusilovsky, A. Kobsa, and W. Nejdl (Eds.): The Adaptive Web, LNCS 4321, pp. 720–762, 2007.

the interface of the final product is efficient to use, easy to learn, and satisfying to use. This includes heuristics and guidelines, expert reviews, and user-centered design methods. The rational of user-centered design (UCD) is to place the user as opposed to the software artifact, at the center of the design process [77]. Users are involved in the development process in very early phases of the software development and in fact throughout the complete development life-cycle. Involving users from the very beginning can help to discover their ideas and expectations about the system (the so-called mental model). Moreover, it can help to identify and analyze tasks, workflows and goals, and in general to validate the developers' assumptions about the users.

As usability engineering and user centered design methods focus on cognitive and ergonomic factors (such as perception, memory, learning, problem-solving, etc.) they seem particularly suitable for the design of user-adaptive systems. The anticipation and the prevention of usability side effects should form an essential part of the iterative design of user-adaptive systems [74]. Many of these methods are described throughout this chapter. Before applying them though, we strongly encourage readers to still consult a "practical" textbook on user needs analysis and evaluation, such as [36].

24.1.2 The User's Perspective for System Evaluation

Evaluation is the systematic determination of merit, worth, and significance of something. In software development, evaluations are used to determine the quality and feasibility of preliminary products such as mock-ups and prototypes as well as of the final system. It also has the advantage of providing useful feedback for subsequent redesigns.

Adaptive systems adapt their behavior to the user and/or the user's context. The construction of a user model usually requires making many assumptions about users' skills, knowledge, needs or preferences, as well as about their behavior and interaction with the system. Empirical evaluation offers a way of testing these assumptions in the real world or under more controlled conditions [154]. Evaluation results can offer valuable insights about the real behavior and preferences of users. They can demonstrate that a certain adaptation technique actually works, i.e., that it is accurate, effective, and efficient. Evaluation studies are an important means to convince users, customers or investors of the usefulness and feasibility of a system. Finally, evaluations are important for scientific advancement as they offer a way to compare different approaches and techniques.

24.1.3 Formative Versus Summative Evaluation

Often evaluation is seen as the final mandatory stage of a project. While the focus of many project proposals is on new theoretical considerations or some innovative features of an adaptive system, a summative evaluation study is often planned in the end as empirical validation of the results. However, when constructing a new adaptive system, the whole development cycle should be covered by various evaluation studies, from the gathering of requirements to the testing of the system under development (see Sec. 24.3.2).

Formative evaluations are aimed at checking the first design choices before actual implementation and getting the clues for revising the design in an iterative design-re-design process.

From this perspective, evaluation can be considered as a *generative method* [43], since it offers contributions during the design phase by providing the means of combining design specification and evaluation into the same framework. Evaluation results can offer insights about the real behavior and the preferences of users, and therefore be adopted in the construction of the user models and system adaptation mechanisms. Expert and real users are a strong source of information for the knowledge base of the system and their real behavior offers insight for the intelligent behavior of the system. Therefore, as will be demonstrated, in adaptive web systems, evaluation is important not only to test usability and functionality, but also because testing methodologies can be a knowledge source for the development of the adaptivity components (e.g., user data acquisition, interface adaptations, inference mechanisms, etc).

The focus of this chapter is on practical issues for carrying out adaptive web-based system evaluation under a usability engineering point of view, suggesting methods and criteria to help researchers and students that are faced with evaluation problems. Since evaluation is still a challenge, we have to promote appropriate testing methodologies and publish empirical results that can be generalized, in order to check the effectiveness of adaptive web systems and put them into practice.

The chapter presents a comprehensive overview of empirical and non-empirical methods focusing on the peculiarities of the web. A detailed list of techniques will be presented, derived from Human Computer Interaction (HCI). In the evaluation of adaptive systems, especially in the final evaluation phase, metrics from information retrieval and information filtering systems are used (e.g., accuracy of recommendations, accuracy of system predictions and/or system preferences, similarity of expert rating and system prediction, inferred domain knowledge in the user model, etc) in order to evaluate the effectiveness of content adaptations. As far as these methodologies are concerned, relevant surveys are already available [19, 28, 55, 132], and Chapter 12 of this book shows examples of recommender systems evaluation [29].

therefore the main focus of the chapter will be on those HCI methods which are used in the iterative design-evaluation process. These are often disregarded in the adaptive web, even if they can contribute to an improvement in the evaluation of adaptive web systems.

When designing empirical studies on adaptive web-based systems a number of typical issues may arise. Section 24.4 provides an overview of these issues and suggests possible solutions or workarounds.

24.2 The Proposed Approach to the Analysis of Evaluation Techniques

In order to produce effective results, evaluation should occur throughout the entire design life cycle and provide feedback for design modifications [109, 159]. Early focus on users and tasks, continual testing of different solution-prototypes, empirical measurement, and integrated and iterative design can help to avoid expensive design mistakes.

All the mentioned principles are also the key-factors of the user-centered design approach [114]: to involve users from the first design decisions of an interactive system and to understand the user's needs and address them in very specific ways. Gould and Lewis [56] originally phrased this principle as follows:

- early focus on users and tasks;
- empirical measurements of product usage;
- iterative design in the production process.

A more direct engagement with final users can help to discover the context in which interaction takes place. This is particularly important both when considering ethnographic approaches (see Sec. 24.3.3) and when investigating the adaptation to the context in adaptive web sites for portable devices.

Since we believe that the usability engineering methodologies and the user-centered approach can become key factors for successful design and evaluation of adaptive web systems, in this chapter the evaluation techniques will be listed according to the life-cycle stage in which they can occur: requirement phase, preliminary evaluation phase, and final evaluation phase.

24.2.1 Classification of Techniques

In order to give some practical suggestions, at the end of every technique we have specified the following dimension: *importance for the adaptive web*. This is intended to help the researcher in the choice of the right technique for a specific situation by summarizing the way in which the method could be especially helpful for adaptive web-based systems.

At the end of every section we have also added a table providing criteria which should be helpful in choosing the most appropriate method to be applied in respect to that particular temporal phase presented in the corresponding section. For these purposes the table classifies the methods according to the following dimensions:

- *Kind of factors*, which highlights the factors the methods are most suited to generate and evaluate.
- *Applicability conditions*, which underline if there are constraints or particular conditions necessary to utilize methodologies.
- *Pros and cons*, which summarize advantages and disadvantages deriving from the application of each method.

24.2.2 Data Collection Methods

Before presenting methods and techniques for evaluation, it is worth describing data collection methods and how they interact together. Evaluation experts can choose between different methods and data collection tools depending on a number of circumstances (e.g., the type of evaluation techniques, the temporal phase, eventual constraints, etc). It is possible, in connection with a particular evaluation technique, to use more than one data collection method (e.g., users can be observed in a controlled experiment and queried at the end by means of a questionnaire). The data collection methods will be examined below.

The Collection of User's Opinion. The collection of user's opinion, also known as *query technique*, is a method that can be used to elicit details about the user's point of view of a system and it can be particularly useful for adaptive web systems in order to collect ideas and details to produce adaptation.

Questionnaires. Questionnaires have pre-defined questions and a set of closed or open answers. The styles of questions can be general, open-ended, scalar, multi-choice, ranked. Questionnaires are less flexible than interviews, but can be administered more easily (for details see [43]). Questionnaires can be used to collect information useful to define the knowledge base of the system for user modeling or system adaptations (especially in the requirement phase, see Section 24.3.1). For instance, questionnaires and scenarios[1] have been used for creating a user modeling component [1].

Since a large number of responses to a questionnaire is required in order to generalize results (which, otherwise, could be biased), existing surveys about the target population (e.g., psycho-graphic and lifestyle surveys, web-users researches, etc) can be exploited for the knowledge base definition, to build stereotype-based user-modeling systems (see for example [49] and [61]), or to inspire the adaptation strategies (see for instance Chapter 16 of this book [57]). Questionnaires (and log files) can also be used to evaluate the accuracy of system recommendations.

In adaptive web systems and their evaluation, questionnaires can further be exploited as:

– *on-line questionnaires*, to collect general users' data and preferences in order to generate recommendations. For instance, they can be used to acquire a user interest profile in collaborative [136] and feature-based recommender systems (see [126] and Chapter 18 of this book [118]). At the beginning, the system can use the user's rating to generate recommendations. Then, the data collected through the questionnaires (and web log files) can also be used to evaluate the accuracy of system recommendations by comparing the system assumptions with the real user choices [35, 102, 5].
– *pre-test questionnaires*, to establish the user's background and place her within the population of interest, and/or to use this information to find a possible correlation after the test session (e.g., computer skilled users could perform better, etc). Pre-test questionnaires can also be useful to gather data in order to classify the user before the experimental session (for instance in a stereotype [5]).
– *post-test questionnaires*, to collect structured information after the experimental session, or after having tried a system for a while. For instance, Matsuo [98] asked the users questions about the system functionality using a 5-point Likert scale[2], Alfonseca and Rodrguezi [2] asked questions concerning usability of their system, Bul [26] used a post-test questionnaire about the potential utility of their system. Besides, post-test questionnaires can be exploited to compare the assumption in the user model to an external test [158].

---

[1] A scenario is aimed at illustrating usage situations by showing step-by-step the possible user's actions and options. It can be represented by textual descriptions, images, videos and it can be employed in different design phases

[2] A Likert scale is a type of survey question where users are asked to evaluate the level at which they agree or disagree with a given sentence

– *pre and post-test questionnaires*, exploited together to collect changes due to real or experimental user-system interaction. For instance, in adaptive web-learning systems, pre and post-test questionnaires can be exploited to register improvements in the student's knowledge after one or more interactions. Pre-test questionnaires can also be used to group students on the basis of their ability [103], their knowledge [138], their motivational factors and their learning strategies [70] and then to test separately the results of the different groups (with post-test questionnaires), or to propose to the different groups solutions adapted to their cognitive profile [60].

Interviews. Interviews are used to collect self-reported opinions and experiences, preferences and behavioral motivations [43, 109]. Interviews are more flexible than questionnaires and they are well suited for exploratory studies (see for instance contextual design, Section 24.3.1). Interviews can be structured, semi-structured, and unstructured. Structured interviews have been exploited in combination with scenarios to identify adaptivity requirements [157]. However, in this experiment, results were not satisfactory to their purpose and they suggested alternative approaches to elicit requirements, such as mock-up prototypes. Unstructured interviews are often used after a test session to gather user's opinion, such as the user's satisfaction with the system [53].

User Observation Methods. This family of methods is based on direct or indirect user's observation. They can be carried out with or without predetermined tasks.

Think Aloud Protocols. Think aloud protocols are methods that make use of the user's thought throughout the experimental session, or simply while the user is performing a task. In think aloud protocols the user is explicitly asked to think out loud when she is performing a task in order to record her spontaneous reactions. The main disadvantage of this method is that it disturbs performance measurements. See for example [121] who have encouraged their users to think aloud while performing experimental tasks for the evaluation of a user modeling system based on the theory of information scent. Another possible protocol is **constructive interaction**, where more users work collaboratively to solve problems at the interface.

User Observation. Observation is a data collection method wherein the user's behavior is observed during an experimental session or in her real environment when she interacts with the system. In the former case, the user's actions are usually quantitatively analyzed and measurements are taken, while in the latter case the user's performance is typically studied from a qualitative [3] point of view. Moreover, as described in Chapter 17 of this book [84] about the evaluation of the GUIDE system, a user study can be based at the same time on direct observation, audio recording and logging data.

---

[3] "The choice between quantitative and qualitative methodologies depends on the point of view of the evaluation: while quantitative research tries to explain the variance of the dependent variable(s) generated through the manipulation of independent variable(s) (variable-based), in qualitative research the object of the study becomes the individual subject (case-based). Qualitative researchers sustain that a subject cannot be reduced to a sum of variables and therefore a deeper knowledge of a fewer group of subjects is more useful than an empirical experiment with a representative sample. Even if the final goals of both approaches are similar (they bothwant to come up with predictive theories to generalize over individual behaviours), they

Table 1. Data Collection Methods

Data Collection Methods	Kind of Factors	Applicability conditions	Pros and cons
Questionnaires and surveys	Demographic data, users' opinions, preferences and attitudes	A sample of target users	+ users involvement; subjective data − data may be biased by non representative sample; questions must be phrased carefully
On-line Questionnaires	User opinions, user satisfaction, user preferences	A sample of target users; web application	+ users involvement; subjective data − data may be biased by non representative sample; questions must be phrased carefully; little control over participation
Pre-test Questionnaires	Classification of users	A sample of target users; an adaptive web-based prototype/system to be tested	+ external user classification − erroneous classification
Post-test Questionnaires	User opinions, user satisfaction	A sample of target users; an adaptive web-based prototype/system to be tested	+ subjective data − data may be biased
Pre and post-test Questionnaires	Learning gain, change in opinion or attitude	A sample of target users; an adaptive educational web-based prototype/system to be tested	+ measuring change or development − sequential effects

are carried out in a different way: while quantitative researchers try to explain the cause-effect relationships between variables and make generalizations on the obtained results (extensive approach), qualitative researchers want to comprehend the subjects under the study by interpreting their points of view and by analyzing the facts in depth (intensive approach) in order to propose new general understanding of the reality." [51]

	Kind of Factors	Applicability conditions	Pros and cons
Interviews	User opinion, user satisfaction	A sample of target users	+ subjective data − time consuming; subjective interpretation
Think aloud protocol	Cognitions; usability of interface adaptations	A sample of target users; an adaptive web-based prototype/system	+ user's spontaneous reactions; information about what the users are thinking/why they do something − interferes with human thought processes, so actions cannot be measured; protocol might bias actions and performance
Users observation	Observation of real user-system interactions	A sample of target users; an (adaptive) web-based prototype/system	+ observing user in action − user may be influenced
Logging use	Real usage data, data for simulation, clustering	A sample of target users; an adaptive web-based prototype/system to be tested	+ large amount of useful information; unobtrusive − many client-side actions not recorded; no information about what the users are thinking/why they do something

Log data have been used to run algorithm with real users data [10, 133], and for simulations such as reproducing Web surfing [85], simulating e-mail usage [101], and calculating the accuracy of the system's predictions. Log file analysis can also suggest the way to design system adaptation on the basis of the behavior of the users (see also "Automatic usability testing and web usage mining" in Sec. 24.3.3). For instance, Herder et al. [63] conducted a long-term client study to investigate the design implication for more personalized browser history support.

Logging Use. The logging use can be considered a kind of indirect observation and consists in the analysis of log files that register all the actions of the users. The log files analysis shows the real behavior of users and is one of the most reliable ways to demonstrate the real effectiveness of user modeling and adaptive solutions [9, 25, 41, 87, 137].

24.3 Phases of Evaluation

The techniques described in this section can be categorized according to the phases of the development cycle they are usually used in, i.e., the requirement phase, the preliminary evaluation phase, and the final evaluation phase.

24.3.1 The Requirement Phase

The requirement phase is usually the first phase in the system design process. It can be defined as a "process of finding out what a client (or a customer) requires from a software system" [125]. During this phase it can be useful to gather data about typical users (features, behavior, actions, needs, environment, etc), the application domain, the system features and goals, etc.

In the case of adaptive web-based systems, the choice of relevant features to model the user (such as goals and plans of the user, the social and physical environment, etc) and consequently adapt the system, may be aided by prior knowledge of the real users of the system, the context of use, and domain experts' opinion. A deeper knowledge of these factors can offer a broader view of the application goals and prevent serious mistakes, especially in the case of innovative systems. As Benyon [17] has underlined, adaptive systems should benefit more than other systems from the requirement analysis before starting any kind of evaluation, because a higher number of features has to be taken into account in the development of these systems. The recognition that an adaptive capability may be desirable leads to a improved system analysis and design.

According to Benyon, five related and interdependent activities need to be considered in the requirement phase of an adaptive system:

- *functional analysis*, aimed at establishing the main functions of the system;
- *data analysis*, concerned with understanding and representing the meaning and structure of data in the application;
- *task knowledge analysis*, focused on the cognitive characteristics required by users of the system such as the user's mental model, cognitive loading, the search strategy required, etc.;

- *user analysis*, that determines the scope of the user population that the system is to respond to. This is concerned with obtaining attributes of users that are relevant for the application such as required intellectual capability, cognitive processing ability, and similar. The target population will be analyzed and classified according to the aspects of the application derived from the point mentioned above;
- *environment analysis*, that covers the environment within which the system is to operate.

The above activities presented by Benyon directly correspond to the following stages of the requirement analysis [125]. In the following we present techniques for gathering requirements highlighting the specific contribution for adaptive web systems, according to Benyon's proposal.

Task Analysis. Task analysis methods are based on breaking down the tasks of potential users into users' actions and users' cognitive processes [43]. In most cases, the tasks to be analyzed are broken down into in sub-tasks (see for instance *Hierarchical Task Analysis* (HTA), [40]). So far, there has been little experience in the application of this method to adaptive web-based system, even if task analysis could be used to deeply investigate users' actions and plans in order to decide in advance in which phase of the interaction the system could propose adaptations. For instance, if the task analysis shows that a goal can be reached faster by proposing some shortcut in the interface, adaptation can be proposed at that point in order to anticipate the user's plans. Task analysis results can also be useful to avoid the well-known cold-start problem[4] of knowledge-based systems by proposing default adaptations at the beginning of the user-system interaction. For instance, if it is possible to identify different kinds of target users of the website (e.g., students, teachers, administration staff, etc), task analysis can investigate the main goals of these typical users (e.g., students want to check course timetables and examination results, teachers want to insert course slides, etc), analyze in depth the tasks to be performed, and proposed possible adaptations.

Importance for the adaptive web: useful for functional, data, and task knowledged analysis of Benyon's classification.

Cognitive and Socio-technical Models. The understanding of the internal cognitive process as a person performs a task, and the representation of knowledge that she needs to do that, is the purpose of the cognitive task models [43, 125]. An example of goal-oriented cognitive model is the GOMS model (Goals, Operators, Methods and Selection) that consists of descriptions of the methods (series of steps consisting of actions performed by the users) needed to accomplish specific goals. For instance, cognitive models have been applied in the development of a mixed-initiative framework [27], by investigating the performance implications of customization decisions by means of a simplified form of GOMS analysis.

Additional methods for requirements analysis also include socio-technical models, which consider social and technical issues and recognize that technology is a part of a

---

[4] Adaptive web-based systems can suffer from cold-start problem, when no initial information about the user is available early on upon which to base adaptations.

wider organizational environment [43]. For instance, the USTM/ CUSTOM [81] model focuses on establishing stakeholder requirements[5]. Even if seldom applied in the adaptive web, both goal-oriented cognitive models and socio-technical models could offer fruitful contributions during the design phase since they are strong generative models [43]. They can help to make predictions respectively about the internal cognitive processes and the social behaviors of users, and therefore be adopted in the construction of the user model knowledge base and the corresponding system adaptations.

Importance for the adaptive web: useful for task knowledge and user analysis of Benyon's classification.

Contextual Design. Contextual design is usually organized as a semi-structured interview (see Sec. 24.2.2) covering the interesting aspects of a system while users are working in their natural work environment on their own work [18, 125]. Often the interview is recorded in order to be elaborated on by both the interviewer and by the interviewee [6]. Contextual design is a qualitative observational methodology that can be applied in the adaptive web in order to gather social and environmental information (such as structure and language used at work; individual and group actions and intentions; the culture affecting the work; explicit and implicit aspects of the work, etc) useful to inspire the design of system adaptations.

Contextual design has been used in Intelligent Tutoring Systems, for instance, through the observations of the strategies employed by teachers [3]. Masthoff [97] has also exploited contextual design together with a variant of Wizard of Oz studies.

Importance for the adaptive web: useful for user and environment analysis of Benyon's classification.

Focus Group. Focus group [58], [109] is an informal technique that can be used to collect user opinions. It is structured as a discussion about specific topics moderated by a trained group leader [58]. A typical focus group session includes from 8 to 12 target users and lasts around for two hours.

Depending on the type of users involved (e.g., final users, domain experts, technicians) focus groups can be exploited to gather functional requirements, data requirements, usability requirements, and environmental requirements to be considered in the design of system adaptations. For instance, during the development of an adaptive web-based system for the local public adminstration, mock-ups have been developed which had been discussed and redesigned after several focus group sessions with experts and final users involved in the project [52]. Focus groups can also be successfully used in combination with other methods that simulate the interaction phases when the system is not yet implemented. For instance, van Barneveld and van Setten [149] use focus groups and creative brainstorming sessions to inspire a recommender systems user interface.

---

[5] A stakeholder is here defined as anyone who is affected by the success or the failure of the system (e.g., who uses the systems, who receive output from it or provide input, etc) [43].

[6] The additional "testing" after-the-fact is also known as **retrospective testing**, and it is usually conducted after a recorded user testing session. Retrospective testing consist in reviewing the tape with the user to ask additional questions and get further clarification.

Table 2. Requirements phase

Requirement phase	Kind of Factors	Applicability conditions	Pros and cons
Task analysis	Fine decomposition of user actions; focus on cognitive processes required by the users	Formalization of the system; presence of expert evaluators	+ find out where adaptation is required − possibly artificial situations
Cognitive models	Discovering of cognitive factors to be considered during the design of system adaptations	Formalization of the system; presence of expert evaluators	+ consideration of cognitive factors in the early phases − expert evaluators required
Socio-technical models	Discovering social factors to be considered during the design of system adaptations	Representative users collaboration; presence of expert evaluators	+ consideration of social factors in early phases − expert evaluators required
Contextual design	Social and environmental information (such as structure and language used at work; individual and group actions and intentions; the culture affecting the work; explicit and implicit aspects of the work, etc)	Existing or similar web sites; representative users collaboration	+ valuable insights into usage context − expert in qualitative evaluation required

Kind of Factors	Applicability conditions	Pros and cons
Focus group	Gathering of heterogeneous requirement data from real users and domain experts for the design of system adaptations	Contact with target and representative users, technicians, domain experts	+ final users involvement; gathering of users opinions and requirements during an informal discussion − not to be confused with final evaluations
Systematic observation	Systematic analysis of features (e.g., interaction patterns, recurring activities, etc) that can be modeled by adaptation strategies	Contact with target and representative users; existing or similar web sites	+ quantification of user activities; observation of users in their natural context − expensive; expert in qualitative evaluation/observation required; time consuming

Importance for the adaptive web: useful for functional, data, user and environment analysis of Benyon's classification.

Systematic Observation. Systematic observation can be defined as a "particular approach to quantifying behavior. This approach is typically concerned with naturally occurring behavior observed in a real context" [6]. The observation is conducted in two phases: First, various forms of behavior, so-called behavioral codes are defined. Secondly, observers are asked to record whenever behavior corresponding to predefined codes occurs. The data can be analysed in two ways: either in the form of non-sequential analysis (subjects are observed for the given time slots during different time intervals) or as sequential analysis (subjects are observed for a given period of time).

In the adaptive web, systematic observation can be used during the requirement phase to systematically analyze significant interactions in order to discover interaction patterns, recurrent and typical behavior, the user's plans (e.g., sequences of user actions-interactions, distribution of user's activities along the time, etc) that can be modelled by the adaptation. For instance, in order to model teaching strategies in an Intelligent Tutoring System, Rizzo et al. [128] recorded the interactions taking place between the tutor and the student in a natural setting or computer-mediated interface. Then the records were systematically analyzed to find teaching patterns useful to inspire adaptation mechanisms.

Importance for the adaptive web: useful for task knowledge, user and environment analysis of Benyon's classification.

24.3.2 Preliminary Evaluation Phase

The preliminary evaluation phase occurs during the system development. It is very important to carry out one or more evaluations during this phase to avoid expensive and complex re-design of the system once it is finished. It can be based on predictive or formative methods.

Predictive evaluations are aimed at making predictions, based on experts' judgement, about the performance of the interactive systems and preventing errors without performing empirical evaluations with users. Formative evaluations are aimed at checking the first design choices before actual implementation and getting the clues for revising the design in an iterative design-re-design process.

Heuristic Evaluation. A heuristic is a general principle or a rule of thumb that can guide a design decision or be used to critique existing decisions. Heuristic evaluation [113] describes a method in which a small set of evaluators examine a user interface and look for problems that violate some of the general principles of good interface design.

Unfortunately, in the adaptive web field a set of recognized and accepted guidelines to follow is still missing. On the one side, this lack can be filled only by publishing statistically significant results that can demonstrate, for instance, that one adaptation strategy is better than another one in a given situation, or that some adaptation technique should be carefully applied. For instance, Sears & Shneiderman [134] performed an evaluation on menu choices sorted on the basis of their usage frequency. Their results reported that the users were disoriented by the menu choices sorted on usage frequency because of the lack of order in the adapted menu. A preferable solution could be the positioning of the most often used choices at the top of the list before all the other ordered items (the so-called *split menu*). Therefore, researchers should be careful in applying this technique. The key point is to carry out evaluations leading to significant results that can be re-used in other research, and promote the development of standard measures that would be able to reasonably evaluate the systems' reliability. To this purpose, Weibelzahl & Weber [156] promoted the development of an online database for studies of empirical evaluations to assist researchers in the evaluation of adaptive systems and to promote the construction of a corpus of guidelines.

On the other side, also general principles have to be considered. For instance, Magoulas et al. [93] proposed an integration of heuristic evaluation in the evaluation of adaptive learning environments. They modified the Nielsen's heuristics [109] to reflect pedagogical consideration and then they collocated their heuristics into the level of adaptation [93]. E.g., the Nielsen's heuristic "Recognition rather than recall" is specified in "instructions and cues that the system provides for users to identify results of adaptations easily". As sketched in Section 24.3.3, Jameson [73] proposed five usability challenges for adaptive interfaces to deal with usability problems that can arise with these systems.

Importance for the adaptive web: making prediction about the usability and the applicability of interface adaptations.

Domain Expert Review. In the first implementation phases of an adaptive web site, the presence of domain experts and human designers can be beneficial. For instance, a domain expert can help defining the dimensions of the user model and domain-relevant features. They can also contribute towards the evaluation of correctness of the inference mechanism [5] and interface adaptations [54]. For instance, an adaptive web site that suggests TV programs can benefit from audience TV experts working in TV advertising that may illustrate habits, behaviors and preferences of homogeneous groups of TV viewers. In this specific case a domain expert review can be beneficial in the requirement phase.

For example, Chapter 1 outlines how experts can contribute to the development of an uncertainty-based user model [23]. Experts can also be asked to pick up a set of relevant documents for a certain query and their judgments are used to check the correctness of system recommendations. For examples of evaluation of a recommender system with the estimation of precision and recall returned to a human advisor proposal see [92]. More metrics for evaluating recommender systems without users are listed in Chapter 3 [105].

Expert review, as well as cognitive walkthrough, scenario-based design and prototypes, can be used to evaluate *parallel designs* [109], which consist of exploring different design alternatives before setting on a single proposal to be developed further. Parallel design can very suitable for systems that have a user model since in this way designers can propose different solutions (what to model) and different interaction strategies (what the user can control) depending on the identified users. Parallel design is a very useful approach since it lets one to explore adaptive solutions and simulate strategies with users before the system is implemented. Design rationale[7] and design space analysis[8] can also be helpful in context of exploring and reasoning among different design alternatives. For details about design rationale see [90], while for design space analysis see [15]. Experts can be involved in **coaching methods**, which are usability testing techniques wherein users are encouraged to ask questions to the expert/coach, who responds with appropriate instruction. Typical user questions help at identifying usability problems.

Importance for the adaptive web: predicting the correctness of inference mechanisms and usability of interface adaptations; simulations of design alternatives.

Cognitive Walkthrough. Cognitive walkthrough is an evaluation method wherein experts play the role of users in order to identify usability problems [124]. Similar to heuristic evaluation, this predictive technique should benefit from a set of guidelines for the adaptive web that should help evaluators to assess not only general HCI mistakes but also recognized errors in the design of adaptations.

Importance for the adaptive web: making prediction about the usability and the reliability of interface adaptations that help the user to accomplish tasks.

Wizard of Oz Prototyping. Wizard of Oz prototyping [109, 125] is a form of prototyping in which the user appears to be interacting with the software when, in fact, the input is transmitted to the wizard (the experimenter) who is responding to user's actions. The user interacts with the emulated system without being aware of the trick.

Wizard of Oz prototyping can be applied in the evaluation of adaptive web systems, for instance, when a real time user-system interaction has to be simulated in the early implementation phases (e.g., speech recognition, interaction with animated agents, etc). For example, a Wizard of Oz interface that enables the tutor to communicate with the student in a computer-mediated environment has been used to model tutorial strategies [128]. Maulsby, Greenberg & Mander [99] used Wizard of Oz to prototype an intelligent agent, and Masthoff [97] applied a variant of Wizard of Oz under a contextual design point of view, making users to take the role of the wizard: humans tend to be good at adaptation, thus, observing them in the role of the wizard may help to design the adaptation.

Importance for the adaptive web: simulation of a real time user-adapted interaction.

Prototyping. Prototypes are artifacts that simulate or animate some but not all features of the intended system [43]. They can be divided in two main categories: static, paper-based prototypes and interactive, software-based prototypes. Testing prototypes is very common, however they should not be considered to be finished products. Prototypes

---

[7] Design rationale "is the information that explains why a computer systems is the way it is, including its structural or architectural description and its functional or behavioral description" [43].

[8] Design space analysis is an "approach to design that encourages the designer to explore alternative design solution" [125]

can also be: *horizontal*, when they contain a shallow layer of the whole surface of the user interface; *vertical*, when they include a small number of deep paths through the interface, but do not include any part of the remaining paths; *scenario-based* when they fully implement some important tasks that cut through the functionality of the prototype. For instance, Gena & Ardissono [52] evaluated an adaptive web prototype in a controlled experiment with real users. The main aims of the test were to discover whether the interface adaptations were visible and effective and whether the content adaptations were consistent and helpful to the task completion. In Chapter 17 of this book [84] is reported a prototype evaluation of the TellMaris system.

As described above for parallel design, scenario based prototypes can be helpful at simulating adaptation strategies and design alternatives with real users and expert before the initial implementations.

Importance for the adaptive web: early evaluation of adaptation strategies; simulations of adaptations strategies and design alternatives.

Card Sorting. Card sorting is a generative method for exploring how people group items and it is particularly useful for defining web site structures [129]. It can be used to discover the latent structure of an unsorted list of categories or ideas. The investigator writes each category on a small index card (e.g., the menu items of a web site), and requests users to groups these cards into clusters (e.g., the main item of the navigational structure). The clusters can be predefined (closed card sorting) or defined by the user herself (open card sorting).

So far, there has been little experience of card sorting in adaptive web systems. Card sorting could be carried out with different groups of representative users for the definition of the information architecture of an adaptive web site. It can inspire different information structures for different groups of users (e.g., how novice and experts see the structure of the web site information space).

Importance for the adaptive web: definition of different information architectures for different group of representative users.

Cooperative Evaluation. An additional methodology that can be carried out during the preliminary evaluation phase is the cooperative evaluation [107], which includes methods wherein the user is encouraged to act as a collaborator in the evaluation to identify usability problems and their solutions. Even if seldom applied, cooperative evaluation is a qualitative technique that could be applied in the evaluation of adaptive web based systems to detect general problems (e.g., usability, reliability of adaptations, etc) in early development phases and to explore the user's point of view to collect design inspiration for the adaptive solutions.

Importance for the adaptive web: detection of general problems concerning adaptations; design inspirations for adaptive solutions.

Table 3. Preliminary evaluation phase

Preliminary evaluation phase	Kind of Factors	Applicability conditions	Pros and cons
Heuristic evaluation	Usability of interface adaptations	A user-adaptive prototype/system	+ making prediction about the interface design without involving users − guidelines for adaptive systems are still missing
Expert review	User, domain and interface knowledge	Only for (adaptive web) knowledge-based system	+ valuable source of information for the system KB − experts may use background and contextual knowledge that are not available to a system
Cognitive walkthrough	Usability of interface adaptations	A user-adaptive prototype/system; presence of expert evaluators	+ make prediction about the design without involving users − guidelines for adaptive systems are still missing; time intensive
Wizard of Oz simulation	Early prototype evaluation	Only for systems that simulate a real time user-system interaction	+ useful when the system is still not completed − the Oz is more intelligent than the system!
Prototypes	Evaluation of vertical or horizontal prototype; design of scenarios	A running user-adaptive prototype	+ evaluation in early phases; simulation of scenarios − limited functionality available

	Kind of Factors	Applicability conditions	Pros and cons
Card sorting	To set the web site structure from the user's point of view	- Top-down information architecture of the web site to be adapted	+ considers the user's mental model – only applicable to web sites that have a categories-based information architecture (top-down)
Cooperative evaluation	Detection of general problems concerning adaptations in the early stages	Final users collaborating during the design-redesign phase	+ direct, immediate user feedback and suggestions – not all the target users are considered
Participative evaluation	Gathering of heterogeneous requirement data from real users and domain experts	Final users involved in the design team	+ direct, immediate user suggestions and inspirations – not all the target users are considered

Participative Evaluation. Another qualitative technique useful in the former evaluation phases is the participative evaluation [109, 125] wherein final users are involved with the design team and participate in design decisions. Participative evaluation is strictly tied to participatory design techniques where users are involved in all the design phases [58, 59]. So far, this methodology is rather disregarded in the adaptive web, however it could be applied to have users directly participating at the design of adaptation strategies.

Importance for the adaptive web: gathering of heterogenous requirement data from real users and domain experts; users and expert participating at the design of adaptation strategies.

24.3.3 Final Evaluation Phase

The final evaluation phase occurs at the end of the system development and it is aimed at evaluating the overall quality of a system with users performing real tasks.

Usability Testing. According to the ISO definition ISO 9241-11:1998 usability is "the extent to which a product can be used by specified users, to achieve specified goals, with effectiveness, efficiency and satisfaction, in a specified context of use" [68] . Based on this definition, the usability of a web site could be measured by how easily and effectively a specific user can browse the web site, to carry out a fixed set of tasks, in a defined set of environments [31].

The core of usability testing [109], [130], [44] is to make the users use the web site and record what happens. In this way it is possible to evaluate the response of a real user rather than to propose interfaces as designed by the designers. In particular, the usability test has four necessary features:

- participants represent real users;
- participants do real tasks;
- users' performances are observed and sometimes recorded (see Sec. 24.2.2);
- users' opinions are collected by means of interviews or questionnaires (see Sec. 24.2.2).

According to [110] a usability test with 4-5 representative users will discover 80% of major usability problems of a web site, while 10 users will discover up to 90% of problems.

One or more usability tests on an adaptive web site should always be performed. The usability of adaptive interfaces has been widely discussed, this will be reported in Section 24.4. Due to inherent problems tied to adaptive interfaces and to the importance of usability in the web, the usability of an adaptive web site should always be tested by taking into account both interface adaptations and general interface solutions. Some examples of usability testing in the adaptive web can be found in [2, 16, 131, 133], while for details on testing procedures see [44, 109, 130].

Jameson [74] pointed out that the anticipation and the prevention of usability side effects should form an essential part of the iterative design of user-adaptive systems.

Jameson [73] proposed five usability challenges for adaptive interfaces: (1) predictability and transparency, (2) controllability, (3) unobtrusiveness, (4) privacy, and (5) breadth of experience. He tried to match usability goals and typical adaptive systems properties to deal with usability problems which these systems can suffer. Transparency and controllability, nevertheless, could imply involving the user in the personalization process and/or adding some adaptability into the system. But sometimes users have difficulty understanding and controlling personalization. For an evaluation of problems with transparency and control of adaptive web systems see [72], [38]. However, there are also examples of learning systems that show systems that expose the user model to the student enhance learning [78],[104]. It is important to notice that usability tests of adaptive web sites can only be applied to evaluate general usability problems at the interface. If one would test the usability of one adaptation technique compared to another one, a controlled experiment should be carried out.

Importance for the adaptive web: usability of the overall web site and of interface adaptations.

Automatic Usability Testing and Web Usage Mining. In recent years, interest in automatic tools able to support the evaluation process has been increasing. The methods for usability evaluation of Web sites has been classified into two types of approaches [117]: *methods based on empirical evaluation*, where user's logs data generated by a web server are analyzed, and *methods based on analytical evaluation*, where various combinations of criteria, guidelines and models are automatically applied.

In the former ones, the analysis of real usage data is considered to be a solution to discover real user-system interaction. For instance, Web usage analysis [106, 139, 120] is a long process of learning to see a website from the perspective of its users. By analyzing Web server log data usage patterns could be discovered (e.g., pages occurring frequently together and in the same order). This may be a signal that many users navigate differently than originally anticipated when the site was designed. The usage mining process can involve the discovery of association rules, sequential patterns, page view clusters, user clusters, or any other pattern discovery methods. After having collected web log data and reached some evidence (confirmed by statistical analysis), the re-design of the interface may be accomplished in two ways [119]:

– by *transformation*, improving the site structure based on interactions with all visitors.
– by *customization*, adapting the site presentation to the needs of each individual visitor based on information about those individuals.

Between these two alternatives, a third solution could be adopted: personalizing a site according to a different cluster of users' behavior (for instance occasional, regular, novice, expert user, etc) emerged from the data mining process. Finally, to help the analysis of this large amount of data, logs of user interactions can be analyzed through graphical tools that visualize the paths followed by the users during the site visit [37].

Analytical methods comprehend automatic tools such as Bobby[9], that verifies the application of accessibility guidelines; WebSat[10], that evaluates usability by analyzing

---

[9] http://www.cast.org/bobby

[10] http://www.research.att.com/conf/hfweb/ proceedings/scholtz/index.html

the HTML code through the application of usability guidelines; or Design Advisor[11], which is based on eye-tracking techniques.

Between analytical and empirical methods are mixed approaches that combine the analysis of browsers logs with usability guidelines and models of user's actions. See for example [117].

Importance for the adaptive web: usability of the overall web site and of interface adaptations; inspiration for the adaptive behavior of the web site.

Accessibility. According to the ISO definition ISO/TS 16071:2003 accessibility is "the usability of a product, service, environment or facility by people with the widest range of capabilities" [69]. This definition strictly correlates accessibility to usability, with the difference that an accessible web site must be usable for every one, also for people with disabilities. There are a variety of tools and approaches for evaluating Web site accessibility. For more details see [150].

Adaptive web sites, which by definition pay more attention to users' needs, should respect accessibility guidelines. Moreover, they could adapt to the specific users with disabilities taking into account their specific problems, since impaired users need their specific requirement. For example, in the AVANTI project, the system adapted the content and the presentation of web pages to each individual user, also taking into account elderly and disabled users [47]. Stephanidis [141] highlighted the potential adaptive techniques have to facilitate both accessibility and high quality interaction, for the broadest possible end-user population.

Importance for the adaptive web: proposing adaptive solutions for different groups of disabled users to increase the accessibility of the web site.

Controlled Experiments. Controlled experiments [79, 80] are one of the most relevant evaluation techniques for the development of the adaptive web, and their impact in user-adapted systems has been largely discussed [32, 51]. Indeed, they are often performed in the evaluation of adaptive systems (mostly for the evaluation of interface adaptations), but sometimes experiments are not properly designed and thus they do not produce significant results to be taken into account. As will be discussed in Section 24.4 significant results are necessary for the growth of the adaptive web, because they can be extended to provide generalizations and guidelines for future works, therefore it is important to correctly carry out every design step and evaluate results with the required statistics.

The general idea underlying a controlled experiment is that by changing one element (the independent variable) in a controlled environment its effects on user's behavior can be measured (on the dependent variable). The aim of a controlled experiment is to empirically support a hypothesis and to verify cause-effect relationships by controlling the experimental variables. Therefore, as described in [73], controlled experiments can be used to evaluate the accuracy of modeling (content layer: e.g. are the system recommendations correct?) and the usability of the adaptive system (interface layer: e.g. do the interface adaptations enhance the quality of the interaction?). The most important criteria to follow in every experiment are:

---

[11] http://www.tri.sbc.com/hfweb/faraday/faraday.htm

- participants have to be credible: they have to be real users of the application under evaluation;
- experimental tasks have to be credible: users have to perform tasks usually performed when they are using the application;
- participants have to be observed during the experiment (see Sec. 24.2.2) and their performance recorded;
- finally, users' opinions are collected by means of interviews or questionnaires (see Sec. 24.2.2).

Empirical evaluation takes place in a laboratory environment. Well equipped laboratory may contain sophisticated audio/video recording facilities, two-way mirrors, and instrumented computers. On the one hand, the lack of context, and the unnatural condition creates an artificial situation, far from the place where the real action takes place. On the other hand, there are some situations where the laboratory observation is the only option, for instance if the location is dangerous and sometimes the experimenters may want to deliberately manipulate the context in order to create unexplored situations [Dix et al. 1998]. The schematic process of a controlled experiment can be summarized in the following steps [80], while a more detailed discussion on problems that can arise will be presented in Sec. 24.4.

Develop research hypothesis. In statistics, usually two hypotheses are considered: the null hypothesis and the alternative hypothesis. The null hypothesis foresees no dependencies between independent and dependent variables and therefore no relationships in the population of interest (e.g., the adaptivity does not cause any effect on user performance). On the contrary, the alternative hypothesis states a dependency between independent and dependent variables: the manipulation of the independent variable(s) causes effects on the dependent variable(s) (e.g., the adaptivity causes some effects on user performance).

Identify the experimental variables. The hypothesis can be verified by manipulating and measuring variables in a controlled situation. In a controlled experiment two kinds of variables can be identified: independent variable(s) (e.g., the presence of adaptive behavior in a web site) and dependent variable(s) (e.g., the task completion time, the number of errors, proportion/qualities of tasks achieved, interaction patterns, learning time/rate, user satisfaction, number of clicks, back button usage, home page visit, cognitive load measured through blood pressure, pupil dilatation, eye-tracking, number of fixations and fixation times, etc). See [75] for an example of how these variables are measured and analyzed during an evaluation of an adaptive web-based system; [13], [34] for eye-tracking in user modeling systems, and [67] for an experimental methodology to evaluate cognitive load in adaptive information filtering.

It is important to notice that it could also be interesting to analyze the correlation between variables that are characteristics naturally occurring in the subject. Statistical correlation (for more details see [79]) tells whether there is a relationship between two variables. In this kind of experiments, namely *correlational studies*, both variables are measured because there are no true independent variables. For example [66] an empirical study of adaptive help system for web-based applications correlated the ACT-value of procedural knowledge with subjective and objective measures of performance. For other examples of correlational studies see [95].

Select the Subjects. The goal of sampling is to collect data from a representative sample drawn from a larger population to make inferences about that population. A common problem of most evaluations in adaptive systems is that often the sample is too narrow to produce significant results. Rules of thumb for the sampling strategies are: i) the number of subjects has to be representative of the target population, ii) they should fit the statistics applied in data analysis, iii) they should fit subjects and resources availability.

Select the Experimental Methods and Conduct the Experiment. The selection of an experimental method consists primarily of collecting the data using a particular experimental design. The simplest design for an experiment is the **single factor design** in which one independent variable is manipulated (e.g., is the adaptive version more successful or the one without adaptations?). When two or more independent variables are manipulated the design is called **factorial design** (e.g., testing the adaptivity and the scrutability of an adaptive web site). Then, subjects are assigned to different treatment conditions. In the simplest procedure, the **between-subjects design**, an experimental group of subjects is assigned to the treatment (e.g., adaptivity), while another group of subjects, the control group, is assigned to the condition consisting of absence of a specific experimental treatment. For example in [91], six users conducted dialogs with the adaptive version of system, and six other users conducted dialogs with the non-adaptive one; while Farzan & Brusilovsky [46] have evaluated a course recommendation system by preparing two different version of the system: one with social navigation support (experimental group) and the other one without (control group).

There may be more than two groups, depending on the number of independent variables and the number of levels each variable can assume.

At the other extreme is the **within-subjects design** in which each subject is assigned to all treatment conditions (e.g., subjects completing tasks using both the application with adaptations and the one without). For instance, in the evaluation of a learning system that adapts the interface to the user's cognitive style, the same subjects used the system under three different treatment conditions [147]. Kumar [86] proposed a within-subject approach categorizing student-concepts as control and test groups instead of the student themselves. In between are designs in which the subjects are serving in some but not all the treatment conditions (**partial, or mixed, within-subjects factorial design**). For example, in [50] the subjects were split into two groups and every group completed the tasks with and without system adaptations (the tasks completed without adaptations by one group were completed with adaptations by the other one, and vice versa).

In an ideal experiment only the independent variable should vary from condition to condition. In reality, other factors are found to vary along with the treatment differences. These unwanted factors are called **confounding variables** (or nuisance variables) and they usually pose serious problems if they influence the behavior under study since it becomes hard to distinguish between the effects of the manipulated variable and the effects due to confounding variables. As indicated by [32], one way to control the potential source of confounding variables is holding them constant, so that they have the same influence on each of the treatment conditions (for instance, the testing environment, the location of the experiment, the instructions given to the participants may be controlled by holding them physically constant). Unfortunately, not all the potential variables can be handled in this way (for instance, reading speed, intelligence, etc). For these nuisance variables, their effect can be neutralized by **randomly assigning** subjects to the different treatment conditions.

Data Analysis and Conclusion. In controlled experiments, data are usually analyzed by means of descriptive and inferential statistics. **Descriptive statistics**, such as mean, variance, standard deviation, are designed to describe or summarize a set of data. In order to report significant results and make inference about the population of interest, the descriptive statistics are not sufficient, but some inferential statistic measure is required. Indeed, **inferential statistics** are used to evaluate the statistical hypotheses. These statistics are designed to make inferences about larger populations. The choice of the right statistics to be used depends on the kind of collected data and the questions to be answered.

Parametric statistics are exploited when data are normally distributed. Example of parametric tests are: ANOVA (ANalysis Of VAriance) calculated by means of F-test or t-test, and linear (or non-linear) regression factor analysis. For instances of the use of F test in adaptive systems see [20, 108], while for examples of t-test see [46, 94, 131].

The *non-parametric statistics* make no assumptions about the distribution of the scores making up a treatment condition. Examples of non-parametric tests are Wilcoxon rank-sum test, rank-sum version of ANOVA, Spearman's rank correlation, Mann-Whitney Test. For examples about the use of non-parametric measures in adaptive systems see [25, 42, 72].

While the above statistics can be applied when the dependent variables to measure are continuous (they can take values as, for instance, time or number of errors, etc), the *Chi square test* (χ^2) instead is the common measure used to evaluate the significant values assumed by categorical data. For example of use of Chi square tests in adaptive systems see [25, 87, 121].

Sensitivity measures should also be calculated. In this context, sensitivity refers to the ability to detect any effects that may exist in the treatments population. The sensitivity of an experiment is given by the effect size and the power. *The effect size or treatment magnitude* (ω^2) measures the strength, or the magnitude, of the treatment effects in the experiment. The *power* of an experiment is the ability to recognize treatment effects. The power can be used for estimating the sample size. Designing the experiments to have a high power rating not only ensures greater repeatability of results, but it makes it more likely to find the desired effects. For an example of sensitivity measures applied to analyze the evaluation results of an adaptive web site see [52], while for details on the importance of sensitivity measures in adaptive and user modeling systems see [32].

Ethnography. Sustainers of qualitative approaches affirm that laboratory conditions are not real world conditions and that only observing users in natural settings can detect the real behavior of the users. From this perspective, a subject cannot be reduced to a sum of variables and therefore a deeper knowledge of a fewer group of subjects is more useful than an empirical experiment with a representative sample. Qualitative methods of research often make use of ethnographic investigations, also known as participant-observation[12].

Preece et al. [125] classify the ethnographic investigations under the umbrella term "interpretative evaluation". The interpretative evaluation can be best summed up as "spending time with users" and it is based on the assumption that small factors that go behind the visible behavior greatly influence outcomes. According to [125], the interpretative evaluation comes in these flavors:

- contextual inquiry (see Sec. 24.3.1);
- cooperative evaluation (see Sec. 24.3.2);
- participative evaluation (see Sec. 24.3.2);
- ethnography.

While the first three techniques have been already described, since they should be used in former evaluation phases, ethnography can be better performed in the final evaluation phase.

Ethnography is a qualitative observational technique that is well established in the field of sociology and anthropology. It involves immersing the researcher in the everyday activities of an organization or in the society for a prolonged period of time. Ethnography provides the kind of information that is impossible to gather from the laboratory, since it is concerned with collecting data about real work circumstances. The ethnographic approach in HCI acknowledges the importance of learning more about the way technology is used in real situations [107].

Qualitative methods are seldom applied in the evaluation of adaptive web-based systems. However, statistical analyses are sometimes false, misleading, and too narrow, while insights and qualitative studies do not suffer from these problems as they strictly rely on the users' observed behavior and reactions [111]. Qualitative methods, such as ethnography, could bring fruitful results, especially in order to discover new phenomena (e.g., by observing the users interacting with a web site in their context, new solutions on how to adapt the site can emerge). In fact, qualitative researchers want to comprehend the subjects under study by interpreting their points of view and by analyzing the facts in depth (intensive approach) in order to propose new general understanding of the reality.

Importance for the adaptive web: collection of data in real situations; exploratory studies; discovering new phenomena.

The Grounded Theory. The Grounded Theory is "a theory derived from data, systematically gathered and analyzed through the research process. In this method, data collection, analysis and eventual theory stand in close relationship to one another. The researcher does not begin a project with a preconceived theory in mind (...). Rather, the researcher begins with an area of study and allows the theory to emerge from the data" [142]. The collected data may be qualitative, quantitative, or a combination of both types, since an interplay between qualitative and quantitative methods is advocated.

---

[12] In social sciences, and in particular in field-study research, participant-observation is a qualitative method of research that requires direct involvement of the researcher with the object of the study. For more details see [140].

Table 4. Final evaluation phase

	Kind of Factors	Applicability conditions	Pros and cons
Final evaluation phase			
Usability test	Usability of interface adaptation	A running user-adaptive prototype/system	+ objective performance measures and subjective user feedback - adaptive systems require also evaluation of the content layer: usability test is necessary, but not sufficient
Automatic usability testing and web usage mining	Automatic detection of usability of interface adaptation; discovering of sequences of user actions and behavioral patterns	A running user-adaptive system; software for the analysis of log data	+ discovering of real usage of the system; unobtrusive - no information about what the users are thinking/why they do something
Accessibility evaluation	Information on accessibility	A running user-adaptive prototype/system	+ accessibility has always to be tested
Experimental evaluation	Interface (and content) adaptations	A running user-adaptive prototype/system	+ when properly designed gives significant results - possibly artificial lab situation
Ethnography	Collection of data in real work situation	A running user-adaptive system	+ users interact with the system in a real situation - time consuming and expensive
Grounded theory	To combine qualitative and quantitative evaluation, to discover new theories	A running user-adaptive prototype/system	+ comprehensive and explorative - results may be subjective; time consuming

See Cena, Gena & Modeo [30] for an application of the Grounded Theory methodology with heterogeneous sources of data (both qualitative and quantitative) in an empirical evaluation aimed at choosing a better way to communicate recommendations to the users in the interface for mobile devices. For the development of a cooperative student model in a multimedia application, Grounded Theory has been applied to understand the many and complex interactions between learners, tutors and learning environment by integrating the range of qualitative and quantitative results collected during the several experimental sessions [8].

Importance for the adaptive web: combined analysis of qualitative and quantitative data; exploratory studies; discovering of new phenomena that can inspire adaptation.

24.4 Key Issues in the Evaluation of Adaptive Systems

Choosing appropriate methods for the evaluation of adaptive web-based systems is crucial. However, when conducting an evaluation study on an adaptive system, a number of issues might arise that are specific for this kind of system. The purpose of this section is to review these issues in order to raise the awareness for the potential problems and to sketch possible counter measures where available. A more in depth discussion of these issues can be found in [154].

24.4.1 Allocation of Resources

Resources required for evaluation studies are frequently underestimated. Set-up, data collection and analysis require a high amount of personnel, organizational and sometimes even financial resources [96]. In some cases, small-scale experiments (i.e., assessing every participant for a short time) are not feasible, when adaptation does not happen on the spot, but takes time. The system needs to gather some information about the user before it actually adapts.

However, there are several ways to either reduce the required resources or to assure the allocation of resources in advance. First of all, as described throughout this chapter, it might be useful to spread the evaluation across the whole development cycle. The summative evaluation would then be only a final validation of previous findings under real world conditions. Experience with empirical research has shown that it is a good idea to plan several small experiments or studies rather than a single large one, because this strategy provides more flexibility and limits the risk of flawed experimental designs. Nevertheless, a project proposal should not underestimate the required resources.

Second, several aspects of the evaluation may also be covered by expert assessment rather than user studies. Several of the methods described in this chapter, for instance, cognitive walkthrough (Section 24.3.1) and heuristic evaluation (Section 24.3.2) have been shown to be an effective and efficient way to detect many frequent usability problems with limited resources. There also exist heuristics for the evaluation of adaptivity [93]. However, it should be pointed out that expert evaluations run the risk of being biased if they are conducted by researchers who evaluate their own system.

Third, simulated users might be considered for testing the inference mechanism [71]. If the system is able to distinguish between groups among these simulated users it can at least be assumed to work in the expected way. However, to improve the ecological validity of this kind of study the users should be based on real empirical data.

In the area of information retrieval testing the adaptive system in terms of accuracy, precision and recall with open data sets is a common research method (e.g., [135]). Obviously, simulated users require less resources than real user studies, because the data can be reused in further improvement cycles and even in the evaluation of other systems. Moreover, the simulation strategy can guarantee that all possible combinations of user characteristics are covered. Therefore, simulated users can be seen as a variant of test cases. However, there are also limitations: simulated users can be used to test the inferences of an adaptive system, but both the user assessment and the effect of the adaptation on the user are excluded. However, if the sample is not based on empirical data, it might deviate from real users in essential aspects. For instance it might contain characteristics or combinations of characteristics that are impossible or that do not exist in the user group.

Finally, cognitive models have been proposed for the evaluation of adaptive systems [96]. A cognitive model is basically a computer program that implements process-oriented specifications of some of the main modules and mechanisms underlying human cognition and social activity [127]. Such a model may interact with an adaptive system and demonstrate important characteristics, e.g., cognitive effort or completion time. The main advantage of this approach is that it facilitates prediction of cognitive processes with variants of the target system without unwanted side effects such as learning, fatigue or reaction. However, adapting a cognitive model to a specific task and environment often requires a lot of effort and expertise even if it is based on an existing cognitive architecture (i.e., a framework for implementing cognitive models).

The last two types of studies (using simulated users and cognitive models) can be categorized as in silico experiments [146], a term that has been coined in biology in order to describe experimental settings that are executed in a virtual environment based on computer models (e.g., [161]). Though there are several threats to the validity of in silico experiments, they are a powerful and cost-effective strategy if used in combination with in vivo (real life) and in vitro (laboratory) experiments.

24.4.2 Specification of Control Conditions

Another problem, that is inherent in the evaluation of adaptive systems, occurs when the control conditions of experimental settings are defined. In many studies the adaptive system is compared to a non-adaptive version of the system with the adaptation mechanism switched off [24]. However, adaptation is often an essential feature of these systems and switching the adaptivity off might result in an absurd or useless system [64, 65]. In some systems, in particular if they are based on machine learning algorithms [82, 122, 123], it might even be impossible to switch off the adaptivity.

A preferred strategy might be to compare a set of different adaptation decisions (as far as applicable). Based on the same inferred user characteristics the system can be adapted in different ways. For instance, an adaptive learning system that adapts to the current knowledge of the learner might use a variety of adaptation strategies, including link annotation, link hiding, or curriculum sequencing. Comparing these variants in terms of relevant criteria sketches a much more complete picture of the adaptation impact than just comparing the standard system with a non-adaptive version. The variants might also include combinations of existing adaptation decisions. However, the variants should be as similar as possible in terms of functionality and layout (often referred to as *ceteris paribus*, all things being equal) in order to be able to trace back the effects to the adaptivity itself. Also matching the context of an experimental setting with real environments seems to be crucial in order to achieve sufficient external validity. Using the example of a recommender system evaluation, Missier & Ricci [39] suggested that it will be necessary to reproduce the real decision environment, i.e., the real system should be tested, with no changes in databases, interface, algorithms, and parameters. Even if this might be a difficult task for some types of adaptation decisions that have an impact on the interaction structure, the interpretability of the results relies a great deal upon these aspects.

24.4.3 Sampling

A proper experimental design requires not only to specify control conditions but also to select adequate samples. On the one hand the sample should be very heterogeneous in order to maximize the effects of the system's adaptivity: the more differences between users, the higher the chances that the system is able to detect these differences and react accordingly. On the other hand, from a statistical point of view, the sample should be very homogeneous in order to minimize the secondary variance and to emphasize the variance of the treatment. It has been reported frequently that too a high variance is a cause of the lack of significance in evaluation studies [21, 96, 104]. For instance, learners in online courses usually differ widely in reading times which might corrupt further comparisons in terms of time savings due to adaptive features. A common strategy to reduce this secondary (undesired) variance is to homogenize or parallelize the sample as much as possible. However, this strategy might be in conflict with the first requirement of sample heterogeneity. The ideal sample would differ widely in terms of the assessed user characteristics but would be homogeneous in terms of all other factors.

A second common strategy to reduce undesired variance is using repeated measurement. The main advantages of this kind of experimental design include: less participants are required, and statistical analysis is based on differences between treatments rather than between groups that are assigned to different treatments. However, this strategy is often not adequate for the evaluation of adaptive systems, because of order effects. If people get used to the first version of the system they might have problems to interact with the second version, because they have built up expectations (a mental model) about the system that are inadequate for the second version. Balancing the order of treatments might alleviate this problem, but the danger of biased results due to unexpected and undesired interactions between the treatments will remain. A third strategy is to control for variables that might have an impact on the results and to include these variables in the analysis. This strategy, sometimes referred to as dicing, might help to explain results that are diluted by the mean values. E.g., the adaptation decision might be correct for one subgroup, but it has a negative effect for the other subgroup. While the mean value would indicate that there is no effect at all, the detailed analysis demonstrates the strengths and weaknesses of the system. Moreover, there are obviously other criteria that have to be considered when selecting the sample in general. In order to generalize the results the sample should either be representative for the target group or at least not differ from the target group in terms of factors that are known to affect the results (e.g., expertise or motivation). Therefore, samples for evaluation studies with adaptive systems need to be selected carefully.

24.4.4 Definition of Criteria

Current evaluation studies use a broad range of different criteria [153]. The diversity of these criteria inhibits a comparison of different modeling approaches.

The criteria usually taken in consideration for evaluation (e.g., task completion time, number of errors, number of viewed pages) sometimes do not fit the aims of the system. For instance, during an evaluation of a recommender system the relevance of the information provided is more important than the time spent to find it. Another good example is reported by a preliminary study on evaluation of an in-vehicle adaptive system [89]. The results showed that adaptivity is beneficial for routine tasks, while performance of infrequent tasks is impaired. Furthermore, lots of applications are designed for long-time interaction and therefore it is hard to correctly evaluate them in a short and controlled test.

A precise specification of the modeling goals is required in the first place, as this is a prerequisite for the definition of the criteria. The criteria might be derived from the abstract system goals for instance by using the Goal-Question-Metric method (GQM) [148], which systematically defines metrics for a set of quality dimensions in products, processes, and resources. Tobar [144] presented a framework that supports the selection of criteria by separating design perspectives.

Many adaptive web-based systems are concerned with some kind of navigation support. Adaptivity might reduce the complexity of the navigation behavior [75, 155]. Accordingly, accepted graph complexity measures might be used for analyzing the users' behavior. However, as argued by Herder [62], the browsing activity is expected to produce more complex navigation than goal-directed interaction. Therefore, the metrics for the evaluation of user navigation should take into account both the site structure and the kind of user's tasks since, depending on these factors, a reduction in the complexity of the interaction is not necessarily caused by the adaptive behavior of the system. However, as claimed by Krug [83], "It doesn't matter how many times I have to click, as long as each click is a mindless, unambigous choice". Therefore, if the web site proposes some kind of adaptation, the adaptive solutions could help the user to disambiguate her choices, reducing the feeling of "being lost in the hyperspace".

Future research should aim at establishing a set of commonly accepted criteria and assessment methods that can be used independently of the actual user model and inference mechanism in order to explore the strength and weaknesses of the different modeling approaches across populations, domains, and context factors. While current evaluation studies usually yield a single data point in the problem space, common criteria would allow integration of the results of different studies for a broader picture.

So-called utility-based evaluation [62] shows how such a comparison across systems could be achieved.

24.4.5 Violation of Accepted Usability Principles

While we argue that the evaluation of adaptive systems must not be seen as being a mere usability testing problem, usability is certainly an important issue. However, several discussions have arisen about the usability of adaptive interfaces [65]. As already sketched in Section 24.3.3 Jameson [73] proposes five usability challenges for adaptive interfaces. These challenges complicate matters for the evaluation of adaptive systems even more, because usability goals and adaptivity goals need to be considered concurrently. For instance, lack of transparency and control can become a threat to the usability of an adaptive system [72, 38]. However, under certain conditions it is possible to match usability and adaptivity goals [74].

24.4.6 Asking for Adaptivity Effects

In many studies the users estimate the effect of adaptivity (e.g., [12]) or rate their satisfaction with the system (e.g., [7, 45, 48] after a certain amount of interaction. However, from a psychological point of view these assessment methods might be inadequate in some situations. Users might have no anchor of what good or bad interaction means for the given task if they do not have any experience with the 'usual' non-adaptive way. They might not even have noticed the adaptivity at all, because adaptive action often flows (or should flow) in the subjective expected way rather than in the static predefined way (i.e., rather than prescribing a certain order of tasks or steps, an adaptive system should do what the user wants to do). Therefore, users might notice and report only those events when the system failed to meet their expectations.

On the other hand, qualitative user feedback can be of high value, in particular in early stages of the development. Therefore, surveys and interviews should definitely be considered when planning the assessment, but in order to avoid interpretation problems they should be accompanied by objective measures such as performance, and number of navigation steps. It is highly recommended to at least informally debrief and converse with participants if possible after the trial both from a ethical point of view in order to detect problems such as design.

24.4.7 Separation of Concerns: Layered Evaluation

Comparing an adaptive version with the non-adaptive version in terms of their effectiveness and efficiency might not be a fair test (see Section 24.4.2). Moreover, this design does not provide insights into why the system is better or not.

When designing evaluation studies, it is fundamental to distinguish the different adaptation constituents, and sometimes it might be necessary to evaluate them separately from the beginning. So-called *layered approaches* [22, 76] have been proposed in the literature to separately evaluate the identified adaptation components (layers) of adaptive systems. The cited approaches identify, at least, two layers: the content layer, and the interface layer. This idea comes from Totterdell and Boyle [145], who first phrased the principle of layered evaluation, *"Two types of assessment were made of the user model: an assessment of the accuracy of the model's inferences about user difficulties; and an assessment of the effectiveness of the changes made at the interface"*. More recent approaches [116, 152, 153] identified several adaptation components and therefore more corresponding evaluation layers, and [115] also proposed specific evaluation techniques to be adopted in every layer. We can see that layered evaluation is one of the peculiarities that characterize the evaluation of adaptive systems, as well as the presence of several *typical users* of the system, to which the system adapts itself. Therefore, groups of significant users should be separately observed across the layers, and the evaluation could underline that adaptive solutions are useful for some users and for others they are not.

24.4.8 Reporting the Results

Even a perfect experimental design will be worthless if the results are not reported in a proper way. In particular statistical data require special care, as the findings might not be interpretable for other researchers if relevant information is skipped. This problem obviously occurs in other disciplines and research areas dealing with empirical findings. Therefore, there are many guidelines and standard procedures for reporting empirical data as suggested or even required by some journals (e.g., [3, 14, 88, 160]. In the special case of adaptive systems, several other things should be reported. First, the inference mechanism should be described in detail, or the reader should at least be referred to a detailed description. Second, the user model should be described in terms of the dimensions or characteristics that are modeled. If applicable the report should contain the theoretically possible values or states of the model as well as the empirically identified states. This is important to characterize both the sample (cf. Section 24.4.3) and the potential impact of the treatment. For instance, if the adaptivity is responsive to user characteristics that occur only once in a while, the impact on the total interaction will be limited. Third, besides statistical standard identifiers (i.e., sample size, means, significance level, confidence interval) the effect size [33] of the treatment is of interest, because it estimates the adaptivity effect in comparison to the total variance and is therefore an indicator of the utility. It enables practitioners to estimate the expected impact of a new technique or approach and facilitates meta-analyses.

24.5 Conclusions

This chapter has presented a review of methods and techniques for design and evaluation of adaptive web-based systems under a usability engineering perspective. Even though improvement has been registered in a number of evaluation studies in the recent years [51], the evaluation of adaptive web systems needs to reach a more rigorous level in terms of subject sampling, statistical analysis, correctness in procedures, experiment settings, etc. Evaluation studies should benefit from the application of qualitative methods of research and from a rigorous and complete application of user-centered design approach in every development phase of these systems.

To conclude, we advocate the importance of evaluation in every design phase of an adaptive web-based system and at different layers of analysis. Significant testing results can lead to more appropriate and successful systems and the user's point of view can be a very inspiring source of information for adaptation strategies. From our point of view, both quantitative and qualitative methodologies of research can offer fruitful contributions and their correct application has to be carried out by the researchers working in this area in every design phase. Finally, since evaluation in adaptive systems is still in a exploratory phase, new approaches are strongly called for and these can include combining together different techniques, exploring new metrics to assess adaptivity, and adapting the evaluation technique to the adaptive systems features.

References

1. Alepis E., Virvou M., 2006. User Modelling: An Empirical Study for Affect Perception Through Keyboard and Speech in a Bi-modal User Interface. In Proceedings of AH2006, LNCS 4018, pp. 338-341.
2. Alfonseca E. and Rodriguezi P., 2003. Modelling Users' Interests and Needs for an Adaptive On-line Information System. In: P. Brusilovsky, A. Corbett and F. De Rosis (Eds.), Lecture Notes in Computer Science n. 2702: User Modeling 2003. Berlin: Springer, pp. 76-82.
3. Altman D., Gore S., Gardner M., and Pocock S., 1983. Statistical Guidelines for Contributors to Medical Journals. British Medical Journal, 286, 1489-1493.
4. Anderson, J. R., Boyle, C. F., and Yost, G., 1985. The Geometry Tutor. 9th International Joint Confenrence on AI, pp 1-7.
5. Ardissono L., Gena C., Torasso P., Bellifemine F., Chiarotto A., Difino A., Negro B., 2004. User Modeling and Recommendation Techniques for Personalized Electronic Program Guides. In L. Ardissono, A. Kobsa and M. Maybury (Eds.), Personalization and user-adaptive interaction in digital tv, Kluwer Academic Publishers, pp. 30-26.
6. Bakeman R. and Gottman J. M., 1986. Observing Behavior: An Introduction to Sequential Analysis. Cambridge: Cambridge University.
7. Bares W. H., and Lester J. C., 1997. Cinematographic User Models for Automated Re altime Camera Control in Dynamic 3D Environments. In A. Jameson, C. Paris, and C. Tasso (Eds.), User modeling: Proceedings of the Sixth International Conference, UM97. Vienna, New York: Springer, pp. 215-226
8. Barker T., Jones S., Britton C., Messer D., 2002. The Use of a Co-operative Student Model of Learner Characteristics to Configure a Multimedia Application. User Modeling and User-adaptive Interaction 12(2), pp. 207-241.
9. Baudisch P., Brueckner L., 2002. TV Scout: Lowering the Entry Barrier to Personalized TV Program Recommendation. In: P. De Bra, P. Brusilovsky and R. Conejo (Eds.), Lecture Notes in Computer Science n. 2347: Adaptive Hypermedia and Adaptive Web-Based Systems. Berlin: Springer, pp. 58-68.
10. Beck J. E., Jia P., Sison J., and Mostow J., 2003. Predicting Student Help-request Behavior in an Intelligent Tutor for Reading. In: P. Brusilovsky, A. Corbett and F. De Rosis (Eds.), Lecture Notes in Computer Science n. 2702: User Modeling 2003. Berlin: Springer, pp. 303-312.
11. Beck J. E., Jia P., Sison J., and Mostow J., 2003. Assessing Student Proficiency in a Reading Tutor That Listens. In: P. Brusilovsky, A. Corbett and F. De Rosis (Eds.), Lecture Notes in Computer Science n. 2702: User Modeling 2003. Berlin, etc.: Springer, pp. 323-327.

12. Beck J. E., Stern M., and Woolf B. P., 1997. Using the Student Model to Control Problem Difficulty. In: A. Jameson, C. Paris, and C. Tasso (Eds.), User modeling: Proceedings of the Sixth International Conference, UM97. Vienna, New York: Springer, pp. 277-288.
13. Bednarik R., 2005. Potentials of Eye-Movement Tracking in Adaptive Systems. In: Weibelzahl, S. Paramythis, A. and Masthoff Judith (eds.). Proceedings of the Fourth Workshop on Empirical Evaluation of Adaptive Systems, held at the 10th International Conference on User Modeling UM2005, Edinburgh, pp. 1-8.
14. Begg C., Cho M., Eastwood S., Horton R., Moher D., Olkin I., Pitkin R., Rennie D., Schultz K., Simel D., and Stroup D., 1996. Improving the Quality of Reporting Randomized Trials (the CONSORT Statement). Journal of the American Medical Association, 276(8), 637-639.
15. Bellotti V. and MacLean A.   Design Space Analysis (DSA).   http://www.mrc-cbu.cam.ac.uk/amodeus/summaries/DSAsummary.html
16. Bental D., Cawsey A., Pearson J. and Jones R., 2003. Does Adapted Information Help Patients with Cancer? In: P. Brusilovsky, A. Corbett and F. De Rosis (Eds.), Lecture Notes in Computer Science n. 2702: User Modeling 2003. Berlin: Springer, pp. 288-291.
17. Benyon D., 1993. Adaptive Systems: A Solution to Usability Problems. User Modeling and User-adaptive Interaction (3), pp. 65-87.
18. Beyer H. and Holtzblatt K., 1998. Contextual Design: Defining Customer-Centered Systems, Morgan Kaufmann Publishers, Inc., San Francisco CA.
19. Billsus, D. and Pazzani, M., 1998. Learning Collaborative Information Filters. In: Proceedings of the International Conference on Machine Learning. Morgan Kaufmann Publishers. Madison, Wisc.
20. Brunstein, A., Jacqueline, W., Naumann, A., and Krems, J.F., 2002. Learning Grammar with Adaptive Hypertexts: Reading or Searching? In: P. De Bra, P. Brusilovsky and R. Conejo (Eds.), Lecture Notes in Computer Science N. 2347: Adaptive Hypermedia and Adaptive Web-Based Systems. Berlin: Springer.
21. Brusilovsky P., and Eklund J., 1998. A study of user-model based link annotation in educational hypermedia. Journal of Universal Computer Science, Special Issue on Assessment Issues for Educational Software, 4(4), 429-448.
22. Brusilovsky P., Karagiannidis C., and Sampson D., 2001. The benefits of layered evaluation of adaptive applications and services. In: S. Weibelzahl, D. N. Chin, & G., Weber (Eds.), Empirical Evaluation of Adaptive Systems. Proceedings of Workshop At the Eighth International Conference on User Modeling, UM2001, Pp. 1-8.
23. Brusilovsky P., and Millán, E. 2006. User Models for Adaptive Hypermedia and Adaptive Educational Systems. In Brusilovsky, P., Kobsa, A., Nejdl, W. (eds.): The Adaptive Web: Methods and Strategies of Web Personalization, Lecture Notes in Computer Science, Vol. 4321. Springer-Verlag, Berlin Heidelberg New York (2007) This Volume.
24. Brusilovsky P., and Pesin L., 1998. Adaptive navigation support in educational hypermedia: An evaluation of the ISIS-tutor. Journal of Computing and Information Technology, 6(1), 27-38.
25. Brusilovsky P., Sosnovsky S., Yudelson M., 2006. Addictive Links: The Motivational Value of Adaptive Link Annotation in Educational Hypermedia In Proceedings of AH2006, LNCS 4018, Pp. 51-60.
26. Bull S., 2003. User Modelling and Mobile Learning. In: P. Brusilovsky, A. Corbett and F. De Rosis (Eds.), Lecture Notes in Computer Science N. 2702: User Modeling 2003. Berlin: Springer, Pp. 383-387.
27. Bunt A., 2005. User Modeling to support user customization. In: Proceedings of UM 2005, LNAI 3538, Pp. 499-501.
28. Burke R., 2002. Hybrid Recommender Systems: Survey and Experiments. User Modeling and User-Adapted Interaction. 12(4), Pages 331-370.

29. Burke R., 2006. Hybrid Web Recommender Systems. In Brusilovsky, P., Kobsa, A., Nejdl, W. (eds.): The Adaptive Web: Methods and Strategies of Web Personalization, Lecture Notes in Computer Science, Vol. 4321. Springer-Verlag, Berlin Heidelberg New York (2007) This Volume.

30. Cena F., Gena C., and Modeo S., 2005. How to communicate recommendations? Evaluation of an adaptive annotation technique. In the Proceedings of the Tenth IFIP TC13 International Conference on Human-Computer Interaction (INTERACT 2005), Pp. 1030-1033.

31. Chapanis, A., 1991. Evaluating usability. In: B. Shackel and S. J.Richardson (Eds.), Human Factors for Informatics Usability, Cambridge: Cambridge University, Pp. 359-395.

32. Chin D.N., 2001. Empirical evaluation of user models and user-adapted systems. *User Modeling and User-Adapted Interaction*, 11(1-2), Pp. 181-194.

33. Cohen J., 1977. Statistical power analysis for the behavioral sciences (revised ed.). New York: Academic Press.

34. Conati C., Merten C. and Muldner K., 2005. Exploring Eye Tracking to Increase Bandwidth in User Modeling In: Proceedings of UM 2005, LNAI 3538, pp. 357-366.

35. Console L., Gena C. and Torre I., 2003. Evaluation of an On-vehicle Adaptive Tourist Service. In: Weibelzahl, S. and Paramythis, A. (eds.). Proceedings of the Second Workshop on Empirical Evaluation of Adaptive Systems, held at the 9th International Conference on User Modeling UM2003, Pittsburgh, pp. 51-60.

36. Courage C., and Baxter K., 2005. Understanding Your Users: A Practical Guide to User Requirements Methods, Tools, and Techniques. Morgan Kaufmann Publishers, San Francisco, CA.

37. Cugini J. Scholtz J., 1999. VISVIP: 3D Visualization of Paths Through Web Sites. In: Proceedings of the International Workshop on Web-Based Information Visualization, pp. 259-263.

38. Czarkowski M., 2005. Evaluating Scrutable Adaptive Hypertext. In: Weibelzahl, S. Paramythis, A. and Masthoff Judith (eds.). Proceedings of the Fourth Workshop on Empirical Evaluation of Adaptive Systems, held at the 10th International Conference on User Modeling UM2005, Edinburgh, pp. 37-46.

39. Del Missier F. and Ricci F., 2003. Understanding Recommender Systems: Experimental Evaluation Challenges. In: Weibelzahl, S. and Paramythis, A. (eds.). Proceedings of the Second Workshop on Empirical Evaluation of Adaptive Systems, held at the 9th International Conference, pp. 31-40.

40. Diaper D. (Ed.). Task Analysis for Human-computer Interaction. Chicester, U.K.: Ellis Horwood, 1989.

41. Dimitrova V., 2003. Using Dialogue Games to Maintain Diagnostic Interactions. In: P. Brusilovsky, A. Corbett and F. De Rosis (Eds.), Lecture Notes in Computer Science n. 2702: User Modeling 2003. Berlin, etc.: Springer, pp. 117-121.

42. Dimitrova V., Self J. and Brna P., 2001. Applying Interactive Open Learner Models to Learning Technical Terminology. In: M. Bauer, P.J. Gmytrasiewicz and J. Vassileva (Eds.), Lecture Notes in Computer Science n. 2109: User Modeling 2001. Berlin, etc.: Springer, pp. 148-157.

43. Dix A., Finlay J., Abowd G. and Beale R., 1998. Human Computer Interaction. Second Edition, Prentice Hall.

44. Dumas J. S. and Redish J. C., 1999. A Practical Guide To Usability Testing. Norwood, N.J. Ablex Publishing Corp.

45. Encarnação L. M., and Stoev S. L., 1999. Application-independent Intelligent User Support System Exploiting Action-sequence Based User Modeling. In: J. Kay (Ed.), User modeling: Proceedings of the Seventh International Conference, UM99. Vienna, New York: Springer, pp. 245-254

46. Farzan R., and Brusilovsky P, 2006 Social Navigation Support in a Course Recommendation System In Proceedings of AH2006, LNCS 4018, pp. 91-100.
47. Fink J., Kobsa A. and Nill A., 1998. Information Provision for All Users, Including Disabled and Elderly People. New Review of Hypermedia and Multimedia, 4, pp. 163-188.
48. Fischer G., and Ye Y., 2001. Personalizing Delivered Information in a Software Reuse Environment. In: M. Bauer, J. Vassileva, and P. Gmytrasiewicz (Eds.), User modeling: Proceedings of the Eighth International Conference, UM2001. Berlin: Springer, pp. 178-187
49. Gena C., 2001. Designing TV Viewer Stereotypes for an Electronic Program Guide. In: M. Bauer, P. J. Gmytrasiewicz and J. Vassileva (Eds.), Lecture Notes in Computer Science n. 2109: User Modeling 2001. Berlin, etc.: Springer, pp. 247-276.
50. Gena C., 2003. Evaluation Methodologies and User Involvement in User Modeling and Adaptive Systems. Unpub. Phd thesis, Università degli Studi di Torino, Italy.
51. Gena C., 2005. Methods and Techniques for the Evaluation of User-adaptive Systems. The Knowledge Engineering Review, Vol 20:1,1-37, 2005.
52. Gena C. and Ardissono L., 2004. Intelligent Support to the Retrieval of Information About Hydric Resources. In: Proceedings of the International Conference on Adaptive Hypermedia and Adaptive Web-based Systems AH2004 , Eindhoven, The Netherlands, Lecture Notes in Computer Science. Pp. 126-135.
53. Gena C., Perna A. and Ravazzi M., 2001. E-tool: a Personalized Prototype for Web Based Applications. In: D. de Waard, K. Brookhuis, J. Moraal and A. Toffetti (Eds.), Human Factors in Transportation, Communication, Health, and the Workplace. Shaker Publishing, pp. 485-496.
54. Gena C. and Torre I., 2004. The Importance of Adaptivity to Provide On-Board Services. A Preliminary Evaluation of an Adaptive Tourist Information Service on Board Vehicles. Special Issue on Mobile A.I. in Applied Artificial Intelligence Journal.
55. Good N., Schafer J.B., Konstan J.A., Botchers A., Sarwar B.M., Herlocker J.L. and Ricdl J., 1999. Combining Collaborative Filtering with Personal Agents for Better Recommendations. In: Proceedings of the Sixteenth National Conference on Artificial Intelligence, pp. 439-446.
56. Gould J. D. and Lewis C., 1983. Designing for Usability – Key Principles and What Designers Think. In: Human Factors in Computing Systems, CHI '83 Proceedings, New York: ACM, pp. 50-53.
57. Goy A., Ardissono L. Petrone G., 2006. Personalization in E-Commerce Applications. In Brusilovsky, P., Kobsa, A., Nejdl, W. (eds.): The Adaptive Web: Methods and Strategies of Web Personalization, Lecture Notes in Computer Science, Vol. 4321. Springer-Verlag, Berlin Heidelberg New York (2007) this volume.
58. Greenbaum J. and Kyng M. (Eds.), 1991. Design At Work: Cooperative Design of Computer Systems. Lawrence Erlbaum Associates, Inc., Hillsdale, NJ.
59. Greenbaum T. L., 1998. The Handbook of Focus Group Research (2nd Edition). Lexington Books: New York, NY.
60. Habieb-Mammar H., Tarpin-Bernard F. and Prévôt P., 2003. Adaptive Presentation of Multimedia Interface Case Study: Brain Story Course. In: P. Brusilovsky, A. Corbett and F. De Rosis (Eds.), Lecture Notes in Computer Science n. 2702: User Modeling 2003. Berlin: Springer, pp. 15-24.
61. Hara Y., Tomomune Y., Shigemori M., 2004. Categorization of Japanese TV Viewers Based on Program Genres They Watch. In: L. Ardissono, A. Kobsa and M. Maybury (Eds.), Personalization and user-adaptive interaction in digital TV, Kluwer Academic Publishers.

62. Herder E., 2003. Utility-Based Evaluation of Adaptive Systems. In: Proceedings of the Second Workshop on Empirical Evaluation of Adaptive Systems, held at the 9th International Conference on User Modeling UM2003, Pittsburgh, pp. 25-30.
63. Herder E., Weinreich H., Obendorf H., Mayer M., 2006. Much to Know About History. In Proceedings of AH2006, LNCS 4018, pp. 283-287.
64. Höök K., 1997. Evaluating the Utility and Usability of an Adaptive Hypermedia System. In: Proceedings of 1997 International Conference on Intelligent User Interfaces, ACM, Orlando, Florida, 179-186.
65. Höök K., 2000. Steps to Take Before IUIs Become Real. Journal of Interacting with Computers, 12(4), 409-426.
66. Iglezakis D., 2005. Is the ACT-value a Valid Estimate for Knowledge? An Empirical Evaluation of the Inference Mechanism of an Adaptive Help System. In: Weibelzahl, S. Paramythis, A. and Masthoff J. (eds.). Proceedings of the Fourth Workshop on Empirical Evaluation of Adaptive Systems, held at the 10th International Conference on User Modeling UM2005, Edinburgh, pp. 19-26.
67. Ikehara S., Chin D.N. and Crosby M. E., 2003. A Model for Integrating an Adaptive Information Filter Utilizing Biosensor Data to Assess Cognitive Load. In: P. Brusilovsky, A. Corbett and F. De Rosis (Eds.), Lecture Notes in Computer Science n. 2702: User Modeling 2003. Berlin: Springer, pp. 208-212.
68. International Organisation for Standardisation ISO 9241-11:1998 Ergonomic Requirements for Office Work with Visual Display Terminals (VDTs), Part 11 Guidance on Usability, 1998. http://www.iso.org/
69. International Organisation for Standardisation ISO/TS 16071:2003 Ergonomics of Human-system Interaction - Guidance on Software Accessibility. Technical Specification, 2003. http://www.iso.org/
70. Jackson T., Mathews E., Lin D., Olney A. and Graesser A., 2003. Modeling Student Performance to Enhance the Pedagogy of AutoTutor. In: P. Brusilovsky, A. Corbett and F. De Rosis (Eds.), Lecture Notes in Computer Science n. 2702: User Modeling 2003. Berlin: Springer, pp. 368-372.
71. Jameson A., Schäfer R., Weis T., Berthold A., and Weyrath T., 1999. Making Systems Sensitive to the User's Changing Resource Limitations. Knowledge-Based Systems, 12(8), 413-425.
72. Jameson A., Schwarzkopf E., 2003, Pros and Cons of Controllability: An Empirical Study. In: P. De Bra, P. Brusilovsky and R. Conejo (Eds.), Proceedings of AH'2002, Adaptive Hypermedia and Adaptive Web-Based Systems, Springer, 193-202.
73. Jameson A., 2003. Adaptive Interfaces and Agents. The Human-Computer Interaction Handbook, Lawrence Erlbaum Associates, New Jersey, pp. 316-318.
74. Jameson A., 2005. User Modeling Meets Usability Goals. In: Proceedings of UM 2005, LNAI 3538, pp. 1-3.
75. Juvina I. and Herder E., 2005. The Impact of Link Suggestions on User Navigation and User Perception. In: Proceedings of UM 2005, LNAI 3538, pp. 483-492.
76. Karagiannidis C. and Sampson D., 2000. Layered Evaluation of Adaptive Applications and Services. In: P. Brusilovsky, O. Stock, C. Strapparava (Eds.), Adaptive Hypermedia and Adaptive Web-Based Systems, Lecture Notes in Computer Science Vol.1892, pp. 343-346.
77. Katz-Haas R., 1998. Ten Guidelines for User-Centered Web Design, Usability Interface, 5(1), pp. 12-13.
78. Kay J., 2001. Learner Control. User Modeling and User-Adapted Interaction, Tenth Anniversary Special Issue, 11(1-2), Kluwer, 111-127.
79. Keppel G., 1991. Design and Analysis: A Researcher's Handbook. Englewood Cliffs, NJ: Prentice-Hall.

80. Keppel G., Saufley W. H. and Tokunaga H., 1998. Introduction to Design and Analysis, A Student's Handbook. Second Edition, Englewood Cliffs, NJ: Prentice-Hall.
81. Kirby M.A.R., 1991. CUSTOM Manual Dpo/std/1.0. Huddersfield: HCI Research Centre, University of Huddersfield.
82. Krogsæter M., Oppermann R. and Thomas C. G., 1994. A User Interface Integrating Adaptability and Adaptivity. In: R. Oppermann (Ed.), Adaptive user support. Hillsdale: Lawrence Erlbaum, pp. 97-125
83. Krug S., 2000. Don't Make Me Think! A Common Sense Approach to Web Usability. Indianapolis, IN: New Riders Publishing.
84. Krüger A., Heckmann D., Kruppa M., Wasinger R., 2006. Web-based Mobile Guides. In Brusilovsky, P., Kobsa, A., Nejdl, W. (eds.): The Adaptive Web: Methods and Strategies of Web Personalization, Lecture Notes in Computer Science, Vol. 4321. Springer-Verlag, Berlin Heidelberg New York (2007) this volume.
85. Kuflik T., Shapira B., Elovici Y. and Maschiach A., 2003. Privacy Preservation Improvement By Learning Optimal Profile Generation Rate. In: P. Brusilovsky, A. Corbett and F. De Rosis (Eds.), Lecture Notes in Computer Science n. 2702: User Modeling 2003. Berlin: Springer, pp. 168-177.
86. Kumar A., 2006. Evaluating Adaptive Generation of Problems in Programming Tutors – Two Studies. In the Proc. of the Fifth Workshop on User-Centred Design and Evaluation of Adaptive Systems, 2006, pp. 450-459.
87. Kurhila J., Miettinen M., Nokelainen P., Tirri H., 2002. EDUCO - A Collaborative Learning Environment Based on Social Navigation. In: P. De Bra, P. Brusilovsky and R. Conejo (Eds.), Lecture Notes in Computer Science n. 2347: Adaptive Hypermedia and Adaptive Web-Based Systems. Berlin: Springer, pp. 242-252.
88. Lang T. and Secic M., 1997. How to Report Statistics in Medicine: Annotated Guidelines for Authors, Editors and Reviewers. Philadelphia, PA: American College of Physicians.
89. Lavie T., J. Meyer, K. Bengler, and J. F. Coughlin, 2005. The Evaluation of In-Vehicle Adaptive Systems. In: Weibelzahl, S. Paramythis, A. and Masthoff J. (eds.). Proceedings of the Fourth Workshop on Empirical Evaluation of Adaptive Systems, held at the 10th International Conference on User Modeling UM2005, Edinburgh, pp. 9-18.
90. Lee J. and Lai K.-Y., 1991. What's in Design Rationale? Human-Computer Interaction special issue on design rationale, 6(3-4), pp. 251-280.
91. Litman D. J., and Pan, S., 2000. Predicting and Adapting to Poor Speech Recognition in a Spoken Dialogue System. In Proceedings of the Seventeenth National Conference on Artificial Intelligence (pp. 722-728). Austin, TX.
92. Magnini B. and Strapparava C., 2001. Improving User Modelling with Content-based Techniques. In: UM2001 User Modeling: Proceedings of 8th International Conference on User Modeling (UM2001), Sonthofen (Germany), July 2001. Springer Verlag.
93. Magoulas G. D., Chen S. Y. and Papanikolaou K. A., 2003. Integrating Layered and Heuristic Evaluation for Adaptive Learning Environments. In: Weibelzahl, S. and Paramythis, A. (eds.). Proceedings of the Second Workshop on Empirical Evaluation of Adaptive Systems, held at the 9th International Conference on User Modeling UM2003, Pittsburgh, pp. 5-14.
94. Martin B. and Mitrovic T., 2002. WETAS: A Web-Based Authoring System for Constraint-Based ITS . In: P. De Bra, P. Brusilovsky and R. Conejo (Eds.), Lecture Notes in Computer Science n. 2347: Adaptive Hypermedia and Adaptive Web-Based Systems. Berlin: Springer, pp. 396-305.
95. Martin B., Mitrovic T., 2006. The Effect of Adapting Feedback Generality in ITS E-Learning and Personalization In Proceedings of AH2006, LNCS 4018, pp. 192-202.
96. Masthoff, J. (2002). The Evaluation of Adaptive Systems. In N. V. Patel (Ed.), Adaptive evolutionary information systems. Idea Group publishing. pp329-347.

97. Masthoff, J. (2006). The User As Wizard. In the Proc. of the Fifth Workshop on User-Centred Design and Evaluation of Adaptive Systems, 2006, pp. 460-469.
98. Matsuo Y., 2003. Word Weighting Based on User's Browsing History. In: P. Brusilovsky, A. Corbett and F. De Rosis (Eds.), Lecture Notes in Computer Science n. 2702: User Modeling 2003. Berlin, etc.: Springer, pp. 35-44.
99. Maulsby, D., Greenberg, S. and Mander, R. (1993) Prototyping an Intelligent Agent Through Wizard of Oz. In ACM SIGCHI Conference on Human Factors in Computing Systems, Amsterdam, The Netherlands, May, p277-284, ACM Press.
100. Maybury M. and Brusilovsky P. (Eds.), 2002. The Adaptive Web, Volume 45. Communications of the ACM.
101. McCreath E. and Kay J., 2003. IEMS : Helping Users Manage Email. In: P. Brusilovsky, A. Corbett and F. De Rosis (Eds.), Lecture Notes in Computer Science n. 2702: User Modeling 2003. Berlin: Springer, pp. 263-272.
102. McNee S. M., Lam S. K, Konstan J. A. and Riedl J., 2003. Interfaces for Eliciting New User Preferences in Recommender Systems. In: P. Brusilovsky, A. Corbett and F. De Rosis (Eds.), Lecture Notes in Computer Science n. 2702: User Modeling 2003. Berlin: Springer, pp. 178-187
103. Mitrovic A., 2001. Investigating Students' Self-assessment Skills. In: M. Bauer, P. J. Gmytrasiewicz and J. Vassileva (Eds.), Lecture Notes in Computer Science n. 2109: User Modeling 2001. Berlin: Springer, pp. 247-250.
104. Mitrovic A. and Martin B. 2002. Evaluating the Effects of Open Student Models on Learning. In: P. DeBra, P. Brusilovsky and R. Conejo (eds), Proceedings of Adaptive Hypermedia and Adaptive Web-Based Systems, Berlin: Springer, pp. 296-305.
105. Mobasher R., 2006. Data Mining for Web Personalisation. In Brusilovsky, P., Kobsa, A., Nejdl, W. (eds.): The Adaptive Web: Methods and Strategies of Web Personalization, Lecture Notes in Computer Science, Vol. 4321. Springer-Verlag, Berlin Heidelberg New York (2007) this volume.
106. Mobasher R., Cooley R., and Srivastava J., 2000. Automatic Personalization Based on Web Usage Mining. Communications of the ACM, (43) 8, 2000.
107. Monk A., Wright P., Haber J. and Davenport L., 1993. Improving Your Human Computer Interface: A Practical Approach. BCS Practitioner Series, Prentice-Hall International, Hemel Hempstead.
108. Müller C., Großmann-Hutter B., Jameson A., Rummer R., Wittig F., 2001. Recognizing Time Pressure and Cognitive Load on the Basis of Speech: An Experimental Study. In: M. Bauer, P. J. Gmytrasiewicz and J. Vassileva (Eds.), Lecture Notes in Computer Science n. 2109: User Modeling 2001. Berlin, etc.: Springer, pp 24-33.
109. Nielsen J., 1993. Usability Engineering. Boston, MA, Academic Press.
110. Nielsen J., 2000. Why You Only Need to Test With 5 Users. In: Alertbox, http://www.useit.com/alertbox/20000319.html.
111. Nielsen J., 2004. Risks of Quantitative Studies. In: Alertbox, http://www.useit.com/alertbox/20040301.html.
112. Nielsen J. and Mack R. L. (Eds.), 1994. Usability Inspection Methods. New York, NY: John Wiley & Sons.
113. Nielsen J. and Molich R., 1990. Heuristic Evaluation of User Interfaces. In: Proceedings of CHI '90, Seattle, Washington, pp. 249-256.
114. Norman D.A. and Draper S.W., 1986. User Centered System Design: New Perspective on HCI. Hillsdale NJ, Lawrence Erlbaum.
115. Paramythis A., Totter A. and Stephanidis C., 2001. A Modular Approach to the Evaluation of Adaptive User Interfaces. In: S. Weibelzahl, D. Chin, and G. Weber (Eds.). Proceedings of the First Workshop on Empirical Evaluation of Adaptive Systems, Sonthofen, Germany, pp. 9-24.

116. Paramythis A., and Weibelzahl S., 2005. A Decomposition Model for the Layered Evaluation of Interactive Adaptive Systems. In: Proceedings of UM 2005, LNAI 3538, pp. 438-442.

117. Paganelli L. and Paterno' F., 2002. Intelligent Analysis of User Interactions with Web Applications. In: Proceedings of the 2002 International Conference on Intelligent User Interfaces. ACM Press.

118. Pazzani M. J. and Billsus D., 2006. Content-based Recommendation Systems. In Brusilovsky, P., Kobsa, A., Nejdl, W. (eds.): The Adaptive Web: Methods and Strategies of Web Personalization, Lecture Notes in Computer Science, Vol. 4321. Springer-Verlag, Berlin Heidelberg New York (2007) this volume.

119. Perkowitz M., and Etzioni O., 2000. Adaptive Web Sites. Communications of the ACM, 43(8), 2000, pp. 152-158.

120. Pierrakos D., Paliouras G., Papatheodorou, C., and Spyropoulos, C.D., 2003. Web Usage Mining As a Tool for Personalization: a Survey. International Journal of User Modeling and User-Adapted Interaction, 13(4), pp. 311-372.

121. Pirolli P. and Fu W. T., 2003. SNIF-ACT: A Model of Information Foraging on the World Wide Web. In: P. Brusilovsky, A. Corbett and F. De Rosis (Eds.), Lecture Notes in Computer Science n. 2702: User Modeling 2003. Berlin: Springer, pp. 45-54.

122. Pohl, W., 1997. LaboUr-machine Learning for User Modeling. In: M. J. Smith, G. Salvendy, and R. J. Koubek (Eds.). Design of computing systems: Social and ergonomic considerations. proceedings of the seventh international conference on human-computer interaction. Amsterdam: Elsevier, Vol. B, pp. 27-30

123. Pohl, W., 1998. User-adapted Interaction, User Modeling, and Machine Learning. In: U. J. Timm and M. R̈ossel (Eds.), Proceedings of the sixth german workshop on adaptivity and user modeling in interactive systems, ABIS98. Erlangen.

124. Polson P.G., Lewis C., Rieman J. and Wharton C., 1992. Cognitive Walkthroughs: A Method for Theory- Based Evaluation of User Interfaces. International Journal of Man-Machine Studies 36, 741-773.

125. Preece J., Rogers Y., Sharp H. and Benyon D., 1994. Human-computer Interaction. Addison-Wesley Pub.

126. Resnick P. and Varian H. R., 1997. Special Issue on Recommender Systems. Communications of the ACM, 40, 1997.

127. Ritter F. E., Shadbolt N., Elliman D., Young R., Gobet F. and Baxter G., 2002. Techniques for Modeling Human and Organizational Behaviour in Synthetic Environments: A Supplementary Review. Wright-Patterson Air Force Base, OH: Human Systems Information Analysis Center.

128. Rizzo P., Lee I.H., Shaw E., Johnson W.L., Wang N. and Mayer R.E., 2005. A Semi-Automated Wizard of Oz Interface for Modeling Tutorial Strategies In: Proceedings of UM 2005, LNAI 3538, pp. 174-178.

129. Rosenfeld L. and Morville P., 1998. Information Archtecture. O'Reilly.

130. Rubin J., 1994. Handbook of Usability Testing: How to Plan, Design, and Conduct Effective Tests. John Wiley & Sons; 1 edition.

131. Santos Jr. E., Nguyen H., Zhao Q. and Pukinskis E., 2003. Empirical Evaluation of Adaptive User Modeling in a Medical Information Retrieval Application. In: P. Brusilovsky, A. Corbett and F. De Rosis (Eds.), Lecture Notes in Computer Science n. 2702: User Modeling 2003. Berlin: Springer, pp. 292-296.

132. Sarwar B. M., Konstan J. A., Borchers A., Herlocker J., Miller B. and Riedl J., 1998. Using Filtering Agents to Improve Prediction Quality in the GroupLens Research Collaborative Filtering System. In: Proceeding of the ACM Conference on Computer Supported Cooperative Work (CSCW), Seattle, WA, pp. 439-446.

133. Scarano V., Barra M., Maglio P. and Negro A., 2002. GAS: Group Adaptive Systems. In: P. De Bra, P. Brusilovsky and R. Conejo (Eds.), Lecture Notes in Computer Science n. 2347: Adaptive Hypermedia and Adaptive Web-Based Systems. Berlin: Springer, pp. 47-57.

134. Sears A. and Shneiderman B., 1994. Split Menus: Effectively Using Selection Frequency to Organize Menus. ACM Transactions on Computer-Human Interaction, 1, 1, pp. 27-51.

135. Semeraro G., Ferilli S., Fanizzi N. and Abbattista F., 2001. Learning Interaction Models in a Digital Library Service. In: M. Bauer, J. Vassileva, and P. Gmytrasiewicz (Eds.), User modeling: Proceedings of the Eighth International Conference, UM2001. Berlin: Springer, pp. 44-53.

136. Shardanand U. and Maes P., 1995. Social Information Filtering for Automating "Word of Mouth". In: Proceedings of CHI-95, Denver, CO, pp. 210-217.

137. Smyth B. and Cotter P., 2002. Personalized Adaptive Navigation for Mobile Portals. In: Proceedings of the 15th European Conference on Artificial Intelligence - Prestigious Applications of Intelligent Systems, Lyons, France, 2002.

138. Specht M. and Kobsa A., 1999. Interaction of Domain Expertise and Interface Design. In: Adaptive Educational Hypermedia, Workshops on Adaptive Systems and User Modeling on the World Wide Web at WWW-8, Toronto, Canada, and Banff, Canada.

139. Spiliopoulou M., 2000. Web Usage Mining for Web Site Evaluation. Communications of the ACM, (43) 8, 2000.

140. Spradley J., 1980. Participant Observation. Wadsworth Publishing.

141. Stephanidis C., 2001. Adaptive Techniques for Universal Access. User Modeling and User-Adapted Interaction, Volume 11, Issue 1-2, pp. 159 - 179.

142. Strauss A. L. and Corbin J. M., 1998. Basics of Qualitative Research: Techniques and Procedures for Developing Grounded Theory. SAGE, Thousand Oaks.

143. Tasso C. and Omero P., 2002. La Personalizzazione Dei Contenuti Web. Franco Angeli, Milano, Italy.

144. Tobar C. M., 2003. Yet Another Evaluation Framework. In: S. Weibelzahl and A. Paramythis (Eds.), Proceedings of the Second Workshop on Empirical Evaluation of Adaptive Systems, held at the 9th International Conference on User Modeling UM2003. Pittsburgh, pp. 15-24

145. Totterdell P., and Boyle E., 1990. The Evaluation of Adaptive Systems. In: D. Browne, P. Totterdell and M. Norman (Eds.), Adaptive User Interfaces. London: Academic Press, pp. 161-194.

146. Travassos G. H., and Barros M. O., 2003. Contributions of In Virtuo and In Silico Experiments for the Future of Empirical Studies in Software Engineering. In: A. Jedlitschka and M. Ciolkowski (Eds.), Proceedings of the Second Workshop on the Future of Empirical Studies in Software Engineering (pp. 109-121). Roman Castles, Italy.

147. Uruchurtu E., MacKinnon L., Rist R., 2005. User Cognitive Style and Interface Design for Personal, Adaptive Learning In: Proceedings of UM2005, LNAI 3538, pp. 154-163.

148. van Solingen R., and Berghout E., 1999. The Goal/question/metric Method: A Practical Guide for Quality Improvement of Software Development. London: McGraw-Hill.

149. van Barneveld, J. and van Setten, M. (2003). Involving users in the design of user interfaces for TV recommender systems. 3rd Workshop on Personalization in Future TV, Associated with UM03, Johnstown, PA

150. Web Accessibility Initiative W3C - Evaluating Web Sites for Accessibility Http://www.w3.org/WAI/eval/

151. Web Accessibility Initiative W3C - Web Content Accessibility Guidelines 1.0 Http://www.w3c.org/TR/WAI-WEBCONTENT

152. Weibelzahl S., 2001. Evaluation of Adaptive Systems. In: M. Bauer, P. J. Gmytrasiewicz and J. Vassileva (Eds.), Lecture Notes in Computer Science N. 2109: User Modeling 2001. Berlin: Springer, Pp. 292-294.

153. Weibelzahl S., 2003. Evaluation of Adaptive Systems. Dissertation. University of Trier, Germany.

154. Weibelzahl S., 2005. Problems and pitfalls in the evaluation of adaptive systems. In: S. Chen and G. Magoulas (Eds.). Adaptable and Adaptive Hypermedia Systems (pp. 285-299). Hershey, PA: IRM Press

155. Weibelzahl S. and Lauer C.U., 2001. Framework for the Evaluation of Adaptive CBRSystems. In: U. Reimer, S. Schmitt, and I. Vollrath (Eds.), Proceedings of the 9th German Workshop on Case-Based Reasoning (GWCBR01), Aachen: Shakerm, Pp. 254-263.

156. Weibelzahl S. and Weber G., 2001. A database of empirical evaluations of adaptive systems. In: R. Klinkenberg, S. Rüping, A. Fick, N. Henze, C. Herzog, R. Molitor, and O. Schröder (Eds.), Proceedings of Workshop Lernen - Lehren - Wissen - Adaptivität (LLWA 01), Research Report in Computer Science Nr. 763, University of Dortmund, Pp. 302-306.

157. Weibelzahl S., Jedlitschka2 A., and Ayari B., 2006. Eliciting Requirements for a Adaptive Decision Support System through Structured User Interviews. In the Proc. of the Fifth Workshop on User-Centred Design and Evaluation of Adaptive Systems, 2006, Pp. 470-478.

158. Weibelzahl S. and Weber, G., 2003. Evaluating the inference mechanism of adaptive learning systems. In: Lecture Notes in Computer Science N. 2702: User Modeling 2003, Berlin: Springer, Pp. 154-162.

159. Whiteside J., Bennett J., and Holtzblatt K., 1988. Usability Engineering: Our Experience and Evolution. In: M. Helander (ed.). Handbook of Human-Computer Interaction, New York: North-Holland, 1988, Pp. 791-817.

160. Wilkinson, L., and Task Force on Statistical Inference, 1999. Statistical methods in psychology journals: Guidelines and explanations. American Psychologist, 54(8), 594-604.

161. Wingender, E. (Ed।)., 1998. In Silico Biology. An international journal on computational molecular biology. IOS Press.

Author Index